TCNA HANDBOOK

TCNA
Tile Council of North America, Inc.

FOR CERAMIC, GLASS, AND STONE TILE INSTALLATION

The Tile Council of North America (TCNA) is a trade association representing North American manufacturers of ceramic tile, tile installation materials, tile equipment, raw materials, and other tile-related products. Established in 1945 as the Tile Council of America (TCA), the Tile Council has a leadership role in facilitating the development of North American and international industry quality standards to benefit tile consumers. Additionally, the Tile Council regularly conducts independent research and product testing, works with regulatory, trade, and other government agencies, offers professional training, and publishes installation guidelines, tile standards, economic reports, and promotional literature.

Our member companies help sustain the publication of this document and other technical literature — look for the "Member of Tile Council of North America" logo on their products.

For more information regarding the Tile Council of North America, or for links to TCNA members, please visit www.TCNAtile.com.

Standards. Science. Education. Advocacy. TCNA.

Generic edition of the *TCNA Handbook*

[inset] In 1919, William Randolph Hearst started construction on his estate, "La Cuesta Encantada" (now commonly referred to as the Hearst Castle) that overlooks San Simeon, CA. In 1947, Hearst had created an estate of 165 rooms, three guesthouses, and 127 acres of terraced gardens, fountains, and pools.

[cover] Hearst was impressed by the beauty of the mosaics in the 5th Century Mausoleum of Galla Placidia, in Ravenna, Italy. When building The Roman Pool at Hearst Castle, he incorporated similar styles with designs created by Camille Solon, a British muralist and ceramist. The indoor pool is decorated from floor to ceiling with 1" square mosaic tiles that are blue, orange, or clear with fused gold inside representing traditional marine monster themes that can be found in ancient Roman baths.

Photo Credits: Trey Ratcliff/Creative Commons (main); Jejim/Shutterstock (inset)

Co-branded editions of the *TCNA Handbook*

Cover: Co-branded editions of the *TCNA Handbook* are marked with the sponsor's logo under the words "This copy sponsored by."

TCNA allows sponsors to customize copies of the *Handbook* they distribute with advertisements showcasing their products, services, or programs on the front and back covers. Inclusion of these advertisements does not constitute any endorsement, recommendation, approval, or evaluation by TCNA, or the TCNA Handbook Committee, of any sponsor's products, services, or programs.

Excluding covers, all editions contain the same content.

TCNA HANDBOOK

TCNA
Tile Council of North America, Inc.

FOR CERAMIC, GLASS, AND STONE TILE INSTALLATION

2021 | Release Date April 2021
Tile Council of North America

2021 TCNA HANDBOOK FOR CERAMIC, GLASS, AND STONE TILE INSTALLATION

The Tile Council of North America (TCNA) provides the *TCNA Handbook for Ceramic, Glass, and Stone Tile Installation* as a guide, not a product, to assist in clarifying and standardizing installation specifications for tile. Each installation method requires a properly designed, constructed, and prepared substructure using materials and construction techniques that meet nationally recognized material and construction standards.

The quick reference details cover most installation methods and conditions. However, some installation methods and materials are not recognized and may not be suitable in some geographical areas because of local trade practices, climatic conditions, or construction methods. Therefore, while the goal of the Conference for the *TCNA Handbook* is to produce accurate guidelines, they should be used only with the independent approval of technically qualified persons. All specifications for tile installations must conform to local building codes, ordinances, and trade standards and practices.

The information presented herein represents a consensus of the national and regional organizations and individuals who serve on the Handbook Committee. The Committee considers revisions on a regular basis to present architects and specification writers with current, accurate recommendations.

Production Staff

Katelyn Simpson, *Handbook Committee Secretary and Technical Content Manager*

Ryan Marino, *Technical Content Manager*

Kathy Meyer, *Project Manager*

Cynthia Faber Smith, *Graphic Design*

Eric Astrachan and Bill Griese, *Technical Content Review and Development*

Roxanne Shiflet, *Prepress Production*

Fernando Lozano Assad, Martha Salazar, and Samantha Montenegro, *Spanish Translations*

TCNA
TILE COUNCIL OF NORTH AMERICA, INC.

100 Clemson Research Blvd.
Anderson, SC 29625
USA

Ave. Batallón de San Patricio #109-627.
Colonia Valle Oriente, San Pedro Garza García
Nuevo León, México. CP 66269

www.TCNAtile.com

(864) 646-8453

52 (81) 8625-3306

Printed in the United States of America.

ISBN: 978-1-7330220-9-5

Library of Congress Control Number: 2021933071

Published by Tile Council of North America.

TECHNICAL CHANGES TO THE 2021 *TCNA HANDBOOK*

Changes to the *TCNA Handbook* are made at the discretion of TCNA Handbook Committee members serving on the committee when such changes are balloted; note that any given revision may or may not have been supported by any individual committee member. See the following page for committee roster as of November 2020.

Technical changes made to this 2021 edition of the *TCNA Handbook* are listed in the order in which they appear, with any global changes to standard (repeated) language listed according to the first appearance of the revised language. Editorial changes and clarifications are not listed.

Revised: Deep Abrasion and Dynamic Coefficient of Friction sections within the "CERAMIC TILE SELECTION GUIDE"

Added: Surface Wear Resistance section, within the "CERAMIC TILE SELECTION GUIDE"

Added: Surface Wear Resistance and Dynamic Coefficient of Friction sections, within the "GLASS TILE SELECTION AND INSTALLATION GUIDE"

Revised: General, Applicable Standards, and Considerations when Selecting Natural Stone Tile sections, within the "NATURAL STONE TILE SELECTION AND INSTALLATION GUIDE"

Revised: Introduction, Neat Cement and Portland Cement Mortar, Dry-Set Cement Mortar, Modified Dry-Set Cement Mortar, Exterior Glue Plywood (EGP) Modified Dry-Set Cement Mortar and Dry-Set Mortar for Large and Heavy Tile (LHT Mortar), Formerly Medium Bed Mortar sections, within the "SETTING MATERIAL SELECTION GUIDE"

Revised: Gauged Porcelain Tiles and Gauged Porcelain Tile Panels/Slabs section, within the "ADDITIONAL PRODUCTS USED IN TILE INSTALLATIONS"

Revised: Specifying Conforming Sustainable Tile Products, Using Publicly Available Data to Understand the Environmental Impacts of Products being Specified, Additional Green Tile and Installation Material Considerations, and Common Green Building Standards and Rating Systems sections, within the "GREEN BUILDING STANDARDS AND GREEN PRODUCT SELECTION GUIDE"

Revised: System Modularity section, within the "GROUT JOINT SIZE, LAYOUTS AND PATTERNS"

Revised: General and Components of an Installation Method sections, within the "USING THE TCNA HANDBOOK FOR SPECIFICATION WRITING"

Revised: General, Advanced Certification for Tile Installers (ACT), and Ceramic Tile Education Foundation (CTEF) Certified Tile Installer Program sections, within the "INSTALLER AND CONTRACTOR QUALIFICATION GUIDE"

Revised: Recommended Uses, and Requirements sections, within Tile and Stone methods "**F101**" and "**F112**"

Revised: Requirements section, within Tile and Stone methods "**F102**", "**F104**", "**F113**", "**F113A**", "**F122**", "**F122A**", "**RH110**", "**RH110A**", "**RH112**", "**RH112A**", "**RH116**" and "**RH116A**"

Revised: Recommended Uses section, within Tile and Stone methods "**F111**", "**F121**", "**RH117**", and "**F141**"

Revised: Recommended Uses section, within Tile methods "**F114**", "**F145**" and "**F185**"

Revised: Requirements, within Tile methods "**F115**", "**F115A**", "**F116E**", "**F116O**", "**F131**", "**F133**", "**F135**", "**RH115**", and "**RH115A**"

Revised: Preparations by Other Trades, within Tile and Stone methods "**F200**", "**F200A**", "**RH111**", and "**RH111A**"

Revised: Drawing and Title, Recommended Uses, Typical Weight of Installation, Requirements, Materials, and Installation Specifications sections, within Tile and Stone method "**F205**"

Revised: Drawing and Title, Recommended Uses, Typical Weight of Installation, Requirements, Materials, and Installation Specifications sections, within Tile and Stone method "**F205A**"

Revised: Limitations and Requirements sections, within Tile and Stone method "**F125**"

Revised: Recommended Uses and Requirements sections, within Tile method "**F132**"

Revised: Drawing, within Tile and Stone methods "**W221**" and "**W222**"

Revised: Movement Joint Type, Other section, within "**EJ171** Movement Joint Guidelines for Ceramic, Glass, and Stone"

Added: Drawing EJ171M, within "**EJ171** Movement Joint Guidelines for Ceramic, Glass, and Stone"

TCNA HANDBOOK COMMITTEE

Chairperson Eric Astrachan

Secretary Katelyn Simpson

Committee Members	Represented by
ANSI ASC A108 Committee Chairman	Chris Walker
APA The Engineered Wood Association	BJ Yeh
Artistic Tile	Josh Levinson
Backer Board Manufacturers Subcommittee	Tony Fuller
Ceramic Tile and Stone Association of Arizona	Sam Hibbs and Ken Milbery
Ceramic Tile Distributors Association	Frank Donahue
Ceramic Tile Education Foundation	Scott Carothers
Ceramic Tile Institute of America	Gray LaFortune and Dave Gobis
Chicago Tile Institute	Gavin Collier
Construction Specifications Institute	Randall Chapple
Deltek	Barbara Larson
Glass Tile Consultants	Scott Fleming
Glass Tile Manufacturers Subcommittee	Brian Fitzgerald
Great Lakes Ceramic Tile Council	Gerald Chioini and Kurt von Koss
Gypsum Association	Michael Schmeida
International Masonry Institute	Scott Conwell and Ron Spezzaferri
International Union of Bricklayers and Allied Craftworkers	Mike Hawthorne
ISO TC-189 Committee	Noah Chitty
Materials and Methods Standards Association	Steve Fine
Membrane Manufacturers Subcommittee	Jim Whitfield
Mortar and Grout Manufacturers Subcommittee	Mike Micalizzi
National Tile Contractors Association	Nyle Wadford
National Tile Contractors Association	James Woelfel
Natural Stone Institute	Chuck Muehlbauer
Porcelain Tile Manufacturers Subcommittee	Jim Neel
Professional Consultants International, LLC	Richard Goldberg
Quarry Tile Manufacturers Subcommittee	Dan Marvin and Guy Renkert
Radiant Panel Association	Tracy Hall
Terrazzo, Tile and Marble Association of Canada	Dale Kempster
Tile and Stone Council Consultants	Kent Klaser
Tile and Stone Council of Northern California	Rich Galliani
Tile Contractors' Association of America–West Coast	Rich Papapietro and Ronald Schwartz
Tile Contractors' Association of America–East Coast	Bradford Trostrud and John Trendell
Tile Council of North America	Bill Griese
Tile Council of North America - Mexico	Fernando Lozano Assad
Underlayment Manufacturers Subcommittee	Greg Schad
United Brotherhood of Carpenters	Eric Kastner

CONTENTS

NOTICE OF DISCLAIMER AND LIMITATION OF LIABILITY

THIS PUBLICATION IS PROVIDED AS A GUIDE, AND ALL OF THE INFORMATION IT CONTAINS IS PROVIDED 'AS IS' WITHOUT WARRANTY OF ANY KIND, WHETHER EXPRESS OR IMPLIED. ALL IMPLIED WARRANTIES, INCLUDING, WITHOUT LIMITATION, IMPLIED WARRANTIES OF MERCHANTABILITY, FITNESS FOR A PARTICULAR PURPOSE, AND NON-INFRINGEMENT, ARE HEREBY EXPRESSLY DISCLAIMED.

PUBLISHER EXPRESSLY DISCLAIMS ANY OBLIGATION TO OBTAIN AND INCLUDE INFORMATION OTHER THAN THAT PRESENTED HEREIN RESULTING FROM THE HANDBOOK COMMITTEE CONSENSUS PROCESS. THIS INFORMATION DOES NOT PURPORT TO ADDRESS SAFETY ISSUES OR APPLICABLE REGULATORY REQUIREMENTS ASSOCIATED WITH ITS USE. IT IS THE RESPONSIBILITY OF THE USER OF THIS INFORMATION TO REVIEW ANY APPLICABLE CODES AND OTHER REGULATIONS AND ANY SITE SPECIFIC CONDITIONS IN CONNECTION WITH THE USE OF THIS INFORMATION. PUBLISHER EXPRESSLY MAKES NO REPRESENTATIONS OR WARRANTIES REGARDING USE OF THIS INFORMATION AND COMPLIANCE WITH ANY APPLICABLE STATUTE, RULE OR REGULATION.

THE READER IS EXPRESSLY WARNED TO CONSIDER AND ADOPT ALL SAFETY PRECAUTIONS APPROPRIATE FOR THE ACTIVITIES HEREIN AND TO AVOID ALL POTENTIAL HAZARDS. SUCH PRECAUTIONS ARE GENERALLY NOT LISTED HEREIN AND ARE OUTSIDE THE SCOPE OF THIS DOCUMENT.

UNDER NO CIRCUMSTANCES WILL PUBLISHER BE LIABLE TO ANY PERSON OR BUSINESS ENTITY FOR ANY DAMAGES, INCLUDING WITHOUT LIMITATION ANY AND ALL DIRECT, INDIRECT, SPECIAL, INCIDENTAL, CONSEQUENTIAL, OR EXEMPLARY DAMAGES, RESULTING, IN WHOLE OR IN PART, FROM ANY USE OF, REFERENCE TO, OR RELIANCE UPON THIS PUBLICATION, EVEN IF ADVISED OF THE POSSIBILITY OF SUCH DAMAGES. THE FOREGOING LIMITATION OF LIABILITY IS A FUNDAMENTAL ELEMENT OF THE USE OF THIS INFORMATION AND THE INFORMATION WOULD NOT BE OFFERED BY THE PUBLISHER WITHOUT SUCH LIMITATION.

2021 | PRODUCT SELECTION GUIDES

[For natural stone tile, see "Natural Stone Tile Selection and Installation Guide"; for glass tile, see "Glass Tile Selection and Installation Guide."]

Ceramic Tile Types

Ceramic tiles suitable for *TCNA Handbook* installation methods are those that meet the specifications outlined in ANSI A137.1 *American National Standard Specifications for Ceramic Tile.* ANSI A137.1 contains performance and aesthetic criteria for the five major types of ceramic tiles: porcelain, pressed floor, mosaic, quarry, and glazed wall tiles.

Porcelain tiles are ceramic tiles for floor and wall applications with a water absorption of 0.5% or less per the ASTM C373 test method and that are generally made by the pressed or extruded method. This category does not include materials with very little or no crystallinity, such as glass tiles. Tiles specifically warranted as porcelain by a manufacturer, and independently tested and confirmed to comply with this water absorption requirement, may be certified by the Porcelain Tile Certification Agency (PTCA).

Pressed floor tiles are ceramic tiles made by pressing, having a facial area equal to or greater than 9 square inches, and although specified by manufacturers as appropriate for floors, are also suitable for use on walls.

Ceramic mosaic tiles are tiles for floor and wall applications having a facial area less than 9 square inches, which are typically mounted on sheets or strips.

Quarry tiles are ceramic tiles for floor and wall applications made by the extrusion process from natural clay or shale.

Glazed wall tiles are nonvitreous ceramic tiles intended for interior use, and are not required nor expected to withstand excessive impact or exposure to freeze/thaw cycling. Tiles in this category are not intended for use on floors.

Water Absorption

Water absorption per the ASTM C373 test method is the determinative factor for whether a tile is considered to be impervious (absorption ≤ 0.5%), vitreous (0.5% < absorption ≤ 3.0%), semivitreous (3.0% < absorption ≤ 7.0%), or nonvitreous (7.0% < absorption ≤ 20.0%).

- Porcelain tiles in compliance with ANSI A137.1 are impervious. Due to their low water absorption, bonding materials specifically appropriate for porcelain tiles are recommended.
- Pressed floor tiles in compliance with ANSI A137.1 may be vitreous, semivitreous, or nonvitreous.
- Mosaic tiles in compliance with ANSI A137.1 may be impervious, vitreous, semivitreous, or nonvitreous.
- Quarry tiles in compliance with ANSI A137.1 may be impervious, vitreous, or semivitreous with an absorption value up to 5.0%.
- Glazed wall tiles in compliance with ANSI A137.1 are nonvitreous.

Grade, Sampling, and Labeling

Grade: Standard grade tiles (also referenced as first grade) shall be in compliance with all testing criteria set forth in ANSI A137.1. Second grade tiles (marked as such by the tile manufacturer) shall be in compliance with all testing criteria set forth in ANSI A137.1, except that the evaluation for facial defects shall be performed from a distance of 10' instead of 3'.

Sampling: Evaluation of a shipment is made based on random sampling and acceptance numbers per ANSI/ASQ Z1.4-2003 (Normal Sampling Plan, Level 2) with an AQL of 2.5%. Table 2 in ANSI A137.1 provides the number of allowable defective tiles per sample with acceptance criteria applied separately for each characteristic under test.

Labeling: All tiles in compliance with ANSI A137.1 shall be shipped in sealed packages with the claim of grade, type of tile, producer identification, and country of origin clearly indicated.

Aesthetic Classifications

Tiles may vary in color, texture, or appearance according to the manufacturer's design for that particular tile series or product line. The following designations classify the range of such variation.

V0 tiles are very uniform in appearance and smooth in texture. When measured by a colorimetric spectrophotometer, they have an overall color difference (delta E) of less than 3 Judds.

V1 tiles have a generally uniform appearance with minimal production run differences among pieces. Some color differences can be observed.

V2 tiles are similarly colored with variations in texture and/or pattern, which are clearly distinguishable.

V3 tiles are moderately variant, and although the colors and/or textures present on a single piece are indicative of the colors and/or textures on another, the amount of colors and/or textures on each piece may vary significantly. It is recommended that the range be viewed before selection and that a mock layout be made.

V4 tiles are substantially variant with random color and/or texture differences from tile to tile, where one tile may have totally different colors and/or textures from another. Because a final installation with V4 tiles will be unique, it is recommended that the range be viewed before selection and that a mock layout be made.

Although checking the range of color and texture for V3 and V4 tiles is specifically recommended, all of the categories above have some range and should be checked and generally installed from several boxes in random fashion to avoid aesthetic issues.

Deep Abrasion

Unglazed tiles are tested for deep abrasion resistance by test method ASTM C1243, in which tiles are quantitatively evaluated for material loss after being subjected to an abrasive disk. For an unglazed tile, the total material loss from abrasion must be below a specified maximum. The maximum value depends on whether the tile was pressed or extruded and its water absorption category: impervious, vitreous, semivitreous, or nonvitreous.

Surface Wear Resistance

ANSI A137.1 specifies that a manufacturer must clearly and definitively state the surface wear classification of a tile. The following classifications defined in ANSI A137.1 are used for manufacturers to declare surface wear.

Class 0 tile should not be exposed to wear/traffic or aggressive maintenance practices and is generally used on walls.

Class I tile may withstand soft-soled foot traffic provided that dirt and/or other abrasives are not present on the wear surface. Class I tile should not be used in areas with direct access to the outside or large amounts of foot traffic.

Class II tile may withstand soft-soled and some normal foot traffic provided the dirt and/or other abrasives are kept to a minimum. Class II tile should not be used in areas with direct access to the outside, large amounts of foot traffic, or areas exposed to abnormal footwear.

Class III tile may withstand normal footwear and regular traffic with some dirt and/or other abrasives present in limited quantities. Tile in this class may be used in light commercial installations with limited foot traffic and with no direct access to the outside, such as hallways with limited traffic from the outside.

Class IV tile may withstand heavier amounts of traffic with greater amounts of dirt and/or other abrasives present, such as areas with regular traffic from the outside.

Class V tile may withstand constant foot traffic with larger amounts of dirt and/or other abrasives, such as areas subject to high volumes of foot traffic and constant traffic from the outside.

The above definitions are a general guideline. Final selection for all applications should take into account expected traffic, conditions, and maintenance. Final selection should also consider whether there is exposure to water (especially standing water), oil, grease, etc., which create slippery conditions. Floor applications with exposure to these elements require extra caution in product selection.

Dimensional Consistency

Tiles in compliance with ANSI A137.1 shall meet the specified criteria for facial dimensions, thickness, warpage, and wedging. Facial dimension criteria specify the maximum allowable variation in size and deviation from the manufacturer-specified nominal size. Thickness criteria specify the acceptable range of thickness variation. Warpage criteria establish allowable deviation from planarity of a tile's surface, and wedging criteria define the allowable extent tiles can be out of square.

Tiles may be rectified (mechanically finished along their edges to achieve more precise facial dimensions), calibrated (sorted to meet a specified caliber range), or "natural" (packaged directly after manufacturing with no rectification or sorting). Depending on the type of tile, dimensional criteria may be given for one, two, or all three categories as detailed in A137.1 Tables 6–10.

See "Grout Joint Size, Layouts, and Patterns" for related information.

Breaking Strength, Bond Strength, Crazing, Thermal Shock

Tiles in compliance with ANSI A137.1 shall have an average breaking strength equal to or greater than 275 lbf for porcelain and quarry tile, 250 lbf for pressed floor and mosaic tiles, and 125 lbf for glazed wall tiles.

When tested per ASTM C482, they shall exhibit a bond strength of at least 50 psi and demonstrate no evidence of crazing when tested per ASTM C424 and no evidence of thermal shock when tested per ASTM C484.

Freeze/Thaw Cycling

When tiles will be used in an environment subject to freeze/thaw cycling, testing per ASTM C1026 can be required to evaluate their potential performance under such conditions.

Dynamic Coefficient of Friction

The method for measuring the dynamic coefficient of friction (DCOF) specified in ANSI A137.1 is ANSI A326.3, *American National Standard Test Method for Measuring Dynamic Coefficient of Friction of Hard Surface Flooring Materials*. Previously, the method for measuring DCOF was included in ANSI A137.1-2012, Section 9.6, but in 2019 the ANSI A137.1 specification was updated to reference ANSI A326.3.

The following is excerpted from ANSI A326.3, section 3.1:

3.1 Unless otherwise specified, hard surface flooring materials suitable for level[1] interior spaces expected to be walked upon when wet with water shall have a wet DCOF of 0.42 or greater when tested using SBR sensor material and SLS solution as per this standard (A326.3). However, hard surface flooring materials with a DCOF of 0.42 or greater are not necessarily suitable for all projects. The specifier shall determine materials appropriate for specific project conditions, considering by way of example, but not in limitation, type of use, traffic, expected contaminants, expected maintenance, expected wear[2], and manufacturers' guidelines and recommendations.

End of Excerpt

Refer to Section 6.2.2.1.10 of ANSI A137.1 and ANSI A326.3 for further explanation and detail.

---

[1] Hard surface flooring materials appropriate for pedestrian inclines shall be chosen for the specific properties and use of the incline and require a wet DCOF greater than 0.42 if the inclined pathway will be used under wet conditions. Specifier shall determine materials appropriate for specific project conditions, considering by way of example, but not in limitation, type of use, traffic, grade of ramp, expected contaminants, expected maintenance, expected wear, and manufacturers' guidelines and recommendations.

[2] The COF of installed hard surface flooring materials can change over time as a result of wear and surface contaminants. In addition to regular cleaning, deep cleaning and traction-enhancing maintenance may be needed periodically to maintain DCOF values.

Stain and Chemical Resistance

While stain and chemical resistance are not required, ceramic tiles generally perform very well in these categories. Testing for stain resistance is performed per ASTM C1378 and testing for chemical resistance is performed per ASTM C650. Depending on the number of stains or chemicals that affect the surface, the tiles are classified from A (no effect) to E, where 4 or more stains or chemicals affected the tile.

Trim Units

Trim units are specified in ANSI A137.1, including line drawings and nomenclature for the 50 most common types.

Mounted Tile

Mounted tile is assembled into units or sheets by suitable material to facilitate handling and installation. Mounted tile assemblies shall have sufficient exposure to bonding surfaces of the tile body to allow for 80% coverage of the bond mortar in dry areas and 95% in wet areas. Tile manufacturers must specify whether their assemblies are suitable for installation in swimming pools, on exteriors, and in other wet areas. Paper back-mounted mosaics are not recommended in wet areas.

Back-mounted and edge-mounted tile assemblies may have perforated paper, fiber mesh, resin, polyurethane, or other bonding material on the back or edges of each tile, which becomes an integral part of the tile installation. Clear film-faced tile is assembled with clear plastic adhesive film on the face, which is removed after final set has occurred. Paper-faced tile has paper applied to the face of the tile with water-soluble adhesives, which should be removed during the installation process by wetting and removing the paper, followed by adjusting the tile prior to its final set.

Specialty Tile

Specialty tiles are designed to meet special physical requirements or to have special appearance characteristics. They are not required to meet all requirements of ANSI A137.1. Consult the manufacturer's specifications. They are sometimes manufactured to create an architectural effect toward the casual. These tiles vary in size, one tile from the other. Variations in plane may be expected. Larger tiles will usually require greater variations in joint width. For each specialty tile being chosen, review installation guidelines supplied by manufacturer/distributor of specialty tiles and/or adhesive manufacturer. Specialty tiles include, but are not limited to, tiles made from nonceramic materials.

TCNA Handbook installation methods that can be used for installing glass tile have glass tile listed in the Materials section of the method. Suitable glass tiles are those that meet the specifications outlined in ANSI A137.2 *American National Standard Specifications for Glass Tile.*

Glass Tile Types

Based on temperature of formation, there are three types of glass tile: cast, fused, and low-temperature-coated. Each type is further categorized, based on size, as: large format, mosaic, or miniature mosaic. Fused and low-temperature-coated glass tiles with any edge longer than 24" and cast glass tiles with any edge longer than 12" are defined as glass panels and are not covered by A137.2 or this *Handbook.*

Cast glass tiles are formed in a liquid state at 1600°F or higher. Many cast glass tiles are wavy and slightly textured, with inherent folds, bubbles, and creases. These unique characteristics are achieved through the casting process.

Fused glass tiles are typically made from sheet glass units that are altered through heat between 1023°F and 1599°F. During manufacturing, different materials and glazes are fused to the sheet glass units, usually in multiple stages, to create a variety of colors and patterns. Fused glass tiles can be smooth, textured, uniform, or nonuniform.

Low-temperature-coated glass tiles are made from sheet glass units that are altered through heat at temperatures less than 1022°F. Typically, these alterations involve heat transference of coatings to the back of transparent sheet glass units. These coatings can contain wide varieties of colors and patterns.

Aesthetic Properties and Considerations

Dimensional Requirements

Glass tiles must meet the criteria for facial dimensions, thickness, warpage, and wedging found in A137.2 Tables 5–7. Facial dimension criteria specify the maximum allowable variation in size and deviation from the manufacturer-specified nominal size. Thickness criteria specify the acceptable range of thickness variation. Warpage criteria establish allowable deviation from planarity of a tile's surface, and wedging criteria define the allowable amount that tiles can be out-of-square.

Glass tiles may be classified by the manufacturer as standard or natural. For the purposes of specifying glass tile, standard glass tiles are those that are manufactured to meet specific sizing tolerances, and natural glass tiles are those that are manufactured with inherent variation and thus not held to specific sizing tolerances. (The term natural, when used to reference a type of glass tile, is not associated with sustainability criteria.)

Aesthetic Classes

Glass tiles may vary in color, texture, or appearance according to the manufacturer's design for a particular tile series or product line. The tile manufacturer uses aesthetic classes V0–V4 to indicate the amount of variation that will be present in the tile. Although checking the range of color and texture for V3 and V4 tiles is specifically recommended, all categories have some range and should be checked and generally installed from several boxes in random fashion to avoid aesthetic issues.

V0 tiles are very uniform in appearance and smooth in texture. They can be measured for small color differences, and conform to manufacturer-reported limits.

V1 tiles have a generally uniform appearance with minimal production run differences among pieces. Some color differences can be observed.

V2 tiles are similarly colored with variations in texture and/or pattern, which are clearly distinguishable.

V3 tiles are moderately variant, and although the colors and/or textures present on a single piece are indicative of the colors and/or textures on another, the amount of colors and/or textures on each piece may vary significantly. It is recommended that the range be viewed before selection and that a mock layout be made.

V4 tiles are substantially variant with random color and/or texture differences from tile to tile, where one tile may have totally different colors and/or textures from another. Because a final installation with V4 tiles will be unique, it is recommended that the range be viewed before selection and that a mock layout be made.

Translucence and Opacity

Opaque glass tiles are defined in ANSI A137.2 as those with an opacity percentage greater than or equal to 95 when tested per section 7.5. Translucent glass tiles are defined as those with an opacity percentage less than 95.

With translucent glass tiles, setting material color consistency and full mortar coverage are essential; white mortar is commonly recommended. Avoid using setting materials from different production batches and check for bag-to-bag color consistency. To achieve uniform coverage on the back of translucent glass tile, back-buttering and/or flattening the trowel ridges is recommended. Regardless of installation technique, mesh backing on translucent tiles may be visible after installation.

Bonding translucent glass tiles directly to membranes or other impervious surfaces is not recommended because any moisture trapped between the tile and membrane would be visible. Membranes should be placed behind or below the tile setting substrate where translucent glass tile will be installed. Select only opaque glass tile if the tile will be bonded to a membrane.

For translucent, shaped pieces (v-cap, quarter round, crown caps, etc.), follow manufacturer recommendations on back-buttering and whether (and how) to fill the cavity. Excess setting material can shrink and cause cracking or ghosting, a slight visible separation of the mortar from the glass tile.

When grouting translucent glass tile, grout joints must be filled uniformly and consistently, as spaces and voids can be visible.

Determination of Aesthetic Defects

Because of the difference in appearance of each glass tile type, resulting from the different methods of manufacture, ANSI A137.2 outlines separate aesthetic criteria for each glass tile type with regard to aesthetic defects. The casting process produces inherent and unique appearance characteristics, such as bubbles below the tile surface, which are not considered defects, as they do not affect the integrity of the tile. In fact, cast glass tiles are typically selected when a less-uniform appearance is desired. At the same time, a bubble at the surface of the tile could leave a chipped or sharp edge and is considered a defect. Fused and low-temperature-coated glass tiles generally have a more uniform appearance and thus, different criteria for determining aesthetic defects. See ANSI A137.2 Section 7 for evaluation procedure and criteria for determining aesthetic defects.

Performance Requirements

Surface wear resistance: ANSI A137.2 specifies that a manufacturer must clearly and definitively state the surface wear classification of a tile. The following classifications defined in ANSI A137.2 are used for manufacturers to declare surface wear.

Class 0 tile should not be exposed to wear/traffic or aggressive maintenance practices and is generally used on walls.

Class I tile may withstand soft-soled foot traffic provided that dirt and/or other abrasives are not present on the wear surface. Class I tile should not be used in areas with direct access to the outside or large amounts of foot traffic.

Class II tile may withstand soft-soled and some normal foot traffic provided the dirt and/or other abrasives are kept to a minimum. Class II tile should not be used in areas with direct access to the outside, large amounts of foot traffic, or areas exposed to abnormal footwear.

Class III tile may withstand normal footwear and regular traffic with some dirt and/or other abrasives present in limited quantities. Tile in this class may be used in light commercial installations with limited foot traffic and with no direct access to the outside, such as hallways with limited traffic from the outside.

Class IV tile may withstand heavier amounts of traffic with greater amounts of dirt and/or other abrasives present, such as areas with regular traffic from the outside.

Class V tile may withstand constant foot traffic with larger amounts of dirt and/or other abrasives, such as areas subject to high volumes of foot traffic and constant traffic from the outside.

The above definitions are a general guideline. Final selection for all applications should take into account expected traffic, conditions, and maintenance. Final selection should also consider whether there is exposure to water (especially standing water), oil, grease, etc., which create slippery conditions. Floor applications with exposure to these elements require extra caution in product selection.

Generally, glass tile is less resistant to abrasion than ceramic tile; select glass tile suitable for expected traffic.

Freeze/thaw cycling resistance: Glass tiles must pass freeze/thaw cycling resistance testing per ASTM C1026.

Crazing: Glass tiles must pass ASTM C424 test for crazing.

Thermal shock resistance: Glass tiles (except those intended for dry interior wall applications) must pass thermal shock cycling when tested per ANSI A137.2 Section 7.9. Glass tile is more sensitive than ceramic tile to thermal shock, and may not be suitable for applications where an instantaneous temperature change of more than 100°F is likely. Consult glass tile manufacturer/distributor for specific written recommendations.

Thermal expansion: Thermal expansion is addressed in A137.2 on an "as reported" basis. Glass tile expands more, when heated, than ceramic tile. Consult manufacturer for minimum distance of the tile from direct heat sources (fireplaces, stoves, etc.).

Dynamic coefficient of friction (DCOF): The method for measuring the dynamic coefficient of friction (DCOF) specified in ANSI A137.2 is ANSI A326.3, *American National Standard Test Method for Measuring Dynamic Coefficient of Friction of Hard Surface Flooring Materials.*

The following is excerpted from ANSI A326.3, section 3.1:

3.1 Unless otherwise specified, hard surface flooring materials suitable for level[1] interior spaces expected to be walked upon when wet with water shall have a wet DCOF of 0.42 or greater when tested using SBR sensor material and SLS solution as per this standard (A326.3). However, hard surface flooring materials with a DCOF of 0.42 or greater are not necessarily suitable for all projects. The specifier shall determine materials appropriate for specific project conditions, considering by way of example, but not in limitation, type of use, traffic, expected contaminants, expected maintenance, expected wear[2], and manufacturers' guidelines and recommendations.

End of Excerpt

Refer to Section 5.2.2.1.11 of ANSI A137.2 and ANSI A326.3 for further explanation and detail.

[1] Hard surface flooring materials appropriate for pedestrian inclines shall be chosen for the specific properties and use of the incline and require a wet DCOF greater than 0.42 if the inclined pathway will be used under wet conditions. Specifier shall determine materials appropriate for specific project conditions, considering by way of example, but not in limitation, type of use, traffic, grade of ramp, expected contaminants, expected maintenance, expected wear, and manufacturers' guidelines and recommendations.

[2] The COF of installed hard surface flooring materials can change over time as a result of wear and surface contaminants. In addition to regular cleaning, deep cleaning and traction-enhancing maintenance may be needed periodically to maintain DCOF values.

Stain and chemical resistance: While stain and chemical resistance are only required on an "as reported" basis in A137.2, glass tiles are generally resistant to a wide range of common chemicals and staining agents. Testing for stain resistance is performed per ASTM C1378 and testing for chemical resistance is performed per ASTM C650. Depending on the number of stains or chemicals that affect the surface, the tiles are classified from A (no effect) to E, where 4 or more stains or chemicals affected the tile. Consult manufacturer for specific chemical exposure recommendations, such as those used in pools and fountains, as applicable.

Breaking strength/modulus of rupture/compressive strength: Since differently sized tiles respond to loading in different ways, ANSI A137.2 assesses glass tile strength through different test methods for each size category. Large format glass tiles must meet modulus of rupture (MOR) criteria; mosaic glass tiles must meet point load criteria; and miniature mosaic glass tiles must meet compressive strength criteria.

Bond strength: Glass tiles must facilitate bond strength greater than or equal to 150 psi when tested in a dry condition and 100 psi when tested in a submerged condition.

Mounting: Glass tiles are generally sold and installed as mounted assemblies, and whether they are back-mounted or face-mounted is important. Some tiles with re-emulsifiable glues and/or insufficient contact area between the tile backing and mortar, can delaminate in submerged applications. It is recommended that glass tiles intended for submerged applications be face-mounted to prevent problems stemming from improper coverage or backing material failure. For back-mounted and edge-mounted products, manufacturers must clearly state for which areas their mounted tiles are suitable.

Grade, Sampling, and Labeling

Standard grade (or first grade) glass tiles must conform to all testing criteria set forth in ANSI A137.2. Conformance of a shipment is based on sampling and acceptance numbers per ANSI/ASQ Z1.4-2003 (Normal Sampling Plan, Level 2) with an AQL of 2.5%. Table 1 in ANSI A137.2 provides the number of allowable defective tiles per sample, with acceptance criteria applied separately for each characteristic under test. Glass tiles in conformance with ANSI A137.2 must be shipped in sealed packages with the claim of grade, tile type, producer identification, and country of origin clearly marked.

Installation Considerations

Substrates

Although the substrates recommended for glass tile installations are similar to ceramic tile installations, extra attention should be given to substrate stability, rigidity, and preparation. Substrates should be sufficiently flat to allow uniform coverage and avoid excessive mortar thickness. Glass tile should not be installed on single-float (one coat) mortar bed walls framed with wood or metal, and glass tile should never be bonded directly to wood substrates. For above-ground slabs, reinforced mortar beds may be required, especially for large-format glass tile. Some manufacturers require mortar bed substrates be cured a minimum of 7 days prior to glass tile installation.

Glass tile is generally more vulnerable to crack propagation than ceramic tile. Where opacity allows (see Translucence and Opacity) the glass tile manufacturer may recommend use of an ANSI A118.12 crack isolation membrane for large format glass tiles.

Setting Materials

Mortars specifically formulated or recommended for glass tile must be used. Some glass tile manufacturers require use of a specific mortar. When installed over an impervious substrate, setting materials may take longer to cure.

Some low-temperature-coated glass tiles are not intended for installation with cementitious materials, as cement mortar alkalinity may deteriorate the low-temperature-applied tile backing and cause delamination, particularly in wet or humid environments. Tiles that cannot be installed with cementitious mortars do not meet ANSI A137.2 and are not covered by this *Handbook.*

Grouting

Grout recommendations for glass tile vary based on aesthetics and joint size. While some glass tile manufacturers specify unsanded grout to prevent scratching, others require sanded grout. Because glass tile is impervious, additional setting time may be necessary to allow the grout to firm prior to cleaning. Manufacturer recommendations vary with regard to whether or not the use of epoxy grout is acceptable.

Movement Joints

Because glass tile expands more than ceramic tile, more movement accommodation is necessary. Also consult tile manufacturer for minimum distance from a direct heat source (fireplace, stove, etc.).

This guide addresses natural stone tiles that are adhered to a substrate. Natural stone tile, also called dimension stone, is stone that has been harvested from its in-situ position in the Earth and cut and machined into tiles without altering the natural fabric of the material. Dimension stone tile has a nominal thickness of less than ¾" (19 mm) and does not have any facial dimension greater than 2' (610 mm). Agglomerates and other engineered products do not meet this definition and therefore are excluded. Natural stone tiles that meet the definition but not the dimensional requirements are also excluded.

Stone tiles are generally supplied in a gauged, or calibrated, thickness, which typically has a tolerance of ±1⁄32" (±0.8 mm) in thickness. Tolerances on face sizes will generally be ±1⁄16" (±1.5 mm), with some tiles available in tighter tolerances.

General

The success of any stone installation requires consideration of the stone's specific properties and how they affect its use as a building material. When dealing with most other construction products, a different color material is only different visually, whereas in natural stones a change in color is a change to a different species of stone with different behavior and performance. This is because every stone is comprised of a variety of minerals that give the stone its behavior and performance in addition to its appearance. The mode of formation, or genesis, of a stone also can be a factor.

Stones, referred to as "rocks" by Earth scientists, are formed by three different methods. Sedimentary rocks are formed by clasts, or particles that settle and are then cemented together by another agent. Limestone, sandstone, travertine, and onyx are examples of sedimentary rock. Metamorphic stones are those that have undergone a change in composition resulting from intense heat and pressure, typically, a crystallization or recrystallization of the original rock fabric. Marble, slate, serpentine, quartzite, gneiss, and soapstone are examples of metamorphic rocks or stones. Igneous stones are formed by solidification, caused by the cooling of molten or semimolten material. Igneous materials can be intrusive, meaning they cooled below the Earth's surface, or extrusive, meaning they cooled above ground. Granite and basalt are examples of igneous rocks or stones.

Of equal importance to the method of formation is the primary chemical makeup of the stone—either calcareous or siliceous. Calcareous rocks, such as marble, travertine, and limestone, are made up primarily of calcium carbonate mineralogy. Because calcium carbonate, or the calcite crystal, can be attacked by relatively mild acids, such stones are predictably vulnerable in acidic environments.

Siliceous rocks are made up primarily of silicates, such as quartz and feldspar. Quartz-based sandstone and granite are examples of siliceous rocks.

	Sedimentary	Metamorphic	Igneous
Calcareous	Onyx Travertine Limestone	Marble	
Siliceous	Sandstone	Slate Serpentine Quartzite Soapstone	Granite

Some generalizations can be made from the table above. Moving from left to right, from sedimentary to igneous, the stone will likely be more dense and thus have higher abrasion resistance and lower absorption. Moving from top to bottom in the table, from calcareous to siliceous, the chemical resistance of the material, particularly resistance to acids, will increase. For more information on the geology of the various dimension stone types, refer to the *Dimension Stone Design Manual* published by the Natural Stone Institute (NSI).

Applicable Standards

Dimension stone products are governed by the following ASTM standards, test methods, and guides. Some of these were drafted for standardization of anchored veneer applications and do not wholly apply to tile and/or adhered systems.

ASTM C97	Test Methods for Absorption and Bulk Specific Gravity of Dimension Stone
ASTM C99	Test Method for Modulus of Rupture of Dimension Stone
ASTM C119	Terminology Relating to Dimension Stone
ASTM C120	Test Method for Flexure Testing of Structural and Roofing Slate
ASTM C170	Test Method for Compressive Strength of Dimension Stone

ASTM C241	Test Method for Abrasion Resistance of Stone Subjected to Foot Traffic
ASTM C503	Specification for Marble Dimension Stone
ASTM C568	Specification for Limestone Dimension Stone
ASTM C615	Specification for Granite Dimension Stone
ASTM C616	Specification for Quartz-Based Dimension Stone
ASTM C629	Specification for Slate Dimension Stone
ASTM C880	Test Method for Flexural Strength of Dimension Stone
ASTM C1201	Test Method for Structural Performance of Exterior Dimension Stone Cladding Systems by Uniform Static Air Pressure Difference
ASTM C1242	Standard Guide for Selection, Design, and Installation of Dimension Stone Attachment Systems
ASTM C1352	Test Method for Flexural Modulus of Elasticity of Dimension Stone
ASTM C1353	Test Method Using the Taber Abraser for Abrasion Resistance of Dimension Stone Subjected to Foot Traffic
ASTM C1354	Test Method for Strength of Individual Stone Anchorages in Dimension Stone
ASTM C1496	Standard Guide for Assessment and Maintenance of Exterior Dimension Stone Masonry Walls and Façades
ASTM C1515	Standard Guide for Cleaning of Exterior Dimension Stone, Vertical and Horizontal Surfaces, New or Existing
ASTM C1526	Specification for Serpentine Dimension Stone
ASTM C1527	Specification for Travertine Dimension Stone
ASTM C1528	Standard Guide for Selection of Dimension Stone
ASTM C1721	Standard Guide for Petrographic Examination of Dimension Stone
ASTM C1799	Standard Guide to Dimension Stone Test Specimen Sampling and Preparation

References for Further Reading

Dimension Stone Design Manual, published by the Natural Stone Institute (NSI)

ANSI *American National Standard Specifications for the Installation of Ceramic Tile*, published by the Tile Council of North America (TCNA)

NTCA Reference Manual, published by the National Tile Contractors Association (NTCA)

Considerations When Selecting Natural Stone Tile

Water-Sensitivity

Many green-colored marbles and serpentines warp, or curl, when exposed to water or a water-based adhesive such as a cementitious mortar. To avoid having stones warp from the moisture in such setting materials, an epoxy adhesive bond coat, which does not contain water, may be required. Because some green marbles and serpentines do not warp, or only mildly exhibit this behavior, research is encouraged on the particular stone species selected. Some can be successfully installed with water-based portland cement mortars. For stones that are mildly water-sensitive, consider mixing the mortar with reduced water volumes as allowed by the mortar manufacturer or wetting the exposed surface of the tiles after installation to provide balanced saturation levels on both surfaces. There are a few stones other than green marbles or serpentines that will warp with unbalanced water exposure. Research is encouraged on any particular stone with which one has no direct experience.

Fiberglass Mesh Reinforced Stone

Many stone products have a mesh reinforcement adhered to the back surface of the tile. The adhesive used to attach the mesh is not standardized, but is commonly an epoxy or polyester-based resin to which portland cement-based mortars will not reliably bond. Such reinforced stone tiles require epoxy bonding mortar.

Filled Voids in Travertine

Travertine typically has voids, or holes, which are commonly filled at the fabrication site with cementitious or resinous filler. Travertine can be supplied in a vein cut, where the slab is sawn perpendicular to the bedding planes to expose the veined character, or in a fleuri cut, also called crosscut, where the slab is sawn parallel to the bedding planes to yield a random, flowery appearance ("fleuri" is the French word for "flowery"). Particularly for fleuri cut tiles, many voids can exist just below the finished surface, concealed only by a slight shell of stone material. This concealment prevents detection and filling of the voids in the factory. Once in service, a concentrated load or point load will fracture the thin shell of stone, exposing the void below. High-heeled shoes and wheeled carts are examples of common causes of this occurrence, and maintenance to fill voids as they appear over time will be required when such stones are used. This issue is normally self-arresting, as eventually all of the voids will be filled.

Pinholes in Marble

On a much smaller scale than the voids found in travertine, many marbles have pinholes visible in the face of the material. Frequency can be as high as 100 or more per square foot. This is not considered a defect, but rather a natural characteristic of the material. Proper sampling is required to be aware of these occurrences prior to commencing with installation.

Viewing/Inspection Distance

As products of nature, dimension stones will have inherent characteristics that may be considered objectionable to some viewers. Inspection of the material should be done at reasonable distances, similar to how the material is viewed by building occupants. It is recommended that samples, range samples, mockups, and finished work be viewed for inspection at a distance of 6½' from a position normal (perpendicular) to the stone face, and with natural lighting. Lighting of reduced angles of incidence, for instance downwashed lights, can accentuate lippage and warpage that is within industry tolerances. Such lighting should be turned off for inspection purposes. See also "Flatness and Lippage."

Variation of Abrasion Resistance in Patterned Floors

When two or more stone species are used together in a floor pattern, it is possible to have different rates of wear when the materials have different resistance to abrasion. If any of the materials have an abrasion resistance less than 20.0 when tested per ASTM C241 (H_a) or C1353 (I_w), the difference between the abrasion resistance of any two stones being used together should not differ in abrasion resistance by more than 5.0.

Acid Sensitivity

Stones resistant to acid are required where acid exposure is likely. Most notably, calcium carbonate and calcite-based stones will react to relatively mild acids like lemon juice, vinegar, and cut tomatoes. The result is not a stain, which could be removed, but an etching of the stone surface, which cannot be removed without refinishing the stone. Acid attack of polished stones will usually manifest itself as a dull spot within the surrounding glossy surface.

Anisotropy

An anisotropic material is one that has different physical or visual properties in different directions. Most stones, particularly sedimentary stones, have some degree of anisotropy. Therefore, it is common that they will have a strong axis and a weak axis. Attention must be paid to this feature, particularly when reviewing test results, to ensure that the direction tested is the same as the direction used.

Delamination

"Rift" is the direction along which a stone is most easily split, and many stone varieties have pronounced bedding planes, along which a predetermined rift exists. Bedding planes can be a region of weakness, and some stones, particularly sandstones and some slates, will spall or delaminate along bedding plane boundaries after installation. Delamination is when a portion of the stone separates from the rest of the stone—typically the exposed surface dislodges from the portion of stone which still remains bonded to the substrate. Differential mineral expansion due to weathering and thermal cycles accelerate this occurrence, and freeze/thaw cycling is an additional factor. Additionally, floor traffic and dynamic floor deformation can cause spalling and delamination. Delamination is not related to any failure in adhesive; and research is required to verify the adequacy of stones with bedding planes for the intended application.

Shading and Variation

As products of nature, dimension stone will have inherent variation in color, shade, and character, including markings and veining, throughout any given lot of material. Variations from piece to piece are characteristic of the material and acceptable. Because of this variation, several pieces should be supplied as a sample, and because the material from a quarry can change over time, ensure samples are current before basing the project specification on them. Depending on the stone, it may be impossible for samples to represent all possible variations in color and character. Variations can be greater than or less than what is expressed by the sample pieces. The range of stones actually supplied should be reviewed and accepted prior to their installation. Dry laying pieces before installation helps ensure a pleasing blend of the tiles prior to permanently installing them, but is not always feasible, nor is doing so required unless specified in writing by the project superintendent or owner.

For viewing and installation, it is recommended that the material be blended from multiple packaging units to distribute the variation throughout the project and to avoid a blocking effect, where an area of one shade is distinctly bordered or surrounded by a different shade. For larger projects, it may not be feasible to view pieces from the full lot being supplied for the project.

Ensure adequate quantities are ordered, including requirements for waste, breakage, and attic stock. While some stone deposits are uniform through hundreds of feet of depth times vast lateral distances, there are also quarries where the inter-block, and even intra-block variability are so great that a given lot of material cannot be matched one month later.

Soundness of Marbles

Marbles are classified according to soundness based on the level of repair or reinforcement commonly required to facilitate their use. The soundness classifications A, B, C, and D are in descending order of soundness and ascending order of reinforcement and/or repair requirements. These classifications have no bearing on the commercial value of the stone. The lesser soundness classifications of C and D include many of the aesthetically prized, decorative marbles. However, these stones typically have too much repair or reinforcement to be used in wet or exterior areas. For more information about marble soundness, refer to the Natural Stone Institute's *Dimension Stone Design Manual.*

Efflorescence

Efflorescence results from migration of soluble salts originating from soils or masonry. These salts are carried to the surface where the water evaporates, leaving a white residue. Natural stones are not generally the source of efflorescence; however, they can sometimes provide efficient paths for the migration. In some high-porosity or open-veined stone varieties, efflorescence can travel through the stone tile in addition to traveling through the grout joint. Cleaning the salt deposit is generally easy and mechanical methods such as brushing are usually effective, but this only removes the symptom. The efflorescence will reoccur unless the source (water infiltration) can be eliminated.

Water Spots

Differing water chemistries may spot a given stone, which is generally most visible on dark, polished stones. This deposit will be on the surface of the stone, and can normally be removed mechanically by buffing with #000 steel wool, but in some cases requires buffing with a polishing powder.

Stone Finishes

A variety of standard stone finishes are available as well as some proprietary finishes. In the interest of pedestrian safety, only textured finishes are recommended for outdoor or wet area applications. While the texture enhances the safety aspect of the walking surface, it may complicate the maintenance by requiring brushes for cleaning, as soft mops may snag by some of the sharper profile finishes. For more information on stone finishes, refer to ASTM C119 Standard Nomenclature for Dimension Stone and the Natural Stone Institute's *Dimension Stone Design Manual.*

Iron Staining

Some stones, slates in particular, contain soluble iron that can cause "bleeding" if installed in a wet area. The use of a sealer will not prevent this from occurring. Yellowing of light-colored marbles may also be due to iron oxide.

Mohs Scale of Relative Hardness

The Mohs scale of relative hardness is based on 10 minerals. Mohs scale is one of relative hardness, not absolute hardness, meaning it is not linear. By example, orthoclase is harder than apatite, which is harder than fluorite, which is harder than calcite. However, orthoclase is not twice as hard as calcite even though 3 doubled is 6. For perspective, a human fingernail would be about 2.5 on the Mohs scale, gold and silver would be 2.5 to 3, a copper penny would be about 3.5, an iron nail would be about 4, window glass would be about 5.5, and aluminum oxide abrasive would be about 8.

Mohs Scale of Relative Hardness	
Hardness	**Mineral**
1	Talc or Mica
2	Gypsum
3	Calcite
4	Fluorite
5	Apatite
6	Orthoclase
7	Quartz

Mohs Scale of Relative Hardness	
Hardness	**Mineral**
8	Topaz
9	Corundum
10	Diamond

Considerations When Selecting Installation Materials

Staining

Light-colored marbles and nearly all onyx and limestone must be installed with white mortars due to their susceptibility to staining from the grey portland cement components in grey bonding mortars.

Although silicone itself does not stain natural stone, silicone sealants can contain plasticizers that can wick into the stone and stain it. This is more problematic for higher-porosity stones, but can occur in dense stones as well. Some silicone manufacturers keep a database of projects where their sealants have been used with various species of stones so compatibility can be verified. In the absence of an exemplar project to view, compatibility can be verified via accelerated laboratory testing.

The use of dyes in grouts, epoxies, polyesters, and any other joint fillers should always be tested to verify that the dyes will not leach into the stone, causing a "picture frame" stain. This is most common when using materials of contrasting color, although dyes of matching color can also cause staining.

Sanded and Unsanded Grout

The sand in sanded grout is generally harder than any of the calcium-carbonate-based stones, such as limestone, travertine, marbles, and onyx. Masking the stone surface or the use of unsanded grout should be considered to avoid scratching the surface finish. This is particularly true of these stones when polished. If unsanded grout is required, the grout joint must be $\frac{1}{16}$" to $\frac{1}{8}$" wide (nominal).

Impregnators

It is a common, although not universal, practice to treat natural stone installations with impregnating repellants, often referred to as sealers, although this is a misnomer. The products typically used on natural stones are not truly sealers, in that they are formulated to allow the transmission of vapor, which maintains breathability of the stone. The use of some impregnators with a particular stone species may create an undesirable color alteration, typically darkening. There are also products marketed as "color enhancers," which are intentionally formulated to darken colors and accentuate natural veining and character of the stone.

SETTING MATERIALS SELECTION GUIDE

Note

Specifications in this *Handbook* for the materials used to adhere ceramic, glass, and stone tile have been developed according to the requirements and procedures of two standards-setting bodies: the American National Standards Institute (ANSI) and the International Organization for Standardization (ISO). These standards were developed with differing criteria and do not correlate with each other; they cannot be cross-referenced or used interchangeably. Both sets of standards are in widespread and common use in North America. Accordingly, each installation method in this *Handbook* provides setting material specifications according to both sets of standards where applicable. For complete material properties and requirements, refer to the applicable ANSI or ISO standard or "Appendix A."

Introduction

The following are the most widely used materials for setting ceramic, glass, and natural stone tiles. Each possesses specific qualities that make it suitable for installing tile over certain backings or under a given set of conditions.

The conventional portland cement mortar method, including the one-coat method, is the only recognized thick-bed method. All other methods are thin-bed methods and are covered by existing trade jurisdictional decisions of record. Dry-set cement mortars and modified dry-set cement mortars can be used in lieu of neat cement as a 1⁄16"-thick bond coat to bond ceramic, glass, and natural stone tiles to a portland cement mortar bed that is still workable. These mortars can also be used on a cured portland cement mortar bed (minimum 3⁄32" thickness after tile is embedded) according to ANSI A108.1B.

A neat cement bond coat can be used only when the portland cement mortar is still workable (A108.1A). Absorptive ceramic and stone tiles must be soaked before setting on a mortar bed that is still workable when using a neat portland cement bond coat. Under normal job conditions, a minimum of 20 hours cure at 70°F is adequate, but longer mortar bed cures of up to 10 days are desirable. When epoxy mortars, epoxy adhesives, furan, or organic adhesives are used, the mortar bed must be dry.

To ensure practical and satisfactory installations, the cement mortar bed to receive the tile, whether left workable or allowed to harden, is to be applied by the tile contractor who must establish all the finished dimensions at the time this bed is applied.

Noncement setting materials such as epoxies and furans offer properties not possible with cement-based mortar (e.g., chemical resistance, quick setting times). However, special skills on the part of the tile setter may be required.

Neat Cement and Portland Cement Mortar

Portland cement mortar is a mixture of portland cement and sand, roughly in proportions of 1:5 for floors, and of portland cement, sand, and lime, in proportions of 1:5:1⁄2 to 1:7:1 for walls.

Portland cement mortar is suitable for most surfaces and ordinary types of installation. A mortar bed, up to 2" in thickness, facilitates accurate slopes or planes in the finished tilework on floors and walls.

The mortar bed can be modified with the inclusion of a latex/redispersible polymer per the manufacturer's directions as part or all of the liquid portion of the mixture to enhance certain performance properties.

There are two equivalent methods recognized for installing ceramic, glass, and natural stone tiles with a portland cement mortar bed on walls, ceilings, and floors. They are: the method covered by ANSI A108.1A, which requires that the tile be set on a mortar bed that is still workable; and the method covered by ANSI A108.1B, which requires that the tile be set on a cured mortar bed with dry-set or modified dry-set cement mortar. Absorptive ceramic and stone tiles must be soaked before setting on a mortar bed that is still workable when using a neat portland cement bond coat. Portland cement mortars can be bonded to concrete floors, backed with membranes and reinforced with wire mesh or metal lath, or applied on metal lath over open studding on walls. They are structurally strong, are not affected by prolonged contact with water, and can be used to plumb and square surfaces installed by others.

Suitable backings when properly prepared are: brick or cement masonry, concrete, wood or steel stud frame, rough wood floors, plywood floors, foam insulation board, gypsum board, and gypsum plaster. The one coat method may be used over masonry, plaster, or other solid backing that provides firm anchorage for metal lath.

Complete installation and material specifications are contained in ANSI A108.1A, A108.1B, and A108.1C. These mortars are not characterized by ISO criteria.

Dry-Set Cement Mortar

Dry-set cement mortar is a mixture of portland cement with sand and additives imparting water retentivity that is used as a bond coat for setting tile.

Dry-set cement mortar is suitable for thin-bed installations of ceramic and natural stone tiles over a variety of surfaces. It is used in one layer nominally between 3⁄32" and 1⁄4" after tiles are embedded, has excellent water and impact resistance, is water-cleanable, nonflammable, good for exterior work, and does not require soaking of tile.

Dry-set cement mortar is available as a factory-sanded mortar to which only water need be added. Cured dry-set cement mortar is not affected by prolonged contact with water, but does not form a water barrier. It is not intended to be used in truing or leveling the work of others.

Suitable backings, when properly prepared, include plumb and true masonry, concrete, gypsum board, cement backer board, fiber-cement backer board, cementitious coated foam backer board, cured portland cement mortar beds, brick, ceramic tile, and dimension stone. Consult membrane and setting material manufacturer for use over uncoupling membranes, waterproof membranes, and crack isolation membranes.

Complete installation and material specifications are contained in ANSI A108.5 and ANSI A118.1. For applicable ISO material specifications, see ISO C criteria.

Modified Dry-Set Cement Mortar

Modified dry-set cement mortar is a mixture of portland cement, sand, and special additives that is used as a bond coat for setting tile.

The uses of modified dry-set cement mortar are similar to those of dry-set mortar. Additives for use in thin-bed dry-set cement mortars are designed to improve adhesion, reduce water absorption, and provide greater bond strength and resistance to shock and impact. These additives allow some latitude in time, working conditions, and temperatures. Modified dry-set cement mortar is required for the installation of porcelain tile.

When modified dry-set cement mortar is used to install ceramic, glass, and natural stone tiles in an area that may not thoroughly dry out in use (e.g., swimming pools and gang showers, etc.) or where initial drying is inhibited (between tile and impervious substrates), it is recommended that the completed installation be allowed to dry out thoroughly before exposure to water. This drying period can range from 14 days to more than 60 days depending upon the temperature, humidity and other climatic conditions, and whether the installation is interior or exterior. Consult setting material manufacturer for minimum set times before grouting tile or allowing traffic, water exposure, or submersion.

When installing 8" × 8" or larger impervious tile over a waterproof or crack isolation membrane, or other impervious substrate, longer curing times may be required. Cure times may also be extended with narrow grout joints and when high performance grouts such as those meeting ANSI A118.3, A118.5, and A118.7 are used. When one or more of these conditions exists, delaying grouting will allow better evaporation of excess moisture.

When longer cure times are required on floors, extend the amount of time before allowing traffic on the floor.

A rapid-setting modified dry-set cement mortar may need to be specified for faster curing. Consult manufacturer for recommendations and requirements. Because polymers vary considerably, the directions of the modified dry-set cement mortar manufacturer must be followed explicitly.

Not all modified dry-set cement mortars are suitable for wet areas. Consult manufacturer for suitability for intermittent and submerged applications.

Complete installation and material specifications are contained in, respectively, ANSI A108.5 and A108.12 installation standards and ANSI A118.4, A118.11, and A118.15 material standards. For applicable ISO material specifications, see ISO C criteria.

Exterior Glue Plywood (EGP) Modified Dry-Set Cement Mortar

EGP mortar is a modified dry-set cement mortar specifically for bonding ceramic, glass, and natural stone tiles to exterior glue plywood. When added in latex form, the polymer is added as a replacement for part or all of the gauging water in accordance with the manufacturer's instructions.

Complete installation and material specifications are contained in ANSI A108.12 and ANSI A118.11. For applicable ISO material specifications, see ISO C, P criteria.

Dry Set Mortar for Large and Heavy Tile (LHT Mortar), formerly *Medium Bed Mortar*

LHT mortar (formerly medium bed mortar) is a thin-set bonding mortar for ceramic and stone tile, formulated by the manufacturer so as to minimize slump and facilitate a thicker bond coat as compared with a bonding mortar that is not labeled as dry set mortar for large and heavy tile, or LHT mortar. It is intended to be used as a bond coat $\frac{3}{32}$" thick (nominal) to $\frac{1}{2}$" thick (nominal) after the tile is embedded.

Complete installation and material specifications are contained in, respectively, ANSI A108.5 installation standards and ANSI A118.1, A118.4, and A118.15 material standards. Products in compliance with these material standards may be identified with an "H" designation (A118.1H, A118.4H, and A118.15H). There are no ISO standards specific to this type of mortar.

The characteristics of LHT mortar make it useful for setting heavy tiles (generally, tiles that are 5 pounds/square foot or heavier) and tiles with ungauged thickness. It is also useful and commonly used for setting large format tiles (tiles with at least one side greater than 15" long) because, when setting such tiles, larger trowels are used and needed to apply enough bonding mortar to achieve mortar coverage requirements. For large format tiles, a thicker bond coat is often required to achieve mortar coverage requirements if there is warpage in the tile, which necessitates additional bonding mortar to eliminate voids under the tile where the tile's curvature creates a larger space between tile and substrate, typically in the center of the tile. Refer to ANSI A137.1 for allowable warpage for ceramic tile.

NOTE TO SPECIFIER: LHT mortar is not intended for truing or leveling substrates or the work of others. Where substrate variation exceeds industry tolerances, LHT mortar cannot be used to remedy such, because the application would exceed the limitations of the mortar. LHT mortar is intended to be used to install tile per ANSI A108.5, the installation standard for installing tile by the thin-bed method. Accordingly, LHT mortar is a product, not an installation method. Project plans and specifications that call for or refer to setting tile by a "medium bed method" or "large and heavy method" or that call for the use of bonding mortar to level, flatten, or fill substrates or to create slopes or transitions between finish floor heights do not conform to tile industry standards or norms.

Epoxy Mortar

Epoxy mortar is a mortar system designed for chemical resistance employing epoxy resin and epoxy hardener portions. Epoxy mortar is suitable for thin-bed installations of ceramic and natural stone tiles where chemical resistance of floors, high bond strength, and high impact resistance are important considerations. High-temperature-resistant formulas are also available. Acceptable substrates, when properly prepared, include concrete, wood and plywood, steel plate, ceramic tile, and stone tile. Application is made in one thin layer. Pot life, adhesion, water-cleanability before cure, and chemical resistance vary with manufacturer. Epoxy grout is also available. See "Grout Selection Guide."

Complete installation and material specifications are contained in ANSI A108.6 and ANSI A118.3. For applicable ISO material specifications, see ISO R criteria.

Modified Epoxy Emulsion Mortar

Modified epoxy emulsion mortar is a mortar system employing emulsified epoxy resin and hardener with portland cement and silica sand. Modified epoxy emulsion mortars are formulated for thin-bed installations of ceramic and natural stone tiles on floors and walls, interior and exterior. Their features include high bond strength, ease of application, little or no shrinkage, and economical epoxy application. They are not designed for chemical resistance.

Recommended uses include residential floors over substrates such as cementitious backer units and concrete. Where complete and firm support under the tiles is mandatory, 95% coverage is required. This material is recommended by most manufacturers as a bond coat or setting material. Some also recommend it for grouting.

Complete installation and material specifications are contained in ANSI A108.9 and ANSI A118.8. For applicable ISO material specifications, see ISO R criteria.

Furan Resin Mortar

Furan resin mortar is a mortar system designed for chemical resistance consisting of furan resin and furan hardener portions. Furan mortar is suitable for thin-bed installations of ceramic tile where chemical resistance of floors is an important consideration. Acceptable subfloors, when properly prepared, include concrete, wood and plywood, steel plate, and ceramic tile. They are typically not used in the installation of glass or stone tiles. Furan grout is also available. See "Grout Selection Guide."

Complete installation and material specifications are contained in ANSI A108.8 and ANSI A118.5. For applicable ISO material specifications, see ISO R criteria.

Epoxy Adhesive

Epoxy adhesive is an adhesive system employing epoxy resin and epoxy hardener portions. Epoxy adhesive is formulated for thin-bed installations of ceramic and stone tile on floors, walls, and counters. It is designed primarily for high bond strength and ease of application and not for optimum chemical resistance. However, its chemical and solvent resistance tends to be better than that of organic adhesives.

Complete installation and material specifications are contained in ANSI A108.4 and ANSI A118.3. For applicable ISO material specifications, see ISO R criteria.

Spot-Bonding Epoxy

Spot-bonding epoxy is a multi-component high-strength epoxy adhesive designed for spot-bonding ceramic tile and stone.

Refer to ANSI A118.3. For applicable ISO material specifications, see ISO R criteria.

Organic Adhesive

Organic adhesive is a prepared organic material for interior use only, ready to use with no further addition of liquid or powder, which cures or sets by evaporation. Organic adhesives are suitable for setting ceramic tile on floors, walls, and countertops, where surfaces are appropriate and properly prepared in accordance with adhesive manufacturer's directions. Adhesives are applied in one thin layer with a trowel, first using the flat edge for continuous coverage and then the notched edge for uniform thickness. Where leveling or truing is required, an underlayment is used.

Adhesives are not suitable for swimming pools, exteriors, or areas exposed to temperatures exceeding 140°F. They supply some flexibility to the tile facing. Bond strength varies greatly among the numerous brands available. Solvents in some adhesives are irritating to some persons, and some adhesives are flammable.

Complete installation and material specifications are contained in ANSI A108.4 and ANSI A136.1. For applicable ISO material specifications, see ISO D criteria.

GROUT SELECTION GUIDE

Note

Specifications in this *Handbook* for the materials used to grout ceramic, glass, and stone tile have been developed according to the requirements and procedures of two standards-setting bodies: The American National Standards Institute (ANSI) and the International Organization for Standardization (ISO). These standards were developed with differing criteria and do not correlate with each other; they cannot be cross-referenced or used interchangeably. Both sets of standards are in widespread and common use in North America. Accordingly, each installation method in this *Handbook* provides grouting material specifications according to both sets of standards where applicable. For complete material properties and requirements, refer to the applicable ANSI or ISO standard or "Appendix A."

Introduction

Grouting materials for ceramic, glass, and stone tiles are available in many forms to meet the requirements of the different kinds of tile and types of exposures. Portland cement is the base for most grouts and is modified to provide specific qualities such as whiteness, mildew resistance, uniformity, hardness, flexibility, and water retentivity. Non-cement-based grouts such as epoxies, furans, and premixed polymer resin grouts offer properties not possible with cement grouts. However, special skills and precautions on the part of the tile setter may be required.

Sand-Portland Cement Grout

An on-the-job mixture of 1 part portland cement to 1 part fine, graded, clean sand (ASTM C144) is used for joints up to ⅛" in width; 1:2 for joints up to ½" in width; and 1:3 for joints wider than ½." Up to ⅕ part lime may be added.

Sand-portland cement grout can be used with ceramic mosaic tiles, quarry tiles, pressed floor tiles, porcelain tiles, some glass tiles, and some stone tiles on floors and walls. Damp curing is necessary. On stone tiles and glass tiles, make sure the surface will not be scratched by the sand in the grout; unsanded grout may be required. Unsanded grout is commonly used with polished calcium carbonate-based stones (see "Natural Stone Tile Selection and Installation Guide"). Some glass tile manufacturers require unsanded grout to prevent scratching on smooth glass tile surfaces (see "Glass Tile Selection and Installation Guide").

Complete installation and material specifications are contained in ANSI A108.10. This type of grout is not characterized by ISO criteria.

Standard Cement Grout

Standard Unsanded Cement Grout: A factory-prepared mixture of cement, fine aggregate, and other ingredients to produce a water-resistant, dense, uniformly-colored material meant for joints 1⁄16" to ⅛" wide (nominal).

Standard Sanded Cement Grout: A factory-prepared mixture of cement, graded sand, and other ingredients to produce a water-resistant, dense, uniformly-colored material meant for joints ⅛" in width or greater.

Grouts in this category are suitable for grouting walls and floors subject to ordinary use. On stone and glass tiles, make sure the surface will not be scratched by sand in the grout; unsanded grout may be required. Unsanded grout is commonly used with polished calcium carbonate-based stones (see "Natural Stone Tile Selection and Installation Guide.") Some glass tile manufacturers require unsanded grout to prevent scratching on smooth glass tile surfaces (see "Glass Tile Selection and Installation Guide").

Complete installation and material specifications are contained in ANSI A108.10 and A118.6. For applicable ISO material specifications, see ISO CG criteria.

High Performance Tile Grout

High Performance Unsanded Tile Grout: A factory-prepared mixture of cement and other ingredients, including a redispersible latex/polymer powder, to which only water is added at the jobsite, or a liquid latex admixture. When added in a latex form, it is added as a replacement for part or all of the mixing water. These grouts are designed for installation in joints 1⁄16" to ⅛" wide (nominal).

High Performance Sanded Tile Grout: A factory-prepared mixture of cement, sand, and other ingredients, including a redispersible latex/polymer powder, to which only water is added at the jobsite, or a liquid latex admixture. When added in a latex form it is added as a replacement for part or all of the mixing water. These grouts are designed for installation in joints ⅛" in width or greater. The maximum allowable joint width is designated by the grout manufacturer.

Grouts in this category provide improved characteristics such as increased bond strengths, flexural strengths, and lower water absorption to resist frost damage. On stone and glass tiles, make sure the surface will not be scratched by sand in the grout; unsanded grout may be required. Unsanded grout is commonly used with polished calcium carbonate-based stones (see "Natural Stone Tile Selection and Installation Guide"). Some glass tile manufacturers require unsanded grout to prevent scratching on smooth glass tile surfaces (see "Glass Tile Selection and Installation Guide").

Complete installation and material specifications are contained in ANSI A108.10 and A118.7. For applicable ISO material specifications, see ISO CG criteria.

Grout Color Selection

Selection of a grout color that contrasts with the tile will emphasize the grout joint and correspondingly any variations in tile size and position. Selection of a similar or complementary grout color will de-emphasize such variations.

When a similar color grout is used, it is often more difficult to see the haze left after the initial grout cleaning and extra care must be taken to remove all haze.

Stain Resistance and Color Consistency of Cementitious Grouts

Color consistency and stain resistance vary depending on several factors: water absorption of the tile (with lower absorption generally producing a lighter shade), grout color chosen, whether polymer or water is used, amount of water used in mixing and cleaning, degree of curing before cleaning, grout compaction, joint depth, joint finishing, humidity, and many other parameters. In-field color can vary significantly from factory-prepared color samples, and many grouts are darker when wet.

While there is currently no industry standard to evaluate color consistency or stain resistance of cementitious grouts, there are grout formulations and/or admixtures available that allow for improved color consistency. Similarly, there are grout formulations, and/or admixtures, and/or sealers that allow for improved stain resistance by reducing the extent or severity of staining. Not all stains are affected similarly; consult grout manufacturers for application suitability and maintenance recommendations.

Regardless of the grout used and whether or not a sealer was applied, tile and grout can generally be cleaned by sweeping first and rinsing with water. In cases where excess dirt cannot be removed, use an alkaline or pH-neutral cleaner at the lowest effective concentration and rinse thoroughly to eliminate residue. Do not use oil-based cleaners, which are difficult to remove and entrap dirt.

Epoxy Grout

Epoxy grout is a grout system employing epoxy resin and hardener portions, often containing coarse silica filler, especially formulated for industrial and commercial installations where chemical resistance is important. These grouts also provide high bond strength, impact resistance, and improved stain resistance. High-temperature, chemical-resistant formulas are also available. Chemical and stain resistance can vary between epoxies; consult grout manufacturers for application suitability, including exterior UV exposure and maintenance recommendations.

Complete installation and material specifications are contained in ANSI A108.6 and A118.3. For applicable ISO material specifications, see ISO RG criteria.

Epoxy Emulsion Grout

Epoxy emulsion grout is a grout system employing epoxy resin and hardener portions, both of which contain additional water to form an emulsion. The coarse silica filler portion also contains portland cement. These grouts are not designed for chemical resistance nor designed to meet ANSI A108.6 or ANSI A118.3.

Complete installation and material specifications are contained in ANSI A108.9 and A118.8. For applicable ISO material specifications, see ISO RG criteria.

Furan Resin Grout

Furan resin grout is a grout system consisting of furan resin and hardener portions. Furan grout is used in industrial and commercial areas requiring chemical resistance. Use of this grout requires waxed tiles, special installation skills, and additional safety precautions when compared to all other grouts. Architects should select the type of furan grout applicable to the specified chemical and temperature exposure. This grout is not used with stone or glass tiles.

Complete installation and material specifications are contained in ANSI A108.8 and ANSI A118.5. For applicable ISO material specifications, see ISO RG criteria.

Premixed Polymer Resin Grout

Premixed polymer resin grout is a one-part liquid ready-to-use grout that requires no mixing with water. These grouts may contain various types of water-based polymer including acrylics or urethanes and fillers that may be sanded or unsanded. The appropriateness of these grouts for commercial, water-exposed, and exterior applications varies widely; review and follow manufacturer recommendations before use.

There are no product or installation standards for this type of grout.

BACKER BOARD SELECTION GUIDE

Cement Backer Board

A backer board designed for use on floors, walls, and ceilings in wet or dry areas, this board is applied directly to wood or metal wall studs or over wood subfloors—referred to in ANSI as cementitious backer units (CBU). Ceramic tile can be bonded to it with dry-set, latex/polymer modified portland cement mortar, or epoxy by following the backer board manufacturer's instructions. It can also be used for glass and stone installations. Complete interior installation and material specifications are contained in ANSI A108.11 and ANSI A118.9 or ASTM C1325.

Coated Glass Mat Water-Resistant Gypsum Backer Board

A backer board conforming to ASTM C1178. Designed for use on floors, walls, and ceilings in wet or dry areas, this board is applied directly to wood or metal wall studs or over wood subfloors. Ceramic tile can be bonded to it with latex/polymer modified portland cement mortar or epoxy by following the backer board manufacturer's instructions. It can also be used for glass and stone installations.

Glass Mat Water-Resistant Gypsum Backer Board

A backer board conforming to ASTM C1658. Designed for use on walls and ceilings in dry areas, this board is applied directly to wood or metal studs and ceiling framing. Ceramic tile can be bonded to it with dry-set, latex/polymer modified portland cement mortar, organic adhesive, or epoxy by following the manufacturer's instructions. It can also be used for glass and stone installations.

Fiber-Cement Backer Board

A dispersed fiber-reinforced cement backer board designed for use on floors, walls, and ceilings in wet or dry areas, this board is applied directly to wood or metal wall studs or over wood subfloors. Ceramic tile can be bonded to it with latex/polymer modified portland cement mortar, organic adhesive, or epoxy by following the backer board manufacturer's instructions. It can also be used for glass and stone installations. General interior installation and material specifications are contained in ANSI A108.11 and ASTM C1288. Consult the manufacturer's written literature for specific application details.

Fiber-Reinforced Water-Resistant Gypsum Backer Board

A backer board conforming to ASTM C1278. Designed for use on floors, walls, and ceilings in dry or wet areas, this board is applied directly to wood or metal wall studs or over wood subfloors. Ceramic tile can be bonded to it with latex/polymer modified portland cement mortar, organic adhesive, or epoxy by following the backer board manufacturer's instructions. It can also be used for glass and stone installations.

Cementitious Coated Extruded Foam Backer Board

A waterproof backer board constructed from extruded polystyrene and coated with a cementitious coating, designed as a substrate for ceramic tile and stone in wet and dry areas. Designed for use on floors, walls, and ceilings in dry or wet areas, this board is applied directly to wood or metal wall studs or over wood subfloors. Ceramic tile can be bonded to it with dry-set, latex/polymer modified portland cement mortar, or epoxy by following the backer board manufacturer's instructions. It can also be used for glass and stone installations. Material specifications are contained in ASTM C578 and ASTM D4068.

Direct Bond to Wood or Gypsum Wall Board—Caution

Wood-based panels such as particle board, composite panels (veneer faces bonded to reconstituted wood cores), nonveneer panels (wafer board, oriented strand board (OSB), and other similar boards), lauan plywood, and softwood plywood expand and contract with changes in moisture content and are not recommended as backing materials for direct bonding of ceramic tile. Plywood manufactured with fully waterproof adhesive and with an exposure durability rating of Exposure 1 or Exterior may be used on residential horizontal surfaces when installed in accordance with ANSI specifications for the installation of ceramic tile (A108.01).

Gypsum wall board, including water-resistant gypsum backing board, may not be used as a backing for direct application of tile in intermittent wet or wet areas.

MEMBRANE SELECTION GUIDE

Membrane Types

Cleavage Membranes

A cleavage membrane is a thin layer of material within a tile assembly that is loose laid (floating) or mechanically attached but not bonded. Cleavage membranes are incorporated below the mortar setting bed in a thick-bed tile installation when the backing or substrate surface can be damaged by water, is not continuous, is cracked, or is dimensionally unstable. The cleavage membrane separates the backing surface from the mortar setting bed and tile allowing the mortar setting bed to be unbonded and free floating, and thereby not subject to the instability of the backing surface, preventing reflective cracking. When a cleavage membrane is incorporated, the setting bed is required to be reinforced with lath or wire.

Cleavage membranes are always materials that are moisture resistant. Cleavage membranes do not necessarily form an impermeable membrane that will hold water, but are intended to provide other materials some protection from moisture and vapor. Typical cleavage membranes include roofing felt, reinforced asphalt paper, asphalt laminated paper, polyethylene sheeting, chlorinated polyethylene (CPE) sheeting, polyvinyl chloride (PVC) membrane, or high-solids cold-liquid-applied membrane. ANSI A108.02-3.8 provides the specific requirements for these various materials. Some of these same materials can be used as a vapor retarder membrane. See Vapor Retarder Membranes.

Vapor Retarder Membranes

No material can completely eliminate water vapor transmission. However, the transmission of vapor can be reduced through the use of a vapor retarder membrane. These membranes are used in wet areas, placed behind mortar bed walls and some types of backer board to retard moisture and vapor transmission into the wall cavity. Examples of vapor retarder membranes include: roofing felt, reinforced asphalt paper, asphalt laminated paper, polyethylene sheeting, chlorinated polyethylene (CPE) sheeting, and polyvinyl chloride (PVC) membrane.

Waterproof Membranes

In addition to built-up membranes, single-ply membranes, and nonmetallic and lead or copper waterproofing, there are also waterproof membranes (ANSI A118.10) available for use with both vertical and horizontal thin-bed and thick-bed installations of tile, which may be installed by tile trades. Among these are single- or multi-component membranes applied in liquid/paste form, which cure into continuous membranes and membranes applied in flexible sheet form. Some of these membranes have integral reinforcing fabrics for tensile strength and minor crack-bridging properties. Others are designed to be used as a combination waterproofing and setting material for the tile.

Depending upon the type of waterproofing membrane, manufacturers may require that tile products be installed on a reinforced portland cement mortar bed applied over the membrane; directly on the membrane with dry-set or latex/polymer modified portland cement mortar; or bonded to the membrane using a troweled application of the waterproofing membrane product.

Low Perm Waterproof Membranes

No material can completely eliminate water vapor transmission. However, the transmission of vapor can be reduced through the use of a low perm waterproof membrane. Low perm waterproof membranes are waterproof membranes (ANSI A118.10) with a water vapor permeance of less than 0.5 perms when tested per ASTM E96 Standard Test Methods for Water Vapor Transmission of Materials, Procedure E (desiccant method at 100°F) and 90% relative humidity. Not all waterproof membranes meeting ANSI A118.10 are low perm waterproof membranes.

Crack Isolation Membranes

Crack isolation membranes (ANSI A118.12) for thin-bed ceramic, glass, and stone installations act to isolate the tile from minor in-plane substrate cracking. Membranes covered by this definition are bonded to a variety of manufacturer-approved substrates covered by ANSI specifications. In some cases, the trowel-applied products can be used as the adhesive for the tile. Other products within the scope of this category are allowed to cure or are applied as sheet goods and are then used as the substrate for the tile.

Uncoupling Membranes

An uncoupling membrane is a plastic membrane system geometrically configured to provide air space between the tile and the substrate to allow independent movement between the two and limit the transfer of stresses. These membranes are not characterized by ANSI or ISO standards. The uncoupling membrane must achieve 50 psi or greater shear bond strength in 7 days per the test method in ANSI A118.12 Section 5.1.3. Consult the manufacturer's written literature for specific application details.

Bonded Sound Reduction Membranes

Bonded sound reduction membranes are intended to reduce floor-to-floor impact insulation class (IIC) sound. The higher the IIC or delta value, the less transmission of impact noise that is allowed by the membrane. Bonded sound reduction membranes must reduce impact noise (increase IIC) by at least 10 IIC points when tested to ASTM E2179, in accordance with ANSI A118.13. The Uniform Building Code requires a minimum IIC of 50 for sound reduction membranes, and local building codes may require even more stringent minimum acceptable IIC values.

Bonded sound reduction membranes may be trowel applied, sheet, or composite membranes that are bonded to a suitable substrate so that tile can be bonded directly to the membrane.

Material specifications are contained in ANSI A118.13.

Considerations When Using Membranes

Membranes installed properly can produce hollow-sounding tiles. This is not indicative of loss of bond without concomitant installation issues.

Where membranes are used, mortar cure times may be extended. See "Setting Material Selection Guide."

Membranes and/or membrane adhesives and primers may be sensitive to naturally occurring moisture and alkalinity when used over cement and gypsum-based substrates. Consult manufacturer for acceptable limits of moisture and alkalinity.

ADDITIONAL PRODUCTS USED IN TILE INSTALLATIONS

Exterior Ceramic Tile Panels

To reduce construction time, these prefabricated, lightweight, custom-built panels are shipped to the construction site and attached to the building by welding or mechanical fasteners.

Panel size, shape, and thickness are determined by building design. Size, spacing, and gauge of the steel stud framework are also dependent on design but may be related to the climatic conditions of the building location.

Manufacturers of tile, mortars, and backing materials and regional contractor associations can supply guide specifications for panel construction. However, the ceramic tile installation is usually done in accordance with W201 (from the membrane out) or W244E (bonding directly to solid backing attached to the studs). Tile to be installed per ANSI specifications.

Integrated Bonding Flange

An integrated bonding flange is designed to provide a large contact area at the top of the drain assembly, which will allow ample surface adhesion to thin, load bearing, bondable waterproof membranes. Drain assemblies of this type are constructed in such a way that the waterproof membrane is bonded to the top of the substrate rather than below it. All drains must comply with local code.

Profiles

Floor profiles, wall profiles, transition profiles, cove profiles, and preformed joint profiles may be incorporated into a tile installation where suitable. Profiles are available in various materials, finishes, colors, and heights. Select the correct material considering traffic and environmental conditions (water exposure, chemical exposure, etc.) and the correct height considering the tile thickness and setting bed thickness.

These products are not characterized by ASTM, ANSI, or ISO product standards. Consult the manufacturer for all performance and installation criteria.

Gauged Porcelain Tiles and Gauged Porcelain Tile Panels/Slabs

Gauged porcelain tiles (GPT) and gauged porcelain tile panels/slabs (GPTPS) are tiles that have been manufactured to a specific nominal thickness, and for which its thickness characteristic is largely associated with how it is installed and/or used, (see ANSI A137.3). Products are categorized based on a nominal thickness that is typically expressed by the manufacturer in millimeters. Characteristics and criteria for products can be found in Tables 4 and 5 of ANSI A137.3.

Depending on the thickness, while typically meeting ISO 13006 modulus of rupture requirements (one measurement of strength), many of these thinner tiles do not meet the ANSI A137.1 breaking strength requirements and require handling and installation that takes the lower breaking strength into consideration. Not all manufacturers recommend their tiles for all substrates. Check with the manufacturer for recommended applications, and whether flooring applications are supported.

Some GPT and GPTPS employ reinforcement on the back changing the physical properties of the tiles, adding impact resistance, raising the breaking strength (although in this category, the breaking strength after reinforcement can still be below the ANSI A137.1 threshold), and reducing crack propagation. Known as back-layered GPT and GPTPS, these products incorporate a material—commonly mesh—that has been attached to the back of the product after firing. There are a variety of technologies employed for back-layering and such reinforcement requires additional consideration when selecting appropriate setting materials.

In general, specialized tools, equipment, thin-bed mortar, and training are required for the successful installation of GPT and GPTPS. With larger tiles, flattening the substrate before installation may be required as stringent substrate and installation requirements apply, especially in flooring applications. Special care may be required to achieve sufficient mortar contact between the tile and substrate, especially near the grout joints.

Complete installation and material specifications are contained in, respectively, ANSI A108.19 and A108.20 installation standards and ANSI A137.3 material standards. Consult manufacturer for all substrate, performance and installation criteria. ANSI A108.19 and A108.20 require the use of pre-qualified installers.

Structural Ribbed Self-Supporting Boards (SRSB)

Structural ribbed self-supporting boards (SRSB) are a pultruded, lightweight board manufactured of a composite dovetailed rib system, integrated and adhered to a fiberglass backing. The boards are engineered to be structural and are attached directly to the wood or steel joist system of an exterior deck or balcony. By eliminating the wood deck boards or a plywood substrate, the deck and the tiled surface are no longer subject to wood swelling and shrinking due to humidity and moisture. The dovetailed cavities are filled with mortar to establish the tile substrate. A waterproof membrane conforming to ANSI A118.10 is required for all installations subject to freeze/thaw conditions.

These products are not characterized by ASTM, ANSI, or ISO product standards. Consult manufacturer for all performance and installation criteria.

Thresholds

By acting as a transitional piece between two different finished floor levels, thresholds permit the use of the conventional mortar method in rooms where it would not otherwise be possible. They also can be used with thin-bed methods.

Ventilated Rainscreen Exterior Wall Systems

Tiles up to 1¼" thick, up to 4' × 4' (and larger) in size, have the strength and characteristics necessary to be used as mechanically anchored exterior wall cladding, similar to traditional mechanically anchored natural stone panels. As an engineered product, tile can be manufactured to have consistent physical characteristics with many color, texture, and surface finish choices. These tiles are well suited for ventilated, back-drained rainscreen wall systems, a relatively new type of exterior wall cladding with functional, environmental, and aesthetic benefits.

These systems, often referred to as ventilated tile façades, consist of tiles attached to a structural aluminum framework attached to a primary back-up wall, creating an open air cavity between the primary wall and the tiles. The tiles form a secondary wall cladding, and the joints between the tiles are left open to induce air circulation within the cavity, allowing for continuous insulation and efficient heat, moisture, and acoustic dissipation. The open cavity creates convection in the insulating air chamber between the tile cladding and the primary wall, effectively insulating the primary wall. The ventilated cavity also minimizes thermal bridging, moisture condensation, and the inefficiencies of traditional internal wall insulation. The backs of the tiles can be reinforced with mesh to reduce accidental breakage, and tiles can also be easily removed and replaced, facilitating access to and maintenance of internal wall components.

Ventilated rainscreen tile wall systems can reduce building energy consumption substantially, reduce damaging water infiltration and condensation, and reduce material and installation costs by use of prefabricated high strength, lightweight components.

GREEN BUILDING STANDARDS AND GREEN PRODUCT SELECTION GUIDE

Environmentally sustainable buildings have many different attributes, and there are many options to consider when "building green". While most green building projects target certification or compliance with green building standards or rating systems, others are initiated for tax incentives, operational cost savings, or solely in the spirit of following environmentally responsible building practices.

Product selection is a major component in green building. Products can impact the environment in different ways, and it is important to understand the variety of contributions by all products. The sustainability of a product involves much more than recycled material content, energy efficiency, or any other single attribute. Multi-attribute sustainability—including environmental and social impacts, and whether the environmental information is transparently reported—should be considered when evaluating a product's true sustainability. Additionally, how products combine into installed product systems is important. For green building projects, use the installation guidelines in this *Handbook* with products meeting the Green Squared® standard to achieve durable, long-lasting assemblies that are truly sustainable. Publicly available North American industry-wide EPDs (environmental product declarations) for tile, mortar, and grout can also be reviewed to obtain in-depth data on the environmental impacts of tile assemblies throughout their life cycle.

Specifying Conforming Sustainable Tile Products

Green Squared, the Multi-Attribute Product Sustainability Standard: The tile industry considered a host of sustainability criteria in creating the wide-ranging, multi-attribute product sustainability standard, ANSI A138.1. This standard establishes requisite and elective criteria for sustainable tiles and installation materials, including social and environmental criteria, and is the first multi-attribute product sustainability standard to encompass a full range of products within an industry. These criteria establish a consistent approach to the evaluation and determination of product sustainability and are in accordance with North American green building practices, expectations, and leading initiatives. Tiles and installation materials in conformance with ANSI A138.1 should be specified for green building projects.

Green Squared Certification: TCNA developed the Green Squared certification program as a way for specifiers and consumers to easily determine which products meet ANSI A138.1. Tile and installation materials labeled with the Green Squared Certified® mark meet ANSI A138.1, as certified by one of two credible and widely recognized certification bodies. The use of Green Squared Certified products helps earn and/or directly contributes toward points in several green building codes, standards and rating systems, including the National Green Building Standard (ICC 700), Leadership in Energy and Environmental Design (LEED), the International Green Construction Code IgCC powered by ANSI/ASHRAE/ICC/USGBC/IES Standard 189.1 (IgCC/Standard 189.1), Green Globes for New Construction, and the General Services Administration's Facilities Standards (GSA P-100). Additionally, Green Squared is included in the *US EPA Recommendations of Specifications, Standards, and Ecolabels,* which aids federal purchasers in identifying and procuring environmentally sustainable products and services. The EPA recommendations provide guidelines and specification targets to tens of thousands of specifiers and contractors nationwide..

For additional information on Green Squared, see "Tile: The Natural Choice" published by TCNA and visit www.GreenSquaredCertified.com.

Using Publicly Available Data to Understand the Environmental Impacts of Products Being Specified

An EPD (environmental product declaration) provides a comprehensive overview of how a product impacts the environment—specifically with regard to global warming, abiotic resource depletion, acidification, smog formation, eutrophication, and ozone depletion. The primary intent of an EPD is transparency and, while developed within a standardized reporting framework, the EPD itself does not indicate conformance to any particular environmental performance threshold(s). Just as nutrition labels inform with respect to food choices, an EPD informs with respect to sustainability.

Tile and related installation materials industry-wide EPDs are publicly available for North American-made ceramic tile, for North American-made cement mortar for tile installation, and for North American-made cement grout for tile installation. Additionally, product-specific (proprietary) EPDs are available.

The combined use of EPDs for tile, mortar, and grout provides specifiers and design professionals data needed to evaluate and understand the environmental footprint of a full assembly. Using products with proprietary or industry-wide EPDs can directly contribute toward points in, or conformance to LEED, Green Globes,

IgCC/Standard 189.1, and GSA P-100. Because most green building standards, codes, and rating systems provide incremental credit for each product used with an EPD, using tile can contribute three times the credit when applicable mortar and grout is also used.

Each of the North American industry-wide EPDs is independently certified by UL Environment and provides 60 or 75-year environmental impacts, based on a full "cradle to grave" LCA (Life Cycle Assessment) per square meter of installed product. Each LCA was conducted based on energy, resource, and emissions data provided by major North American manufacturing companies toward the development of a LCI (life cycle inventory) associated with product manufacturing and transportation. Further, industry-wide information on installation, use, and disposal were considered in the development of a LCIA (life cycle impact analysis) to calculate the full-life environmental impacts of tile, mortar, and grout. The final "cradle to grave" LCA, comprised of the LCI and LCIA, was critically reviewed and approved by a stakeholder panel comprised of environmental experts prior to being certified. For additional information see "Tile: The Natural Choice" published by TCNA and visit www.TCNAtile.com to download copies of the EPDs.

Additional Green Tile and Installation Material Considerations

Recycled Content: Tile and installation material manufacturers offer wide varieties of products with pre- and post-consumer recycled content. Recycled content in tile products can contribute to overall building recycled content and can help achieve compliance with the recycled content criteria in LEED, ICC 700, Green Globes, and IgCC/Standard 189.1. Consult with manufacturers to ensure that materials labeled as having recycled content meet project-specific requirements. Often, green building projects require that recycled content fits the following ISO and FTC definitions:

- **Post-consumer material:** Waste material generated by households or by commercial, industrial, and institutional facilities in their role as end-users of the product, which can no longer be used for its intended purpose.
- **Pre-consumer material:** Material diverted from the waste stream during the manufacturing process. Reutilization of materials is excluded (i.e., rework, regrind, or scrap generated in a process and capable of being reclaimed within the same process that generated it).

Additionally, high levels of responsibly-recovered waste, including dust, powder, unfired scrap and water, are commonly reincorporated into the manufacturing process. Such materials, when reincorporated into the same processes that created them, are not typically considered pre-consumer recycled content by ISO and FTC definitions. However, waste reclamation is a vital component to minimizing waste and maximizing resources, and is becoming more widely acknowledged in green building. In fact, many factories are so efficient with waste reclamation that they are effectively closed-loop facilities, reducing waste to zero and fully utilizing all inputs.

Regional Availability: Tile products manufactured with indigenous raw materials and within close proximity of building sites can help reduce overall energy consumption and air pollution. Manufacturers can provide the necessary paperwork regarding production locations, quantities of indigenous raw materials, and modes of product/raw material transportation in order for their products to contribute toward points in, or conformance to LEED, ICC 700, Green Globes, and IgCC/Standard 189.1 projects.

Indoor Air Quality: Building materials with low or no volatile organic compound emissions (VOCs) are necessary for good indoor air quality. Ceramic and porcelain tiles have zero VOCs (because they are manufactured at high temperatures where organic matter in raw materials is completely burned-off) and inherently meet the product emission requirements established by the California Department of Public Health (commonly referred to as California Section 01350). Indication of such compliance for ceramic tile is not required for LEED, ICC 700, Green Globes, or IgCC/Standard 189.1 projects, as tile is exempted from VOC emission testing. Low VOC tile adhesives and other installation materials in compliance with California public health requirements or SCAQMD Rule 1168 can also contribute toward points in, or conformance to LEED, ICC 700, Green Globes, and IgCC/Standard 189.1 projects.

Exterior Contribution: Light-colored tiles used for site hardscaping (as opposed to traditional paving materials) can lower a site's heat absorption or heat island effect. Using tile this way can contribute toward compliance in LEED, ICC 700, Green Globes, and IgCC/Standard 189.1 projects when tile products have a solar reflective index (SRI) of 29 or greater. Manufacturers should be contacted for more information on such products.

Cleaning and Maintenance: Ceramic tile's easy maintenance generally eliminates the need for harsh cleaning chemicals and the associated environmental concerns with some chemical ingredients. Tile and grout manufacturers should be contacted to provide cleaning and maintenance recommendations for products used in green building projects. When sealers are used, consult with sealer manufacturers to confirm compliance with the VOC criteria in California public health requirements or SCAQMD Rule 1168 for products being utilized in LEED, ICC 700, Green Globes, or IgCC/Standard 189.1 projects.

Cleanliness and Sterility: Ceramic tiles are inherently inhospitable to mold, germs, and bacteria and tile does not harbor dust mites, making tile ideal for people with allergies or asthma. Additionally, some tiles and installation materials have innovative components that inhibit microbial growth.

Durability: The durability of tile products contributes to the inherent sustainability of tile installations, reducing the frequency of replacement and the associated environmental impact. When building green, it is important to use products that meet or exceed the industry durability criteria referenced in the Green Squared standard.

Life Cycle Performance: LCAs conducted in conformance with ISO 14040 and ISO 14044 are the basis for EPDs, which are standardized under ISO 14025 and ISO 21930. Products that have undergone such evaluation in conformance with these standards are eligible for contribution to LEED, ICC 700, Green Globes, and IgCC/Standard 189.1 projects. Furthermore, various green building standardization efforts are bringing more attention to the subject of life cycle evaluation, especially as it pertains to building and material service life and performance.

Energy Reduction: Tile finishes can contribute to the energy efficiency of a building in many ways. By their nature, tile products have exceptional thermal mass that allows them to store and release heat. In the summer, tiled surfaces capture and store heat from interior environments without significantly changing temperature, keeping interiors cool during the hottest parts of the day. Similarly, during the winter, tiled surfaces radiate heat back to an interior environment when the ambient temperature is below the temperature of the tile. Additionally, there are several innovative technologies, such as radiant heating systems, ventilated façades, and photovoltaic products, which can help improve energy efficiency to lower a building's energy footprint, reduce operational costs, and help meet energy reduction targets in LEED, ICC 700, Green Globes, and IgCC/Standard 189.1 projects.

Innovation: Manufacturers of tile and installation materials continue to create new products that lower energy usage, reduce material consumption, improve human health, and decrease environmental burdens. The use of products with innovative environmental technologies can contribute toward innovation credits in LEED if approved by the relevant project authorities.

Common Green Building Standards and Rating Systems

The most common green building standards and rating systems are described in this section. While there are several others—some consensus-based and some written exclusively for certification by various third-party entities—many have product selection criteria similar to these most commonly used green standards and rating systems:

LEED (Leadership in Energy and Environmental Design): LEED is a suite of green building rating systems developed by the U.S. Green Building Council (USGBC) that allows builders, owners, and tenants to evaluate the impact of their design on energy and environmental issues. LEED certification of a project can be earned based on points received for choices made during the design, construction, and operation of a new or existing building or major renovation.

ICC 700: Through the American National Standards Institute (ANSI) consensus process, and based on initiatives of the National Association of Home Builders (NAHB), the International Code Council (ICC) published the National Green Building Standard (ICC 700), a voluntary green building standard dedicated to residential construction. Similar to LEED, points are awarded for choices made during the design, construction, and operational phases of a project.

Green Globes: Green Globes is a rating system and certification program by the Green Building Initiative (GBI) to provide an alternative to LEED for building environmental design and management. New or existing commercial buildings can be submitted for assessment per the standard ANSI/GBI 01 and points are awarded toward four different levels of certification, from "One Globes" to "Four Globes."

IgCC/Standard 189.1: IgCC (International Green Construction Code) was created by the International Code Council (ICC) in partnership with the American Institute of Architects (AIA); ASTM International; the American Society of Heating, Refrigerating, and Air-Conditioning Engineers (ASHRAE); the United States Green Building Council (USGBC); and the Illuminating Engineering Society (IES) as a "model code" providing a regulatory framework for adoption by governing jurisdictions. ASHRAE 189.1 was created as an ANSI-approved standard for designing, building, and operating high-performance green buildings and was initially recognized by IgCC as an alternative compliance path. IgCC and ASHRAE Standard 189.1 are now integrated into a single model code and standard for high-performance buildings—IgCC powered by ANSI/ASHRAE/ICC/USGBC/IES Standard 189.1 requires adoption by a municipality before becoming law for new and existing commercial buildings.

GSA P-100 Facilities Standards: GSA P-100 incorporates sustainability criteria into design standards for the construction of government-funded buildings. Section 3.4 (Interior Performance) sets forth several sustainability provisions for ceramic tile, including Green Squared Certification and EPDs.

2021 | FIELD AND INSTALLATION REQUIREMENTS

SUBSTRATE REQUIREMENTS

Maximum Allowable Deflection for Floor Systems and Substrates

Floor systems, whether wood framed or concrete, over which the tile will be installed using the appropriate TCNA method, according to the "Floor Tiling Installation Guide," shall be in conformance with the International Residential Code (IRC) for residential applications, the International Building Code (IBC) for commercial applications, or applicable building codes. For ceramic tile installations maximum allowable floor member live load and concentrated load deflection for framed floor systems shall not exceed L/360, where "L" is the clear span length of the supporting member per applicable building code. For natural stone tile installations, maximum allowable floor member live load and concentrated load deflection for wood framed floor systems shall not exceed L/720, where "L" is the clear span length of the supporting member, per applicable building code.

The owner should communicate in writing to the project design professional and general contractor the intended uses of the tile installation, including in-service loads or information to allow a project design professional to calculate such. Project design professional and general contractor must make necessary allowances for the expected live load, concentrated loads, impact loads, and dead loads, including maximum allowable loads during construction and maintenance. When concentrated loads such as scissor lifts, pallet jacks, automobiles, forklifts, etc., will be utilized on a tile or stone floor, the project design professional shall include their use in the determination of the appropriate substrate. For the weight of the tile and setting bed (contribution to dead load), see Typical Weight of Tile Installation in the method being specified.

The tile contractor shall not be responsible for problems resulting from any structural subfloor installation not compliant with applicable building codes, unless structural subfloor was designed and installed by tile contractor, nor for problems from overloading. As tile is a finish applied to and relying upon the underlying structure, an inadequate substructure can cause a tile failure. In many cases, problems in the substructure may not be obvious, and the tile contractor cannot be expected to discover such and tile contractor shall not be responsible for designing flooring assembly, unless specifically engaged to do so in writing. Tile contractor cannot determine possibility of an overloaded condition.

In addition to deflection considerations, above-ground installations are inherently more susceptible to vibration. Consult grout, mortar, and membrane manufacturers to determine appropriate installation materials for above-ground installations. Crack isolation or uncoupling membranes and higher quality setting materials can increase the performance capabilities of above-ground applications. However, these upgraded materials cannot mitigate structural deficiencies including floors not meeting code requirements and/or overloading or other abuse of the installation in excess of design parameters.

Natural Stone Tile Installations on Post-Tensioned Concrete

Because dynamic movements of post-tensioned slabs have proven problematic for directly adhered natural stone, F111 is the only method of installing natural stone tile over post-tensioned slabs, on-ground or above-ground.

Proper Spacing for Wood Subflooring and Wood Underlayments

Plywood subflooring, OSB subflooring, and plywood underlayment shall be installed with proper spacing between the sheets (typically ⅛", except if specified otherwise by the wood manufacturer). If the subfloor or underlayment is installed without proper spacing, this condition cannot be corrected by the tile installer. It is the responsibility of the project owner (or owner's subflooring or underlayment installer) to ensure proper spacing is used as failure to do so may not be obvious, and the tile installer cannot be expected to discover such. If expansion takes place in wood subflooring or underlayment installed without proper spacing, the tile installation could fail.

Natural Stone Tile Installations Over Wood Substrates

Two layers of structural wood panels are required on floors to receive stone tile when backer board will be used as the tile substrate. The MIA prohibits installation of stone tile over single-layer wood floor systems under backer board because of the discontinuity of the system at seams between the subfloor panels. If an unbonded mortar bed will be installed as the tile substrate, a single layer of wood subflooring is permitted.

Natural Stone Tile Installation Over Frame Construction

Strongbacks, bridging, or other load-sharing members may be required within a wood framed system to reduce differential deflection between adjacent framing members; project design professional is responsible for determining the necessity of such. Without the use of such load-sharing members, deflection may differ significantly between adjoining members. Accordingly, for framed construction, an L/720 requirement has been specified (see Maximum Allowable Deflection for Floor Systems and Substrates) while an L/360 requirement can be accepted for concrete/masonry substrates.

Moisture-Affected Materials—Caution

The performance of a properly installed ceramic tile installation is dependent upon the durability and dimensional stability of the substrate. Some substrate materials used in wet areas are subject to deterioration from moisture penetration. (Reference ANSI A108.01-2.4.) Therefore, while the goal of the conference for the *TCNA Handbook* is to produce accurate guidelines, they should be used only with the independent approval of technically qualified persons.

Substrate Tolerances

To facilitate a tile installation that will meet ANSI finish flatness requirements or a stone installation that will meet MIA finish flatness requirements (see "Flatness and Lippage"), the installation substrate must meet the following flatness tolerances per ANSI A108.02 and MIA's DSDM.

Substrate Tolerances for Mortar Bed Methods and Self-Leveling Methods

For thick-bed (mortar bed) ceramic and stone tile installations and self-leveling methods: maximum allowable variation in the substrate prior to the installation of the mortar bed or self-leveling material—¼" in 10' from the required plane.

Substrate Tolerances for Thin-Bed Methods

For thin-bed ceramic tile installations when a cementitious bonding material will be used, including mortar for large and heavy tile (formerly medium bed mortar): maximum allowable variation in the tile substrate—for tiles with all edges shorter than 15", maximum allowable variation is ¼" in 10' from the required plane, with no more than 1/16" variation in 12" when measured from the high points in the surface. For tiles with at least one edge 15" in length or longer, maximum allowable variation is ⅛" in 10' from the required plane, with no more than 1/16" variation in 24" when measured from the high points in the surface. For modular substrate units, such as plywood panels or adjacent concrete masonry units, adjacent edges cannot exceed 1/32" difference in height.

For thin-bed ceramic tile installations when an organic adhesive or epoxy adhesive will be used: maximum allowable variation in the tile substrate is 1/16" in 3' from the required plane with no abrupt irregularities greater than 1/32".

Thin-bed stone tile installations: maximum allowable variation in the tile substrate—⅛" in 10' from the required plane.

Project specifications shall include a specific and separate requirement to bring the substrate into compliance if a thin-bed method is specified but substrate does not meet the flatness requirements.

See also: Disparity Between Concrete Flatness Tolerances Based on F-Numbers and the 10-Foot Straightedge Method and Substrate Tolerances and Large Tile.

Substrate Tolerances and Large Tile

As tile size increases, the negative effect of substrate irregularities is compounded. If specifying a thin-bed method, project specifications should include a separate specification and requirement (such as a pourable underlayment) to bring the substrate into compliance if the substrate does not meet the required flatness tolerance. Alternatively, when specifying tile with any edge longer than 15" consider specifying a recessed installation substrate and a mortar bed (thick-set) method to produce a tile substrate that meets the more stringent flatness requirement for large format tiles. There is no "medium bed installation method" that can be used to flatten the substrate while installing the tile, as mortar (including LHT mortar) is not intended for truing or leveling substrates or the work of others.

See also: Disparity Between Concrete Flatness Tolerances Based on F-Numbers and the 10-Foot Straightedge Method and Dry Set Mortar For Large and Heavy Tile (LHT Mortar), Formerly *Medium Bed Mortar.*

Substrate Tolerance When More Stringent Finish Tolerance is Desired

Should the architect/designer require a more stringent finish tolerance (e.g., ⅛" in 10'), the substrate specification must reflect that tolerance, or the tile specification must include a specific and separate requirement to bring the substrate into compliance with the desired tolerance.

Disparity Between Concrete Flatness Tolerances Based on F-Numbers and the 10-Foot Straightedge Method

Division 3 project specifications typically require concrete substrates to comply with floor flatness (F_F) tolerances based on the ASTM E1155 Standard Test Method to Determine Floor Flatness and Floor Levelness Numbers. A floor's F_F value reflects the aggregate of many measurements and is an indication of the overall flatness of the concrete at the time of measurement. A specified overall flatness SOF_F of 20 is considered "conventional," while an SOF_F of 35 is considered "flat." To establish if the concrete substrate meets the specified F_F the ASTM E1155 test is performed.

By contrast, Division 9 project specifications typically incorporate ANSI A108.02 criteria and as such require areas to receive tile to meet flatness tolerances based on measurements taken with a 10-foot straightedge, where the tolerance is measured as a gap between the straightedge and the floor, e.g., ¼" in 10'. Unlike an F_F number, where results are aggregated, straightedge measurements are used individually, and each area exhibiting a gap greater than the allowed tolerance is out of compliance with the required substrate flatness.

Because an F_F value is derived from many measurements, concrete that meets a specified F_F of 35 will typically have areas that do not meet the required flatness tolerance for tile. This is particularly true when larger tiles will be installed because of the tighter tolerance that applies—the maximum allowable variation for any areas where tiles with at least one edge 15" in length or longer will be installed is ⅛" in 10' from the required plane, with no more than 1/16" variation in 24" when measured from the high points in the surface. For tiles with all edges shorter than 15" the maximum allowable variation is ¼" in 10' from the required plane, with no more than 1/16" variation in 12" when measured from the high points in the surface. If these tolerances as specified in ANSI A108.02 are not met, the tile contractor cannot provide an installation that complies with tile industry tolerances for flatness and lippage without additional effort and compensation for first bringing the substrate into compliance.

In addition to the lack of correlation between the two methods for measuring concrete flatness, there are other provisions of the ASTM E1155 method that are problematic in consideration of the flatness requirements for tile. The method requires measurements be taken within 72 hours of concrete placement, before concrete curling and shrinkage resulting from the curing process has culminated. While the concrete may meet the tolerance at the time of measurement, the flatness can change significantly afterward, generally becoming less flat, not more. Additionally, no measurements are taken at construction, isolation, or control joints; at column block-outs; or within 2 feet of the perimeter of the slab; these are typically the areas with the greatest amount of variation due to curling.

Therefore, a concrete substrate that meets specified Division 3 F_F criteria will likely not be suitably flat to receive tile as defined in ANSI A108.02 and Division 9. Accordingly, a separate specification and requirement is often necessary to bring the substrate into tolerance for tile, with the amount of substrate flattening required dependent on how close the concrete is to meeting the A108.02 flatness criteria at the time of tile installation.

Including a separate requirement helps avoid change orders. If such substrate preparation is not separately and specifically required and included in bid proposals, the tile contractor assumes a suitably flat substrate will be provided. Industry standards for tile finish flatness and lippage do not apply if the project owner does not provide a substrate that meets required ANSI A108.02 substrate flatness tolerances and chooses not to correct substrate flatness issues. See also: Substrate Tolerances and Large Tile.

Substrate Flatness Tolerance and Framed Wall Construction

In framed wall construction, where framing supports or straps are needed for mounted fixtures (grab bars, mirrors, towel dispensers, etc.), such must be located between studs. When supports or straps are installed over studs, the thickness of the supports and the fasteners used to attach them will cause flatness or plumb deviations, and studs must be shimmed out by the framing contractor sufficiently to compensate for the thickness of the supports and anchors.

Alternately, project specifications can include a separate requirement for the tile contractor to bring out-of-tolerance walls into tolerance. However such is not included unless clearly and additionally specified and included in bid proposals. Industry requirements for tile finish flatness and lippage do not apply if the project owner does not provide a substrate that meets minimum flatness requirements and chooses not to correct substrate flatness issues. See also: Substrate Tolerances and Large Tile.

Maximum Allowable Deflection for Framed Walls

Walls intended for ceramic tile and stone installations shall meet applicable building code requirements.

"Equivalent Gauge" Steel Framing for Drywall (Commonly Referred to as "EQ" Drywall Studs)

For tile and stone installations on cementitious backer board fastened to steel studs, American National Standard Specifications for the Interior Installation of Cementitious Backer Units (A108.11) call for 20 gauge structural steel studs. Similar criteria are used for gypsum board and other backer units in tile and stone applications. According to steel stud industry standards, 20 gauge structural (load-bearing) steel studs have a minimum base metal thickness of 0.0329" and 20 gauge drywall (nonstructural, non-load-bearing wall) studs have a minimum base metal thickness of 0.0296". Some gypsum board and backer board manufacturers require the use of 20 gauge structural studs for tile and stone installations as specified by ANSI A108.11, while others allow the use of 20 gauge drywall studs.

On interior walls with a tile or stone finish, gypsum board and backer board manufacturers commonly do not support the use of steel studs with less than 0.0296" base metal thickness. This includes the use of "equivalent gauge" steel drywall framing, also known as "EQ" drywall studs, typically marketed and engineered for interior drywall applications where painted drywall is the intended finished surface.

Although referred to as "equivalent," the base metal thickness for "EQ" drywall studs is 0.019" to 0.024" (approximately two-thirds the thickness of 20 gauge structural and 20 gauge drywall studs.) Use of "EQ" drywall studs in tile or stone applications can lead to problems with wall deflection and overall flatness, screw spin-out and stripping of the framing, torsional effects, stud flange bending, and stud warpage.

If considering "EQ" drywall studs for use in a tile or stone installation, work with a project engineer, structural engineer, and/or design professional to ensure that appropriate testing and engineering analysis has been performed to confirm the suitability of the "EQ" drywall studs for the specific application, and confirm with backer board manufacturer. If such an installation is pursued, consider increased stud frequency, additional bracing, lighter weight tiling products, counter balancing, and/or use of special fasteners. Additionally, consider performing a small system mockup prior to board installation to evaluate that the fastening tools are properly adjusted such that fasteners are installed in a manner to avoid fastener spin out and overdriving, yet snug enough to remove gaps between connected parts. See *Specifications for the Application and Finishing of Gypsum Panel Products*, published by the Gypsum Association, for stud thickness criteria for abuse-resistant and impact-resistant gypsum panels, as the inherent density and weight of such products present similar challenges to those of tile and stone walls.

As tile is a finish applied to and relying upon the underlying structure, an inadequate substructure can cause a tile failure. In many cases, problems in the substructure may not be obvious, and the tile contractor cannot be expected to discover such. Accordingly, the installing tile contractor shall not be responsible for problems resulting from any structural installation not compliant with applicable building codes or not sufficient for the intended use or application.

LIGHTING AND TILE INSTALLATIONS

Use of wall-washer and cove-type lighting, where the lights are located either at the wall/ceiling interface or mounted directly on the wall, are popular techniques for producing dramatic room lighting effects. However, when proper backing surfaces, installation materials and methods, and location of light sources are not carefully coordinated these lighting techniques produce shadows and undesirable effects with ceramic tiles, particularly when light strikes the tile surface at a low or nearly flat angle (i.e., nearly parallel to the tile surface.) Similar shadows can be created from natural light from windows and doors side-lighting interior walls and floors, and when light shines at a low angle on exterior walls and floors, as commonly occurs in the early morning and evening.

Because such low angle lighting highlights and exaggerates normal and acceptable inconsistencies in tile and tilework, the shadow-producing light is often referred to as "critical light" or "critical lighting." Some of the allowable and acceptable characteristics of tile and tilework that can be highlighted or exaggerated by critical lighting include:

- Die release lines on the edges of tiles
- Difference in appearance between a factory edge of a tile versus an edge that has been field-cut by scoring, grinding, or wet cutting
- Allowable warpage in the tile (see ANSI A137.1 for definition and allowances)
- Allowable lippage (see ANSI A108.02 for definition and allowances)
- Allowable variation in height between field tile and trim pieces and/or accent tiles, whether such are or are not part of the same tile line

In addition, critical lighting can worsen the appearance of tilework that does not meet the finish flatness and lippage tolerances set forth in ANSI A108.02, which is considered acceptable tile workmanship under certain specified conditions:

- Tiled floors sloping to drains (see ANSI A108.02)
- When the project owner does not provide a substrate that meets required flatness tolerances for tile nor contract the tile contractor to correct substrate flatness defects (see ANSI A108.01 and Disparity Between Concrete Flatness Tolerances Based on F-Numbers and the 10-Foot Straightedge Method in this *Handbook*)

To minimize the undesirable effects of critical lighting:

- Place lighting such that it will decrease or eliminate the undesirable effects of critical lighting, and install permanent lights prior to tile installation to provide the installer the maximum opportunity to reduce undesirable shadows.
- Reduce the amount of inherent or allowable lippage by ensuring substrates meet required flatness tolerances for ceramic tile, for example by: specifying a mortar bed, pourable underlayment, or other tile substrate that facilitates a flat tile installation; including adequate allowance in the tile specification for substrate preparation; and specifying the trade responsible for required alterations to a substrate that does not meet flatness tolerances for tile.
- Reduce the amount of inherent or allowable lippage by specifying tile with minimal warpage, such as rectified tile per ANSI A137.1.
- Minimize the effect of lippage due to warpage by specifying wider grout joints, cushioned or beveled edge tile, and tile sizes and patterns that minimize lippage due to warpage, for example avoid offset or brick-joint patterns or specify a maximum 25% or 33% offset (see ANSI A108.02).

MORTAR APPLICATION AND COVERAGE

Mortar Coverage for Ceramic Tile

Average contact area for dry areas is 80% and for wet areas is 95%. Mortar coverage is to be evenly distributed to support edges and corners.

It is not possible or practical to achieve 100% coverage consistently and such should not be specified.

Mortar Coverage for Natural Stone Tile

Mortar coverage must be sufficient to prevent cracks in the stone resulting from voids in the setting bed. In dry and wet areas, the minimum coverage is 95% with no voids exceeding 2 square inches and no voids within 2" of tile corners. All corners and edges of the stone tiles must be fully supported, and back-parging, or back-buttering, is recommended in all areas. Coating the back of the tile, however, does not constitute coverage, which is the area where the mortar makes contact with the tile and the substrate.

It is not possible or practical to achieve 100% coverage consistently and such should not be specified.

Directional Troweling

To ensure proper coverage of the bonding surface of 8" × 8" and larger tiles and to provide full support of edges and corners, select a notched trowel sized to facilitate the proper coverage. Key the mortar into the substrate with the flat side of the trowel. Comb with the notched side of the trowel in one direction. Firmly press tiles into the mortar and move them perpendicularly across the ridges forward and back to flatten the ridges and fill the valleys. This method can produce maximum coverage, with the corners and edges fully supported, without back-buttering or beat-in. Periodically remove and check a tile to ensure proper coverage is being attained.

Excessive or Nonuniform Depth of Mortar

Many stone varieties, particularly the softer marbles, limestones, and travertines, are extremely vulnerable to nonuniform shrinkage of cement mortar. Uneven mortar shrinkage during curing can produce a fine but visible crack in the stone. This is also true for larger glass tiles. To avoid this, the mortar must be applied in uniform thickness, and within the minimum and maximum thickness allowed by the manufacturer. Substrate preparation, or flattening, is required if the substrate does not meet the substrate tolerances. Mortar bed and self-leveling methods may be a good choice when larger glass tiles or such stones have been selected.

High-Porosity Stone Tile

Installing high-porosity stones in hot, dry climates may require that the stone be presoaked prior to installation to prevent the stone from wicking excessive amounts of water out of the mortar and inhibiting hydration of the cement needed for bonding. Alternately, rapid-setting mortar can be used.

FLATNESS AND LIPPAGE

Flatness for Ceramic and Stone Tile Installations

The finish flatness requirement for ceramic tile installations is ¼" in 10' from the required plane according to ANSI A108.02. The finish flatness requirement for stone tile installations is ⅛" in 10' from the required plane according to MIA.

Lippage in Natural Stone Tile Installations

Lippage is a condition where one edge of a tile is higher than an adjacent tile, giving the finished surface an uneven appearance. Lippage in stone tile flooring is accentuated from what would be perceived in a ceramic tile installation for several reasons. First, natural stones tend to be installed with tighter joints, particularly because some stones require narrower joints to allow for the use of unsanded grout. Second, natural stones typically have slight or no chamfer at their edges. Third, many natural stones have a high-gloss, polished finish.

Industry standards limit lippage between adjacent units of smooth-finished natural stone flooring to 1/32". When this is not achieved, there is the option of grinding the stone tiles in place to eliminate the lippage. This work must be done by skilled, experienced technicians, with care taken to avoid a wavy-looking finished floor. In some cases, installing contractors plan to grind the finished floor regardless of the lippage amount, usually per the requirement of the client. Lippage should still be minimized to the extent possible during installation to avoid extremely thin stone sections after grinding.

Lippage in Ceramic Tile Installations

Lippage is a condition where one edge of a tile is higher than an adjacent tile, giving the finished surface an uneven appearance. Lippage is inherent in all ceramic installation methods and may also be unavoidable due to the tile tolerances, in accordance with ANSI A137.1.

The following is excerpted from ANSI A108.02—2019, Section 4.3.7:

> **4.3.7 Lippage—guidelines, explanation, and caution:** Lippage refers to differences in elevation between edges of adjacent tile modules. These differences or perception thereof are influenced by many factors such as:
>
> A) The allowable thickness variation of the tile modules when judged in accordance with manufacturing standards.
>
> B) The allowable warpage of the tile modules.
>
> C) The spacing or separation of each tile module, which would influence a gradual or abrupt change in elevation.
>
> D) Angle of natural or manufactured light accentuating otherwise acceptable variance in modules.
>
> E) Highly reflective surfaces of tile modules accentuating otherwise acceptable variance in modules.

The following is a guideline for identifying acceptable lippage. The total allowable lippage shall be calculated by adding the actual warpage of the tile supplied to the allowable lippage number in the chart below (1/16 in. or 1/32 in. depending on the tile type, size, and grout joint width).

Example (not applicable to Gauged Porcelain Tile (GPT) or GPT Panels/Slabs): If the tile being evaluated for lippage is a porcelain tile with an 1/8 in. grout joint, and it has an actual warpage of 1/32 in., then the allowable lippage is 1/32 in. (warpage in tile) + 1/32 in. (value from chart) or 2/32 in. (1/16 in.).

These guidelines only apply to tiles manufactured in accordance with ANSI A137.1 or ANSI A137.3.

Tile Type	Tile Size (in.)	Joint Width (in.)	Allowable Lippage (in.)
Glazed Wall/Mosaics	1 × 1 to 6 × 6	1/16 to 1/8	1/32
Quarry	6 × 6 to 8 × 8	¼ or greater	1/16
Pressed Floor and Porcelain Tiles	All	1/16 to less than ¼	1/32
Pressed Floor and Porcelain Tiles	All	¼ or greater	1/16
Gauged Porcelain Tiles (GPT) and GPT Panels/Slabs	All	All	1/32*

***NOTE:** *Due to the lack of warpage in Gauged Porcelain Tiles (GPT) and GPT Panels/Slabs, only the value in the chart shall be used as a guideline for identifying acceptable lippage. No additional calculation is needed when evaluating GPT or GPT Panels/Slabs.*

CAUTION: *This chart does not apply to tiled floors sloping to drains. Lippage will be present when using tiles 6in.×6in. and larger over interior and exterior conical surfaces sloped to drains. The larger the tile unit surface area, the greater the lippage. Cutting the individual units can reduce the amount of lippage but may not eliminate lippage. Using smaller units in sloping areas will reduce lippage.*

CAUTION: *Tiles with a warpage exceeding the maximum allowable in ANSI A137.1 can potentially become a trip hazard in flooring applications.*

End of Excerpt

GROUT JOINT SIZE, LAYOUTS, AND PATTERNS

Minimum Grout Joint Width

The minimum required joint width for ceramic tile and natural stone tile is 1⁄16". Setting ceramic or stone without a grout joint of at least 1⁄16", often referred to as a butt joint, does not provide sufficient accommodation for dynamic building movement, differential thermal expansion, or allowable variation in fabrication or manufacturing.

ANSI Grout Joint and Offset Pattern Requirements for Ceramic Tile Installations

The following criteria are excerpted from ANSI A108.02—2019, Section 4.3.8 regarding grout joint size, particularly in relation to the tile size, dimensional precision, and offset pattern:

> **4.3.8 Grout joint size:** To accommodate the range in facial dimensions of the tile supplied for a specific project, the actual grout joint size may, of necessity, vary from the grout joint size specified. The actual grout joint size shall be at least three times the actual variation of facial dimensions of the tile supplied. Example: for tile having a total variation of 1/16in. in facial dimensions, a minimum of 3/16 in. grout joint shall be used. Nominal centerline of all joints shall be straight with due allowances for hand-molded or rustic tiles. In no circumstance shall the grout joint be less than 1/16 in.
>
> **4.3.8.1 Running bond/brick joint patterns:** For running bond/brick joint patterns utilizing tiles (square or rectangular) with any side greater than 15 in., the grout joint shall be, on average, a minimum of 1⁄8 in. wide for rectified tiles and, on average, a minimum of 3⁄16 in. wide for calibrated (non-rectified) tiles. The grout joint width shall be increased over the minimum requirement by the amount of edge warpage on the longest edge of the actual tiles being installed. For example, for a rectified tile exhibiting 1⁄32 in. edge warpage on the longest edge, the minimum grout joint for a running bond/brick joint pattern will be 1⁄8 in. + 1⁄32 in. or 5⁄32 in., on average. Of necessity, in any installation, some grout joints will be less and some more than the average minimum dimension to accommodate the specific tiles being installed.
>
> **4.3.8.2 Running bond/brick joint and any offset pattern:** For running bond/brick joint and any offset patterns (i.e. non-continuous grout joints) utilizing tiles (square and/or rectangular) where the side being offset is greater than 15 in. (nominal dimension), only patterns with an offset of 33% or less shall be specified. If a pattern with an offset greater than 33% is desired, the specifier and owner must approve a mock-up and the resulting lippage.
>
> Examples of offset patterns where the offset is greater than 33%: Standard Herringbone or Versailles. (See Figure 13.0 for examples).
>
> Examples of offset patterns where offset is 33% or less: Blunt Herringbone or Large Pinwheel. (See Figure 13.0 for examples).
>
> **NOTE:** *This restriction on bonding pattern does not apply to Gauged Porcelain Tile and Gauged Porcelain Tile Panels/Slabs which can typically be set with a 50% offset.*
>
> **End of Excerpt**

Tile Layout Considerations

Basic provisions for tile layout are given in ANSI A108.02 Section 4.3, however such may not be practical or applicable under the following conditions:

- The size and configuration of the room and the size of the tile make it impossible to center the layout in all aspects or avoid cut tiles less than ½ tile.
- The tile layout is continuous, flowing into multiple spaces, making it impossible to center the layout in all aspects or avoid cut tiles less than ½ tile.
- The overall layout and/or aesthetics are improved by laying out the tiles such that the layout is not centered in all aspects and/or not all cut tiles are at least ½ tile, for example to place a full tile or larger cut tile, or decorative element, in a high visibility location.
- Tile waste can be dramatically reduced using a layout that is not centered in all aspects and/or does not have any cut tiles less than ½ tile.

The layout preferences of the design professional should be well defined for each space prior to bidding. If a layout is not provided by the design professional prior to bidding, the tile contractor may, of necessity, determine the layout based on the space, tiles, and pattern specified.

Given allowable size variation between tiles (per ANSI A137.1), spacers shall not be relied on exclusively for, or in lieu of, proper layout and squaring of the installation.

System Modularity

Nominal sizes only provide a general idea of tile size (see ANSI A137.1) and cannot be relied upon as an indicator of size compatibility or pattern compatibility with other tiles having the same nominal size, including for tiles from the same tile line. Grout joint sizing for patterns can only be determined based on actual tiles ready for installation.

Per ANSI A137.1 system modularity is defined as follows: "tiles of various nominal dimensions are sized so that they may be installed together in patterns with a common specified grout joint width." System modularity is not a required or standardized tile property. Although various trade names exist for common modular patterns (e.g., basketweave, ashlar, versailles, etc.—see ANSI A108.02 Figure 13.0), pattern names and designs vary from manufacturer to manufacturer. While tiles manufactured to be installed in a modular pattern will have both the pattern and suggested grout joint width provided by the manufacturer, the specified grout joint width may not be appropriate for the site conditions or the size variations in the actual tiles supplied for installation. Adjusting the grout joint width may be required to accommodate the actual tiles, and, because of site conditions and size variations in the tiles, some grout joints will be wider than others. This is a necessary fact of the overall modular layout.

Similarly, when the design professional creates a modular pattern using tiles that are not specifically manufactured for a pattern, some grout joints will of necessity be wider than others. Verifying system modularity and specifying the achievable grout joint for the pattern and actual tiles to be installed is the responsibility of the design professional.

If grout joint alignment from floor to wall is desired (i.e., carrying grout joints continuously from a floor onto adjoining walls), the design professional must confirm suitability of such based on actual tiles to be installed. If grout joint alignment cannot be accommodated with the actual tiles, the tile contractor may, by necessity, reasonably adjust the pattern, for example by "breaking the joint" at the floor/wall interface. Tiles should not be field-cut to size to accomplish modular patterns or to align grout joints, as field-cut edges will be dissimilar from factory edges and cannot be held to the same squareness tolerance.

Protecting New Tilework

To avoid damage to finished tilework, schedule floor installations to begin only after all structural work, building enclosure, and overhead finishing work, such as ceilings, painting, mechanical, and electrical work, are completed. Keep all traffic off finished tile floors until they have fully cured. Builder shall provide up to ¾"-thick plywood or OSB protection over nonstaining kraft paper to protect floors after installation materials have cured. Covering the floor with polyethylene or plywood in direct contact with the floor may adversely affect the curing process of grout and latex/polymer modified portland cement mortar.

Maintenance

All tile installations (especially exterior installations, which include the movement joint sealant) require periodic inspection and maintenance by the owner. Consult material manufacturers and maintenance product manufacturers for recommended procedures.

Visual Inspection of Tilework

All components of a tile installation, including but not limited to the substrate, the tile, the installation materials, and the workmanship have tolerances, allowances, and industry best practices that govern the determination of acceptable work and the visual and aesthetic appearance of the finished installation. All of these factors shall be taken into consideration when compiling punch lists or deficiency lists.

When visually inspecting finished ceramic, stone, and glass tile installations, do so without magnification under the permanent intended lighting (artificial and/or natural) and without the use of additional lighting such as flashlights, spot lights, or temporary lights. View the installation 36" from walls and 60" or normal standing height from floors. Recognizing the hand-built aspect of tile installations, any aesthetic concerns not visible at these distances (but apparent at closer distances) are acceptable under industry best practices.

Refer to the following industry-recognized publications and standards for details on tolerances and allowances in tile and tilework:

- **Lippage:** Allowances are provided in ANSI A108.02; see also "Substrate Requirements" and "Flatness and Lippage" in this *Handbook.*
- **Finished flatness:** Allowances are provided in ANSI A108.02; see also "Substrate Requirements" and "Flatness and Lippage" in this *Handbook.*
- **Allowable tile warpage, wedging, and facial dimensions:** Allowances are provided in ANSI A137.1 and ANSI A137.2.
- **Effects of lighting on installation appearance:** See "Lighting and Tile Installations" in this *Handbook.*
- **Workmanship, cutting, and fitting:** Allowances are provided in ANSI A108.02.
- **Tile layout, system modularity:** See "Grout Joint Size, Layouts, and Patterns" in this *Handbook.*
- **Grout joint size:** Minimum size and related allowances are provided in ANSI A108.02; see also "Grout Joint Size, Layouts, and Patterns" in this *Handbook.*
- **Shade variation in ceramic tile:** Allowances and related information are provided in ANSI A137.1; see also "Ceramic Tile Selection Guide" in this *Handbook.*
- **Shade variation in natural stone tile:** Provisions are given in the *Dimension Stone Design Manual* published by MIA; see also "Natural Stone Tile Selection and Installation Guide" in this *Handbook.*
- **Facial and structural defects in glass tile:** Allowances and related information are provided in ANSI A137.2; see also "Glass Tile Selection and Installation Guide" in this *Handbook.*
- **Mortar coverage:** Allowances are provided in ANSI A108.02; see also "Mortar Application and Coverage" in this *Handbook.*
- **Hollow-sounding tile:** See "Membrane Selection Guide" in this *Handbook.*

When ceramic tile is used as the flooring surface, design professionals should consider the following, based on ANSI A108.01 and A108.02, where accessibility is a primary consideration.

Changes in Level

Changes in level up to ¼" may be vertical and without edge treatment. Changes in level between ¼" and ½" shall be beveled with a slope no greater than 1:2. Changes in level greater than ½" shall be accomplished by means of a ramp. The maximum slope of a ramp in new construction shall be no greater than 1:12. Ramps where space limitations prohibit this may have slopes and rises as follows: a slope between 1:10 and 1:12 is allowed for a maximum rise of 6", and a slope between 1:8 and 1:10 is allowed for a maximum rise of 3". A slope steeper than 1:8 is not allowed.

Flatness and Lippage

With regard to flatness, the amount of substrate variation generally is reflected in the finished tile installation. For any application, a tiled floor should comply with the flatness requirements in ANSI A108.02: no variations exceeding ¼" in 10' from the required plane. Conformance to this standard requires that substrates conform to the following: For tiles with all edges shorter than 15", maximum allowable variation is ¼" in 10' from the required plane, with no more than ¹⁄₁₆" variation in 12" when measured from the high points in the surface. For tiles with at least one edge 15" in length, maximum allowable variation is ⅛" in 10' from the required plane, with no more than ¹⁄₁₆" variation in 24" when measured from the high points in the surface. For modular substrate units, such as plywood panels, adjacent edges cannot exceed ¹⁄₃₂" difference in height. Additionally, the effect from irregularities in the substrate increases as the tile size increases. A substrate tolerance of ⅛" in 10' may be required.

Because the flatness of wood and concrete substrates can change over time, it is recommended that the designer make provisions for evaluating substrate flatness just before installation of the tile. Project specifications should make clear which trade is responsible for the required alterations if the subfloor is found not to be in compliance with the flatness requirements. Alternately, the designer may choose to specify a mortar bed method or a pourable underlayment installed by the tile contractor to ensure substrate flatness sufficient to facilitate a flat tile installation.

Lippage is most significantly influenced by substrate flatness and tile warpage. Allowable lippage is calculated by adding the actual warpage of the tile supplied, plus either ¹⁄₃₂" or ¹⁄₁₆" based on tile type and size and grout joint width (see "Flatness and Lippage"). Specifying wider grout joints allows for more gradual changes. To minimize lippage due to warpage, specify tile that meets the dimensional requirements for rectified tile according to ANSI A137.1, and use a larger grout joint. Some patterns, such as a 50% offset (brick-joint) pattern, accentuate the effects of warpage and result in more lippage than other patterns would. Cushioned or beveled-edge tiles can minimize the effects of lippage.

In addition to taking measures to ensure a flat substrate, designers should consult with the tile manufacturer to discuss grout joint size and tile and pattern selections that will minimize issues relating to flatness and lippage. See also "Grout Joint Size, Layouts, and Patterns."

The installation methods in this *Handbook* are rated for water exposure suitability—these ratings are listed under the Environmental Exposure Classifications subhead within each method as well as in the Environmental Exposure Classifications chart. However, this rating system cannot predict the actual amount of water any installation will be exposed to, nor compensate for water exposure exceeding the method's rating.

The design professional must choose an installation method suitable for the amount of water the installation will be subjected to; damage can result from water exposure exceeding the method's Environmental Exposure Classifications rating. Where the use of a waterproof membrane is optional, the rating assumes the optional membrane will not be used. If use of a waterproof membrane is intended in these applications, to provide a more water-resistant installation, it must be clearly specified.

When selecting a method, in addition to the use of the tiled area, consider maintenance practices that will be employed. For example, commercial restrooms and locker rooms typically do not require waterproofing in order to be suitable for their intended uses and normal maintenance; typically these are limited water exposure (Res2/Com2) applications. However, if such areas are to be hosed down or otherwise saturated, specify a wet area (Res3/Com3) method. Wet area installation methods typically incorporate waterproofing to contain and evacuate water and to protect building materials. The two general categories of waterproof installation methods are:

- Use of an unbonded water-containment membrane (referenced in this *Handbook* as an unbonded shower pan membrane).
- Use of a bonded waterproof membrane meeting ANSI A118.10.

A shower pan membrane is a loose laid, or unbonded, liner that is placed below a mortar bed, with the mortar bed receiving the tile. Acceptable shower pan membrane materials are listed in ANSI A108.01 Section 3.6 and include plastics such as polyvinyl chloride (PVC) and chlorinated polyethylene (CPE), metals such as lead and copper, and hot-mop systems, which employ layers of asphalt or coal-tar saturated roofing felt. Shower pan membranes connect into drains at the clamping ring level. Refer to B414, B415, B420, B426, B431, and F121 for shower pan membrane assemblies.

Bonded waterproof membranes can be sheet materials, or roll-on or trowel-on materials that dry/cure to form a waterproof membrane. These membranes also connect to drains, either at the clamping ring level, or just below the tile when an integrated bonding flange is used. Because these membranes can be bonded to various substrates, and tile is bonded to the membrane, they allow for thin-bed tile installation. Refer to B421 and B422 for bonded waterproof membrane assembly specifications.

Both systems manage water. The following are the essential requirements and considerations for designing wet areas.

Incorporation of a Drain/Slope to Drain

To fully evacuate water, shower pan membranes and bonded waterproof membranes must slope to and connect with a drain. Plumbing code typically requires membranes to be sloped a minimum of ¼" per foot and extend at least 3" above the height of the curb or threshold. Account for the perimeter floor height required to form adequate slopes. Membranes must be installed over other horizontal surfaces in wet areas subject to deterioration, like shower seats. They must be sloped and configured so as to direct water to the membrane connected to the drain.

Open Weep Holes

The weep holes of clamping drains enable water to pass from the membrane into the plumbing system. Crushed stone or tile or other positive weep protectors placed around/over weep holes prevent their blockage.

Membrane Connection with Drain or Integrated Bonding Flange

To form a watertight seal, membranes must have adequate contact with the clamping ring of the drain or with the bonding area of an integrated bonding flange.

Membrane Cuts and Penetrations

Membranes must be protected to prevent punctures resulting from traffic on the membrane before the mortar bed is installed (for shower pan membranes) or before the tile is installed (for bonded waterproof membranes). For punctures that do occur, the membrane must be replaced or repaired according to the membrane manufacturer's directions for repairs. Ensure the integrity of any repairs by water testing the repaired membrane.

Backer board cannot be fastened to studs lower than 3" above the finished curb height, nor fastened to the top or the inside of a curb. Backer board on shower seats must be topically waterproofed due to the use of fasteners.

In-Corners, Out-Corners, and Seams

Shower pan membrane in-corners should be folded not cut. For out-corners, such as where the shower curb meets the jamb, membrane manufacturers typically offer preformed out-corners to better enable wrapping of the membrane at the curb/jamb interface. For sheet-type bonded waterproof membranes applied topically, pre-made in-corners and out-corners enable waterproofing of corners without excessive material thickness that would result from folding. Sheet membranes in large areas are seamed, bonded, or otherwise welded together to form a continuous membrane.

Liquid-Applied and Trowel-Applied Bonded Waterproof Membranes

These membrane types are manufactured in the field by the installer who applies the waterproofing material. These products require a minimum wet film thickness and have specified cure/dry time requirements. Many membranes of this variety incorporate a mesh that is embedded in the wet material during installation. Mesh may be required over the entire surface to be waterproofed or only in corners and/or joints.

Configuration of Shower Receptor Components

When a shower pan membrane system is employed, some backer board types must be installed with the board held out of the mortar bed due to the saturation that occurs below this level. Vapor retarder membranes fastened to studs must weather-lap the shower pan membrane or flange of the tub or prefabricated shower receptor.

Whether a shower pan membrane or a bonded waterproof membrane system is used, the membrane must completely wrap the curb, and the jamb must be waterproofed to its outside edge a minimum of 3" above the curb. Curb and jamb waterproofing must be seamed together without breach to form a continuous barrier.

Performing a Water Test

Where complete waterproofing is required such as in showers, water testing of the membrane, by the installing contractor, is recommended and may be required by applicable plumbing code.

FLOOR TILING INSTALLATION GUIDE

Performance Level Requirement Guide and Selection Table

Based on results from ASTM Test Method C627 Standard Test Method for Evaluating Ceramic Floor Tile Installation Systems Using the Robinson Type Floor Tester, all methods are material dependent—performance rating should not exceed rating of weakest component—consult each material manufacturer for individual component rating. Tests to determine performance levels utilized representative products meeting recognized industry standards.

Consideration must also be given to wear properties of surface of tile selected, tile size, and coefficient of friction. Unglazed standard grade tile will give satisfactory wear or abrasion resistance in installations listed. Glazed tile or soft-body, decorative, unglazed tile should have the manufacturer's approval for intended use. Color, pattern, surface texture, and glaze hardness must be considered in determining tile acceptability on a particular floor.

Those methods also applicable to stone tile installation will carry a similar rating when stone tiles with adequate compressive strength, flexural strength, and resistance to abrasion are used.

SERVICE REQUIREMENTS	INSTALLATION SUBSTRATE	
Find required performance level and choose installation method that meets or exceeds it. Performance results are based on ceramic tile meeting ANSI A137.1 or tile designated by tile manufacturer.	**CONCRETE**	**WOOD**
EXTRA HEAVY Extra heavy and high-impact use in food plants, dairies, breweries, and kitchens. Requires quarry tile, packing house tile, or tile designated by tile manufacturer. (Passes ASTM C627 cycles 1 through 14.)	F101 F114 F132 F102 F115 F133 F111 F115A F134 F112 F116E F205 F113 F128[e] F205A F113A F131 RH117	
HEAVY Shopping malls, stores, commercial kitchens, work areas, laboratories, auto showrooms and service areas, shipping/receiving, and exterior decks. (Passes ASTM C627 cycles 1 through 12).	F103 F103B[f] F104 F121	
MODERATE Normal commercial and light institutional use in public space of restaurants and hospitals. (Passes ASTM C627 cycles 1 through 10.)	F122 F122A F200 F200A RH110 RH110A RH111 RH111A RH112 RH112A RH115 RH115A RH116 RH116A	
LIGHT Light commercial use in office space, reception areas, kitchens, and bathrooms. (Passes ASTM C627 cycles 1 through 6.)	F128[e]	F105 F141 F145 F143[d] F144[b,c] F146 F150[d] F160 F170 F175 F180 F250[d] RH122 RH123 RH130[d] RH135[b,c] RH141
RESIDENTIAL Kitchens, bathrooms, and foyers. (Passes ASTM C627 cycles 1 through 3.)	F116O F135 F136 TR711[a]	F142 F143[d] F144[b,c] F147 F148 F149 F150[d] F151 F152 F155 F185 F250[d] RH130[d] RH135[b,c] RH140

a. Tile bonded to existing resilient flooring with epoxy adhesive.

b. Minimum 7/16"-thick cement backer board or minimum 1/4"-thick fiber-cement backer board tested.

c. Minimum 1/4"-thick cement backer board can be used for residential applications over minimum 19/32"-thick subfloor; minimum 1/4"-thick cement backer board can be used for light commercial applications over minimum 23/32"-thick subfloor.

d. Requires minimum 19/32" exterior glue plywood underlayment for light rating; 15/32" exterior glue plywood underlayment may be used for residential rating.

e. Rated light for mosaic tile and extra heavy for 12" × 12" and larger porcelain tiles.

f. Not all drainage mat systems will receive heavy rating. Consult drainage mat manufacturer.

ENVIRONMENTAL EXPOSURE CLASSIFICATIONS

The end user significantly affects the amount of water and vapor an installation will be exposed to. Examples provided below and the classifications assigned to the individual methods are guidelines only and are not meant to be all inclusive. Base installation method selection on actual exposure levels, and consult with product manufacturers and their specifications. In methods where inclusion of a waterproof membrane is optional, the design professional must clearly specify that a membrane is desired. Optional membranes are not included unless clearly specified.

Res1 (Residential Dry): Tile surfaces that will not be exposed to moisture or liquid, except for cleaning purposes. Includes areas adjacent to R2 areas. Examples: Floors in rooms with no direct access to the outdoors and no wet utility function, such as living rooms, dining rooms, and bedrooms; dry area ceilings, soffits, decorative/accent walls, fireplaces, some backsplashes and some wainscots.

Res2 (Residential Limited Water Exposure): Tile surfaces that are subjected to moisture or liquids but do not become soaked or saturated due to the system design or time exposure. If waterproofing is desired, it must be clearly specified. Includes areas adjacent to R3 areas. Examples: Floors in bathrooms, kitchens, mudrooms, laundry, and foyers, where water exposure is limited and/or water is removed; some backsplashes, some wainscots, some countertops.

Res3 (Residential Wet): Tile surfaces that are soaked, saturated, or regularly and frequently subjected to moisture or liquids. Examples: Shower floors; floors and other horizontal surfaces where water is not removed or drained, such as some countertops; tub walls, shower walls, and enclosed pool area walls.

Res4 (Residential High Humidity, Heavy Moisture Exposure): Tile surfaces that are subject to continuous high humidity or heavy moisture exposure. Examples: Intermittent-use steam shower walls, ceilings, and floors.

Res5 (Residential High Temperature ≥125°F): Tile surfaces frequently subjected to water or vapor equal to or greater than 125°F. Examples: Furnace and boiler areas.

Res6 (Residential Exterior): Tile surfaces exposed to exterior conditions. When designing such installations, consider local climate and conditions including temperature and temperature fluctuations, humidity and humidity fluctuations, and freeze/thaw cycling. If waterproofing is desired, it must be clearly specified. Examples: Exterior walls, balconies, decks.

Res7 (Residential Submerged): Tile surfaces exposed to continuous water submersion in interior or exterior conditions. Examples: Swimming pools, water features, and fountains.

Com1 (Commercial Dry): Tile surfaces that will not be exposed to moisture or liquid, except for cleaning purposes. Commercial cleaning and maintenance practices typically generate greater water exposure than residential practices. Includes areas adjacent to C2 areas. Examples: Floors in areas with no direct access to the outdoors and no wet utility function, such as hallways; dry area ceilings; soffits; decorative/accent walls; corridor walls.

Com2 (Commercial Limited Water Exposure): Tile surfaces that are subjected to moisture or liquids but do not become soaked or saturated due to the system design or time exposure. If waterproofing is desired, it must be clearly specified. Includes areas adjacent to C3 areas. Examples: Floors in bathrooms and locker rooms; some backsplashes and other walls, such as bathroom walls and wainscots where water exposure is limited and/or water is removed.

Com3 (Commercial Wet): Tile surfaces that are soaked, saturated, or regularly and frequently subjected to moisture or liquids. Includes areas adjacent to C4 areas. Examples: Tub walls, shower walls and floors, enclosed pool areas, natatoriums, gang showers, and some commercial kitchen floors and walls.

Com4 (Commercial High Humidity, Heavy Moisture Exposure): Tile surfaces that are subject to continuous high humidity or heavy moisture exposure, especially in enclosed areas. Examples: Continuous use steam shower/ steam room walls and ceilings.

Com5 (Commercial High Temperature ≥125°F): Tile surfaces that are frequently subjected to water or vapor equal to or greater than 125°F. Examples: Commercial saunas, furnace and boiler areas, and some commercial kitchen floors and walls.

Com6 (Commercial Exterior): Tile surfaces exposed to exterior conditions. When designing such installations, consider local climate and conditions including temperature and temperature fluctuations, humidity and humidity fluctuations, and freeze/thaw cycling. If waterproofing is desired, it must be clearly specified. Examples: Exterior walls, balconies, decks.

Com7 (Commercial Submerged): Tile surfaces exposed to continuous water submersion in interior or exterior conditions. Examples: Swimming pools, water features, and fountains.

	Installation Method	Method Type T = Tile S = Stone	Residential (Res)							Commercial (Com)						
			1	2	3	4	5	6	7	1	2	3	4	5	6	7
FLOORS	F101	T, S						•							•	
	F102	T, S						•							•	
	F103	T, S						•							•	
	F103B	T, S						•							•	
	F104	T, S						•							•	
	F105	T, S						•							•	
	F111	T, S	•	•	•		•			•	•	•		•		
	F112	T, S	•	•	•		•			•	•	•		•		
	F113	T, S	•	•	•		•			•	•	•		•		
	F113A	T, S	•	•	•		•			•	•	•		•		
	F114	T	•	•	•	•	•			•	•	•	•	•		
	F115	T	•	•	•	•	•			•	•	•	•	•		
	F115A	T	•	•	•	•	•			•	•	•	•	•		
	F116E	T	•	•	•	•	•			•	•	•	•	•		
	F116O	T	•													
	F121	T, S	•	•	•		•			•	•	•		•		
	F122	T, S	•	•	•		•			•	•	•		•		
	F122A	T, S	•	•	•		•			•	•	•		•		
	F128	T	•	•	•	•	•			•	•	•	•	•		
	F131	T	•	•	•	•	•			•	•	•	•	•		
	F132	T	•	•	•	•	•			•	•	•	•	•		
	F133	T	•	•	•	•	•			•	•	•	•	•		
	F134	T	•	•	•	•	•			•	•	•	•	•		
	F135	T	•													
	F136	T	•													
	F141	T, S	•	•						•	•					
	F142	T	•													
	F143	T	•	•						•	•					
	F144	T	•	•						•	•					
	F145	T	•	•						•	•					
	F146	T	•	•						•	•					
	F147	T	•	•												
	F148	T	•	•												
	F149	T	•													
	F150	T	•							•						
	F151	T	•	•												
	F152	T	•	•												
	F155	T	•													
	F160	T	•							•						

	Installation Method	Method Type T = Tile S = Stone	Residential (Res)							Commercial (Com)						
			1	2	3	4	5	6	7	1	2	3	4	5	6	7
FLOORS	F170	T	•	•						•	•					
FLOORS	F175	T	•	•						•	•					
FLOORS	F180	T	•							•						
FLOORS	F185	T	•	•												
FLOORS	F200	T, S	•							•						
FLOORS	F200A	T, S	•							•						
FLOORS	F205	T, S	•							•						
FLOORS	F205A	T, S	•							•						
FLOORS	F250	S	•	•						•	•					
FLOORS	RH110	T, S	•	•	•		•			•	•	•		•		
FLOORS	RH110A	T, S	•	•	•		•			•	•	•		•		
FLOORS	RH111	T, S	•							•						
FLOORS	RH111A	T, S	•							•						
FLOORS	RH112	T, S	•							•						
FLOORS	RH112A	T, S	•							•						
FLOORS	RH115	T	•	•						•	•					
FLOORS	RH115A	T	•	•						•	•					
FLOORS	RH116	T, S	•							•						
FLOORS	RH116A	T, S	•							•						
FLOORS	RH117	T, S	•	•	•		•			•	•	•		•		
FLOORS	RH122	T	•							•						
FLOORS	RH123	T	•							•						
FLOORS	RH130	T	•							•						
FLOORS	RH135	T	•	•						•	•					
FLOORS	RH140	T	•							•						
FLOORS	RH141	T, S	•	•						•	•					
WALLS	W201	T, S						•							•	
WALLS	W202E	T, S						•							•	
WALLS	W202I	T, S	•	•	•		•			•	•	•		•		
WALLS	W211	T, S	•	•	•		•			•	•	•		•		
WALLS	W215	T, S	•							•						
WALLS	W221	T, S	•	•	•		•			•	•	•		•		
WALLS	W222	T, S	•	•			•			•	•			•		
WALLS	W223	T	•	•						•	•					
WALLS	W231/W241	T, S	•	•			•			•	•			•		
WALLS	W242	T	•							•						
WALLS	W243	T, S	•							•						
WALLS	W244C	T, S	•	•			•			•	•			•		
WALLS	W244E	T, S						•							•	

	Installation Method	Method Type T = Tile S = Stone	Residential (Res)							Commercial (Com)						
			1	2	3	4	5	6	7	1	2	3	4	5	6	7
WALLS	W244F	T, S	•	•			•			•	•			•		
	W245	T, S	•	•						•	•					
	W246	T, S	•	•						•	•					
	W247	T, S	•	•						•	•					
	W248	T, S	•							•						
	W260	T, S	•							•						
BATHTUBS AND SHOWERS	B411	T, S	•	•	•		•			•	•	•		•		
	B412	T, S	•	•	•		•			•	•	•		•		
	B413	T, S	•							•						
	B414	T, S	•	•	•		•			•	•	•		•		
	B415	T, S	•	•	•		•			•	•	•		•		
	B419	T, S	•	•	•					•	•	•				
	B420	T, S	•	•	•					•	•	•				
	B421	T, S	•	•	•					•	•					
	B421C	T	•	•	•					•	•					
	B422	T, S	•	•	•					•	•					
	B422C	T	•	•	•					•	•					
	B425	T, S	•	•	•					•	•	•				
	B426	T, S	•	•	•					•	•	•				
	B430	T, S	•	•	•					•	•					
	B431	T, S	•	•	•					•	•					
	B440	T, S	•	•	•		•			•	•	•		•		
	B441	T, S	•	•	•		•			•	•	•		•		
	SR613	T				•							•			
	SR614	T				•										
CEILINGS	C311	T, S	•	•	•		•			•	•	•		•		
	C312	T, S	•							•						
	C315	T, S	•	•	•					•	•					
COUNTERTOPS	C511	T	•	•						•	•					
	C512	T	•	•						•	•					
	C513	T	•	•						•	•					
POOLS & WATER FEATURES	P601MB	T			•	•		•	•			•	•		•	•
	P601TB	T			•	•		•	•			•	•		•	•
	P602	T			•	•		•	•			•	•		•	•

USING THE *TCNA HANDBOOK* FOR SPECIFICATION WRITING

The quick-reference details in the *TCNA Handbook* provide a means of simplifying and standardizing installation specifications for ceramic, glass, and stone tile. In addition to referencing *Handbook* methods, architects and specifiers should specify tile and installation materials, including showing specific locations of movement joints on drawings. Use American National Standards Institute (ANSI) standards, where applicable, in project specifications. Additionally, project specifications shall conform to applicable building codes, ordinances, trade standards and practices, and climatic conditions. Use of materials and methods not specifically designated by the architect, will, of necessity, be determined by the tile contractor.

General

Select an installation method (found in the sections for Ceramic and Glass Tile Installation Methods or Natural Stone Tile Installation Methods) for the given backing or substrate with the appropriate Environmental Exposure Classification, and for floors, having the required Service Rating.

The "Product Selection Guide" section is provided to aid in the selection of the individual components of an installation: tile (ceramic, glass, stone), setting materials, grout, backer boards, and membranes. The "Field and Installation Requirements" section outlines many of the important conditions and considerations that must be provided for when specifying and installing tile.

Special installation procedures or proprietary installation methods or materials should be specified in accordance with manufacturer recommendations.

Components of an Installation Method

Drawing: *Handbook* illustrations show the necessary components of the given tile installation method, including the installation substrate. Glass tile can be substituted for ceramic tile in the assembly if glass tile is listed as an option under Materials.

The illustrations are intentionally not to scale in order for each material and its location within the assembly to be clearly visible. Consult manufacturers for the actual dimensions of any material.

Recommended Uses: Provides typical areas or applications for the installation method.

Service Rating: Provided for floor methods only. Indicates the expected level of use (service rating) based on Robinson Floor Test (ASTM C627) results.

Environmental Exposure Classifications: Indicates the conditions in which the installation is generally expected to perform.

Typical Weight of Tile Installation: Provided for floor methods only. Indicates estimated weight the tile installation contributes to the dead load.

Limitations: Provides special considerations and restrictions.

Membrane Options: Lists the possible membrane types that can be specified.

This section and the *Handbook* illustrations indicate when use of a membrane is optional, and what types may be used. They also indicate when an installation method requires the use of a membrane and what types may be used.

If using a membrane is optional for the installation method specified, and one is not designated by the architect as part of the execution criteria of the specification (e.g., Part 3 – Execution, commonly found in specifications written using CSI MasterFormat®) or on the drawings, the contractor is not required to install a membrane. If using a membrane is required for the installation method specified, and one is not designated by the architect, the tile contractor may choose any listed membrane option.

Requirements: Lists important conditions and considerations.

Materials: This section lists tile and installation material (e.g., setting material, grout) options that can be used and the minimum criteria the materials must meet—such as ANSI, ISO, or ASTM specifications.

Note: When more than one type of material is listed, for example cementitious grout and epoxy grout, the job specification must clearly indicate which material is desired for a specific tile installation. If such an indication is not included, the tile contractor by necessity may choose any listed material option.

Preparation by Other Trades: Indicates the materials and/or preparation work required of other trades preceding the work of the tile contractor. ANSI A108.01 and A108.02 provide particular specifications for related trades regarding preparation for tilework.

Preparation by Tile Trade: Indicates the materials, preparations, or other requirements typically expected of the tile contractor.

Preparation by Backer Board Installers: Indicates the materials, preparations, or other requirements expected of the contractor that installs the backer board.

Movement Joint: Refers user to EJ171 movement joint guidelines.

Installation Specifications: Lists applicable ANSI installation specifications.

Notes: Lists additional considerations that may be important in the specification or installation of the tile and installation materials for the given method.

Design Considerations When Specifying Tile

Numerous tile industry standards, guidelines, and best practices exist, which define parameters for acceptable tilework, many of which are impacted by the tiles and designs selected. Through familiarity with tile industry standards and best practices, building design professionals can factor such parameters into selection and design decisions to specify products, patterns, and installation details that will produce the desired end result.

When variation from existing standards and guidelines is desired, such must be clearly defined prior to bidding and installation and noted within project specifications as a variance from industry standards. Such variances are discouraged because they will not always be possible to meet, with the given materials and job conditions. In such cases, established tile industry ANSI standards and industry guidelines apply.

The following are the commonly referenced industry standards, guidelines, and best practices that pertain to finished tilework and serve as the defining criteria for evaluating quality of installation and workmanship unless project specifications define otherwise:

Base/cove alternatives: Appropriate trim shapes and details must be specified by the design professional or left to the discretion of the tile contractor. All details in this *Handbook* are stylized cross-sectional drawings of typical installations and are not intended to show specific treatments of the transition between vertical and horizontal tile surfaces.

Finished flatness: Tolerances for finish flatness are provided in ANSI A108.02; see also "Substrate Requirements" and "Flatness and Lippage" in this *Handbook*.

Grout joint size: Tolerances for grout joint size are provided in ANSI A108.02; see also "Grout Joint Size, Layouts, and Patterns" in this *Handbook*.

Tile layout: Workmanship with regard to layout is addressed in ANSI A108.02; see also "Grout Joint Size, Layouts, and Patterns" in this *Handbook*.

Tile patterns and modularity: See "Grout Joint Size, Layouts, and Patterns" in this *Handbook*.

Mortar coverage: Required mortar coverage is provided in ANSI A108.02; see also "Mortar Application and Coverage" in this *Handbook*.

Effect of lighting on tilework: Natural and artificial light can dramatically affect the appearance of a tile installation; see "Lighting and Tile Installations" in this *Handbook*.

Lippage: Allowances are provided in ANSI A108.02; see also ANSI A137.1 for allowable tile warpage.

Tile warpage, wedging, and facial dimensions: Allowances are provided in ANSI A137.1.

Shade variation in ceramic tile: Allowances and related information are provided in ANSI A137.1; see also "Ceramic Tile Selection Guide" in this *Handbook*.

Shade variation in natural stone tile: Provisions are given in the Dimension Stone Design Manual published by MIA; see also Shading and Variation section within the "Natural Stone Tile Selection and Installation Guide," in this *Handbook*.

Facial and structural defects in glass tile: Allowances and related information are provided in ANSI A137.2.

ANSI Installation Specifications

A108.01 General Requirements: Subsurfaces and Preparations by Other Trades

A108.02 General Requirements: Materials, Environmental, and Workmanship

A108.1A Installation of Ceramic Tile in the Wet-Set Method with Portland Cement Mortar

A108.1B Installation of Ceramic Tile on a Cured Portland Cement Mortar Setting Bed with Dry-Set or Latex-Portland Cement Mortar

A108.1C Contractor's Option: Installation of Ceramic Tile in the Wet-Set Method with Portland Cement Mortar or Installation of Ceramic Tile on a Cured Portland Cement Mortar Setting Bed with Dry-Set or Latex-Portland Cement Mortar

A108.4 Installation of Ceramic Tile with Organic Adhesives or Water Cleanable Tile-Setting Epoxy Adhesive

A108.5	Installation of Ceramic Tile with Dry-Set Portland Cement Mortar or Latex-Portland Cement Mortar
A108.6	Installation of Ceramic Tile with Chemical Resistant, Water Cleanable Tile-Setting and -Grouting Epoxy
A108.8	Installation of Ceramic Tile with Chemical Resistant Furan Resin Mortar and Grout
A108.9	Installation of Ceramic Tile with Modified Epoxy Emulsion Mortar/Grout
A108.10	Installation of Grout in Tilework
A108.11	Interior Installation of Cementitious Backer Units
A108.12	Installation of Ceramic Tile with EGP (Exterior Glue Plywood) Latex-Portland Cement Mortar
A108.13	Installation of Load Bearing, Bonded, Waterproof Membranes for Thin-Set Ceramic Tile and Dimension Stone
A108.14	Installation of Paper-Faced Glass Mosaic Tile
A108.15	Alternative Method: Installation of Paper-Faced Mounted Glass Mosaic Tile
A108.16	Installation of Paper-Faced, Back-Mounted, or Clear Film Face-Mounted Glass Mosaic Tile
A108.17	Installation of Crack Isolation Membranes for Thin-Set Ceramic Tile and Dimension Stone

Mockups

In addition to tile industry standards and guidelines, design professionals can utilize mock ups, especially in-situ mock-ups, to see the specified products and designs executed under the given job conditions.

INSTALLER AND CONTRACTOR QUALIFICATIONS GUIDE

General

A home or any building today is one of the few things still made entirely "by hand" and every aspect of a tile installation relies on the tile contracting company and its installers. How good the finished installation looks, how well it performs, and how long it lasts are in their hands. It is for this reason that the Handbook Committee and the Tile Council of North America strongly recommend hiring tile installation professionals who have demonstrated their commitment to their craft and taken the time to stay current with the latest materials and methods. Because tile is a permanent finish, the lowest bid should not be the driving factor, but rather who is the most qualified to perform the scope of the work specified.

Requiring a portfolio and references reflecting the installer's/contractor's experience, along with a bid or estimate, is a good way to ensure work of similar size, scope, and complexity has been completed. Pools, exterior facades, mortar beds, shower pans, steam showers, and other potentially complex installations, require different skills. Matching installer ability to the project at hand requires close evaluation of their experience, training, state licensing (where applicable), and certifications/credentials (where applicable). The Ceramic Tile Education Foundation (CTEF) provides a Contractor Questionnaire that can be used to aid in evaluating and comparing contractors (www.ceramictilefoundation.org).

Various programs administered by associations, nonprofit educational organizations, unions, and private companies serve the tile industry by providing education, hands-on training, and evaluation of the skills and competency of installers and contractors. It is important to distinguish between the many programs available:

- Classroom and/or online knowledge training
- Hands-on installation skills training
- Evaluation/certification of contractor qualifications
- Evaluation/certification of knowledge
- Evaluation/certification of hands-on skills

As with all programs, the rigor and credibility of the program must also be considered.

The following nonprofit programs are well-established and recognized by the Handbook Committee (listed alphabetically):

Advanced Certifications for Tile Installers (ACT): ACT is an evaluation and certification program for tile installers in advanced, specific skills areas: large format tile installation and substrate preparation, mortar (mud) beds for walls, mortar (mud) beds for floors, shower receptor (waterproof shower pan membrane, mortar bed and mortar curb) installation, membrane (both sheet and liquid) applications, grouting, and installation of gauged porcelain tiles (GPT) and gauged porcelain tile panels/slabs. To pass an ACT test module, an installer must demonstrate hands-on proficiency and specific knowledge. Only installers who have already passed the Certified Tile Installer (CTI) test or those who are recognized by the U.S. Department of Labor as Journeyman Tile Layers may register for ACT testing. ACT tests were developed collaboratively by product manufacturers and industry associations including the Ceramic Tile Education Foundation (CTEF), Tile Contractors Association of America (TCAA), International Union of Bricklayers and Allied Craftworkers (IUBAC), International Masonry Institute (IMI), National Tile Contractors Association (NTCA), and Tile Council of North America (TCNA). Certification testing is administered by CTEF and IMI. For more information see www.tilecertifications.com. See also: Sample Language for Including Installer and Contractor Qualifications in Specifications.

Ceramic Tile Education Foundation (CTEF) Certified Tile Installer Program: CTEF tests hands-on installation skills and knowledge. Installers must achieve the minimum required score on both tests to earn the "CTEF Certified Tile Installer" (CTI) designation. Contractors that employ CTEF Certified Tile Installers are listed on the CTEF website. See www.ceramictilefoundation.org for more information. See also: Sample Language for Including Installer and Contractor Qualifications in Specifications.

International Masonry Institute (IMI) Contractor College Program: IMI conducts professional and technical courses for union masonry and tile contractors, which lead to certification in installation and project supervision. See www.imiweb.org for more information. See also: Sample Language for Including Installer and Contractor Qualifications in Specifications.

Journeyman Tile Layer Apprenticeship Programs: Installers recognized by the U.S. Department of Labor (DOL) as Journeyman Tile Layers are required to fulfill and document several years of training and on-the-job experience as apprentices to become Journeymen. The majority of these setters earn their Journeyman status through registered apprenticeship programs that are jointly-trusteed by union and management trustees, although some non-union tile contractors administer their own DOL-recognized apprenticeship programs and employ journeyman tile layers. Contractors that employ union Journeyman Tile Setters can be found through the union locals that list their signatory contractors, primarily the Bricklayer and Allied Craftworkers (BAC) and the United Brotherhood of Carpenters (UBC). See www.bacweb.org and www.carpenters.org for more information. See also: Sample Language for Including Installer and Contractor Qualifications in Specifications.

Natural Stone Institute (NSI) Accreditation for Natural Stone Tile Installation Contractors: The Natural Stone Institute is an internationally recognized trade association that operates an industry accreditation program for natural stone tile installation contractors. NSI Accredited companies are required to: have been in business for at least three years and have completed at least 200 projects in that timeframe, agree to a code of ethics, and have passed the NSI Accreditation exam. NSI Accredited companies are evaluated and tested with regard to best practices in installation and safety. Visit www.naturalstoneinstitute.org/gettheseal for more information.

National Tile Contractors Association (NTCA) Five Star Contractor Program: NTCA is a tile contractors association, with membership open to all tile contractors. Their Five Star program is a peer review program to recognize NTCA members who have demonstrated a track record of providing successful installations. Earning the Five Star designation requires recommendations from customers, suppliers, and peers as well as participation in continuing education, training, and safety programs. See www.tile-assn.com for more information. See also: Sample Language for Including Installer and Contractor Qualifications in Specifications.

Tile Contractors' Association of America (TCAA) Trowel of Excellence Program: TCAA is a contractors association for BAC signatory contractors. Its Trowel of Excellence program is a peer review program to recognize TCAA members who have demonstrated a track record of providing successful installations. Earning the Trowel of Excellence designation requires letters of reference, submittal of a detailed project description and photos, employee participation in educational programming, and proof of financial responsibility. See www.tcaainc.org for more information. See also: Sample Language for Including Installer and Contractor Qualifications in Specifications.

Other Programs: The *Handbook* Committee recognizes many qualified tile installers and contractors can be found that have not participated in the programs listed above. They are encouraged to do so, in order that consumers may more easily identify them. Similarly, the *Handbook* Committee recognizes there are useful training and evaluation programs for contractors and installers not listed in this guide, such as consumer-review-based rating systems, programs offered by for-profit educational firms, and government-based programs. See also: Sample Language for Including Installer and Contractor Qualifications in Specifications.

Sample Language for Including Installer and Contractor Qualifications in Specifications

Where inclusion of installer and contractor qualifications in project specifications is desired, include the qualifications under a quality assurance section and require proof of qualifications under a submittals section. The language below may be used for the purpose of including such requirements.

Qualifications/Quality Assurance

Installing contractor is [a five-star member of the National Tile Contractors Association] [and] [or] [a Trowel of Excellence member of the Tile Contractors' Association of America].

Installing contractor's foreman or superintendent for the Project holds the International Masonry Institute's Foreman Certification.[1]

Installing contractor employs [Ceramic Tile Education Foundation Certified Installers] [and] [or] [installers recognized by the U.S. Department of Labor as Journeyman Tile Layers].

Individual installer(s) working for installing contractor are certified through Advanced Certifications for Tile Installers (ACT) for installation of <Insert one or more ACT certification requirements>: [Large Format Tile and Substrate Preparation] [and] [or] [Membranes] [and] [or] [Mud (Mortar Bed) Floors] [and] [or] [Mud (Mortar) Walls] [and] [or] [Shower Receptors] [and] [or] [Grouts] [and] [or] gauged porcelain tiles (GPT) and gauged porcelain tile panels/slabs.

Installer meets the requirements of a program identified and approved by the architect with the criteria for such

[1] The International Masonry Institute's Foreman Certification is available only to members of the International Union of Bricklayers and Allied Craftworkers; accordingly, only IUBAC signatory contractors can meet this requirement.

program similar to or exceeding [Advanced Certifications for Tile Installers (ACT) for installation of Large Format Tile and Substrate Preparation] [and] [or] [Advanced Certifications for Tile Installers (ACT) for installation of Membranes] [and] [or] [Advanced Certifications for Tile Installers (ACT) for installation of Mud (Mortar Bed) Floors] [and] [or] [Advanced Certifications for Tile Installers (ACT) for installation of Mud (Mortar) Walls] [and] [or] [Advanced Certifications for Tile Installers (ACT) for installation of Shower Receptors] [and] [or] [Advanced Certifications for Tile Installers (ACT) for installation of Grout] [and] [or] [Advanced Certifications for Tile Installers (ACT) for installation of gauged porcelain tiles (GPT) and gauged porcelain tile panels/ slabs]; [and] [or] [Ceramic Tile Education Foundation (CTEF) Certified Tile Installer Program] [and] [or] [International Masonry Institute (IMI) Contractor College Program] [and] [or] [Department of Labor recognized Journeyman Tile Layer Apprenticeship Programs].

Installing contractor meets the requirements of a program identified and approved by the architect with the criteria for such program similar to or exceeding [National Tile Contractors Association (NTCA) Five Star Contractor Program] [and] [or] [Tile Contractors' Association of America (TCAA) Trowel of Excellence Program].

Submittals

Submit certificates and other qualification data for Installer(s).

Submit certificates and other qualification data for Installing Contractor(s).

NOTES

NOTES

NOTES

TCNA

Tile Council of North America, Inc.

TILE INITIATIVE

2021

Handbook for Ceramic, Glass, and Stone Tile Installation
2021
ANSI
American National Standard Specifications for the Installation of Ceramic Tile
American National Standard Specifications for Ceramic Tile
GREEN SQUARED
AMERICAN NATIONAL STANDARD SPECIFICATIONS FOR SUSTAINABLE CERAMIC TILES, GLASS TILES, AND TILE INSTALLATION MATERIALS

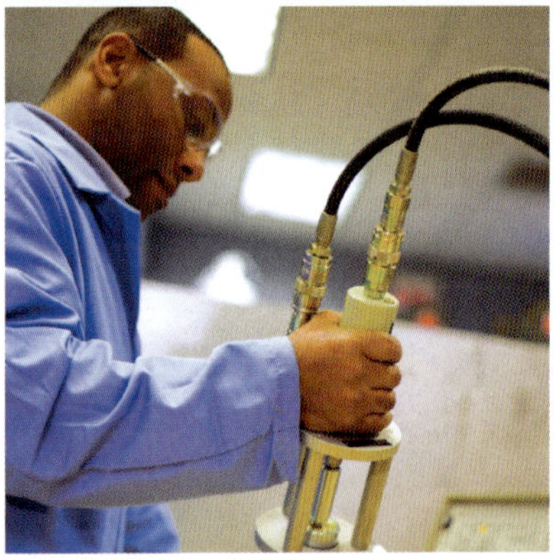

CONTENTS

ABOUT THE ADVERTISERS

The advertisements in this issue come from the Elite, Signature, and Platinum level sponsors of the Ceramic Tile Education Foundation (CTEF). Their generous donations enable the administration and growth of CTEF's educational and certification programs for tile installers and other tile industry professionals. These programs benefit the construction industry and tile-consuming public by facilitating a higher level of craftsmanship in tile installation and a means of identifying qualified installers in the trade. For more information, visit www.tilecareer.org. All of the funds received go directly to CTEF with the cost of printing donated by the Tile Council of North America (TCNA), publisher of the *TCNA Handbook*.

TCNA BULLETIN
ANSI A326.3 UNIFIES MEASUREMENT OF SLIP RESISTANCE FOR HARD SURFACES

The standard for DCOF, ANSI A326.3, unifies the measurement of slip resistance for hard surface flooring materials.

Since 2012, the DCOF AcuTest® has been the ceramic tile industry standard for testing dynamic coefficient of friction (DCOF). The DCOF AcuTest® is the method used by the tile industry for assessing a floor's relative "slipperiness," whether in-situ after installation or during manufacturing. Although initially developed for ceramic tile floors and specified in ANSI 137.1[1] since 2012, the methodology has been equally applicable and useful with other hard surface flooring, where its use has been "gaining traction" every year.

With facility managers, inspectors, cleaning professionals, and forensic engineers using the method increasingly in the field, ANSI A326.3, *Test Method for Dynamic Coefficient of Friction of Hard Surface Flooring Materials*, was developed based on the DCOF AcuTest®.

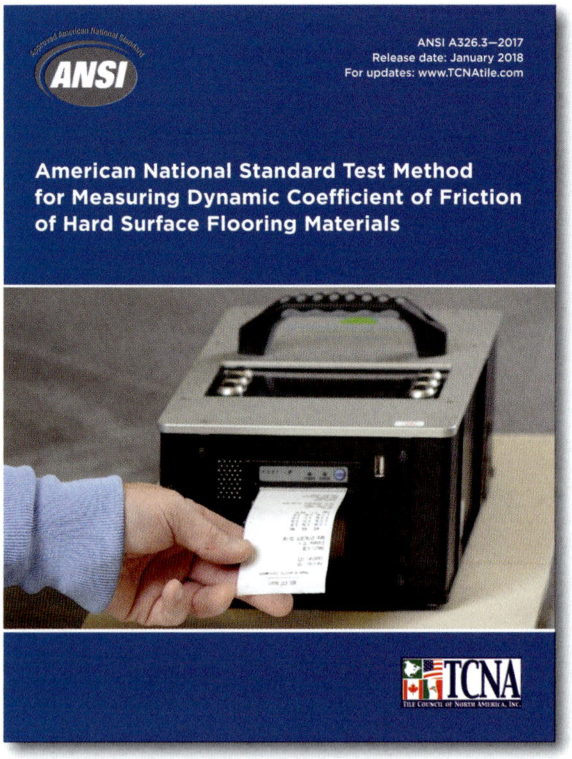

Approved in March 2017 by a broad consensus of hard surface flooring stakeholders, ANSI A326.3[2] incorporates the well-known method from ANSI A137.1 into a separate, stand-alone DCOF standard for hard surface flooring materials. A field testing section is provided for in-situ testing, including criteria and guidance for testing an installed floor under prevailing conditions, as would be the case when evaluating cleaning/maintenance procedures. Additionally, the field testing section addresses testing after cleaning, including the possible use of stronger cleaning agents than the cleaner referenced in the laboratory method. This evaluation is useful for testing the flooring surface after installation and use, without contaminants, to check for wear and to compare to the manufacturer's reported DCOF.

[1] ANSI A137.1 is the American National Standard Specification for Ceramic Tile

[2] As a simple mnemonic, 326.3 spells "DCOF" on a touchtone phone.

With the adoption of ANSI A326.3, the ANSI A137.1 and A137.2[3] standards have been updated to reference ANSI A326.3 as the stand-alone DCOF test method.

Additional Guidance for the Specifier

The ANSI A137.1 specification released in 2012 included a minimum threshold of 0.42 for tiles intended to be walked upon when wet with water, and guidance to the specifier about important factors relevant to slip resistance. ANSI A326.3 has the same specification but additionally provides further guidance helpful for the specifier. Examples of the additional guidance given in ANSI A326.3 include:

- While specifying products with higher COF for use under contaminated conditions can be considered, surfaces with higher COF can lead to maintenance/cleanliness issues and hard to remove contaminants and films, which can cause hazardous and unfavorable conditions. In addition to maintenance issues, a surface with a high COF can create a difficult walking condition for that subset of the elderly and disabled who slide their feet on the floor.

- Hard surface flooring materials with a wet DCOF less than 0.42 are often used in areas such as shopping malls (outside the food court), hotel lobbies, office buildings, etc. where appearance and ease of cleaning are highly desired and measures are in place to keep the floor dry when walked upon.

- Hard surface flooring materials that have a coating applied shall only be used in areas that can be kept dry, unless otherwise specified by the coating's manufacturer. If testing data is required after a coating is applied, use the test method specified by the manufacturer, or the dry testing procedure in ANSI A326.3 if no test method is suggested by the manufacturer of the coating.

The ANSI A326.3 standard identifies many factors that must be considered when determining the suitability of a hard surface flooring material for a particular

[3] ANSI A137.2 is the American National Standard Specification for Glass Tile.

TCNA BULLETIN
ANSI A326.3 UNIFIES MEASUREMENT OF SLIP RESISTANCE FOR HARD SURFACES

application. For example, in exterior applications, the suitability of a flooring material depends significantly on drainage of the assembly, physical structure of the flooring material, expected footwear, intended use, and the variety of contaminants present, among many other factors. This is why a single DCOF limit value is not provided for exterior applications, interior ramps/ inclines, and areas expected to be contaminated with something other than water.

Putting the Standard into Practice

Tile manufacturers already report DCOF according to ANSI A137.1/A326.3, with some including the information directly on packaging and others providing the information in their product literature. While the ANSI specification for ceramic tile does not require a minimum or maximum DCOF value independent of how the tile is used, reporting the value per the DCOF AcuTest® is required. Other hard surface flooring industries are similarly following suit.

However, it is important to understand that DCOF is only one factor in determining the suitability of a flooring material for a particular application. As the standard correctly notes, there are many factors that can affect the possibility of a slip occurring, including by way of example, but not in limitation, the following: the material of the shoe sole and the degree of its wear; the presence and nature of surface contaminants; the speed and length of stride at the time of a slip; the physical and mental condition of the individual at the time of a slip; whether the floor is flat or inclined; how the hard surface flooring material is used and maintained; and the COF of the material, how the flooring surface is structured, and how drainage takes place if liquids are involved. Specifying flooring materials appropriately requires knowledge of how the space will be used

and maintained; often the better the communication regarding this early on, the better the flooring selections.

Considering slip resistance is especially important at entrances and exits where flooring can become unintentionally contaminated, for example, from water and soil tracked inside, or from fire-fighting efforts under emergency conditions. All building chokepoints deserve extra attention, as well as spots where water and other liquids can be anticipated, such as by drinking fountains, inside and around restrooms and kitchens, etc. Also, areas where the floor may be worn or polished deserve attention and periodic monitoring. Remedial treatments are always possible when DCOF monitoring indicates the need for such.

Above all, the easy measurement of in-situ flooring through the ANSI A326.3 method allows building owners/facility managers a means to monitor their spaces for essential flooring safety. Such monitoring, sometimes called DCOF auditing, can be used to keep maintenance providers accountable. After all, if a floor is not maintaining its intended DCOF value under actual use conditions, an accident can be anticipated. When floors are properly specified and maintained, safer walkways result, benefitting all.

As a public service, the Tile Council of North America, publisher of the ANSI A326.3 standard, offers the standard without charge on the TCNA website, TCNAtile.com/DCOF.

TCNA
Tile Council of North America, Inc.
www.tcnatile.com
PREPARE FOR
SUCCESS
ANSI A137.3–2017, A108.19–2020, A108.20–2020
Release date: January 2021
For updates: www.TCNAtile.com
ANSI
American National Standard Specifications for Gauged Porcelain Tiles and Gauged Porcelain Tile Panels/Slabs
Material and Installation Standards
ANSI A137.3, A108.19 and A108.20
In addition to having a unique product standard (A137.3), gauged porcelain tiles and gauged porcelain tile panels/slabs require unique installation procedures and workmanship standards (A108.19 and A108.20).
ORDER YOUR COPY TODAY

Perform with porcelain.
Create without compromise.
Our 100% porcelain tiles are completely resistant to stains, scratches, frost, and fire. They're manufactured here in the United States to ensure jobsite success, and they provide long-lasting durability so you can avoid costly callbacks. For you and your clients, it means you'll never compromise on creative solutions again.
Get your next sample on us at CrossvilleInc.com.
CROSSVILLE
Floor: Alaska by Crossville
Proud Sponsor of the Ceramic Tile Education Foundation
CTEF

Stand Out
from the Crowd.
Never before has a pre-mixed grout offered the superior strength and performance of an epoxy, now found in SPECTRALOCK® 1.
LATICRETE
SPECTRALOCK 1
Make the move.
For more about the SPECTRALOCK® 1 story, visit SPECTRALOCK 1.com or scan QR code
Pre-Mixed Grout with Epoxy Performance – SPECTRALOCK® 1 available now.
10X
LOW VOC/ODOR
Patent Pending Technology
www.laticrete.com | 1.800.243.4788
Proud sponsor of the Ceramic Tile Education Foundation
LATICRETE
A-9040-1120 ©2020 LATICRETE International, Inc. All trademarks shown are the intellectual properties of their respective owners.

INTRODUCING

TILE TAKEOUT

HERE'S HOW IT WORKS –

STAY IN YOUR RIDE

When you arrive, **call to let us know you're here** and we will process your order (book ahead to get in and out)!

SKIP THE LINE

We will **deliver your order**, curbside, to keep you in your seat. Open accounts will be billed directly, **OR** prepay by credit card.

FOR FASTER SERVICE...

SCHEDULE PICK-UP

Use **ProExchange** to schedule your pick-up time online (at least 1 hour in advance) or **call us** to coordinate.

PRO EXCHANGE™

SIGN UP FOR PROEXCHANGE AT:

WWW.TRADEPROEXCHANGE.COM

daltile® — daltile.com

AMERICAN OLEAN® — americanolean.com

MARAZZI® — marazzitile.com

A DIVISION OF MOHAWK INDUSTRIES

Proud sponsor of the Ceramic Tile Education Foundation

TCNA BULLETIN

NEW, MEASURABLE CODE CRITERIA FOR SLIP RESISTANT SURFACES AROUND POOLS, SPAS, AND HOT TUBS

In 2021, the Uniform Swimming Pool, Spa & Hot Tub Code (USPSHTC), published by the International Association of Plumbing and Mechanical Officials (IAPMO), established for the first time, measurable criteria for slip resistant surfaces around pools, spas, and hot tubs.

Introduction to the USPSHTC

According to IAPMO, the USPSHTC establishes "minimum requirements and standards for the protection of the public health, safety, and welfare." It is relevant to "the design, construction or installation, repair or alterations of swimming pools, spas, and hot tubs," and "is available for adoption and use by jurisdictions in the United States and Internationally. Its use within a governmental jurisdiction is accomplished through adoption by reference in accordance with applicable jurisdictional laws." Prior to the publication of the 2021 edition, the USPSHTC did not include any quantitative threshold related to slip resistance, nor did it provide code officials, inspectors, specifiers, or other users with any criteria to appropriately assess walkway surfaces that are intended to be slip-resistant.

Changes to the USPSHTC

In the summer of 2019, a Slip-Resistance Task Group comprised of technical experts representing a variety of industries and fields was formed to draft recommended code revisions to standardize the specification of "slip-resistant" walkway surfaces and address the applicability of such surfaces in swimming pool, spa, and hot tub applications. The group, chaired by TCNA's Director of Standards Development and Sustainability Initiatives, developed updates to the code language that were accepted by the IAPMO Technical Committee in 2020 and subsequently approved for inclusion in the 2021 USPSHTC.

The new USPSHTC criteria for slip-resistant walkway surfaces contain the following provisions:

1. A required wet dynamic coefficient of friction (DCOF) value of no less than 0.42, for level walkway surfaces intended to be walked upon when wet, as determined in accordance with the current ANSI A326.3 standardized test method.

2. A corrected minimum DCOF requirement for inclined/sloped walkways, which require greater available friction to prevent slipping.

3. For three-dimensionally patterned or profiled surfaces, testing shall be conducted on a nominally flat section (i.e., a non-profiled, non-patterned section) of such walkways. Where that is not possible, the specifier must provide documentation substantiating the product choice.

An appendix is also provided in the code with additional information regarding the determination of DCOF values for hard surface walkways intended to be slip-resistant.

Background on ANSI A326.3

Published in 2017, ANSI A326.3 is the *American National Standard Test Method for Measuring Dynamic Coefficient of Friction of Hard Surface Flooring Materials* that is now referenced by the 2021 USPSHTC. It was developed by the ANSI Accredited Standards Committee (ASC) A108, which includes 65 members representing a broad range of stakeholders across a variety of hard surface flooring industries.

A326.3 includes a standardized test method for the measurement of DCOF between a test foot and hard surface flooring under specified conditions; it can be performed in the laboratory and in the field. The standard defines a specified minimum threshold value of 0.42 for surfaces intended to be walked upon when wet with water, and additionally provides further guidance for specifiers throughout the standard.

Product choice substantiation where DCOF measurements are not possible

To meet the new code criteria for slip-resistant three-dimensional or textured surfaces where nominally flat sections are not present, and thus where specifiers must provide documentation substantiating their product choice, TCNA recommends the internationally-recognized DIN 51130 standard, commonly referred to as the "German Ramp," as a means to characterize the slip resistance properties of such surfaces. The method requires that human subjects, wearing standardized footwear and a safety harness, walk on an increasingly inclined plane until a slip occurs. Products are classified

NEW, MEASURABLE CODE CRITERIA FOR SLIP RESISTANT SURFACES AROUND POOLS, SPAS, AND HOT TUBS

using "R" values, which are derived from the angle at which the operators slipped while conducting the test. These values are often utilized to assess product suitability for various applications, and they help illustrate a product's slip resistance characteristics. In general, products with a higher R value, or where a steeper angle of slip occurs, will have a higher degree of slip resistance.

While an understanding of the application of DIN 51130 is helpful in characterizing product slip resistance, it is important to note that the method can only be conducted in laboratory settings; field testing is not possible. Also, as a reminder, the code states that product substantiation for three-dimensional and textured surfaces is only necessary when nominally flat sections of such surfaces are not present for A326.3 testing.

DIN 51130 Slip-resistance classification chart

6° up to 10°	R9
Over 10° up to 19°	R10
Over 19° up to 27°	R11
Over 27° up to 35°	R12
Over 35°	R13

Anticipated updates to ANSI A326.3

Currently, ASC A108 is developing updates to ANSI A326.3, including a product use classification system expected to be published in 2021. At the time of publishing this bulletin, this system has been proposed to address dry areas, two categories of interior wet areas, exterior areas, and areas exposed to oils/greases. Product classification is accomplished through reference to DCOF threshold values in ANSI A326.3 and/or manufacturer declarations. For the latter, manufacturing parameters, internal quality control criteria, and manufacturer experience with similar surfaces, form the basis for making such declarations, with manufacturers defining their internal product selection criteria. These criteria may include DCOF limit values established using ANSI A326.3 or other test methods, internal reference standards and practices, and/or the presence of abrasive grain and/or surface structure. The intent of the classification system is to facilitate increased communication between manufacturers, specifiers, designers, suppliers, and consumers regarding the types of areas where products can be appropriately installed and used, ultimately providing greater consumer protection.

Although this and possible future revisions to ANSI A326.3 are not automatically adopted into the 2021 USPSHTC, but rather will be considered in future code cycles, the product use classifications can be used for further product choice substantiation.

Going beyond the code . . . other important slip resistance factors

To further address the complicated issue of slip-resistance, it is important that specifiers are aware of many different factors not addressed by standardized product testing that will affect the slip resistance an individual experiences. Referencing ANSI A326.3:

> *"There are many factors that affect the possibility of a slip occurring on a surface, including, by way of*

example, but not in limitation, the following: the material of the shoe sole and the degree of its wear; the presence and nature of surface contaminants; the speed and length of stride at the time of a slip; the physical and mental condition of the individual at the time of a slip; whether the floor is flat or inclined; how the hard surface flooring material is used and maintained; ... how the flooring surface is structured, and how drainage takes place if liquids are involved. Because many variables affect the risk of a slip occurring, the COF shall not be the only factor in determining the appropriateness of a hard surface flooring material for a particular application."

Simply stated, understanding how a space will be used and maintained, in conjunction with the code criteria in the 2021 USPSHTC and specifications contained in ANSI A326.3, allows the specifier to choose appropriate flooring for the intended use.

For more information on the code language on slip-resistant walkway surfaces, consult Section 314.0 of the USPSHTC 2021 edition on the IAPMO website.

For more information on the ANSI A326.3 standard, its publisher (TCNA) offers the standard free of charge on the TCNA website.

For parties interested in DIN 51130 "German Ramp" testing and ANSI A326.3 "DCOF" testing, please contact TCNA Product Performance Testing Laboratory, which offers both tests as part of its slip resistance testing services.

TCNA BULLETIN

INTERNATIONAL BUILDING CODE (IBC) UPDATES REQUIREMENTS FOR EXTERIOR ADHERED PORCELAIN TILE

Submitted by Scott Conwell, International Masonry Institute

For the first time since the introduction of porcelain tile to the International Building Code (IBC) in 2009, the newest edition of IBC published this year includes major changes to the requirements for exterior adhered porcelain tile. The new criteria for exterior tile were examined and developed by the International Masonry Institute (IMI), Tile Council of North America (TCNA), and manufacturers of tile and setting materials. Once consensus was reached, IMI and TCNA made a formal proposal to the International Code Council (ICC), publisher of the IBC, and testified at ICC's public hearing on the new code criteria. ICC unanimously approved the proposal and has implemented its language into the 2021 IBC. The updated requirements are in response to the increased availability of larger, thinner porcelain tiles, improved technologies in bonding mortar, and the tile industry's focus on qualified labor. These new code requirements will allow safe and more liberal use of large format adhered porcelain tile on exterior walls and will create unprecedented opportunities to use ultra-large format tile on building exteriors. The updated section from the 2021 IBC reads as follows in its entirety:

> **1404.10.2 Exterior adhered masonry veneers—porcelain tile.** Adhered units weighing more than 3.5 pounds per square foot (0.7 kN/m^2) shall not exceed 48 inches (1219 mm) in any face dimension nor more than 9 square feet (0.8 m^2) in total face area and shall not weigh more than 6 pounds per square foot (0.29 kN/m^2). Adhered units weighing less than or equal to 3.5 pounds per square foot (0.17 kN/m^2) shall not exceed 72 inches (1829 mm) in any face dimension nor more than 17.5 square feet (1.6 m^2) in total face area. Porcelain tile shall be adhered to an approved backing system.

Code requirements from 2009 through 2018

In 2009, porcelain tiles meeting ANSI A137.1 *American National Standard Specifications for Ceramic Tile* were typically manufactured to a thickness of 9/32–7/16 in. (7-11 mm), and were adhered with mortar manufactured in compliance with ANSI A118.4 *American National Standard Specifications for Modified Dry-Set Cement Mortar*. Due to safety concerns and recommendations of the tile industry at that time, the IBC limited the size of these relatively thick and heavy tiles to a maximum of 24 in. (610 mm) in any dimension, and a maximum area of 3 ft^2 (0.28 m^2) for any one tile. Moreover, the IBC imposed the restriction that any one tile shall not weigh more than 9 pounds per square foot (0.43 kN/m^2). By today's standards these requirements seem overly restrictive, but they served the tile industry well in the last decade.

What factors have sparked the code changes?

The past ten years have seen major developments in the tile industry leading to the safe use of larger adhered porcelain tiles: the advent of larger and thinner tiles, the development of stronger mortars, and the increased focus on installer training and certification programs.

Larger, thinner tiles

Manufacturers of porcelain tile in North America and worldwide now have the technology to produce tiles that are significantly thinner and larger than tile manufactured in previous decades. Porcelain tiles manufactured to ANSI A137.3 *American National Standard Specifications for Gauged Porcelain Tiles and Gauged Porcelain Tile Panels/Slabs* are commonly 1/8–7/32 in. (3.5–5 mm) in thickness. Gauged porcelain tile (GPT) panels also frequently have facial dimensions nearly as large as 6 ft. x 12 ft. (1800 mm x 3600 mm in actual size). The thinner profile of these tiles results in units that remain relatively lightweight despite their ultra-large format.

Stronger mortar

Just as tile manufacturers were developing technology that would test the limits of tile size and thickness, setting material manufacturers were improving their products to provide more tenacious bond strength. ANSI A118.15 *American National Specifications for Improved Modified Dry-Set Cement Mortars* was approved as a new standard in 2014, establishing more rigid criteria for resistance to shear forces. For example, the required 28-day shear strength of mortar meeting ANSI A118.15, the standard recommended by most GPT panel manufacturers, is double the requirement for mortars meeting ANSI A118.4, at 400 psi (2.76 MPa) and 200 psi (1.38 MPa) respectively.

TCNA BULLETIN

INTERNATIONAL BUILDING CODE (IBC) UPDATES REQUIREMENTS FOR EXTERIOR ADHERED PORCELAIN TILE

Industry's focus on qualified labor

Finally, even with these tremendous improvements in materials, the tile setters, finishers, and contractors installing them must be qualified. Trained, skilled craftworkers and contractors are paramount to the success of any tile project, and when the public's health, safety, and welfare are at stake, qualified labor is critical. The tile industry has made great strides in the last decade in expanding the delivery of longstanding programs like the apprenticeship and training program of the International Masonry Training and Education Foundation (IMTEF), which is the training arm of the IMI for union installers, and the Certified Tile Installer (CTI) program of the Ceramic Tile Education Foundation (CTEF) for union and non-union installers.

Further, the Tile Contractors' Association of America's (TCAA) Trowel of Excellence certification and the National Tile Contractor Association's (NTCA) Five Star certification establish credentials for best practice tile contractors. And for installers who aspire to be the most technically proficient in their field, the Advanced Certifications for Tile Installers (ACT) allow for installers to be assessed and certified in seven critical skill areas including the installation of large format tile (LFT) and GPT. Each of these qualified labor training and certification programs works closely with manufacturers of tile and setting materials to ensure that the skilled labor force is constantly kept up to date with the latest material and installation technologies.

The new code requirements in 2021

So how large a porcelain tile can now be adhered to a facade? To simplify the new code requirements, the criteria in the 2021 IBC may be summarized as follows:

> For tiles greater than 1/4 in. (6 mm) thick, the maximum tile size is 9 ft^2 (0.84 m^2), which in common tile sizes is 36 x 36 in. (915 x 915 mm) or 24 x 48 in. (610 x 1220 mm). Tiles of this thickness may not have any face longer than 48 in. (1220 mm).
>
> For tiles less than or equal to 1/4 in. (6 mm) thick, the maximum tile size is 17.5 ft^2 (1.6 m^2), which translates to 48 x 48 in. (1220 x 1220 mm) or 36 x 72 in. nominally (900 x 1800 mm in actual size). Tiles of this thickness may not have any face longer than 72 in.

Finally, while this update applies to large and ultra-large format tiles, it is important to remember that ceramic and porcelain tiles in traditional sizes have always been and continue to be allowed under the IBC for exteriors. With this code change, a design professional or owner can now specify these larger size tiles without submitting any special requests to the local building code official. If a design professional or owner wishes to install tiles that exceed these requirements, it may still be possible to do so by applying for a code variance with the authority having jurisdiction.

What's next?

As states, provinces, and municipalities adopt the 2021 IBC, we encourage design professionals across North America to consider ceramic and porcelain tiles in all sizes for their benefits on exteriors as well as interiors. Their health benefits, durability, low maintenance, sustainability, and unlimited design aesthetics make ceramic and porcelain tiles the perfect choice for building facades, and it is now easier than ever to meet building code requirements and deliver a beautiful, healthy building.

Project: Perry's Steakhouse

Architect: Aria Group

General Contractor: International Construction Inc.

Tile Contractor: Ready Tile

Craftworkers: BAC Local 21 Illinois

Photo credits: International Masonry Institute

PORCELAIN TILE CERTIFICATION
AND THE PORCELAIN TILE CERTIFICATION AGENCY (PTCA)

In 2007 Tile Council of North America (TCNA) partnered with the Ceramic Tile Distributors Association (CTDA) to create the Porcelain Tile Certification Agency (PTCA) to certify porcelain tile. This program was formed at the initiative of distributors and manufacturers who were concerned with the amount of tile being sold in the United States that was marked as porcelain but which was in fact not porcelain.

CERTIFIED PORCELAIN® TILE
0.5% or less water absorption

What is porcelain tile, and why does it matter whether a tile is porcelain?

As defined by the ANSI A137.1 ceramic tile standard, porcelain tile is a ceramic tile with a very low water absorption (0.5% or less), as tested per ASTM C373. Porcelain tile is denser and has a lower water absorption than other types of ceramic tile. When non-porcelain tiles are unknowingly substituted, freeze/thaw and expansion failures can result from unexpected moisture absorption.

Why was this porcelain tile certification program created?

It is well known that some non-porcelain tiles made overseas are knowingly mislabeled as porcelain, with exporters and importers choosing to ignore the relevant North American standard (ANSI A137.1). While the criteria for porcelain tiles have been well-defined for several decades in North America, this practice of mislabeling tiles began when the term porcelain was undefined in international standards.

Today the term is well defined and the ISO and ASTM water absorption methods are harmonized, but historically the method used internationally for measuring water absorption (ISO) was less rigorous than the ASTM C373 method used in North America.

Simply stated, this means that some tiles classified as

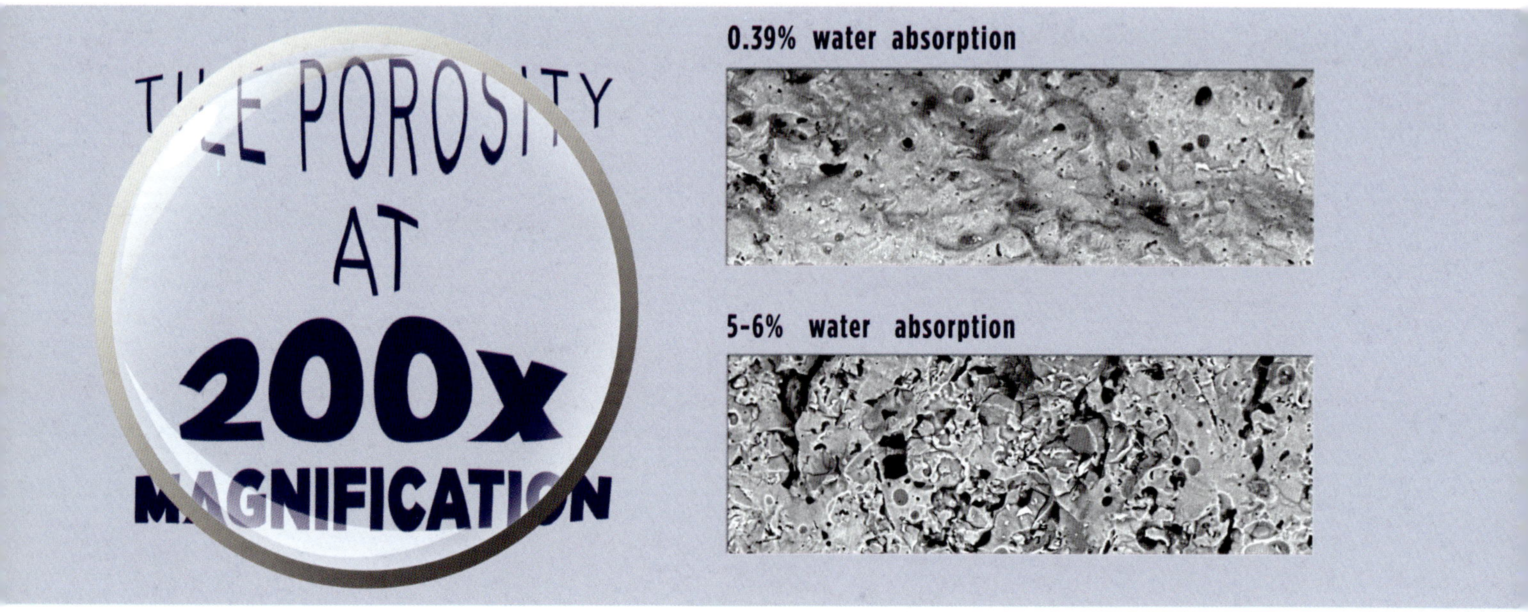

porcelain overseas haven't met the stricter and more demanding water absorption criteria used in North America. In freeze/thaw and wet environments, that can be important. Given that approximately 70% of the tiles sold in the United States are imported, PTCA certification was developed to protect the consumer from tiles either intentionally mislabeled or mislabeled due to differences in testing.

Through the PTCA program the need for porcelain certification has become even more evident, as 360 tile series failed as of November 2019.

Can only manufacturers sign up for PTCA certification?

No. The program is open to both manufacturers and sellers of porcelain tiles. Either can be a program participant.

PTCA certification: What does it mean?

Recognizing that the extent of this false labeling issue only applies to whether or not tiles meet the water absorption criteria of the ANSI A137.1 standard, PTCA certification was developed only to independently evaluate if the program participant understands North American water absorption criteria and can meet such. Tiles are not checked to see whether or not they meet all the other relevant properties for porcelain tiles in the ANSI A137.1 standard; variance from those properties has not been an issue in general, and the criteria are well understood. For each series being evaluated, five commercially available samples (selected by the participant) are sent once every three years by manufacturing participants and annually by non-manufacturing participants.

Passing the initial testing establishes that the participant understands and can meet North American water absorption criteria. For more details on the PTCA program, the PTCA Program Participation Agreement is publicly available and can be found at www.ptcaonline.org/PTCA_Participation_Agreement.pdf.

PORCELAIN TILE CERTIFICATION
AND THE PORCELAIN TILE CERTIFICATION AGENCY (PTCA)

If a box of tiles has the PTCA certification mark on it, is PTCA stating that those tiles meet ANSI A137.1 water absorption criteria?

No. PTCA establishes that the program participant understands North American water absorption criteria and is able to meet such.

The quality of the tiles being sold is exclusively controlled by the actual manufacturer.

If a box of tiles has the PTCA certification mark on it, is the program participant stating that those tiles meet all ANSI A137.1 criteria?

While the program participant may independently claim compliance with all ANSI A137.1 porcelain tile criteria, that is not required by PTCA of program participants.

By participating in the PTCA certification program, the program participant (i.e., the manufacturer and/or seller) is stating that the tiles it produces or sells labeled with the PTCA mark meet the ANSI A137.1 porcelain tile water absorption requirements.

Non-manufacturing participants are further required to obtain a written assurance from the actual manufacturer that it will immediately notify the participant of any changes in the conforming porcelain tiles or any manufacturing variances that may affect the certification.

To further ensure the program's effectiveness, participants have agreed not to use the PTCA mark in any way misleading or confusing to buyers, including displaying the certification mark in a way that would imply non-certified products are certified. Participants also are not allowed to transfer use of the mark to any other person or entity.

As noted above, PTCA certification does not mean the tiles tested met all ANSI A137.1 or ISO 13006 criteria, which would require testing for other physical properties such as dimensions, warpage, breaking strength, etc. That assurance would need to come from the manufacturer or via a third-party lab.

If a box of tiles has the PTCA certification mark on it, can those tiles be used in freeze/thaw and wet environments without concern?

While the tiles may be perfectly appropriate for such use, the PTCA certification mark does not suggest that. The suitability of any tiles for specific applications requires an analysis of the project conditions by a qualified individual and proper installation. The certification mark does not assure fitness for any particular purpose.

What are the benefits of the PTCA program?

The PTCA program is designed to directly benefit consumers purchasing porcelain tiles and, indirectly, everyone involved in the supply chain.

Participants benefit by being able to independently confirm to customers that what they are producing or selling is truly porcelain, and by being able to differentiate their products from falsely-labeled porcelain products.

Producing porcelain tiles can be a more intensive and costly process than producing non-porcelain tiles, so certification is a good way for manufacturers and sellers to confirm that investment to the market.

For distributors PTCA certification helps differentiate real porcelain tiles from those that are falsely-labeled as porcelain.

Who polices PTCA-certified tile?

The marketplace does. If a question arises about whether a tile sold as PTCA-certified truly meets the water absorption criteria for porcelain tiles, PTCA is authorized to acquire further samples and test such for compliance. The board of PTCA then reviews the available data and relevant actions taken by the program participant to decide whether to withdraw PTCA certification and use of the mark.

Anyone who suspects a non-porcelain tile is being sold as PTCA-certified tile is encouraged to notify PTCA at 630-942-6588 or info@ptcaonline.org.

TCNA BULLETIN
WHAT IS TRUE PORCELAIN?

CERTIFIED PORCELAIN® TILE
0.5% or less water absorption

The difference between real and false porcelain cannot be detected by eye.... Suppliers of falsely-labeled porcelain are defrauding the consumer and benefitting from the popularity and market value of genuine porcelain.

Porcelain tile has become increasingly popular over the past decade. The American National Standard Specifications for Ceramic Tile (ANSI A137.1) require tile to have a water absorption of 0.5% or less to be classified as porcelain when tested per ASTM C373.

Manufacturing tile that meets this standard—true porcelain—requires porcelain-grade clays and other unique raw materials, plus precision milling processes and kilns set to extremely high firing temperatures (2100°F to 2500°F). The required raw materials, energy, and manufacturing equipment needed to produce such low porosity, high density tile are why real porcelain is typically more expensive than non-porcelain tile.

The difference between real and false porcelain cannot be detected by eye—the only way to know is to have a laboratory verify the tile's water absorption is 0.5% or less. Through its lab, Tile Council has identified 360 series that did not meet the PTCA water absorption criteria necessary to be certified as porcelain.

Suppliers of falsely-labeled porcelain are defrauding the consumer and benefitting from the popularity and market value of genuine porcelain. This is particularly true for imported tile, and considering that approximately 70 percent of the tile sold in the United States is imported, much of the "porcelain" being sold may be falsely-labeled.

ASTM C373 Water Absorption Test

For ceramic tile, water absorption refers to the maximum amount of water that a tile can be made to absorb. In the lab test ASTM C373, water is drawn into the deepest pores of the tile by a strong vacuum.

So, measuring water absorption can also be looked at as measuring available tile porosity—the more water that can be absorbed, the more porous (less dense) the tile.

STEP 1

Tile sample is dried in an oven to ensure accurate dry weight.

STEP 2

Dried tile sample is weighed using a digital scale accurate to 0.001 gram.

STEP 3

Water is drawn into the tile sample by a vacuum and soaking it.

STEP 4

Saturated tile sample is weighed to determine the amount of weight gain due to absorption of water.

STEP 5

Water absorption is calculated.

The change in weight is expressed as the percentage of the tile's dry weight.

WHEN QUALITY IS THE BOTTOM LINE
MATCH ACT-CERTIFIED INSTALLERS TO THE JOB AT HAND

ACT
ADVANCED CERTIFICATIONS FOR TILE INSTALLERS

Tile setting has become a more and more specialized trade, yet it remains largely unregulated when it comes to requirements for installers, whether for training or for proven adherence to best practices and industry standards. The easy entry into tile setting means a contractor may have seasoned, skilled craftworkers or untrained installers with little experience under their belts. And, without an established skills baseline, the contractors that don't invest in installer training and education have a competitive edge, if the only consideration for choosing from a pool of tile contractors is which one has submitted the lowest bid, the norm for the vast majority of commercial work today.

This is the system for awarding tile jobs—too often to unqualified companies—that ACT (Advanced Certifications for Tile Installers) seeks to improve by establishing a skills baseline that allows consumers to compare costs and qualifications.

Launched in 2014, ACT is a program of written and hands-on testing for defined skill sets, like large format tile installation. While other training and certification programs are available to tile installers, ACT has garnered wide support from the tile industry because it is standards-based and highly demanding.

Construction and building design professionals are encouraged to integrate installer qualifications as requirements for bidding contractors, under "quality assurance" sections of their specs.

Requiring evidence of program completion or certifications under "submittals" is also recommended to help ensure the specified requirements for installers are met.

For ready-to-use boilerplate spec language, see the Installer and Contractor Qualifications Guide in the TCNA Handbook, or, for easy copy and paste, visit TCNAtile.com.

ACT tests are not show-up-for-a-demonstration-and-get-your-certificate events. A percentage of installers fail, which differentiates ACT as a meaningful certification, not an educational session. The tests have strictly enforced time limits, and installers' hands-on work is evaluated and scored in-person, by approved evaluators only. Upon completion of the hands-on component by the installer, the evaluator literally tears it apart. By prying up tiles and probing fresh mortar beds, ACT evaluators judge what's below the surface, a crucial component of the program, as so much of what is required for a successful tile installation lies below the finished tile work.

ACT tests are administered by the Ceramic Tile Education Foundation (CTEF) and the International Masonry Institute (IMI), which collaborated to develop the program, with support from product manufacturers and industry organizations including the National Tile Contractors Association (NTCA), Tile Contractors Association of America (TCAA), Tile Council of North America (TCNA), and the International Union of Bricklayers and Allied Craftworkers (IUBAC).

ACT
ADVANCED CERTIFICATIONS FOR TILE INSTALLERS
GROUTS

ACT Certification: GROUTS

Specify ACT GROUTS certification on every job where cementitious grout, epoxy grout, or modified epoxy emulsion grout will be used.

Critical Installation Skills Tested: Proper mixing, installation, and curing of cementitious grout, epoxy grout, and modified epoxy emulsion grout

ACT
ADVANCED CERTIFICATIONS FOR TILE INSTALLERS
LARGE FORMAT TILE

ACT Certification: LARGE FORMAT TILE / SUBSTRATE PREP

Specify ACT LARGE FORMAT TILE certification when tile larger than 15" long will be installed by a thin-bed method.

Critical Installation Skills Tested: Flattening a substrate to receive large tile and installing large tile within industry tolerances for coverage, flatness, and lippage

ACT
ADVANCED CERTIFICATIONS FOR TILE INSTALLERS
MEMBRANES

ACT Certification: MEMBRANES

Specify ACT MEMBRANES certification when a sheet or liquid membrane will be used for waterproofing or crack isolation.

Critical Installation Skills Tested: Application of sheet and liquid membranes with emphasis on avoiding installation errors that affect waterproofness

ACT
ADVANCED CERTIFICATIONS FOR TILE INSTALLERS
SHOWERS

ACT Certification: SHOWERS

Specify ACT SHOWERS certification when designing showers with a mortar bed and tile floor over a shower-pan membrane.

Critical Installation Skills Tested: Creating a watertight (leak-proof) shower base that effectively evacuates water

ACT
ADVANCED CERTIFICATIONS FOR TILE INSTALLERS
MUD WALLS

ACT Certification: MUD WALLS

Specify ACT MUD WALLS certification when a mortar bed has been selected as the substrate for tiling walls.

Critical Installation Skills Tested: Installing wall mud to ANSI standards, with emphasis on proper materials and precision of finished work (flat, plumb, level, square)

ACT
ADVANCED CERTIFICATIONS FOR TILE INSTALLERS
MUD FLOORS

ACT Certification: MUD FLOORS

Specify ACT MUD FLOORS certification when a mortar bed has been selected as the substrate for tiling floors.

Critical Installation Skills Tested: Installing floor mud to ANSI standards, with emphasis on proper materials and precision of finished work (flat, level)

TCNA BULLETIN
CHOOSING YOUR TILE CONTRACTOR

TILE. It's the go-to finish when you're looking for high fashion and high function. But you might not get either if you leave it to just anyone to install. Unlike plumbing, electrical, and structural masonry trades, tile installers and the tile contractors that employ them are not generally required to meet minimum trade craft criteria to be in business.

The difference between trained, experienced installers and inexperienced installers is noticeably reflected in their work, and the difference between a quality contractor and a deficient one is reflected in their service and business operations.

Together, contractor and installer transform your concept into reality. Whether you're a design/build professional selecting tile contractors on a regular basis or a homeowner with a single tile project, it's just not possible to overestimate the importance of finding qualified contractors and installers.

The Reputable Tile Contractor

- ☑ **Operates a legitimate business,** with responsible business practices and a policy of standing behind their work.
- ☑ **Invests in continuing education** necessary to stay up-to-date on current building codes, regulations, standards, and best practices. On-the-job training is the most popular way to learn a construction trade, but formalized training is a must for ensuring correct installation methods are being taught to and used by installers on your project.
- ☑ **Carries all required business licenses and insurances,** and doesn't push liabilities for property damages or worker injuries onto others.
- ☑ **Does not misclassify workers** to avoid paying into social security, unemployment, workers' compensation, and other employee programs.
- ☑ **Has a traceable business location** so customers can be sure post-installation questions and issues are addressed and resolved.
- ☑ **Has a track record for quality and service:** Good contractors can easily produce references and verifiable documentation of their commitment to quality and service.

Architects and Specifiers

Include language in job specifications requiring qualified labor and enforce it with the GC. See the TCNA Handbook for a list of industry recognized prequalification programs for installers and contractors such as the CTEF Certified Tile Installer Program, the ACT (Advanced Certifications for Tile Installers) Program, the NTCA 5-Star Contractor Program, and the TCAA Trowel of Excellence Program.

General Contractors

Deliver a quality tile installation by fulfilling contractor qualification requirements in job specifications. When not included, utilize internally developed qualifications. Require proof of qualifications to be included with all project bids. Thoroughly compare estimates from bidding contractors before awarding contracts. Often, higher estimates reflect better materials and additional necessary components and tasks, like substrate preparation and movement joints.

Homeowners

Don't hesitate to ask contractors for proof of insurance, their license (where required), and their installation qualifications. Thoroughly interview bidding contractors and check several references. Utilize consumer resources available from your state on the internet and from the Ceramic Tile Education Foundation.

"Because tile is a permanent finish, the lowest bid should not be the driving factor, but rather who is the most qualified to perform the scope of the work specified."

— *TCNA Handbook*

Call CTEF at 864-222-2131 or visit CeramicTileFoundation.org for assistance finding or specifying a quality contractor.

Schluter Systems
Schluter®-EDUCATION
So Many Ways to Upgrade Your Trade!
CONTINUING EDUCATION AIA
IDCEC INTERIOR DESIGN CONTINUING EDUCATION COUNCIL
Seminars
Live Online Training
Workshops
Schluter®-Systems
www.schluter.com/education
Proud Sponsor of the Ceramic Tile Education Foundation (CTEF)

ENDURING DESIGNS NEED SOLID FOUNDATIONS
American National Standard Specifications for Ceramic Tile
American National Standard Specifications for the Installation of Ceramic Tile
Material and Installation Standards
TCNA Handbook for Ceramic, Glass, and Stone Tile Installation
2021
ANSI A137
product performance—
ceramic and glass tile
ANSI A108, A118, A136
installation and
workmanship; product
performance—mortars,
grouts, and membranes
TCNA Handbook
installation systems—
ceramic, glass, and
stone tile
START WITH STANDARDS
Now available in digital or print editions at www.TCNAtile.com
TCNA
TILE COUNCIL OF NORTH AMERICA, INC.
Guastavino tile circa 1909, Bridgemarket, New York City
Photo: Michael Freeman

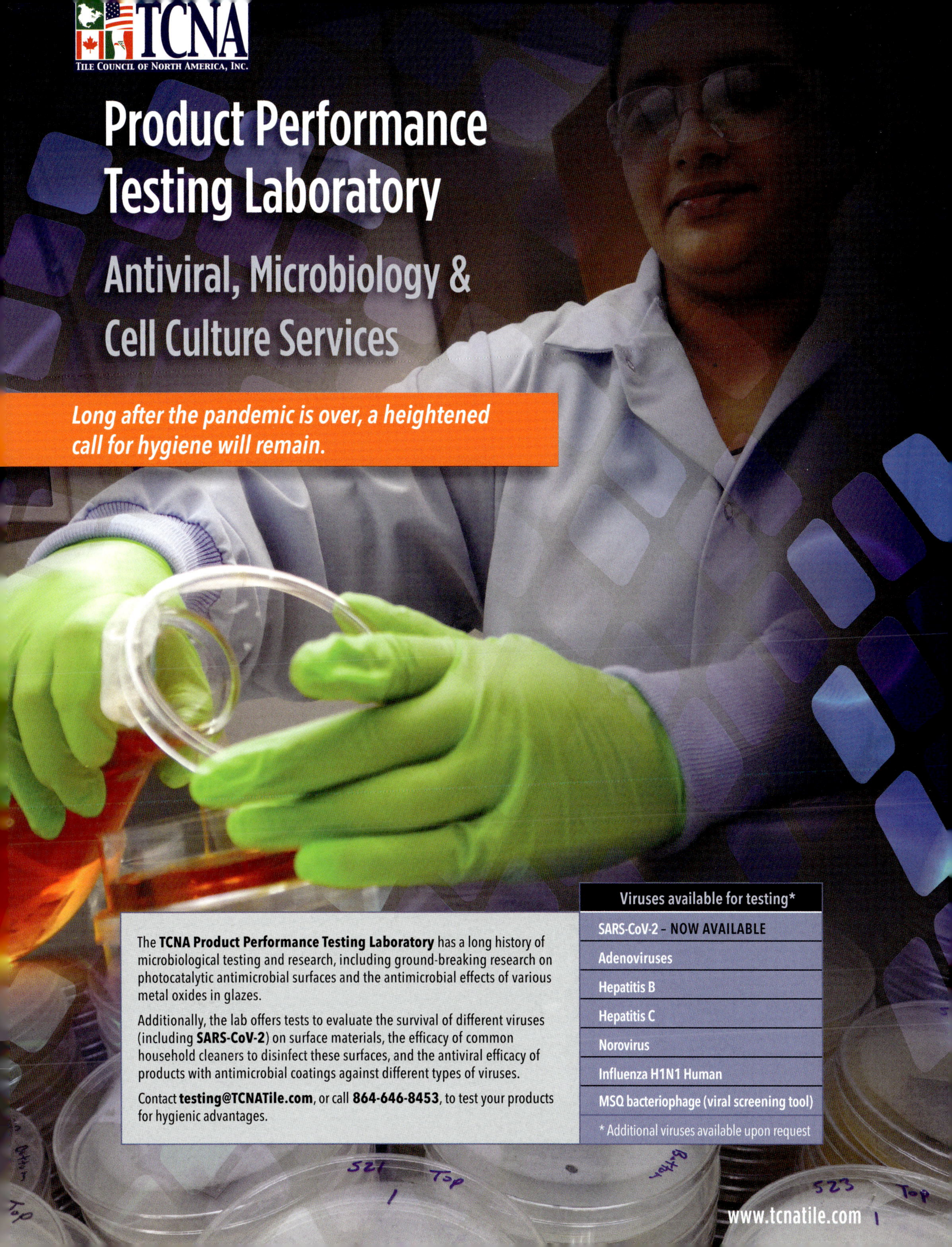
TCNA
Tile Council of North America, Inc.
Product Performance Testing Laboratory
Antiviral, Microbiology & Cell Culture Services
Long after the pandemic is over, a heightened call for hygiene will remain.
The TCNA Product Performance Testing Laboratory has a long history of microbiological testing and research, including ground-breaking research on photocatalytic antimicrobial surfaces and the antimicrobial effects of various metal oxides in glazes.
Additionally, the lab offers tests to evaluate the survival of different viruses (including SARS-CoV-2) on surface materials, the efficacy of common household cleaners to disinfect these surfaces, and the antiviral efficacy of products with antimicrobial coatings against different types of viruses.
Contact testing@TCNATile.com, or call 864-646-8453, to test your products for hygienic advantages.
Viruses available for testing*
SARS-CoV-2 – NOW AVAILABLE
Adenoviruses
Hepatitis B
Hepatitis C
Norovirus
Influenza H1N1 Human
MSQ bacteriophage (viral screening tool)
* Additional viruses available upon request
www.tcnatile.com

TCNA BULLETIN

CERAMIC TILE TESTING AND RESEARCH IN A PANDEMIC WORLD

Throughout the COVID-19 pandemic, the words "unprecedented" and "new normal" have been used more than anyone could have imagined. It should come as no surprise that the TCNA Product Performance Testing Laboratory has experienced a significant shift in daily testing and the types of testing and research inquiries received. The TCNA lab, historically focused on testing and research toward the development of standards and technical advancement of the tile industry, has expanded over the years to become a trusted source for third party product performance testing of tile, related installation materials, and many other products used in construction. For over 10 years, the lab has also performed microbiological testing and research, including ground-breaking research on photocatalytic antimicrobial surfaces and the antimicrobial effects of various metal oxides in glazes.

Pandemic-driven demands have called for more microbiological information about surfaces with which building occupants regularly come into contact. Because of the lab's extensive experience, microbiological and antimicrobial testing capabilities, and expertise testing a variety of surfaces and products, TCNA's lab is uniquely positioned to provide state-of-the-art research and testing on materials used in public and residential spaces to meet these new, critical needs.

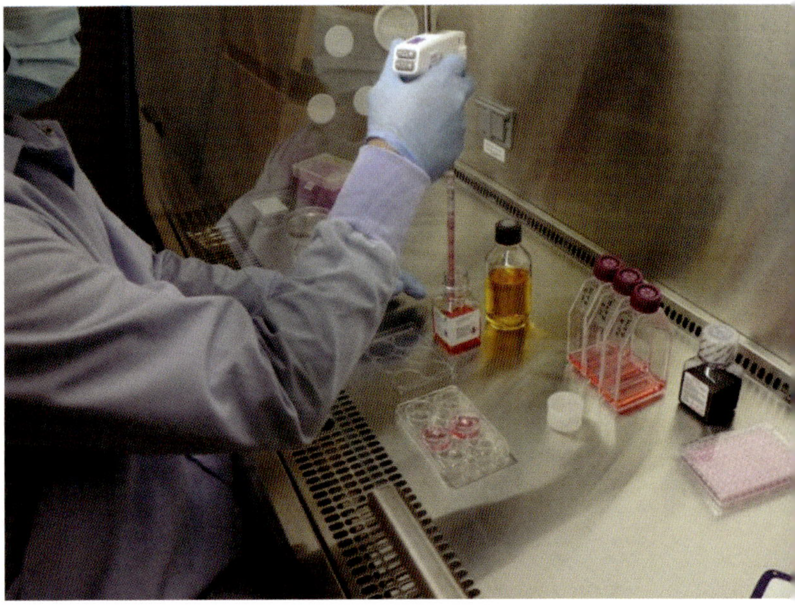

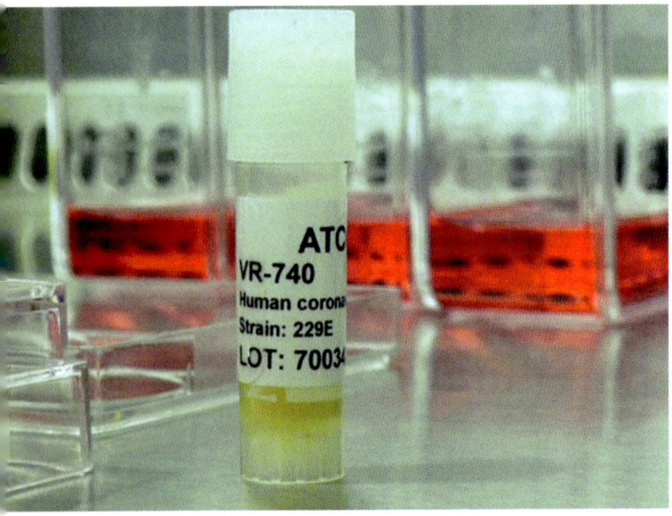

As the lab focuses more attention on antiviral testing, such testing will contribute to the race to protect individuals against contracting and/or spreading viruses from contaminated interior surfaces. This data can also form a basis for reference when specifying interior products—a heightened consideration that will prevail long after the pandemic has subsided.

Bridging the Disconnect between Construction Experts and Testing Experts

Prior to COVID-19, there was a disconnect between building construction experts and microbiologists. This disconnect is now being bridged through work at TCNA and by those companies bringing antimicrobial solutions to the market. The TCNA Product Performance Testing Laboratory is the only laboratory in the U.S. specializing in microbiological testing of floor, wall, and countertop surfaces. In 2020, the lab expanded testing to serve a broader spectrum of materials, such as sanitaryware, textiles, plastic films, copper composites, transportation interiors, food prep areas, disinfectants, and cleaning agents.

Increased testing through the TCNA lab has provided information about virus survivability on various surfaces, an area where there was previously very little information. There is currently a lack of standards for evaluating the efficacy of antiviral surfaces and agents and virus survival rates on surfaces. The work being done at TCNA is leading the development of internationally recognized test methods to standardize this type of testing as emphasis grows on the importance of antimicrobial properties.

Encouraging Product-Specific Discoveries

Thanks to increased testing, information is actively being collected and communicated but there is still a great deal of work to be done. There is encouraging news in some of the product-specific insights discovered in the lab thus far. As examples: There are certain antimicrobial coatings and various glaze ingredients that can be effective against the survival of SARS-CoV-2. Copper additives can also be effective, although oxidation on any metals can be a concern.

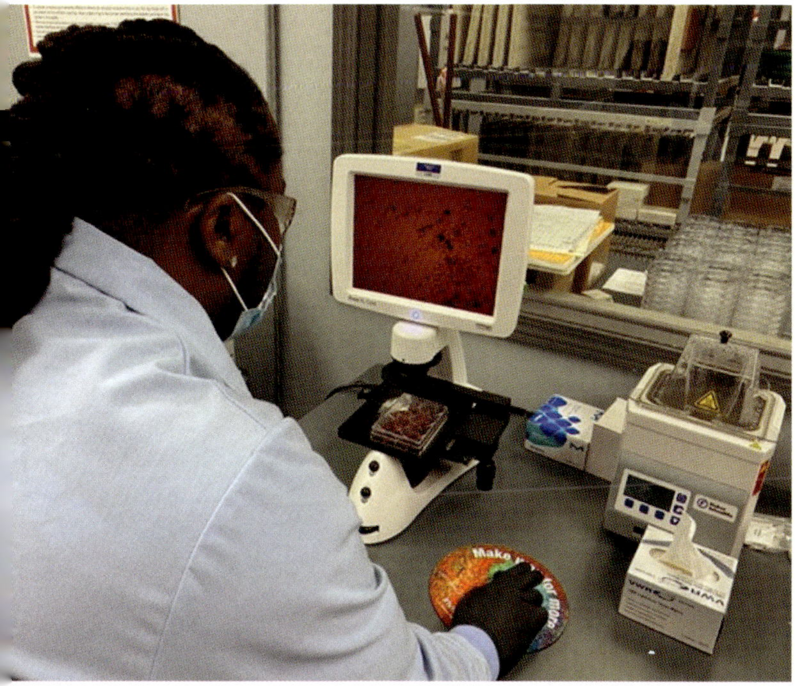

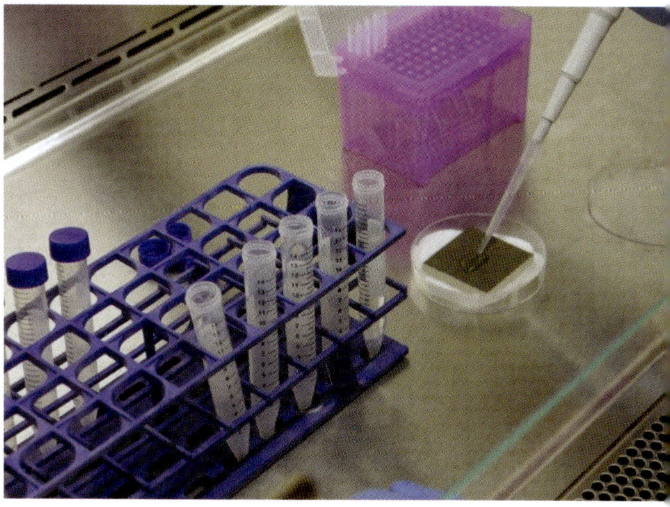

The TCNA lab continues to evaluate the duration of virus survival on ceramics and other non-porous surfaces. Additionally, the lab is studying the antimicrobial performance of porous surfaces, including treated/non-treated textiles and plastics.

The COVID-19 pandemic has changed how people look at the spaces around them; consumers and designers are seeking hygienic finishes and furnishings. Public health will be at the forefront of our world for the foreseeable future and products will continue to be developed to meet specific needs. The call for antimicrobial testing will continue to grow as standards evolve. Ongoing microbiological work at TCNA's lab provides necessary insights into the performance of finishes and furnishings, information that will prove beneficial to building occupants for many years to come.

Tile Heritage Foundation: Guardian of American Tile History

Tile Heritage Foundation was established in 1987 as a member supported, not-for-profit organization whose sole purpose is to protect and preserve the history of the American Tile Industry.

Tile Heritage is dedicated to promoting an awareness and appreciation of ceramic surfaces in the United States.

The Foundation is a repository, an archive, which embraces all aspects of the industry from its inception in the 1870s through to the present time, validating its significance for posterity.

The Foundation's archives include an estimated 40,000 documents and an equal number of images, both historic and contemporary. Manufacturing, distribution and installation history are represented.

The body of information on hand, coupled with expertise resulting from over 70 years of combined experience in the field and access to a network of experts worldwide, provides assurance of both helpful and accurate answers to questions and solutions to problems.

Above, Ross Purdy scrapes away excess clay from a tile mold at Orton Hall, OSU (1903). Photo courtesy of the Ohio State University archives. At left, a patented process for manufacturing mosaics (No. 537703 dated Apr. 16, 1895) provided Mosaic Tile Co. with a potential means of competing with its neighbor, American Encaustic Tiling Co., by making multilayered or multicolored tiles faster and less expensively.

Unless otherwise specified, all imagery courtesy of the Tile Heritage Digital Library.

Manufacturing represents the cornerstone of the tile industry.

Mosaic Tile Company, Zanesville, Ohio, circa 1910

After seven years modeling for American Encaustic, Herman Mueller along with his fellow ceramist Karl Langenbeck left in 1894 to establish the Mosaic Tile Company, also in Zanesville. Within a year, in April of 1895, Mueller was granted a patent for the "Process of and Apparatus for Manufacturing Mosaics," an encaustic tile that imitated fine mosaics.

Although Mueller referred to the tile as "mosaic," the decorative patterns were executed in small tesserae shapes of clay one-eighth inch square that extended through the clay body. Along with a series of individual motifs, large mosaic murals were produced like the pictorial mural at the Moerlin Bottling Department in Cincinnati, the facade of St. Nicholas Catholic Church in Zanesville, the "Eureka" panels throughout the main floor of the California State Capitol Building in Sacramento (replaced in recent years with Heath tile), and the "Presidential Seal" in the center of the floor at the National Building Museum in Washington DC (see previous page).

Tympanum, St. Nicholas Catholic Church, 925 E. Main St., Zanesville.

Herman C. Mueller, circa 1890

The primary goal of the Tile Heritage Foundation is to assist in the preservation of ceramic surfaces, which includes its legendary history, significant installations, as well as the objects themselves. By providing pertinent information, unbiased consultation and specific recommendations when needed, the Foundation serves both the industry and the public at large as no other agency can.

Marketing, sales, and distribution have served as an essential realm within the tile industry since tiles were first made to sell.

Tile showroom of J. A. Finn Co., 31 Madison Ave., New York City, exclusively featuring Mosaic Tile Company's products, circa 1930. Designed by Paul A. Faire. Tile Contractor: M. F. Welch.

During the last half of the 1920s and into the 1930s a number of tile manufacturers provided exclusive representation to independent showrooms to sell their products sometimes in addition to hosting showrooms themselves. There were competitive advantages for tile companies to do so. No doubt, Mosaic Tile was well aware of their Zanesville rival, American Encaustic, bankrolling its exotic and successful showroom in Manhattan just ten blocks to the north. Photos gift of Julie Mackall, Gilbertson Collection.

Office of J. A. Finn Co. immediately adjacent to the showroom.

Honoring the work and artistry of tile installers through the archiving of their accomplishments validates tiles for posterity.

Robert Howden, without question a "self-made man," chose tile setting for what became an illustrious career. Born in Scotland in 1863, the youngest of eleven children, he worked half time in the woolen mills from age 9, leaving school entirely at 13 to work full time. During this time he worked in six different mills while attending seven different schools!

The Howden Building at 17th and Webster Streets in Oakland, CA.

Robert with two of his three grandchildren, Ed and Betty, broke ground for the Howden Building in 1925.

In 1882 at age 19 Robert immigrated to the U.S. settling in Minneapolis and took a job as a helper in a mantel and tile business; a year later he became a "full-fledged" (his word) setter. He married in 1887 and soon moved to Tacoma before discovering the Bay Area on his return from the Chicago World's Fair in 1893. Oakland became his home where he rented a store, fitted up a display in front, and lived in the rear with his wife and three children. Soon he bought a house and moved with his family there. Next he purchased property for cash and built a 2-story brick building for his tile business, completing the work just three months before the 1906 earthquake.

Despite his losses business was booming during the reconstruction period following the quake. By the time Robert was breaking ground for the Howden Building at 17th and Webster, in his own words: "I had made the name Robert Howden a household word"!

Note: The original tile showroom in the Howden Building remains virtually intact today, featuring Batchelder, Claycraft and Solon & Schemmel tiles as well as housing a popular cafe.

Identification encourages both maintenance and restoration of historic tile installations.

Homeowner in Dallas, Texas was told that his 1926 fireplace was from the Rookwood Pottery. Could it be true? No, the tiles were produced by California Clay Products Co., popularly known as "Calco," located in South Gate, California and referred to as "Mayan Art" in the company's 1930 catalog.

The Tile Heritage Foundation offers tile identification services to the public at no charge. Simply email foundation@tileheritage.org with clear, low res images of individual tiles or tile installations along with whatever relevant information is readily available: site (city/state), size, date (approximate), architect/ designer if known. If the experts at Tile Heritage are not able to identify the work, your email will be forwarded to others who are likely to know.

The archives held by THF are of national importance—they tell our story. We are all part of this heritage!

The owner of this early 1920s Tudor Revival home in Birmingham, Alabama suspected that her tiles were Batchelder's but needed confirmation. She contacted Tile Heritage to find out and received the cleaning recommendations she asked for along with the verification of her suspicions.

Preservation of significant installations is of paramount importance.

Original ceramic tile mural (11′ x 45′) designed and produced by Guillermo Wagner Granizo for the City of Monterey for the façade of the city's Conference Center.

In 1983 the City of Monterey (California) commissioned Guillermo Wagner Granizo to design a tile mural (11′ x 45′) depicting the history of this coastal community for the exterior of the city's Conference Center. The city, needing to renovate the Center, made the decision to have the mural removed in 2015, all 633 pieces of 9″ x 12″ tile. The arduous task was awarded to Architectural Resources Group in San Francisco. In the process sixty-five of the tiles were broken, most in multiple pieces.

Imagine the care involved in just securing each of the tiny pieces, keeping the pieces for each broken tile separate, reconfiguring what goes where once back in the studio, and finally the skill involved in adhering the pieces together and masking the repairs! The mural was successfully reinstalled at the Center by C. L. Frost Inc. in November 2018.

Before restoration ***After restoration***

Pietro "Pete" Ferrante (1867-1954), considered by many the "father" of sardine fisheries in Monterey Bay, inspired the success of the canning industry on Cannery Row.

The restored and re-installed historic ceramic tile mural (11′ x 45′) designed and produced by Guillermo Wagner Granizo for the City of Monterey Conference Center. Photo courtesy of Ron Wagner.

To be most effective, education necessitates direct "hands-on" communication.

Between 1991 and 2005 Tile Heritage presented annual symposiums in different cities around the United States, partnering with local organizations, to bring like minds together and to raise a community's consciousness about the significance of local tile installations within these diverse venues. The program, most often 4-5 days, included workshops, tours and lectures on both historic and contemporary tile-related subjects.

Tile contractor demonstrates the proper use of a trowel.

Publishing is central to the Foundation's outreach. Between 1988 and 2003 Tile Heritage published 42 issues of "Flash Point" (ISSN 1078-5647). Since 2004 "E-News" and "Shards & Snippets" have served as the Foundation's principal means of communication. Since 2016 THF has resumed the annual publication of *Tile Heritage: A Review of American Tile History* (ISSN 1978-5655) featuring articles by the leading tile historians in America.

From their earliest age children have a natural attraction to soft clay.

Communicate is what we do!

Email provides our principal means of communication today, both from the office and when we're "on the road," responding to the daily inquiries from throughout the United States.

Adults too find clay irresistible!
Photos on this page courtesy of Carter Sietsema Photography .

Workshops hosted or attended by Tile Heritage augment the educational objectives of the Foundation.

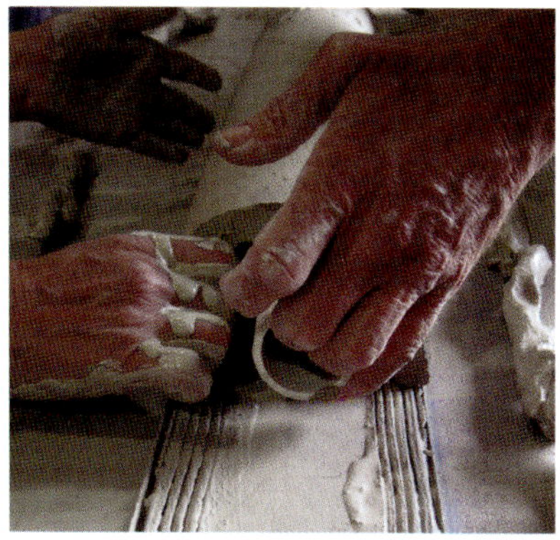

Internationally recognized architectural ceramists present a workshop in Seattle.

Tile Heritage joins with others for a week "playing" with mosaics at "Heaven on Earth" in Todos Santos on the Baja Peninsula, Mexico.

A mosaic workshop in Davis, California (left) resulted in four decorative tile pillars supporting a palapa in Palm Desert, California.

"Keeping the Craft Alive," an integral part of the Foundation's mission, brings the importance of tile's heritage into studios throughout the United States and beyond.

The library at Tile Heritage contains file folders on over 700 contemporary tile artists and artisans in the U.S., the folders representing only those who have sent their information to the Foundation with intention. No doubt there are hundreds of additional studios in the country whose work is currently not represented. Since the Foundation's inception in 1987, there has evolved a sense of community that assists in keeping the craft alive.

Preparing a custom job in Park City, Utah.

Sculpting clay for a commercial project in San Diego.

Visiting a manufacturer's showroom in Portland, OR.

Applying glaze to a lofted mural in Torrance, California.

Removing excess clay from a molded tile in Milanville, PA.

Tile Heritage tours to Europe provide opportunities once we return with new friends and broader perspectives.

Leaving the Jackfield Tile Museum at Ironbridge Gorge (above), the tour group gathers in Nottingham (left).

First stop in the Netherlands is the tile museum in Otterlo followed by a tour and demonstrations at the Makkum Tile Factory where traditional Dutch tiles are produced (right).

After the better part of a day at the tile museum in Lisbon (right), this group enjoys the spectacular display of Rococo *azulejaria* throughout the gardens at *Quinta dos Azulejos* in Lumiar, resting upon the benches in the semicircular pergola.

The Foundation is now engaged in maintaining the industry's history as a living archive through a publicly accessible Finding-aid Index.

Tile Heritage Foundation's archived tile collections, available by appointment.

"Where art and architecture meld and merge in the world of tile, Tile Heritage Foundation is there, preserving and documenting to educate the future of our industry."

Eric Astrachan
Tile Council of North America

The Foundation's tile collection alone contains over 4000 different glazed and decorative samples from scores of American companies dating to the late 19th century. All of the tiles in the collection have been donated; Tile Heritage does not buy or sell historic material.

The present time is of critical importance as we strive to enhance the accessibility of the Tile Heritage archives and collections for industry-wide and public use*. Our goal is to keep the archives "alive" with our ongoing development of the online Finding-aid Index. It is imperative that this work be completed. Expanding our industry partnership is essential to its success.*

The Tile Heritage Library contains hundreds of books and over 40,000 documents. The collections include over 700 original company catalogs and more than 90 tile-related periodicals dating back to the 1880s.

"Tile Heritage represents the 'soul' of the industry in America."

Donato Grosser

Tile Heritage Foundation's archived periodical collections.

A broad funding base is essential for the long-term sustainability of Tile Heritage. Membership and sponsorship have always provided the core of the Foundation's financial stability.

For the past 34 years Tile Heritage has received substantial support from sponsors in the tile industry. Diverse membership within and beyond the industry has also played a major role. The Foundation has benefited from substantial grant support as well.

We invite you to partner with the Tile Heritage Foundation by becoming an Industry Sponsor protecting tile history today, validating that history for tomorrow!

Tile enthusiasts tour installations in Philadelphia.

TEAM UP WITH TILE HERITAGE!

Email: foundation@tileheritage.org

www.tileheritage.org

Tile Council of North America (TCNA) has embraced Tile Heritage for many years recognizing the importance of maintaining the historic tile industry archives and collections. TCNA advocacy and inclusiveness has contributed greatly to the Foundation's validation and visibility.

Tile Heritage Archives has been accepted as a contributor by the Online Archive of California (OAC), providing access to the Tile Heritage Archives Index nationwide and beyond!

"It is very important that we, as an industry, promote an appreciation of tiles - to know what came before. Individually we are not always able to do that, but by supporting the Tile Heritage Foundation we can preserve the history. As an industry we should support that work."

Svend Hovmand

YOU ARE INVITED!

The Tile Heritage Foundation encourages manufacturers, studios, distributors, dealers, contractors, installers, the design community and other tile-related support industries to establish a **THF Archive File** where materials related to your company's history including photography, promotional materials, correspondence and other important items of interest will be kept safe for posterity. Digital images of documented installations are also encouraged.

This **THF Archive File**, a Tile Heritage service, is open to the greater tile community—THF members and non-members alike, at no charge. Currently we have over 450 historic tile companies and over 700 files in the Contemporary Archive, which includes manufacturers, studios, dealers and installers. Utilize the free resources of the Tile Heritage Foundation to hold your valuable company records for posterity.

To CONTRIBUTE to the SUPPORT
and continuing DEVELOPMENT
of the Foundation's **THF Archive Files**
contact foundation@tileheritage.org

TILE HERITAGE RESOURCE DIRECTORY

Take Advantage of a Free Listing

The Tile Heritage Foundation is presently compiling information for the next edition of the Resource Directory. We need updated information from you. Unique in its content, breadth and format, the THF Directory is designed to be used as a resource by architects and designers, preservationists, historians, collectors, dealers, installers as well as tile artists, artisans and manufacturers, among others.

Directions: Please circle the code below that best represents your primary occupation (or preoccupation). Circle the code next that represents your secondary interest and put a "2" beside the circle. You may make a third choice as well: put a "3" beside it. Return this form with your code(s) marked, and we will enter your information to the master database or make the necessary changes in your present entry. We urge you to act now as the deadline is near. Take just a moment and complete this form. Copy and enclose with your renewal or mail to Tile Heritage, P. O. Box 1850, Healdsburg, CA 95448.

All applicants receive an initial listing free of charge. The listing will include name, company (when applicable), address, phone, fax, email and website. Your primary ("1") code choice will determine the placement of your listing. **NOTE:** THF membership alone will not place you in the Directory; you must choose one or more codes to be listed.

Specialty Description: You are encouraged to submit a specialty description; there is no charge. **IMPORTANT:** limit your description to **170 characters** or less (otherwise we must do the editing).

Cost: The Tile Heritage Resource Directory is priced at $25 ($20 for THF members). Members who contribute $60 or more each year will be offered the Directory for a nominal charge of $5 to help defray production and mailing expenses.

CODES: Circle first ("1"), second ("2") and third ("3") choices as needed.

AB Builder/General Contractor
AC Commercial architect
AD Decorator
AH Residential architect
AI Interior designer
AK Kitchen/bathroom designer
AR Restoration architect

CC Tile contractor
CI Tile setter
CN Conservator
CR Specialist in tile restoration

DA Dealer of antique tiles
DC Dealer of commercial tiles
DD Tile distributor
DI Tile importer
DR Manufacturer or artist rep
DS Dealer of custom tiles

EC Educator/teacher
ET Tile enthusiast

HA Architectural historian
HC Collector of contemporary tiles
HH Collector of historic tiles
HS Appraiser/scholar/historian

MN Museum without a collection
MT Museum with tile collection

OA Art-related organization
OH Historical organization
OL Library or ceramic dept.
OT Tile-related organization

PA Publisher: art
PH Publisher: history
PT Publisher: tile-related

SC Supplier of clay, glazes, etc.
SG Supplier of setting materials
SM Supplier of machinery/tools
SP Supplier of maintenance products

TA Tile artist/designer
TC Architectural ceramist
TF Tile manufacturer
TM Tile maker (studio)
TP Potter
TR Specialist in historic reproduction
TS Specialist in tile production
TT Mosaic artists, suppliers, setters

Other ______________________

Complete and return this form to Tile Heritage.

Name ______________________

Company ______________________

Address ______________________

City ______________ State ______ Zip ______________

Work phone ______________ Fax ______________

Email ______________ www. ______________

Describe your specialty:

TILE HERITAGE FOUNDATION RESOURCE DIRECTORY

2021 - 2022

"Plus Ultra" by Mandy Baker
Jamison, Pennsylvania

"Plus Ultra," approx. 8" x 8" x 1/2".
Tile Heritage Award
presented to Cleota Reed in May 2017.

The THF Resource Directory 2021-2022
Email: foundation@tileheritage.org for availability.

Thirteenth Edition

TILE HAS MANY FACES.
Beauty like this doesn't have to be fleeting... when your clients choose tile, that is. A properly installed ceramic tile floor will outperform and outlast any other floor covering product created for the same application, even outdoors in coastal conditions. In fact, neither mold, mildew, bacteria, nor the side effects of Man's Best Friend will leave tile in ruins: its hard surfaces stand up to the test of time. Smart and beautiful — that's why tile.
WHY TILE .COM
WhyTile.com

Release date: April 2021

TCNA
TILE COUNCIL OF NORTH AMERICA, INC.

100 Clemson Research Blvd. Anderson, SC 29625 | 864.646.8453 | www.TCNAtile.com

TCNA
TILE COUNCIL OF NORTH AMERICA, INC.

TILE
THE NATURAL CHOICE

2021

O'Hare
Board here
Forest Park

CONTENTS

TILE
THE NATURAL CHOICE

Stands the Test of Time

Ceramic tile has been the preferred choice around the world for centuries for its inherent beauty and durability, and because the natural materials used to make tile are so abundant. And the endless design options can make each installation a unique work of art to be enjoyed for generations.

Inspires Design

Explore unlimited design and customization possibilities. Mix colors, textures, patterns, shapes and styles to make every installation unique. Get the look of wood, natural stone, wallpaper, even delicate fabrics, but with the durability of ceramic tile. And with tiles following the latest color trends, color matching is a breeze. You can even take your design to the next level. Find a manufacturer that specializes in custom mosaic murals, waterjet cutting, or hand painting.

Provides Unequaled Versatility

- Hypoallergenic
- Easy to sterilize
- Fire resistant
- Chemical resistant
- Enhanced traction products available
- Stain resistant, easy to clean
- Exteriors, including freeze/thaw climates
- Dry, wet, and submerged applications
- Durable: withstands heavy foot and wheel traffic

Delivers Robust Sustainability

Recycled Content and Waste Reclamation

Many factories are effectively closed loop facilities with zero waste, high levels of recycling, and efficient resource management.

Exceptional Life-Cycle Cost = Best Value

Ceramic tile has an exceptional life cycle. See the life-cycle cost study in this guide for data on how ceramic tile outperforms other floor finishes in terms of cost per square foot over time.

Regional Manufacturing and Raw Materials

With tile manufacturers in many regions of North America, tile and installation products made with local raw materials are widely available, greatly reducing the energy consumption and emissions of long distance shipping. In many cases, the raw materials are mined within 500 miles of both the manufacturing facility and the job site. Plus, the clays and other materials used to make tile are plentiful.

Best Indoor Air Quality and Zero VOCs

The high firing temperature of ceramic tile—generally more than 2000°F—burns off all organic compounds. Plus, ceramic tile is inhospitable to bacteria, fungi, mold, and other irritants that contribute to allergies and asthma. All this allows ceramic and glass tiles to easily meet low-emitting material requirements, contribute to pre-occupancy air quality plans, and provide superior indoor air quality for building occupants.

Lower Energy Needs

Tile can reduce the amount of energy needed for heating and cooling because of its exceptional thermal mass.

Reduced Heat Island Effect

Using light-colored tiles instead of traditional paving materials can lower an area's heat absorption, or heat island effect. Green buildings in urban areas in particular are required to reduce the potential heat island effect.

Building Material Reuse

Tile finishes are among the few surfaces that can be salvaged in a major renovation.

Eco-Friendly Maintenance

No harsh chemicals required!

Ceramic tile is an ideal choice for ventilated façade projects.

Tile can provide an aesthetic advantage, due to the wide range of colors, surfaces and sizes, allowing for virtually unlimited design opportunities.

A ventilated tile façade can reduce a building's energy consumption 20-30% due to the natural chimney effect it creates between the structure and the cladding. Hot air is evacuated in summer, and insulation is more effective in winter.

TILE AND GREEN BUILDING
CREDITS AND REQUIREMENTS

	Leadership In Energy and Environmental Design (LEED v4.1)	2020* International Green Construction Code (Powered by ASHRAE Standard 189.1) *Anticipated
MULTI-ATTRIBUTE SUSTAINABILITY: CONFORMANCE TO GREEN SQUARED	Pilot Credit—Certified Multi-attribute Products and Materials	901.4.1.4.3 (9.4.1.4.3) Third-Party Multi-attribute Certification
LCA OR EPD	Building Product Disclosure and Optimization—Environmental Product Declarations	901.4.1.4.1 (9.4.1.4.1) Industry-Wide Declaration 901.4.1.4.2 (9.4.1.4.2) Product-Specific Declaration 901.4.1.4.4 (9.4.1.4.4) Product Life Cycle
RECYCLED CONTENT	Building Product Disclosure and Optimization—Sourcing of Raw Materials	901.4.1.1.1 (9.4.1.1.1) Recycled Content
PRODUCT REUSABILITY/ RECYCLABILITY	Building Product Disclosure and Optimization—Sourcing of Raw Materials	901.4.1.1.2 (9.4.1.1.2) Salvaged Material Content
REGIONAL MATERIALS	Building Product Disclosure and Optimization—Sourcing of Raw Materials	901.4.1.2 (9.4.1.2) Regional Materials
CONSTRUCTION/ DESIGN MANAGEMENT	Integrative Process Construction Indoor Air Quality Management Plan Indoor Air Quality Assessment Construction and Demolition Waste Management	901.3.1 (9.3.1) Construction Waste Management 1001.7.2 (10.7.2) IAQ Construction Management
INDOOR ENVIRONMENTAL QUALITY	Low-Emitting Materials Thermal Comfort Interior Lighting	801.4.2.1 (8.4.2.1) Adhesives and Sealants 801.4.2.3 (8.4.2.3) Floor Covering Materials 801.4.2.6 (8.4.2.6) Ceiling and Wall Assemblies and Systems
HEAT ISLAND EFFECT	Heat Island Reduction	501.3.5.1 (5.3.5.1) Site Hardscape
ENERGY EFFICIENCY	Optimize Energy Performance	701.4.2 (7.4.2) 701.5 (7.5) Building Envelope Performance Option
INHERENT DURABILITY	Building Life-Cycle Impact Reduction	1001.10 (10.10) Service Life Plan
OTHER PRODUCT ATTRIBUTES OR MANUFACTURER ENGAGEMENT INITIATIVES	Open Space Building Product Disclosure and Optimization—Sourcing of Raw Materials Building Product Disclosure and Optimization—Material Ingredients Innovation	105.4 Innovative Approaches and Alternative Materials, Design, and Methods of Construction and Equipment 1001.9.5 (10.9.5) Building Green Cleaning Plan

NAHB National Green Building Standard (ICC/ASHRAE 700-2020*) *Anticipated	ANSI/GBI 01-2019 Green Globes	
612.2 Sustainable Products	10.2.1.1 Third-Party Multiple-Attribute Product Certification	**MULTI-ATTRIBUTE SUSTAINABILITY: CONFORMANCE TO GREEN SQUARED**
611.1 Product Declarations	10.2.1.1 Cradle-to-Gate EPD 10.2.1.2 Cradle-to-Grave EPD	**LCA OR EPD**
604.1 Recycled Content	10.4.1 Sustainable Material Attributes	**RECYCLED CONTENT**
603.1 Reuse of Existing Building 603.2 Salvaged Materials 605.3 Onsite Recycling 605.4 Offsite Recycling	10.5.1 Reuse Structural/Non-Structural Elements 10.5.2 Reused, Refurbished, and Salvaged Materials	**PRODUCT REUSABILITY/ RECYCLABILITY**
609.1 Regional Materials	N/A	**REGIONAL MATERIALS**
605.2 Construction Waste Management Plan	6.1.2 Integrated Design Process 6.2 Environmental Mgmt. During Construction	**CONSTRUCTION/ DESIGN MANAGEMENT**
901.7 Hard-Surface Flooring 901.8 Wall Coverings 901.10 Adhesives & Sealants	11.2.1 Volatile Organic Compounds	**INDOOR ENVIRONMENTAL QUALITY**
505.2 Heat Island Mitigation	7.3.4 Heat Island Effect	**HEAT ISLAND EFFECT**
702.2.2 Energy Performance Levels 703.2 Building Envelope	8.3.1 Building Envelope	**ENERGY EFFICIENCY**
602.1.6 Termite-Resistant Materials 602.1.11 Tile Backing Materials 901.6 Wall-to-wall Carpeting Not Installed	6.3.2 Building Service Life Plan	**INHERENT DURABILITY**
601.3 Building Dimensions and Layouts 601.7 Pre-Finished Materials	10.3 Product Risk Assessment	**OTHER PRODUCT ATTRIBUTES OR MANUFACTURER ENGAGEMENT INITIATIVES**

TCNA BULLETIN
CERAMIC TILE FOR A SAFE AND HEALTHY HOME

Ceramic tile is not only beautiful, it's also healthy for your home. It contains none of the chemicals that have been in the news associated with other flooring products, and it's long-lasting, easy to clean, fire-safe, and offers many slip-resistant choices.

Made of Clay and other Naturally-Occurring Minerals: Free of Formaldehyde, VOCs, and PVC

Ceramic tile* does not contain the chemicals in other floor coverings that are receiving increased attention for their possible adverse health effects, such as formaldehyde,[1-4] VOCs (volatile organic compounds)[5-9] or PVC (polyvinyl chloride).[10-11]

Zero VOCs. Some VOCs emitted by non-ceramic building products can contribute to a wide variety of health problems and "Sick Building Syndrome," according to various health experts.[12-15] Because ceramic tile is fired at extremely high temperatures, producing an inorganic material, it has zero VOCs.*

Formaldehyde-Free. Ceramic tile* contains none of the formaldehyde found in some other floor coverings. Formaldehyde is a chemical long associated with respiratory disorders, and exposure is a particular concern for children and the elderly, according to the Centers for Disease Control (CDC).[16-17] In fact, formaldehyde emissions for several wood-based building and flooring products have been federally restricted since 2011 under the Formaldehyde Standards for Composite Wood Products Act.[18-21] As a fired product, ceramic tile is formaldehyde-free and not subject to this or any other such restriction.

PVC-Free. Ceramic tile* is also free from PVC, a resin used in other types of floor coverings, and regularly a subject of concern and discussion among health experts.[22-33]

The easy solution for flooring? Use ceramic tile, which is free of VOCs, formaldehyde, and PVC!

Occupant Safety

Of the many flooring choices available, ceramic tile stands out when slip/fall safety is a consideration because of the thousands of floor tile choices that are slip-resistant when wet. This is in stark contrast to floorings that, according to their manufacturers, must be kept dry in order to be slip-resistant.

Additionally, ceramic tile* is non-flammable and does not produce smoke in a fire, meeting the flame spread and smoke development requirements of Section 803 of the International Building Code (IBC) for interior wall and ceiling materials.

Environment

The UL-certified Environmental Product Declaration (EPD) for North American-made ceramic tile shows that North American-made ceramic tile has the lowest environmental impact across all impact categories, when compared to other floorings with generic EPDs[34-35] evaluated under the same product category rules.[36] With a 60-year service life[34] this makes ceramic tile both cost-effective and the best choice for reducing negative environmental impacts, resource use, and demolition waste, when compared to flooring products that need to be replaced more frequently.

Disclaimer: The above information is general in nature and is not specific to any one building product. Ceramic tile products and their properties can vary among manufacturers, products lines, and even among tiles in the same product line. Likewise, non-ceramic tile products have their own properties and can also vary among manufacturers and products. Products that contain any of the chemicals referenced above may contain them in manners that are not known to be harmful and/or that comply with applicable health and safety regulations. Contact the manufacturer of a specific building product to learn about its particular characteristics. See opposite for notes and numbered references.

References

1. Godish, T. "Residential formaldehyde: Increased Exposure Levels Aggravate Adverse Health Effects." Journal of Environment and Health 53, no. 3 (1990): 34-37.

2. Böhm, Martin, Mohamed ZM Salem, and Jaromír Srba. "Formaldehyde Emission Monitoring from a Variety of Solid Wood, Plywood, Blockboard and Flooring Products Manufactured for Building and Furnishing Materials." Journal of Hazardous Materials 221 (2012): 68-79.

3. Cox, Steven S., John C. Little, and Alfred T. Hodgson. "Predicting the Emission Rate of Volatile Organic Compounds from Vinyl Flooring." Environmental Science & Technology 36, no. 4 (2002): 709-714.

4. Salem, Mohamed ZM, Martin Böhm, Jaromír Srba, and Jitka Beránková. "Evaluation of Formaldehyde Emission` from Different Types of Wood-Based Panels and Flooring Materials Using Different Standard Test Methods." Building and Environment 49 (2012): 86-96.

5. Franck, Ulrich, Annegret Weller, Stefan W. Röder, Gunda Herberth, Kristin M. Junge, Tibor Kohajda, Martin von Bergen et al. "Prenatal VOC Exposure and Redecoration Are Related to Wheezing in Early Infancy." Environment International 73 (2014): 393-401.

6. Hodgson, A. T., A. F. Rudd, D. Beal, and S. Chandra. "Volatile Organic Compound Concentrations and Emission Rates in New Manufactured and Site-Built Houses." Indoor Air 10, no. 3 (2000): 178-192.

7. Risholm-Sundman, M., M. Lundgren, E. Vestin, and P. Herder. "Emissions of Acetic Acid and Other Volatile Organic Compounds from Different Species of Solid Wood." Holz als Roh-und Werkstoff 56, no. 2 (1998): 125-129.

8. Weschler, Charles J. "Changes in Indoor Pollutants since the 1950s." Atmospheric Environment 43, no. 1 (2009): 153-169.

9. Won, Doyun, and Mike Rynes. "Sorptive Interactions between VOCs and indoor materials." Indoor Air 11, no. 4 (2001): 246-256.

10. Allsopp, Michelle, David Santillo, and Paul Johnston. "Hazardous chemicals in PVC flooring." Greenpeace Research Laboratories Technical Note 14/00 (2000): 10.

11. Bornehag, C.-G., Lundgren, B., Weschler, C.J., Sigsgaard, T., Hagerhed-Engman, L., Sundell, J. "Phthalates in Indoor Dust and Their Association with Building Characteristics." Environmental Health Perspectives 113 (2005): 1399–1404.

12. Reiser, R., A. Meile, C. Hofer, and R. Knutti. "Indoor Air Pollution by Volatile Organic Compounds (VOC) Emitted from Flooring Material in a Technical University in Switzerland." Proceedings: Indoor Air (2002): 1004-1009.

13. Ryan, Christopher M., and Lisa A. Morrow. "Dysfunctional Buildings or Dysfunctional People: An Examination of the Sick Building Syndrome and Allied Disorders." Journal of Consulting and Clinical Psychology 60, no. 2 (1992): 220.

14. Nakaoka, Hiroko, Emiko Todaka, Hiroshi Seto, Ikue Saito, Masamichi Hanazato, Masahiro Watanabe, and Chisato Mori. "Correlating the Symptoms of Sick-Building Syndrome to Indoor VOCs Concentration Levels and Odour." Indoor and Built Environment 23, no. 6 (October 2014): 804-813.

15. Liang, Huang-Min, and Chung-Min Liao. "Modeling VOC-Odor Exposure Risk in Livestock Buildings." Chemosphere 68, no. 4 (2007): 781-789.

16. http://www.atsdr.cdc.gov/phs/phs. asp?id=218&tid=39#bookmark06

17. Formaldehyde, http://www.cdc.gov/niosh/idlh/50000.html

18. http://www.cdc.gov/nceh/formaldehyde/

19. http://www.cpsc.gov/PageFiles/121919/AN%20UPDATE%20ON%20FORMALDEHYDE%20final%200113.pdf

20. http://www2.epa.gov/formaldehyde/questions-and-answers-regarding-laminate-flooring-0

21. Formaldehyde Standards for Composite Wood Products Act, http://www.gpo.gov/fdsys/pkg/BILLS-111s1660enr/pdf/BILLS-111s1660enr.pdf

22. Carlstedt, Fredrik, B. A. G. Jönsson, and C. G. Bornehag. "PVC Flooring Is Related to Human Uptake of Phthalates in Infants." Indoor Air 23, no. 1 (2013): 32-39.

23. Xu, Ying, Elaine A. Cohen Hubal, Per A. Clausen, and John C. Little. "Predicting Residential Exposure to Phthalate Plasticizer Emitted from Vinyl Flooring: A Mechanistic Analysis." Environmental Science & Technology 43, no. 7 (2009): 2374-2380.

24. Bornehag, Carl-Gustaf, Jan Sundell, Charles J. Weschler, Torben Sigsgaard, Björn Lundgren, Mikael Hasselgren, and Linda Hägerhed-Engman. "The Association between Asthma and Allergic Symptoms in Children and Phthalates in House Dust: A Nested Case-Control Study." Environmental Health Perspectives (2004): 1393-1397.

25. Jaakkola, J. J., Leif Oie, Per Nafstad, Grete Botten, Sven Ove Samuelsen, and Per Magnus. "Interior Surface Materials in the Home and the Development of Bronchial Obstruction in Young Children in Oslo, Norway." American Journal of Public Health 89, no. 2 (1999): 188-192.

26. Allsopp, Michelle, David Santillo, and Paul Johnston. "Hazardous Chemicals in PVC Flooring." Greenpeace Research Laboratories Technical Note 14/00 (2000): 10.

27. Casals-Casas, Cristina, J. N. Feige, and Béatrice Desvergne. "Interference of Pollutants with PPARs: Endocrine Disruption Meets Metabolism." International Journal of Obesity 32 (2008): S53-S61.

28. Casals-Casas, Cristina, and Béatrice Desvergne. "Endocrine Disruptors: From Endocrine to Metabolic Disruption." Annual Review of Physiology 73 (2011): 135-162.

29. Grün, Felix, and Bruce Blumberg. "Endocrine Disrupters as Obesogens." Molecular and Cellular Endocrinology 304, no. 1 (2009): 19-29.

30. Gray, L. Earl, Joseph Ostby, Johnathan Furr, Matthew Price, DN Rao Veeramachaneni, and Louise Parks. "Perinatal Exposure to the Phthalates DEHP, BBP, and DINP, but not DEP, DMP, or DOTP, Alters Sexual Differentiation of the Male Rat." Toxicological Sciences 58, no. 2 (2000): 350-365.

31. Diamanti-Kandarakis, Evanthia, Jean-Pierre Bourguignon, Linda C. Giudice, Russ Hauser, Gail S. Prins, Ana M. Soto, R. Thomas Zoeller, and Andrea C. Gore. "Endocrine-Disrupting Chemicals: An Endocrine Society Scientific Statement." Endocrine Reviews 30, no. 4 (2009): 293-342.

32. Mylchreest, Eve, Madhabananda Sar, Russell C. Cattley, and Paul MD Foster. "Disruption of Androgen-Regulated Male Reproductive Development by Di (n-butyl) Phthalate during Late Gestation in Rats Is Different from Flutamide." Toxicology and Applied Pharmacology 156, no. 2 (1999): 81-95.

33. Grün, Felix, Hajime Watanabe, Zamaneh Zamanian, Lauren Maeda, Kayo Arima, Ryan Cubacha, David M. Gardiner, Jun Kanno, Taisen Iguchi, and Bruce Blumberg. "Endocrine-Disrupting Organotin Compounds Are Potent Inducers of Adipogenesis in Vertebrates." Molecular Endocrinology 20, no. 9 (2006): 2141-2155.

34. http://www.TCNAtile.com/images/pdfs/EPD-for-Ceramic-Tile-Made-in-North-America.pdf

35. http://www.rfci.com/environmental-product-declaration/

36. http://www.nsf.org/newsroom_pdf/flooring_pcr-new.pdf

Notes

*Applies to fired ceramics free of any mesh backing or topical sealer. Not all mesh backings and topical sealers produce volatile organic compounds or contribute to flame spread. Consult manufacturer for specific product information if desired.

†Based on independent studies performed at TCNA Laboratory on tiles without mesh backing or topical sealers.

‡More than 95% of a fired ceramic tile is made from clay and other raw materials found in nature.

§Per UL-Certified Environmental Product Declaration (EPD) for North American-made Ceramic Tile.

TCNA GREEN REPORT
LIFE-CYCLE COST STUDY

Because a tile installation will last as long as the building it's installed in, its environmental impacts are minimal when compared to other surface coverings that would have to be replaced numerous times over the life of the building. This also means tile costs less per square foot over the long haul, whereas carpet, vinyl, and other floor coverings are significantly more expensive due to their replacement, maintenance, and refinishing requirements.

Construction cost consulting firm Scharf-Godfrey compared the life-cycle costs of various flooring types. The per year cost for each flooring type includes installation, maintenance, and removal costs. Their comparison, provided below, shows why ceramic tile is the natural choice for sustainable design and long-term value.

Life-Cycle Costs for Floor Finishes (per square foot)

Floor Finish	Installed Cost	Life Cycle Cost	Expected Life (y)	Cost Per Year
Quarry Tile	$6.83	$16.13	50	$0.32
Glazed Ceramic Floor Tile	$7.00	$16.30	50	$0.33
Glazed Porcelain Tile	$8.34	$17.64	50	$0.35
Mosaic Tile	$8.20	$17.50	50	$0.35
Unglazed Porcelain	$8.30	$17.60	50	$0.35
Natural Hardwood	$9.31	$20.80	50	$0.42
Travertine-Turkish	$12.50	$21.80	50	$0.44
Marble	$21.00	$30.30	50	$0.61
Laminate	$8.84	$17.77	25	$0.71
Man-Made Hardwood	$9.58	$18.51	25	$0.74
Portland Cement Terrazzo	$14.88	$24.27	30	$0.81
Stained Concrete	$12.40	$24.60	25	$0.98
Carpet	$3.22	$ 6.50	6	$1.08
Resin Terrazzo	$8.50	$16.53	15	$1.10
Sheet Vinyl	$6.90	$13.90	10	$1.39
Poured Epoxy	$8.18	$15.18	10	$1.52
VCT	$3.91	$18.35	10	$1.83

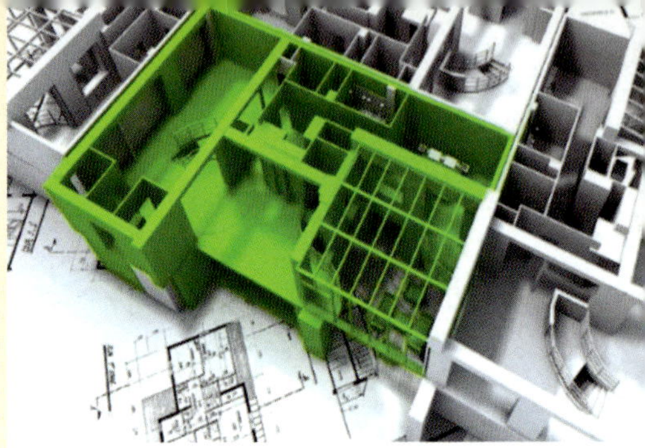

Guide to Green Squared®:

The Tile Industry's Standard and Certification Program for Product Sustainability

2021

CONTENTS

WHAT IS
GREEN SQUARED®?

Green Squared® is the world's only multi-attribute sustainability standard and certification program for tile and the associated products needed to install it.

When you see the Green Squared Certified® mark, it means a product meets the broad array of rigorous sustainability requirements of the American National Standard Specifications for Sustainable Ceramic Tiles, Glass Tiles and Tile Installation Materials (ANSI A138.1).

Including raw material extraction, end of product life management, and all life cycle stages in-between, Green Squared is a true cradle-to-grave performance standard that eliminates the need to cross evaluate single-attribute sustainability claims.

Products independently verified as conforming to the Green Squared standard may bear the Green Squared Certified mark. To be in conformance, products must meet a battery of requirements and electives, as set forth in the standard.

Qualities of the Green Squared Standard

Thorough Green Squared incorporates a full range of sustainability issues into one standard, one mark — allowing for easy marketplace recognition of sustainable tiles and tile installation materials.

Standardized Criteria Green Squared establishes performance-based thresholds, setting the bar for technical specification of sustainable tiles and installation materials.

Clarity As the tile industry's own sustainable product certification program, Green Squared facilitates increased uniformity and decreased confusion in green marketing.

Opportunity By passing rigorous third-party evaluation criteria, certification to the Green Squared standard confers a high level of credibility and a smooth, clear, and responsible path for products to be specified in green building projects.

Sustainable Systems Green Squared addresses the components of a tile installation, allowing the specification and installation of systems of certified conforming products.

THE STANDARD

Developed by an ANSI-accredited committee of green building professionals, manufacturers, distributors, installers, and other stakeholders, the Green Squared standard, ANSI A138.1, contains a spectrum of multi-attribute environmental and social criteria.

PRODUCTS COVERED

Green Squared covers a variety of products, including ceramic and glass tiles, mortars, grouts, backerboards, membranes, and other installation materials.

THIRD PARTY CERTIFICATION

Manufacturers seeking to have their product(s) certified can apply to either of two internationally-recognized Green Squared certification bodies.

SCSglobal SERVICES

UL Environment

PRODUCT LABELING

Once a product is third-party certified as meeting the ANSI A138.1 standard, the certification body awards use of the Green Squared Certified® logo on products, packaging and/or associated literature.

Multi-Attribute Sustainability
(condensed from standard)

Environmental Product Characteristics	Recycled Content/Reclaimed Waste Content
	Indigenous Raw Goods
	Environmental Packaging
	Durability
	Low Emissions
	Environmental Cleaning & Maintenance
	Solar Reflectance Index (SRI)
	Light Reflectance Value (LRV)
	Sound Abatement
	Third Party LCA & EPD
	Participation in LCA Database Initiatives
Environmental Product Manufacturing and Raw Material Extraction	Particulate Emissions
	Combustion and Fuel Usage
	Raw Goods Sourcing & Extraction
	Outsourced Packaging & Manufacturing Services
	Environmental Management Plans & Systems
	Utility Usage
	Renewable Energy
	Manufacturer Waste Diversion/Minimization
	Shipping Material Waste Minimization
End of Life Product Mgmt.	Manufacturer Guidelines on Clean Fill Usage
	Post-Life Material Waste Minimization
Progressive Corporate Governance	Social Responsibility Strategy
	Labor Law Compliance
	Environmental Regulation Compliance
	Health & Safety Regulation Compliance
	Voluntary Participation in Health/Safety Programs
	FTC Green Guides Compliance
	Continuous Community Involvement
	Public Disclosure
	Sustainability Reports
	Certified "Green" Facilities
Innovation	"Above & Beyond" Standardized Criteria
	Innovative Sustainability Technologies
	Carbon Footprint Awareness/Reduction Strategy

THE GREEN SQUARED® STANDARD
ANSI A138.1

Environmental Product Characteristics

The first section of ANSI A138.1, Environmental Product Characteristics, standardizes product attributes such as the amount of recycled content, level of volatile organic emissions, amount of indigenous raw materials, etc.

Three levels of recycled content and reclaimed waste are defined in the standard. Level one is mandatory for conformance, and levels two and three are elective. Thresholds vary depending on product type.

There are also three standardized levels of indigenous raw material content. Recognizing the environmental burden from transportation, indigenous raw materials are those extracted within 500 miles of the manufacturing site. If transported by rail or sea, the limit is 2000 miles. All three levels are elective.

There are five standard attributes for sustainable packaging. All five are elective, but it is mandatory that at least one be achieved:

- Minimal packaging (packaging cannot exceed a specified percentage of the product weight)
- Recyclable packaging (a specified percentage of the packaging is recyclable)
- On-site reusable packaging (100% reusable on site)
- Biodegradable or compostable packaging (meets specific ASTM biodegradability criteria)
- Recycled content packaging (contains a specified percentage of recycled content)

Mandatory for conformance to ANSI A138.1, a product must meet industry standards for durability and performance. For example, tile must comply with ANSI A137.1. Also mandatory for conformance, a product shall have low (for installation products) or no (for tile) volatile organic compound (VOC) emissions. Environmentally-friendly maintenance instructions must also be available.

Finally, as elective options for conformance to A138.1, criteria for solar reflectance index (SRI), light reflectance value (LRV), and sound abatement are all established. Also, electives for life cycle assessment (LCA) and/or environmental product declaration (EPD) are available.

Environmental Product Manufacturing and Raw Material Extraction

The second section of ANSI A138.1, Environmental Product Manufacturing and Raw Material Extraction, standardizes sustainable production practices.

Mandatory for a product to conform to the standard, buildings containing process equipment shall not allow visible particulate matter (PM) emissions. Additionally, there are four standardized PM pollution control levels based on grain size; the first is mandatory and other three are elective.

It is mandatory that the only types of fuels used are natural gas, LP gas, landfill generated methane, or bio-based fuel. The usage of landfill generated methane or bio-based fuel can earn elective credits.

Elective credits can also be achieved through the use of low nitrogen oxide (NOx) burners and acid gas controls.

Also mandatory for a product to conform, the manufacturer of that product must have a written procurement policy for its raw material suppliers addressing social issues, environmental issues, and applicable requirements for mining laws. Additionally, the manufacturer is expected to maintain a list of all materials used to manufacture the product.

As an elective, the manufacturer may choose to enter into a sustainability-based buying agreement with the raw material suppliers contributing, in aggregate, at least 90% or more of the product by weight.

For outsourced services, such as packaging or additional manufacturing services, manufacturers must require that their service providers have written sustainability policies.

Manufacturers are required to have an environmental management plan that addresses waste minimization, lighting efficiency, heating fuel usage, electricity consumption, and water conservation. There are elective options for minimal water discharge, renewable energy usage, cogeneration, heat recovery or combined heat and power integration, ASHRAE auditing, and ISO 14001 environmental plan registration.

Finally, as electives, several innovative criteria for waste diversion are established. These include incorporation of waste into other manufacturers' products or beneficial reuse projects, donation of finished products to charitable organizations in lieu of disposal, and recycling/ reuse of incoming shipping materials.

End of Product Life Management

This section of the standard opens with the following preface:

Inherently, tile products are durable, inert, and intended to have life spans as long as the buildings in which they are installed. They are engineered to serve as permanent finishes capable of outlasting multiple generations of building occupants. Tile product end of life management is pertinent to building demolition waste and small quantities of waste generated during construction.

It is perhaps self-evident—but worth noting—that a product with a long lifetime is more sustainable than a similar product with a short lifespan. When considering the environmental, social, and economic sustainability of a product, all relevant impacts are repeated each time that product is replaced. Ideally, a product's expected service life is at least as long as the building in which it is installed, in which case it's impacts are considered only once.

Although a tiled finish is inherently durable and typically desirable for a lifetime, there are instances in which end of product life must be addressed. Thus, there are end of product life management electives in ANSI A138.1 that are intended for instances where buildings are demolished, scrap waste is generated during construction, or a remodel occurs.

The first end of product life management elective option requires that the product be eligible for use as clean fill. To satisfy this elective, a manufacturer shall provide documentation verifying that the product is inert and solid such that it can potentially be considered along with other eligible construction and demolition debris for state and local Clean Fill acquisition initiatives.

A second elective requires an end of product life collection plans. To satisfy this elective, the manufacturer shall establish and implement a plan that addresses the collection, processing, and recycling or re-tasking of a product for other purposes once the product's useful life is completed.

Progressive Corporate Governance

The fourth section of ANSI A138.1 specifies criteria for Progressive Corporate Governance.

Mandatory for conformance to the standard, a manufacturer shall have a written and implemented social responsibility strategy which addresses at least the following: labor law compliance, forced labor prohibitions, child labor prohibitions, environmental regulation compliance, health and safety regulation compliance, and community involvement.

As an elective, the manufacturer may choose to participate in a voluntary safety program such as OSHA Safety Consultation, Voluntary Protection Program (VPP), or OHSAS 18001.

It is mandatory that all green marketing claims made by the manufacturer be in compliance with the Federal Trade Commission (FTC) Fair Packaging and Labeling Act (FPLA) Green Guides (publicly available) that indicate how the FTC applies Section 5 of the FPLA, prohibiting unfair or deceptive acts or practices in environmental claims.

As an elective, the manufacturer may choose to regularly engage in its community, building upon the community involvement plan established in its mandatory social responsibility strategy.

Also, as an elective, the manufacturer may publicly disclose on an annual basis one of the following: utilities consumption, registered Environmental Management System (EMS) data, or Life Cycle Assessment (LCA) data.

An elective credit is also available if a manufacturer provides a detailed sustainability report each year, conforms to the requirements of the Global Reporting Initiative (GRI), or is selected for inclusion in the Dow Jones Sustainability Index (DJSI).

If a manufacturer has at least one facility with LEED® or Green Globes certification, an elective credit is also available for that.

Additionally, manufacturers are required to have a program in place that demonstrates continued conformance to the specified criteria of ANSI A138.1/Green Squared for all pertinent products.

THE GREEN SQUARED® STANDARD
ANSI A138.1

Innovation

Technological advancement and outstanding achievement, well beyond what is required, can be key to the development of sustainable products and operations. ANSI A138.1 allows products to achieve conformance, in part, through innovation. This may involve exceptional performance well above the requirements set forth in other sections of the standard, and/or innovative performance recognized by the ANSI ASC A108 Committee in categories not otherwise addressed by the standard.

A product may earn up to two electives through exceptional conformance if quantitative criteria already addressed by the standard are greatly exceeded. Either, the most stringent threshold already established for a specific criterion must be exceeded by one and a half times, or certain specific requirements defined in the standard's Appendix C must be met.

For sustainability attributes not directly addressed by the standard, an evolving list of approved innovations is managed by the ANSI ASC A108 Committee. Innovations not included in this list can be added if they are submitted to and approved by the Committee. This can include ecological attributes or processes not otherwise addressed by the standard or innovative corporate governance.

A product may also earn a fourth innovation elective if the product's carbon footprint is provided and prescriptive measures were taken to reduce either the product's or the manufacturing organization's greenhouse gases.

WHY TILE .COM
SMART.
BEAUTIFUL.
RESOURCE.
If clean air is important to you and your family, tile is the natural choice.
Endless finishes await, from timeless elegance to top trends.
Consumer Information
Searchable Image Database
Commercial Guide and Case Studies
13 Style Innovators and Their Tile Inspirations
Trending Articles
Ceramic Tile's Answer to the Plastic Problem

THIRD PARTY CERTIFICATION
GREEN SQUARED CERTIFIED®

Background on Green Squared Certification

Green Squared® certification was developed by TCNA for independent third-party certifiers to assess if products are in conformance with ANSI A138.1. Products certified under Green Squared may be labeled using the Green Squared Certified mark. This easily-recognizable mark helps architects, designers, and end users choose products, while resting assured that the products they choose meet the industry's broad range of sustainability criteria.

The Green Squared framework for product evaluation and certification is in accordance with ISO Type 1 environmental labeling and declaration requirements (ISO 14024). With so many different brands of green today, Green Squared certification provides authenticity and conveys that the product certified is the real deal. In a marketplace wary of unsubstantiated product sustainability claims, Green Squared third-party validation is provided by well-known and widely-recognized certification bodies.

The Green Squared Certification Process

A manufacturer seeking Green Squared certification for various products they make applies directly to a certification body approved under the Green Squared program. The certification body then makes necessary arrangements with the manufacturer to evaluate if the products meet the ANSI A138.1 standard. This evaluation, which is in accordance with the Green Squared Certification Program Criteria, includes a comprehensive and objective review of the products and the associated manufacturing facility. Based on this evaluation, for those products that meet ANSI A138.1, the certification body authorizes the use of the Green Squared Certified mark.

Global Applicability

The ANSI A138.1 standard, and certification of conformance to that standard (i.e., Green Squared certification) were developed specifically to be relevant to products no matter where in the world they are produced. Developed to meet the sustainability needs of the North American marketplace, the Green Squared standard and certification program provide all producers, foreign and domestic, with a clear benchmark for designing sustainable products that can be accepted by North American green building programs.

Six Steps of Certification

Here's what to expect when pursuing Green Squared® certification of your product(s):

1 APPLICATION PROCESS

Manufacturer interested in having a product certified applies directly to an authorized Green Squared certification agency.

Certification Agencies: SCSglobal SERVICES UL Environment

2 EVALUATION

The certification process involves the following in determination of ANSI A138.1 conformance:

- General evaluation of the applicant organization
- Product evaluation
- On-site facility evaluation

3 CONFIDENTIALITY

All proprietary information remains confidential between the certification agency and the manufacturer.

4 CERTIFICATION

- Green Squared Certified® mark awarded by certification agency
- Scope of certified products determined by certification agency
- Certifications valid for a five-year period
- Surveillance audits each year
- Re-certification after five years

5 PRODUCT LISTING

- Agency lists all certified products
- Certified products also listed on GreenSquaredCertified.com
- Certified product data syndicated by Ecomedes to Fulcrum, as well as to many public and private A&D purchasing libraries

6 USING THE MARK

- Directly on certified products
- Packaging of certified products
- Promotional literature published about certified products

G2 GREEN SQUARED CERTIFIED®

SPECIFYING GREEN SQUARED®
GREEN BUILDING STANDARDS, CODES, AND RATING SYSTEMS

LEED v4.1

Under the LEED pilot credit, *Certified Multi-attribute Products and Materials*, a point toward LEED certification can be earned by using products from manufacturers who have confirmed multiple environmental attributes through an independent, third-party certification, such as Green Squared®.

In order to achieve this point, at least 25%, by cost, of the total value of permanently installed products in a project must have earned certification under any of the following programs: Level® certification for furniture, NSF 140 certification for carpet, NSF 332 certification for resilient flooring, Green Squared certification for tiles and tile installation materials, UL 100 certification for wall board, NSC 373 certification for natural stone, NSF 343 for wallcoverings, and NSF 347 for roofing membranes.

Additionally, the product literature must disclose which sections of the relevant multi-attribute sustainability standard were met, and an LCA (life-cycle assessment) conforming to ISO 14044 must be available, either as a component of or in addition to the product's multi-attribute certification.

GREEN SQUARED CERTIFIED. Tiles or installation materials that are Green Squared Certified® and for which an LCA is available (or which meet section 3.8.1 of Green Squared®/ANSI A138.1) can contribute:

- 50% of the total product cost for baseline certification

- 75% of the total product cost if ANSI 138.1 Level 2 criteria for recycled/reclaimed content and/or Level 2 criteria for indigenous raw materials were met as part of the certification
- 100% of the total product cost if ANSI A138.1 Level 3 criteria for recycled/reclaimed content and/or Level 3 criteria for indigenous raw materials were met as part of the certification

Sample calculation for a project with $1,000,000 in permanently installed products, including $80,000 of Green Squared Certified tile and $20,000 of Green Squared Certified mortar that meets ANSI A138.1 Level 3 criteria for indigenous raw materials.

Tile installation contribution to the 25%, by cost, of the total value of permanently installed products in the project

$$\frac{(0.5)\ (\$80\text{k tile value}) + (1.0)\ (\$20\text{k mortar value})}{\$1\text{M (all materials value)}} = 6\%$$

In the example above, the contribution of Green Squared Certified® tiles and installation materials toward the 25% minimum would be 6%.

Green Globes ANSI/GBI 01-2019

Under the provisions of Section 10.2, *ANSI/GBI 01-2019 Product Life Cycle,* using Green Squared Certified® products can contribute toward the 19 available points for using products with EPDs or that are third-party certified to multi-attribute standards. Points are awarded for using products with:

EPDs: Generic or Proprietary AND/OR Third-Party Verified Proprietary Life-Cycle Assesments AND/OR Third-Party Certifications to Multiple Attribute Standards

Points are awarded based on the number of products used, as follows:

# of Products	Points	# of Products	Points
30+ products	19 pts	24 products	9 pts
29 products	14 pts	23 products	8 pts
28 products	13 pts	22 products	7 pts
27 products	12 pts	21 products	6 pts
26 products	11 pts	20 products	5 pts
25 products	10 pts	<20 products	No pts

Without limit, using Green Squared Certified tiles or installation materials will contribute to this requirement. Products that also have an LCA or EPD can contribute double.

Sample calculation for a project with 6 Green Squared Certified tiles, 6 Green Squared Certified grouts, 1 Green Squared Certified mortar and 9 other products certified under their respective industry programs:

13 total tiles/tile installation products + 9 other compliant products = 22 total products (7 points)

Using the above example, tile's contribution is 13 out of 22 products, or 59%.

Sample calculation for the same project, but with tile that is Green Squared Certified and has an LCA or EPD:

[(13 Green Squared Certified products with LCA or EPD) x 2]
\+ 9 other compliant products
35 total products (19 points)

Using the above example, although 35 total products are achieved, only 30 are required to obtain the maximum points tier. Toward the threshold, tile's contribution is 26 out of 30 products, or 87%.

SPECIFYING GREEN SQUARED®

GREEN BUILDING STANDARDS, CODES, AND RATING SYSTEMS

International Green Construction Code (IgCC) Powered by ASHRAE Standard 189.1

About IgCC: An initiative of the International Code Council (ICC), IgCC was developed as model code language for states and municipalities to establish baseline sustainability requirements for new and existing commercial buildings.

About ASHRAE 189.1: An ANSI-accredited standard, ASHRAE 189.1 was written to be incorporated into governing jurisdictions' codes and ordinances for commercial buildings and enforced by building officials and inspectors.

IgCC and ASHRAE Standard 189.1 are now integrated into a single model code and standard for high-performance buildings.

For compliance with the IgCC, specifically Section 901.4.1.4.3 (ASHRAE 189.1 Section 9.4.1.4.3), *Multiple-Attribute Product Declaration or Certification*, at least 10 different products installed in the building at the time of occupancy shall have a generic EPD, have a proprietary EPD, conform to a third-party multi-attribute certification, or have a third-party reviewed life cycle assessment (LCA).

Each Green Squared Certified® tile or installation material used on a project counts as two products toward the 10-product threshold, as the Green Squared® standard is one of the multi-attribute certification programs specified by the standard.

NAHB National Green Building Standard

Commonly referred to as ICC 700, NAHB's National Green Building Standard is an ANSI-accredited specification for residential construction. Green Squared Certified® tiles and installation materials can help achieve points under section 612.2, Sustainable Products.

Three points are awarded if 50% or more, based on square footage, of all tiles installed in a project have been Green Squared Certified. In order for such tiles to qualify, tile must make up at least 30% of the floor or wall area of the project.

GREEN SQUARED CERTIFIED.

Three points are awarded if 50% or more, based on square footage, of all tile installed in a project have been Green Squared Certified. In order for such tiles to qualify tile in general must make up at least 30% of the floor or wall area of the project.

GSA P-100 Facilities Standards

This standard incorporates sustainability criteria into design standards for the construction of government-funded buildings. Section 3.4 (Interior Performance) sets forth several provisions for ceramic tile:

The "Baseline" and Tier 1 High Performance provisions specify that tiles conform to ANSI A137.1, the ANSI standard for ceramic tile, and be Green Squared Certified.

The Tier 2 High Performance and Tier 3 High Performance provisions specify that tiles conform to ANSI A137.1, the ANSI standard for ceramic tile, be Green Squared Certified, and have a publicly-available EPD.

GREEN SQUARED CERTIFIED.

Green Squared Certified tiles and installation materials satisfy GSA's baseline and Tier 1 High Performance tiling requirements. Green Squared Certified products that also have an EPD are eligible for use on Tier 2 and Tier 3 federal building projects.

GREEN SQUARED® IN THE FIELD
PROJECTS WITH GREEN SQUARED CERTIFIED® PRODUCTS

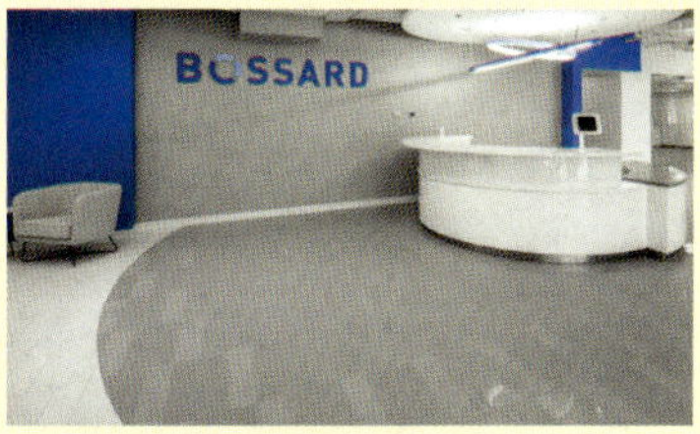

The inherent eco-friendliness of tile and tile installation products is amply demonstrated by Green Squared Certified® products.

Green Squared® is a multi-attribute, industry-wide standard that reflects the North American tile industry's commitment to sustainable manufacturing. The standard provides architects, specifiers, and consumers with a consistent set of criteria for evaluating the sustainability of tile and tile installation products.

Products bearing the Green Squared Certified logo are manufactured according to rigorous sustainability criteria that address a broad range of considerations, from the extraction of raw materials, to reuse or disposal at the end of a product's life.

Manufacturers of Green Squared Certified products invest considerable resources to ensure their facilities and manufacturing practices are socially and environmentally responsible, and to obtain third-party certification.

Besides being robustly sustainable, Green Squared Certified products are also beautiful, as the photos on the following pages illustrate.

The tile and installation products pictured here represent only a handful of the Green Squared Certified products that are currently available. Hundreds of products have been certified, and still more are in the certification process.

PROJECT SITE: Pensacola International Airport
Pensacola, Florida

PROJECT SCOPE: N/A

GREEN SQUARED CERTIFIED PRODUCTS:

- Moonstruck Collection 18" x 36" porcelain planks by Crossville

Pensacola International Airport welcomes nearly two million travelers annually. When it came time to update the flooring throughout the center concourse—the highest traffic area of the facility—airport leaders searched for the most efficient way to renovate the space. The design team planned a seamless renovation by scheduling the tile installation to take place after hours, thus minimizing disruption for travelers and airport staff.

The project specifications required that the flooring offer exceptional durability and low maintenance, while also providing a smooth surface for equipment, luggage, and wheelchairs. The team chose 18" x 36" Green Squared Certified porcelain tile planks for the center concourse. The material's natural look and sustainable features made the decision to choose Green Squared Certified tile simple for an area active in environmental preservation. With its sandstone texture and coastal aesthetic, the tiles evoke the feeling of the beach. To greet travelers in true Pensacola style, the design team artfully incorporated waterjet-cut designs of pelicans, seagulls, and waves into the flooring. Pensacola is recognized as Pelican City thanks to the natural presence of the iconic bird throughout the region. The waterjet-cut pelican designs throughout the concourse are a stunning nod to Pensacola's avian ambassadors.

With the renovation completed, travelers can now experience the first waves of tranquility straight from the plane—well before they reach the beach.

PROJECT SITE: Beckett Park by Cortland Walnut Hill Apartments
Irving, Texas

PROJECT SCOPE: 10,000 square ft

GREEN SQUARED CERTIFIED PRODUCTS:

- Frontier20 Limestone 24x24 porcelain paver, Frontier20 Simply Grey 24x24 porcelain paver by Landmark

Beckett Park by Cortland Apartments is an upscale apartment complex located minutes away from downtown Dallas, Texas. The complex offers several amenities, such as a fitness studio, a community clubhouse, a playground, and a newly renovated resort-style pool and sun deck.

The modernized pool area required a durable, slip resistant, and fade-resistant flooring material to serve Beckett Parks' many residents. With sustainability in mind, Landmark's Frontier20 Green Squared Certified porcelain tile collection was chosen to cover the sun deck, the area surrounding the pool, and the pool trim.

The design team decided to go with a checkered pattern across the entire pool surround using limestone-look and stone-look porcelain pavers. Residents are now welcomed to enjoy the outdoor oasis by lounging by the pool or relaxing under the sun deck gazebos.

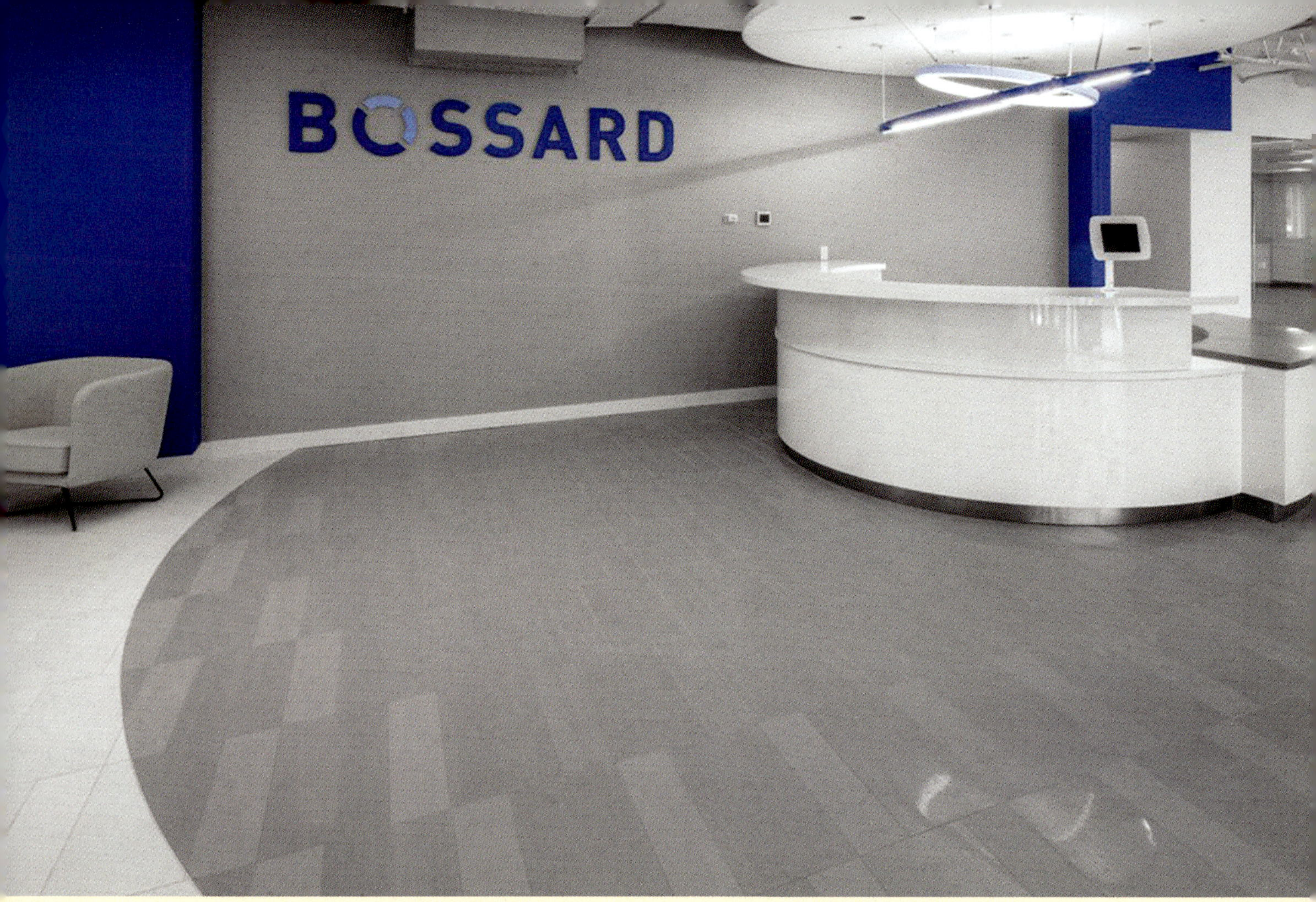

PROJECT SITE: Bossard Group Customer Experience Zone
Stroughton, Maryland

PROJECT SCOPE: N/A

GREEN SQUARED CERTIFIED PRODUCTS:

- Simpatico collection and Shades 6" x 24" and 12" x 24" collection by Crossville

Bossard Group, the worldwide leader in modern fastening technology and production, decided they wanted to use the ample space of a former warehouse to showcase the company's full range of fastening elements in a retail-focused environment. Using Bossard's logo as the foundation for the design, the designers envisioned the renovated facility with a classic, uncomplicated aesthetic that also conveys a feeling of modern technology. The designers' attention to detail resulted in an inviting, multi-purpose customer showroom featuring beautifully configured areas decked in surfacing solutions and materials that align with Bossard's reputation for quality and innovation.

The "Customer Experience Zone" is part of the manufacturing warehouse, so the design team recognized the need for durable materials that would be easy to maintain. They agreed ceramic tile was the best choice for the renovated warehouse. Tile from Crossville's Shade collection creates an aesthetically balanced space with a neutral color scheme that will withstand the high traffic demands of a warehouse environment. Green Squared certified tiles made from recycled materials were chosen to complement Bossard Group's high sustainability standards.

By varying the size and finish of the tile, the designers created a subtle take on the segmented 'O' in the Bossard logo. The pattern continues throughout the showroom and around the reception desk area. To complement the showroom, the design team extended the tile to the customer restrooms, while also adding dimensional tile from Crossville's Simpatico collection on the restroom walls.

The materials and design create an exceptional —and uniquely branded— customer experience.

PROJECT SITE: First Church of Christ, Scientist
Winter Park, Florida

PROJECT SCOPE: 6,500 square ft

GREEN SQUARED CERTIFIED PRODUCTS:

- Shades 12" x 24" exterior cladding by Crossville

The First Church of Christ, Scientist in Winter Park, Florida, was facing the dilemma of a deteriorating facility. Eager to continue serving, the church congregation searched for the best option for redesigning the aging church property. It was decided that a new church would be built on the property to meet the fiscal, social, architectural, and spiritual goals the church community wanted to achieve.

Designers decided to go with a modern approach when designing the new facility, while keeping the look of the original building in mind. To create a building that would stand the test of time, the team selected porcelain stone-look tile for the exterior cladding of the new facility. The earth-toned tiles were installed in a horizontal pattern giving the illusion of an elongated building on the property.

"Early in the conceptual phase of the project, we asked the church congregation to provide us with a list of spiritual principles that helped to shape their faith, explained Wade Miller, ACi's senior project designer and project director. "We found that in many ways, selecting Crossville tile, through its material honesty, lasting design integrity, low-maintenance nature, and recycled content composition, matched the guiding tenants of the congregation."

With these tenets in mind, the design team specified porcelain tile with a minimum of 20% recycled content that was Green Squared certified.

PROJECT SITE: Eastland-Fairfield Career and Technical Schools
Groveport, Ohio

PROJECT SCOPE: 1,000+ square ft

GREEN SQUARED CERTIFIED PRODUCTS:

- Metropolitan Ceramics' #507 Puritan Gray Quarry Tile by Ironrock

Eastland-Fairfield Career and Technical Schools serve 16 school districts in Franklin, Fairfield, and Pickaway, Ohio counties. Covering 700 square miles, it is one of the most extensive career and technical programs in the state, with nearly 40 in-demand programs, including studies in the culinary arts.

The school's culinary program includes a full-service commercial kitchen where students can create all types of restaurant cuisine. When designing the culinary arts space, the school knew the kitchen needed a floor that was made to last and would be easy to maintain as future chefs honed their skills in the kitchen classroom.

It was agreed ceramic quarry tile would be the material of choice for the kitchen floor due to quarry tile's durability, longevity, slip resistant options, and hygienic advantages. The school selected Ironrock's #507 Puritan Gray quarry tile for the entire floor of the learning space. Because all Ironrock quarry tile options are Green Squared Certified, the kitchen floor gained the added bonus of being environmentally friendly and sustainable.

PROJECT SITE: Issaquah Salmon Hatchery
Issaquah, Washington

PROJECT SCOPE: 76 square ft

GREEN SQUARED CERTIFIED PRODUCTS:

- Flexcolor® 3D grout in Crystal Moon by MAPEI

Over 100 volunteers gathered to bring Cheryl Smith's artistic design to life with the installation of a mosaic wall at the Issaquah Salmon Hatchery. The Friends of Issaquah Salmon Hatchery (FISH) reached out to Cheryl to be the lead designer of the community wall project after she taught a mini mosaic class for FISH the previous year.

The mosaic wall depicts a variety of salmon swimming through a stream of elaborate bubble designs. Each volunteer was given mosaic tile and a template of a salmon with the freedom to design their own fish.

A key priority of the FISH mosaic wall project was to minimize the impact on the environment, specifically the ecosystem of Pacific Northwest salmon. The wall was solely constructed using recycled mosaic glass and Flexcolor 3D grout by MAPEI. The Green Squared Certified specialty translucent grout complemented the goal of using recycled products, and with its iridescent-effect finish, complemented the aesthetics of the glass tile.

This unique aquatic mural will continue to grace the Issaquah community for many years to come.

PROJECT SITE: T.A. Brown Elementary
Austin, Texas

PROJECT SCOPE: N/A

GREEN SQUARED CERTIFIED PRODUCTS:

- Norway 7" x 36" and Kaleido 12" x 24" by Interceramic

Austin Independent School District's T.A. Brown Elementary School is a state-of-the-art, STEAM (Sciences, Technology, Engineering, Arts, and Mathematics) learning facility for 522 students. The new campus was built to incorporate spaces that allow collaborative, interdisciplinary, and project-based instruction in a flexible learning environment. Designers kept sustainability at the forefront of the project, with goals of achieving a LEED Silver rating and at a minimum, a two-star rating with the Austin Energy Green Building program.

The adaptable learning environment and sustainable systems throughout the building ensure this school will positively impact the lives of the students, teachers, faculty, and community. The school was completed on-time in January 2020, which allowed students to begin the spring semester in the new building.

According to interior designer Roberto Pruneda of ERO Architects, "When designing for high-traffic areas, we like to choose materials that are easy to maintain, durable, and long-lasting like ceramic tile. Ceramic tile is also non-toxic and free of harsh chemicals, which is ideal for areas continuously filled with students. We also wanted to use locally sourced materials that contributed to LEED certification and the building's overall sustainability, so selecting Green Squared Certified ceramic tile was the obvious choice."

Release date: April 2021

Green building standards and rating systems are living documents, and the consensus regarding their use and interpretation is ever-evolving. Check for updates to *Tile: The Natural Choice* at www.TCNAtile.com.

100 Clemson Research Blvd. Anderson, SC 29625 | 864.646.8453 | www.TCNAtile.com

TCNA
TILE COUNCIL OF NORTH AMERICA, INC.

UL-Certified
ISO STANDARDIZED
PCRs
Green Building Credits
Independently Verified
CRADLE-TO-GRAVE
EPD
Environmental Product Declaration
TILE MORTAR GROUT
CREDIBILITY
Natural Resources
Tile Council
ATMOSPHERE
LOW ENVIRONMENTAL IMPACT
Transparency
ENERGY DEMAND
Carbon Footprint
LCA
60-YEAR IMPACT ASSESSMENT

Guide to EPDs for Ceramic Tile, Mortar, and Grout Made in North America

2021

CONTENTS

TOTAL TRANSPARENCY
EPDs FOR NORTH AMERICAN-MADE CERAMIC TILE, MORTAR, AND GROUT

The environmental product declarations (EPDs) for North American-made ceramic tile, mortar, and grout provide a comprehensive overview of how these products impact the environment. The combination of the three provides environmental impact data, per square meter, of the main materials used to set tile.

The EPDs address the most important environmental considerations affecting the well-being of the planet and those who call it home—specifically, global warming, fossil fuel resource depletion, acidification, smog formation, eutrophication, and ozone depletion.

And, because EPDs and product environmental transparency are increasingly required in green construction standards and rating systems, specifying ceramic tiles, mortars, and grouts covered by the EPDs can help achieve green building points and meet transparency requirements. In fact, as most green building standards, codes and rating systems provide incremental credit for the use of each product covered by an EPD, using tile in a project, with the accompanying mortar and grout, means that a single tile installation using materials covered by North American EPDs could potentially contribute "three times (3X)."

The Credibility Process

When selecting construction materials, it's important to know decisions are supported by reliable, life cycle-based sustainability information. For the North American-made Ceramic Tile EPD, the life-cycle assessments were independently performed by WAP Sustainability Consulting, and for Mortars and Grouts,

by Sphera (formerly Thinkstep), all per the ISO standard 14040/44, *LCA Principles, Framework, Requirements, and Guidelines.* Additionally, UL Environment, as Program Operator, verified the validity of the LCAs and certified the requirements of ISO 14025, *Principles and Procedures for EPDs*, and ISO 21930, *Core Rules for EPDs of Construction Products and Services,* were met. Conformance to these standards, plus certification of the EPDs by a globally recognized sustainability leader, ensures the EPDs are accurate, not unsubstantiated marketing claims or other attempts at "green washing."

To produce the EPDs, the participating manufacturers provided extensive data on their materials and operations and participated in a cradle-to-grave evaluation of their products, from raw material sourcing/extraction, through manufacturing, delivery, installation, use, and end of life.

For each EPD, the specific environmental impacts measured, and the methodology for measuring those, are dictated by a Product Category Rule (PCR). For the tile EPD, the North American PCR, UL 10010-7 Part B: Flooring EPD Requirements v2.0, was followed. Using product category rules developed for all flooring allows tile covered by the EPD to be compared to other flooring products evaluated under the same PCR.

For the mortar and grout EPDs, international PCRs for factory-made mortar were adapted for North American usage per an addendum published by UL Environment. This allowed the North American PCR to build on criteria that was previously used in the development of EPDs for other types of cementitious construction products. This provides valuable references when analyzing and interpreting EPDs in a global context.

The standardized reporting of tile, mortar and grout

DATA COLLECTION

North American ceramic tile, mortar and grout manufacturers submit extensive data on their materials and operations, covering everything from raw material sourcing and extraction through end of life.

DATA ANALYSIS AND LIFE-CYCLE ASSESSMENT (LCA)

Sustainability leaders WAP and Sphera analyze the data to determine the environmental impacts and life-cycle assessments of North American-made ceramic tile, mortar and grout.

REVIEW OF DRAFT REPORT

Sustainability leader UL Environment independently evaluates and certifies that PCRs were followed, as well as ISO 14025, ISO 21930, and ISO 14040/44 standards for EPDs and LCAs.

EXTERNAL VERIFICATION

A panel of independent stakeholders verifies conformance to standards, applicability of the PCR, and that the information presented is accurate, not unsubstantiated marketing claims or other attempts at green washing.

EPD CERTIFICATION AND REGISTRATION

UL Environment certifies the EPDs and registers them in its online Sustainable Product Database, SPOT, a publicly available resource for authentic and credible environmental product information.

EPDs FOR NORTH AMERICAN-MADE CERAMIC TILE, MORTAR, AND GROUT

required by the PCRs additionally provides transparency and prevents green washing, as all parameters and impacts in the PCR must be reported, not just those areas in which a product performs well.

Products Covered by the EPDs

Because the EPDs are based on aggregated data from all participating manufacturers, they are "generic" or "non-proprietary" EPDs. As most major North American tile industry manufacturers participated, over 85% of North American-made products are covered: 3.4 billion square feet of tile and 2.25 billion kilograms of mortar and grout.

To find out if a specific product line is covered by a certified EPD, contact the participating manufacturers.

Participating Tile Manufacturers

American Wonder Porcelain
Arto
Crossville
Dal-Tile Corporation
Del Conca USA
Florida Tile
Interceramic
Ironrock
Porcelanite Lamosa
Portobello America
Quarry Tile Company
StonePeak Ceramics
Vitromex de Norteamérica

Participating Grout and Mortar Manufacturers

Ardex
Bexel
Bostik
Cemix/Texrite
Crest
Custom Building Products
HB Fuller/Tec
Interceramic
Laticrete
Mapei

The EPDs were initiated by Tile Council of North America (TCNA) and its participating members, with data analysis and modeling by WAP Sustainability and Sphera, and third-party, independent certification by UL Environment.

Download the EPDs at TCNAtile.com

Just as nutrition labels inform the calorie conscious on food choices, an EPD informs with respect to sustainability. When using the North American-made Tile EPD alongside other flooring products' generic EPDs, one thing is clear: Overall ceramic tile has the lowest 75-year environmental impact per square meter. Similarly, the industry-wide EPDs for North American-made mortar and grout report very low 75-year environmental impacts per installed square meter.

Environmental Facts

Functional unit: 1 m^2 of Ceramic Tile Floor Covering
Reference Service Life (RSL): 75 Years

75 YEAR IMPACT ASSESSMENT	Total
Global Warming Potential (kg CO_2 eq)	19.6
Acidification Potential (kg SO_2 eq)	0.037
Ozone Depletion Potential (kg R11 eq)	6.7E-10
Smog Potential (kg O_3 eq)	0.8
Eutrophication Potential (kg N eq)	0.003
Fossil Resource Depletion (MJ)	32
OTHER INFORMATION	
Zero VOCs	
Boundaries	Cradle to Grave
Recycled Content	Varies
Green Squared Certification®	Some Tiles

Ceramic Tile Ingredients: Clay (69.8%), Sand (6.2%), Talc (1.1%), Feldspar (11.3%), Scrap (6.5%), Frit (0.4%), Calcium Carbonate (1.5%), Ash (1.4%), Additives (0.7%), Ink (0.2%), Glaze (1.1%)

Visit **www.TCNAtile.com** for further information.

Environmental Facts

Functional Unit: 1 m^2 of Installed Grout
(Application Rate, 0.212 kg/m^2)
Reference Service Life (RSL): 75 Years

75 YEAR IMPACT ASSESSMENT*	Total
*Assumes linear extrapolation from 60-year use phase impacts reported in EPD, and equivalence to 60-year impacts for all other phases	
Global Warming Potential (kg CO_2 eq)	0.206
Acidification Potential (kg SO_2 eq)	0.000823
Ozone Depletion Potential (kg R11 eq)	3.75E-10
Smog Potential (kg O_3 eq)	1.24E-02
Eutrophication Potential (kg N eq)	6.78E-05
Abiotic Depletion Potential-Fossil (MJ)*	2.1
*Per CML 2001-2016 LCIA methodology due to unavailability of Fossil Resource Depletion per TRACI 2.1 LCIA methodology	
OTHER INFORMATION	
VOC Emissions-CDPH Section 01350	Meets
Recycled Content	Varies
Green Squared® Certification	Some Grout
ANSI A118.6, ANSI A118.7	Meets

Tile Grout Ingredients: Sand, White Cement, Limestone, Grey Cement, Calcium Aluminate, Calcium Formate, Other Additives

Visit **www.TCNAtile.com** for further information.

Environmental Facts

Functional Unit: 1 m^2 of Installed Tile Mortar
(Application Rate, 4.07 kg/m^2)
Reference Service Life (RSL): 75 Years

75 YEAR IMPACT ASSESSMENT*	Total
*Assumes equivalence to 60-year impacts reported in EPD	
Global Warming Potential (kg CO_2 eq)	2.88
Acidification Potential (kg SO_2 eq)	0.0122
Ozone Depletion Potential (kg R11 eq)	5.69E-09
Smog Potential (kg O_3 eq)	0.201
Eutrophication Potential (kg N eq)	5.62E-04
Abiotic Depletion Potential-Fossil (MJ*)	30.9
*Per CML 2001-2016 LCIA methodology due to unavailability of Fossil Resource Depletion per TRACI 2.1 LCIA methodology	
OTHER INFORMATION	
VOC Emissions-CDPH Section 01350	Meets
Recycled Content	Varies
Green Squared® Certification	Some Mortar
ANSI A118.1, ANSI A118.4, ANSI A118.11, ANSI A118.15, ISO 13007	Meets

Tile Mortar Ingredients: Sand, Calcium Carbonate, Grey Cement, White Cement, Ethylene Vinyl Acetate, Admixture, Cellulose Ether, Other Additives

Visit **www.TCNAtile.com** for further information.

CREDITS AND COMPLIANCE
NORTH AMERICAN-MADE CERAMIC TILE, MORTAR, AND GROUT

LEED v4.1

Under the credit category Building Product Disclosure and Optimization–Environmental Product Declarations, up to 2 points toward LEED certification can be earned by using products with EPDs, such as North American-made ceramic tile, mortar and grout.

Option 1 (1 Point) Environmental Product Declaration

AND/OR

One point can be earned for using at least 20 different permanently installed "products," which are construction materials that meet one of the specified disclosure criteria and are therefore eligible for the valuation associated with that disclosure criteria. Valuations for such materials are awarded on a per-material basis, and this LEED point is earned by using enough materials that their combined valuations total at least 20 "products."

Disclosure Criteria	Valuation
Proprietary LCA (Internally Reviewed)	1 product
Proprietary EPD (Internally Reviewed)	1 product
Generic EPD (Third-Party Certified)	1 product
Proprietary EPD (Third-Party Certified)	1.5 products

Materials that meet multiple criteria are eligible only for the highest valuation available; the valuations cannot be combined. Also, materials must be sourced from at least five different manufacturers.

EPD Products covered by the North American-made Ceramic Tile, Mortar, and Grout EPDs can each contribute at least ½ product toward the 20-product requirement. For example, for a project with 8 different tiles installed with the same mortar but different grouts (all covered by EPDs), the 17 products (8 tiles, 8 grouts, 1 mortar) would contribute 8.5 "products" toward the 20-product requirement (Example A). In the same scenario, if the mortar, grout, and 1 of the 8 tiles also have a proprietary EPD, the same 17 tiles and related materials would contribute 13.5 "products" (Example B).

Example A

(8 tiles + 1 mortar + 8 grouts) × .5 products = 8.5

Example B

(1 tile × 1 product) + (1 mortar × 1 product)
+ (8 grouts × 1 product) + (7 tiles × .5 products) = 13.5

Option 2 (1 Point) Multi-Attribute Optimization

One point can be earned by using products with improved, or plans for improved environmental life-cycle impacts when at least 10 of those products are used or those products comprise at least 10% of the total value of permanently installed products in the project. Valuations for such materials are awarded on a per-material basis, and this LEED point is earned by using enough materials that their combined valuations total at least 10 "products" or 10% of the total value of the project.

Optimization Criteria	Valuation
Life-Cycle Impact Reduction Plan	½ product or 50% cost
Reduced Global Warming Potential (GWP)	1 product or 100% cost
10% Reduced GWP	1.5 products or 150% cost
20% Reduced GWP & 5% Reduction in Two Other Impact Categories	2 products or 200% cost

EPD For this LEED point, generic EPDs provide a convenient baseline for comparison, and manufacturers of tiles, mortars, or grouts included in generic EPDs that also have one or more proprietary EPDs have the option to show below-industry-average impacts or an impact reduction plan, either of which can contribute to meeting the 10-product or 10% value threshold.

Example calculation 1: For a project with \$1,000,000 in permanently installed products, including \$20,000 of mortar and grout with GWP 10% lower than industry average, and \$80,000 of ceramic tiles for which the manufacturer has included a publicly available action plan to reduce life-cycle environmental impacts:

$$\frac{[\$20K \text{ (mortar and grout value) x } 150\%] + [\$80K \text{ (tile value) x } 50\%]}{\$1M \text{ (all materials value)}} = 7\%$$

In the previous example using the cost calculation method, the contribution of the tile and related installation materials toward the 10% minimum would be 7% (70% of the requirement).

Example calculation 2: For a project with 2 different tiles with GWP lower than industry average, 1 tile with a life-cycle impact reduction plan, 3 different grouts with life-cycle environmental impact reduction plans, and 1 mortar with GWP 10% lower than industry average:

(2 tiles x 1 product) + (1 tile x .5 product) +
(3 grout x .5 product) + (1 mortar x 1.5 products) = 5.5

In the above example using the product calculation method, the 7 tiles and related installation materials would contribute 5.5 "products" toward the 10-product minimum (55% of the requirement).

Green Globes ANSI/GBI 01-2019

Under the provisions of Section 10.2, *Product Life Cycle*, using North American-made ceramic tiles, mortars, and/or grouts can contribute toward the 29 available points for using products with EPDs.

19 points are awarded for using products with **cradle-to-gate** life-cycle evaluations based on:				
EPDs: Generic or Proprietary	AND / OR	Third-Party Verified Proprietary Life-Cycle Assessment	AND / OR	Third-Party Certification to Multiple Attribute Standards
10 points are awarded for using products with **cradle-to-grave** life-cycle evaluations based on:				
EPDs: Generic or Proprietary	AND / OR	Third-Party Verified Proprietary Life-Cycle Assessment		

Products recognized as having cradle-to-grave life cycle evaluations are additionally recognized for cradle-to-gate.

Points are awarded based on the number of products used. Products with EPDs are valued the same, regardless of whether their EPDs are generic or proprietary; both contribute equally toward the product total.

	# Products	Points
Cradle-to-Gate Life-Cycle Evaluations	30+	19
	29	14
	28	13
	27	12
	26	11
	25	10
	24	9
	23	8
	22	7
	21	6
	20	5
Cradle-to-Grave Life-Cycle Evaluations	10	10
	9	9
	8	8
	7	7
	6	6
	5	5

EPD Without limit, using North American-made ceramic tiles, mortars, and grouts covered by EPDs will contribute to this requirement on a cradle-to-gate and cradle-to-grave basis. Products that are also Green Squared Certified® will earn double points under the cradle-to-gate provisions for also meeting the requirement for products with third-party certifications based on multiple attribute standards.

Sample calculation for a project with 10 tile products, 2 mortars, and 10 grouts covered by the EPD:

7 Points (22 products with cradle-to-gate, covered by EPD)
+ 10 Points (22 products with cradle-to-grave, also covered by EPD)
= 17 Total Points

Sample calculation for the same project, but with 3 of the tile products, 1 mortar and 2 grouts additionally having Green Squared® Certification:

13 Points (28 products; 22 with cradle-to-gate, covered by EPD, 6 of which are additionally recognized as Green Squared Certified®)
+ 10 Points (22 products with cradle-to-grave, also covered by EPD)
= 23 Total Points

CREDITS AND COMPLIANCE
NORTH AMERICAN-MADE CERAMIC TILE, MORTAR, AND GROUT

NAHB National Green Building Standard

NAHB National Green Building Standard is commonly referred to as ICC 700. North American-made ceramic tile can help achieve points under Section 611.4, Product Declarations, in this ANSI-approved standard and rating system for residential construction. In this section of the standard, 5 points are awarded if EPDs are submitted for 10 different products installed in the building project. Each building material with an industry-wide EPD contributes a full 'product' toward the 10-product goal. Each building material with a product-specific EPD contributes two 'products' toward the 10-product goal.

Additionally, up to 10 points can be awarded under Section 610, Life Cycle Analysis, for using products with lower environmental impacts, compared to products intended for the same use. To garner points under Section 610, the selected product must improve upon multiple environmental impacts by an average of 15%, with the number of points awarded tied to the number of impact categories in which the selected product performs better.

Number of Impact Categories	Points
4	2
5	3

Applicable Impact Categories:

- Fossil fuel consumption
- Global warming potential
- Acidification potential
- Eutrophication potential
- Ozone depletion potential

The standard requires that the LCA tool used to determine the environmental impacts must be based on a minimum reference service life for a building of 60 years. It allows comparisons to be made between different types of products with the same use. It also allows comparison of individual product impacts to the industry averages for the same product type.

Each tiling product covered by the North American-made tile, mortar and/or grout EPDs contributes a full product toward the 10-product threshold specified by Section 611.4 of this standard. Those which have product-specific EPDs count double, expanding opportunities for garnering the previously discussed 5 points and creating the possibility of satisfying the 10-product requirement solely with tile installations alone!

Furthermore, the North American-made Ceramic Tile, Mortar, and Grout EPDs report environmental impacts over 75 years, as analyzed by GaBi LCA software, according to the requirements of the standard, fulfilling the 60-year minimum building service life with no requirement to account for product replacements.

When choosing between tile and an alternate flooring material, using ceramic tile covered by the EPD is advantageous when building to this standard, because the EPD shows very low environmental impacts in all categories. In fact, tile covered by the EPD has the lowest impact in all six impact categories, as compared to the impacts reported for other common flooring options in their publicly available, generic EPDs. Refer to the 75-year impact analysis graphs included in this guide for details.

When choosing between specific tiles, mortars, and/or grouts, products covered by generic EPDs will additionally contribute when such products also have proprietary EPDs showing impacts that are on average 15% lower, as compared to the industry average reported in the generic EPD, in at least 4 categories.

Additionally, products that are Green Squared Certified® can also contribute points as detailed elsewhere in the NAHB Green Building Standard. Refer to the Tile and Green Building chart in *Tile: The Natural Choice* for details.

International Green Construction Code (IgCC) Powered by ASHRAE Standard 189.1

About IgCC: An initiative of the International Code Council (ICC). IgCC was developed as model code language for states and municipalities to establish baseline sustainability requirements for new and existing commercial buildings.

About ASHRAE 189.1: An ANSI-approved standard, ASHRAE 189.1 was designed to be incorporated into governing jurisdictions' codes and ordinances for commercial buildings and enforced by building officials and inspectors.

IgCC and ASHRAE Standard 189.1 are now integrated into a single model code and standard for high-performance buildings.

In IgCC Section 901.4.1.4 (ASHRAE 189.1 Section 9.4.1.4.2), *Multiple-Attribute Product Declaration or Certification*, at least 10 different products installed in the building by the time occupancy is allowed shall have a generic EPD, have a proprietary EPD, conform to a third-party multi-attribute certification, or have a third-party reviewed life cycle assessment (LCA).

EPD Each tiling product used, if covered by one or more of the North American made Ceramic Tile, Mortar, and/or Grout EPDs, counts as one product toward the ten-product threshold.

CREDITS AND COMPLIANCE

NORTH AMERICAN-MADE CERAMIC TILE, MORTAR, AND GROUT

Collaborative for High Performance Schools (CHPS)

Under Section MW 6.1, Environmental Product Declarations, 2 points are available for using products with EPDs. To earn them, the standard requires selection of at least 10 products with third party certified EPDs conforming to ISO 14025 and/or ISO 21930. Products are valued the same, whether their EPDs are generic or proprietary.

EPD

North American-made ceramic tile, mortar and grout can contribute toward earning these points, as the EPDs have been third-party certified by UL Environment as conforming to ISO standards for EPDs.

CALGreen

This standard is part of the California Building Standards Code and is commonly adopted as model language by developers of green building standards and rating systems in other parts of the country.

Section A5.409.3 for Non-Residential Construction requires at least 50% use of materials or assemblies based on life-cycle assessment (LCA) of their global warming potential (climate change/greenhouse gases) and two more environmental impacts from the list below:

- Fossil fuel depletion
- Stratospheric ozone depletion
- Acidification of land and water resources
- Eutrophication
- Photochemical oxidants (smog)

EPD Tiling products covered by the North American-made Ceramic Tile, Mortar and/or Grout EPDs are eligible for consideration under this section of CALGreen, as the EPDs provides LCA data and address the referenced environmental impacts.

GSA P-100 Facilities Standards

This standard incorporates sustainability criteria into design standards for the construction of government-funded buildings. Section 3.4 (Interior Performance) sets forth several provisions for ceramic tile:

The "Baseline" and Tier 1 High Performance provisions specify that tiles conform to ANSI A137.1, the ANSI standard for ceramic tile, and be Green Squared Certified.®

The Tier 2 High Performance and Tier 3 High Performance provisions specify that tiles conform to ANSI A137.1, the ANSI standard for ceramic tile, be Green Squared Certified,® and have a publicly-available EPD.

EPD Tiles covered by the North American-made Ceramic Tile EPD are eligible for use on Tier 2 and Tier 3 federal building projects, provided they are also Green Squared Certified® and meet the tile performance standards given in ANSI A137.1.

*Note: At the time of this publication, there are no GSA P-100 provisions for tile mortar or tile grout.

ENVIRONMENTAL FOOTPRINT: TILE, MORTAR AND GROUT
A 75-YEAR IMPACT ANALYSIS

Today, North American construction products are increasingly being evaluated based on their 75-year environmental footprint, a time frame chosen for the average service life of a building. Accordingly, the EPDs for North American-made Ceramic Tile, Mortar, and Grout provide long-term environmental impact data. In fact, the North American flooring Product Category Rule (PCR), UL 10010-7, requires EPDs to provide a life-cycle assessment (LCA) with a 75-year impact analysis. This appropriately includes the environmental impacts associated with acquiring a product's raw materials, manufacturing the product, and then delivering, installing, maintaining, and disposing of it, multiplied by the theoretical number of times the product would wear out and would have to be reinstalled (i.e., 75 years divided by the product's service life, in years).

In all six standard LCA environmental impact assessment categories, as included in the EPD for North American-made Ceramic Tile, Mortar, and Grout, the environmental benefits resulting from tile's inherent durability, long life, and low maintenance are clearly realized.

The EPDs for North American-made tiles, mortars, and grouts all report very low 75-year environmental impacts per installed square meter. In fact, when the EPD for North American-made Tile is compared to the generic EPDs for other types of flooring, developed under the same PCR and using the same building service life, North American-made ceramic tile has the lowest overall environmental impact. This includes the lowest Global Warming Potential and the lowest Fossil Fuel Resource Depletion.

Global Warming Potential (GWP)

GWP, commonly referred to as "carbon footprint," is measured by emissions of carbon-equivalent gases and relates to climate change.

The 75-year GWP for North American-made Ceramic Tile, Mortar and Grout, based on their EPDs, is 19.6 kg CO_2 equivalent (eq.) per square meter for tile, 2.8 kg per m^2 for mortar, and 0.2 kg per m^2 for grout.

GWP: Tile vs Other Types of Flooring

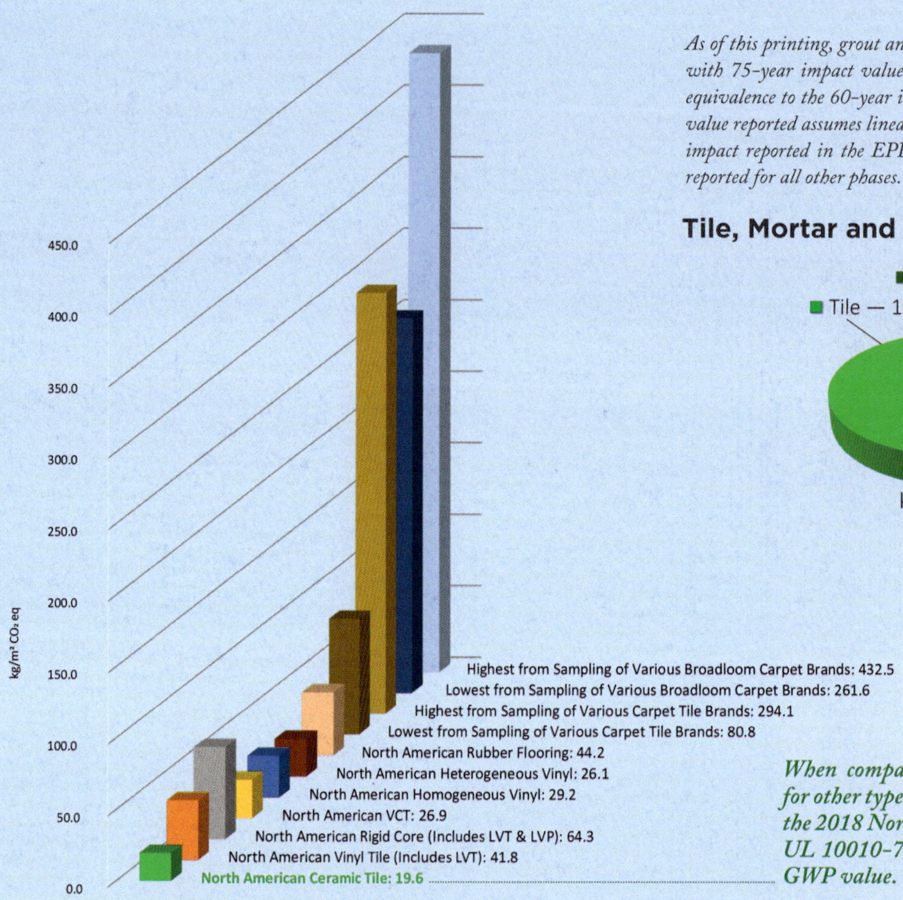

As of this printing, grout and mortar EPDs have not yet been updated with 75-year impact values. For mortar, the value reported assumes equivalence to the 60-year impact reported in the EPD. For grout, the value reported assumes linear extrapolation from the 60-year use phase impact reported in the EPD, and equivalence to the 60-year impact reported for all other phases.

Tile, Mortar and Grout: Combined GWP

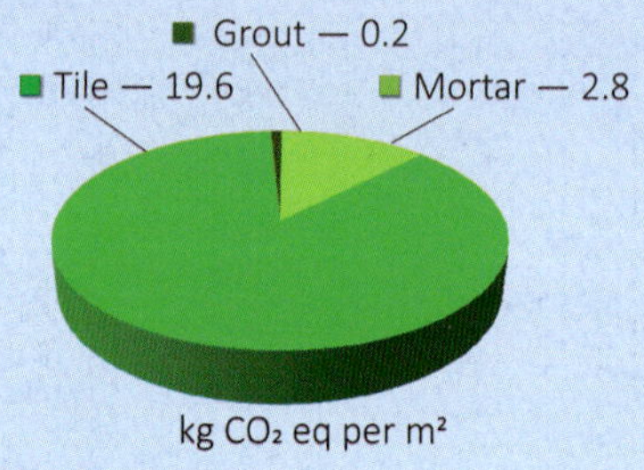

When compared to UL Certified EPDs for other types of flooring developed under the 2018 North American Flooring PCR, UL 10010-7, tile has the lowest 75-year GWP value.

ENVIRONMENTAL FOOTPRINT: TILE, MORTAR AND GROUT
A 75-YEAR IMPACT ANALYSIS

Photochemical Oxidant Creation Potential (POCP)

POCP is directly linked to smog formation. Smog is harmful to humans and animals, most notably for aggravating bronchial-related diseases; it also inhibits plant growth due to decreased UV radiation, and can discolor ground-level materials like building exteriors.

The 75-year POCP for North American-made Ceramic Tile, Mortar, and Grout, based on their EPDs, is 0.8 kg O_3 equivalent (eq.) per square meter for tile, 0.2 kg per m^2 for mortar, and 0.01 kg per m^2 for grout.

POCP: Tile vs Other Types of Flooring

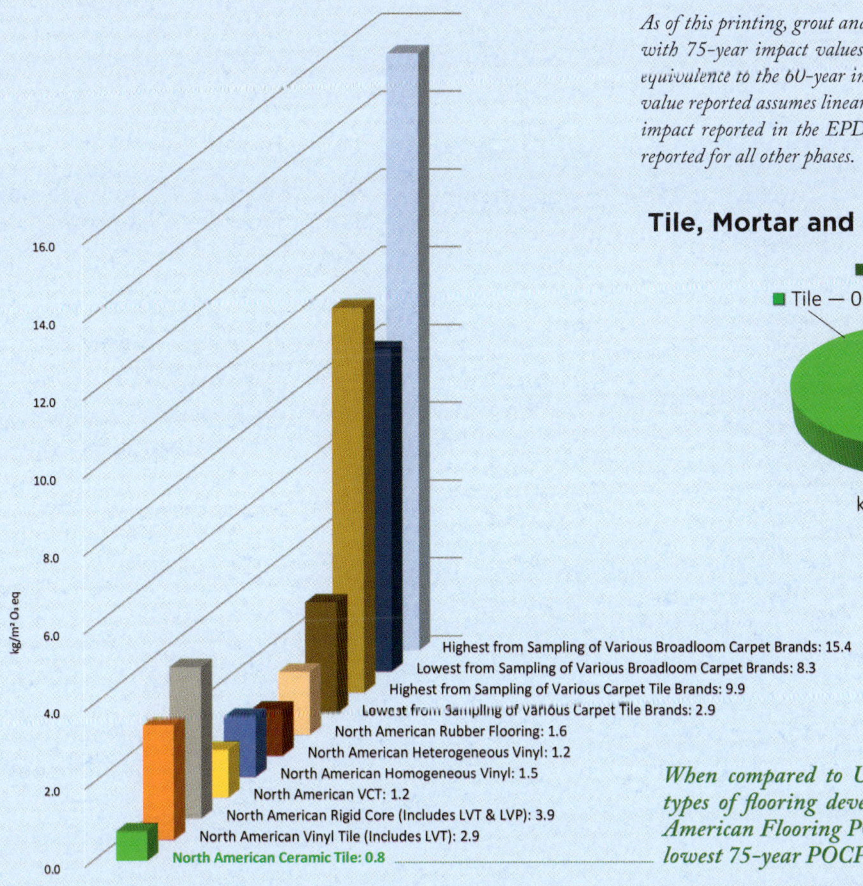

As of this printing, grout and mortar EPDs have not yet been updated with 75-year impact values. For mortar, the value reported assumes equivalence to the 60-year impact reported in the EPD. For grout, the value reported assumes linear extrapolation from the 60-year use phase impact reported in the EPD, and equivalence to the 60-year impact reported for all other phases.

Tile, Mortar and Grout: Combined POCP

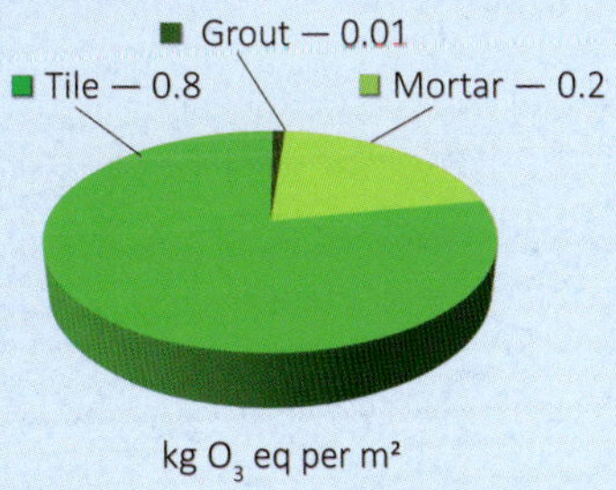

When compared to UL Certified EPDs for other types of flooring developed under the 2018 North American Flooring PCR, UL 10010-7, tile has the lowest 75-year POCP value.

Ozone Depletion Potential (ODP)

Ozone depletion occurs when chlorofluorocarbons (CFCs) and hydrofluorocarbons (HCFCs) reach the stratosphere, react with the sun, and break down the ozone layer. Decreased ozone can lead to an increase in the amount of UV-B radiation that reaches Earth's surface, having harmful effects on human health, animal health, terrestrial and aquatic ecosystems, biochemical cycles, and materials.

The 75-year ODP for North American-made Ceramic Tile, Mortar, and Grout, based on their EPDs, is 0.00000000067 kg R11 equivalent (eq.) per square meter for tile, 0.0000000057 kg per m^2 for mortar, and 0.00000000038 kg per m^2 for grout.

ODP: Tile vs Other Types of Flooring

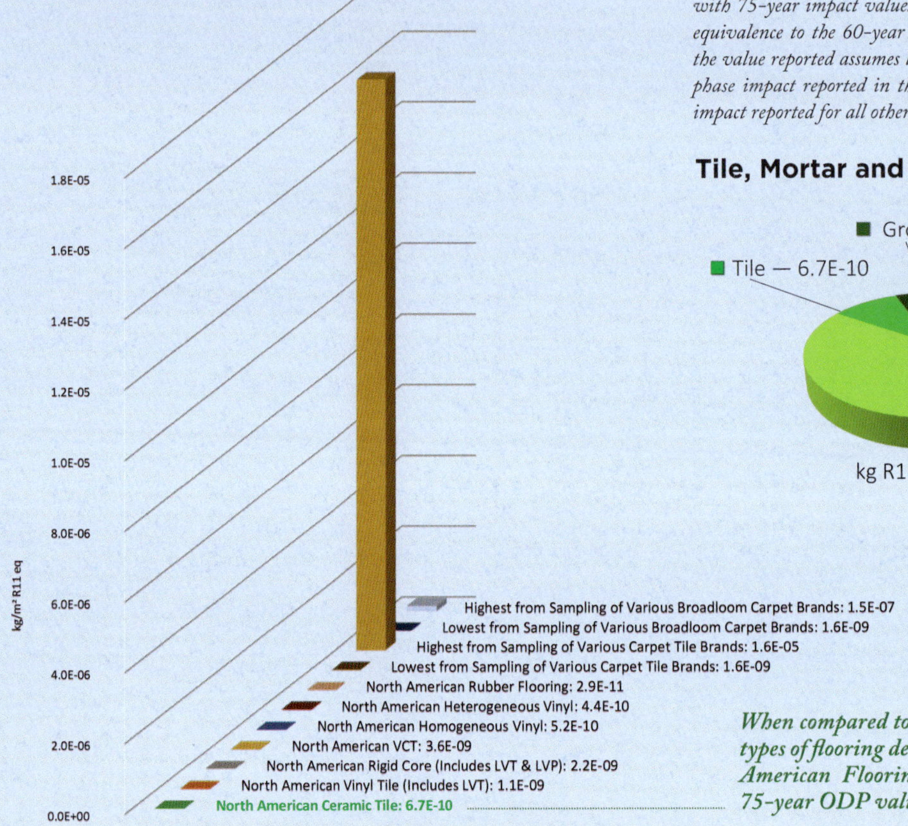

When compared to UL Certified EPDs for other types of flooring developed under the 2018 North American Flooring PCR, UL 10010-7, tile's 75-year ODP value is among the lowest.

As of this printing, grout and mortar EPDs have not yet been updated with 75-year impact values. For mortar, the value reported assumes equivalence to the 60-year impact reported in the EPD. For grout, the value reported assumes linear extrapolation from the 60-year use phase impact reported in the EPD, and equivalence to the 60-year impact reported for all other phases.

Tile, Mortar and Grout: Combined ODP

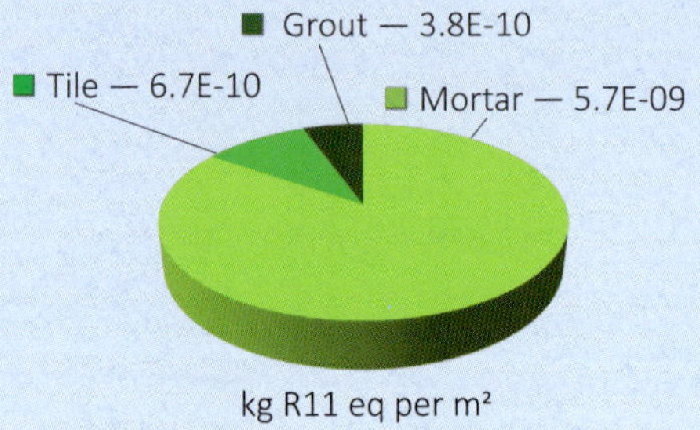

ENVIRONMENTAL FOOTPRINT: TILE, MORTAR AND GROUT
A 75-YEAR IMPACT ANALYSIS

Acidification Potential (AP)

Acidification occurs when soil, groundwater, surface water and atmospheric moisture pH levels are lowered. This can have a harmful impact on organisms, ecosystems, and man-made materials, including buildings.

The 75-year AP for North American-made Ceramic Tile, Mortar and Grout, based on their EPDs, is 0.037kg SO_2 equivalent (eq.) per square meter for tile, 0.012 kg per m^2 for mortar, and 0.00082 kg per m^2 for grout.

AP: Tile vs Other Types of Flooring

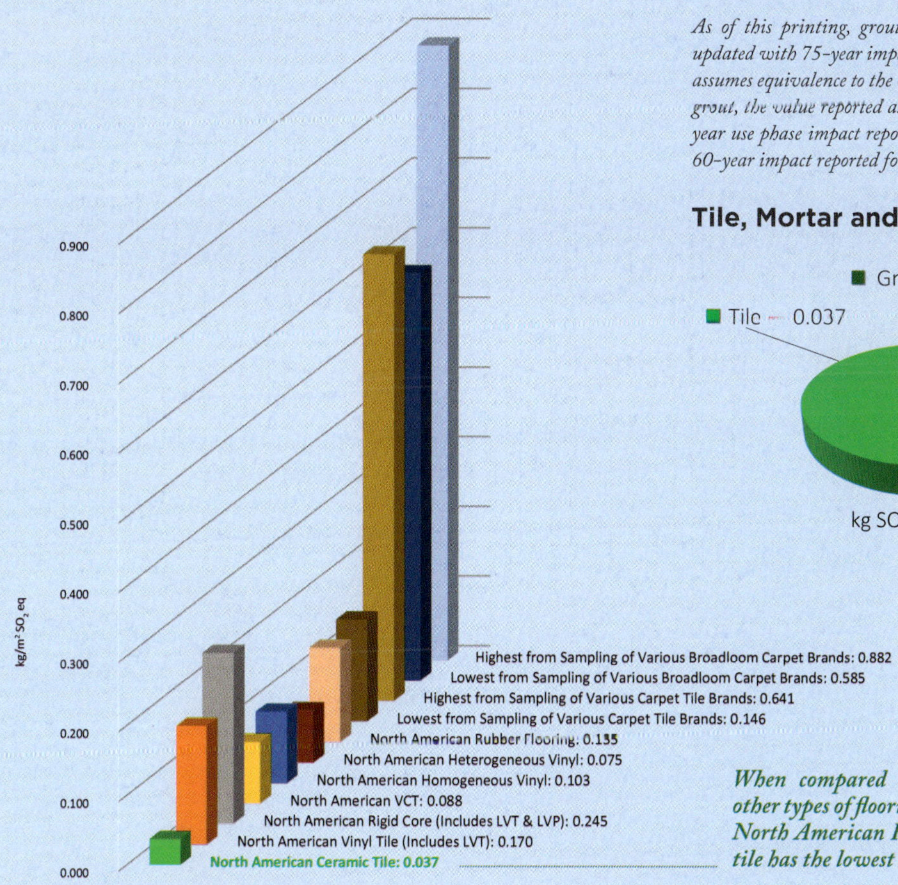

As of this printing, grout and mortar EPDs have not yet been updated with 75-year impact values. For mortar, the value reported assumes equivalence to the 60-year impact reported in the EPD. For grout, the value reported assumes linear extrapolation from the 60-year use phase impact reported in the EPD, and equivalence to the 60-year impact reported for all other phases.

Tile, Mortar and Grout: Combined AP

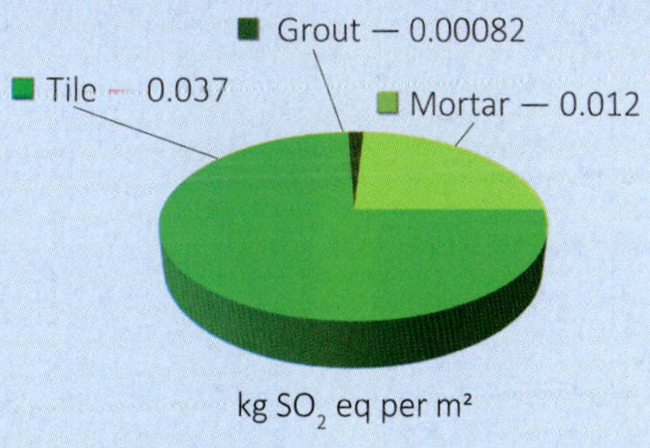

When compared to UL Certified EPDs for other types of flooring developed under the 2018 North American Flooring PCR, UL 10010-7, tile has the lowest 75-year AP value.

Eutrophication Potential (EP)

EP: Tile vs Other Types of Flooring

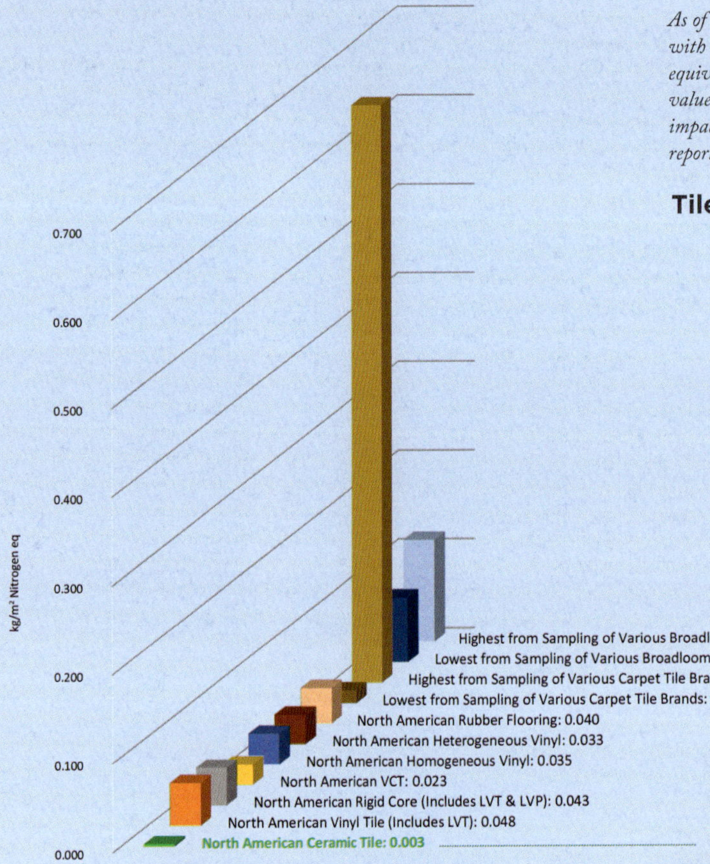

As of this printing, grout and mortar EPDs have not yet been updated with 75-year impact values. For mortar, the value reported assumes equivalence to the 60-year impact reported in the EPD. For grout, the value reported assumes linear extrapolation from the 60-year use phase impact reported in the EPD, and equivalence to the 60-year impact reported for all other phases.

Tile, Mortar and Grout: Combined EP

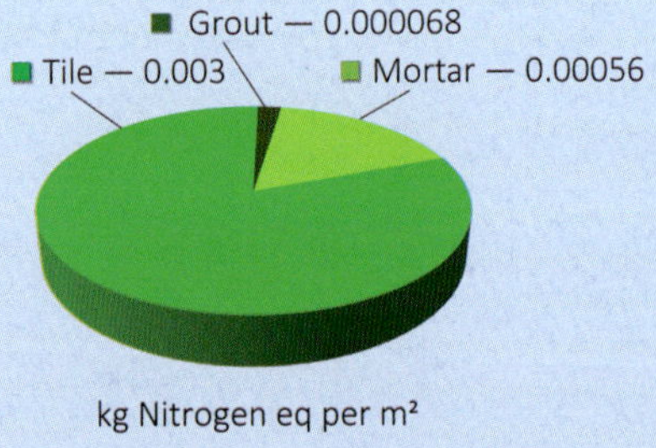

When compared to UL Certified EPDs for other types of flooring developed under the 2018 North American Flooring PCR, UL 10010-7, tile has the lowest 75-year EP value.

Eutrophication is the enrichment of a body of water (or ecosystem) with nutrients needed for photosynthesis, such as carbon dioxide and nutrient compounds containing nitrogen and/or phosphorus, commonly from fertilizer runoff and sewage. It is a significant cause of oxygen depletion in bodies of water, resulting from excess plant and algal growth, causing hypoxic conditions in which marine life cannot be sustained.

The 75-year EP for North American-made Ceramic Tile, Mortar, and Grout, based on their EPDs, is 0.003 kg Nitrogen equivalent (eq.) per square meter for tile, 0.00056 kg per m^2 for mortar, and 0.000068 kg per m^2 for grout.

Fossil Fuel Resource Depletion

The utilization of fossil fuel resources throughout the life cycle of a product includes primary non-renewable energy resources such as petroleum, coal, and natural gas used as raw materials and as fuel in the manufacture, installation, maintenance, and disposal of the product. Efforts among environmental life cycle experts have been made to consider depletion of other types of resources, including land and water, but such evaluations are not often included in product LCAs due to difficulties in quantification at the site level.

The Fossil Fuel Resource Depletion over 75 years for North American-made Ceramic Tile, based on its EPD, is 32 MJ per square meter using TRACI 2.1 life cycle impact analysis (LCIA) methodology. Fossil Abiotic Resource Depletion (ADP-f) over 75 years for North American-made Mortar and Grout, based on their EPDs, is 31 MJ per m^2 for mortar and 2 MJ per m^2 for grout using CML 2001-2016 LCIA methodology.

TRACI LCIA methodology is generally preferable over CML in North America based on strong correlations with other impact categories, including Global Warming Potential, Acidification, and Smog Formation.

Fossil Fuel Resource Depletion: Tile vs Other Types of Flooring

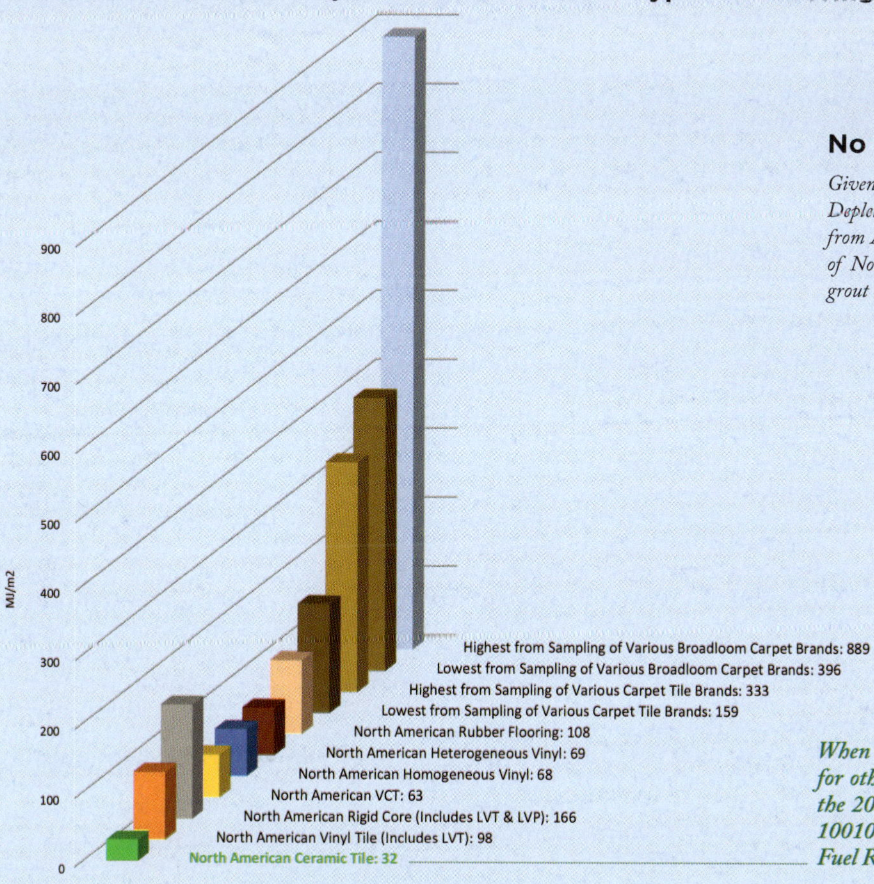

No Combined Pie Chart

Given the unavailability of Fossil Fuel Resource Depletion data per TRACI, and inherent differences from ADP-f data per CML, a graphical comparison of North American-made ceramic tile, mortar and grout is not possible.

When compared to UL Certified EPDs for other types of flooring developed under the 2018 North American Flooring PCR, 10010-7, tile has the lowest 75-year Fossil Fuel Resource Depletion.

Now Available in Print and Digital Formats
2021
TCNA Handbook
and ANSI Standards
ANSI A137.2
American National Standard Specifications for Glass Tile
ANSI A137.1
American National Standard Specifications for Ceramic Tile
ANSI A108, A118, and A136
American National Standard Specifications for the Installation of Ceramic Tile
Material and Installation Standards
ANSI A137.3, A108.19 and A108.20
American National Standard Specifications for Gauged Porcelain Tiles and Gauged Porcelain Tile Panels/Slabs
Material and Installation Standards
PLUS
The Tile Initiative
EPD and G2 Guides
Tile: The Natural Choice
TCNA Bulletins
TCNA
Tile Council of North America, Inc.
Handbook for Ceramic, Glass, and Stone Tile Installation
Hearst Castle
San Simeon, California
2021
Update your reference library today with the standards information you need for a successful project.
TCNA
Tile Council of North America, Inc.
www.tcnatile.com

EPD

TRANSPARENCY SUMMARY: TILE

COMPANY NAME	Industry-Wide
PRODUCT NAME	North America Ceramic Tile: Porcelain, Pressed Floor, Mosaic,Quarry, Glazed Wall
PRODUCT DESCRIPTION	Floor or wall covering which is inert, fire resistant, non-combustible, durable, easy to maintain and made of mineral-based natural materials. Manufacturers include American Wonder Porcelain, Arto, Crossville, Dal-Tile, Del Conca USA, Florida Tile, Interceramic, Ironrock, Porcelanite-Lamosa, Portobello America, Quarry Tile Company, StonePeak Ceramics, and Vitromex
PRODUCT CATEGORY RULE (PCR)+ VERSION	UL PCR Part B: Flooring, 10010-7, Version 2.0, September 2018
CERTIFICATION PERIOD	April 1, 2020 to April 1, 2025
DECLARATION NUMBER	4789101745.101.1
EPD TYPE	☐ PRODUCT SPECIFIC ☑ INDUSTRY AVERAGE
DECLARED/ FUNCTIONAL UNIT	Functional Unit: 1 meter squared

CERTIFIED ENVIRONMENTAL PRODUCT DECLARATION UL.COM/EPD

GREEN BUILDING QUALIFICATIONS

LEED v4 Building Product Disclosure and Optimization - EPDs, Option 1
ASHRAE 189.1 Material Compliance
IgCC Material Compliance
Green Globes 3.5.1.2.1
NAHB Material Selection

REFERENCE SERVICE LIFE (IF APPLICABLE)	75 Years
LCA SOFTWARE + VERSION	GaBi Database Version 9.2.0.58
IMPACT ASSESSMENT METHOD + VERSION	TRACI 2.1 & CML 2001-2016

LIFECYCLE IMPACT CATEGORIES

The environmental impacts listed below were assessed through the product's lifecycle over 75 years (cradle to grave impacts)—including production, construction, use (including use, maintenance, refurbishment and replacement), and end of life.

	ATMOSPHERE			WATER		EARTH	
	Global Warming Potential refers to long-term changes in global weather patterns that are caused by increased concentrations of greenhouse gases in the atmosphere.	**Ozone Depletion Potential** is the destruction of the stratospheric ozone layer, which shields the earth from ultraviolet radiation that's harmful to life, caused by human-made air pollution.	**Photochemical Ozone Creation Potential** happens when sunlight reacts with hydrocarbons, nitrogen oxides, and volatile organic compounds, to produce air pollution known as smog.	**Acidification Potential** is the result of human-made emissions and refers to the decrease in pH and increase in acidity of oceans, lakes, rivers, and streams – polluting groundwater and harming aquatic life.	**Eutrophication Potential** occurs when excessive nutrients cause increased algae growth in lakes, blocking the underwater penetration of sunlight needed to produce oxygen and resulting in the loss of aquatic life.	**Depletion of Abiotic Resources (Elements)** refers to the reduction of available non-renewable resources, such as metals, that are found on the periodic table of elements, due to human activity.	**Depletion of Abiotic Resources (Fossil Fuels)** refers to the decreasing availability of non-renewable carbon-based compounds, such as oil and coal, due to human activity.
TRACI	19.6 kg CO_2-Equiv.	6.7E-10 kg CFC 11-Equiv.	0.757 kg O_3-Equiv.	0.0371 kg SO_2-Equiv.	0.00268 kg N-Equiv.	N/A kg Sb-Equiv.	N/A MJ
CML	19.7 kg CO_2-Equiv.	5.72E-10 kg R11-Equiv.	0.00176 kg Ethene-Equiv.	0.0323 kg SO_2-Equiv.	0.00536 kg PO_4-Equiv.	1.34E-05 kg Sb-Equiv.	242 MJ

Environment

MATERIAL CONTENT

Material content measured to 1%.

COMPONENT	MATERIAL	AVAILABILITY	MASS%	ORIGIN
Body	Clay	Mineral perpetual	69.78	US, Mexico, Europe
	Sand	Mineral perpetual	6.22	US, Mexico, Europe
	Talc	Mineral perpetual	1.12	US, Mexico, Europe
	Feldspar	Mineral perpetual	11.26	US, Mexico, Europe
	Internal Scrap	Pre-consumer	6.48	US, Mexico, Europe
	Frit	Mineral perpetual	0.41	US, Mexico, Europe
	Calcium Carbonate	Mineral perpetual	1.46	US, Mexico, Europe
	Ash	Mineral perpetual	1.35	US, Mexico, Europe
	Additives	Mineral perpetual	0.66	US, Mexico, Europe
Surface	Ink	Mineral perpetual	0.15	US, Mexico, Europe
	Glaze	Mineral perpetual	1.06	US, Mexico, Europe

ADDITIONAL ENVIRONMENTAL INFORMATION

PRE-CONSUMER RECYCLED CONTENT	VARIES %
POST-CONSUMER RECYCLED CONTENT	VARIES %
VOC EMISSIONS	ZERO/INORGANIC
WATER CONSUMPTION	0.0242 CUBIC METERS

RECYCLING OR REUSE

Ceramic tile can outlast multiple generations of building occupants and is commonly reused in an existing building or salvaged for use in a new building. Solid and inert, it can also be used in a variety of post-life applications such as clean fill, roadway paving, and raw materials used to manufacture new products. To represent a "worst case" scenario, this EPD reports environmental impacts based on 100% of all tile removal waste being disposed of in a landfill.

ENERGY

RENEWABLE ENERGY	5.12 %	13.7 MJ
NON-RENEWABLE ENERGY	94.88 %	254 MJ

MANUFACTURER CONTACT INFO

NAME	Tile Council of North America (TCNA)
PHONE	864-646-8453
EMAIL	
WEBSITE	www.TCNAtile.com

STANDARDS

ANSI A137.1 American National Standard Specifications for Ceramic Tile

ISO 13006 International Organization for Standardization Specifications for Ceramic Tile

ANSI A138.1/Green Squared® American National Standard Specifications for Sustainable Ceramic Tiles, Glass Tiles, and Tile Installation Materials

CERTIFICATIONS

GREEN SQUARED CERTIFIED®

EPD
TRANSPARENCY SUMMARY: GROUT

COMPANY NAME	**Industry-Wide**
PRODUCT TYPE	**ANSI A118.6, A118.7, and ISO 13007**
PRODUCT NAME	Cement Grout for Tile Installation: Made in North America
PRODUCT DEFINITION	Factory-prepared mixture of cement, aggregate and other ingredients used to fill joints, or the spaces between tiles. Once cured, it is durable, fire-resistant, and moisture insensitive. Manufacturers include Ardex, Bexel, Bostik, Crest, Custom, HB Fuller/TEC, Interceramic, Laticrete, MAPEI, and Cemix/Texrite
PRODUCT CATEGORY RULE (PCR)	IBU Part A & B for Mineral Factory-made Mortar, 07.2014, with UL Addendum
CERTIFICATION PERIOD	**September 30, 2016 to September 30, 2021**
DECLARATION NUMBER	**4787109018.101.1**

CERTIFIED
ENVIRONMENTAL PRODUCT DECLARATION
UL.COM/EPD

LIFECYCLE IMPACT CATEGORIES

The environmental impacts listed below were assessed throughout the product's lifecycle – including raw material extraction, transportation, manufacturing, packaging, use, and disposal at end of life.

	ATMOSPHERE			WATER		EARTH	
	Global Warming Potential refers to long-term changes in global weather patterns – including temperature and precipitation – that are caused by increased concentrations of greenhouse gases in the atmosphere.	**Ozone Depletion Potential** is the destruction of the stratospheric ozone layer, which shields the earth from ultraviolet radiation that's harmful to life, caused by human-made air pollution.	**Photochemical Ozone Creation Potential** happens when sunlight reacts with hydrocarbons, nitrogen oxides, and volatile organic compounds, to produce a type of air pollution known as smog.	**Acidification Potential** is the result of human-made emissions and refers to the decrease in pH and increase in acidity of oceans, lakes, rivers, and streams – a phenomenon that pollutes groundwater and harms aquatic life.	**Eutrophication Potential** occurs when excessive nutrients cause increased algae growth in lakes, blocking the underwater penetration of sunlight needed to produce oxygen and resulting in the loss of aquatic life.	**Depletion of Abiotic Resources (Elements)** refers to the reduction of available non-renewable resources, such as metals and gases, that are found on the periodic table of elements, due to human activity.	**Depletion of Abiotic Resources (Fossil Fuels)** refers to the decreasing availability of non-renewable carbon-based compounds, such as oil and coal, due to human activity.
TRACI	**0.203** kg CO2-Equiv.	**3.73E-10** kg CFC 11-Equiv.	**0.0123** kg O3-Equiv.	**8.13E-04** kg SO2-Equiv.	**6.14E-05** kg N-Equiv.	**N/A**	**N/A**
CML	**0.203** kg CO2-Equiv.	**2.94E-10** kg CFC 11-Equiv.	**6.21E-05** kg Ethene-Equiv.	**7.89E-04** kg SO2-Equiv.	**9.40E-05** kg Phosphate-Equiv.	**2.34E-07** kg SB-Equiv.	**2.07** MJ

FUNCTIONAL UNIT One square meter (1 sqm) of installed grout. Life cycle impact results are expressed for 60 years. Assumed application rate is 0.212 kilograms per square meter. Divide results by 0.212 to calculate impact per kilogram of grout.

Environment

MATERIAL CONTENT

Material content measured to 1%.

COMPONENT	MATERIAL	AVAILABILITY	MASS%	ORIGIN
Sand	Sand	Mineral perpetual	53	Locally sourced US, Canada, Mexico
Cement	White cement	Mineral manufactured	27	US, Canada, Mexico, Europe
Limestone	Limestone	Mineral perpetual	10	Locally sourced US, Canada, Mexico
Cement	Grey cement	Mineral manufactured	5	Locally sourced US, Canada, Mexico
Additive	Calcium aluminate	Mineral perpetual	2	Locally sourced US, Canada, Mexico
Additive	Calcium formate	Mineral perpetual	1	Locally sourced US, Canada, Mexico
Additive	Other additives	Various	2	US, Canada, Europe, Asia

ADDITIONAL ENVIRONMENTAL INFORMATION

PRE-CONSUMER RECYCLED CONTENT	VARIES %
POST-CONSUMER RECYCLED CONTENT	VARIES %
VOC EMISSIONS	CDPH 01350 COMPLIANT
WATER CONSUMPTION	0.51 L

ENERGY

RENEWABLE ENERGY	9 %	0.21 MJ
NON-RENEWABLE ENERGY	91 %	2.3 MJ

MANUFACTURER CONTACT INFO

NAME	Tile Council of North America (TCNA)
PHONE	864-646-8453
EMAIL	
WEBSITE	www.TCNAtile.com

RECYCLING OR REUSE

Tile installations with cement grout can last as long as buildings and are often reused for multiple generations of building occupants. Although tile and cement grout can be used in a variety of post-life applications (e.g., clean fill, roadway paving, aggregate for new products, etc.), this analysis adopts a conservative approach and assumes that 100% of tile and associated waste are disposed in a landfill.

STANDARDS

ANSI A118.6 American National Standard Specifications, Standard Cement Grouts for Tile Installation
ANSI A118.7 American National Standard Specifications, High Performance Cement Grouts for Tile Installation
ISO 13007 Terms, Definitions and Specifications, Ceramic Tile Grouts and Adhesives
ANSI A138.1/Green Squared® American National Standard Specifications, Sustainable Ceramic Tiles, Glass Tiles and Tiles, and Tile Installation Materials

CERTIFICATIONS

www.UL.com/environment | environment@ul.com

EPD
TRANSPARENCY SUMMARY: MORTAR

COMPANY NAME **Industry-Wide**

PRODUCT TYPE **ANSI A118.1, A118.4/11, A118.15, and ISO 13007**

PRODUCT NAME Cement Mortar for Tile Installation: Made in North America

PRODUCT DEFINITION Blend of cement, sand, water retention compounds and other additives used to adhere tile to a substrate. Once cured, it is durable, fire-resistant, moisture insensitive, and maintenance-free. Manufacturers include Ardex, Bexel, Bostik, Crest, Custom, HB Fuller/TEC, Interceramic, Laticrete, MAPEI, and Cemix/Texrite

PRODUCT CATEGORY RULE (PCR) IBU Part A & B for Mineral Factory-made Mortar, 07.2014, with UL Addendum

CERTIFICATION PERIOD **September 30, 2016 to September 30, 2021**

DECLARATION NUMBER **4787109018.102.1**

CERTIFIED ENVIRONMENTAL PRODUCT DECLARATION UL.COM/EPD

LIFECYCLE IMPACT CATEGORIES

The environmental impacts listed below were assessed throughout the product's lifecycle – including raw material extraction, transportation, manufacturing, packaging, use, and disposal at end of life.

	ATMOSPHERE			WATER		EARTH	
	Global Warming Potential refers to long-term changes in global weather patterns – including temperature and precipitation – that are caused by increased concentrations of greenhouse gases in the atmosphere.	**Ozone Depletion Potential** is the destruction of the stratospheric ozone layer, which shields the earth from ultraviolet radiation that's harmful to life, caused by human-made air pollution.	**Photochemical Ozone Creation Potential** happens when sunlight reacts with hydrocarbons, nitrogen oxides, and volatile organic compounds, to produce a type of air pollution known as smog.	**Acidification Potential** is the result of human-made emissions and refers to the decrease in pH and increase in acidity of oceans, lakes, rivers, and streams – a phenomenon that pollutes groundwater and harms aquatic life.	**Eutrophication Potential** occurs when excessive nutrients cause increased algae growth in lakes, blocking the underwater penetration of sunlight needed to produce oxygen and resulting in the loss of aquatic life.	**Depletion of Abiotic Resources (Elements)** refers to the reduction of available non-renewable resources, such as metals and gases, that are found on the periodic table of elements, due to human activity.	**Depletion of Abiotic Resources (Fossil Fuels)** refers to the decreasing availability of non-renewable carbon-based compounds, such as oil and coal, due to human activity.
TRACI	**2.88** kg CO2-Equiv.	**5.65E-09** kg CFC 11-Equiv.	**0.200** kg O3-Equiv.	**0.0122** kg SO2-Equiv.	**5.62E-04** kg N-Equiv.	**N/A**	**N/A**
CML	**2.89** kg CO2-Equiv.	**4.43E-09** kg CFC 11-Equiv.	**9.66E-04** kg Ethene-Equiv.	**0.0118** kg SO2-Equiv.	**1.21E-03** kg Phosphate-Equiv.	**3.51E-06** kg SB-Equiv	**30.8** MJ

FUNCTIONAL UNIT One square meter (1 sqm) of installed mortar. Life cycle impact results are expressed for 60 years of use. Assumed application rate is 4.07 kilograms per square meter. Divide results by 4.07 to calculate impact per kilogram of mortar.

Environment

MATERIAL CONTENT

Material content measured to 1%.

COMPONENT	MATERIAL	AVAILABILITY	MASS%	ORIGIN
Sand	Sand	Mineral perpetual	37	Locally sourced US, Canada, Mexico
Calcium carbonate	Calcium carbonate	Mineral perpetual	30	Locally sourced US, Canada, Mexico
Cement	Grey cement	Mineral manufactured	15	Locally sourced US, Canada, Mexico
Cement	White cement	Mineral manufactured	13	US, Canada, Mexico, Europe
Additive	E/VA	Polymer manufactured	1	US, Europe
Additive	Admixture	Mineral perpetual	1	US, Mexico, Europe
Additive	Cellulose ether	Organic rapidly renewable	1	US, Europe
Additive	Other additives	Various	1	US, Canada, Europe, Asia

ADDITIONAL ENVIRONMENTAL INFORMATION

PRE-CONSUMER RECYCLED CONTENT	VARIES %
POST-CONSUMER RECYCLED CONTENT	VARIES %
VOC EMISSIONS	CDPH 01350 COMPLIANT
WATER CONSUMPTION	6.5 L

ENERGY

RENEWABLE ENERGY	6 %	2.11 MJ
NON-RENEWABLE ENERGY	94 %	32.5 MJ

MANUFACTURER CONTACT INFO

NAME	Tile Council of North America (TCNA)
PHONE	864-646-8453
EMAIL	
WEBSITE	www.TCNAtile.com

RECYCLING OR REUSE

Tile installations with cement mortar can last as long as buildings and are often reused for multiple generations of building occupants. Although tile and cement mortar can be used in a variety of post-life applications (e.g., clean fill, roadway paving, aggregate for new products, etc.), this analysis adopts a conservative approach and assumes that 100% of tile and associated waste are disposed in a landfill.

STANDARDS

ANSI A118.1 American National Standard Specifications, Dry-Set Cement Mortar
ANSI A118.4/11 American National Standard Specifications, Modified Dry-Set Cement Mortar
ANSI A118.15 American National Standard Specifications, Improved Modified Dry-Set Cement Mortar
ISO 13007 Terms, Definitions and Specifications, Ceramic Tile Grouts and Adhesives
ANSI A138.1/Green Squared® American National Standard Specifications, Sustainable Ceramic Tiles, Glass Tiles and Tiles, and Tile Installation Materials

CERTIFICATIONS

Release date: April 2021

Green building standards and rating systems are living documents, and the consensus regarding their use and interpretation is ever-evolving. Check for updates to *Tile: The Natural Choice* at www.TCNAtile.com.

TCNA Tile Council of North America, Inc.

U.S. Green Building Council Member

100 Clemson Research Blvd. Anderson, SC 29625 | 864.646.8453 | www.TCNAtile.com

Release date: April 2021

Green building standards and rating systems are living documents, and the consensus regarding their use and interpretation is ever-evolving. Check for updates to *Tile: The Natural Choice* at www.TCNAtile.com.

TCNA
TILE COUNCIL OF NORTH AMERICA, INC.

U.S. GREEN BUILDING COUNCIL MEMBER

100 Clemson Research Blvd. Anderson, SC 29625 | 864.646.8453 | www.TCNAtile.com

2021 | CERAMIC AND GLASS TILE INSTALLATION METHODS

F101-21

- **On-Ground Concrete**
- **Bonded Mortar Bed**
- **Ceramic Tile, Glass Tile**

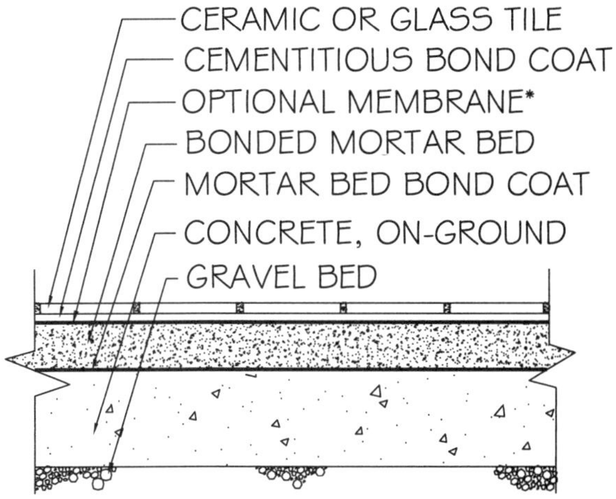

*USE OF A MEMBRANE IS OPTIONAL. SEE MEMBRANE OPTIONS.

Recommended Uses

- For exterior slab-on-ground construction where no bending stresses occur and positive drainage below slab is provided, for example, floors, decks, or patios.
- For areas where leveling, flattening, or contouring of finish floor height is required.
- For areas where floor flatness is critical, such as when tiles with any edge longer than 15" are specified or where the finished tile assembly is required to achieve in-plane transitions (as may be necessary where accessibility is a concern).

Service Rating

- Extra heavy.
- When glass tile is used, service rating may be lower.

Environmental Exposure Classifications

- Res6; Com6.
- For installations that may be exposed to staining, specify tile and grout suitable for exposure. Consult product manufacturers; see also "Product Selection Guides."

Typical Weight of Tile Installation

- 15 pounds/square foot with ¾" mortar bed. Add 3 pounds/square foot for each additional ¼" of mortar bed.
- Does not include weight of substrate. See "Appendix B" for assumptions, included materials, and their individual weights.

Limitations

- Where the installation will be subjected to freeze-thaw cycles, snow and ice accumulation, and/or snow melting chemicals, degradation can occur over time.

Membrane Options

- A waterproof membrane (A118.10) may be specified to prevent moisture intrusion and reduce the need for below-slab drainage. Specifier shall indicate if complete waterproofing is required, including if/how membrane connects to drain assembly, if base flashing is required, and treatment at other termination points.
- Check with membrane manufacturer for suitability for applicable conditions, as not all membranes are suitable for steam, high temperature, and/or chemical exposure, exterior use, use over above-ground structural slabs, use over pourable underlayments, use with radiant heat, or use over concrete with excessive moisture vapor transmission and/or alkalinity. Membrane may also affect service rating.
- When glass tile is used, consult glass tile manufacturer for membrane options and recommendations.

Requirements

- Slab to be well cured, dimensionally stable, and free of cracks, surface laitance, waxy or oily films, curing compounds, and other surface bond-inhibiting contaminants.
- When bond coat is portland cement paste, cover completed tilework and allow to cure.
- Seal edges of mortar bed with a trowelable membrane/ sealant when water intrusion is expected.
- Mortar beds in excess of 2" thick shall be detailed by the architect.
- Mortar bed thickness—¾" minimum to 2" maximum.

Materials

- **Multiple options exist for membranes, mortars, grouts, and other materials and *must be clearly specified* to be included. If not specifically indicated, optional materials are not included and mortar/grout choice defaults to minimum performance specification indicated. Consider each system component and intended use to determine minimum requirements and to specify options.**
- Ceramic tile—ANSI A137.1.
- Glass tile, when used—ANSI A137.2; see also "Glass Tile Selection and Installation Guide," and consult tile manufacturer for service rating and environmental exposure classification recommendations. Not all glass tiles are suitable.
- Cementitious grout—ANSI A118.6 or better or ISO CG1 or better. When glass tile is used, specify grout designated by tile and grout manufacturers.

- Cementitious bond coat—portland cement paste on a mortar bed that is still workable. For a cured mortar bed, follow recommendations below to select appropriate bonding mortar:
  - When a waterproof membrane is not used—ANSI A118.4 or better or ISO C2 or better.
  - When a waterproof membrane is used—ANSI A118.4 or better or ISO C2S1 or better unless ANSI A118.1 or ISO C1 is recommended by membrane manufacturer.
  - When glass tile is used, specify mortar designated by tile and mortar manufacturers. Bond coat color will impact the final appearance of translucent glass tile. Specifier shall confirm bond coat color is acceptable.
- Waterproof membrane, when used—ANSI A118.10.
- Mortar bed—ANSI A108.1A.
- Mortar bed bond coat—portland cement slurry.

Materials for Green/Sustainable Design

- See "Green Building Standards and Green Product Selection Guide" and consult manufacturers and suppliers for product sustainability and contribution to green building design.
- Consider specifying tile and installation materials that meet ANSI A138.1, the *American National Standard Specifications for Sustainable Ceramic Tiles, Glass Tiles, and Tile Installation Materials.*

Preparation by Other Trades

- Gravel bed or other means of drainage below slab.
- Slope slab for complete drainage.
- Slab to have steel trowel and fine broom finish with no curing compounds used. When used, mechanical scarifying is necessary.
- Maximum allowable variation in the installation substrate (concrete)—¼" in 10' from the required plane.

Movement Joint (architect must specify type of joint and show location and details on drawings)

- Movement joints—mandatory according to EJ171.
- When glass tile is used, adhere to more frequent placement recommendations within the ranges listed in EJ171.

Installation Specifications

- Tile—ANSI A108.1A, .1B, or .1C. A108.1B required if membrane to be used.
- Glass tile—ANSI A108.14, .15, .16, or manufacturer's directions.
- Cementitious grout—ANSI A108.10.
- Waterproof membrane—ANSI A108.13.
- Movement Joints—EJ171 and ASTM C1193.

Notes

- Not all bonding mortars are suitable for exterior use.
- Protection of installation may be required to prevent premature exposure of setting materials to moisture.
- If a waterproof membrane is not used, positive drainage below slab must be provided.
- When glass tile is used, see "Glass Tile Selection and Installation Guide," and consult manufacturer for recommendations and requirements.

F102-21

- **On-Ground Concrete**
- **Ceramic Tile, Glass Tile**

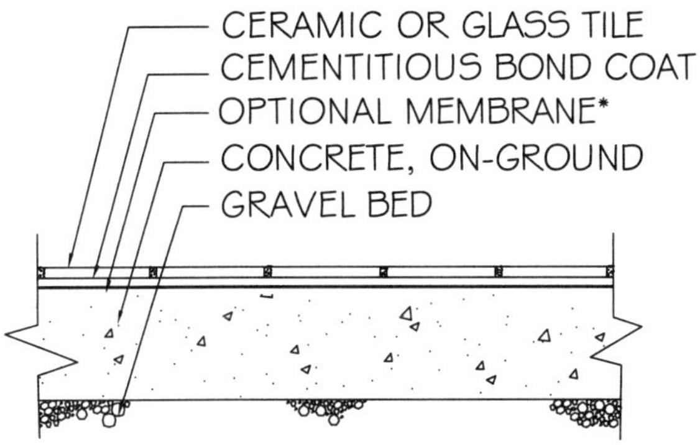

*USE OF A MEMBRANE IS OPTIONAL. SEE MEMBRANE OPTIONS.

Recommended Uses

- For exterior slab-on-ground construction where no bending stresses occur, positive drainage below the slab is provided, and thin-bed installation of tile is desired, for example, floors, decks, or patios.

Service Rating

- Extra heavy.
- When glass tile is used, service rating may be lower.

Environmental Exposure Classifications

- Res6; Com6.
- For installations that may be exposed to staining, specify tile and grout suitable for exposure. Consult product manufacturers; see also "Product Selection Guides."

Typical Weight of Tile Installation

- 5 pounds/square foot.
- Does not include weight of substrate. See "Appendix B" for assumptions, included materials, and their individual weights.

Limitations

- Where the installation will be subjected to freeze-thaw cycles, snow and ice accumulation, and/or snow melting chemicals, degradation can occur over time.

Membrane Options

- A waterproof membrane (A118.10) may be specified to prevent moisture intrusion and reduce the need for below-slab drainage. Specifier shall indicate if complete waterproofing is required, including if/how membrane connects to drain assembly, if base flashing is required, and treatment at other termination points.
- A crack isolation membrane (A118.12) may be specified to treat existing in-plane cracks (F125-partial) or for protection against future in-plane cracking (F125-full). See F125 for more information.
- Check with membrane manufacturer for suitability for applicable conditions, as not all membranes are suitable for steam, high temperature, and/or chemical exposure, exterior use, use over above-ground structural slabs, use over pourable underlayments, use with radiant heat, or use over concrete with excessive moisture vapor transmission and/or alkalinity. Membrane may also affect service rating.
- When glass tile is used, consult glass tile manufacturer for membrane options and recommendations.

Requirements

- Slab to be well cured, dimensionally stable, and free of cracks, surface laitance, waxy or oily films, curing compounds, and other surface bond-inhibiting contaminants.

Materials

- **Multiple options exist for membranes, mortars, grouts, and other materials and *must be clearly specified* to be included. If not specifically indicated, optional materials are not included and mortar/grout choice defaults to minimum performance specification indicated. Consider each system component and intended use to determine minimum requirements and to specify options.**
- Ceramic tile—ANSI A137.1.
- Glass tile, when used—ANSI A137.2; see also "Glass Tile Selection and Installation Guide," and consult tile manufacturer for service rating and environmental exposure classification recommendations. Not all glass tiles are suitable.
- Cementitious grout—ANSI A118.6 or better or ISO CG1 or better. When glass tile is used, specify grout designated by tile and grout manufacturers.
- Cementitious bond coat:
  - When a membrane is not used—ANSI A118.4 or better or ISO C2 or better.
  - When a crack isolation membrane is used—ANSI A118.4 or better or ISO C2S1 or better.
  - When a waterproof membrane is used—ANSI A118.4 or better or ISO C2S1 or better unless ANSI A118.1 or ISO C1 is recommended by membrane manufacturer.
  - When glass tile is used, specify mortar designated by tile and mortar manufacturers. Bond coat color will impact the final appearance of translucent glass tile. Specifier shall confirm bond coat color is acceptable.
- Waterproof membrane, when used—ANSI A118.10.
- Crack isolation membrane, when used—ANSI A118.12.

Materials for Green/Sustainable Design

- See "Green Building Standards and Green Product Selection Guide" and consult manufacturers and suppliers for product sustainability and contribution to green building design.
- Consider specifying tile and installation materials that meet ANSI A138.1, the *American National Standard Specifications for Sustainable Ceramic Tiles, Glass Tiles, and Tile Installation Materials.*

Preparation by Other Trades

- Gravel bed or other means of drainage below slab.
- Slope slab for complete drainage.
- Slab to have steel trowel and fine broom finish with no curing compounds used. When used, mechanical scarifying is necessary.
- Maximum allowable variation in the tile substrate—for tiles with all edges shorter than 15", maximum allowable variation is ¼" in 10' from the required plane, with no more than ¹⁄₁₆" variation in 12" when measured from the high points in the surface. For tiles with at least one edge 15" in length, maximum allowable variation is ⅛" in 10' from the required plane, with no more than ¹⁄₁₆" variation in 24" when measured from the high points in the surface.

Movement Joint (architect must specify type of joint and show location and details on drawings)

- Movement joints—mandatory according to EJ171.
- When glass tile is used, adhere to more frequent placement recommendations within the ranges listed in EJ171.

Installation Specifications

- Tile—ANSI A108.5.
- Glass tile—manufacturer's directions.
- Cementitious grout—ANSI A108.10.
- Waterproof membrane—ANSI A108.13.
- Crack isolation membrane—ANSI A108.17.
- Movement Joints—EJ171 and ASTM C1193.

Notes

- Not all bonding mortars are suitable for exterior use.
- Protection of installation may be required to prevent premature exposure of setting materials to moisture.
- If a waterproof membrane is not used, positive drainage below slab must be provided.
- When glass tile is used, see "Glass Tile Selection and Installation Guide," and consult manufacturer for recommendations and requirements.

F103-21

- **Roof Deck or Balcony**
- **Drainage Layer**
- **Mortar Bed**
- **Ceramic Tile, Glass Tile**

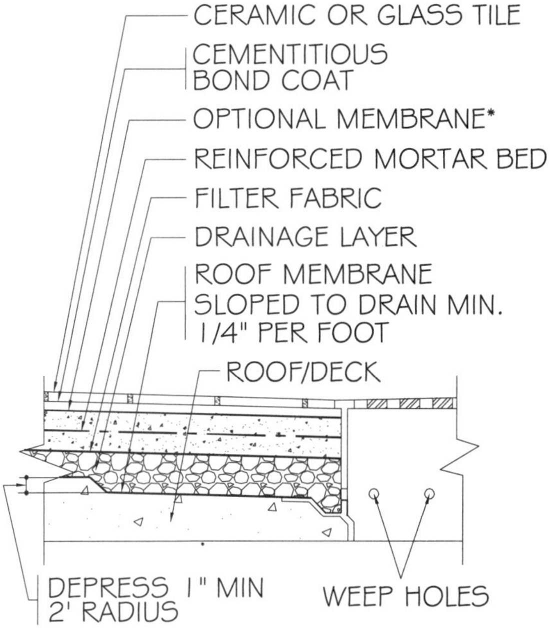

*USE OF A MEMBRANE IS OPTIONAL. SEE MEMBRANE OPTIONS.

Recommended Uses

- For exterior roofs or decks of concrete, steel, or wood where a waterproof roof membrane is used and sloped minimum ¼" per foot to a drain.

Service Rating

- Heavy.
- When glass tile is used, service rating may be lower.

Environmental Exposure Classifications

- Res6; Com6.
- For installations that may be exposed to staining, specify tile and grout suitable for exposure. Consult product manufacturers; see also "Product Selection Guides."

Typical Weight of Tile Installation

- 31 pounds/square foot with 1¼" mortar bed. Add 3 pounds/square foot for each additional ¼" of mortar bed.
- Does not include weight of substrate. See "Appendix B" for assumptions, included materials, and their individual weights.

Limitations

- Where the installation will be subjected to freeze-thaw cycles, snow and ice accumulation, and/or snow melting chemicals, degradation can occur over time.
- Requires additional consideration by design professional to accommodate movement and/or deflection. Setting materials with improved bond strength and deformability are required.

Membrane Options

- A waterproof membrane (A118.10) may be specified to prevent moisture intrusion. Specifier shall indicate if complete waterproofing is required, including if/how membrane connects to drain assembly, if base flashing is required, and treatment at other termination points.
- Check with membrane manufacturer for suitability for applicable conditions, as not all membranes are suitable for steam, high temperature, and/or chemical exposure, exterior use, use over above-ground structural slabs, use over pourable underlayments, use with radiant heat, or use over concrete with excessive moisture vapor transmission and/or alkalinity. Membrane may also affect service rating.
- When glass tile is used, consult glass tile manufacturer for membrane options and recommendations.

Requirements

- Mortar beds in excess of 2" thick shall be detailed by the architect.
- Reinforcing mesh mandatory.
- Protect weep holes to prevent blockage.
- When bond coat is portland cement paste, cover completed tilework and allow to cure.
- Mortar bed thickness—1¼" minimum to 2" maximum.
- Above-ground installations are inherently more susceptible to vibration and deflection. Grout and mortar manufacturers to warrant product suitability.

Materials

- **Multiple options exist for membranes, mortars, grouts, and other materials and *must be clearly specified* to be included. If not specifically indicated, optional materials are not included and mortar/grout choice defaults to minimum performance specification indicated. Consider each system component and intended use to determine minimum requirements and to specify options.**
- Ceramic tile—ANSI A137.1.
- Glass tile, when used—ANSI A137.2; see also "Glass Tile Selection and Installation Guide," and consult tile manufacturer for service rating and environmental exposure classification recommendations. Not all glass tiles are suitable.
- Cementitious grout—ANSI A118.6 or better or ISO CG1 or better. When glass tile is used, specify grout designated by tile and grout manufacturers.

- Cementitious bond coat—portland cement paste on a mortar bed that is still workable. For a cured mortar bed, follow recommendations below to select appropriate bonding mortar:
  - Must be recommended by manufacturer for above-ground use.
  - When a waterproof membrane is not used—ANSI A118.15 or better or ISO C2S1 or better.
  - When a waterproof membrane is used—ANSI A118.15 or better or ISO C2S1 or better unless ANSI A118.1 or ISO C1 is recommended by membrane manufacturer.
  - When glass tile is used, specify mortar designated by tile and mortar manufacturers. Bond coat color will impact the final appearance of translucent glass tile. Specifier shall confirm bond coat color is acceptable.
- Waterproof membrane, when used—ANSI A118.10.
- Mortar bed and reinforcing—ANSI A108.1A.
- Drainage layer—minimum 1" thick bed of crushed stone (½" maximum size).
- Filter fabric—burlap or closely woven cheesecloth.

Materials for Green/Sustainable Design

- See "Green Building Standards and Green Product Selection Guide" and consult manufacturers and suppliers for product sustainability and contribution to green building design.
- Consider specifying tile and installation materials that meet ANSI A138.1, the *American National Standard Specifications for Sustainable Ceramic Tiles, Glass Tiles, and Tile Installation Materials.*

Preparation by Other Trades

- Floor systems over which tile will be installed shall be in conformance with the IRC for residential applications, the IBC for commercial applications, or applicable building codes. Maximum allowable substrate deflection under live load not to exceed l/360. See also "Substrate Requirements."
- When concentrated loads (scissor lifts, pallet jacks, automobiles, forklifts, etc.) will be used on an above-ground tiled floor, the engineer and/or specifier shall specify a substrate to accommodate the concentrated loads. Owner/specifier is responsible for protecting the tilework from damage, including allowing sufficient time for installed materials to cure properly.
- Roof drains and membrane by other trades—provide complete drainage at roof membrane level by use of weep holes as shown or other methods. Flat deck with poor or no drainage will not perform well.
- Maximum allowable variation in the roof/deck—¼" in 10' from the required plane.

Movement Joint (architect must specify type of joint and show location and details on drawings)

- Movement joints—mandatory according to EJ171. For above-ground installations, additional movement joints are required.
- Movement joints should not go through the gravel bed; they should extend only to the bottom of the setting bed.
- When glass tile is used, adhere to more frequent placement recommendations within the ranges listed in EJ171.

Installation Specifications

- Tile—ANSI A108.1A, .1B, or .1C. A108.1B required if membrane to be used.
- Glass tile—ANSI A108.14, .15, .16, or manufacturer's directions.
- Cementitious grout—ANSI A108.10.
- Waterproof membrane—ANSI A108.13.
- Movement Joints—EJ171 and ASTM C1193.

Notes

- Not all bonding mortars are suitable for exterior use.
- Not all mortars, grouts, and membranes are suitable for above-ground use. Check manufacturer recommendations.
- Protection of installation may be required to prevent premature exposure of setting materials to moisture.
- Some substrate materials used in wet areas are subject to deterioration from moisture. See ANSI A108.01-2.4.
- When glass tile is used, see "Glass Tile Selection and Installation Guide," and consult manufacturer for recommendations and requirements.

F103B-21

- **Roof Deck or Balcony**
- **Drainage Mat**
- **Mortar Bed**
- **Ceramic Tile, Glass Tile**

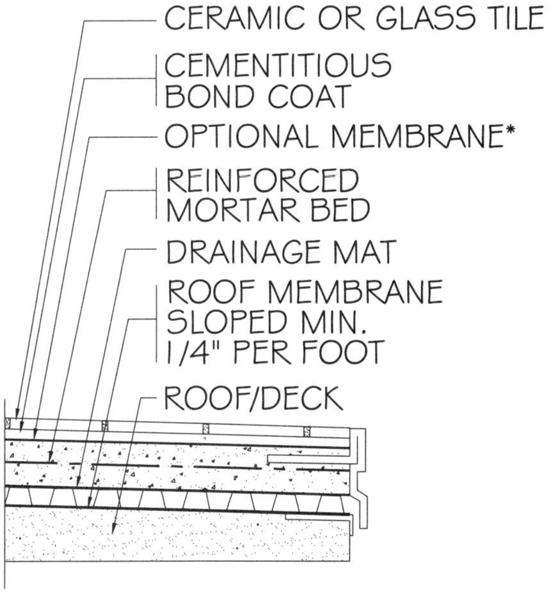

*USE OF MEMBRANE IS OPTIONAL. SEE MEMBRANE OPTIONS.

Recommended Uses

- For exterior roofs or decks of concrete, steel, or wood where a waterproof roof membrane is used and sloped minimum ¼" per foot.

Service Rating

- Consult drainage mat manufacturer.
- When glass tile is used, service rating may be lower.

Environmental Exposure Classifications

- Res6; Com6.
- For installations that may be exposed to staining, specify tile and grout suitable for exposure. Consult product manufacturers; see also "Product Selection Guides."

Typical Weight of Tile Installation

- 21 pounds/square foot with 1¼" mortar bed. Add 3 pounds/square foot for each additional ¼" of mortar bed.
- Does not include weight of substrate. See "Appendix B" for assumptions, included materials, and their individual weights.

Limitations

- Where the installation will be subjected to freeze-thaw cycles, snow and ice accumulation, and/or snow melting chemicals, degradation can occur over time.
- Requires additional consideration by design professional to accommodate movement and/or deflection. Setting materials with improved bond strength and deformability are required.

Membrane Options

- A waterproof membrane (A118.10) may be specified to prevent moisture intrusion. Specifier shall indicate if complete waterproofing is required, including if/how membrane connects to drain assembly, if base flashing is required, and treatment at other termination points.
- Check with membrane manufacturer for suitability for applicable conditions, as not all membranes are suitable for steam, high temperature, and/or chemical exposure, exterior use, use over above-ground structural slabs, use over pourable underlayments, use with radiant heat, or use over concrete with excessive moisture vapor transmission and/or alkalinity. Membrane may also affect service rating.
- When glass tile is used, consult glass tile manufacturer for membrane options and recommendations.

Requirements

- Mortar beds in excess of 2" thick shall be detailed by the architect.
- Reinforcing mesh mandatory.
- When bond coat is portland cement paste, cover completed tilework and allow to cure.
- Mortar bed thickness— 1¼" minimum to 2" maximum.
- Above-ground installations are inherently more susceptible to vibration and deflection. Grout and mortar manufacturers to warrant product suitability.

Materials

- **Multiple options exist for membranes, mortars, grouts, and other materials and *must be clearly specified* to be included. If not specifically indicated, optional materials are not included and mortar/grout choice defaults to minimum performance specification indicated. Consider each system component and intended use to determine minimum requirements and to specify options.**
- Ceramic tile—ANSI A137.1.
- Glass tile, when used—ANSI A137.2; see also "Glass Tile Selection and Installation Guide," and consult tile manufacturer for service rating and environmental exposure classification recommendations. Not all glass tiles are suitable.
- Cementitious grout—ANSI A118.6 or better or ISO CG1 or better. When glass tile is used, specify grout designated by tile and grout manufacturers.
- Cementitious bond coat—portland cement paste on a mortar bed that is still workable. For a cured mortar bed, follow recommendations below to select appropriate bonding mortar:

- Must be recommended by manufacturer for above-ground use.
- When a waterproof membrane is not used—ANSI A118.15 or better or ISO C2S1 or better.
- When a waterproof membrane is used—ANSI A118.15 or better or ISO C2S1 or better unless ANSI A118.1 or ISO C1 is recommended by membrane manufacturer.
- When glass tile is used, specify mortar designated by tile and mortar manufacturers. Bond coat color will impact the final appearance of translucent glass tile. Specifier shall confirm bond coat color is acceptable.

- Waterproof membrane, when used—ANSI A118.10.
- Mortar bed and reinforcing—ANSI A108.1A.
- Drainage mat—recommended by manufacturer for intended use or application.

Materials for Green/Sustainable Design

- See "Green Building Standards and Green Product Selection Guide" and consult manufacturers and suppliers for product sustainability and contribution to green building design.
- Consider specifying tile and installation materials that meet ANSI A138.1, the *American National Standard Specifications for Sustainable Ceramic Tiles, Glass Tiles, and Tile Installation Materials.*

Preparation by Other Trades

- Floor systems over which tile will be installed shall be in conformance with the IRC for residential applications, the IBC for commercial applications, or applicable building codes. Maximum allowable substrate deflection under live load not to exceed l/360. See also "Substrate Requirements."
- When concentrated loads (scissor lifts, pallet jacks, automobiles, forklifts, etc.) will be used on an above-ground tiled floor, the engineer and/or specifier shall specify a substrate to accommodate the concentrated loads. Owner/specifier is responsible for protecting the tilework from damage, including allowing sufficient time for installed materials to cure properly.
- Roof membrane must be sloped and provide complete protection from moisture. Tile over flat deck with poor or no drainage will not perform well.
- Maximum allowable variation in the roof/deck—¼" in 10' from the required plane.

Movement Joint (architect must specify type of joint and show location and details on drawings)

- Movement joints—mandatory according to EJ171. For above-ground installations, additional movement joints are required.
- When glass tile is used, adhere to more frequent placement recommendations within the ranges listed in EJ171.

Installation Specifications

- Tile—ANSI A108.1A, .1B, or .1C. A108.1B required if membrane to be used.
- Glass tile—ANSI A108.14, .15, .16, or manufacturer's directions.
- Cementitious grout—ANSI A108.10.
- Waterproof membrane—ANSI A108.13.
- Drainage mat—manufacturer's directions.
- Movement Joints—EJ171 and ASTM C1193.

Notes

- Not all bonding mortars are suitable for exterior use.
- Not all mortars, grouts, and membranes are suitable for above-ground use. Check manufacturer recommendations.
- Protection of installation may be required to prevent premature exposure of setting materials to moisture.
- Some substrate materials used in wet areas are subject to deterioration from moisture. See ANSI A108.01-2.4.
- When glass tile is used, see "Glass Tile Selection and Installation Guide," and consult manufacturer for recommendations and requirements.

F104-21

- **Roof Deck or Balcony**
- **Ceramic Tile, Glass Tile**

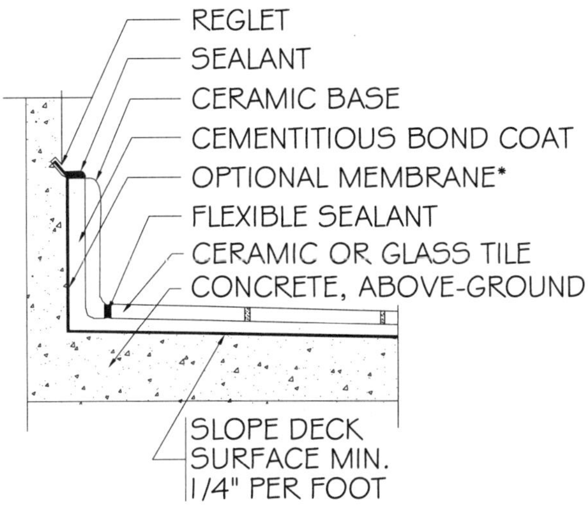

*USE OF A MEMBRANE IS OPTIONAL.

Recommended Uses

- For exterior balconies or decks of concrete over unoccupied space where waterproof roof membrane is not used, where slab is sloped minimum ¼" per foot, and where thin-bed installation of tile is desired.

Service Rating

- Heavy.
- When glass tile is used, service rating may be lower.

Environmental Exposure Classifications

- Res6; Com6.
- For installations that may be exposed to staining, specify tile and grout suitable for exposure. Consult product manufacturers; see also "Product Selection Guides."

Typical Weight of Tile Installation

- 5 pounds/square foot.
- Does not include weight of substrate. See "Appendix B" for assumptions, included materials, and their individual weights.

Limitations

- Where the installation will be subjected to freeze-thaw cycles, snow and ice accumulation, and/or snow melting chemicals, degradation can occur over time.
- Requires additional consideration by design professional to accommodate movement and/or deflection. Setting materials with improved bond strength and deformability are required.

Membrane Options

- A waterproof membrane (A118.10) may be specified to prevent moisture intrusion. Specifier shall indicate if complete waterproofing is required, including if/how membrane connects to drain assembly, if base flashing is required, and treatment at other termination points.
- A crack isolation membrane (A118.12) may be specified to treat existing in-plane cracks (F125-partial) or for protection against future in-plane cracking (F125-full). See F125 for more information.
- Check with membrane manufacturer for suitability for applicable conditions, as not all membranes are suitable for steam, high temperature, and/or chemical exposure, exterior use, use over above-ground structural slabs, use over pourable underlayments, use with radiant heat, or use over concrete with excessive moisture vapor transmission and/or alkalinity. Membrane may also affect service rating.
- When glass tile is used, consult glass tile manufacturer for membrane options and recommendations.

Requirements

- Slab to be well cured, dimensionally stable, and free of cracks, surface laitance, waxy or oily films, curing compounds, and other surface bond-inhibiting contaminants.
- Above-ground installations are inherently more susceptible to vibration and deflection. Grout and mortar manufacturers to warrant product suitability.

Materials

- **Multiple options exist for membranes, mortars, grouts, and other materials and *must be clearly specified* to be included. If not specifically indicated, optional materials are not included and mortar/grout choice defaults to minimum performance specification indicated. Consider each system component and intended use to determine minimum requirements and to specify options.**
- Ceramic tile—ANSI A137.1.
- Glass tile, when used—ANSI A137.2; see also "Glass Tile Selection and Installation Guide," and consult tile manufacturer for service rating and environmental exposure classification recommendations. Not all glass tiles are suitable.
- Cementitious grout—ANSI A118.6 or better or ISO CG1 or better. When glass tile is used, specify grout designated by tile and grout manufacturers.
- Cementitious bond coat:.
  - Must be recommended by manufacturer for above-ground use.
  - When a membrane is not used—ANSI A118.15 or better or ISO C2S1 or better.
  - When a crack isolation membrane is used—ANSI

A118.15 or better or ISO C2S1 or better.

  - When a waterproof membrane is used—ANSI A118.15 or better or ISO C2S1 or better unless ANSI A118.1 or ISO C1 is recommended by membrane manufacturer.
  - When glass tile is used, specify mortar designated by tile and mortar manufacturers. Bond coat color will impact the final appearance of translucent glass tile. Specifier shall confirm bond coat color is acceptable.
- Waterproof membrane, when used—ANSI A118.10.
- Crack isolation membrane, when used—ANSI A118.12.

Materials for Green/Sustainable Design

- See "Green Building Standards and Green Product Selection Guide" and consult manufacturers and suppliers for product sustainability and contribution to green building design.
- Consider specifying tile and installation materials that meet ANSI A138.1, the *American National Standard Specifications for Sustainable Ceramic Tiles, Glass Tiles, and Tile Installation Materials.*

Preparation by Other Trades

- Floor systems over which tile will be installed shall be in conformance with the IRC for residential applications, the IBC for commercial applications, or applicable building codes. Maximum allowable substrate deflection under live load not to exceed l/360. See also "Substrate Requirements."
- When concentrated loads (scissor lifts, pallet jacks, automobiles, forklifts, etc.) will be used on an above-ground tiled floor, the engineer and/or specifier shall specify a substrate to accommodate the concentrated loads. Owner/specifier is responsible for protecting the tilework from damage, including allowing sufficient time for installed materials to cure properly.
- Maximum allowable variation in the tile substrate—for tiles with all edges shorter than 15," maximum allowable variation is ¼" in 10' from the required plane, with no more than 1/16" variation in 12" when measured from the high points in the surface. For tiles with at least one edge 15" in length, maximum allowable variation is ⅛" in 10' from the required plane, with no more than 1/16" variation in 24" when measured from the high points in the surface.
- Slope slab minimum ¼" per foot. Flat deck with poor or no drainage will not perform well.
- Slab to have steel trowel finish with no curing compounds used. When used, mechanical scarification is necessary.

Movement Joint (architect must specify type of joint and show location and details on drawings)

- Movement joints—mandatory in accordance with EJ171. For above-ground installations, additional movement joints are required.
- When glass tile is used, adhere to more frequent placement recommendations within the ranges listed in EJ171.

Installation Specifications

- Tile—ANSI A108.5.
- Glass tile—manufacturer's directions.
- Cementitious grout—ANSI A108.10.
- Waterproof membrane—ANSI A108.13.
- Crack isolation membrane—ANSI A108.17.
- Movement Joints—EJ171 and ASTM C1193.

Notes

- Not all bonding mortars are suitable for exterior use.
- Not all mortars, grouts, and membranes are suitable for above-ground use. Check manufacturer recommendations.
- Protection of installation may be required to prevent premature exposure of setting materials to moisture.
- When glass tile is used, see "Glass Tile Selection and Installation Guide," and consult manufacturer for recommendations and requirements.

F105-21

- **Roof Deck or Balcony**
- **Structural Ribbed Self-Supporting Boards (SRSB) Filled with Portland Cement Mortar**
- **Ceramic Tile**

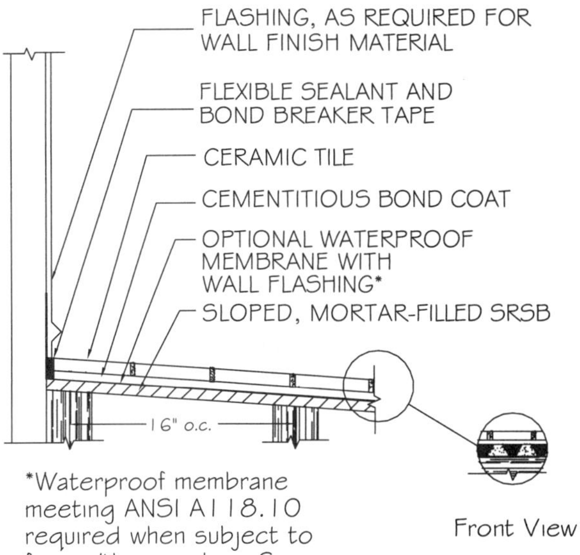

*Waterproof membrane meeting ANSI A118.10 required when subject to freeze/thaw cycling. See Membrane Options.

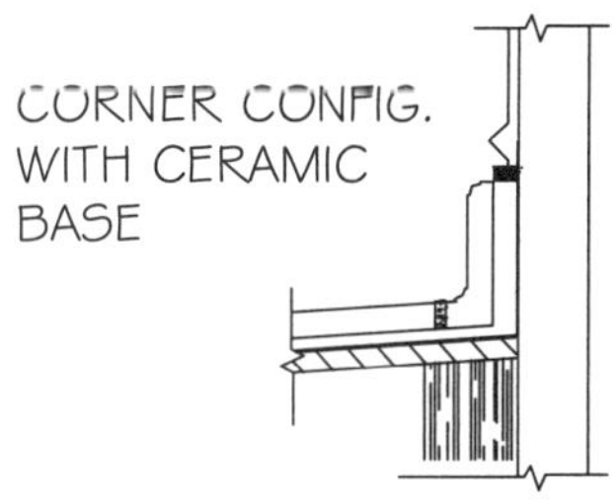

Recommended Uses

- For residential balconies and decks over unoccupied space, constructed of a joist system sloped min. ¼" per foot.

Service Rating

- Light.

Environmental Classifications

- Res6; Com6.
- For installations that may be exposed to staining, specify tile and grout suitable for exposure. Consult product manufacturers; see also "Product Selection Guides."

Typical Weight of Tile Installation

- 5 pounds/square foot.
- Does not include weight of substrate (mortar-filled SRSB). See "Appendix B" for assumptions, included materials, and their individual weights.

Limitations

- 12" × 12" and larger tile only.
- Maximum joist spacing 16" on center.
- Where the installation will be subjected to freeze-thaw cycles, snow and ice accumulation, and/or snow melting chemicals, degradation can occur over time.
- Requires additional consideration by design professional to accommodate movement and/or deflection. Setting materials with improved bond strength and deformability are required.

Membrane Options

- A waterproof membrane (A118.10) is required for installations that will be exposed to freeze/thaw cycling. When used, membrane flashing is required.
- Specifier shall indicate if complete waterproofing is required, including if/how membrane connects to drain assembly, if base flashing is required, and treatment at other termination points.
- Check with membrane manufacturers for suitability for applicable conditions, as not all membranes are suitable for steam, high temperature, and/or chemical exposure, exterior use, use over above-ground structural slabs, use over pourable underlayments, use with radiant heat, or use over concrete with excessive moisture vapor transmission and/or alkalinity. Membrane may also affect service rating.

Requirements

- Above-ground installations are inherently more susceptible to vibration and deflection. Grout and mortar manufacturers to warrant product suitability.
- Mortar bed thickness—screed fill channels of SRSB boards flush to top of ribs and allow to cure 24 hours.

Preparation by SRSB Installers

- The ends of SRSB must be fully supported. Cantilevering is not permitted.
- Do not extend movement joints through SRSB boards.
- Fasteners—1⅝" pan head screws, non-corrosive and non-oxidizing.

Materials

- **Multiple options exist for membranes, mortars, grouts, and other materials and *must be clearly specified* to be included. If not specifically indicated, optional materials are not included and mortar/grout choice defaults to minimum performance specification indicated. Consider each system component and intended use to determine minimum requirements and to specify options.**
- Ceramic tile—ANSI A137.1.
- Cementitious grout—ANSI A118.7 or better or ISO CG2 or better.
- Cementitious bond coat:
  - Must be recommended by manufacturer for above-ground use.

- When a membrane is not used—ANSI A118.15 or better or ISO C2S1 or better.
- When a waterproof membrane is used—ANSI A118.15 or better or ISO C2S1 or better unless ANSI A118.1 or ISO C1 is recommended by membrane manufacturer.

- Waterproof membrane, when used—ANSI A118.10.
- Waterproof membrane flashing, when used—per applicable building codes.
- Mortar bed—ANSI A108.1A.
- Structural ribbed self-supporting boards—recommended by manufacturer for intended use or application and meeting the following requirements:
    - Water absorption < 3% when tested per ASTM D570.
    - Passes 300 cycles when tested per ASTM C67, Section 9.
    - Flexural strength ≥ 2,600 psf when tested per ASTM D790.
    - Direct screw withdrawal resistance (using SRSB manufacturer's recommended fastener) ≥ 250 psf when tested per ASTM D1037, Section 16.
    - Passes Robinson Floor Test (ASTM C627) cycles 1-10 when tested with standard 12" × 12" ANSI A137.1 porcelain tile, ANSI A118.6 grout, and ANSI A118.4 mortar; and joists spaced 16" o.c.
    - Passes Robinson Floor Test (ASTM C627) cycles 1-10 when tested with standard 12" × 12" ANSI A137.1 porcelain tile, ANSI A118.6 grout, and ANSI A118.4 mortar; joists spaced 16" o.c.; and modified to accommodate a 14' suspended floor assembly.
    - Receives flame spread and smoke development indices of 0-0-0-0 when the assembly is exposed to fire, when tested per ASTM E84.

Materials for Green/Sustainable Design

- See "Green Building Standards and Green Product Selection Guide" and consult manufacturers and suppliers for product sustainability and contribution to green building design.
- Consider specifying tile and installation materials that meet ANSI A138.1, the *American National Standard Specifications for Sustainable Ceramic Tiles, Glass Tiles, and Tile Installation Materials.*

Preparation by Other Trades

- For decks with 8 foot and longer joists, cross bracing may be needed between joists to prevent racking. Consult SRSB manufacturer.
- Floor systems over which tile will be installed shall be in conformance with the IRC for residential applications, the IBC for commercial applications, or applicable building codes. Maximum allowable substrate deflection under live load not to exceed l/360. See also "Substrate Requirements."
- When concentrated loads (scissor lifts, pallet jacks, automobiles, forklifts, etc.) will be used on an above-ground tiled floor, the engineer and/or specifier shall specify a substrate to accommodate the concentrated loads. Owner/specifier is responsible for protecting the tilework from damage, including allowing sufficient time for installed materials to cure properly.
- Maximum allowable variation in the joist system—¼" in 10' and ¹⁄₁₆" in 1' from the required plane.
- Slope joist system min. ¼" per foot. Flat deck with poor or no drainage will not perform well.

Movement Joint (architect must specify type of joint and show location and details on drawing)

- Movement joints—mandatory according to EJ171. For above-ground installations additional movement joints are required.

Installation Specifications

- Tile—ANSI A108.5.
- Cementitious grout—ANSI A108.10.
- Waterproof membrane—ANSI A108.13.
- Mortar bed—ANSI A108.1B.
- SRSB—manufacturer's directions.
- Movement Joints—EJ171 and ASTM C1193.

Notes

- State and local building codes for exterior deck construction, membrane requirements, and membrane flashing requirements vary; check applicable codes for requirements.
- SRSB are structural panels. Specifier to ensure compliance with building codes and expected load. Consult with SRSB manufacturer for design criteria.
- Not all bonding mortars are suitable for exterior use.
- Not all mortars, grouts, and membranes are suitable for above-ground use. Check manufacturer recommendations.
- Protection of installation may be required to prevent premature exposure of setting materials to moisture.
- Some substrate materials used in wet areas are subject to deterioration from moisture. See ANSI A108.01-2.4.

F111-21

- **On-Ground or Above-Ground Concrete**
- **Unbonded Mortar Bed**
- **Ceramic Tile, Glass Tile**

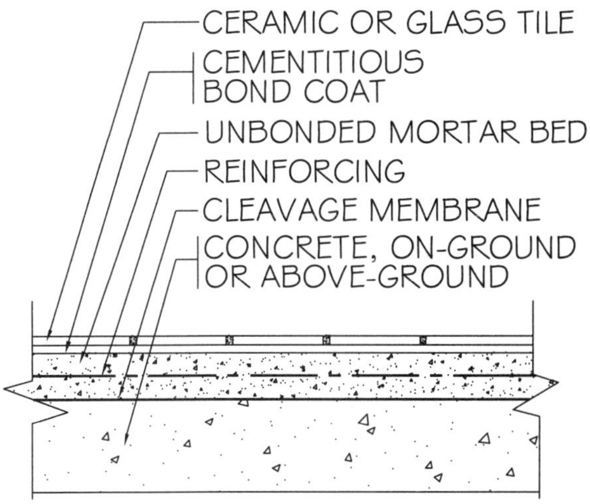

Recommended Uses

- For above-ground structural slabs and for slab-on-ground construction.
- For isolating the tile installation from the concrete substrate.
- For areas where leveling, flattening, or contouring of finish floor height is required, such as commercial kitchens and gang showers.
- For areas where floor flatness is critical, such as when tiles with any edge longer than 15" are specified or where the finished tile assembly is required to achieve in-plane transitions (as may be necessary where accessibility is a concern).
- For concrete substrates that are cracked or present bonding issues.
- For areas with in-slab hydronic tubing.

Service Rating

- Extra heavy.
- When glass tile is used, service rating may be lower.

Environmental Exposure Classifications

- Res1, 2, 3, 5; Com1, 2, 3, 5.
- May be suitable for exterior applications in areas not subject to freeze/thaw cycling when appropriate precautions are taken, including expansion joint placement, proper slope, waterproofing, and material selection.
- For installations that may be exposed to staining, specify tile and grout suitable for exposure. Consult product manufacturers; see also "Product Selection Guides."
- For installations that may be exposed to staining and/or chemical attack, see also F114, F132, and F134.

Typical Weight of Tile Installation

- 21 pounds/square foot with 1¼" mortar bed. Add 3 pounds/square foot for each additional ¼" of mortar bed.
- Does not include weight of substrate. See "Appendix B" for assumptions, included materials, and their individual weights.

Limitations

- Above-ground structural slabs and other floors subject to movement and/or deflection—requires additional consideration by design professional to accommodate movement and/or deflection. Setting materials with improved bond strength and deformability are required.

Requirements

- Reinforcing mesh mandatory.
- Mortar bed thickness—1¼" minimum to 2" maximum.
- Mortar beds in excess of 2" thick shall be detailed by the architect.
- Above-ground—above-ground installations are inherently more susceptible to vibration and deflection. Grout and mortar manufacturers to warrant product suitability.

Materials

- **Multiple options exist for membranes, mortars, grouts, and other materials and *must be clearly specified* to be included. If not specifically indicated, optional materials are not included and mortar/grout choice defaults to minimum performance specification indicated. Consider each system component and intended use to determine minimum requirements and to specify options.**
- Ceramic tile—ANSI A137.1.
- Glass tile, when used—ANSI A137.2; see also "Glass Tile Selection and Installation Guide," and consult tile manufacturer for service rating and environmental exposure classification recommendations. Not all glass tiles are suitable.
- Cementitious grout—ANSI A118.6 or better or ISO CG1 or better. When glass tile is used, specify grout designated by tile and grout manufacturers.
- Cementitious bond coat (on-ground)—portland cement paste on a mortar bed that is still workable. For a cured mortar bed, follow recommendations below to select appropriate bonding mortar:
  - ANSI A118.1 or better or ISO C1 or better.
  - When porcelain tile is used—ANSI A118.4 or better or ISO C2 or better.
  - When glass tile is used, specify mortar designated by tile and mortar manufacturers. Bond coat color will impact the final appearance of translucent glass tile. Specifier shall confirm bond coat color is acceptable.

- Cementitious bond coat (above-ground)—portland cement paste on a mortar bed that is still workable. For a cured mortar bed, follow recommendations below to select appropriate bonding mortar:
  - Must be recommended by manufacturer for above-ground use.
  - ANSI A118.15 or better or ISO C2S1 or better.
  - When glass tile is used, specify mortar designated by tile and mortar manufacturers. Bond coat color will impact the final appearance of translucent glass tile. Specifier shall confirm bond coat color is acceptable.
- Mortar bed, reinforcing, and cleavage membrane—ANSI A108.1A.

Materials for Green/Sustainable Design

- See "Green Building Standards and Green Product Selection Guide" and consult manufacturers and suppliers for product sustainability and contribution to green building design.
- Consider specifying tile and installation materials that meet ANSI A138.1, the *American National Standard Specifications for Sustainable Ceramic Tiles, Glass Tiles, and Tile Installation Materials.*

Preparation by Other Trades

- Floor systems over which tile will be installed shall be in conformance with the IRC for residential applications, the IBC for commercial applications, or applicable building codes. Maximum allowable substrate deflection under live load not to exceed l/360. See also "Substrate Requirements."
- When concentrated loads (scissor lifts, pallet jacks, automobiles, forklifts, etc.) will be used on an above-ground tiled floor, the engineer and/or specifier shall specify a substrate to accommodate the concentrated loads. Owner/specifier is responsible for protecting the tilework from damage, including allowing sufficient time for installed materials to cure properly.
- Where slab is depressed to accommodate mortar bed thickness, slab depression to be accurate with steel trowel finish.
- Slope, when required, to be in subfloor.
- Maximum allowable variation in the installation substrate (concrete)—¼" in 10' from the required plane.

Movement Joint (architect must specify type of joint and show location and details on drawings)

- Movement joints—mandatory according to EJ171. For above-ground installations, additional movement joints are required.
- When glass tile is used, adhere to more frequent placement recommendations within the ranges listed in EJ171.

Installation Specifications

- Tile—ANSI A108.1A, .1B, or .1C.
- Glass tile—ANSI A108.14, .15, .16, or manufacturer's directions.
- Cementitious grout—ANSI A108.10.
- Movement Joints—EJ171 and ASTM C1193.

Notes

- Above-ground—not all mortars and grouts are suitable for above-ground use. Check manufacturer recommendations.
- For waterproof application when using a mortar bed, see F121 or F112.
- For hydronic tubing on top of the slab when using a mortar bed, see RH117.
- When glass tile is used, see "Glass Tile Selection and Installation Guide," and consult manufacturer for recommendations and requirements.

F112-21

- **On-Ground or Above-Ground Concrete**
- **Bonded Mortar Bed**
- **Ceramic Tile, Glass Tile**

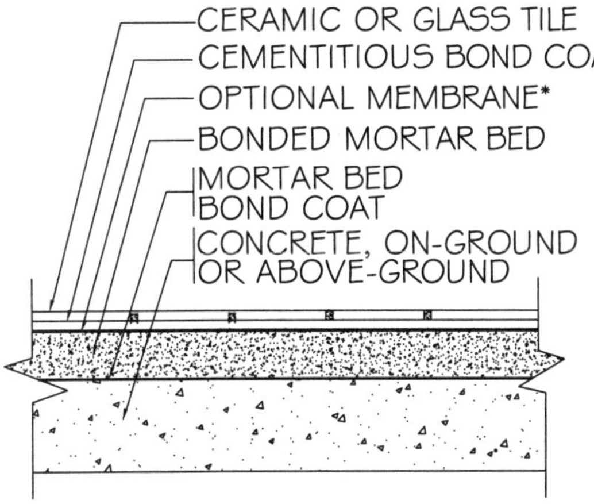

*USE OF A MEMBRANE IS OPTIONAL. SEE MEMBRANE OPTIONS.

Recommended Uses

- For slab-on-ground construction where no bending stresses occur and for above-ground structural slabs where tile installation is 100 square feet or less.
- For areas where leveling, flattening, or contouring of finish floor height is required, such as commercial kitchens and gang showers.
- For areas where floor flatness is critical, such as when tiles with any edge longer than 15" are specified or where the finished tile assembly is required to achieve in-plane transitions (as may be necessary where accessibility is a concern).
- For areas with in-slab hydronic tubing.

Service Rating

- Extra heavy.
- When glass tile is used, service rating may be lower.

Environmental Exposure Classifications

- Res1, 2, 3, 5; Com1, 2, 3, 5.
- May be suitable for exterior applications in areas not subject to freeze/thaw cycling when appropriate precautions are taken, including expansion joint placement, proper slope, waterproofing, and material selection.
- For installations that may be exposed to staining, specify tile and grout suitable for exposure. Consult product manufacturers; see also "Product Selection Guides."
- For installations that may be exposed to staining and/or chemical attack, see also F114, F132, and F134.

Typical Weight of Tile Installation

- 15 pounds/square foot with ¾" mortar bed. Add 3 pounds/square foot for each additional ¼" of mortar bed.
- Does not include weight of substrate. See "Appendix B" for assumptions, included materials, and their individual weights.

Limitations

- Not for use over above-ground structural slabs or other floors subject to movement and/or deflection except where tile installation is 100 square feet or less. See Notes.
- Above-ground structural slabs and other floors subject to movement and/or deflection—requires additional consideration by design professional to accommodate movement and/or deflection. Setting materials with improved bond strength and deformability are required.

Membrane Options

- A waterproof membrane (A118.10) may be specified to prevent moisture intrusion. Specifier shall indicate if complete waterproofing is required, including if/how membrane connects to drain assembly, if base flashing is required, and treatment at other termination points.
- Check with membrane manufacturers for suitability for applicable conditions, as not all membranes are suitable for steam, high temperature, and/or chemical exposure, exterior use, use over above-ground structural slabs, use over pourable underlayments, use with radiant heat, or use over concrete with excessive moisture vapor transmission and/or alkalinity. Membrane may also affect service rating.
- When glass tile is used, consult glass tile manufacturer for membrane options and recommendations.

Requirements

- Mortar bed thickness—¾" minimum to 2" maximum.
- Slab to be well cured, dimensionally stable, and free of cracks, surface laitance, waxy or oily films, curing compounds, and other surface bond-inhibiting contaminants.
- Mortar beds in excess of 2" thick shall be detailed by architect.
- Above-ground—above-ground installations are inherently more susceptible to vibration and deflection. Grout and mortar manufacturers to warrant product suitability.

Materials

- **Multiple options exist for membranes, mortars, grouts, and other materials and *must be clearly specified* to be included. If not specifically indicated, optional materials are not included and mortar/grout choice defaults to minimum performance specification indicated. Consider each system component and intended use to determine minimum requirements and to specify options.**

- Ceramic tile—ANSI A137.1.
- Glass tile, when used—ANSI A137.2; see also "Glass Tile Selection and Installation Guide," and consult tile manufacturer for service rating and environmental exposure classification recommendations. Not all glass tiles are suitable.
- Cementitious grout—ANSI A118.6 or better or ISO CG1 or better. When glass tile is used, specify grout designated by tile and grout manufacturers.
- Cementitious bond coat (on-ground)—portland cement paste on a mortar bed that is still workable. For a cured mortar bed, follow recommendations below to select appropriate bonding mortar:
  - When a waterproof membrane is not used—ANSI A118.1 or better or ISO C1 or better.
  - When a waterproof membrane is used—ANSI A118.4 or better or ISO C2S1 or better unless ANSI A118.1 or ISO C1 is recommended by membrane manufacturer.
  - When porcelain tile is used—ANSI A118.4 or better or ISO C2 or better.
  - When glass tile is used, specify mortar designated by tile and mortar manufacturers. Bond coat color will impact the final appearance of translucent glass tile. Specifier shall confirm bond coat color is acceptable.
- Cementitious bond coat (above-ground)—portland cement paste on a mortar bed that is still workable. For a cured mortar bed, follow recommendations below to select appropriate bonding mortar:
  - Must be recommended by manufacturer for above-ground use.
  - When a waterproof membrane is not used—ANSI A118.15 or better or ISO C2S1 or better.
  - When a waterproof membrane is used—ANSI A118.4 or better or ISO C2S1 or better unless ANSI A118.1 or ISO C1 is recommended by membrane manufacturer.
  - When glass tile is used, specify mortar designated by tile and mortar manufacturers. Bond coat color will impact the final appearance of translucent glass tile. Specifier shall confirm bond coat color is acceptable.
- Waterproof membrane, when used—ANSI A118.10.
- Mortar bed—ANSI A108.1A.
- Mortar bed bond coat—portland cement slurry.

Materials for Green/Sustainable Design

- See "Green Building Standards and Green Product Selection Guide" and consult manufacturers and suppliers for product sustainability and contribution to green building design.
- Consider specifying tile and installation materials that meet ANSI A138.1, the *American National Standard Specifications for Sustainable Ceramic Tiles, Glass Tiles, and Tile Installation Materials.*

Preparation by Other Trades

- Floor systems over which tile will be installed shall be in conformance with the IRC for residential applications, the IBC for commercial applications, or applicable building codes. Maximum allowable substrate deflection under live load not to exceed l/360. See also "Substrate Requirements."
- When concentrated loads (scissor lifts, pallet jacks, automobiles, forklifts, etc.) will be used on an above-ground tiled floor, the engineer and/or specifier shall specify a substrate to accommodate the concentrated loads. Owner/specifier is responsible for protecting the tilework from damage, including allowing sufficient time for installed materials to cure properly.
- Slab to have steel trowel and fine broom finish with no curing compounds used. When used, mechanical scarifying is necessary.
- Slope, when required, to be in subfloor.
- Maximum allowable variation in the installation substrate (concrete)—¼" in 10' from the required plane.

Movement Joint (architect must specify type of joint and show location and details on drawings)

- Movement joints—mandatory according to EJ171. For above-ground installations, additional movement joints are required.
- When glass tile is used, adhere to more frequent placement recommendations within the ranges listed in EJ171.

Installation Specifications

- Tile—ANSI A108.1A, .1B, or .1C. A108.1B required if membrane to be used.
- Glass tile—ANSI A108.14, .15, .16, or manufacturer's directions.
- Cementitious grout—ANSI A108.10.
- Waterproof membrane—ANSI A108.13.
- Movement Joints—EJ171 and ASTM C1193.

Notes

- F111 is preferred over above-ground structural slabs and other floors subject to movement and/or deflection.
- Above-ground—not all mortars, grouts, and membranes are suitable for above-ground use. Check manufacturer recommendations.
- For hydronic tubing on top of the slab when using a mortar bed, see RH117.
- For waterproofing below mortar bed (unbonded), see F121.
- When glass tile is used, see "Glass Tile Selection and Installation Guide," and consult manufacturer for recommendations and requirements.

F113-21

- **On-Ground Concrete**
- **Ceramic Tile, Glass Tile**

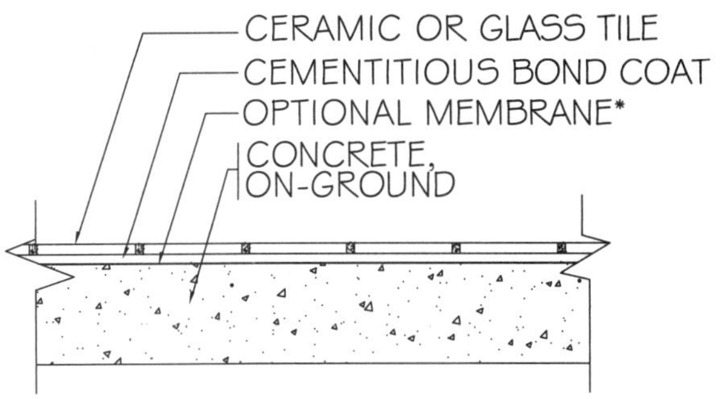

*USE OF A MEMBRANE IS OPTIONAL.
SEE MEMBRANE OPTIONS.

Recommended Uses

- For slab-on-ground construction where no bending stresses occur and thin-bed installation of tile is desired.

Service Rating

- Extra heavy.
- When glass tile is used, service rating may be lower.

Environmental Exposure Classifications

- Res1, 2, 3, 5; Com1, 2, 3, 5.
- May be suitable for exterior applications in areas not subject to freeze/thaw cycling when appropriate precautions are taken, including expansion joint placement, proper slope, waterproofing, and material selection.
- For installations that may be exposed to staining, specify tile and grout suitable for exposure. Consult product manufacturers; see also "Product Selection Guides."
- For installations that may be exposed to staining and/or chemical attack, see also F115, F131, and F133.

Typical Weight of Tile Installation

- 5 pounds/square foot.
- Does not include weight of substrate. See "Appendix B" for assumptions, included materials, and their individual weights.

Limitations

- Not for use over above-ground structural slabs and other floors subject to movement and/or deflection. See F113A.

Membrane Options

- A crack isolation membrane may be specified to treat existing in-plane cracks (F125-partial) or for protection against future in-plane cracking (F-125 full). See F125 for more information.
- An uncoupling membrane may be specified to accommodate subfloor movement. Check with membrane manufacturer for recommendations and requirements.
- Check with membrane manufacturers for suitability for applicable conditions, as not all membranes are suitable for steam, high temperature, and/or chemical exposure, exterior use, use over above-ground structural slabs, use over pourable underlayments, use with radiant heat, or use over concrete with excessive moisture vapor transmission and/or alkalinity. Membrane may also affect service rating.
- When glass tile is used, consult glass tile manufacturer for membrane options and recommendations.

Requirements

- Slab to be well cured, dimensionally stable, and free of cracks, surface laitance, waxy or oily films, curing compounds, and other surface bond-inhibiting contaminants.

Materials

- **Multiple options exist for membranes, mortars, grouts, and other materials and *must be clearly specified* to be included. If not specifically indicated, optional materials are not included and mortar/grout choice defaults to minimum performance specification indicated. Consider each system component and intended use to determine minimum requirements and to specify options.**
- Ceramic tile—ANSI A137.1.
- Glass tile, when used—ANSI A137.2; see also "Glass Tile Selection and Installation Guide," and consult tile manufacturer for service rating and environmental exposure classification recommendations. Not all glass tiles are suitable.
- Cementitious grout—ANSI A118.6 or better or ISO CG1 or better. When glass tile is used, specify grout designated by tile and grout manufacturers.
- Cementitious bond coat:
  - When a membrane is not used—ANSI A118.1 or better or ISO C1 or better.
  - When a crack isolation membrane is used—ANSI A118.4 or better or ISO C2S1 or better.
  - When an uncoupling membrane is used, check with membrane manufacturer.
  - When porcelain tile is used—ANSI A118.4 or better or ISO C2 or better.
  - When glass tile is used, specify mortar designated by tile and mortar manufacturers. Bond coat color will impact the final appearance of translucent glass tile. Specifier shall confirm bond coat color is acceptable.
- Crack isolation membrane, when used—ANSI A118.12.
- Uncoupling membrane, when used—recommended by manufacturer; must achieve 50 psi or greater shear bond strength in 7 days per the test method in ANSI A118.12 Section 5.1.3.

Materials for Green/Sustainable Design

- See "Green Building Standards and Green Product Selection Guide" and consult manufacturers and suppliers for product sustainability and contribution to green building design.
- Consider specifying tile and installation materials that meet ANSI A138.1, the *American National Standard Specifications for Sustainable Ceramic Tiles, Glass Tiles, and Tile Installation Materials.*

Preparation by Other Trades

- Slab to have steel trowel and fine broom finish free of curing compounds. When used, mechanical scarifying is necessary.
- Slope, when required, to be in subfloor.
- Maximum allowable variation in the tile substrate—for tiles with all edges shorter than 15", maximum allowable variation is ¼" in 10' from the required plane, with no more than ¹⁄₁₆" variation in 12" when measured from the high points in the surface. For tiles with at least one edge 15" in length, maximum allowable variation is ⅛" in 10' from the required plane, with no more than ¹⁄₁₆" variation in 24" when measured from the high points in the surface.

Movement Joint (architect must specify type of joint and show location and details on drawings)

- Movement joints—mandatory according to EJ171.
- When glass tile is used, adhere to more frequent placement recommendations within the ranges listed in EJ171.

Installation Specifications

- Tile—ANSI A108.5.
- Glass tile—manufacturer's directions.
- Cementitious grout—ANSI A108.10.
- Crack isolation membrane—ANSI A108.17.
- Uncoupling membrane—manufacturer's directions.
- Movement Joints—EJ171 and ASTM C1193.

Notes

- For waterproof application where thin-bed installation is desired, see F122.
- For electric radiant in-floor heat application, where thin-bed installation is desired, see RH115.
- For in-slab hydronic tubing, see RH110.
- For young concrete, see F128.
- When glass tile is used, see "Glass Tile Selection and Installation Guide," and consult manufacturer for recommendations and requirements.

F113A-21

- **Above-Ground Concrete**
- **Ceramic Tile, Glass Tile**

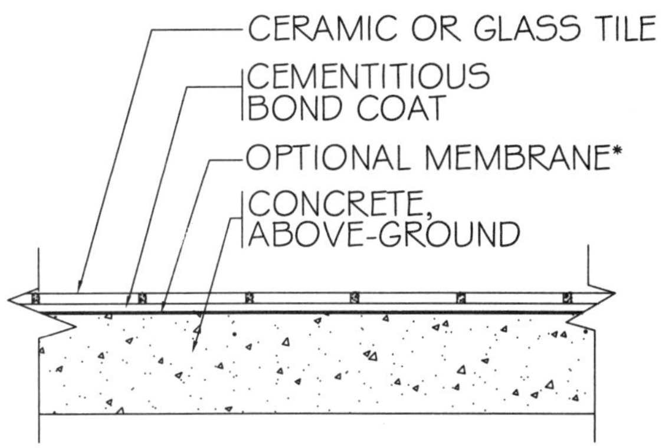

*USE OF A MEMBRANE IS OPTIONAL. SEE MEMBRANE OPTIONS.

Recommended Uses

- For above-ground structural slabs and other floors subject to movement and/or deflection where thin-bed installation of tile is desired.

Service Rating

- Extra heavy.
- When glass tile is used, service rating may be lower.

Environmental Exposure Classifications

- Res1, 2, 3, 5; Com1, 2, 3, 5.
- May be suitable for exterior applications in areas not subject to freeze/thaw cycling when appropriate precautions are taken, including expansion joint placement, proper slope, waterproofing, and material selection.
- For installations that may be exposed to staining, specify tile and grout suitable for exposure. Consult product manufacturers; see also "Product Selection Guides."
- For installations that may be exposed to staining and/or chemical attack, see also F115A, F131, and F133.

Typical Weight of Tile Installation

- 5 pounds/square foot.
- Does not include weight of substrate. See "Appendix B" for assumptions, included materials, and their individual weights.

Limitations

- Requires additional consideration by design professional to accommodate movement and/or deflection. Setting materials with improved bond strength and deformability are required.

Membrane Options

- A crack isolation membrane (A118.12) may be specified to treat existing in-plane cracks (F125 partial) or for protection against future in-plane cracking (F125-full). See F125 for more information.
- An uncoupling membrane may be specified to accommodate subfloor movement. Check with membrane manufacturer for recommendations and requirements.
- Check with membrane manufacturer for suitability for applicable conditions, as not all membranes are suitable for steam, high temperature, and/or chemical exposure, exterior use, use over above-ground structural slabs, use over pourable underlayments, use with radiant heat, or use over concrete with excessive moisture vapor transmission and/or alkalinity. Membrane may also affect service rating.
- When glass tile is used, consult glass tile manufacturer for membrane options and recommendations.

Requirements

- Slab to be well cured, dimensionally stable, and free of cracks, surface laitance, waxy or oily films, curing compounds, and other surface bond-inhibiting contaminants.
- Above-ground installations are inherently more susceptible to vibration and deflection. Grout and mortar manufacturers to warrant product suitability.

Materials

- **Multiple options exist for membranes, mortars, grouts, and other materials and *must be clearly specified* to be included. If not specifically indicated, optional materials are not included and mortar/grout choice defaults to minimum performance specification indicated. Consider each system component and intended use to determine minimum requirements and to specify options.**
- Ceramic tile—ANSI A137.1.
- Glass tile, when used—ANSI A137.2; see also "Glass Tile Selection and Installation Guide," and consult tile manufacturer for service rating and environmental exposure classification recommendations. Not all glass tiles are suitable.
- Cementitious grout—ANSI A118.6 or better or ISO CG1 or better. When glass tile is used, specify grout designated by tile and grout manufacturers.
- Cementitious bond coat:
  - Must be recommended by manufacturer for above-ground use.
  - When a membrane is not used—ANSI A118.15 or better or ISO C2S1 or better.
  - When a crack isolation membrane is used—ANSI A118.4 or better or ISO C2S1 or better.
  - When an uncoupling membrane is used, check with membrane manufacturer.
  - When glass tile is used, specify mortar designated by tile and mortar manufacturers. Bond coat color will

impact the final appearance of translucent glass tile. Specifier shall confirm bond coat color is acceptable.

- Crack isolation membrane, when used—A118.12.
- Uncoupling membrane, when used—recommended by manufacturer; must achieve 50 psi or greater shear bond strength in 7 days per the test method in ANSI A118.12 Section 5.1.3.

Materials for Green/Sustainable Design

- See "Green Building Standards and Green Product Selection Guide" and consult manufacturers and suppliers for product sustainability and contribution to green building design.
- Consider specifying tile and installation materials that meet ANSI A138.1, the *American National Standard Specifications for Sustainable Ceramic Tiles, Glass Tiles, and Tile Installation Materials.*

Preparation by Other Trades

- Floor systems over which tile will be installed shall be in conformance with the IRC for residential applications, the IBC for commercial applications, or applicable building codes. Maximum allowable substrate deflection under live load not to exceed l/360. See also "Substrate Requirements."
- When concentrated loads (scissor lifts, pallet jacks, automobiles, forklifts, etc.) will be used on an above-ground tiled floor, the engineer and/or specifier shall specify a substrate to accommodate the concentrated loads. Owner/specifier is responsible for protecting the tilework from damage, including allowing sufficient time for installed materials to cure properly.
- Slab to have steel trowel and fine broom finish free of curing compounds. When used, mechanical scarifying is necessary.
- Slope, when required, to be in subfloor.
- Maximum allowable variation in the tile substrate—for tiles with all edges shorter than 15", maximum allowable variation is ¼" in 10' from the required plane, with no more than ¹⁄₁₆" variation in 12" when measured from the high points in the surface. For tiles with at least one edge 15" in length, maximum allowable variation is ⅛" in 10' from the required plane, with no more than ¹⁄₁₆" variation in 24" when measured from the high points in the surface.

Movement Joint (architect must specify type of joint and show location and details on drawings)

- Movement joints—mandatory according to EJ171. For above-ground installations, additional movement joints are required.
- When glass tile is used, adhere to more frequent placement recommendations within the ranges listed in EJ171.

Installation Specifications

- Tile—ANSI A108.5.
- Glass tile—manufacturer's directions.
- Cementitious grout—ANSI A108.10.
- Crack isolation membrane—A108.17.
- Uncoupling membrane—manufacturer's directions.
- Movement Joints—EJ171 and ASTM C1193.

Notes

- F111 is preferred over above-ground structural slabs and other floors subject to movement and/or deflection.
- Not all mortars, grouts, and membranes are suitable for above-ground installations. Check manufacturer recommendations.
- For waterproof application where thin-bed installation is desired, see F122A.
- For electric radiant in-floor heat application, where thin-bed installation is desired, see RH115A.
- For in-slab hydronic tubing, see RH110A.
- When glass tile is used, see "Glass Tile Selection and Installation Guide," and consult manufacturer for recommendations and requirements.

F114-21

- **On-Ground or Above-Ground Concrete**
- **Unbonded Mortar Bed**
- **Ceramic Tile**
- **Epoxy or Furan Grout**

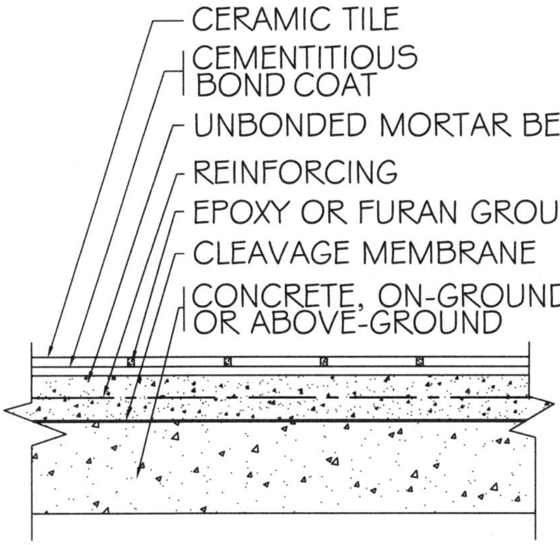

Recommended Uses

- For above-ground structural slabs and for slab-on-ground construction where good stain resistance and mild chemical resistance are preferred, for example, in commercial dining areas, dark rooms, public toilets, public foyers, etc.
- For areas where leveling, flattening, or contouring of finish floor height is required, such as commercial kitchens and gang showers, where good stain resistance and mild chemical resistance are preferred.
- For areas where floor flatness is critical, such as when tiles with any edge longer than 15" are specified or where the finished tile assembly is required to achieve in-plane transitions (as may be necessary where accessibility is a concern) and good stain resistance and mild chemical resistance are preferred.
- For concrete substrates that are cracked or present bonding issues and good stain resistance and mild chemical resistance are preferred.
- For areas with in-slab hydronic tubing where good stain resistance and mild chemical resistance are preferred.

Service Rating

- Extra heavy.

Environmental Exposure Classifications

- Res1, 2, 3, 4, 5; Com1, 2, 3, 4, 5.
- For increased chemical resistance, see also F132 and F134.

Typical Weight of Tile Installation

- 21 pounds/square foot with 1¼" mortar bed. Add 3 pounds/square foot for each additional ¼" of mortar bed
- Does not include weight of substrate. See "Appendix B" for assumptions, included materials, and their individual weights.

Limitations

- Above-ground structural slabs and other floors subject to movement and/or deflection—requires additional consideration by design professional to accommodate movement and/or deflection. Setting materials with improved bond strength and deformability are required.
- Not for severe chemical resistance such as may be needed in commercial kitchens, dairies, breweries, and food processing plants. See Environmental Exposure Classifications in this method.
- Extreme heat or improper steam cleaning will soften epoxy grouts and wash them out of joints. High-temperature-resistant formulas are available.

Requirements

- Specify tile and grout suitable for stain and/or chemical exposure. Consult product manufacturers. See also "Product Selection Guides."
- Tile surface must be waxed before grouting when furan resin grout is used.
- Reinforcing mesh mandatory.
- Mortar bed thickness—1¼" minimum to 2" maximum.
- Mortar beds in excess of 2" thick shall be detailed by the architect.
- Joints must be clean and completely filled with epoxy or furan. Partial filling with sand or mortar is unacceptable.
- Above-ground—above-ground installations are inherently more susceptible to vibration and deflection. Grout and mortar manufacturers to warrant product suitability.

Materials

- **Multiple options exist for membranes, mortars, grouts, and other materials and *must be clearly specified* to be included. If not specifically indicated, optional materials are not included and mortar/grout choice defaults to minimum performance specification indicated. Consider each system component and intended use to determine minimum requirements and to specify options.**
- Ceramic tile—ANSI A137.1.
- Epoxy or furan grout—ANSI A118.3, A118.5 or ISO RG.
- Cementitious bond coat (on-ground)—portland cement paste on a mortar bed that is still workable. For a cured mortar bed, follow recommendations below to select appropriate bonding mortar:
  - ANSI A118.1 or better or ISO C1 or better.
  - When porcelain tile is used—ANSI A118.4 or better or ISO C2 or better.

- Cementitious bond coat (above-ground)—portland cement paste on a mortar bed that is still workable. For a cured mortar bed, follow recommendations below to select appropriate bonding mortar:
  - Must be recommended by manufacturer for above-ground use.
  - ANSI A118.15 or better or ISO C2S1 or better.
- Mortar bed, reinforcing, and cleavage membrane—A108.1A.

Materials for Green/Sustainable Design

- See "Green Building Standards and Green Product Selection Guide" and consult manufacturers and suppliers for product sustainability and contribution to green building design.
- Consider specifying tile and installation materials that meet ANSI A138.1, the *American National Standard Specifications for Sustainable Ceramic Tiles, Glass Tiles, and Tile Installation Materials.*

Preparation by Other Trades

- Floor systems over which tile will be installed shall be in conformance with the IRC for residential applications, the IBC for commercial applications, or applicable building codes. Maximum allowable substrate deflection under live load not to exceed l/360. See also "Substrate Requirements."
- When concentrated loads (scissor lifts, pallet jacks, automobiles, forklifts, etc.) will be used on an above-ground tiled floor, the engineer and/or specifier shall specify a substrate to accommodate the concentrated loads. Owner/specifier is responsible for protecting the tilework from damage, including allowing sufficient time for installed materials to cure properly.
- Where slab is depressed to accommodate mortar bed thickness, slab depression to be accurate with steel trowel finish.
- Slope, when required, to be in subfloor.
- Maximum allowable variation in the installation substrate (concrete)—¼" in 10' from the required plane.

Movement Joint (architect must specify type of joint and show location and details on drawings)

- Movement joints—mandatory according to EJ171. For above-ground installations, additional movement joints are required.

Installation Specifications

- Tile—ANSI A108.1B.
- Epoxy grout—ANSI A108.6.
- Furan grout—ANSI A108.8
- Movement Joints—EJ171 and ASTM C1193.

Notes

- Above-ground—not all mortars and grouts are suitable for above-ground use. Check manufacturer recommendations.
- For hydronic tubing on top of the slab when using a mortar bed, see RH117.
- For waterproof application when using a mortar bed, see F121 or F112.

F115-21

- **On-Ground Concrete**
- **Ceramic Tile**
- **Epoxy or Furan Grout**

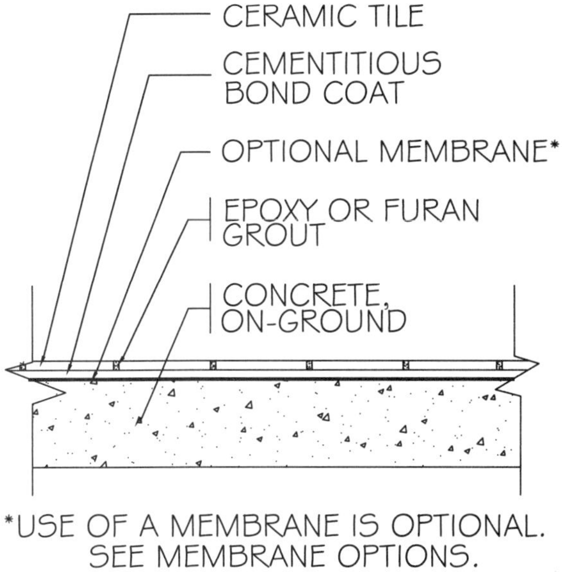

Recommended Uses

- For slab-on-ground construction where no bending stresses occur, thin-bed installation of tile is desired, and good stain resistance and mild chemical resistance are preferred, for example, in commercial dining areas, dark rooms, public toilets, public foyers, etc.

Service Rating

- Extra heavy.

Environmental Exposure Classifications

- Res1, 2, 3, 4, 5; Com1, 2, 3, 4, 5.
- For increased chemical resistance, see also F131 and F133.

Typical Weight of Tile Installation

- 5 pounds/square foot.
- Does not include weight of substrate. See "Appendix B" for assumptions, included materials, and their individual weights.

Limitations

- Not for severe chemical resistance such as may be needed in commercial kitchens, dairies, breweries, and food processing plants. See Environmental Exposure Classifications in this method.
- Extreme heat or improper steam cleaning will soften epoxy grouts and wash them out of joints. High-temperature-resistant formulas are available.
- Not for use over above-ground structural slabs and other floors subject to movement and/or deflection. See F115A.

Membrane Options

- A crack isolation membrane may be specified to treat existing in-plane cracks (F125-partial) or for protection against future in-plane cracking (F-125 full). See F125 for more information.
- Check with membrane manufacturers for suitability for applicable conditions, as not all membranes are suitable for steam, high temperature, and/or chemical exposure, exterior use, use over above-ground structural slabs, use over pourable underlayments, use with radiant heat, or use over concrete with excessive moisture vapor transmission and/or alkalinity. Membrane may also affect service rating.

Requirements

- Specify tile and grout suitable for stain and/or chemical exposure. Consult product manufacturers. See also "Product Selection Guides."
- Tile surface must be waxed before grouting when furan resin grout is used.
- Slab to be well cured, dimensionally stable, and free of cracks, surface laitance, waxy or oily films, curing compounds, and other surface bond-inhibiting contaminants.
- Joints must be clean and completely filled with epoxy or furan. Partial filling with sand or mortar is unacceptable.

Materials

- **Multiple options exist for membranes, mortars, grouts, and other materials and *must be clearly specified* to be included. If not specifically indicated, optional materials are not included and mortar/grout choice defaults to minimum performance specification indicated. Consider each system component and intended use to determine minimum requirements and to specify options.**
- Ceramic tile—ANSI A137.1.
- Epoxy or furan grout—ANSI A118.3, A118.5, or ISO RG.
- Cementitious bond coat:
  - When a membrane is not used—ANSI A118.1 or better or ISO C1 or better.
  - When a crack isolation membrane is used—ANSI A118.4 or better or ISO C2S1 or better.
  - When porcelain tile is used—ANSI A118.4 or better or ISO C2 or better.
- Crack isolation membrane, when used—ANSI A118.12.

Materials for Green/Sustainable Design

- See "Green Building Standards and Green Product Selection Guide" and consult manufacturers and suppliers for product sustainability and contribution to green building design.
- Consider specifying tile and installation materials that meet ANSI A138.1, the *American National Standard*

Specifications for Sustainable Ceramic Tiles, Glass Tiles, and Tile Installation Materials.

Preparation by Other Trades

- Slab to have steel trowel and fine broom finish free of curing compounds. When used, mechanical scarifying is necessary.
- Slope, when required, to be in subfloor.
- Maximum allowable variation in the tile substrate—for tiles with all edges shorter than 15", maximum allowable variation is ¼" in 10' from the required plane, with no more than 1/16" variation in 12" when measured from the high points in the surface. For tiles with at least one edge 15" in length, maximum allowable variation is ⅛" in 10' from the required plane, with no more than 1/16" variation in 24" when measured from the high points in the surface.

Movement Joint (architect must specify type of joint and show location and details on drawings)

- Movement joints—mandatory according to EJ171.

Installation Specifications

- Tile—ANSI A108.5.
- Epoxy grout—ANSI A108.6.
- Furan grout—ANSI A108.8.
- Crack isolation membrane—ANSI A108.17.
- Movement Joints—EJ171 and ASTM C1193.

Notes

- For waterproof application where thin-bed installation is desired, see F122.

F115A-21

- **Above-Ground Concrete**
- **Ceramic Tile**
- **Epoxy or Furan Grout**

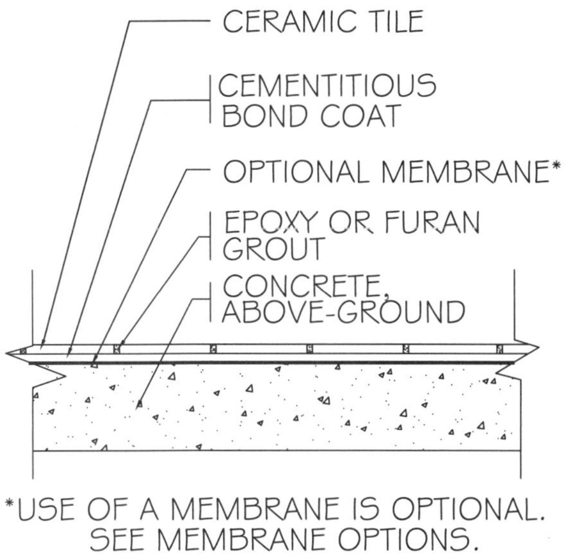

Recommended Uses

- For above-ground structural slabs where thin-bed installation of tile is desired and good stain resistance and mild chemical resistance are preferred, for example, in commercial dining areas, dark rooms, public toilets, public foyers, etc.

Service Rating

- Extra heavy.

Environmental Exposure Classifications

- Res1, 2, 3, 4, 5; Com1, 2, 3, 4, 5.
- For increased chemical resistance, see also F131 and F133.

Typical Weight of Tile Installation

- 5 pounds/square foot.
- Does not include weight of substrate. See "Appendix B" for assumptions, included materials, and their individual weights.

Limitations

- Not for severe chemical resistance such as may be needed in commercial kitchens, dairies, breweries, and food processing plants. See Environmental Exposure Classifications in this method.
- Requires additional consideration by design professional to accommodate movement and/or deflection. Setting materials with improved bond strength and deformability are required.
- Extreme heat or improper steam cleaning will soften epoxy grouts and wash them out of joints. High-temperature-resistant formulas are available.

Membrane Options

- A crack isolation membrane may be specified to treat existing in-plane cracks (F125-partial) or for protection against future in-plane cracking (F-125 full). See F125 for more information.
- Check with membrane manufacturers for suitability for applicable conditions, as not all membranes are suitable for steam, high temperature, and/or chemical exposure, exterior use, use over above-ground structural slabs, use over pourable underlayments, use with radiant heat, or use over concrete with excessive moisture vapor transmission and/or alkalinity. Membrane may also affect service rating.

Requirements

- Specify tile and grout suitable for stain and/or chemical exposure. Consult product manufacturers. See also "Product Selection Guides."
- Slab to be well cured, dimensionally stable, and free of cracks, surface laitance, waxy or oily films, curing compounds, and other surface bond-inhibiting contaminants.
- Tile surfaces must be waxed before grouting when furan resin grout is used.
- Joints must be clean and completely filled with epoxy or furan. Partial filling with sand or mortar is unacceptable.
- Above-ground installations are inherently more susceptible to vibration and deflection. Grout and mortar manufacturers to warrant product suitability.

Materials

- **Multiple options exist for membranes, mortars, grouts, and other materials and *must be clearly specified* to be included. If not specifically indicated, optional materials are not included and mortar/grout choice defaults to minimum performance specification indicated. Consider each system component and intended use to determine minimum requirements and to specify options.**
- Ceramic tile—ANSI A137.1.
- Epoxy or furan grout—ANSI A118.3, A118.5, or ISO RG.
- Cementitious bond coat:
  - Must be recommended by manufacturer for above-ground use.
  - When a membrane is not used—ANSI A118.15 or better or ISO C2S1 or better.
  - When a crack isolation membrane is used—ANSI A118.4 or better or ISO C2S1 or better.
- Crack isolation membrane, when used—ANSI A118.12.

Materials for Green/Sustainable Design

- See "Green Building Standards and Green Product Selection Guide" and consult manufacturers and suppliers for product sustainability and contribution to green building design.
- Consider specifying tile and installation materials that meet ANSI A138.1, the *American National Standard Specifications for Sustainable Ceramic Tiles, Glass Tiles, and Tile Installation Materials.*

Preparation by Other Trades

- Floor systems over which tile will be installed shall be in conformance with the IRC for residential applications, the IBC for commercial applications, or applicable building codes. Maximum allowable substrate deflection under live load not to exceed l/360. See also "Substrate Requirements."
- When concentrated loads (scissor lifts, pallet jacks, automobiles, forklifts, etc.) will be used on an above-ground tiled floor, the engineer and/or specifier shall specify a substrate to accommodate the concentrated loads. Owner/specifier is responsible for protecting the tilework from damage, including allowing sufficient time for installed materials to cure properly.
- Slab to have steel trowel and fine broom finish free of curing compounds. When used, mechanical scarifying is necessary.
- Slope, when required, to be in subfloor.
- Maximum allowable variation in the tile substrate—for tiles with all edges shorter than 15", maximum allowable variation is ¼" in 10' from the required plane, with no more than ¹⁄₁₆" variation in 12" when measured from the high points in the surface. For tiles with at least one edge 15" in length, maximum allowable variation is ⅛" in 10' from the required plane, with no more than ¹⁄₁₆" variation in 24" when measured from the high points in the surface.

Movement Joint (architect must specify type of joint and show location and details on drawings)

- Movement joints—mandatory according to EJ171. For above-ground installations, additional movement joints are required.

Installation Specifications

- Tile—ANSI A108.5.
- Epoxy grout—ANSI A108.6.
- Furan grout—ANSI A108.8
- Crack isolation membrane—A108.17.
- Movement Joints—EJ171 and ASTM C1193.

Notes

- F111 is preferred over above-ground structural slabs and other floors subject to movement and/or deflection.
- Not all mortars, grouts, and membranes are suitable for above-ground installations. Check manufacturer recommendations.
- For waterproof application where thin-bed installation is desired, see F122A.

F116E-21

- **On-Ground or Above-Ground Concrete**
- **Epoxy Adhesive**
- **Ceramic Tile**

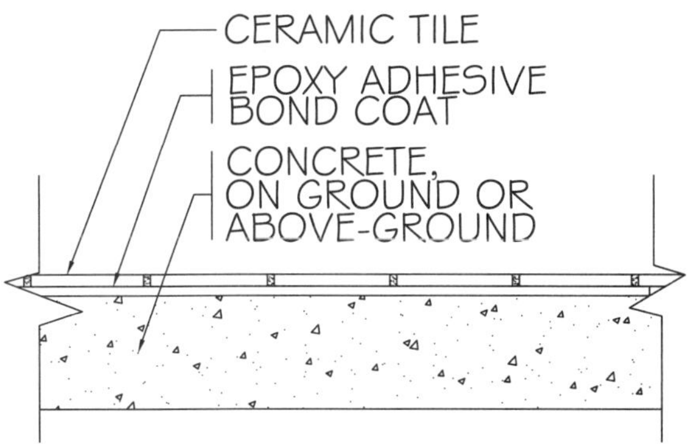

Recommended Uses

- For above-ground structural slabs and for slab-on-ground construction where thin-bed installation of tile is desired and high bond strength and/or high impact strength is desired.

Service Rating

- Extra heavy.

Environmental Exposure Classifications

- Res1, 2, 3, 4, 5; Com1, 2, 3, 4, 5.
- For installations that may be exposed to staining, specify tile and grout suitable for exposure. Consult product manufacturers; see also "Product Selection Guides."
- For installations that may be exposed to staining and/or chemical attack, see also F131 and F133.

Typical Weight of Tile Installation

- 6 pounds/square foot.
- Does not include weight of substrate. See "Appendix B" for assumptions, included materials, and their individual weights.

Limitations

- Above-ground structural slabs and other floors subject to movement and/or deflection—requires additional consideration by design professional to accommodate movement and/or deflection. Setting materials with improved bond strength and deformability are required.

Requirements

- Slab to be well cured, dimensionally stable, and free of cracks, surface laitance, waxy or oily films, curing compounds, and other surface bond-inhibiting contaminants.
- Above-ground—above-ground installations are inherently more susceptible to vibration and deflection. Grout and mortar manufacturers to warrant product suitability.

Materials

- **Multiple options exist for membranes, mortars, grouts, and other materials and *must be clearly specified* to be included. If not specifically indicated, optional materials are not included and mortar/grout choice defaults to minimum performance specification indicated. Consider each system component and intended use to determine minimum requirements and to specify options.**
- Ceramic tile—ANSI A137.1.
- Cementitious grout—ANSI A118.6 or better or ISO CG1 or better.
- Epoxy or furan grout, when used—ANSI A118.3, A118.5, or ISO RG.
- Epoxy adhesive—ANSI A118.3 or ISO R1 or better. For above-ground use, must also be recommended by manufacturer.

Materials for Green/Sustainable Design

- See "Green Building Standards and Green Product Selection Guide" and consult manufacturers and suppliers for product sustainability and contribution to green building design.
- Consider specifying tile and installation materials that meet ANSI A138.1, the *American National Standard Specifications for Sustainable Ceramic Tiles, Glass Tiles, and Tile Installation Materials.*

Preparation by Other Trades

- Floor systems over which tile will be installed shall be in conformance with the IRC for residential applications, the IBC for commercial applications, or applicable building codes. Maximum allowable substrate deflection under live load not to exceed l/360. See also "Substrate Requirements."
- When concentrated loads (scissor lifts, pallet jacks, automobiles, forklifts, etc.) will be used on an above-ground tiled floor, the engineer and/or specifier shall specify a substrate to accommodate the concentrated loads. Owner/specifier is responsible for protecting the tilework from damage, including allowing sufficient time for installed materials to cure properly.
- Slab to have steel trowel and fine broom finish free of curing compounds. When used, mechanical scarifying is necessary.
- Slope, when required, to be in subfloor.
- Maximum allowable variation in the tile substrate—$\frac{1}{16}$" in 3' with no abrupt irregularities greater than $\frac{1}{32}$."

Movement Joint (architect must specify type of joint and show location and details on drawings)

- Movement joints—mandatory according to EJ171.

For above-ground installations, additional movement joints are required.

Installation Specifications

- Tile—ANSI A108.4.
- Cementitious grout—ANSI A108.10.
- Epoxy grout—ANSI A108.6.
- Furan grout—ANSI A108.8.
- Movement Joints—EJ171 and ASTM C1193.

Notes

- As the tile size increases, there is less tolerance for variation in the subfloor from the required plane. Epoxy adhesive thickness must be thin and uniform; therefore, subfloor flattening may be required.
- Above-ground—not all mortars and grouts are suitable for above-ground use. Check manufacturer recommendations.
- F111 is preferred over above-ground structural slabs and other floors subject to movement and/or deflection.
- For waterproof application where thin-bed installation is desired, see F122 and F122A.

F116O-21

- **On-Ground or Above-Ground Concrete**
- **Organic Adhesive**
- **Ceramic Tile**

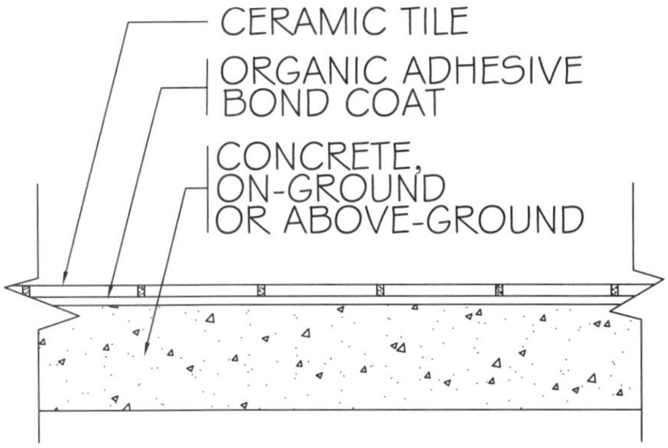

Recommended Uses

- For above-ground structural slabs and for slab-on-ground construction where thin-bed installation of tile is desired in residential dry areas with light traffic and will not be subjected to high impact, wheel loads, or point loads.

Service Rating

- Residential.

Environmental Exposure Classifications

- Res1.
- For installations that may be exposed to staining, specify tile and grout suitable for exposure. Consult product manufacturers; see also "Product Selection Guides."

Typical Weight of Tile Installation

- 5 pounds/square foot.
- Does not include weight of substrate. See "Appendix B" for assumptions, included materials, and their individual weights.

Limitations

- Will not withstand high impact, wheel loads, or point loads.
- Not for below-ground installations.
- Not for use where moisture vapor transmission exceeds recommendation of organic adhesive manufacturer.
- Not for areas exposed to temperatures exceeding 140°F.
- Above-ground structural slabs and other floors subject to movement and/or deflection—requires additional consideration by design professional to accommodate movement and/or deflection. Setting materials with improved bond strength and deformability are required.
- Maximum tile size 8" × 8" unless organic adhesive manufacturer allows larger tile size. See Notes.

Requirements

- Consult the adhesive manufacturer for maximum moisture vapor transmission limitations prior to tile setting.
- Slab to be well cured, dimensionally stable, and free of cracks, surface laitance, waxy or oily films, curing compounds, and other surface bond-inhibiting contaminants.
- Above-ground—above-ground installations are inherently more susceptible to vibration and deflection. Grout and mortar manufacturers to warrant product suitability.

Materials

- **Multiple options exist for membranes, mortars, grouts, and other materials and *must be clearly specified* to be included. If not specifically indicated, optional materials are not included and mortar/grout choice defaults to minimum performance specification indicated. Consider each system component and intended use to determine minimum requirements and to specify options.**
- Ceramic tile—ANSI A137.1.
- Cementitious grout—ANSI A118.6 or better or ISO CG1 or better.
- Organic adhesive bond coat—ANSI A136.1 Type I or ISO D2. For above-ground use, must also be recommended by manufacturer.

Materials for Green/Sustainable Design

- See "Green Building Standards and Green Product Selection Guide" and consult manufacturers and suppliers for product sustainability and contribution to green building design.
- Consider specifying tile and installation materials that meet ANSI A138.1, the *American National Standard Specifications for Sustainable Ceramic Tiles, Glass Tiles, and Tile Installation Materials*.

Preparation by Other Trades

- Floor systems over which tile will be installed shall be in conformance with the IRC for residential applications, the IBC for commercial applications, or applicable building codes. Maximum allowable substrate deflection under live load not to exceed l/360. See also "Substrate Requirements."
- When concentrated loads (scissor lifts, pallet jacks, automobiles, forklifts, etc.) will be used on an above-ground tiled floor, the engineer and/or specifier shall specify a substrate to accommodate the concentrated loads. Owner/specifier is responsible for protecting the tilework from damage, including allowing sufficient time for installed materials to cure properly.
- Slab to have steel trowel and fine broom finish free of curing compounds. When used, mechanical scarifying is necessary.

- Slope, when required, to be in subfloor.
- Maximum allowable variation in the tile substrate—1⁄16" in 3' with no abrupt irregularities greater than 1⁄32".

Movement Joint (architect must specify type of joint and show location and details on drawings)

- Movement joints—mandatory according to EJ171. For above-ground installations, additional movement joints are required.

Installation Specifications

- Tile—ANSI A108.4.
- Cementitious grout—ANSI A108.10.
- Movement Joints—EJ171 and ASTM C1193.

Notes

- As the tile size increases, there is less tolerance for variation in the subfloor from the required plane. Organic adhesive thickness must be thin and uniform; therefore, subfloor flattening may be required.
- Above-ground—not all organic adhesives and grouts are suitable for above-ground installations. Check manufacturer recommendations.
- F111 is preferred over above-ground structural slabs and other floors subject to movement and/or deflection.
- Do not use organic adhesives to bond tile to a membrane or other impervious substrate.
- Extended dry time may be required for larger tiles and tiles with low water absorption when organic adhesive is used.

F200-21

- **On-Ground Concrete**
- **Poured Gypsum Underlayment**
- **Bonded Membrane**
- **Ceramic Tile, Glass Tile**

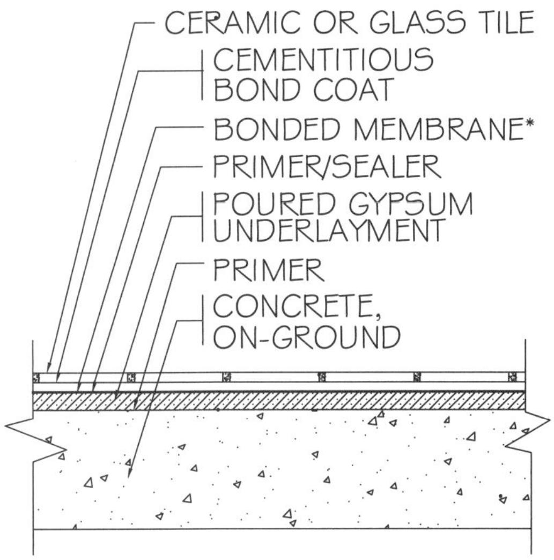

*USE OF A MEMBRANE IS REQUIRED. SEE MEMBRANE OPTIONS.

Recommended Uses

- For slab-on-ground construction where no bending stresses occur and with a poured gypsum underlayment.

Service Rating

- Moderate.
- When glass tile is used, service rating may be lower.

Environmental Exposure Classifications

- Res1; Com1.
- May be suitable for limited water exposure areas (Res2; Com2). See Membrane Options.
- For installations that may be exposed to staining, specify tile and grout suitable for exposure. Consult product manufacturers; see also "Product Selection Guides."
- For installations that may be exposed to mild chemical attack, specify epoxy grout and tile suitable for exposure. Consult product manufacturers; see also "Product Selection Guides."

Typical Weight of Tile Installation

- 12 pounds/square foot with ¾" poured gypsum. Add 2¼ pounds/square foot for each additional ¼" of poured gypsum.
- Does not include weight of substrate. See "Appendix B" for assumptions, included materials, and their individual weights.

Limitations

- Not for below-ground installation.
- Not for use over above-ground structural slabs and other floors subject to movement and/or deflection. See F200A.
- Not for use where moisture vapor transmission rate exceeds gypsum underlayment manufacturer's limitations.

Membrane Options

- Requires use of crack isolation (A118.12) and/or waterproof (A118.10) and/or uncoupling membrane.
- Membrane must meet ANSI A118.10 for limited water exposure areas (Res2; Com2). Consult underlayment manufacturer for requirements.
- Specifier shall indicate if complete waterproofing is required, including if/how membrane connects to drain assembly, if base flashing is required, and treatment at other termination points.
- Check with membrane manufacturers for suitability for applicable conditions, as not all membranes are suitable for steam, high temperature, and/or chemical exposure, exterior use, use over above-ground structural slabs, use over pourable underlayments, use with radiant heat, or use over concrete with excessive moisture vapor transmission and/or alkalinity. Membrane may also affect service rating.
- When glass tile is used, consult glass tile manufacturer for membrane options and recommendations.

Requirements

- Proper curing/drying of gypsum underlayments prior to application of tile is critical for proper performance. Consult the underlayment manufacturer for specific instructions.

Materials

- **Multiple options exist for membranes, mortars, grouts, and other materials and *must be clearly specified* to be included. If not specifically indicated, optional materials are not included and mortar/grout choice defaults to minimum performance specification indicated. Consider each system component and intended use to determine minimum requirements and to specify options.**
- Ceramic tile—ANSI A137.1.
- Glass tile, when used—ANSI A137.2; see also "Glass Tile Selection and Installation Guide," and consult tile manufacturer for service rating and environmental exposure classification recommendations. Not all glass tiles are suitable.
- Cementitious grout—ANSI A118.6 or better or ISO CG1 or better. When glass tile is used, specify grout designated by tile and grout manufacturers.
- Epoxy grout, when used—ANSI A118.3 or ISO RG.
- Cementitious bond coat:
  - When a crack isolation membrane is used—ANSI A118.4 or better or ISO C2S1 or better.

- When a waterproof membrane is used—ANSI A118.4 or better or ISO C2S1 or better unless ANSI A118.1 or ISO C1 is recommended by membrane manufacturer.
- When an uncoupling membrane is used, check with membrane manufacturer.
- When glass tile is used, specify mortar designated by tile and mortar manufacturers. Bond coat color will impact the final appearance of translucent glass tile. Specifier shall confirm bond coat color is acceptable.

- Crack isolation membrane, when used—ANSI A118.12.
- Waterproof membrane, when used—ANSI A118.10.
- Uncoupling membrane, when used—recommended by uncoupling membrane manufacturer; must achieve 50 psi or greater shear bond strength in 7 days per the test method in ANSI A118.12 Section 5.1.3.

Materials for Green/Sustainable Design

- See "Green Building Standards and Green Product Selection Guide" and consult manufacturers and suppliers for product sustainability and contribution to green building design.
- Consider specifying tile and installation materials that meet ANSI A138.1, the *American National Standard Specifications for Sustainable Ceramic Tiles, Glass Tiles, and Tile Installation Materials.*

Preparation by Other Trades

- Poured gypsum underlayment—minimum compressive strength of 2000 psi when tested per ASTM C472 modified and meeting performance requirements of ASTM C627 for the anticipated service level designated by the specifier or intended use.
- Consult the manufacturer for maximum moisture vapor transmission limitations prior to application of underlayment.
- Maximum allowable variation in the tile substrate—for tiles with all edges shorter than 15", maximum allowable variation is ¼" in 10' from the required plane, with no more than $^{1}/_{16}$" variation in 12" when measured from the high points in the surface. For tiles with at least one edge 15" in length, maximum allowable variation is ⅛" in 10' from the required plane, with no more than $^{1}/_{16}$" variation in 24" when measured from the high points in the surface.
- Slab to be well cured, dimensionally stable, and free of cracks, surface laitance, waxy or oily films, curing compounds, and other surface bond-inhibiting contaminants.
- Poured gypsum underlayment thickness and application varies, consult the manufacturer for specific recommendations.
- Surfaces must be prepared and primed according to the underlayment manufacturer's instructions.
- Poured gypsum underlayment—installed only by a manufacturer-approved applicator in accordance with manufacturer's recommendations.

Movement Joint (architect must specify type of joint and show location and details on drawings)

- Movement joints—mandatory in accordance with EJ171.
- When glass tile is used, adhere to more frequent placement recommendations within the ranges listed in EJ171.

Installation Specifications

- Tile—ANSI A108.5.
- Glass tile—manufacturer's directions.
- Cementitious grout—ANSI A108.10.
- Epoxy grout—ANSI A108.6.
- Crack isolation membrane—ANSI A108.17.
- Waterproof membrane—ANSI A108.13.
- Uncoupling membrane—manufacturer's directions.
- Movement Joints—EJ171 and ASTM C1193.

Notes

- Some manufacturers require special primers prior to the application of the membrane. Follow manufacturer's directions.
- When glass tile is used, see "Glass Tile Selection and Installation Guide," and consult manufacturer for recommendations and requirements.

F200A-21

- **Above-Ground Concrete**
- **Poured Gypsum Underlayment**
- **Bonded Membrane**
- **Ceramic Tile, Glass Tile**

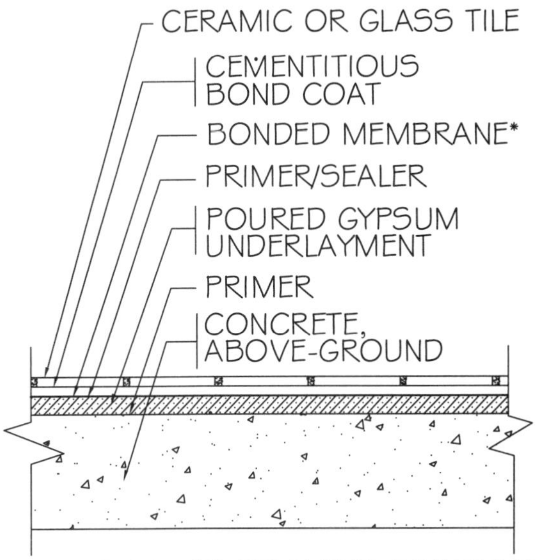

*USE OF A MEMBRANE IS REQUIRED. SEE MEMBRANE OPTIONS.

Recommended Uses

- For above-ground structural slabs with a poured gypsum underlayment.

Service Rating

- Moderate.
- When glass tile is used, service rating may be lower.

Environmental Exposure Classifications

- Res1; Com1.
- May be suitable for limited water exposure areas (Res2; Com2). See Membrane Options.
- For installations that may be exposed to staining, specify tile and grout suitable for exposure. Consult product manufacturers; see also "Product Selection Guides."
- For installations that may be exposed to mild chemical attack, specify epoxy grout and tile suitable for exposure. Consult product manufacturers; see also "Product Selection Guides."

Typical Weight of Tile Installation

- 12 pounds/square foot with ¾" poured gypsum. Add 2¼ pounds/square foot for each additional ¼" of poured gypsum.
- Does not include weight of substrate. See "Appendix B" for assumptions, included materials, and their individual weights.

Limitations

- Not for use where moisture vapor transmission rate exceeds gypsum underlayment manufacturer's limitations.
- Requires additional consideration by design professional to accommodate movement and/or deflection. Setting materials with improved bond strength and deformability are required.

Membrane Options

- Requires use of crack isolation (A118.12) and/or waterproof (A118.10) and/or uncoupling membrane.
- Membrane must meet ANSI A118.10 for limited water exposure areas (Res2; Com2). Consult underlayment manufacturer for requirements.
- Specifier shall indicate if complete waterproofing is required, including if/how membrane connects to drain assembly, if base flashing is required, and treatment at other termination points.
- Check with membrane manufacturers for suitability for applicable conditions, as not all membranes are suitable for steam, high temperature, and/or chemical exposure, exterior use, use over above-ground structural slabs, use over pourable underlayments, use with radiant heat, or use over concrete with excessive moisture vapor transmission and/or alkalinity. Membrane may also affect service rating.
- When glass tile is used, consult glass tile manufacturer for membrane options and recommendations.

Requirements

- Proper curing/drying of underlayments prior to application of tile is critical for proper performance. Consult the underlayment manufacturer for specific instructions.
- Above-ground installations are inherently more susceptible to vibration and deflection. Grout and mortar manufacturers to warrant product suitability.

Materials

- **Multiple options exist for membranes, mortars, grouts, and other materials and *must be clearly specified* to be included. If not specifically indicated, optional materials are not included and mortar/grout choice defaults to minimum performance specification indicated. Consider each system component and intended use to determine minimum requirements and to specify options.**
- Ceramic tile—ANSI A137.1.
- Glass tile, when used—ANSI A137.2; see also "Glass Tile Selection and Installation Guide," and consult tile manufacturer for service rating and environmental exposure classification recommendations. Not all glass tiles are suitable.
- Cementitious grout—ANSI A118.6 or better or ISO CG1 or better. When glass tile is used, specify grout

designated by tile and grout manufacturers.

- Epoxy grout, when used—ANSI A118.3 or ISO RG.
- Cementitious bond coat:
  - Must be recommended by manufacturer for above-ground use.
  - When a crack isolation membrane is used—ANSI A118.4 or better or ISO C2S1 or better.
  - When a waterproof membrane is used—ANSI A118.4 or better or ISO C2S1 or better unless ANSI A118.1 or ISO C1 is recommended by membrane manufacturer.
  - When an uncoupling membrane is used, check with membrane manufacturer.
  - When glass tile is used, specify mortar designated by tile and mortar manufacturers. Bond coat color will impact the final appearance of translucent glass tile. Specifier shall confirm bond coat color is acceptable.
- Crack isolation membrane, when used—ANSI A118.12.
- Waterproof membrane, when used—ANSI A118.10.
- Uncoupling membrane, when used—recommended by uncoupling membrane manufacturer; must achieve 50 psi or greater shear bond strength in 7 days per the test method in ANSI A118.12 Section 5.1.3.

Materials for Green/Sustainable Design

- See "Green Building Standards and Green Product Selection Guide" and consult manufacturers and suppliers for product sustainability and contribution to green building design.
- Consider specifying tile and installation materials that meet ANSI A138.1, the *American National Standard Specifications for Sustainable Ceramic Tiles, Glass Tiles, and Tile Installation Materials.*

Preparation by Other Trades

- Floor systems over which tile will be installed shall be in conformance with the IRC for residential applications, the IBC for commercial applications, or applicable building codes. Maximum allowable substrate deflection under live load not to exceed l/360. See also "Substrate Requirements."
- When concentrated loads (scissor lifts, pallet jacks, automobiles, forklifts, etc.) will be used on an above-ground tiled floor, the engineer and/or specifier shall specify a substrate to accommodate the concentrated loads. Owner/specifier is responsible for protecting the tilework from damage, including allowing sufficient time for installed materials to cure properly.
- Slab to be well cured, dimensionally stable, and free of cracks, surface laitance, waxy or oily films, curing compounds, and other surface bond-inhibiting contaminants.
- Consult the manufacturer for maximum moisture vapor transmission limitations prior to application of underlayment.
- Maximum allowable variation in the tile substrate—for tiles with all edges shorter than 15", maximum allowable variation is ¼" in 10' from the required plane, with no more than ¹⁄₁₆" variation in 12" when measured from the high points in the surface. For tiles with at least one edge 15" in length, maximum allowable variation is ⅛" in 10' from the required plane, with no more than ¹⁄₁₆" variation in 24" when measured from the high points in the surface.
- Poured gypsum underlayment thickness and application varies, consult the manufacturer for specific recommendations.
- Surfaces must be prepared and primed according to the underlayment manufacturer's instructions.
- Poured gypsum underlayment—minimum compressive strength of 2000 psi when tested per ASTM C472 modified and meeting performance requirements of ASTM C627 for the anticipated service level designated by the specifier or intended use.
- Poured gypsum underlayment—installed only by a manufacturer-approved applicator in accordance with manufacturer's recommendations.

Movement Joint (architect must specify type of joint and show location and details on drawings)

- Movement joints—mandatory according to EJ171. For above-ground installations, additional movement joints are required.
- When glass tile is used, adhere to more frequent placement recommendations within the ranges listed in EJ171.

Installation Specifications

- Tile—ANSI A108.5.
- Glass tile—manufacturer's directions.
- Cementitious grout—A108.10.
- Epoxy grout—ANSI A108.6.
- Crack isolation membrane—ANSI A108.17.
- Waterproof membrane—ANSI A108.13.
- Uncoupling membrane—manufacturer's directions.
- Movement Joints—EJ171 and ASTM C1193.

Notes

- Some manufacturers require special primers prior to the application of the membrane. Follow manufacturer's directions.
- Not all mortars, grouts, and membranes are suitable for above-ground use. Check manufacturer recommendations.
- F111 is preferred over above-ground structural slabs and other floors subject to movement and/or deflection.
- When glass tile is used, see "Glass Tile Selection and Installation Guide," and consult manufacturer for recommendations and requirements.

F205-21

- **On-Ground Concrete**
- **Flowable Hydraulic Cement Underlayment/ Cementitious Self-Leveling Underlayment**
- **Ceramic Tile, Glass Tile**

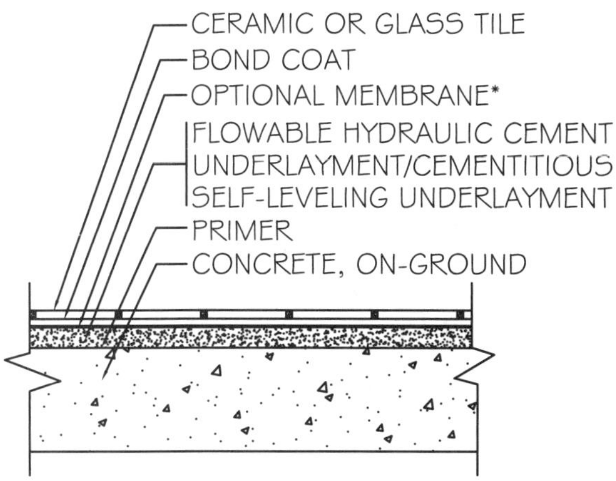

*USE OF A MEMBRANE IS OPTIONAL. SEE MEMBRANE OPTIONS.

Recommended Uses

- For areas where leveling, flattening, or raising the finished floor height is required.
- For slab-on-ground construction where no bending stresses occur and floor flatness is critical, such as when tiles with any edge longer than 15" are specified or where the finished tile assembly is required to achieve in-plane transitions (as may be necessary where accessibility is a concern).

Service Rating

- Extra heavy.
- When glass tile is used, service rating may be lower.

Environmental Exposure Classifications

- Res1; Com1.
- May be suitable for increased water exposure. See Membrane Options.
- For installations that may be exposed to staining, specify tile and grout suitable for exposure. Consult product manufacturers; see also "Product Selection Guides."
- For installations that may be exposed to mild chemical attack, specify epoxy grout and tile suitable for exposure. For greater resistance to chemical exposure, also specify an epoxy bonding material. Consult product manufacturers; see also "Product Selection Guides."

Typical Weight of Tile Installation

- 11 pounds/square foot with ½" flowable hydraulic cement underlayment/cementitious self-leveling underlayment. Add 2¾ pounds/square foot for each additional ¼" of underlayment.
- Does not include weight of substrate. See "Appendix B" for assumptions, included materials, and their individual weights.

Limitations

- Not for use where moisture vapor transmission rate exceeds underlayment manufacturer's limitations.
- Not for use over above-ground structural slabs and other floors subject to movement and/or deflection. See F205A.
- Not for below-ground installation.

Membrane Options

- A waterproof membrane (A118.10) may be required for areas exposed to moisture. Consult underlayment manufacturer for requirements.
- Specifier shall indicate if complete waterproofing is required, including if/how membrane connects to drain assembly, if base flashing is required, and treatment at other termination points.
- A crack isolation membrane may be specified to treat existing in-plane cracks (F125-partial) or for protection against future in-plane cracking (F-125 full). See F125 for more information.
- Check with membrane manufacturers for suitability for applicable conditions, as not all membranes are suitable for steam, high temperature, and/or chemical exposure, exterior use, use over above-ground structural slabs, use over pourable underlayments, use with radiant heat, or use over concrete with excessive moisture vapor transmission and/or alkalinity. Membrane may also affect service rating.
- When glass tile is used, consult glass tile manufacturer for membrane options and recommendations.

Requirements

- Consult the manufacturer for maximum moisture vapor transmission limitations prior to application of underlayment.
- Slab to be well cured, dimensionally stable, and free of cracks, surface laitance, waxy or oily films, curing compounds, and other surface bond-inhibiting contaminants.
- Flowable hydraulic cement underlayment/cementitious self-leveling underlayment thickness and application varies, consult the manufacturer for specific recommendations.
- When used to flatten the substrate and/or raise the finished tile assembly, the height of the finished tile assembly should be indicated by the specifier/design professional.
  - The field conditions must be evaluated by the installation professional prior to determining the amount of material necessary.

- Surfaces must be prepared and primed according to manufacturer's instructions.
- Proper curing/drying of underlayments prior to application of tile is critical for proper performance. Consult the underlayment manufacturer for specific instructions.

Materials

- **Multiple options exist for membranes, mortars, grouts, and other materials and *must be clearly specified* to be included. If not specifically indicated, optional materials are not included and mortar/grout choice defaults to minimum performance specification indicated. Consider each system component and intended use to determine minimum requirements and to specify options.**
- Ceramic tile—ANSI A137.1.
- Glass tile, when used—ANSI A137.2; see also "Glass Tile Selection and Installation Guide," and consult tile manufacturer for service rating and environmental exposure classification recommendations. Not all glass tiles are suitable.
- Cementitious grout—ANSI A118.6 or better or ISO CG1 or better. When glass tile is used, specify grout designated by tile and grout manufacturers.
- Epoxy grout, when used—ANSI A118.3 or ISO RG.
- Cementitious bond coat:
  - When a membrane is not used—ANSI A118.4 or better or ISO C2 or better.
  - When a crack isolation membrane is used—ANSI A118.4 or better or ISO C2S1 or better.
  - When a waterproof membrane is used—ANSI A118.4 or better or ISO C2S1 or better unless ANSI A118.1 or ISO C1 is recommended by membrane manufacturer.
  - When glass tile is used, specify mortar designated by tile and mortar manufacturers. Bond coat color will impact the final appearance of translucent glass tile. Specifier shall confirm bond coat color is acceptable.
- Epoxy bond coat, when used—ANSI A118.3 or ISO R1 or better.
  - When glass tile is used, specify epoxy bond coat designated by tile and bond coat manufacturers. Bond coat color will impact the final appearance of translucent glass tile. Specifier shall confirm bond coat color is acceptable.
- Crack isolation membrane, when used—ANSI A118.12.
- Waterproof membrane, when used—ANSI A118.10.
- Flowable hydraulic cement underlayment/cementitious self-leveling underlayment—meeting performance requirements of ASTM C627 for the anticipated service level designated by the specifier or intended use.

Materials for Green/Sustainable Design

- See "Green Building Standards and Green Product Selection Guide" and consult manufacturers and suppliers for product sustainability and contribution to green building design.
- Consider specifying tile and installation materials that meet ANSI A138.1, the *American National Standard Specifications for Sustainable Ceramic Tiles, Glass Tiles, and Tile Installation Materials.*

Preparation by Other Trades

- Slab to have a fine broom finish with no curing compounds used. When used, mechanical scarifying is necessary.
- Maximum allowable variation in the installation substrate (concrete)—¼" in 10' from the required plane.
- Maximum allowable variation in the tile substrate—for tiles with all edges shorter than 15", maximum allowable variation is ¼" in 10' from the required plane, with no more than ¹⁄₁₆" variation in 12" when measured from the high points in the surface. For tiles with at least one edge 15" in length, maximum allowable variation is ⅛" in 10' from the required plane, with no more than ¹⁄₁₆" variation in 24" when measured from the high points in the surface.

Movement Joint (architect must specify type of joint and show location and details on drawings)

- Movement joints—mandatory according to EJ171.
- When glass tile is used, adhere to more frequent placement recommendations within the ranges listed in EJ171.

Installation Specifications

- Tile—ANSI A108.5.
- Glass tile—manufacturer's directions.
- Cementitious grout—ANSI A108.10.
- Epoxy mortar/grout—ANSI A108.6.
- Flowable hydraulic cement underlayment/cementitious self-leveling underlayment—manufacturer's directions.
- Crack isolation membrane—ANSI A108.17.
- Waterproof membrane—ANSI A108.13.
- Movement Joints—EJ171 and ASTM C1193.

Notes

- Some manufacturers require special primers prior to the application of the setting materials or membrane. Follow manufacturer's directions.
- When glass tile is used, see "Glass Tile Selection and Installation Guide," and consult manufacturer for recommendations and requirements.

F205A-21

- **Above-Ground Concrete**
- **Flowable Hydraulic Cement Underlayment/ Cementitious Self-Leveling Underlayment**
- **Ceramic Tile, Glass Tile**

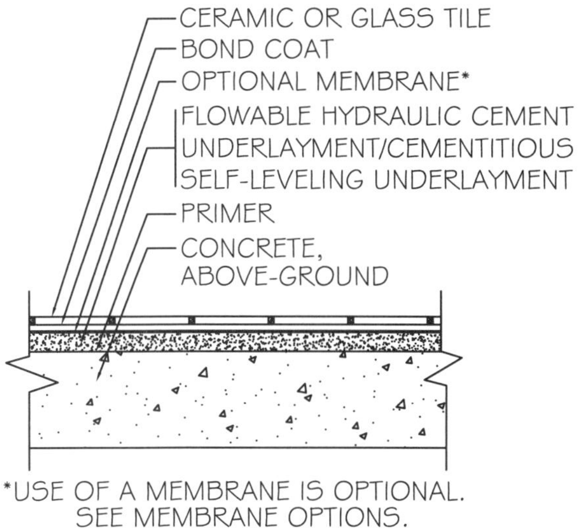

Recommended Uses

- For areas where leveling, flattening, or raising the finished floor height is required.
- For above-ground structural slabs and other floors subject to movement and/or deflection where floor flatness is critical, such as when tiles with any edge longer than 15" are specified or where the finished tile assembly is required to achieve in-plane transitions (as may be necessary where accessibility is a concern).

Service Rating

- Extra heavy.
- When glass tile is used, service rating may be lower.

Environmental Exposure Classifications

- Res1; Com1.
- May be suitable for increased water exposure. See Membrane Options.
- For installations that may be exposed to staining, specify tile and grout suitable for exposure. Consult product manufacturers; see also "Product Selection Guides."
- For installations that may be exposed to mild chemical attack, specify epoxy grout and tile suitable for exposure. For greater resistance to chemical exposure, also specify an epoxy bonding material. Consult product manufacturers; see also "Product Selection Guides."

Typical Weight of Tile Installation

- 11 pounds/square foot with ½" flowable hydraulic cement underlayment/cementitious self-leveling underlayment. Add 2¾ pounds/square foot for each additional ¼" of underlayment.
- Does not include weight of substrate. See "Appendix B" for assumptions, included materials, and their individual weights.

Limitations

- Not for use where moisture vapor transmission rate exceeds underlayment manufacturer's limitations.
- Requires additional consideration by design professional to accommodate movement and/or deflection. Setting materials with improved bond strength and deformability are required.

Membrane Options

- A waterproof membrane (A118.10) may be required for areas exposed to moisture. Consult underlayment manufacturer for requirements.
- Specifier shall indicate if complete waterproofing is required, including if/how membrane connects to drain assembly, if base flashing is required, and treatment at other termination points.
- A crack isolation membrane may be specified to treat existing in-plane cracks (F125-partial) or for protection against future in-plane cracking (F-125 full). See F125 for more information.
- Check with membrane manufacturers for suitability for applicable conditions, as not all membranes are suitable for steam, high temperature, and/or chemical exposure, exterior use, use over above-ground structural slabs, use over pourable underlayments, use with radiant heat, or use over concrete with excessive moisture vapor transmission and/or alkalinity. Membrane may also affect service rating.
- When glass tile is used, consult glass tile manufacturer for membrane options and recommendations.

Requirements

- Consult the manufacturer for maximum moisture vapor transmission limitations prior to application of underlayment.
- Slab to be well cured, dimensionally stable, and free of cracks, surface laitance, waxy or oily films, curing compounds, and other surface bond-inhibiting contaminants.
- Flowable hydraulic cement underlayment/cementitious self-leveling underlayment thickness and application varies, consult the manufacturer for specific recommendations.
- When used to flatten the substrate and/or raise the finished tile assembly, the height of the finished tile assembly should be indicated by the specifier/design professional.
  - The field conditions must be evaluated by the installation professional prior to determining the amount of material necessary.
- Surfaces must be prepared and primed according to manufacturer's instructions.
- Proper curing/drying of underlayments prior to application of tile is critical for proper performance. Consult the underlayment manufacturer for specific instructions.
- Above-ground installations are inherently more susceptible to vibration and deflection. Grout and mortar manufacturers to warrant product suitability.

Materials

- **Multiple options exist for membranes, mortars, grouts, and other materials and *must be clearly specified* to be included. If not specifically indicated, optional materials are not included and mortar/grout choice defaults to minimum performance specification indicated. Consider each system component and intended use to determine minimum requirements and to specify options.**
- Ceramic tile—ANSI A137.1.
- Glass tile, when used—ANSI A137.2; see also "Glass Tile Selection and Installation Guide," and consult tile manufacturer for service rating and environmental exposure classification recommendations. Not all glass tiles are suitable.
- Cementitious grout—ANSI A118.6 or better or ISO CG1 or better. When glass tile is used, specify grout designated by tile and grout manufacturers.
- Epoxy grout, when used—ANSI A118.3 or ISO RG.
- Cementitious bond coat:
  - Must be recommended by manufacturer for above-ground use.
  - When a membrane is not used—ANSI A118.15 or better or ISO C2S1 or better.
  - When a crack isolation membrane is used—ANSI A118.4 or better or ISO C2S1 or better.
  - When a waterproof membrane is used—ANSI A118.4 or better or ISO C2S1 or better unless ANSI A118.1 or ISO C1 is recommended by membrane manufacturer.
  - When glass tile is used, specify mortar designated by tile and mortar manufacturers. Bond coat color will impact the final appearance of translucent glass tile. Specifier shall confirm bond coat color is acceptable.
- Epoxy bond coat, when used—ANSI A118.3 or ISO R1 or better. Must also be recommended by manufacturer for above-ground use.
  - When glass tile is used, specify epoxy bond coat designated by tile and bond coat manufacturers. Bond coat color will impact the final appearance of translucent glass tile. Specifier shall confirm bond coat color is acceptable.
- Crack isolation membrane, when used—ANSI A118.12.
- Waterproof membrane, when used—ANSI A118.10.
- Flowable hydraulic cement underlayment/cementitious self-leveling underlayment—meeting performance requirements of ASTM C627 for the anticipated service level designated by the specifier or intended use.

Materials for Green/Sustainable Design

- See "Green Building Standards and Green Product Selection Guide" and consult manufacturers and suppliers for product sustainability and contribution to green building design.
- Consider specifying tile and installation materials that meet ANSI A138.1, the *American National Standard Specifications for Sustainable Ceramic Tiles, Glass Tiles, and Tile Installation Materials.*

Preparation by Other Trades

- Floor systems over which tile will be installed shall be in conformance with the IRC for residential applications, the IBC for commercial applications, or applicable building codes. Maximum allowable substrate deflection under live load not to exceed l/360. See also "Substrate Requirements."
- When concentrated loads (scissor lifts, pallet jacks, automobiles, forklifts, etc.) will be used on an above-ground tiled floor, the engineer and/or specifier shall specify a substrate to accommodate the concentrated loads. Owner/specifier is responsible for protecting the tilework from damage, including allowing sufficient time for installed materials to cure properly.
- Slab to have a fine broom finish with no curing compounds used. When used, mechanical scarifying is necessary.
- Maximum allowable variation in the installation substrate (concrete)—¼" in 10' from the required plane.
- Maximum allowable variation in the tile substrate—for tiles with all edges shorter than 15,"maximum allowable variation is ¼" in 10' from the required plane, with no more than 1/16" variation in 12" when measured from the high points in the surface. For tiles with at least one edge 15" in length, maximum allowable variation is ⅛" in 10' from the required plane, with no more than 1/16" variation in 24" when measured from the high points in the surface.

Movement Joint (architect must specify type of joint and show location and details on drawings)

- Movement joints—mandatory according to EJ171. For above-ground installations, additional movement joints are required.
- When glass tile is used, adhere to more frequent placement recommendations within the ranges listed in EJ171.

Installation Specifications

- Tile—ANSI A108.5.
- Glass tile—manufacturer's directions.
- Cementitious grout—ANSI A108.10.
- Epoxy mortar/grout—ANSI A108.6.
- Flowable hydraulic cement underlayment/cementitious self-leveling underlayment—manufacturer's directions.
- Crack isolation membrane—ANSI A108.17.
- Waterproof membrane—ANSI A108.13.
- Movement Joints—EJ171 and ASTM C1193.

Notes

- F111 is preferred over above-ground structural slabs and other floors subject to movement and/or deflection.
- Some manufacturers require special primers prior to the application of the setting materials or membrane. Follow manufacturer's directions.
- Not all mortars, grouts, and membranes are suitable for above-ground use. Check manufacturer recommendations.
- When glass tile is used, see "Glass Tile Selection and Installation Guide," and consult manufacturer for recommendations and requirements.

F121-21

- **On-Ground or Above-Ground Concrete**
- **Unbonded Mortar Bed**
- **Waterproof Membrane**
- **Ceramic Tile, Glass Tile**

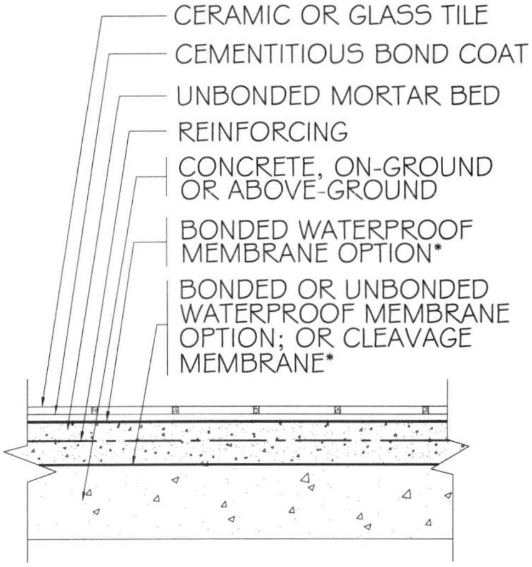

*USE OF A MEMBRANE IS REQUIRED. SEE MEMBRANE OPTIONS.

Recommended Uses

- For above-ground structural slabs and for slab-on-ground construction where a waterproof floor is required.
- For areas where leveling, flattening, or contouring of finish floor height is required, such as commercial kitchens and gang showers, and a waterproof floor is required.
- For areas where floor flatness is critical, such as when tiles with any edge longer than 15" are specified or where the finished tile assembly is required to achieve in-plane transitions (as may be necessary where accessibility is a concern), and a waterproof floor is required.
- For concrete substrates that are cracked or present bonding issues and a waterproof floor is required.

Service Rating

- Heavy. Consult membrane manufacturer to verify.
- When glass tile is used, service rating may be lower.

Environmental Exposure Classifications

- Res1, 2, 3, 5; Com1, 2, 3, 5.
- May be suitable for exterior applications in areas not subject to freeze/thaw cycling when appropriate precautions are taken, including expansion joint placement, proper slope, waterproofing, and material selection.
- For installations that may be exposed to staining, specify tile and grout suitable for exposure. Consult product manufacturers; see also "Product Selection Guides."
- For installations that may be exposed to mild chemical attack, specify epoxy grout and tile suitable for exposure. Consult product manufacturers; see also "Product Selection Guides."
- For installations that may be exposed to staining and/or chemical attack, see also F114, F132, and F134.

Typical Weight of Tile Installation

- 21 pounds/square foot with 1¼" mortar bed. Add 3 pounds/square foot for each additional ¼" of mortar bed.
- Does not include weight of substrate. See "Appendix B" for assumptions, included materials, and their individual weights.

Limitations

- Not for severe chemical exposure. See Environmental Exposure Classifications in this method.
- Above-ground structural slabs and other floors subject to movement and/or deflection—requires additional consideration by design professional to accommodate movement and/or deflection. Setting materials with improved bond strength and deformability are required.

Membrane Options

- A waterproof membrane and a cleavage membrane are required.
- If a bonded waterproof membrane (ANSI A118.10) is used, it may be placed above or below the mortar bed. If a shower pan membrane (ASTM D4551 or D4068) is used, it must be placed below the mortar bed.
- The waterproof membrane can also function as the cleavage membrane if loose laid (not bonded to concrete) below the mortar bed. Check with membrane manufacturer if membrane may be loose laid.
- If waterproof membrane is bonded to the top of the mortar bed, a cleavage membrane must be placed below the mortar bed.
- Specifier shall indicate if complete waterproofing is required, including if/how membrane connects to drain assembly, if base flashing is required, and treatment at other termination points.
- Check with membrane manufacturer for suitability for applicable conditions, as not all membranes are suitable for steam, high temperature, and/or chemical exposure, exterior use, use over above-ground structural slabs, use over pourable underlayments, use with radiant heat, or use over concrete with excessive moisture vapor transmission and/or alkalinity. Membrane may also affect service rating.
- When glass tile is used, consult glass tile manufacturer for membrane options and recommendations.

Requirements

- Mortar bed thickness—1¼" minimum to 2" maximum.
- Mortar beds in excess of 2" thick shall be detailed by the architect.
- Reinforcing mesh mandatory.
- Above-ground—above-ground installations are inherently more susceptible to vibration and deflection. Grout and mortar manufacturers to warrant product suitability.

Materials

- **Multiple options exist for membranes, mortars, grouts, and other materials and *must be clearly specified* to be included. If not specifically indicated, optional materials are not included and mortar/grout choice defaults to minimum performance**

specification indicated. Consider each system component and intended use to determine minimum requirements and to specify options.

- Ceramic tile—ANSI A137.1.
- Glass tile, when used—ANSI A137.2; see also "Glass Tile Selection and Installation Guide," and consult tile manufacturer for service rating and environmental exposure classification recommendations. Not all glass tiles are suitable.
- Cementitious grout—ANSI A118.6 or better or ISO CG1 or better. When glass tile is used, specify grout designated by tile and grout manufacturers.
- Epoxy grout, when used—ANSI A118.3 or ISO RG.
- Cementitious bond coat (on-ground)—portland cement paste on a mortar bed that is still workable. For a cured mortar bed, follow recommendations below to select appropriate bonding mortar:
  - When waterproof membrane is placed below mortar bed—ANSI A118.1 or better or ISO C1 or better.
  - When waterproof membrane is placed above mortar bed—ANSI A118.4 or better or ISO C2S1 or better unless ANSI A118.1 or ISO C1 is recommended by membrane manufacturer.
  - When porcelain tile is used—ANSI A118.4 or better or ISO C2 or better.
  - When glass tile is used, specify mortar designated by tile and mortar manufacturers. Bond coat color will impact the final appearance of translucent glass tile. Specifier shall confirm bond coat color is acceptable.
- Cementitious bond coat (above-ground)—portland cement paste on a mortar bed that is still workable. For a cured mortar bed, follow recommendations below to select appropriate bonding mortar:
  - Must be recommended by manufacturer for above-ground use.
  - When waterproof membrane is placed below mortar bed—ANSI A118.15 or better or ISO C2S1 or better.
  - When waterproof membrane is placed above mortar bed—ANSI A118.4 or better or ISO C2S1 or better unless ANSI A118.1 or ISO C1 is recommended by membrane manufacturer.
  - When glass tile is used, specify mortar designated by tile and mortar manufacturers. Bond coat color will impact the final appearance of translucent glass tile. Specifier shall confirm bond coat color is acceptable.
- Mortar bed, reinforcing, and cleavage membrane—ANSI A108.1A.
- Waterproof membrane, when used—ANSI A118.10.
- Shower pan membrane, when used—ASTM D4551 or D4068.

Materials for Green/Sustainable Design

- See "Green Building Standards and Green Product Selection Guide" and consult manufacturers and suppliers for product sustainability and contribution to green building design.
- Consider specifying tile and installation materials that meet ANSI A138.1, the *American National Standard Specifications for Sustainable Ceramic Tiles, Glass Tiles, and Tile Installation Materials.*

Preparation by Other Trades

- Floor systems over which tile will be installed shall be in conformance with the IRC for residential applications, the IBC for commercial applications, or applicable building codes. Maximum allowable substrate deflection under live load not to exceed l/360. See also "Substrate Requirements."
- When concentrated loads (scissor lifts, pallet jacks, automobiles, forklifts, etc.) will be used on an above-ground tiled floor, the engineer and/or specifier shall specify a substrate to accommodate the concentrated loads. Owner/specifier is responsible for protecting the tilework from damage, including allowing sufficient time for installed materials to cure properly.
- Maximum allowable variation in the installation substrate (concrete)—¼" in 10' from the required plane.
- Where slab is depressed to accommodate mortar bed thickness, slab depression to be accurate, with steel trowel finish.
- Slope subfloor ¼" per foot to drain.

Movement Joint (architect must specify type of joint and show location and details on drawings)

- Movement joints—mandatory according to EJ171. For above-ground installations, additional movement joints are required.
- When glass tile is used, adhere to more frequent placement recommendations within the ranges listed in EJ171.

Installation Specifications

- Tile—ANSI A108.1A, .1B, or .1C. A108.1B required if membrane to be used on top of the mortar bed.
- Glass tile—ANSI A108.14, .15, .16, or manufacturer's directions.
- Cementitious grout—ANSI A108.10.
- Epoxy grout—ANSI A108.6.
- Waterproof membrane—ANSI A108.13 or manufacturer's directions. Comply with applicable plumbing and building codes.
- Shower pan membrane—manufacturer's directions. Comply with applicable plumbing and building codes.
- Movement Joints—EJ171 and ASTM C1193.

Notes

- For hydronic tubing on top of the slab when using a mortar bed, see RH117.
- See shower receptor details B414, B421, and B422 for drain connections.
- For a waterproof application, but with a bonded mortar bed, see F112.
- Above-ground—not all mortars, grouts, and membranes are suitable for above-ground use. Check manufacturer recommendations.
- When glass tile is used, see "Glass Tile Selection and Installation Guide," and consult manufacturer for recommendations and requirements.

F122-21

- **On-Ground Concrete**
- **Waterproof Membrane**
- **Ceramic Tile, Glass Tile**

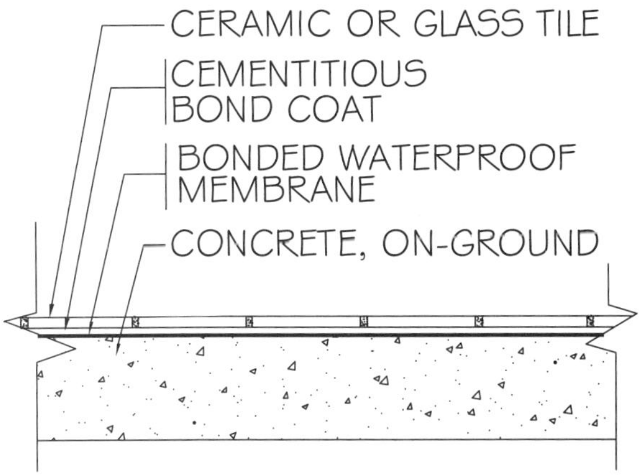

Recommended Uses

- For slab-on-ground construction where no bending stresses occur, where thin-bed installation of tile is desired, and a waterproof floor is required.

Service Rating

- Moderate. Consult membrane manufacturer to verify.
- When glass tile is used, service rating may be lower.

Environmental Exposure Classifications

- Res1, 2, 3, 5; Com1, 2, 3, 5.
- May be suitable for exterior applications in areas not subject to freeze/thaw cycling when appropriate precautions are taken, including expansion joint placement, proper slope, waterproofing, and material selection.
- For installations that may be exposed to staining, specify tile and grout suitable for exposure. Consult product manufacturers; see also "Product Selection Guides."
- For installations that may be exposed to mild chemical attack, specify epoxy grout and tile suitable for exposure. Consult product manufacturers; see also "Product Selection Guides."
- For installations that may be exposed to staining and/or chemical attack, see also F115, F131, and F133.

Typical Weight of Tile Installation

- 6 pounds/square foot.
- Does not include weight of substrate. See "Appendix B" for assumptions, included materials, and their individual weights.

Limitations

- Not for severe chemical exposure. See Environmental Exposure Classifications in this method.
- Not for use over above-ground structural slabs and other floors subject to movement and/or deflection. See F122A.

Requirements

- Slab to be well cured, dimensionally stable, and free of cracks, surface laitance, waxy or oily films, curing compounds, and other surface bond-inhibiting contaminants.
- Check with membrane manufacturers for suitability for applicable conditions, as not all membranes are suitable for steam, high temperature, and/or chemical exposure, exterior use, use over above-ground structural slabs, use over pourable underlayments, use with radiant heat, or use over concrete with excessive moisture vapor transmission and/or alkalinity. Membrane may also affect service rating.
- Specifier shall indicate if complete waterproofing is required, including if/how membrane connects to drain assembly, if base flashing is required, and treatment at other termination points.
- When glass tile is used, consult glass tile manufacturer for membrane options and recommendations.

Materials

- **Multiple options exist for membranes, mortars, grouts, and other materials and *must be clearly specified* to be included. If not specifically indicated, optional materials are not included and mortar/grout choice defaults to minimum performance specification indicated. Consider each system component and intended use to determine minimum requirements and to specify options.**
- Ceramic tile—ANSI A137.1.
- Glass tile, when used—ANSI A137.2; see also "Glass Tile Selection and Installation Guide," and consult tile manufacturer for service rating and environmental exposure classification recommendations. Not all glass tiles are suitable.
- Cementitious grout—ANSI A118.6 or better or ISO CG1 or better. When glass tile is used, specify grout designated by tile and grout manufacturers.
- Epoxy grout, when used—ANSI A118.3 or ISO RG.
- Cementitious bond coat:
  - ANSI A118.4 or better or ISO C2S1 or better unless ANSI A118.1 or ISO C1 is recommended by membrane manufacturer.
  - When glass tile is used, specify mortar designated by tile and mortar manufacturers. Bond coat color will impact the final appearance of translucent glass tile. Specifier shall confirm bond coat color is acceptable.
- Waterproof membrane—ANSI A118.10.

Materials for Green/Sustainable Design

- See "Green Building Standards and Green Product Selection Guide" and consult manufacturers and suppliers for product sustainability and contribution to green building design.
- Consider specifying tile and installation materials that meet ANSI A138.1, the *American National Standard Specifications for Sustainable Ceramic Tiles, Glass Tiles, and Tile Installation Materials.*

Preparation by Other Trades

- Slab to have steel trowel finish free of curing compounds. When used, mechanical scarification is necessary.
- Maximum allowable variation in the tile substrate—for tiles with all edges shorter than 15", maximum allowable variation is ¼" in 10' from the required plane, with no more than 1⁄16" variation in 12" when measured from the high points in the surface. For tiles with at least one edge 15" in length, maximum allowable variation is ⅛" in 10' from the required plane, with no more than 1⁄16" variation in 24" when measured from the high points in the surface.
- Slope subfloor ¼" per foot to drain.

Movement Joint (architect must specify type of joint and show location and details on drawings)

- Movement joints—mandatory according to EJ171.
- When glass tile is used, adhere to more frequent placement recommendations within the ranges listed in EJ171.

Installation Specifications

- Tile—ANSI A108.5.
- Glass tile—manufacturer's directions.
- Cementitious grout—ANSI A108.10.
- Epoxy grout—ANSI A108.6.
- Waterproof membrane—ANSI A108.13 or manufacturer's directions. Comply with applicable plumbing and building codes.
- Movement Joints—EJ171 and ASTM C1193.

Notes

- See shower receptor details B421 and B422 for drain connections.
- When glass tile is used, see "Glass Tile Selection and Installation Guide," and consult manufacturer for recommendations and requirements.

F122A-21

- **Above-Ground Concrete**
- **Waterproof Membrane**
- **Ceramic Tile, Glass Tile**

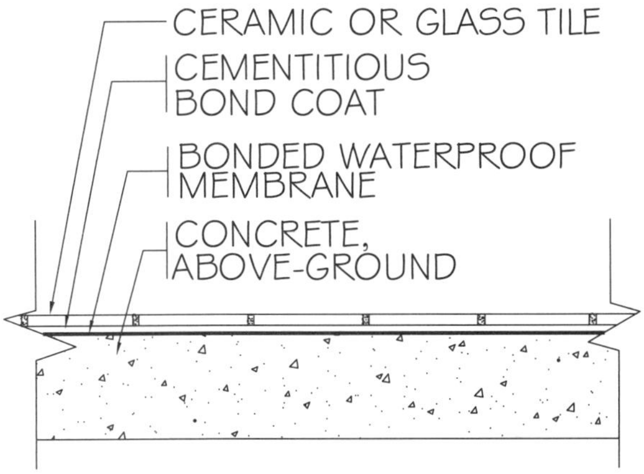

Recommended Uses

- For above-ground structural slabs and other floors subject to movement and/or deflection where thin-bed installation of tile is desired and a waterproof floor is required.

Service Rating

- Moderate. Consult membrane manufacturer to verify.
- When glass tile is used, service rating may be lower.

Environmental Exposure Classifications

- Res1, 2, 3, 5; Com1, 2, 3, 5.
- May be suitable for exterior applications in areas not subject to freeze/thaw cycling when appropriate precautions are taken, including expansion joint placement, proper slope, waterproofing and material selection.
- For installations that may be exposed to staining, specify tile and grout suitable for exposure. Consult product manufacturers; see also "Product Selection Guides."
- For installations that may be exposed to mild chemical attack, specify epoxy grout and tile suitable for exposure. Consult product manufacturers; see also "Product Selection Guides."
- For installations that may be exposed to staining and/or chemical attack, see also F115A, F131, and F133.

Typical Weight of Tile Installation

- 6 pounds/square foot.
- Does not include weight of substrate. See "Appendix B" for assumptions, included materials, and their individual weights.

Limitations

- Requires additional consideration by design professional to accommodate movement and/or deflection. Setting materials with improved bond strength and deformability are required.
- Not for severe chemical exposure. See Environmental Exposure Classifications in this method.

Requirements

- Slab to be well cured, dimensionally stable, and free of cracks, surface laitance, waxy or oily films, curing compounds, and other surface bond-inhibiting contaminants.
- Check with membrane manufacturers for suitability for applicable conditions, as not all membranes are suitable for steam, high temperature, and/or chemical exposure, exterior use, use over above-ground structural slabs, use over pourable underlayments, use with radiant heat, or use over concrete with excessive moisture vapor transmission and/or alkalinity. Membrane may also affect service rating.
- Specifier shall indicate if complete waterproofing is required, including if/how membrane connects to drain assembly, if base flashing is required, and treatment at other termination points.
- When glass tile is used, consult glass tile manufacturer for membrane options and recommendations.
- Above-ground installations are inherently more susceptible to vibration and deflection. Grout and mortar manufacturers to warrant product suitability.

Materials

- **Multiple options exist for membranes, mortars, grouts, and other materials and *must be clearly specified* to be included. If not specifically indicated, optional materials are not included and mortar/grout choice defaults to minimum performance specification indicated. Consider each system component and intended use to determine minimum requirements and to specify options.**
- Ceramic tile—ANSI A137.1.
- Glass tile, when used—ANSI A137.2; see also "Glass Tile Selection and Installation Guide," and consult tile manufacturer for service rating and environmental exposure classification recommendations. Not all glass tiles are suitable.
- Cementitious grout—ANSI A118.6 or better or ISO CG1 or better. When glass tile is used, specify grout designated by tile and grout manufacturers.
- Epoxy grout, when used—ANSI A118.3 or ISO RG.
- Cementitious bond coat:
  - Must be recommended by manufacturer for above-ground use.
  - ANSI A118.4 or better or ISO C2S1 or better

unless ANSI A118.1 or ISO C1 is recommended by membrane manufacturer.

- When glass tile is used, specify mortar designated by tile and mortar manufacturers. Bond coat color will impact the final appearance of translucent glass tile. Specifier shall confirm bond coat color is acceptable.

- Waterproof membrane—ANSI A118.10.

Materials for Green/Sustainable Design

- See Green Building Standards and Green Product Selection Guide and consult manufacturers and suppliers for product sustainability and contribution to green building design.
- Consider specifying tile and installation materials that meet ANSI A138.1, the *American National Standard Specifications for Sustainable Ceramic Tiles, Glass Tiles, and Tile Installation Materials.*

Preparation by Other Trades

- Floor systems over which tile will be installed shall be in conformance with the IRC for residential applications, the IBC for commercial applications, or applicable building codes. Maximum allowable substrate deflection under live load not to exceed l/360. See also "Substrate Requirements."
- When concentrated loads (scissor lifts, pallet jacks, automobiles, forklifts, etc.) will be used on an above-ground tiled floor, the engineer and/or specifier shall specify a substrate to accommodate the concentrated loads. Owner/specifier is responsible for protecting the tilework from damage, including allowing sufficient time for installed materials to cure properly.
- Slab to have steel trowel finish free of curing compounds. When used, mechanical scarification is necessary.
- Maximum allowable variation in the tile substrate—for tiles with all edges shorter than 15," maximum allowable variation is ¼" in 10' from the required plane, with no more than $\frac{1}{16}$" variation in 12" when measured from the high points in the surface. For tiles with at least one edge 15" in length, maximum allowable variation is ⅛" in 10' from the required plane, with no more than $\frac{1}{16}$" variation in 24" when measured from the high points in the surface.
- Slope subfloor ¼" per foot to drain.

Movement Joint (architect must specify type of joint and show location and details on drawings)

- Movement joints—mandatory according to EJ171. For above-ground installations, additional movement joints are required.
- When glass tile is used, adhere to more frequent placement recommendations within the ranges listed in EJ171.

Installation Specifications

- Tile—ANSI A108.5.
- Glass tile—manufacturer's directions.
- Cementitious grout—ANSI A108.10.
- Epoxy grout—ANSI A108.6.
- Waterproof membrane—ANSI A108.13 or manufacturer's directions. Comply with applicable plumbing and building codes.
- Movement Joints—EJ171 and ASTM C1193.

Notes

- F111 is preferred over above-ground structural slabs and other floors subject to movement and/or deflection.
- Not all mortars, grouts, and membranes are suitable for above-ground installations. Check manufacturer recommendations.
- See shower receptor details B421 and B422 for drain connections.
- When glass tile is used, see "Glass Tile Selection and Installation Guide," and consult manufacturer for recommendations and requirements.

F125-Partial-21

- **On- or Above-Ground Concrete**
- **Crack Isolation Membrane (Partial Coverage)**
- **Ceramic Tile**

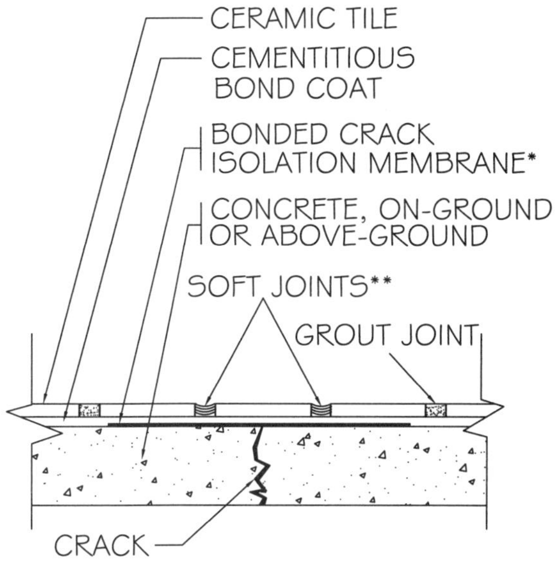

*CONSULT MANUFACTURER FOR MEMBRANE WIDTH

**ONE OR TWO SOFT JOINTS MAY BE REQUIRED (CONSULT MANUFACTURER)

F125-Full-21

- **On- or Above-Ground Concrete**
- **Crack Isolation Membrane (Full Coverage)**
- **Ceramic Tile**

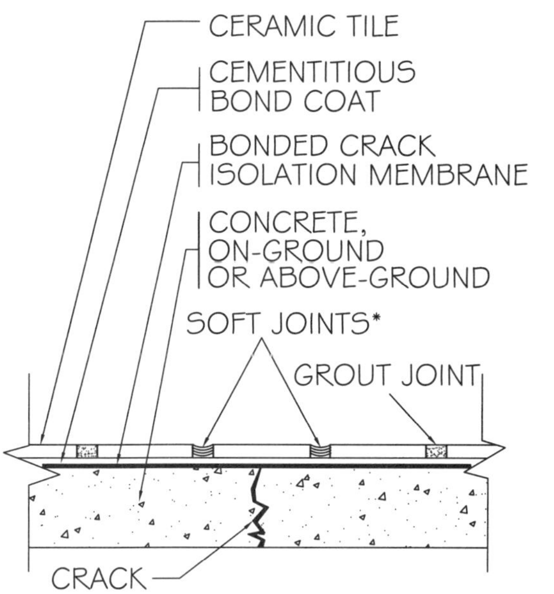

*ONE OR TWO SOFT JOINTS MAY BE REQUIRED (CONSULT MANUFACTURER)

Recommended Uses

- For use in conjunction with other concrete substrate installation methods.
- For slab-on-ground construction and above-ground structural slabs where thin-bed installation of tile is desired and crack isolation is required to treat existing in-plane cracks (F125-Partial).
- For slab-on-ground construction and above-ground structural slabs where thin-bed installation of tile is desired and protection against future in-plane cracking is desired, full-coverage is recommended and must be additionally specified (F125-Full).

Service Rating

- Different crack isolation membranes will produce different service ratings when tested per ASTM C627. Consult membrane manufacturer to verify service rating.

Environmental Exposure Classifications

- Refer to appropriate installation method.

Typical Weight of Tile Installation

- 6 pounds/square foot.
- Does not include weight of substrate. See "Appendix B" for assumptions, included materials, and their individual weights.

Limitations

- Above-ground structural slabs and other floors subject to movement and/or deflection—requires additional consideration by design professional to accommodate movement and/or deflection. Setting materials with improved bond strength and deformability are required.
- May not always be 100% effective in preventing reflective cracking.
- For lateral (in-plane) movement only.
- Standard performance membranes can accommodate up to 1/16" in-plane movement. Consult manufacturer for membrane performance rating.
- High performance membranes can accommodate up to 1/8" in-plane movement. Consult manufacturer for membrane performance rating.
- Not for cracking associated with deflection of above-ground installations.

Requirements

- For existing cracks, soft joints may be required adjacent to cracks per manufacturer's instructions.
- Specifier to identify areas of coverage (partial or full). Unless full coverage is clearly specified, installing contractor's responsibility is limited to partial coverage.
- Slab to be well cured, dimensionally stable, and free of cracks, surface laitance, waxy or oily films, curing compounds, and other surface bond-inhibiting contaminants.

- Check with membrane manufacturers for suitability for applicable conditions, as not all membranes are suitable for steam, high temperature, and/or chemical exposure, exterior use, use over above-ground structural slabs, use over pourable underlayments, use with radiant heat, or use over concrete with excessive moisture vapor transmission and/or alkalinity. Membrane may also affect service rating.
- Above-ground—above-ground installations are inherently more susceptible to vibration and deflection. Grout and mortar manufacturers to warrant product suitability.

Materials

- **Multiple options exist for membranes, mortars, grouts, and other materials and *must be clearly specified* to be included. If not specifically indicated, optional materials are not included and mortar/grout choice defaults to minimum performance specification indicated. Consider each system component and intended use to determine minimum requirements and to specify options. Unless full coverage of crack isolation membrane is clearly specified, installing contractor's responsibility is limited to partial coverage.**
- Refer to appropriate installation method.
- Sealant—ASTM C920.
- Crack isolation membrane—ANSI A118.12.

Materials for Green/Sustainable Design

- See "Green Building Standards and Green Product Selection Guide" and consult manufacturers and suppliers for product sustainability and contribution to green building design.
- Consider specifying tile and installation materials that meet ANSI A138.1, the *American National Standard Specifications for Sustainable Ceramic Tiles, Glass Tiles, and Tile Installation Materials.*

Preparation by Other Trades

- Refer to appropriate installation method.
- Floor systems over which tile will be installed shall be in conformance with the IRC for residential applications, the IBC for commercial applications, or applicable building codes. Maximum allowable substrate deflection under live load not to exceed l/360. See also "Substrate Requirements."
- When concentrated loads (scissor lifts, pallet jacks, automobiles, forklifts, etc.) will be used on an above-ground tiled floor, the engineer and/or specifier shall specify a substrate to accommodate the concentrated loads. Owner/specifier is responsible for protecting the tilework from damage, including allowing sufficient time for installed materials to cure properly.
- Maximum allowable variation in the tile substrate—for tiles with all edges shorter than 15", maximum allowable variation is ¼" in 10' from the required plane, with no more than $^1/_{16}$" variation in 12" when measured from the high points in the surface. For tiles with at least one edge 15" in length, maximum allowable variation is ⅛" in 10' from the required plane, with no more than $^1/_{16}$" variation in 24" when measured from the high points in the surface.
- Slab to have steel trowel and fine broom finish free of curing compounds. When used, mechanical scarifying is necessary.
- Slope, when required, to be in subfloor.

Installation Specifications

- Refer to appropriate installation method.
- Crack isolation membrane—ANSI A108.17.
- Movement Joints—EJ171 and ASTM C1193.

Movement Joint (architect must specify type of joint and show location and details on drawings)

- Movement joints—mandatory according to EJ171. For above-ground installations, additional movement joints are required.

Notes

- Tiles that do not meet the breaking strength requirement of ANSI A137.1, such as Saltillo tiles, can be subject to breakage from strain induced by membrane movement or deflection over softer, compressible membrane material. Consult membrane and tile manufacturers for suitability. Testing per ANSI A118.12, modified by conducting the test using the actual tile can be performed to evaluate its performance under test as compared with the tile meeting ANSI A137.1 specified by the test.

F128-21

- **On-Ground Young Concrete**
- **Uncoupling Membrane**
- **Ceramic Tile**

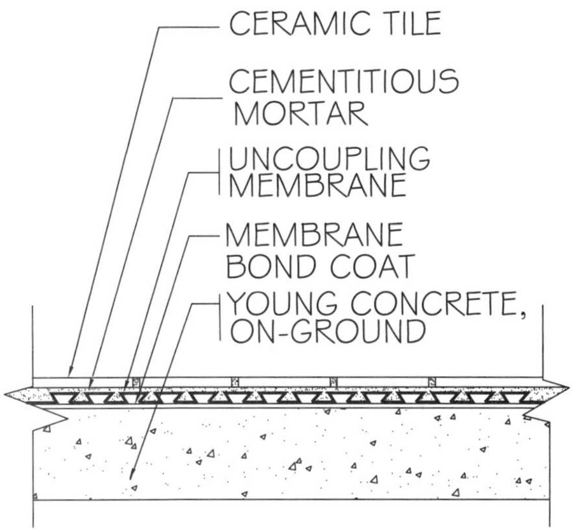

Recommended Uses

- For tiling over young concrete on slab-on-ground construction where no bending stresses occur.

Service Rating

- Light commercial with 2" × 2" and larger tiles.
- Extra heavy with 12" × 12" and larger porcelain tiles.

Environmental Exposure Classifications

- Res1, 2, 3, 4, 5; Com1, 2, 3, 4, 5.
- For installations that may be exposed to staining, specify tile and grout suitable for exposure. Consult product manufacturers; see also "Product Selection Guides."
- For installations that may be exposed to mild chemical attack, specify epoxy grout and tile suitable for exposure. Consult product manufacturers; see also "Product Selection Guides."

Typical Weight of Tile Installation

- 6 pounds/square foot.
- Does not include weight of substrate. See "Appendix B" for assumptions, included materials, and their individual weights.

Limitations

- Not for use over above-ground structural slabs and other floors subject to movement and/or deflection.

Requirements

- Slab to be free of waxy or oily films and curing compounds.
- No standing water on slab during uncoupling membrane installation.
- Concrete must be cured sufficiently to support tile installation traffic as determined by the project design professional, construction manager, or general contractor.
- Check with membrane manufacturers for suitability for applicable conditions, as not all membranes are suitable for steam, high temperature, and/or chemical exposure, exterior use, use over above-ground structural slabs, use over pourable underlayments, use with radiant heat, or use over concrete with excessive moisture vapor transmission and/or alkalinity. Membrane may also affect service rating.

Materials

- **Multiple options exist for membranes, mortars, grouts, and other materials and *must be clearly specified* to be included. If not specifically indicated, optional materials are not included and mortar/grout choice defaults to minimum performance specification indicated. Consider each system component and intended use to determine minimum requirements and to specify options.**
- Ceramic tile—ANSI A137.1.
- Cementitious grout—ANSI A118.6 or better or ISO CG1 or better.
- Epoxy grout, when used—ANSI A118.3 or ISO RG.
- Cementitious mortar—as recommended by membrane manufacturer.
- Uncoupling membrane—recommended by manufacturer for use on young concrete; must achieve 50 psi or greater shear bond strength in 7 days per the test method in ANSI A118.12 Section 5.1.3.

Materials for Green/Sustainable Design

- See "Green Building Standards and Green Product Selection Guide" and consult manufacturers and suppliers for product sustainability and contribution to green building design.
- Consider specifying tile and installation materials that meet ANSI A138.1, the *American National Standard Specifications for Sustainable Ceramic Tiles, Glass Tiles, and Tile Installation Materials.*

Preparation by Other Trades

- Slab to have steel trowel and fine broom finish free of curing compounds. When used, mechanical scarifying is necessary.
- Maximum allowable variation in the installation substrate (concrete)—for tiles with all edges shorter than 15", maximum allowable variation is ¼" in 10' from the required plane, with no more than 1/16" variation in 12" when measured from the high points in the surface. For tiles with at least one edge 15" in length, maximum allowable variation is ⅛" in 10' from the required plane, with no more than 1/16" variation in 24" when measured from the high points in the surface.
- Slope, when required, to be in subfloor.

Movement Joint (architect must specify type of joint and show location and details on drawings)

- Movement joints—mandatory in accordance with EJ171.

Installation Specifications

- Tile—ANSI A108.5.
- Cementitious grout—ANSI A108.10.
- Epoxy grout—ANSI A108.6.
- Uncoupling membrane—follow manufacturer's directions.
- Movement Joints—EJ171 and ASTM C1193.

Notes

- Keep traffic off finished tile floors until fully cured.

F131-21

- **On-Ground or Above-Ground Concrete**
- **Epoxy Mortar**
- **Ceramic Tile**
- **Epoxy Grout**

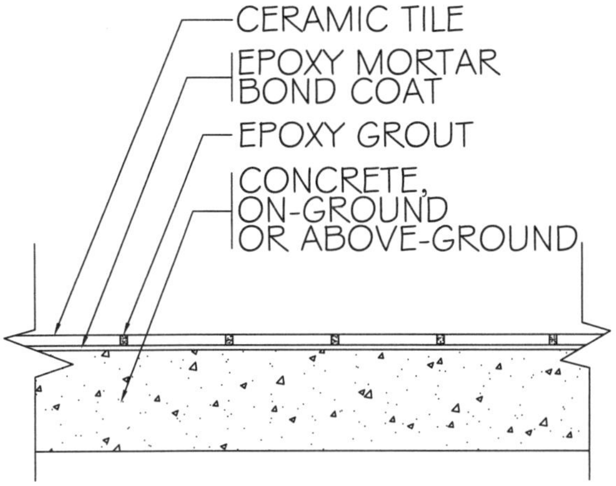

Recommended Uses

- For above-ground structural slabs and for slab-on-ground construction where thin-bed installation of tile is desired and resistance to moderate chemical exposure and severe cleaning methods is desired, for example, in commercial kitchens, dairies, breweries, food processing plants, etc.

Service Rating

- Extra heavy.

Environmental Exposure Classifications

- Res1, 2, 3, 4, 5; Com1, 2, 3, 4, 5.
- For increased chemical and heat resistance, see also F133.

Typical Weight of Tile Installation

- 6 pounds/square foot.
- Does not include weight of substrate. See "Appendix B" for assumptions, included materials, and their individual weights.

Limitations

- Not for severe chemical exposure or extreme heat. See Environmental Exposure Classifications in this method.
- Extreme heat or improper steam cleaning will soften epoxy grouts and wash them out of joints. High-temperature-resistant formulas are available.
- Above-ground structural slabs and other floors subject to movement and/or deflection—requires additional consideration by design professional to accommodate movement and/or deflection. Setting materials with improved bond strength and deformability are required.

Requirements

- Specify tile and grout suitable for stain and/or chemical exposure. Consult product manufacturers. See also "Product Selection Guides."
- Subfloor carefully finished to proper elevation and slope. See Notes.
- Slab to be well cured, dimensionally stable, and free of cracks, surface laitance, waxy or oily films, curing compounds, and other surface bond-inhibiting contaminants.
- Joints must be clean and completely filled with epoxy. Partial filling with sand is unacceptable.
- Consult grout manufacturer for special precautions when chemical exposure is severe or at prolonged temperatures above 140°F.
- Above-ground—above-ground installations are inherently more susceptible to vibration and deflection. Grout and mortar manufacturers to warrant product suitability.

Materials

- **Multiple options exist for membranes, mortars, grouts, and other materials and *must be clearly specified* to be included. If not specifically indicated, optional materials are not included and mortar/grout choice defaults to minimum performance specification indicated. Consider each system component and intended use to determine minimum requirements and to specify options.**
- Ceramic tile—ANSI A137.1.
- Epoxy grout—ANSI A118.3 or ISO RG. For tilework exposed to prolonged high temperatures, use high-temperature, chemical resistant epoxy grout certified by manufacturer for intended use.
- Epoxy mortar—ANSI A118.3 or ISO R1 or better. For above-ground use, must also be recommended by manufacturer. For tilework exposed to prolonged high temperatures, use high-temperature, chemical-resistant epoxy mortar certified by manufacturer for intended use.

Materials for Green/Sustainable Design

- See "Green Building Standards and Green Product Selection Guide" and consult manufacturers and suppliers for product sustainability and contribution to green building design.
- Consider specifying tile and installation materials that meet ANSI A138.1, the *American National Standard Specifications for Sustainable Ceramic Tiles, Glass Tiles, and Tile Installation Materials.*

Preparation by Other Trades

- Floor systems over which tile will be installed shall be in conformance with the IRC for residential applications, the IBC for commercial applications, or applicable building codes. Maximum allowable substrate deflection under live load not to exceed l/360. See also "Substrate Requirements."
- When concentrated loads (scissor lifts, pallet jacks, automobiles, forklifts, etc.) will be used on an above-ground tiled floor, the engineer and/or specifier shall specify a substrate to accommodate the concentrated loads. Owner/specifier is responsible for protecting the tilework from damage, including allowing sufficient time for installed materials to cure properly.
- Slab to have steel trowel and fine broom finish free of curing compounds. When used, mechanical scarifying is necessary.
- Slope, when required, to be in subfloor.
- Maximum allowable variation in the tile substrate—for tiles with all edges shorter than 15", maximum allowable variation is ¼" in 10' from the required plane, with no more than ¹⁄₁₆" variation in 12" when measured from the high points in the surface. For tiles with at least one edge 15" in length, maximum allowable variation is ⅛" in 10' from the required plane, with no more than ¹⁄₁₆" variation in 24" when measured from the high points in the surface.

Movement Joint (architect must specify type of joint and show location and details on drawings)

- Movement joints—mandatory according to EJ171. For above-ground installations, additional movement joints are required.

Installation Specifications

- Tile and grout—ANSI A108.6.
- Movement Joints—EJ171 and ASTM C1193.

Notes

- As tile size increases, there is less tolerance for variation in the subfloor from the required plane. Epoxy adhesive thickness must be thin and uniform; therefore, subfloor flattening may be required.
- Above-ground—not all mortars and grouts are suitable for above-ground use. Check manufacturer recommendations.
- F111 is preferred over above-ground structural slabs and other floors subject to movement and/or deflection.

F132-21

- **On-Ground or Above-Ground Concrete**
- **Bonded Mortar Bed**
- **Epoxy Mortar**
- **Ceramic Tile**
- **Epoxy Grout**

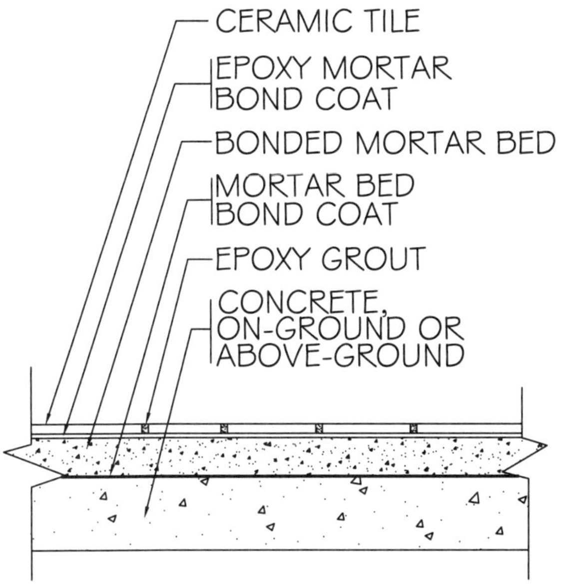

Recommended Uses

- For slab-on-ground construction and for above-ground structural slabs where tile installation is 100 square feet or less, where resistance to moderate chemical exposure and severe cleaning methods is desired, for example commercial kitchens, dairies, breweries, food processing plants, etc.
- For areas where leveling, flattening, or contouring of finish floor height is required, such as commercial kitchens and gang showers, and resistance to moderate chemical exposure and severe cleaning methods is desired.
- For areas where floor flatness is critical, such as when tiles with any edge longer than 15" are specified or where the finished tile assembly is required to achieve in-plane transitions (as may be necessary where accessibility is a concern) and resistance to moderate chemical exposure and severe cleaning methods is desired.

Service Rating

- Extra heavy.

Environmental Exposure Classifications

- Res1, 2, 3, 4, 5; Com1, 2, 3, 4, 5.
- For increased chemical and heat resistance, see also F134.

Typical Weight of Tile Installation

- 15 pounds/square foot with ¾" mortar bed. Add 3 pounds/square foot for each additional ¼" of mortar bed.
- Does not include weight of substrate. See "Appendix B" for assumptions, included materials, and their individual weights.

Limitations

- Not for severe chemical exposures or extreme heat. See Environmental Exposure Classifications in this method.
- Not for use over above-ground structural slabs or other floors subject to movement and/or deflection except where tile installation is 100 square feet or less. See Notes.
- Above-ground structural slabs and other floors subject to movement and/or deflection—requires additional consideration by design professional to accommodate movement and/or deflection. Setting materials with improved bond strength and deformability are required.
- Extreme heat or improper steam cleaning will soften epoxy grouts and wash them out of joints. High-temperature-resistant formulas are available.

Requirements

- Specify tile and grout suitable for stain and/or chemical exposure. Consult product manufacturers. See also "Product Selection Guides."
- Slab to be well cured, dimensionally stable, and free of cracks, surface laitance, waxy or oily films, curing compounds, and other surface bond-inhibiting contaminants.
- Joints must be clean and completely filled with epoxy. Partial filling with sand is unacceptable.
- Consult grout manufacturer for special precautions when chemical exposure is severe or at prolonged temperatures above 140°F.
- Mortar bed thickness—¾" minimum to 2" maximum.
- Mortar beds in excess of 2" thick shall be detailed by architect.
- Above-ground—above-ground installations are inherently more susceptible to vibration and deflection. Grout and mortar manufacturers to warrant product suitability.

Materials

- **Multiple options exist for membranes, mortars, grouts, and other materials and *must be clearly specified* to be included. If not specifically indicated, optional materials are not included and mortar/grout choice defaults to minimum performance specification indicated. Consider each system component and intended use to determine minimum requirements and to specify options.**
- Ceramic tile—ANSI A137.1.

- Epoxy grout—ANSI A118.3 or ISO RG. For tilework exposed to prolonged high temperatures, use high-temperature, chemical-resistant epoxy grout certified by manufacturer for intended use.
- Epoxy mortar—ANSI A118.3 or ISO R1 or better. For above-ground use, must also be recommended by manufacturer. For tilework exposed to prolonged high temperatures, use high-temperature, chemical-resistant epoxy mortar certified by manufacturer for intended use.
- Mortar bed—ANSI A108.1A.
- Mortar bed bond coat—portland cement slurry.

Materials for Green/Sustainable Design

- See "Green Building Standards and Green Product Selection Guide" and consult manufacturers and suppliers for product sustainability and contribution to green building design.
- Consider specifying tile and installation materials that meet ANSI A138.1, the *American National Standard Specifications for Sustainable Ceramic Tiles, Glass Tiles, and Tile Installation Materials.*

Preparation by Other Trades

- Floor systems over which tile will be installed shall be in conformance with the IRC for residential applications, the IBC for commercial applications, or applicable building codes. Maximum allowable substrate deflection under live load not to exceed l/360. See also "Substrate Requirements."
- When concentrated loads (scissor lifts, pallet jacks, automobiles, forklifts, etc.) will be used on an above-ground tiled floor, the engineer and/or specifier shall specify a substrate to accommodate the concentrated loads. Owner/specifier is responsible for protecting the tilework from damage, including allowing sufficient time for installed materials to cure properly.
- Slab to have steel trowel and fine broom finish free of curing compounds. When used, mechanical scarifying is necessary.
- Slope, when required, to be in subfloor.
- Maximum allowable variation in the installation substrate (concrete)—¼" in 10' from the required plane.

Movement Joint (architect must specify type of joint and show location and details on drawings)

- Movement joints—mandatory according to EJ171. For above-ground installations, additional movement joints are required.

Installation Specifications

- Tile and grout—ANSI A108.6.
- Mortar bed—ANSI A108.1B.
- Movement Joints—EJ171 and ASTM C1193.

Notes

- An unbonded mortar bed, per F114, may be specified and is preferred over above-ground structural slabs and other floors subject to movement and deflection. Modify by specifying epoxy mortar and grout as shown here for chemical resistance.
- Above-ground—not all mortars and grouts are suitable for above-ground use. Check manufacturer recommendations.

F133-21

- **On-Ground or Above-Ground Concrete**
- **Furan Resin Mortar**
- **Ceramic Tile (Waxed)**
- **Furan Grout**

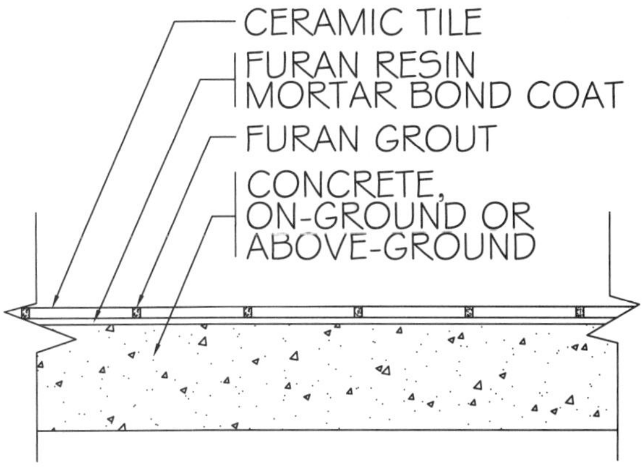

Recommended Uses

- For above-ground structural slabs and for slab-on-ground construction where thin-bed installation of tile is desired and moderate resistance to chemical exposure and severe cleaning methods is preferred, for example, in commercial kitchens, dairies, breweries, food processing plants, etc.

Service Rating

- Extra heavy.

Environmental Exposure Classifications

- Res1, 2, 3, 4, 5; Com1, 2, 3, 4, 5.
- For increased chemical and heat resistance, see also F134.

Typical Weight of Tile Installation

- 5 pounds/square foot.
- Does not include weight of substrate. See "Appendix B" for assumptions, included materials, and their individual weights.

Limitations

- Above-ground structural slabs and other floors subject to movement and/or deflection—requires additional consideration by design professional to accommodate movement and/or deflection. Setting materials with improved bond strength and deformability are required.

Requirements

- Specify tile and grout suitable for stain and/or chemical exposure. Consult product manufacturers. See also "Product Selection Guides."
- Subfloor carefully finished to proper elevation. See Notes.
- Slab to be well cured, dimensionally stable, and free of cracks, surface laitance, waxy or oily films, curing compounds, and other surface bond-inhibiting contaminants.
- Tile surface must be waxed before grouting with furan resin.
- Joints must be clean and completely filled with furan. Partial filling with sand is unacceptable.
- Above-ground—above-ground installations are inherently more susceptible to vibration and deflection. Grout and mortar manufacturers to warrant product suitability.

Materials

- **Multiple options exist for membranes, mortars, grouts, and other materials and *must be clearly specified* to be included. If not specifically indicated, optional materials are not included and mortar/grout choice defaults to minimum performance specification indicated. Consider each system component and intended use to determine minimum requirements and to specify options.**
- Ceramic tile—ANSI A137.1.
- Furan grout—ANSI A118.5.
- Furan mortar bond coat—ANSI A118.5. For above-ground use, must also be recommended by manufacturer.

Materials for Green/Sustainable Design

- See "Green Building Standards and Green Product Selection Guide" and consult manufacturers and suppliers for product sustainability and contribution to green building design.
- Consider specifying tile and installation materials that meet ANSI A138.1, the *American National Standard Specifications for Sustainable Ceramic Tiles, Glass Tiles, and Tile Installation Materials.*

Preparation by Other Trades

- Floor systems over which tile will be installed shall be in conformance with the IRC for residential applications, the IBC for commercial applications, or applicable building codes. Maximum allowable substrate deflection under live load not to exceed l/360. See also "Substrate Requirements."
- When concentrated loads (scissor lifts, pallet jacks, automobiles, forklifts, etc.) will be used on an above-ground tiled floor, the engineer and/or specifier shall specify a substrate to accommodate the concentrated loads. Owner/specifier is responsible for protecting the tilework from damage, including allowing sufficient time for installed materials to cure properly.
- Slab to have steel trowel and fine broom finish free of curing compounds. When used, mechanical scarifying is necessary.
- Slope, when required, to be in subfloor.

- Maximum allowable variation in the tile substrate—for tiles with all edges shorter than 15", maximum allowable variation is 1/4" in 10' from the required plane, with no more than 1/16" variation in 12" when measured from the high points in the surface. For tiles with at least one edge 15" in length, maximum allowable variation is 1/8" in 10' from the required plane, with no more than 1/16" variation in 24" when measured from the high points in the surface.

Movement Joint (architect must specify type of joint and show location and details on drawings)

- Movement joints—mandatory according to EJ171. For above-ground installations, additional movement joints are required.

Installation Specifications

- Tile and grout—ANSI A108.8.
- Movement Joints—EJ171 and ASTM C1193.

Notes

- As the tile size increases, there is less tolerance for variation in the subfloor from the required plane. Furan adhesive thickness must be a thin and uniform bond coat; therefore, subfloor flattening may be required.
- Above-ground—not all mortars and grouts are suitable for above-ground use. Check manufacturer recommendations.

F134-21

- **Concrete**
- **Bonded Mortar Bed**
- **Chemical-Resistant Membrane**
- **Chemical-Resistant Mortar**
- **Packing House Tile**
- **Chemical-Resistant Grout**

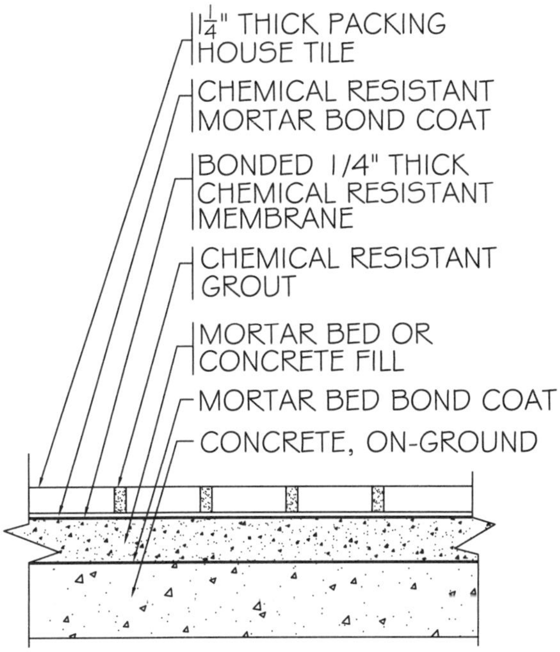

Recommended Use

- For setting 1¼"-thick packing house tiles in areas of continuous or severe chemical exposure, where special protection against leakage or damage to concrete subfloor is required.

Service Rating

- Extra heavy.

Environmental Exposure Classifications

- Res1, 2, 3, 4, 5; Com1, 2, 3, 4, 5.

Typical Weight of Tile Installation

- 26 pounds/square foot with ¾" mortar bed. Add 3 pounds/square foot for each additional ¼" of mortar bed.
- Does not include weight of substrate. See "Appendix B" for assumptions, included materials, and their individual weights.

Requirements

- Requires chemical-resistant membrane.
- Structurally sound subfloor, carefully finished to proper elevation and slope.

Materials

- **Multiple options exist for membranes, mortars, grouts, and other materials and *must be clearly specified* to be included. If not specifically indicated, optional materials are not included and mortar/grout choice defaults to minimum performance specification indicated. Consider each system component and intended use to determine minimum requirements and to specify options.**
- Ceramic tile—ANSI A137.1 and recommended by manufacturer.
- Epoxy or furan grout—ANSI A118.3, A118.5, or ISO RG.
- Epoxy or furan mortar bond coat—ANSI A118.3, A118.5, or ISO R1 or better.
- Chemical-resistant membrane—recommended by membrane manufacturer for intended use.
- Mortar bed—ANSI A108.1A.
- Mortar bed bond coat—portland cement slurry.

Materials for Green/Sustainable Design

- See "Green Building Standards and Green Product Selection Guide" and consult manufacturers and suppliers for product sustainability and contribution to green building design.
- Consider specifying tile and installation materials that meet ANSI A138.1, the *American National Standard Specifications for Sustainable Ceramic Tiles, Glass Tiles, and Tile Installation Materials.*

Preparation by Other Trades

- Slab to have steel trowel and fine broom finish with no curing compounds used. When used, mechanical scarifying is necessary.
- Maximum variation in the concrete fill or mortar bed ¼" in 10' from the required plane.
- Chemical-resistant membrane may be installed by other trades separate from tilework.

Movement Joint (architect must specify type of joint and show location and details on drawings)

- None in subfloor beneath continuous membrane unless special installation method is designed to accommodate them.

Installation Specifications

- Tile—ANSI A108.5.
- Epoxy grout and/or bond coat—ANSI A108.6.
- Furan grout and/or bond coat—ANSI A108.8.
- Mortar bed—ANSI A108.B.
- Membrane—manufacturer's directions.
- Movement Joints—EJ171 and ASTM C1193.

Notes

- The entire floor system is usually installed by a specialty ceramic tile flooring contractor and should be so specified.
- Joints must be clean and completely filled with epoxy or furan. Partial filling with sand or mortar is unacceptable.
- Consult resin manufacturer for special precautions when chemical exposure is severe or at high temperature.

F135-21

- **On-Ground or Above-Ground Concrete**
- **½" Cork Underlayment**
- **Ceramic Tile**
- **Epoxy Grout**

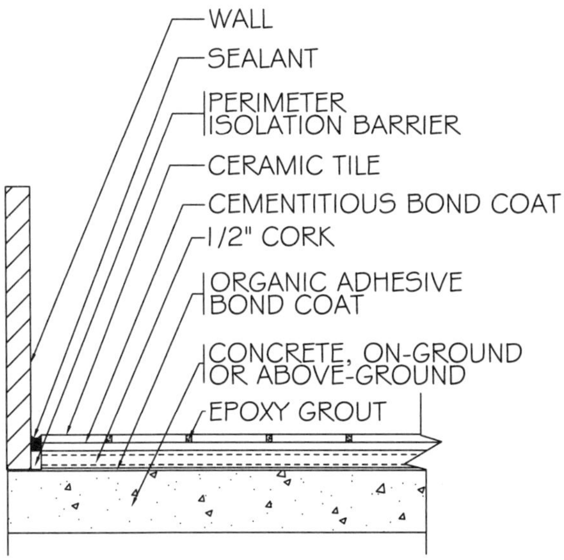

Recommended Uses

- For on-ground and above-ground structural slabs in residential dry areas where a cork underlayment is desired.

Service Rating

- Residential.

Environmental Exposure Classifications

- Res1.

Typical Weight of Tile Installation

- 6 pounds/square foot.
- Does not include weight of substrate. See "Appendix B" for assumptions, included materials, and their individual weights.

Limitations

- For 8" × 8" and larger floor tiles only.
- Requires additional consideration by design professional to accommodate movement and/or deflection. Setting materials with improved bond strength and deformability are required.

Requirements

- Minimum 6"-thick concrete subfloor.
- Slab to be well cured, dimensionally stable, and free of cracks, surface laitance, waxy or oily films, curing compounds, and other surface bond-inhibiting contaminants.
- Above-ground installations are inherently more susceptible to vibration and deflection. Grout and mortar manufacturers to warrant product suitability.

Materials

- **Multiple options exist for membranes, mortars, grouts, and other materials and *must be clearly specified* to be included. If not specifically indicated, optional materials are not included and mortar/grout choice defaults to minimum performance specification indicated. Consider each system component and intended use to determine minimum requirements and to specify options.**
- Ceramic tile—ANSI A137.1.
- Epoxy grout—ANSI A118.3 or ISO RG.
- Cementitious bond coat (on ground):
  - ANSI A118.4 or better or ISO C2S1 or better.
- Cementitious bond coat (above-ground):
  - Must be recommended by manufacturer for above-ground use.
  - ANSI A118.15 or better or ISO C2S1 or better.
- Cork—½"-thick (+/- .010") underlayment sheets consisting of a mix of pure cork granules combined with a polyurethane binder, having a density of between 11.8 and 13.6 pounds/cubic foot.
- Organic adhesive bond coat (for cork underlayment)—ANSI A136.1 Type I or ISO D2.

Materials for Green/Sustainable Design

- See "Green Building Standards and Green Product Selection Guide" and consult manufacturers and suppliers for product sustainability and contribution to green building design.
- Consider specifying tile and installation materials that meet ANSI A138.1, the *American National Standard Specifications for Sustainable Ceramic Tiles, Glass Tiles, and Tile Installation Materials.*

Preparation by Other Trades

- Floor systems over which tile will be installed shall be in conformance with the IRC for residential applications, the IBC for commercial applications, or applicable building codes. Maximum allowable substrate deflection under live load not to exceed l/360. See also "Substrate Requirements."
- When concentrated loads (scissor lifts, pallet jacks, automobiles, forklifts, etc.) will be used on an above-ground tiled floor, the engineer and/or specifier shall specify a substrate to accommodate the concentrated loads. Owner/specifier is responsible for protecting the tilework from damage, including allowing sufficient time for installed materials to cure properly.
- Slab—steel trowel or comparable level of finish.

- Slope, when required, to be in the subfloor.
- Maximum allowable variation in the installation substrate (concrete)—for tiles with all edges shorter than 15", maximum allowable variation is ¼" in 10' from the required plane, with no more than ¹⁄₁₆" variation in 12" when measured from the high points in the surface. For tiles with at least one edge 15" in length, maximum allowable variation is ⅛" in 10' from the required plane, with no more than ¹⁄₁₆" variation in 24" when measured from the high points in the surface.

Movement Joint (architect must specify type of joint and show location and details on drawings)

- Movement joints—mandatory according to EJ171. For above-ground installations, additional movement joints are required.

Installation Specifications

- Tile—ANSI A108.5.
- Epoxy grout—ANSI A108.6.
- Cork—manufacturer's directions.
- Movement Joints—EJ171 and ASTM C1193.

Notes

- Not all mortars and grouts are suitable for above-ground installations. Check manufacturer recommendations.
- F111 is preferred over above-ground structural slabs and other floors subject to movement and/or deflection.

F136-21

- **Above-Ground Concrete**
- **Bonded Sound Reduction Membrane**
- **Ceramic Tile**

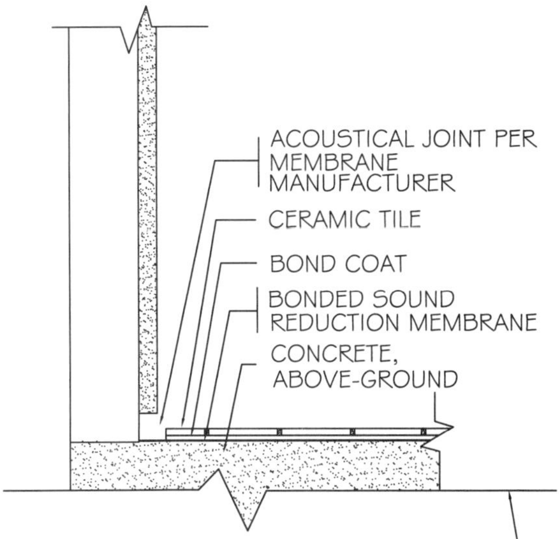

Recommended Uses

- For above-ground structural slabs and other floors subject to movement and/or deflection where thin-bed installation of tile is desired and reduction of sound transference to area below the tile installation is desired.

Service Rating

- Different sound reduction membranes will produce different service ratings when tested per ASTM C627. Consult membrane manufacturer to verify service rating with tile and installation materials that will be used.

Environmental Exposure Classifications

- Res1.
- May be suitable for Res2, Com1, and/or Com2. Consult membrane manufacturer.
- For installations that may be exposed to staining, specify tile and grout suitable for exposure. Consult product manufacturers; see also "Product Selection Guides."
- For installations that may be exposed to mild chemical attack, specify epoxy grout and tile suitable for exposure. Consult product manufacturers; see also "Product Selection Guides."

Typical Weight of Tile Installation

- 6 pounds/square foot.
- Does not include weight of substrate. See "Appendix B" for assumptions, included materials, and their individual weights.

Limitations

- Requires additional consideration by design professional to accommodate movement and/or deflection. Setting materials with improved bond strength and deformability are required.

Requirements

- Slab to be well cured, dimensionally stable, and free of cracks, waxy or oily films, and curing compounds.
- Check with membrane manufacturers for suitability for applicable conditions, as not all membranes are suitable for steam, high temperature, and/or chemical exposure, exterior use, use over above-ground structural slabs, use over pourable underlayments, use with radiant heat, or use over concrete with excessive moisture vapor transmission and/or alkalinity. Membrane may also affect service rating.
- Acoustical joints are required at the perimeter of the floor area being treated and at all penetrations and retaining surfaces to minimize flanking. Proper installation at the wall trim/flooring interface is critical. Observe membrane manufacturer's acoustical joint or perimeter isolation joint details.
- Above-ground installations are inherently more susceptible to vibration and deflection. Grout and mortar manufacturers to warrant product suitability.
- Sound reduction membranes meeting ANSI A118.13 must meet a minimum laboratory Δ IIC rating of 10 when tested per ASTM E2179. When installed in the field, the membrane may yield higher or lower sound reduction. Field installation variables such as slab thickness; slab density and porosity; ceiling heights and assemblies; floor penetrations; tile type and size; adhesive type; grout type and joint width; room size; atmospheric pressure; framing member type and arrangement; fastener locations; and other factors affect field impact (FIIC) sound transmission measurements. Design professional is responsible for ensuring system components are specified and designed to achieve the desired FIIC sound reduction. A suspended ceiling should be specified for greatest sound reduction.
- Sound reduction membranes that meet ANSI A118.13 must meet a minimum laboratory Δ IIC rating of 10 when tested per ASTM E2179. A higher Δ IIC rating may be required to produce a code compliant installation.

Materials

- **Multiple options exist for membranes, mortars, grouts, and other materials and *must be clearly specified* to be included. If not specifically indicated, optional materials are not included and mortar/grout choice defaults to minimum performance specification indicated. Consider each system component and intended use to determine minimum requirements and to specify options.**
- Ceramic tile—ANSI A137.1.

- Cementitious grout—ANSI A118.6 or better or ISO CG1 or better.
- Epoxy grout, when used—ANSI A118.3 or ISO RG.
- Bond coat (for tile)—use manufacturer-recommended material.
- Bond coat (for membrane)—use manufacturer-recommended material.
- Sound reduction membrane—ANSI A118.13.
- Sealant, when used—ASTM C919 and recommended by membrane manufacturer.

Materials for Green/Sustainable Design

- See "Green Building Standards and Green Product Selection Guide" and consult manufacturers and suppliers for product sustainability and contribution to green building design.
- Consider specifying tile and installation materials that meet ANSI A138.1, the *American National Standard Specifications for Sustainable Ceramic Tiles, Glass Tiles, and Tile Installation Materials.*

Preparation by Other Trades

- Floor systems over which tile will be installed shall be in conformance with the IRC for residential applications, the IBC for commercial applications, or applicable building codes. Maximum allowable substrate deflection under live load not to exceed l/360. See also "Substrate Requirements."
- When concentrated loads (scissor lifts, pallet jacks, automobiles, forklifts, etc.) will be used on an above-ground tiled floor, the engineer and/or specifier shall specify a substrate to accommodate the concentrated loads. Owner/specifier is responsible for protecting the tilework from damage, including allowing sufficient time for installed materials to cure properly.
- Slab to have steel trowel finish free of curing compounds. When used, mechanical scarification is necessary.
- Maximum allowable variation in the tile substrate—for tiles with all edges shorter than 15", maximum allowable variation is ¼" in 10' from the required plane, with no more than ¹⁄₁₆" variation in 12" when measured from the high points in the surface. For tiles with at least one edge 15" in length, maximum allowable variation is ⅛" in 10' from the required plane, with no more than ¹⁄₁₆" variation in 24" when measured from the high points in the surface.

Movement Joint (architect must specify type of joint and show location and details on drawings)

- Movement joints—mandatory according to EJ171. For above-ground installations, additional movement joints are required.
- Perimeter joints must be properly located and left open or filled per membrane manufacturer's instructions.

Installation Specifications

- Tile—ANSI A108.5.
- Cementitious grout—ANSI A108.10.
- Epoxy grout—ANSI A108.6.
- Sound reduction membrane—manufacturer's directions.
- Acoustical Sealant, when used—ASTM C919.
- Movement Joints—EJ171 and ASTM C1193.

Notes

- Not all mortars, grouts, and membranes are suitable for above-ground installations. Check manufacturer recommendations.
- In addition to being tested per ANSI A118.13 and meeting all requirements of that standard for sound reduction membranes, sound reduction membranes may be additionally tested using exact or more similar materials and assemblies to those that will be used in the field. Such testing may provide a more accurate indication of actual sound reduction and service rating that could be expected.
- Consult individual manufacturers for specific instructions, application, performance levels, and limitations concerning their materials. Follow the individual manufacturer's written instructions precisely.
- See "Sound Rated Floors Guide" for more information.
- The tile contractor is not responsible for the design or testing of the sound reduction system.

RH110-21

- **On-Ground Concrete Encapsulating Hydronic Tubing**
- **Ceramic Tile, Glass Tile**

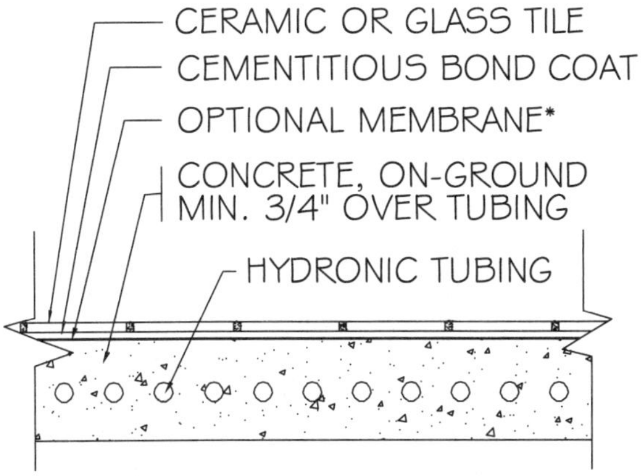

*USE OF A MEMBRANE IS OPTIONAL. SEE MEMBRANE OPTIONS.

Recommended Uses

- For slab-on-ground construction where no bending stresses occur, thin-bed installation of tile is desired, and hydronic tubing is encapsulated within the slab.

Service Rating

- Moderate.
- When glass tile is used, service rating may be lower.

Environmental Exposure Classifications

- Res1, 2, 3, 5; Com1, 2, 3, 5.
- May be suitable for exterior applications in areas not subject to freeze/thaw cycling when appropriate precautions are taken, including expansion joint placement, proper slope, waterproofing, and material selection.
- For installations that may be exposed to staining, specify tile and grout suitable for exposure. Consult product manufacturers; see also "Product Selection Guides."
- For installations that may be exposed to mild chemical attack, specify epoxy grout and tile suitable for exposure. Consult product manufacturers; see also "Product Selection Guides."

Typical Weight of Tile Installation

- 5 pounds/square foot.
- Does not include weight of substrate. See "Appendix B" for assumptions, included materials, and their individual weights.

Limitations

- Not for use on above-ground structural slabs and other floors subject to movement and/or deflection. See RH110A.

Membrane Options

- A crack isolation membrane (A118.12) may be required. Check with heat system and setting material manufacturers.
- A crack isolation membrane may be specified to treat existing in-plane cracks (F125-partial) or for protection against future in-plane cracking (F-125 full). See F125 for more information.
- Crack isolation membrane may also provide waterproofing. Consult membrane manufacturer.
- Check with membrane manufacturers for suitability for applicable conditions, as not all membranes are suitable for steam, high temperature, and/or chemical exposure, exterior use, use over above-ground structural slabs, use over pourable underlayments, use with radiant heat, or use over concrete with excessive moisture vapor transmission and/or alkalinity. Membrane may also affect service rating.
- When glass tile is used, consult glass tile manufacturer for membrane options and recommendations.

Requirements

- Slab to be well cured, dimensionally stable, and free of cracks, surface laitance, waxy or oily films, curing compounds, and other surface bond-inhibiting contaminants.

Materials

- **Multiple options exist for membranes, mortars, grouts, and other materials and *must be clearly specified* to be included. If not specifically indicated, optional materials are not included and mortar/grout choice defaults to minimum performance specification indicated. Consider each system component and intended use to determine minimum requirements and to specify options.**
- Ceramic tile—ANSI A137.1.
- Glass tile, when used—ANSI A137.2; see also "Glass Tile Selection and Installation Guide," and consult tile manufacturer for service rating and environmental exposure classification recommendations. Not all glass tiles are suitable.
- Cementitious grout—ANSI A118.6 or better or ISO CG1 or better. When glass tile is used, specify grout designated by tile and grout manufacturers.
- Epoxy grout, when used—ANSI A118.3 or ISO RG.
- Cementitious bond coat:
  - When a crack isolation membrane is not used—ANSI A118.15 or better or ISO C2S1 or better.
  - When a crack isolation membrane is used—ANSI A118.4 or better or ISO C2S1 or better.
  - When glass tile is used, specify mortar designated by tile and mortar manufacturers. Bond coat color will

impact the final appearance of translucent glass tile. Specifier shall confirm bond coat color is acceptable.

- Crack isolation membrane, when used—ANSI A118.12.

Materials for Green/Sustainable Design

- See "Green Building Standards and Green Product Selection Guide" and consult manufacturers and suppliers for product sustainability and contribution to green building design.
- Consider specifying tile and installation materials that meet ANSI A138.1, the *American National Standard Specifications for Sustainable Ceramic Tiles, Glass Tiles, and Tile Installation Materials.*

Preparation by Other Trades

- Slab to have steel trowel and fine broom finish free of curing compounds. When used, mechanical scarifying is necessary.
- Slope, when required, to be in subfloor.
- Maximum allowable variation in the tile substrate—for tiles with all edges shorter than 15", maximum allowable variation is ¼" in 10' from the required plane, with no more than ¹⁄₁₆" variation in 12" when measured from the high points in the surface. For tiles with at least one edge 15" in length, maximum allowable variation is ⅛" in 10' from the required plane, with no more than ¹⁄₁₆" variation in 24" when measured from the high points in the surface.
- Radiant tubing by others.
- Portland cement concrete poured minimum ¾" over top of hydronic tubing.

Movement Joint (architect must specify type of joint and show location and details on drawings)

- Movement joints—mandatory according to EJ171.
- When glass tile is used, adhere to more frequent placement recommendations within the ranges listed in EJ171.

Installation Specifications

- Tile—ANSI A108.5.
- Glass tile—manufacturer's directions.
- Cementitious grout—A108.10.
- Epoxy grout—A108.6.
- Crack isolation membrane—ANSI A108.17.
- Movement Joints—EJ171 and ASTM C1193.

Notes

- For waterproof application, see F122.
- When glass tile is used, see "Glass Tile Selection and Installation Guide," and consult manufacturer for recommendations and requirements.

RH110A-21

- **Above-Ground Concrete Encapsulating Hydronic Tubing**
- **Ceramic Tile, Glass Tile**

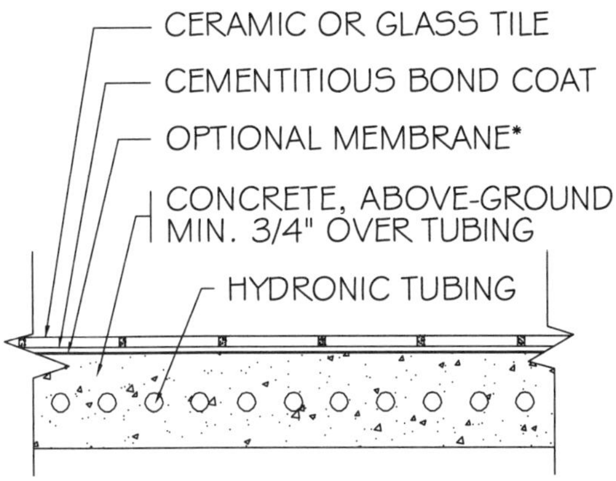

*USE OF A MEMBRANE IS OPTIONAL. SEE MEMBRANE OPTIONS.

Recommended Uses

- For above-ground structural slabs where thin-bed installation of tile is desired and hydronic tubing is encapsulated within the slab.

Service Rating

- Moderate.
- When glass tile is used, service rating may be lower.

Environmental Exposure Classifications

- Res1, 2, 3, 5; Com1, 2, 3, 5.
- May be suitable for exterior applications in areas not subject to freeze/thaw cycling when appropriate precautions are taken, including expansion joint placement, proper slope, waterproofing, and material selection.
- For installations that may be exposed to staining, specify tile and grout suitable for exposure. Consult product manufacturers; see also "Product Selection Guides."
- For installations that may be exposed to mild chemical attack, specify epoxy grout and tile suitable for exposure. Consult product manufacturers; see also "Product Selection Guides."

Typical Weight of Tile Installation

- 5 pounds/square foot.
- Does not include weight of substrate. See "Appendix B" for assumptions, included materials, and their individual weights.

Limitations

- Requires additional consideration by design professional to accommodate movement and/or deflection. Setting materials with improved bond strength and deformability are required.

Membrane Options

- A crack isolation membrane (A118.12) may be required. Check with heat system and setting material manufacturers.
- A crack isolation membrane may be specified to treat existing in-plane cracks (F125-partial) or for protection against future in-plane cracking (F-125 full). See F125 for more information.
- Crack isolation membrane may also provide waterproofing. Consult membrane manufacturer.
- Check with membrane manufacturers for suitability for applicable conditions, as not all membranes are suitable for steam, high temperature, and/or chemical exposure, exterior use, use over above-ground structural slabs, use over pourable underlayments, use with radiant heat, or use over concrete with excessive moisture vapor transmission and/or alkalinity. Membrane may also affect service rating.
- When glass tile is used, consult glass tile manufacturer for membrane options and recommendations.

Requirements

- Slab to be well cured, dimensionally stable, and free of cracks, surface laitance, waxy or oily films, curing compounds, and other surface bond-inhibiting contaminants.
- Above-ground installations are inherently more susceptible to vibration and deflection. Grout and mortar manufacturers to warrant product suitability.

Materials

- **Multiple options exist for membranes, mortars, grouts, and other materials and *must be clearly specified* to be included. If not specifically indicated, optional materials are not included and mortar/grout choice defaults to minimum performance specification indicated. Consider each system component and intended use to determine minimum requirements and to specify options.**
- Ceramic tile—ANSI A137.1.
- Glass tile, when used—ANSI A137.2; see also "Glass Tile Selection and Installation Guide," and consult tile manufacturer for service rating and environmental exposure classification recommendations. Not all glass tiles are suitable.
- Cementitious grout—ANSI A118.6 or better or ISO CG1 or better. When glass tile is used, specify grout designated by tile and grout manufacturers.
- Epoxy grout, when used—ANSI A118.3 or ISO RG.
- Cementitious bond coat:
  - Must be recommended by manufacturer for above-ground use.
  - When a crack isolation membrane is not used—ANSI A118.15 or better or ISO C2S1 or better.

- When a crack isolation membrane is used—ANSI A118.4 or better or ISO C2S1 or better.
- When glass tile is used, specify mortar designated by tile and mortar manufacturers. Bond coat color will impact the final appearance of translucent glass tile. Specifier shall confirm bond coat color is acceptable.

- Crack isolation membrane, when used—ANSI A118.12.

Materials for Green/Sustainable Design

- See "Green Building Standards and Green Product Selection Guide" and consult manufacturers and suppliers for product sustainability and contribution to green building design.
- Consider specifying tile and installation materials that meet ANSI A138.1, the *American National Standard Specifications for Sustainable Ceramic Tiles, Glass Tiles, and Tile Installation Materials.*

Preparation by Other Trades

- Floor systems over which tile will be installed shall be in conformance with the IRC for residential applications, the IBC for commercial applications, or applicable building codes. Maximum allowable substrate deflection under live load not to exceed l/360. See also "Substrate Requirements."
- When concentrated loads (scissor lifts, pallet jacks, automobiles, forklifts, etc.) will be used on an above-ground tiled floor, the engineer and/or specifier shall specify a substrate to accommodate the concentrated loads. Owner/specifier is responsible for protecting the tilework from damage, including allowing sufficient time for installed materials to cure properly.
- Slab to have steel trowel and fine broom finish free of curing compounds. When used, mechanical scarifying is necessary.
- Slope, when required, to be in subfloor.
- Maximum allowable variation in the tile substrate—for tiles with all edges shorter than 15", maximum allowable variation is ¼" in 10' from the required plane, with no more than ¹⁄₁₆" variation in 12" when measured from the high points in the surface. For tiles with at least one edge 15" in length, maximum allowable variation is ⅛" in 10' from the required plane, with no more than ¹⁄₁₆" variation in 24" when measured from the high points in the surface.
- Radiant tubing by others.
- Portland cement concrete poured minimum ¾" over top of hydronic tubing.

Movement Joint (architect must specify type of joint and show location and details on drawings)

- Movement joints—mandatory according to EJ171. For above-ground installations, additional movement joints are required.
- When glass tile is used, adhere to more frequent placement recommendations within the ranges listed in EJ171.

Installation Specifications

- Tile—ANSI A108.5.
- Glass tile—manufacturer's directions.
- Cementitious grout—A108.10.
- Epoxy grout—A108.6.
- Crack isolation membrane—A108.17.
- Movement Joints—EJ171 and ASTM C1193.

Notes

- F111 is preferred over above-ground structural slabs and other floors subject to movement and/or deflection.
- Not all mortars, grouts, and membranes are suitable for above-ground installations. Check manufacturer recommendations.
- For waterproof application, see F122A.
- When glass tile is used, see "Glass Tile Selection and Installation Guide," and consult manufacturer for recommendations and requirements.

RH117-21

- **On-Ground or Above-Ground Concrete**
- **Unbonded Mortar Bed Encapsulating Hydronic Tubing**
- **Ceramic Tile, Glass Tile**

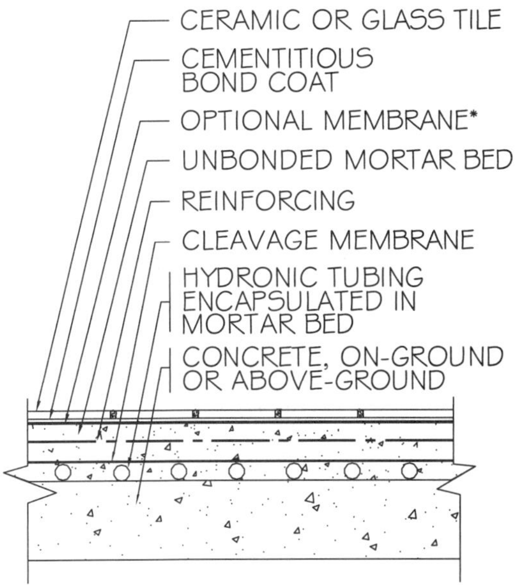

*USE OF A MEMBRANE IS OPTIONAL
SEE MEMBRANE OPTIONS

Recommended Uses

- For above-ground structural slabs and for slab-on-ground construction where hydronic tubing will be placed over the slab.
- For areas where leveling, flattening, or contouring of finish floor height is required where hydronic tubing will be placed over the slab.
- For areas where floor flatness is critical, such as when tiles with any edge longer than 15" are specified or where the finished tile assembly is required to achieve in-plane transitions (as may be necessary where accessibility is a concern), and where hydronic tubing will be placed over the slab.
- For concrete substrates that are cracked or present bonding issues and hydronic tubing will be placed over the slab.

Service Rating

- Extra heavy.
- When glass tile is used, service rating may be lower.

Environmental Exposure Classifications

- Res1, 2, 3, 5; Com1, 2, 3, 5.
- May be suitable for exterior applications in areas not subject to freeze/thaw cycling when appropriate precautions are taken, including expansion joint placement, proper slope, waterproofing, and material selection.
- For installations that may be exposed to staining, specify tile and grout suitable for exposure. Consult product manufacturers; see also "Product Selection Guides."
- For installations that may be exposed to mild chemical attack, specify epoxy grout and tile suitable for exposure. Consult product manufacturers; see also "Product Selection Guides."

Typical Weight of Tile Installation

- 30 pounds/square foot with 1¼" mortar bed over hydronic tubing. Add 3 pounds/square foot for each additional ¼" of mortar bed.
- Does not include weight of substrate. See "Appendix B" for assumptions, included materials, and their individual weights.

Limitations

- Above-ground structural slabs and other floors subject to movement and/or deflection—requires additional consideration by design professional to accommodate movement and/or deflection. Setting materials with improved bond strength and deformability are required.

Membrane Options

- A crack isolation membrane (A118.12) may be required. Check with heat system and setting material manufacturers.
- Crack isolation membrane may also provide waterproofing. Consult membrane manufacturer.
- A waterproof membrane (A118.10) may be specified to prevent moisture intrusion. Specifier shall indicate if complete waterproofing is required, including if/how membrane connects to drain assembly, if base flashing is required, and treatment at other termination points.
- Check with membrane manufacturers for suitability for applicable conditions, as not all membranes are suitable for steam, high temperature, and/or chemical exposure, exterior use, use over above-ground structural slabs, use over pourable underlayments, use with radiant heat, or use over concrete with excessive moisture vapor transmission and/or alkalinity. Membrane may also affect service rating.
- When glass tile is used, consult glass tile manufacturer for membrane options and recommendations.

Requirements

- Reinforcing mesh mandatory.
- Mortar bed thickness—1¼" minimum to 2" maximum.
- Mortar beds in excess of 2" thick shall be detailed by the architect.
- Screed fill flush to top of radiant tubing before placing cleavage membrane and mortar bed.
- Above-ground—above-ground installations are inherently more susceptible to vibration and deflection. Grout and mortar manufacturers to warrant product suitability.

Materials

- **Multiple options exist for membranes, mortars, grouts, and other materials and *must be clearly specified* to be included. If not specifically indicated,**

optional materials are not included and mortar/grout choice defaults to minimum performance specification indicated. Consider each system component and intended use to determine minimum requirements and to specify options.

- Ceramic tile—ANSI A137.1.
- Glass tile, when used—ANSI A137.2; see also "Glass Tile Selection and Installation Guide," and consult tile manufacturer for service rating and environmental exposure classification recommendations. Not all glass tiles are suitable.
- Cementitious grout—ANSI A118.6 or better or ISO CG1 or better. When glass tile is used, specify grout designated by tile and grout manufacturers.
- Epoxy grout, when used—ANSI A118.3 or ISO RG.
- Cementitious bond coat (on ground)—portland cement paste on a mortar bed that is still workable. For a cured mortar bed, follow recommendations below to select appropriate bonding mortar:
  - When a membrane is not used—ANSI A118.15 or better or ISO C2S1 or better.
  - When a crack isolation membrane is used—ANSI A118.4 or better or ISO C2S1 or better.
  - When a waterproof membrane is used—ANSI A118.4 or better or ISO C2S1 or better unless ANSI A118.1 or ISO C1 is recommended by membrane manufacturer.
  - When glass tile is used, specify mortar designated by tile and mortar manufacturers. Bond coat color will impact the final appearance of translucent glass tile. Specifier shall confirm bond coat color is acceptable.
- Cementitious bond coat (above-ground)—portland cement paste on a mortar bed that is still workable. For a cured mortar bed, follow recommendations below to select appropriate bonding mortar:
  - Must be recommended by manufacturer for above-ground use.
  - When a membrane is not used—ANSI A118.15 or better or ISO C2S1 or better.
  - When a crack isolation membrane is used—ANSI A118.4 or better or ISO C2S1 or better.
  - When a waterproof membrane is used—ANSI A118.4 or better or ISO C2S1 or better unless ANSI A118.1 or ISO C1 is recommended by membrane manufacturer.
  - When glass tile is used, specify mortar designated by tile and mortar manufacturers. Bond coat color will impact the final appearance of translucent glass tile. Specifier shall confirm bond coat color is acceptable.
- Crack isolation membrane, when used—ANSI A118.12.
- Waterproof membrane, when used—ANSI A118.10.
- Mortar bed, reinforcing, and cleavage membrane—ANSI A108.1A.

Materials for Green/Sustainable Design

- See "Green Building Standards and Green Product Selection Guide" and consult manufacturers and suppliers for product sustainability and contribution to green building design.
- Consider specifying tile and installation materials that meet ANSI A138.1, the *American National Standard Specifications for Sustainable Ceramic Tiles, Glass Tiles, and Tile Installation Materials.*

Preparation by Other Trades

- Floor systems over which tile will be installed shall be in conformance with the IRC for residential applications, the IBC for commercial applications, or applicable building codes. Maximum allowable substrate deflection under live load not to exceed l/360. See also "Substrate Requirements."
- When concentrated loads (scissor lifts, pallet jacks, automobiles, forklifts, etc.) will be used on an above-ground tiled floor, the engineer and/or specifier shall specify a substrate to accommodate the concentrated loads. Owner/specifier is responsible for protecting the tilework from damage, including allowing sufficient time for installed materials to cure properly.
- Where slab is depressed to accommodate mortar bed thickness, slab depression to be accurate with steel trowel finish.
- Slope, when required, to be in subfloor.
- Maximum allowable variation in the installation substrate (concrete)—1⁄4" in 10' from the required plane.
- Radiant tubing by others.

Movement Joint (architect must specify type of joint and show location and details on drawings)

- Movement joints—mandatory according to EJ171. For above-ground installations, additional movement joints are required.
- When glass tile is used, adhere to more frequent placement recommendations within the ranges listed in EJ171.

Installation Specifications

- Tile—ANSI A108.1A, .1B, or .1C. A108.1B required if membrane or epoxy grout to be used.
- Glass tile—ANSI A108.14, .15, .16, or manufacturer's directions.
- Cementitious grout—A108.10.
- Epoxy grout—A108.6.
- Crack isolation membrane—ANSI A108.17.
- Waterproof Membrane—ANSI A108.13.
- Movement Joints—EJ171 and ASTM C1193.

Notes

- For in-slab hydronic tubing, see F111.
- Above-ground—not all mortars, grouts, and membranes are suitable for above-ground use. Check manufacturer recommendations.
- When glass tile is used, see "Glass Tile Selection and Installation Guide," and consult manufacturer for recommendations and requirements.

RH111-21

- **On-Ground Concrete**
- **Poured Gypsum Underlayment Encapsulating Hydronic Tubing**
- **Bonded Membrane**
- **Ceramic Tile, Glass Tile**

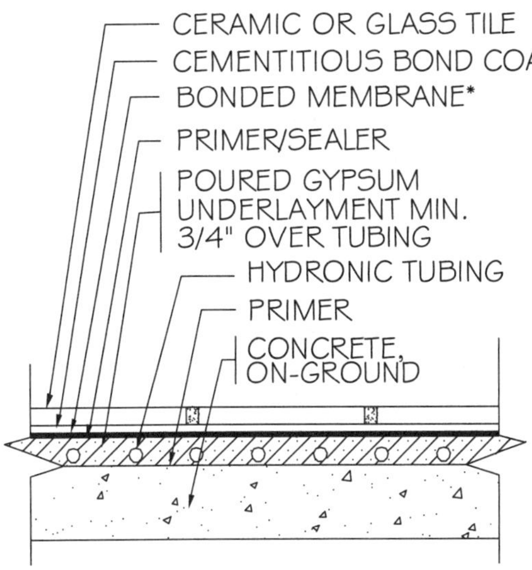

*USE OF A MEMBRANE IS REQUIRED. SEE MEMBRANE OPTIONS.

Recommended Uses

- For slab-on-ground construction where no bending stresses occur and with a poured gypsum underlayment encapsulating hydronic tubing.

Service Rating

- Moderate.
- When glass tile is used, service rating may be lower.

Environmental Exposure Classifications

- Res1; Com1.
- May be suitable for limited water exposure areas (Res2; Com2). See Membrane Options.
- For installations that may be exposed to staining, specify tile and grout suitable for exposure. Consult product manufacturers; see also "Product Selection Guides."
- For installations that may be exposed to mild chemical attack, specify epoxy grout and tile suitable for exposure. Consult product manufacturers; see also "Product Selection Guides."

Typical Weight of Tile Installation

- 19 pounds/square foot with ¾" poured gypsum over hydronic tubing. Add 2¼ pounds/square foot for each additional ¼" of poured gypsum.
- Does not include weight of substrate. See "Appendix B" for assumptions, included materials, and their individual weights.

Limitations

- Not for below-ground installation.
- Not for use over above-ground structural slabs and other floors subject to movement and/or deflection. See RH111A.
- Not for use where moisture vapor transmission rate exceeds gypsum underlayment manufacturer's limitations.

Membrane Options

- Requires use of crack isolation (A118.12) and/or waterproof (A118.10) and/or uncoupling membrane.
- Membrane must meet ANSI A118.10 for limited water exposure areas (Res2; Com2). Consult underlayment manufacturer for requirements.
- Specifier shall indicate if complete waterproofing is required, including if/how membrane connects to drain assembly, if base flashing is required, and treatment at other termination points.
- Check with membrane manufacturers for suitability for applicable conditions, as not all membranes are suitable for steam, high temperature, and/or chemical exposure, exterior use, use over above-ground structural slabs, use over pourable underlayments, use with radiant heat, or use over concrete with excessive moisture vapor transmission and/or alkalinity. Membrane may also affect service rating.
- When glass tile is used, consult glass tile manufacturer for membrane options and recommendations.

Requirements

- Proper curing/drying of gypsum underlayments prior to application of tile is critical for proper performance. Consult the underlayment manufacturer for specific instructions.

Materials

- **Multiple options exist for membranes, mortars, grouts, and other materials and *must be clearly specified* to be included. If not specifically indicated, optional materials are not included and mortar/grout choice defaults to minimum performance specification indicated. Consider each system component and intended use to determine minimum requirements and to specify options.**
- Ceramic tile—ANSI A137.1.
- Glass tile, when used—ANSI A137.2; see also "Glass Tile Selection and Installation Guide," and consult tile manufacturer for service rating and environmental exposure classification recommendations. Not all glass tiles are suitable.
- Cementitious grout—ANSI A118.6 or better or ISO CG1 or better. When glass tile is used, specify grout designated by tile and grout manufacturers.
- Epoxy grout, when used—ANSI A118.3 or ISO RG.

- Cementitious bond coat:
  - When a crack isolation membrane is used—ANSI A118.4 or better or ISO C2S1 or better.
  - When a waterproof membrane is used—ANSI A118.4 or better or ISO C2S1 or better, unless ANSI A118.1 or ISO C1 is recommended by membrane manufacturer.
  - When an uncoupling membrane is used, check with membrane manufacturer.
  - When glass tile is used, specify mortar designated by tile and mortar manufacturers. Bond coat color will impact the final appearance of translucent glass tile. Specifier shall confirm bond coat color is acceptable.
- Crack isolation membrane, when used—ANSI A118.12.
- Waterproof membrane, when used—ANSI A118.10.
- Uncoupling membrane, when used—recommended by uncoupling membrane manufacturer; must achieve 50 psi or greater shear bond strength in 7 days per the test method in ANSI A118.12 Section 5.1.3.

Materials for Green/Sustainable Design

- See "Green Building Standards and Green Product Selection Guide" and consult manufacturers and suppliers for product sustainability and contribution to green building design.
- Consider specifying tile and installation materials that meet ANSI A138.1, the *American National Standard Specifications for Sustainable Ceramic Tiles, Glass Tiles, and Tile Installation Materials.*

Preparation by Other Trades

- Consult the manufacturer for maximum moisture vapor transmission limitations prior to application of underlayment.
- Maximum allowable variation in the tile substrate—for tiles with all edges shorter than 15", maximum allowable variation is ¼" in 10' from the required plane, with no more than ¹⁄₁₆" variation in 12" when measured from the high points in the surface. For tiles with at least one edge 15" in length, maximum allowable variation is ⅛" in 10' from the required plane, with no more than ¹⁄₁₆" variation in 24" when measured from the high points in the surface.
- Slab to be well cured, dimensionally stable, and free of cracks, surface laitance, waxy or oily films, curing compounds, and other surface bond-inhibiting contaminants.
- Poured gypsum underlayment thickness and application varies, consult the manufacturer for specific recommendations when used with floor warming systems. Minimum thickness is ¾" above hydronic tubing.
- Surfaces must be prepared and primed according to the underlayment manufacturer's instructions.
- Radiant tubing by others.
- Poured gypsum underlayment—minimum compressive strength of 2000 psi when tested per ASTM C472 modified and meeting performance requirements of ASTM C627 for the anticipated service level designated by the specifier or intended use.
- Poured gypsum underlayment—installed only by a manufacturer-approved applicator in accordance with manufacturer's recommendations.

Movement Joint (architect must specify type of joint and show location and details on drawings)

- Movement joints—mandatory in accordance with EJ171.
- When glass tile is used, adhere to more frequent placement recommendations within the ranges listed in EJ171.

Installation Specifications

- Tile—ANSI A108.5.
- Glass tile—manufacturer's directions.
- Cementitious grout—ANSI A108.10.
- Epoxy grout—ANSI A108.6.
- Crack isolation membrane—ANSI A108.17.
- Waterproof membrane—ANSI A108.13.
- Uncoupling membrane—manufacturer's directions.
- Movement Joints—EJ171 and ASTM C1193.

Notes

- Consult underlayment manufacturer for minimum cure/dry time of installation before radiant heat system is activated.
- Some manufacturers require special primers prior to the application of the membrane. Follow manufacturer's directions.
- When glass tile is used, see "Glass Tile Selection and Installation Guide," and consult manufacturer for recommendations and requirements.

RH111A-21

- **Above-Ground Concrete**
- **Poured Gypsum Underlayment Encapsulating Hydronic Tubing**
- **Bonded Membrane**
- **Ceramic Tile, Glass Tile**

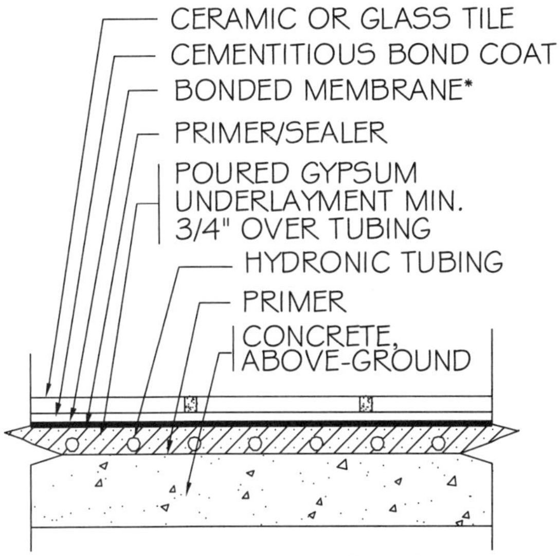

*USE OF A MEMBRANE IS REQUIRED. SEE MEMBRANE OPTIONS.

Recommended Uses

- For above-ground structural slabs with a poured gypsum underlayment encapsulating hydronic tubing.

Service Rating

- Moderate.
- When glass tile is used, service rating may be lower.

Environmental Exposure Classifications

- Res1; Com1.
- May be suitable for limited water exposure areas (Res2; Com2). See Membrane Options.
- For installations that may be exposed to staining, specify tile and grout suitable for exposure. Consult product manufacturers; see also "Product Selection Guides."
- For installations that may be exposed to mild chemical attack, specify epoxy grout and tile suitable for exposure. Consult product manufacturers; see also "Product Selection Guides."

Typical Weight of Tile Installation

- 19 pounds/square foot with ¾" poured gypsum over hydronic tubing. Add 2¼ pounds/square foot for each additional ¼" of poured gypsum.
- Does not include weight of substrate. See "Appendix B" for assumptions, included materials, and their individual weights.

Limitations

- Requires additional consideration by design professional to accommodate movement and/or deflection. Setting materials with improved bond strength and deformability are required.
- Not for use where moisture vapor transmission rate exceeds gypsum underlayment manufacturer's limitations.

Membrane Options

- Requires use of crack isolation (A118.12) and/or waterproof (A118.10) and/or uncoupling membrane.
- Membrane must meet ANSI A118.10 for limited water exposure areas (Res2; Com2). Consult underlayment manufacturer for requirements.
- Specifier shall indicate if complete waterproofing is required, including if/how membrane connects to drain assembly, if base flashing is required, and treatment at other termination points.
- Check with membrane manufacturers for suitability for applicable conditions, as not all membranes are suitable for steam, high temperature, and/or chemical exposure, exterior use, use over above-ground structural slabs, use over pourable underlayments, use with radiant heat, or use over concrete with excessive moisture vapor transmission and/or alkalinity. Membrane may also affect service rating.
- When glass tile is used, consult glass tile manufacturer for membrane options and recommendations.

Requirements

- Proper curing/drying of underlayment prior to application of tile is critical for proper performance. Consult the underlayment manufacturer for specific instructions.
- Above-ground installations are inherently more susceptible to vibration and deflection. Grout and mortar manufacturers to warrant product suitability.

Materials

- **Multiple options exist for membranes, mortars, grouts, and other materials and *must be clearly specified* to be included. If not specifically indicated, optional materials are not included and mortar/grout choice defaults to minimum performance specification indicated. Consider each system component and intended use to determine minimum requirements and to specify options.**
- Ceramic tile—ANSI A137.1.
- Glass tile, when used—ANSI A137.2; see also "Glass Tile Selection and Installation Guide," and consult tile manufacturer for service rating and environmental exposure classification recommendations. Not all glass tiles are suitable.
- Cementitious grout—ANSI A118.6 or better or ISO CG1 or better. When glass tile is used, specify grout designated by tile and grout manufacturers.
- Epoxy grout, when used—ANSI A118.3 or ISO RG.

- Cementitious bond coat:
  - Must be recommended by manufacturer for above-ground use.
  - When a crack isolation membrane is used—ANSI A118.4 or better or ISO C2S1 or better.
  - When a waterproof membrane is used—ANSI A118.4 or better or ISO C2S1 or better, unless ANSI A118.1 or ISO C1 is recommended by membrane manufacturer.
  - When an uncoupling membrane is used, check with membrane manufacturer.
  - When glass tile is used, specify mortar designated by tile and mortar manufacturers. Bond coat color will impact the final appearance of translucent glass tile. Specifier shall confirm bond coat color is acceptable.
- Crack isolation membrane, when used—ANSI A118.12.
- Waterproof membrane, when used—ANSI A118.10.
- Uncoupling membrane, when used—recommended by uncoupling membrane manufacturer; must achieve 50 psi or greater shear bond strength in 7 days per the test method in ANSI A118.12 Section 5.1.3.

Materials for Green/Sustainable Design

- See "Green Building Standards and Green Product Selection Guide" and consult manufacturers and suppliers for product sustainability and contribution to green building design.
- Consider specifying tile and installation materials that meet ANSI A138.1, the *American National Standard Specifications for Sustainable Ceramic Tiles, Glass Tiles, and Tile Installation Materials.*

Preparation by Other Trades

- Floor systems over which tile will be installed shall be in conformance with the IRC for residential applications, the IBC for commercial applications, or applicable building codes. Maximum allowable substrate deflection under live load not to exceed l/360. See also "Substrate Requirements."
- When concentrated loads (scissor lifts, pallet jacks, automobiles, forklifts, etc.) will be used on an above-ground tiled floor, the engineer and/or specifier shall specify a substrate to accommodate the concentrated loads. Owner/specifier is responsible for protecting the tilework from damage, including allowing sufficient time for installed materials to cure properly.
- Slab to be well cured, dimensionally stable, and free of cracks, surface laitance, waxy or oily films, curing compounds, and other surface bond-inhibiting contaminants.
- Consult the manufacturer for maximum moisture vapor transmission limitations prior to application of underlayment.
- Maximum allowable variation in the tile substrate—for tiles with all edges shorter than 15," maximum allowable variation is ¼" in 10' from the required plane, with no more than ¹⁄₁₆" variation in 12" when measured from the high points in the surface. For tiles with at least one edge 15" in length, maximum allowable variation is ⅛" in 10' from the required plane, with no more than ¹⁄₁₆" variation in 24" when measured from the high points in the surface.
- Radiant tubing by others.
- Poured gypsum underlayment thickness and application varies, consult the manufacturer for specific recommendations when used with floor warming systems. Minimum thickness is ¾" above hydronic tubing.
- Surfaces must be prepared and primed according to the underlayment manufacturer's instructions.
- Poured gypsum underlayment—minimum compressive strength of 2000 psi when tested per ASTM C472 modified and meeting performance requirements of ASTM C627 for the anticipated service level designated by the specifier or intended use.
- Poured gypsum underlayment—installed only by a manufacturer-approved applicator in accordance with manufacturer's recommendations.

Movement Joint (architect must specify type of joint and show location and details on drawings)

- Movement joints—mandatory according to EJ171. For above-ground installations, additional movement joints are required.
- When glass tile is used, adhere to more frequent placement recommendations within the ranges listed in EJ171.

Installation Specifications

- Tile—ANSI A108.5.
- Glass tile—manufacturer's directions.
- Cementitious grout—ANSI A108.10.
- Epoxy grout—ANSI A108.6.
- Crack isolation membrane—ANSI A108.17.
- Waterproof membrane—ANSI A108.13.
- Uncoupling membrane—manufacturer's directions.
- Movement Joints—EJ171 and ASTM C1193.

Notes

- Consult underlayment manufacturer for minimum cure/dry time of installation before radiant heat system is activated.
- Some manufacturers require special primers prior to the application of the membrane. Follow manufacturer's directions.
- RH117 is preferred over above-ground structural slabs and other floors subject to movement and/or deflection.
- Not all mortars, grouts, and membranes are suitable for above-ground use. Check manufacturer recommendations.
- When glass tile is used, see "Glass Tile Selection and Installation Guide," and consult manufacturer for recommendations and requirements.

RH112-21

- **On-Ground Concrete**
- **Cementitious Self-Leveling Underlayment Encapsulating Hydronic Tubing**
- **Ceramic Tile, Glass Tile**

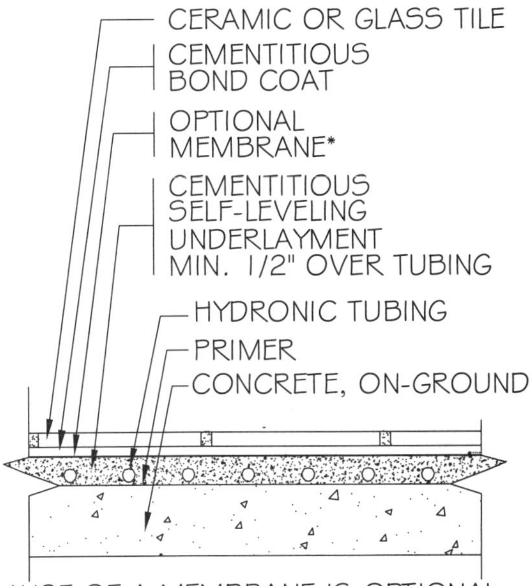

*USE OF A MEMBRANE IS OPTIONAL. SEE MEMBRANE OPTIONS.

Recommended Uses

- For slab-on-ground construction where no bending stresses occur and where hydronic tubing will be placed on the slab.

Service Rating

- Moderate.
- When glass tile is used, service rating may be lower.

Environmental Exposure Classifications

- Res1; Com1.
- May be suitable for increased water exposure. See Membrane Options.
- For installations that may be exposed to staining, specify tile and grout suitable for exposure. Consult product manufacturers; see also "Product Selection Guides."
- For installations that may be exposed to mild chemical attack, specify epoxy grout and tile suitable for exposure. Consult product manufacturers; see also "Product Selection Guides."

Typical Weight of Tile Installation

- 19 pounds/square foot with 1⁄2" cementitious self-leveling underlayment over hydronic tubing. Add 23⁄4 pounds/square foot for each additional 1⁄4" of underlayment.
- Does not include weight of substrate. See "Appendix B" for assumptions, included materials, and their individual weights.

Limitations

- Not for below-ground installations.
- Not for use where moisture vapor transmission rate exceeds underlayment manufacturer's limitations.
- Not for use over above-ground structural slabs and other floors subject to movement and/or deflection. See RH112A.

Membrane Options

- A crack isolation membrane (A118.12) may be required. Check with heat system and setting material manufacturers.
- Crack isolation membrane may also provide waterproofing. Consult membrane manufacturer.
- A waterproof membrane (A118.10) may be required for areas exposed to moisture. Consult underlayment manufacturer for requirements.
- Specifier shall indicate if complete waterproofing is required, including if/how membrane connects to drain assembly, if base flashing is required, and treatment at other termination points.
- A crack isolation membrane may be specified to treat existing in-plane cracks (F125-partial) or for protection against future in plane cracking (F 125 full). See F125 for more information.
- An uncoupling membrane may be specified to accommodate subfloor movement. Check with membrane manufacturer for recommendations and requirements.
- Check with membrane manufacturers for suitability for applicable conditions, as not all membranes are suitable for steam, high temperature, and/or chemical exposure, exterior use, use over above-ground structural slabs, use over pourable underlayments, use with radiant heat, or use over concrete with excessive moisture vapor transmission and/or alkalinity. Membrane may also affect service rating.
- When glass tile is used, consult glass tile manufacturer for membrane options and recommendations.

Requirements

- Consult the manufacturer for maximum moisture vapor transmission limitations prior to application of underlayment.
- Slab to be well cured, dimensionally stable, and free of cracks, surface laitance, waxy or oily films, curing compounds, and other surface bond-inhibiting contaminants.
- Cementitious self-leveling underlayment thickness and application varies, consult the manufacturer for specific recommendations when used with floor warming systems. Minimum thickness is ½" above hydronic tubing.

- Surfaces must be prepared and primed according to manufacturer's instructions.
- Proper curing/drying of underlayments prior to application of tile is critical for proper performance. Consult the underlayment manufacturer for specific instructions.

Materials

- **Multiple options exist for membranes, mortars, grouts, and other materials and *must be clearly specified* to be included. If not specifically indicated, optional materials are not included and mortar/grout choice defaults to minimum performance specification indicated. Consider each system component and intended use to determine minimum requirements and to specify options.**
- Ceramic tile—ANSI A137.1.
- Glass tile, when used—ANSI A137.2; see also "Glass Tile Selection and Installation Guide," and consult tile manufacturer for service rating and environmental exposure classification recommendations. Not all glass tiles are suitable.
- Cementitious grout—ANSI A118.6 or better or ISO CG1 or better. When glass tile is used, specify grout designated by tile and grout manufacturers.
- Epoxy grout, when used—ANSI A118.3 or ISO RG.
- Cementitious bond coat:
  - When a membrane is not used—ANSI A118.15 or better or ISO C2S1 or better.
  - When a crack isolation membrane is used—ANSI A118.4 or better or ISO C2S1 or better.
  - When a waterproof membrane is used—ANSI A118.4 or better or ISO C2S1 or better, unless ANSI A118.1 or ISO C1 is recommended by membrane manufacturer.
  - When an uncoupling membrane is used, check with membrane manufacturer.
  - When glass tile is used, specify mortar designated by tile and mortar manufacturers. Bond coat color will impact the final appearance of translucent glass tile. Specifier shall confirm bond coat color is acceptable.
- Crack isolation membrane, when used—ANSI A118.12.
- Waterproof membrane, when used—ANSI A118.10.
- Uncoupling membrane, when used—recommended by uncoupling membrane manufacturer; must achieve 50 psi or greater shear bond strength in 7 days per the test method in ANSI A118.12 Section 5.1.3.
- Self-leveling underlayment—meeting performance requirements of ASTM C627 for the anticipated service level designated by the specifier or intended use.

Materials for Green/Sustainable Design

- See "Green Building Standards and Green Product Selection Guide" and consult manufacturers and suppliers for product sustainability and contribution to green building design.
- Consider specifying tile and installation materials that meet ANSI A138.1, the *American National Standard Specifications for Sustainable Ceramic Tiles, Glass Tiles, and Tile Installation Materials.*

Preparation by Other Trades

- Slab to have fine broom finish with no curing compounds. When used, mechanical scarification is necessary.
- Maximum allowable variation in the installation substrate (concrete)—¼" in 10' from the required plane.
- Maximum allowable variation in the tile substrate—for tiles with all edges shorter than 15", maximum allowable variation is ¼" in 10' from the required plane, with no more than ¹⁄₁₆" variation in 12" when measured from the high points in the surface. For tiles with at least one edge 15" in length, maximum allowable variation is ⅛" in 10' from the required plane, with no more than ¹⁄₁₆" variation in 24" when measured from the high points in the surface.
- Radiant tubing by others.

Movement Joint (architect must specify type of joint and show location and details on drawings)

- Movement joints—mandatory in accordance with EJ171.
- When glass tile is used, adhere to more frequent placement recommendations within the ranges listed in EJ171.

Installation Specifications

- Tile—ANSI A108.5.
- Glass tile—manufacturer's directions.
- Cementitious grout—ANSI A108.10.
- Epoxy grout—ANSI A108.6.
- Cementitious self-leveling underlayment—manufacturer's directions.
- Crack isolation membrane—ANSI A108.17.
- Waterproof membrane—ANSI A108.13.
- Uncoupling membrane—manufacturer's directions.
- Movement Joints—EJ171 and ASTM C1193.

Notes

- Consult underlayment manufacturer for minimum cure/dry time of installation before radiant heat system is activated.
- Some manufacturers require special primers prior to the application of the setting materials or membrane. Follow manufacturer's directions.
- When glass tile is used, see "Glass Tile Selection and Installation Guide," and consult manufacturer for recommendations and requirements.

RH112A-21

- **Above-Ground Concrete**
- **Cementitious Self-Leveling Underlayment Encapsulating Hydronic Tubing**
- **Ceramic Tile, Glass Tile**

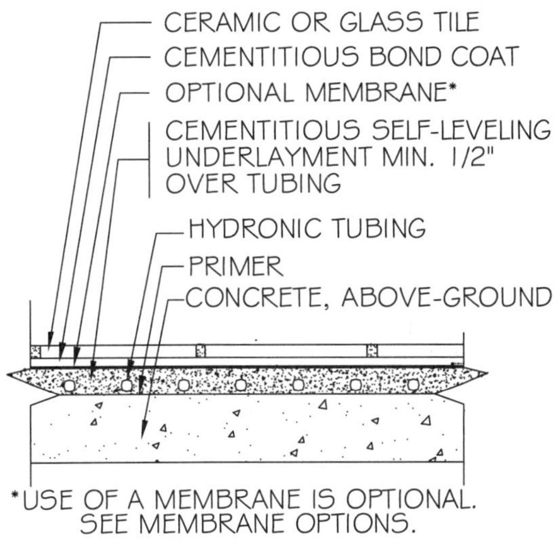

Recommended Uses

- For above-ground structural slabs where hydronic tubing will be placed on the slab.

Service Rating

- Moderate.
- When glass tile is used, service rating may be lower.

Environmental Exposure Classifications

- Res1; Com1.
- May be suitable for increased water exposure. See Membrane Options.
- For installations that may be exposed to staining, specify tile and grout suitable for exposure. Consult product manufacturers; see also "Product Selection Guides."
- For installations that may be exposed to mild chemical attack, specify epoxy grout and tile suitable for exposure. Consult product manufacturers; see also "Product Selection Guides."

Typical Weight of Tile Installation

- 19 pounds/square foot with ½" cementitious self-leveling underlayment over hydronic tubing. Add 2¾ pounds/square foot for each additional ¼" of underlayment.
- Does not include weight of substrate. See "Appendix B" for assumptions, included materials, and their individual weights.

Limitations

- Requires additional consideration by design professional to accommodate movement and/or deflection.Setting materials with improved bond strength and deformability are required.
- Not for use where moisture vapor transmission rate exceeds underlayment manufacturer's limitations.

Membrane Options

- A crack isolation membrane (A118.12) may be required. Check with heat system and setting material manufacturers.
- Crack isolation membrane may also provide waterproofing. Consult membrane manufacturer.
- A waterproof membrane (A118.10) may be required for areas exposed to moisture. Consult underlayment manufacturer for requirements.
- Specifier shall indicate if complete waterproofing is required, including if/how membrane connects to drain assembly, if base flashing is required, and treatment at other termination points.
- A crack isolation membrane may be specified to treat existing in-plane cracks (F125-partial) or for protection against future in-plane cracking (F-125 full). See F125 for more information.
- An uncoupling membrane may be specified to accommodate subfloor movement. Check with membrane manufacturer for recommendations and requirements.
- Check with membrane manufacturers for suitability for applicable conditions, as not all membranes are suitable for steam, high temperature, and/or chemical exposure, exterior use, use over above-ground structural slabs, use over pourable underlayments, use with radiant heat, or use over concrete with excessive moisture vapor transmission and/or alkalinity. Membrane may also affect service rating.
- When glass tile is used, consult glass tile manufacturer for membrane options and recommendations.

Requirements

- Slab to be well cured, dimensionally stable, and free of cracks, surface laitance, waxy or oily films, curing compounds, and other surface bond-inhibiting contaminants.
- Cementitious self-leveling underlayment thickness and application varies, consult the manufacturer for specific recommendations when used with floor warming systems. Minimum thickness is ½" above hydronic tubing.
- Surfaces must be prepared and primed according to the manufacturer's instructions.
- Proper curing/drying of underlayment prior to application of tile is critical for proper performance. Consult the underlayment manufacturer for specific instructions.
- Consult the manufacturer for maximum moisture vapor transmission limitations prior to application of underlayment.
- Above-ground installations are inherently more susceptible to vibration and deflection. Grout and mortar manufacturers to warrant product suitability.

Materials

- **Multiple options exist for membranes, mortars, grouts, and other materials and *must be clearly specified* to be included. If not specifically indicated,**

optional materials are not included and mortar/grout choice defaults to minimum performance specification indicated. Consider each system component and intended use to determine minimum requirements and to specify options.

- Ceramic tile—ANSI A137.1.
- Glass tile, when used—ANSI A137.2; see also "Glass Tile Selection and Installation Guide," and consult tile manufacturer for service rating and environmental exposure classification recommendations. Not all glass tiles are suitable.
- Cementitious grout—ANSI A118.6 or better or ISO CG1 or better. When glass tile is used, specify grout designated by tile and grout manufacturers.
- Epoxy grout, when used—ANSI A118.3 or ISO RG.
- Cementitious bond coat:
  - Must be recommended by manufacturer for above-ground use.
  - When a membrane is not used—ANSI A118.15 or better or ISO C2S1 or better.
  - When a crack isolation membrane is used—ANSI A118.4 or better or ISO C2S1 or better.
  - When a waterproof membrane is used—ANSI A118.4 or better or ISO C2S1 or better, unless ANSI A118.1 or ISO C1 is recommended by membrane manufacturer.
  - When an uncoupling membrane is used, check with membrane manufacturer.
  - When glass tile is used, specify mortar designated by tile and mortar manufacturers. Bond coat color will impact the final appearance of translucent glass tile. Specifier shall confirm bond coat color is acceptable.
- Crack isolation membrane, when used—ANSI A118.12.
- Waterproof membrane, when used—ANSI A118.10.
- Uncoupling membrane, when used—recommended by uncoupling membrane manufacturer; must achieve 50 psi or greater shear bond strength in 7 days per the test method in ANSI A118.12 Section 5.1.3.
- Self-leveling underlayment—meeting performance requirements of ASTM C627 for the anticipated service level designated by the specifier or intended use.

Materials for Green/Sustainable Design

- See "Green Building Standards and Green Product Selection Guide" and consult manufacturers and suppliers for product sustainability and contribution to green building design.
- Consider specifying tile and installation materials that meet ANSI A138.1, the *American National Standard Specifications for Sustainable Ceramic Tiles, Glass Tiles, and Tile Installation Materials.*

Preparation by Other Trades

- Floor systems over which tile will be installed shall be in conformance with the IRC for residential applications, the IBC for commercial applications, or applicable building codes. Maximum allowable substrate deflection under live load not to exceed l/360. See also "Substrate Requirements."
- When concentrated loads (scissor lifts, pallet jacks, automobiles, forklifts, etc.) will be used on an above-ground tiled floor, the engineer and/or specifier shall specify a substrate to accommodate the concentrated loads. Owner/specifier is responsible for protecting the tilework from damage, including allowing sufficient time for installed materials to cure properly.
- Slab to have fine broom finish with no curing compounds used. When used, mechanical scarification is necessary.
- Maximum allowable variation in the installation substrate (concrete)—¼" in 10' from the required plane.
- Maximum allowable variation in the tile substrate—for tiles with all edges shorter than 15", maximum allowable variation is ¼" in 10' from the required plane, with no more than ¹⁄₁₆" variation in 12" when measured from the high points in the surface. For tiles with at least one edge 15" in length, maximum allowable variation is ⅛" in 10' from the required plane, with no more than ¹⁄₁₆" variation in 24" when measured from the high points in the surface.
- Radiant tubing by others.

Movement Joint (architect must specify type of joint and show location and details on drawings)

- Movement joints—mandatory according to EJ171. For above-ground installations, additional movement joints are required.
- When glass tile is used, adhere to more frequent placement recommendations within the ranges listed in EJ171.

Installation Specifications

- Tile—ANSI A108.5.
- Glass tile—manufacturer's directions.
- Cementitious grout—ANSI A108.10.
- Epoxy grout—ANSI A108.6.
- Cementitious self-leveling underlayment—manufacturer's directions.
- Crack isolation membrane—ANSI A108.17.
- Waterproof membrane—ANSI A108.13.
- Uncoupling membrane—manufacturer's directions.
- Movement Joints—EJ171 and ASTM C1193.

Notes

- RH117 is preferred over above-ground structural slabs and other floors subject to movement and/or deflection.
- Some manufacturers require special primers prior to the application of the setting materials or membrane. Follow manufacturer's directions.
- Consult underlayment manufacturer for minimum cure/dry time of installation before radiant heat system is activated.
- Not all mortars, grouts, and membranes are suitable for above-ground use. Check manufacturer recommendations.
- When glass tile is used, see "Glass Tile Selection and Installation Guide," and consult manufacturer for recommendations and requirements.

RH115-21

- **On-Ground Concrete**
- **Electric Radiant Heat System Encapsulated in Portland Cement Mortar**
- **Ceramic Tile**

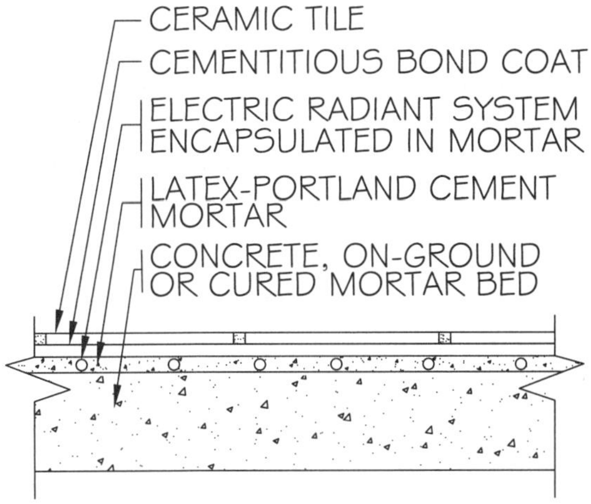

USE OF A MEMBRANE IS OPTIONAL.
SEE MEMBRANE OPTIONS.

Recommended Uses

- For slab-on-ground construction where no bending stresses occur and electric radiant in-floor heat is desired.
- For snow and ice melting applications, as determined by heat system manufacturer.

Service Rating

- Moderate.

Environmental Exposure Classifications

- Res1, 2; Com1, 2.
- May be suitable for exterior applications when appropriate precautions are taken, including expansion joint placement, proper slope, waterproofing, and material selection. Consult heat system manufacturer.
- For installations that may be exposed to staining, specify tile and grout suitable for exposure. Consult product manufacturers; see also "Product Selection Guides."
- For installations that may be exposed to mild chemical attack, specify epoxy grout and tile suitable for exposure. Consult product manufacturers; see also "Product Selection Guides."

Typical Weight of Tile Installation

- 7 pounds/square foot.
- Does not include weight of substrate. See "Appendix B" for assumptions, included materials, and their individual weights.

Limitations

- Not for use over above-ground structural slabs and other floors subject to movement and/or deflection. See RH115A.

Membrane Options

- Consult heat system manufacturer and setting material manufacturer for requirements and options.
- If a membrane is used, follow manufacturer's instructions for membrane type, installation, placement, and setting materials required.

Requirements

- Slab to be well cured, dimensionally stable, and free of cracks, surface laitance, waxy or oily films, curing compounds, and other surface bond-inhibiting contaminants.
- Electric radiant system should not be installed over building expansion joints.

Materials

- **Multiple options exist for membranes, mortars, grouts, and other materials and *must be clearly specified* to be included. If not specifically indicated, optional materials are not included and mortar/grout choice defaults to minimum performance specification indicated. Consider each system component and intended use to determine minimum requirements and to specify options.**
- Ceramic tile—ANSI A137.1.
- Cementitious grout—ANSI A118.6 or better or ISO CG1 or better.
- Epoxy grout, when used—ANSI A118.3 or ISO RG.
- Cementitious bond coat:
  - When a membrane is not used on top of mortar-encapsulated radiant system—ANSI A118.15 or better or ISO C2S1 or better.
  - When a membrane is used on top of mortar-encapsulated radiant system—ANSI A118.4 or better or ISO C2S1 or better, unless ANSI A118.1 or ISO C1 is recommended by membrane manufacturer.

Materials for Green/Sustainable Design

- See "Green Building Standards and Green Product Selection Guide" and consult manufacturers and suppliers for product sustainability and contribution to green building design.
- Consider specifying tile and installation materials that meet ANSI A138.1, the *American National Standard Specifications for Sustainable Ceramic Tiles, Glass Tiles, and Tile Installation Materials.*

Preparation by Other Trades

- Slab to have steel trowel and fine broom finish free of curing compounds. When used, mechanical scarifying is necessary.
- Slope, when required, to be in subfloor.
- Maximum allowable variation in the installation substrate (concrete)—for tiles with all edges shorter than 15", maximum allowable variation is ¼" in 10' from the required plane, with no more than $\frac{1}{16}$" variation in 12" when measured from the high points in the surface. For tiles with at least one edge 15" in length, maximum allowable variation is ⅛" in 10' from the required plane, with no more than $\frac{1}{16}$" variation in 24" when measured from the high points in the surface.
- Electrician or qualified contractor to wire electric radiant system to power source.

Movement Joint (architect must specify type of joint and show location and details on drawings)

- Movement joints—mandatory according to EJ171.

Installation Specifications

- Tile—ANSI A108.5.
- Cementitious grout—ANSI A108.10.
- Epoxy grout—ANSI A108.6.
- Electric radiant system—UL (US) 1693; UL (CAN/CSA) C22.2 #217; NEC Article 424.
- Movement Joints—EJ171 and ASTM C1193.

Notes

- Bond coat thickness and application varies, consult heat system manufacturer for recommendations.
- If latex-portland cement mortar is to be applied over $\frac{3}{16}$" in thickness, consult setting material manufacturer for additional installation instructions.

RH115A-21

- **Above-Ground Concrete**
- **Electric Radiant Heat System Encapsulated in Portland Cement Mortar**
- **Ceramic Tile**

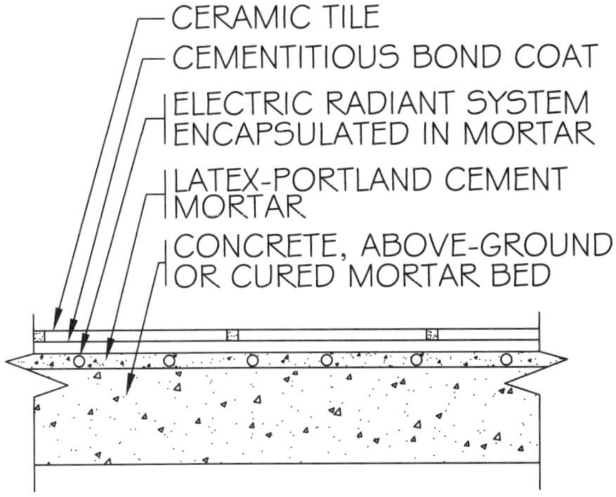

USE OF A MEMBRANE IS OPTIONAL. SEE MEMBRANE OPTIONS.

Recommended Uses

- For above-ground structural slabs where electric radiant in-floor heat is desired.
- For snow and ice melting applications, as determined by heat system manufacturer.

Service Rating

- Moderate.

Environmental Exposure Classifications

- Res1, 2; Com1, 2.
- May be suitable for exterior applications when appropriate precautions are taken, including expansion joint placement, proper slope, waterproofing and material selection. Consult heat system manufacturer.
- For installations that may be exposed to staining, specify tile and grout suitable for exposure. Consult product manufacturers; see also "Product Selection Guides."
- For installations that may be exposed to mild chemical attack, specify epoxy grout and tile suitable for exposure. Consult product manufacturers; see also "Product Selection Guides."

Typical Weight of Tile Installation

- 7 pounds/square foot.
- Does not include weight of substrate. See "Appendix B" for assumptions, included materials, and their individual weights.

Limitations

- Requires additional consideration by design professional to accommodate movement and/or deflection. Setting materials with improved bond strength and deformability are required.

Membrane Options

- Consult heat system manufacturer and setting material manufacturer for requirements and options.
- If a membrane is used, follow manufacturer's instructions for membrane type, installation, placement, and setting materials required.

Requirements

- Slab to be well cured, dimensionally stable, and free of cracks, surface laitance, waxy or oily films, curing compounds, and other surface bond-inhibiting contaminants.
- Electric radiant system should not be installed over building expansion joints.
- Above-ground installations are inherently more susceptible to vibration and deflection. Grout and mortar manufacturers to warrant product suitability.

Materials

- **Multiple options exist for membranes, mortars, grouts, and other materials and *must be clearly specified* to be included. If not specifically indicated, optional materials are not included and mortar/grout choice defaults to minimum performance specification indicated. Consider each system component and intended use to determine minimum requirements and to specify options.**
- Ceramic tile—ANSI A137.1.
- Cementitious grout—ANSI A118.6 or better or ISO CG1 or better.
- Epoxy grout, when used—ANSI A118.3 or ISO RG.
- Cementitious bond coat:
  - Must be recommended by manufacturer for above-ground use.
  - When a membrane is not used on top of mortar-encapsulated radiant system—ANSI A118.15 or better or ISO C2S1 or better.
  - When a membrane is used on top of mortar-encapsulated radiant system—ANSI A118.4 or better or ISO C2S1 or better, unless ANSI A118.1 or ISO C1 is recommended by membrane manufacturer.

Materials for Green/Sustainable Design

- See "Green Building Standards and Green Product Selection Guide" and consult manufacturers and suppliers for product sustainability and contribution to green building design.
- Consider specifying tile and installation materials that meet ANSI A138.1, the *American National Standard*

Specifications for Sustainable Ceramic Tiles, Glass Tiles, and Tile Installation Materials.

Preparation by Other Trades

- Floor systems over which tile will be installed shall be in conformance with the IRC for residential applications, the IBC for commercial applications, or applicable building codes. Maximum allowable substrate deflection under live load not to exceed l/360. See also "Substrate Requirements."
- When concentrated loads (scissor lifts, pallet jacks, automobiles, forklifts, etc.) will be used on an above-ground tiled floor, the engineer and/or specifier shall specify a substrate to accommodate the concentrated loads. Owner/specifier is responsible for protecting the tilework from damage, including allowing sufficient time for installed materials to cure properly.
- Slab to have steel trowel and fine broom finish free of curing compounds. When used, mechanical scarifying is necessary.
- Slope, when required, to be in subfloor.
- Maximum allowable variation in the installation substrate (concrete)—for tiles with all edges shorter than 15", maximum allowable variation is ¼" in 10' from the required plane, with no more than ¹⁄₁₆" variation in 12" when measured from the high points in the surface. For tiles with at least one edge 15" in length, maximum allowable variation is ⅛" in 10' from the required plane, with no more than ¹⁄₁₆" variation in 24" when measured from the high points in the surface.
- Electrician or qualified contractor to wire electric radiant system to power source.

Movement Joint (architect must specify type of joint and show location and details on drawings)

- Movement joints—mandatory according to EJ171. For above-ground installations, additional movement joints are required.

Installation Specifications

- Tile—ANSI A108.5.
- Cementitious grout—ANSI A108.10.
- Epoxy grout—ANSI A108.6.
- Electric radiant system—UL (US) 1693; UL (CAN/CSA) C22.2 #217; NEC Article 424.
- Movement Joints—EJ171 and ASTM C1193.

Notes

- Bond coat thickness and application varies, consult heat system manufacturer for recommendations.
- If latex-portland cement mortar is to be applied over ³⁄₁₆" in thickness, consult setting material manufacturer for additional installation instructions.
- F111 is preferred over above-ground structural slabs and other floors subject to movement and/or deflection.
- Not all mortars, grouts, and membranes are suitable for above-ground installations. Check manufacturer recommendations.

RH116-21

- **On-Ground Concrete**
- **Electric Radiant Heat System Encapsulated in Cementitious Self-Leveling Underlayment**
- **Ceramic Tile**

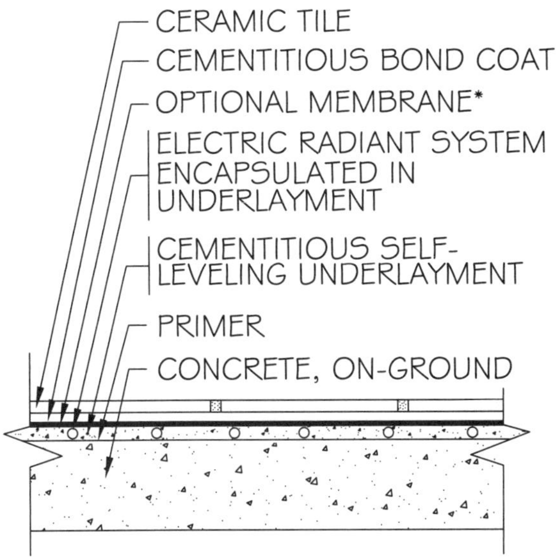

*USE OF A MEMBRANE IS OPTIONAL. SEE MEMBRANE OPTIONS.

Recommended Uses

- For slab-on-ground construction where no bending stresses occur and where floor leveling and electric radiant in-floor heat are desired.

Service Rating

- Moderate.

Environmental Exposure Classifications

- Res1; Com1.
- May be suitable for increased water exposure. See Membrane Options.
- For installations that may be exposed to staining, specify tile and grout suitable for exposure. Consult product manufacturers; see also "Product Selection Guides."
- For installations that may be exposed to mild chemical attack, specify epoxy grout and tile suitable for exposure. Consult product manufacturers; see also "Product Selection Guides."

Typical Weight of Tile Installation

- 11 pounds/square foot with ½" cementitious self-leveling underlayment. Add 2¾ pounds/square foot for each additional ¼" of underlayment.
- Does not include weight of substrate. See "Appendix B" for assumptions, included materials, and their individual weights.

Limitations

- Not for below-ground installations.
- Not for use where moisture vapor transmission rate exceeds underlayment manufacturer's limitations.
- Not for use over above-ground structural slabs and other floors subject to movement and/or deflection. See RH116A.

Membrane Options

- A crack isolation membrane (A118.12) may be required. Check with heat system and setting material manufacturers.
- Crack isolation membrane may also provide waterproofing. Consult membrane manufacturer.
- A waterproof membrane (A118.10) may be required for areas exposed to moisture. Consult underlayment manufacturer for requirements.
- Specifier shall indicate if complete waterproofing is required, including if/how membrane connects to drain assembly, if base flashing is required, and treatment at other termination points.
- A crack isolation membrane may be specified to treat existing in-plane cracks (F125-partial) or for protection against future in-plane cracking (F-125 full). See F125 for more information.
- Check with membrane manufacturers for suitability for applicable conditions, as not all membranes are suitable for steam, high temperature, and/or chemical exposure, exterior use, use over above-ground structural slabs, use over pourable underlayments, use with radiant heat, or use over concrete with excessive moisture vapor transmission and/or alkalinity. Membrane may also affect service rating.

Requirements

- Consult the manufacturer for maximum moisture vapor transmission limitations prior to application of underlayment.
- Slab to be well cured, dimensionally stable, and free of cracks, surface laitance, waxy or oily films, curing compounds, and other surface bond-inhibiting contaminants.
- Cementitious self-leveling underlayment thickness and application varies, consult the manufacturer for specific recommendations.
- Surfaces must be prepared and primed according to manufacturer's instructions.
- Proper curing/drying of underlayments prior to application of tile is critical for proper performance. Consult the underlayment manufacturer for specific instructions.
- Electric radiant system should not be installed over building expansion joints.

Materials

- **Multiple options exist for membranes, mortars, grouts, and other materials and *must be clearly specified* to be included. If not specifically indicated, optional materials are not included and mortar/grout choice defaults to minimum performance specification indicated. Consider each system component and intended use to determine minimum requirements and to specify options.**
- Ceramic tile—ANSI A137.1.
- Cementitious grout—ANSI A118.6 or better or ISO CG1 or better.
- Epoxy grout, when used—ANSI A118.3 or ISO RG.
- Cementitious bond coat:
  - When a membrane is not used—ANSI A118.15 or better or ISO C2S1 or better.
  - When a crack isolation membrane is used—ANSI A118.4 or better or ISO C2S1 or better.
  - When a waterproof membrane is used—ANSI A118.4 or better or ISO C2S1 or better, unless ANSI A118.1 or ISO C1 is recommended by membrane manufacturer.
- Crack isolation membrane, when used—ANSI A118.12.
- Waterproof membrane, when used—ANSI A118.10.
- Self-leveling underlayment—meeting performance requirements of ASTM C627 for the anticipated service level designated by the specifier or intended use.

Materials for Green/Sustainable Design

- See "Green Building Standards and Green Product Selection Guide" and consult manufacturers and suppliers for product sustainability and contribution to green building design.
- Consider specifying tile and installation materials that meet ANSI A138.1, the *American National Standard Specifications for Sustainable Ceramic Tiles, Glass Tiles, and Tile Installation Materials.*

Preparation by Other Trades

- Slab to have a fine broom finish with no curing compounds used. When used, mechanical scarifying is necessary.
- Maximum allowable variation in the installation substrate (concrete)—¼" in 10' from the required plane.
- Maximum allowable variation in the tile substrate—for tiles with all edges shorter than 15", maximum allowable variation is ¼" in 10' from the required plane, with no more than 1/16" variation in 12" when measured from the high points in the surface. For tiles with at least one edge 15" in length, maximum allowable variation is ⅛" in 10' from the required plane, with no more than 1/16" variation in 24" when measured from the high points in the surface.
- Electrician or qualified contractor to wire electric radiant system to power source.

Movement Joint (architect must specify type of joint and show location and details on drawings)

- Movement joints—mandatory according to EJ171.

Installation Specifications

- Tile—ANSI A108.5.
- Cementitious grout—ANSI A108.10.
- Epoxy grout—ANSI A108.6.
- Cementitious self-leveling underlayment—manufacturer's directions.
- Crack isolation membrane—ANSI A108.17.
- Waterproof membrane—ANSI A108.13.
- Electric radiant system—UL (US) 1693; UL (CAN/CSA) C22.2 #217; NEC Article 424.
- Movement Joints—EJ171 and ASTM C1193.

Notes

- Consult underlayment manufacturer for minimum cure/dry time of installation before radiant heat system is activated.
- Some manufacturers require special primers prior to the application of the setting materials or membrane. Follow manufacturer's directions.

RH116A-21

- **Above-Ground Concrete**
- **Electric Radiant Heat System Encapsulated in Cementitious Self-Leveling Underlayment**
- **Ceramic Tile**

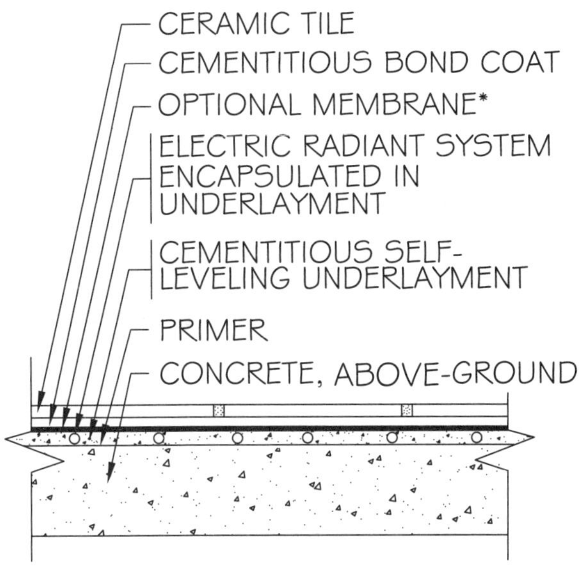

*USE OF A MEMBRANE IS OPTIONAL. SEE MEMBRANE OPTIONS.

Recommended Uses

- For above-ground structural slabs where floor leveling and electric radiant in-floor heat are desired.

Service Rating

- Moderate.

Environmental Exposure Classifications

- Res1; Com1.
- May be suitable for increased water exposure. See Membrane Options.
- For installations that may be exposed to staining, specify tile and grout suitable for exposure. Consult product manufacturers; see also "Product Selection Guides."
- For installations that may be exposed to mild chemical attack, specify epoxy grout and tile suitable for exposure. Consult product manufacturers; see also "Product Selection Guides."

Typical Weight of Tile Installation

- 11 pounds/square foot with ½" cementitious self-leveling underlayment. Add 2¾ pounds/square foot for each additional ¼" of underlayment.
- Does not include weight of substrate. See "Appendix B" for assumptions, included materials, and their individual weights.

Limitations

- Requires additional consideration by design professional to accommodate movement and/or deflection. Setting materials with improved bond strength and deformability are required.
- Not for use where moisture vapor transmission rate exceeds underlayment manufacturer's limitations.

Membrane Options

- A crack isolation membrane (A118.12) may be required. Check with heat system and setting material manufacturers.
- Crack isolation membrane may also provide waterproofing. Consult membrane manufacturer.
- A waterproof membrane (A118.10) may be required for areas exposed to moisture. Consult underlayment manufacturer for requirements.
- Specifier shall indicate if complete waterproofing is required, including if/how membrane connects to drain assembly, if base flashing is required, and treatment at other termination points.
- A crack isolation membrane may be specified to treat existing in-plane cracks (F125-partial) or for protection against future in-plane cracking (F-125 full). See F125 for more information.
- Check with membrane manufacturers for suitability for applicable conditions, as not all membranes are suitable for steam, high temperature, and/or chemical exposure, exterior use, use over above-ground structural slabs, use over pourable underlayments, use with radiant heat, or use over concrete with excessive moisture vapor transmission and/or alkalinity. Membrane may also affect service rating.

Requirements

- Consult the manufacturer for maximum moisture vapor transmission limitations prior to application of underlayment.
- Slab to be well cured, dimensionally stable, and free of cracks, surface laitance, waxy or oily films, curing compounds, and other surface bond-inhibiting contaminants.
- Cementitious self-leveling underlayment thickness and application varies, consult the manufacturer for specific recommendations.
- Surfaces must be prepared and primed according to the underlayment manufacturer's instructions.
- Proper curing/drying of underlayments prior to application of tile is critical for proper performance. Consult the underlayment manufacturer for specific instructions.
- Electric radiant system should not be installed over building expansion joints.
- Above-ground installations are inherently more

susceptible to vibration and deflection. Grout and mortar manufacturers to warrant product suitability.

Materials

- **Multiple options exist for membranes, mortars, grouts, and other materials and *must be clearly specified* to be included. If not specifically indicated, optional materials are not included and mortar/grout choice defaults to minimum performance specification indicated. Consider each system component and intended use to determine minimum requirements and to specify options.**
- Ceramic tile—ANSI A137.1.
- Cementitious grout—ANSI A118.6 or better or ISO CG1 or better.
- Epoxy grout, when used—ANSI A118.3 or ISO RG.
- Cementitious bond coat:
  - Must be recommended by manufacturer for above-ground use.
  - When a membrane is not used—ANSI A118.15 or better or ISO C2S1 or better.
  - When a crack isolation membrane is used—ANSI A118.4 or better or ISO C2S1 or better.
  - When a waterproof membrane is used—ANSI A118.4 or better or ISO C2S1 or better, unless ANSI A118.1 or ISO C1 is recommended by membrane manufacturer.
- Crack isolation membrane, when used—ANSI A118.12.
- Waterproof membrane, when used—ANSI A118.10.
- Self-leveling underlayment—meeting performance requirements of ASTM C627 for the anticipated service level designated by the specifier or intended use.

Materials for Green/Sustainable Design

- See "Green Building Standards and Green Product Selection Guide" and consult manufacturers and suppliers for product sustainability and contribution to green building design.
- Consider specifying tile and installation materials that meet ANSI A138.1, the *American National Standard Specifications for Sustainable Ceramic Tiles, Glass Tiles, and Tile Installation Materials.*

Preparation by Other Trades

- Floor systems over which tile will be installed shall be in conformance with the IRC for residential applications, the IBC for commercial applications, or applicable building codes. Maximum allowable substrate deflection under live load not to exceed l/360. See also "Substrate Requirements."
- When concentrated loads (scissor lifts, pallet jacks, automobiles, forklifts, etc.) will be used on an above-ground tiled floor, the engineer and/or specifier shall specify a substrate to accommodate the concentrated loads. Owner/specifier is responsible for protecting the tilework from damage, including allowing sufficient time for installed materials to cure properly.
- Slab to have a fine broom finish with no curing compounds used. When used, mechanical scarifying is necessary.
- Maximum allowable variation in the installation substrate (concrete)—¼" in 10' from the required plane.
- Maximum allowable variation in the tile substrate—for tiles with all edges shorter than 15", maximum allowable variation is ¼" in 10' from the required plane, with no more than 1⁄16" variation in 12" when measured from the high points in the surface. For tiles with at least one edge 15" in length, maximum allowable variation is ⅛" in 10' from the required plane, with no more than 1⁄16" variation in 24" when measured from the high points in the surface.
- Electrician or qualified contractor to wire electric radiant system to power source.

Movement Joint (architect must specify type of joint and show location and details on drawings)

- Movement joints—mandatory according to EJ171. For above-ground installations, additional movement joints are required.

Installation Specifications

- Tile—ANSI A108.5.
- Cementitious grout—ANSI A108.10.
- Epoxy grout—ANSI A108.6.
- Cementitious self-leveling underlayment—manufacturer's directions.
- Crack isolation membrane—ANSI A108.17.
- Waterproof membrane—ANSI A108.13.
- Electric radiant system—UL (US) 1693; UL (CAN/CSA) C22.2 #217; NEC Article 424.
- Movement Joints—EJ171 and ASTM C1193.

Notes

- Consult underlayment manufacturer for minimum cure/dry time of installation before radiant heat system is activated.
- F111 is preferred over above-ground structural slabs and other floors subject to movement and/or deflection.
- Some manufacturers require special primers prior to the application of the setting materials or membrane. Follow manufacturer's directions.
- Not all mortars, grouts, and membranes are suitable for above-ground use. Check manufacturer recommendations.

RH122-21

- **Joists max. 16" o.c./Plywood Subfloor**
- **Poured Gypsum Underlayment Encapsulating Hydronic Tubing**
- **Bonded Membrane**
- **Ceramic Tile, Glass Tile**

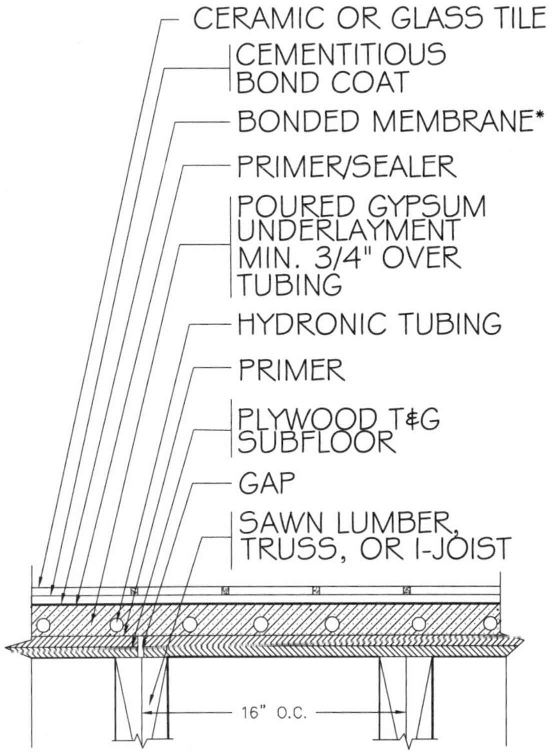

*USE OF A MEMBRANE IS REQUIRED. SEE MEMBRANE OPTIONS.

Recommended Uses

- For wood substrates with a poured gypsum underlayment encapsulating hydronic tubing.

Service Rating

- Light commercial.
- When glass tile is used, service rating may be lower.

Environmental Exposure Classifications

- Res1; Com1.
- May be suitable for limited water exposure areas (Res2; Com2). See Membrane Options.
- For installations that may be exposed to staining, specify tile and grout suitable for exposure. Consult product manufacturers; see also "Product Selection Guides."
- For installations that may be exposed to mild chemical attack, specify epoxy grout and tile suitable for exposure. Consult product manufacturers; see also "Product Selection Guides."

Typical Weight of Tile Installation

- 19 pounds/square foot with ¾" poured gypsum over hydronic tubing. Add 2¼ pounds/square foot for each additional ¼" of poured gypsum.
- Does not include weight of substrate. See "Appendix B" for assumptions, included materials, and their individual weights.

Limitations

- Maximum joist spacing 16" on center.

Membrane Options

- Requires use of crack isolation (A118.12) and/or waterproof (A118.10) and/or uncoupling membrane.
- Membrane must meet ANSI A118.10 for limited water exposure areas (Res2; Com2). Consult underlayment manufacturer for requirements.
- Specifier shall indicate if complete waterproofing is required, including if/how membrane connects to drain assembly, if base flashing is required, and treatment at other termination points.
- Check with membrane manufacturers for suitability for applicable conditions, as not all membranes are suitable for steam, high temperature, and/or chemical exposure, exterior use, use over above-ground structural slabs, use over pourable underlayments, use with radiant heat, or use over concrete with excessive moisture vapor transmission and/or alkalinity. Membrane may also affect service rating.
- When glass tile is used, consult glass tile manufacturer for membrane options and recommendations.

Requirements

- Proper curing/drying of underlayment prior to application of tile is critical for proper performance. Consult the underlayment manufacturer for specific instructions.

Materials

- **Multiple options exist for membranes, mortars, grouts, and other materials and *must be clearly specified* to be included. If not specifically indicated, optional materials are not included and mortar/grout choice defaults to minimum performance specification indicated. Consider each system component and intended use to determine minimum requirements and to specify options.**
- Ceramic tile—ANSI A137.1.
- Glass tile, when used—ANSI A137.2; see also "Glass Tile Selection and Installation Guide," and consult tile manufacturer for service rating and environmental exposure classification recommendations. Not all glass tiles are suitable.
- Cementitious grout—ANSI A118.6 or better or ISO CG1 or better. When glass tile is used, specify grout designated by tile and grout manufacturers.
- Epoxy grout, when used—ANSI A118.3 or ISO RG.

- Cementitious bond coat:
  - When a crack isolation membrane is used—ANSI A118.4 or better or ISO C2S1 or better.
  - When a waterproof membrane is used—ANSI A118.4 or better or ISO C2S1 or better, unless ANSI A118.1 or ISO C1 is recommended by membrane manufacturer.
  - When an uncoupling membrane is used, check with membrane manufacturer.
  - When glass tile is used, specify mortar designated by tile and mortar manufacturers. Bond coat color will impact the final appearance of translucent glass tile. Specifier shall confirm bond coat color is acceptable.
- Crack isolation membrane, when used—ANSI A118.12.
- Waterproof membrane, when used—ANSI A118.10.
- Uncoupling membrane, when used—recommended by uncoupling membrane manufacturer; must achieve 50 psi or greater shear bond strength in 7 days per the test method in ANSI A118.12 Section 5.1.3.

Materials for Green/Sustainable Design

- See "Green Building Standards and Green Product Selection Guide" and consult manufacturers and suppliers for product sustainability and contribution to green building design.
- Consider specifying tile and installation materials that meet ANSI A138.1, the *American National Standard Specifications for Sustainable Ceramic Tiles, Glass Tiles, and Tile Installation Materials.*

Preparation by Other Trades

- Floor systems, including the framing system and subfloor panels, over which tile will be installed shall be in conformance with the IRC for residential applications, the IBC for commercial applications, or applicable building codes. Maximum allowable floor member live load and concentrated load deflection shall not exceed l/360. See also "Substrate Requirements."
- When concentrated loads (scissor lifts, pallet jacks, automobiles, forklifts, etc.) will be used on an aboveground tiled floor, the engineer and/or specifier shall specify a substrate to accommodate the concentrated loads. Owner/specifier is responsible for protecting the tilework from damage, including allowing sufficient time for installed materials to cure properly.
- Poured gypsum underlayment—minimum compressive strength of 2000 psi when tested per ASTM C472 modified and meeting performance requirements of ASTM C627 for the anticipated service level designated by the specifier or intended use.
- Poured gypsum underlayment thickness and application varies, consult the manufacturer for specific recommendations when used with floor warming systems. Minimum thickness is ¾" above hydronic tubing.
- Exterior grade plywood subfloor to be clean and free of dirt, dust, paint, and oily film.
- Face grain of plywood shall run perpendicular to joists.
- Surfaces must be prepared and primed according to the underlayment manufacturer's instructions.
- Subfloor—minimum 23⁄32" tongue and groove exterior glue plywood.
- Maximum allowable variation in the tile substrate—for tiles with all edges shorter than 15", maximum allowable variation is ¼" in 10' from the required plane, with no more than 1⁄16" variation in 12" when measured from the high points in the surface. For tiles with at least one edge 15" in length, maximum allowable variation is ⅛" in 10' from the required plane, with no more than 1⁄16" variation in 24" when measured from the high points in the surface.
- Radiant tubing by others.
- Some manufacturers of poured gypsum underlayments require use of plastic lath. Follow manufacturer's recommendations for installation and use of the lath.
- Poured gypsum underlayment—installed only by a manufacturer-approved applicator in accordance with manufacturer's recommendations.

Movement Joint (architect must specify type of joint and show location and details on drawings)

- Movement joints—mandatory in accordance with EJ171.
- When glass tile is used, adhere to more frequent placement recommendations within the ranges listed in EJ171.

Installation Specifications

- Tile—ANSI A108.5.
- Glass tile—manufacturer's directions.
- Cementitious grout—ANSI A108.10.
- Epoxy grout—ANSI A108.6.
- Crack isolation membrane—ANSI A108.17.
- Waterproof membrane—ANSI A108.13.
- Uncoupling membrane—manufacturer's directions.
- Movement Joints—EJ171 and ASTM C1193.

Notes

- Consult underlayment manufacturer for minimum cure/dry time of installation before radiant heat system is activated.
- Some manufacturers require special primers prior to the application of the membrane. Follow manufacturer's directions.
- When glass tile is used, see "Glass Tile Selection and Installation Guide," and consult manufacturer for recommendations and requirements.

RH123-21

- **Joists max. 16" o.c./Plywood Subfloor**
- **Cementitious Self-Leveling Underlayment Encapsulating Hydronic Tubing**
- **Ceramic Tile, Glass Tile**

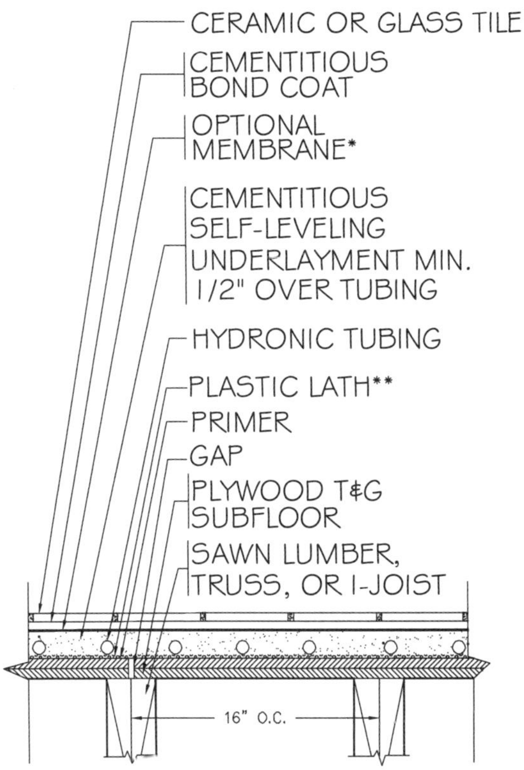

*USE OF A MEMBRANE IS OPTIONAL. SEE MEMBRANE OPTIONS.

**IF NECESSARY, PLASTIC LATH MAY BE INSTALLED ON TOP OF HYDRONIC TUBING. CONTACT MANUFACTURER FOR DIRECTIONS.

Recommended Uses

- For wood substrates where hydronic tubing will be placed.

Service Rating

- Light commercial.
- When glass tile is used, service rating may be lower.

Environmental Exposure Classifications

- Res1; Com1.
- May be suitable for increased water exposure. See Membrane Options.
- For installations that may be exposed to staining, specify tile and grout suitable for exposure. Consult product manufacturers; see also "Product Selection Guides."
- For installations that may be exposed to mild chemical attack, specify epoxy grout and tile suitable for exposure. Consult product manufacturers; see also "Product Selection Guides."

Typical Weight of Tile Installation

- 19 pounds/square foot with ½" cementitious self-leveling underlayment over hydronic tubing. Add 2¾ pounds/square foot for each additional ¼" of underlayment.
- Does not include weight of substrate. See "Appendix B" for assumptions, included materials, and their individual weights.

Limitations

- Maximum joist spacing 16" on center.

Membrane Options

- A crack isolation membrane (A118.12) may be required. Check with heat system and setting material manufacturers.
- Crack isolation membrane may also provide waterproofing. Consult membrane manufacturer.
- A waterproof membrane (A118.10) may be required for areas exposed to moisture. Consult underlayment manufacturer for requirements.
- Specifier shall indicate if complete waterproofing is required, including if/how membrane connects to drain assembly, if base flashing is required, and treatment at other termination points.
- A crack isolation membrane may be specified to treat existing in-plane cracks (F125-partial) or for protection against future in-plane cracking (F-125 full). See F125 for more information.
- An uncoupling membrane may be specified to accommodate subfloor movement. Check with membrane manufacturer for recommendations and requirements.
- Check with membrane manufacturers for suitability for applicable conditions, as not all membranes are suitable for steam, high temperature, and/or chemical exposure, exterior use, use over above-ground structural slabs, use over pourable underlayments, use with radiant heat, or use over concrete with excessive moisture vapor transmission and/or alkalinity. Membrane may also affect service rating.
- When glass tile is used, consult glass tile manufacturer for membrane options and recommendations.

Requirements

- Cementitious self-leveling underlayment thickness and application varies, consult the manufacturer for specific recommendations when used with floor warming systems. Minimum thickness is ½" above hydronic tubing.
- Plywood subfloor to be clean and free of dirt, dust, paint, and oily film.
- Surfaces must be prepared and primed according to the underlayment manufacturer's instructions.
- Use plastic lath designed for this purpose, nailed or stapled in accordance with the underlayment manufacturer's instructions.
- Proper curing/drying of underlayments prior to application of tile is critical for proper performance. Consult the underlayment manufacturer for specific instructions.

Materials

- **Multiple options exist for membranes, mortars, grouts, and other materials and *must be clearly specified* to be included. If not specifically indicated, optional materials are not included and mortar/grout choice defaults to minimum performance specification indicated. Consider each system component and intended use to determine minimum requirements and to specify options.**
- Ceramic tile—ANSI A137.1.
- Glass tile, when used—ANSI A137.2; see also "Glass Tile Selection and Installation Guide," and consult tile manufacturer for service rating and environmental exposure classification recommendations. Not all glass tiles are suitable.
- Cementitious grout—ANSI A118.6 or better or ISO CG1 or better. When glass tile is used, specify grout designated by tile and grout manufacturers.
- Epoxy grout, when used—ANSI A118.3 or ISO RG.
- Cementitious bond coat:
  - When a membrane is not used—ANSI A118.15 or better or ISO C2S1 or better.
  - When a crack isolation membrane is used—ANSI A118.4 or better or ISO C2S1 or better.
  - When a waterproof membrane is used—ANSI A118.4 or better or ISO C2S1 or better, unless ANSI A118.1 or ISO C1 is recommended by membrane manufacturer.
  - When an uncoupling membrane is used, check with membrane manufacturer.
  - When glass tile is used, specify mortar designated by tile and mortar manufacturers. Bond coat color will impact the final appearance of translucent glass tile. Specifier shall confirm bond coat color is acceptable.
- Crack isolation membrane, when used—ANSI A118.12.
- Waterproof membrane, when used—ANSI A118.10.
- Uncoupling membrane, when used—recommended by uncoupling membrane manufacturer; must achieve 50 psi or greater shear bond strength in 7 days per the test method in ANSI A118.12 Section 5.1.3.
- Self-leveling underlayment—meeting performance requirements of ASTM C627 for the anticipated service level designated by the specifier or intended use.

Materials for Green/Sustainable Design

- See "Green Building Standards and Green Product Selection Guide" and consult manufacturers and suppliers for product sustainability and contribution to green building design.
- Consider specifying tile and installation materials that meet ANSI A138.1, the *American National Standard Specifications for Sustainable Ceramic Tiles, Glass Tiles, and Tile Installation Materials.*

Preparation by Other Trades

- Floor systems, including the framing system and subfloor panels, over which tile will be installed shall be in conformance with the IRC for residential applications, the IBC for commercial applications, or applicable building codes. Maximum allowable floor member live load and concentrated load deflection shall not exceed l/360. See also "Substrate Requirements."
- When concentrated loads (scissor lifts, pallet jacks, automobiles, forklifts, etc.) will be used on an above-ground tiled floor, the engineer and/or specifier shall specify a substrate to accommodate the concentrated loads. Owner/specifier is responsible for protecting the tilework from damage, including allowing sufficient time for installed materials to cure properly.
- Face grain of plywood shall run perpendicular to joists.
- Subfloor—minimum 23⁄32" tongue and groove exterior glue plywood.
- Maximum allowable variation in the installation substrate (plywood)—¼" in 10' from the required plane.
- Maximum allowable variation in the tile substrate—for tiles with all edges shorter than 15", maximum allowable variation is ¼" in 10' from the required plane, with no more than 1⁄16" variation in 12" when measured from the high points in the surface. For tiles with at least one edge 15" in length, maximum allowable variation is ⅛" in 10' from the required plane, with no more than 1⁄16" variation in 24" when measured from the high points in the surface.
- Radiant tubing by others.

Movement Joint (architect must specify type of joint and show location and details on drawings)

- Movement joints—mandatory in accordance with EJ171.
- When glass tile is used, adhere to more frequent placement recommendations within the ranges listed in EJ171.

Installation Specifications

- Tile—ANSI A108.5.
- Glass tile—manufacturer's directions.
- Cementitious grout—ANSI A108.10.
- Epoxy grout—ANSI A108.6.
- Cementitious self-leveling underlayment—manufacturer's directions.
- Crack isolation membrane—ANSI A108.17.
- Waterproof membrane—ANSI A108.13.
- Uncoupling membrane—manufacturer's directions.
- Movement Joints—EJ171 and ASTM C1193.

Notes

- Consult underlayment manufacturer for minimum cure/dry time of installation before radiant heat system is activated.
- Some manufacturers require special primers prior to the application of the setting materials or membrane. Follow manufacturer's directions.
- When glass tile is used, see "Glass Tile Selection and Installation Guide," and consult manufacturer for recommendations and requirements.

RH141-21

- **Joists max. 16" o.c./Plywood Subfloor**
- **Unbonded Mortar Bed Encapsulating Hydronic Tubing**
- **Ceramic Tile, Glass Tile**

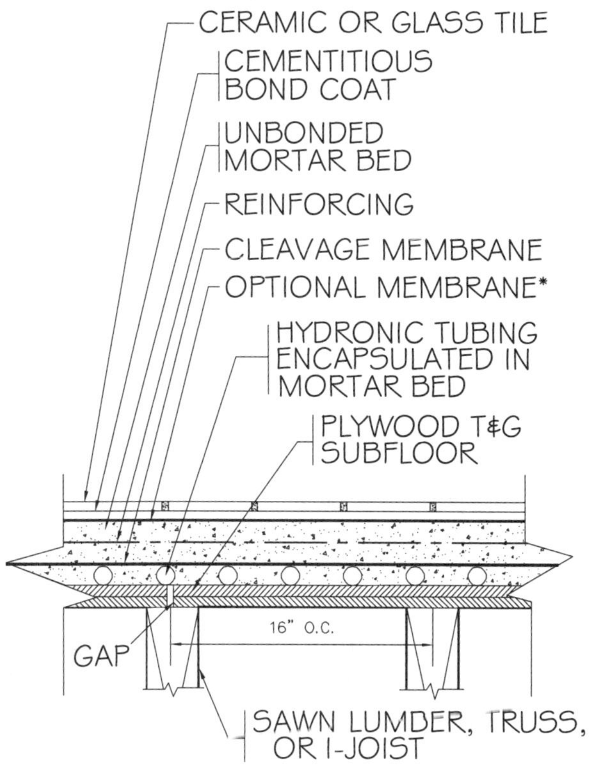

Recommended Uses

- For floor leveling or flattening, especially where multiple finish floor heights must be accommodated or slopes are required and where hydronic tubing will be placed.
- For installations that must be at least 1½" thick (nominal, including tile) and where hydronic tubing will be placed.
- For isolating the tile installation from the floor framing where hydronic tubing will be placed.

Service Rating

- Light commercial.
- When glass tile is used, service rating may be lower.

Environmental Exposure Classifications

- Res1, 2; Com1, 2.
- May be suitable for increased water exposure. See Membrane Options.
- May be suitable for exterior applications in areas not subject to freeze/thaw cycling when appropriate precautions are taken, including expansion joint placement, proper slope, waterproofing, and material selection.
- For installations that may be exposed to staining, specify tile and grout suitable for exposure. Consult product manufacturers; see also "Product Selection Guides."
- For installations that may be exposed to mild chemical attack, specify epoxy grout and tile suitable for exposure. Consult product manufacturers; see also "Product Selection Guides."

Typical Weight of Tile Installation

- 30 pounds/square foot with 1¼" mortar bed over hydronic tubing. Add 3 pounds/square foot for each additional ¼" of mortar bed.
- Does not include weight of substrate. See "Appendix B" for assumptions, included materials, and their individual weights.

Limitations

- Maximum joist spacing 16" on center.

Membrane Options

- A crack isolation membrane (A118.12) may be required. Check with heat system and setting material manufacturers.
- Crack isolation membrane may also provide waterproofing. Consult membrane manufacturer.
- A waterproof membrane (A118.10) may be specified to prevent moisture intrusion and protect adjacent walls and building materials. Base flashing should be used for maximum effectiveness.
- Specifier shall indicate if complete waterproofing is required, including if/how membrane connects to drain assembly, if base flashing is required, and treatment at other termination points.
- Check with membrane manufacturers for suitability for applicable conditions, as not all membranes are suitable for steam, high temperature, and/or chemical exposure, exterior use, use over above-ground structural slabs, use over pourable underlayments, use with radiant heat, or use over concrete with excessive moisture vapor transmission and/or alkalinity. Membrane may also affect service rating.
- When glass tile is used, consult glass tile manufacturer for membrane options and recommendations.

Requirements

- Screed fill flush to top of hydronic tubing before placing a cleavage membrane and reinforced mortar bed.
- Reinforcing mesh mandatory.
- Mortar bed thickness—1¼" minimum to 2" maximum.
- Mortar beds in excess of 2" thick shall be detailed by the architect.

Materials

- **Multiple options exist for membranes, mortars, grouts, and other materials and *must be clearly specified* to be included. If not specifically indicated, optional materials are not included and mortar/grout choice defaults to minimum performance specification indicated. Consider each system component and intended use to determine minimum requirements and to specify options.**
- Ceramic tile—ANSI A137.1.
- Glass tile, when used—ANSI A137.2; see also "Glass Tile Selection and Installation Guide," and consult tile manufacturer for service rating and environmental exposure classification recommendations. Not all glass tiles are suitable.
- Cementitious grout—ANSI A118.6 or better or ISO CG1 or better. When glass tile is used, specify grout designated by tile and grout manufacturers.
- Epoxy grout, when used—ANSI A118.3 or ISO RG.
- Cementitious bond coat—portland cement paste on a mortar bed that is still workable. For a cured mortar bed, follow recommendations below to select appropriate bonding mortar:
  - When a membrane is not used on top of the mortar bed—ANSI A118.15 or better or ISO C2S1 or better.
  - When a crack isolation membrane is used on top of the mortar bed—ANSI A118.4 or better or ISO C2S1 or better.
  - When a waterproof membrane is used on top of the mortar bed—ANSI A118.4 or better or ISO C2S1 or better, unless ANSI A118.1 or ISO C1 is recommended by membrane manufacturer.
  - When glass tile is used, specify mortar designated by tile and mortar manufacturers. Bond coat color will impact the final appearance of translucent glass tile. Specifier shall confirm bond coat color is acceptable.
- Mortar bed, reinforcing, and cleavage membrane—ANSI A108.1A.
- Crack isolation membrane, when used—ANSI A118.12.
- Waterproof membrane, when used—ANSI A118.10.

Materials for Green/Sustainable Design

- See "Green Building Standards and Green Product Selection Guide" and consult manufacturers and suppliers for product sustainability and contribution to green building design.
- Consider specifying tile and installation materials that meet ANSI A138.1, the *American National Standard Specifications for Sustainable Ceramic Tiles, Glass Tiles, and Tile Installation Materials.*

Preparation by Other Trades

- Floor systems, including the framing system and subfloor panels, over which tile will be installed shall be in conformance with the IRC for residential applications, the IBC for commercial applications, or applicable building codes. Maximum allowable floor member live load and concentrated load deflection shall not exceed l/360. See also "Substrate Requirements."
- When concentrated loads (scissor lifts, pallet jacks, automobiles, forklifts, etc.) will be used on an above-ground tiled floor, the engineer and/or specifier shall specify a substrate to accommodate the concentrated loads. Owner/specifier is responsible for protecting the tilework from damage, including allowing sufficient time for installed materials to cure properly.
- Face grain of plywood shall run perpendicular to joists.
- Subfloor—minimum 19⁄32" tongue and groove plywood with 1⁄8" gap between sheets or 1" nominal boards.
- Depressing subfloor between joists on ledger strips permissible in residential use.
- Maximum allowable variation in the installation substrate (plywood)—1⁄4" in 10' from the required plane.

Movement Joint (architect must specify type of joint and show location and details on drawings)

- Movement joints—mandatory according to EJ171.
- When glass tile is used, adhere to more frequent placement recommendations within the ranges listed in EJ171.

Installation Specifications

- Tile—ANSI A108.1A,.1B, or .1C. A108.1B required if membrane to be used.
- Glass tile—ANSI A108.14, .15, .16, or manufacturer's directions.
- Cementitious grout—ANSI A108.10.
- Epoxy grout—ANSI A108.6.
- Crack isolation membrane—A108.17.
- Waterproof membrane—A108.13.
- Movement Joints—EJ171 and ASTM C1193.

Notes

- When glass tile is used, see "Glass Tile Selection and Installation Guide," and consult manufacturer for recommendations and requirements.

RH130-21

- **Joists max. 16" o.c./Plywood Subfloor**
- **Plywood Underlayment**
- **Electric Radiant Heat System Encapsulated in Mortar**
- **Ceramic Tile**

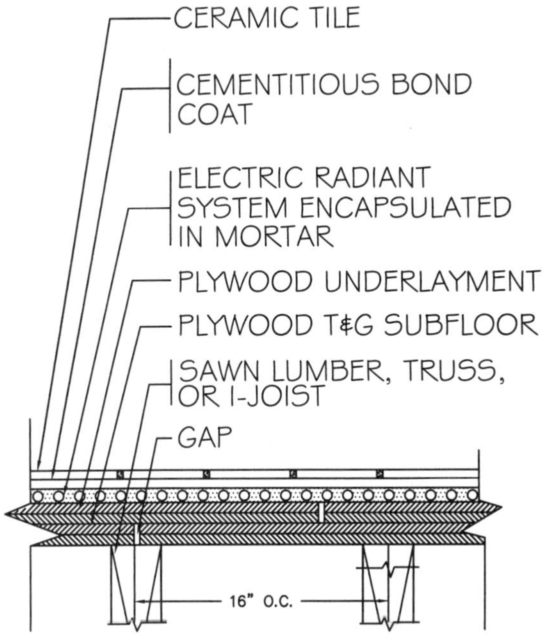

USE OF A MEMBRANE IS OPTIONAL.
SEE MEMBRANE OPTIONS.

Recommended Uses

- For wood substrates in dry areas where electric radiant in-floor heat is desired.

Service Rating

- Residential with minimum 15⁄32"-thick plywood underlayment.
- Light commercial with minimum 19⁄32"-thick plywood underlayment.

Environmental Exposure Classifications

- Res1; Com1.
- For installations that may be exposed to staining, specify tile and grout suitable for exposure. Consult product manufacturers; see also "Product Selection Guides."
- For installations that may be exposed to mild chemical attack, specify epoxy grout and tile suitable for exposure. Consult product manufacturers; see also "Product Selection Guides."

Typical Weight of Tile Installation

- 7 pounds/square foot.
- Does not include weight of substrate. See "Appendix B" for assumptions, included materials, and their individual weights.

Limitations

- Maximum joist spacing 16" on center.

Membrane Options

- Consult heat system manufacturer and setting material manufacturer for requirements and options.
- If a membrane is used, follow manufacturer's instructions for membrane type, installation, placement, and setting materials required.

Requirements

- Plywood underlayment—clean and free of dirt, dust, and oily film.
- Gaps between plywood underlayment sheets to be treated per setting material manufacturer's recommendations.
- Electric radiant system should not be installed over building expansion joints.

Materials

- **Multiple options exist for membranes, mortars, grouts, and other materials and *must be clearly specified* to be included. If not specifically indicated, optional materials are not included and mortar/grout choice defaults to minimum performance specification indicated. Consider each system component and intended use to determine minimum requirements and to specify options.**
- Ceramic tile—ANSI A137.1.
- Cementitious grout—ANSI A118.6 or better or ISO CG1 or better.
- Epoxy grout, when used—ANSI A118.3 or ISO RG.
- Cementitious bond coat:
  - Mortar for encapsulating electric radiant system—ANSI A118.11 or ISO C2S1P1.
  - Mortar for bonding tile, when a membrane is not used—ANSI A118.15 or better or ISO C2S1 or better.
  - Mortar for bonding tile, when a membrane is used—ANSI A118.4 or better or ISO C2S1 or better, unless ANSI A118.1 or ISO C1 is recommended by membrane manufacturer.

Materials for Green/Sustainable Design

- See "Green Building Standards and Green Product Selection Guide" and consult manufacturers and suppliers for product sustainability and contribution to green building design.
- Consider specifying tile and installation materials that meet ANSI A138.1, the *American National Standard Specifications for Sustainable Ceramic Tiles, Glass Tiles, and Tile Installation Materials.*

Preparation by Other Trades

- Floor systems, including the framing system and subfloor panels, over which tile will be installed shall be in conformance with the IRC for residential applications,

the IBC for commercial applications, or applicable building codes. Maximum allowable floor member live load and concentrated load deflection shall not exceed l/360. See also "Substrate Requirements."

- When concentrated loads (scissor lifts, pallet jacks, automobiles, forklifts, etc.) will be used on an above-ground tiled floor, the engineer and/or specifier shall specify a substrate to accommodate the concentrated loads. Owner/specifier is responsible for protecting the tilework from damage, including allowing sufficient time for installed materials to cure properly.
- Plywood subfloor—minimum 19⁄32" exterior-glue tongue and groove plywood.
- Plywood underlayment—minimum 15⁄32" exterior glue plywood with 1⁄8" gaps between sheets. See also Service Rating.
- Maximum allowable variation in the tile substrate—for tiles with all edges shorter than 15", maximum allowable variation is 1⁄4" in 10' from the required plane, with no more than 1⁄16" variation in 12" when measured from the high points in the surface. For tiles with at least one edge 15" in length, maximum allowable variation is 1⁄8" in 10' from the required plane, with no more than 1⁄16" variation in 24" when measured from the high points in the surface. Adjacent edges of plywood sheets not to exceed 1⁄32" difference in height.
- Face grain of plywood subfloor and underlayment shall run perpendicular to joists.
- Electrician or qualified contractor to wire electric radiant system to power source.

Movement Joint (architect must specify type of joint and show location and details on drawings)

- Movement joints—mandatory in accordance with EJ171.

Installation Specifications

- Tile—ANSI A108.5.
- Cementitious grout—ANSI A108.10.
- Epoxy grout—ANSI A108.6.
- Electric radiant system—UL (US) 1693; UL (CAN/CSA) C22.2 #217; NEC Article 424.
- Movement Joints—EJ171 and ASTM C1193.

Notes

- Bond coat thickness and application varies, consult floor warming manufacturer for recommendations.
- If latex-portland cement mortar is to be applied over 3⁄16" in thickness, consult setting material manufacturer for additional installation instructions.
- Underlayment fasteners should not penetrate joists below.

RH135-21

- **Joists max. 16" o.c./Plywood Subfloor**
- **Backer Board**
- **Electric Radiant Heat System Encapsulated in Mortar**
- **Ceramic Tile**

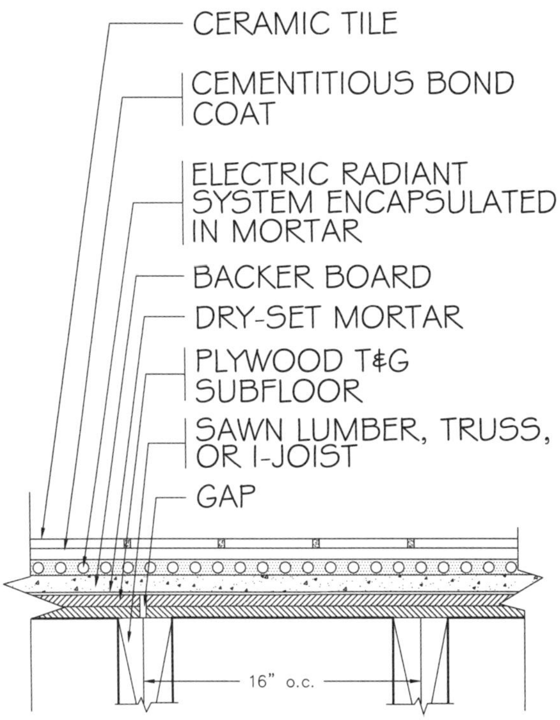

USE OF A MEMBRANE IS OPTIONAL. SEE MEMBRANE OPTIONS.

Recommended Uses

- For wood substrates where electric radiant in-floor heat and water resistance are desired.

Service Rating

- Residential with minimum $1\frac{9}{32}$"-thick plywood subfloor.
- Light commercial with minimum $\frac{23}{32}$"-thick plywood subfloor.

Environmental Exposure Classifications

- Res1, 2; Com1, 2.
- For installations that may be exposed to staining, specify tile and grout suitable for exposure. Consult product manufacturers; see also "Product Selection Guides."
- For installations that may be exposed to mild chemical attack, specify epoxy grout and tile suitable for exposure. Consult product manufacturers; see also "Product Selection Guides."

Typical Weight of Tile Installation

- 8 pounds/square foot if ¼" or ½" cementitious coated foam backer board used.
- 9 pounds/square foot if ¼" fiber-reinforced water-resistant gypsum or ¼" coated glass mat water-resistant gypsum backer board used.
- 10 pounds/square foot if ¼" cement backer board, ¼" or $\frac{7}{16}$" fiber cement backer board, ½" coated glass mat water-resistant gypsum backer board, or ⅜" fiber-reinforced water-resistant gypsum backer board used.
- 11 pounds/square foot if ½" fiber-reinforced water-resistant gypsum backer board used.
- 12 pounds/square foot if ½" cement backer board used.
- Does not include weight of substrate. See "Appendix B" for assumptions, included materials, and their individual weights.

Limitations

- Maximum joist spacing 16" on center.

Membrane Options

- Consult heat system manufacturer and setting material manufacturer for requirements and options.
- If a membrane is used, follow manufacturer's instructions for membrane type, installation, placement, required cure time based on backer board type, and setting materials required.

Requirements

- Treat joints between backer units per manufacturer's directions.
- Surface of backer units—clean and free of dirt, dust, or oily film.
- Electric radiant system should not be installed over building expansion joints.
- Use a sufficient amount of portland cement mortar under the backer board to establish a supporting plane and eliminate voids.
- Fasten backer units with corrosion-resistant fasteners per manufacturer's directions.

Materials

- **Multiple options exist for membranes, mortars, grouts, and other materials and *must be clearly specified* to be included. If not specifically indicated, optional materials are not included and mortar/grout choice defaults to minimum performance specification indicated. Consider each system component and intended use to determine minimum requirements and to specify options.**
- Ceramic tile—ANSI A137.1.
- Cementitious grout—ANSI A118.6 or better or ISO CG1 or better.
- Epoxy grout, when used—ANSI A118.3 or ISO RG.
- Cementitious bond coat:
  - When a membrane is not used—ANSI A118.15 or better or ISO C2S1 or better.

    - When a membrane is used—ANSI A118.4 or better or ISO C2S1 or better unless ANSI A118.1 or ISO C1 is recommended by membrane manufacturer.
- Cement backer board, when used—ANSI A118.9 or ASTM C1325.
- Cementitious-coated foam backer board, when used—ASTM C578 (minimum Type VI).
- Coated-glass mat water-resistant gypsum backer board, when used—ASTM C1178.
- Fiber-cement backer board, when used—ASTM C1288.
- Fiber-reinforced water-resistant gypsum backer board, when used—ASTM C1278.
- Mortar under backer board—ANSI A118.1 or better or ISO C1 or better.

Materials for Green/Sustainable Design

- See "Green Building Standards and Green Product Selection Guide" and consult manufacturers and suppliers for product sustainability and contribution to green building design.
- Consider specifying tile and installation materials that meet ANSI A138.1, the *American National Standard Specifications for Sustainable Ceramic Tiles, Glass Tiles, and Tile Installation Materials.*

Preparation by Other Trades

- Floor systems, including the framing system and subfloor panels, over which tile will be installed shall be in conformance with the IRC for residential applications, the IBC for commercial applications, or applicable building codes. Maximum allowable floor member live load and concentrated load deflection shall not exceed l/360. See also "Substrate Requirements."
- When concentrated loads (scissor lifts, pallet jacks, automobiles, forklifts, etc.) will be used on an aboveground tiled floor, the engineer and/or specifier shall specify a substrate to accommodate the concentrated loads. Owner/specifier is responsible for protecting the tilework from damage, including allowing sufficient time for installed materials to cure properly.
- Face grain of plywood shall run perpendicular to joists.
- Subfloor—minimum 19⁄32" exterior-glue tongue and groove plywood. See also Service Rating.
- Maximum allowable variation in the tile substrate—for tiles with all edges shorter than 15," maximum allowable variation is 1⁄4" in 10' from the required plane, with no more than 1⁄16" variation in 12" when measured from the high points in the surface. For tiles with at least one edge 15" in length, maximum allowable variation is 1⁄8" in 10' from the required plane, with no more than 1⁄16" variation in 24" when measured from the high points in the surface. Adjacent edges of plywood sheets not to exceed 1⁄32" difference in height.
- Electrician or qualified contractor to wire electric radiant system to power source.

Movement Joint (architect must specify type of joint and show location and details on drawings)

- Movement joints—mandatory in accordance with EJ171.

Installation Specifications

- Tile—ANSI A108.5.
- Cementitious grout—ANSI A108.10.
- Epoxy grout—ANSI A108.6.
- Backer board—ANSI A108.11 or manufacturer's directions.
- Electric radiant system—UL (US) 1693; UL (CAN/CSA) C22.2 #217; NEC Article 424.
- Movement Joints—EJ171 and ASTM C1193.

Notes

- Bond coat thickness and application varies, consult floor warming manufacturer for recommendations.
- If latex-portland cement mortar is to be applied over 3⁄16" in thickness, consult setting material manufacturer for additional installation instructions.

RH140-21

- **Joists max. 19.2" o.c./Plywood Subfloor**
- **Electric Radiant Heat System Encapsulated in Cementitious Self-Leveling Underlayment**
- **Ceramic Tile**

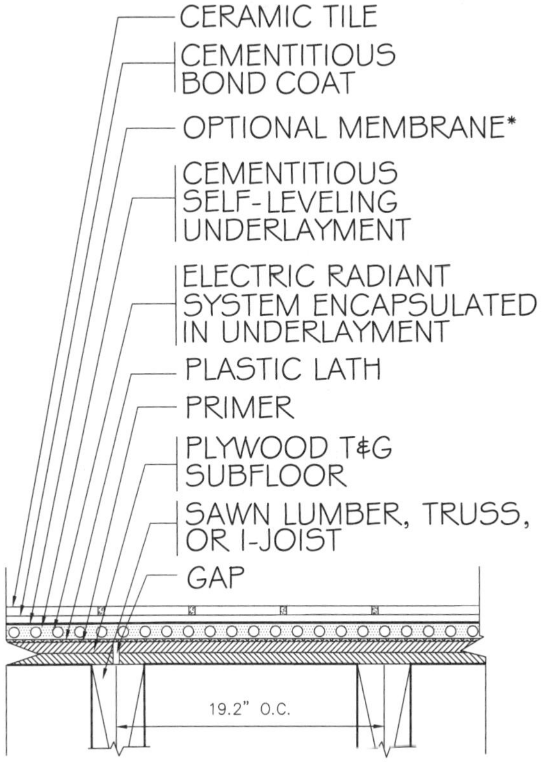

*USE OF A MEMBRANE IS OPTIONAL. SEE MEMBRANE OPTIONS.

Recommended Uses

- For wood substrates where joists are spaced 19.2" o.c. or less and electric radiant in-floor heat is desired.

Service Rating

- Residential.

Environmental Exposure Classifications

- Res1; Com1.
- May be suitable for increased water exposure. See Membrane Options.
- For installations that may be exposed to staining, specify tile and grout suitable for exposure. Consult product manufacturers; see also "Product Selection Guides."
- For installations that may be exposed to mild chemical attack, specify epoxy grout and tile suitable for exposure. Consult product manufacturers; see also "Product Selection Guides."

Typical Weight of Tile Installation

- 11 pounds/square foot with ½" cementitious self-leveling underlayment. Add 2¾ pounds/square foot for each additional ¼" of underlayment.
- Does not include weight of substrate. See "Appendix B" for assumptions, included materials, and their individual weights.

Limitations

- Maximum joist spacing 19.2" on center.

Membrane Options

- A crack isolation membrane (A118.12) may be required. Check with heat system and setting material manufacturers.
- Crack isolation membrane may also provide waterproofing. Consult membrane manufacturer.
- A waterproof membrane (A118.10) may be required for areas exposed to moisture. Consult underlayment manufacturer for requirements.
- Specifier shall indicate if complete waterproofing is required, including if/how membrane connects to drain assembly, if base flashing is required, and treatment at other termination points.
- A crack isolation membrane may be specified to treat existing in-plane cracks (F125-partial) or for protection against future in-plane cracking (F-125 full). See F125 for more information.
- Check with membrane manufacturers for suitability for applicable conditions, as not all membranes are suitable for steam, high temperature, and/or chemical exposure, exterior use, use over above-ground structural slabs, use over pourable underlayments, use with radiant heat, or use over concrete with excessive moisture vapor transmission and/or alkalinity. Membrane may also affect service rating.

Requirements

- Plywood subfloor—clean and free of dirt, dust, paint, and oily film.
- Surfaces must be prepared and primed according to the underlayment manufacturer's instructions.
- Use plastic lath designed for this purpose, nailed or stapled in accordance with the underlayment manufacturer's instructions.
- Self-leveling underlayment thickness and application varies, consult the manufacturer for specific recommendations when used with electric radiant systems.
- Proper curing/drying of underlayments prior to the application of tile is critical for proper performance. Consult the underlayment manufacturer for specific instructions.
- Electric radiant system should not be installed over building expansion joints.

Materials

- **Multiple options exist for membranes, mortars, grouts, and other materials and *must be clearly specified* to be included. If not specifically indicated, optional materials are not included and mortar/grout**

choice defaults to minimum performance specification indicated. Consider each system component and intended use to determine minimum requirements and to specify options.

- Ceramic tile—ANSI A137.1.
- Cementitious grout—ANSI A118.6 or better or ISO CG1 or better.
- Epoxy grout, when used—ANSI A118.3 or ISO RG.
- Cementitious bond coat:
  - When a membrane is not used—ANSI A118.15 or better or ISO C2S1 or better.
  - When a crack isolation membrane is used—ANSI A118.4 or better or ISO C2S1 or better.
  - When a waterproof membrane is used—ANSI A118.4 or better or ISO C2S1 or better, unless ANSI A118.1 or ISO C1 is recommended by membrane manufacturer.
- Waterproof membrane, when used—ANSI A118.10.
- Crack isolation membrane, when used—ANSI A118.12.
- Self-leveling underlayment—meeting performance requirements of ASTM C627 for the anticipated service level designated by the specifier or intended use.

Materials for Green/Sustainable Design

- See "Green Building Standards and Green Product Selection Guide" and consult manufacturers and suppliers for product sustainability and contribution to green building design.
- Consider specifying tile and installation materials that meet ANSI A138.1, the *American National Standard Specifications for Sustainable Ceramic Tiles, Glass Tiles, and Tile Installation Materials*.

Preparation by Other Trades

- Floor systems, including the framing system and subfloor panels, over which tile will be installed shall be in conformance with the IRC for residential applications, the IBC for commercial applications, or applicable building codes. Maximum allowable floor member live load and concentrated load deflection shall not exceed l/360. See also "Substrate Requirements."
- When concentrated loads (scissor lifts, pallet jacks, automobiles, forklifts, etc.) will be used on an above-ground tiled floor, the engineer and/or specifier shall specify a substrate to accommodate the concentrated loads. Owner/specifier is responsible for protecting the tilework from damage, including allowing sufficient time for installed materials to cure properly.
- Face grain of plywood shall run perpendicular to joists.
- Subfloor—minimum $^{23}/_{32}$" tongue and groove exterior-glue plywood with $^{1}/_{8}$" gap between sheets.
- Maximum allowable variation in the installation substrate (plywood)—$^{1}/_{4}$" in 10' from the required plane.
- Maximum allowable variation in the tile substrate—for tiles with all edges shorter than 15", maximum allowable variation is $^{1}/_{4}$" in 10' from the required plane, with no more than $^{1}/_{16}$" variation in 12" when measured from the high points in the surface. For tiles with at least one edge 15" in length, maximum allowable variation is $^{1}/_{8}$" in 10' from the required plane, with no more than $^{1}/_{16}$" variation in 24" when measured from the high points in the surface.
- Electrician or qualified contractor to wire electric radiant system to power source.

Movement Joint (architect must specify type of joint and show location and details on drawings)

- Movement joints—mandatory in accordance with EJ171.

Installation Specifications

- Tile—ANSI A108.5.
- Cementitious grout—ANSI A108.10.
- Epoxy grout—ANSI A108.6.
- Electric radiant systems—UL (US) 1693; UL (CAN/CSA) C22.2 #217; NEC Article 424.
- Cementitious self-leveling underlayment—manufacturer's directions.
- Crack isolation membrane—ANSI A108.17.
- Waterproof membrane—ANSI A108.13.
- Movement Joints—EJ171 and ASTM C1193.

Notes

- Consult underlayment manufacturer for minimum cure time of installation before radiant heat system is activated.
- Some manufacturers require special primers prior to the application of the setting materials or membrane. Follow manufacturer's directions.

F141-21

- **Joists max. 16" o.c./Plywood Subfloor**
- **Unbonded Mortar Bed**
- **Ceramic Tile, Glass Tile**

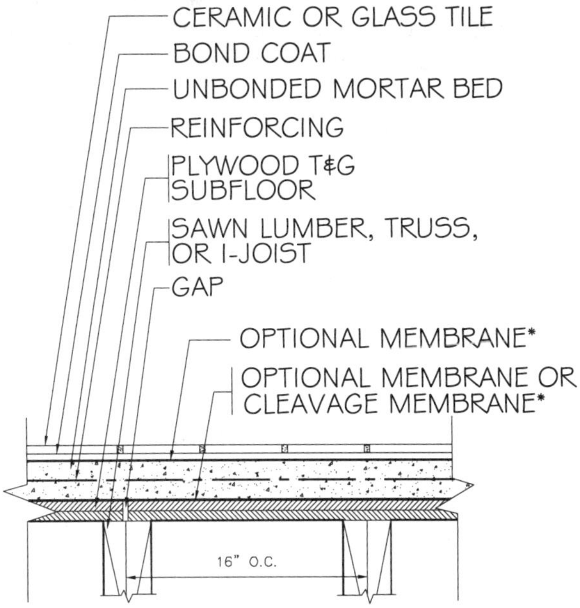

*USE OF A MEMBRANE IS OPTIONAL. SEE MEMBRANE OPTIONS.

Recommended Uses

- For wood substrates where leveling, flattening, or contouring of finish floor height is required, such as when finishing to adjacent floors or where slopes are required.
- For areas where floor flatness is critical, such as when tiles with any edge longer than 15" are specified or where the finished tile assembly is required to achieve in-plane transitions (as may be necessary where accessibility is a concern).

Service Rating

- Light commercial.
- When glass tile is used, service rating may be lower.

Environmental Exposure Classifications

- Res1, 2; Com1, 2.
- May be suitable for increased water exposure. See Membrane Options.
- May be suitable for exterior applications in areas not subject to freeze/thaw cycling when appropriate precautions are taken, including expansion joint placement, proper slope, waterproofing, and material selection.
- For installations that may be exposed to staining, specify tile and grout suitable for exposure. Consult product manufacturers; see also "Product Selection Guides."
- For installations that may be exposed to mild chemical attack, specify epoxy grout and tile suitable for exposure. For greater resistance to chemical exposure, also specify an epoxy bonding material. Consult product manufacturers; see also "Product Selection Guides."

Typical Weight of Tile Installation

- 21 pounds/square foot with 1¼" mortar bed. Add 3 pounds/square foot for each additional ¼" of mortar bed.
- Does not include weight of substrate. See "Appendix B" for assumptions, included materials, and their individual weights.

Limitations

- Maximum joist spacing 16" on center.

Membrane Options

- A waterproof membrane (A118.10) may be specified to prevent moisture intrusion and protect adjacent walls and building materials. Base flashing should be used for maximum effectiveness.
- Specifier shall indicate if complete waterproofing is required, including if/how membrane connects to drain assembly, if base flashing is required, and treatment at other termination points.
- A cleavage membrane is required.
- If a bonded waterproof membrane (ANSI A118.10) is used, it may be placed above or below the mortar bed. If a shower pan membrane (ASTM D4551 or D4068) is used, it must be placed below the mortar bed.
- The waterproof membrane can also function as the cleavage membrane if loose laid (not bonded to concrete) below the mortar bed. Check with membrane manufacturer if membrane may be loose laid.
- If waterproof membrane is bonded to the top of the mortar bed, a cleavage membrane must be placed below the mortar bed.
- Check with membrane manufacturers for suitability for applicable conditions, as not all membranes are suitable for steam, high temperature, and/or chemical exposure, exterior use, use over above-ground structural slabs, use over pourable underlayments, use with radiant heat, or use over concrete with excessive moisture vapor transmission and/or alkalinity. Membrane may also affect service rating.
- When glass tile is used, consult glass tile manufacturer for membrane options and recommendations.

Requirements

- Reinforcing mesh mandatory.
- Mortar bed thickness—1¼" minimum to 2" maximum.
- Mortar beds in excess of 2" thick shall be detailed by the architect.

Materials

- **Multiple options exist for membranes, mortars, grouts, and other materials and *must be clearly specified* to be included. If not specifically indicated, optional materials are not included and mortar/grout choice defaults to minimum performance specification indicated. Consider each system component and intended use to determine minimum requirements and to specify options.**
- Ceramic tile—ANSI A137.1.
- Glass tile, when used—ANSI A137.2; see also "Glass Tile Selection and Installation Guide," and consult tile manufacturer for service rating and environmental exposure classification recommendations. Not all glass tiles are suitable.
- Cementitious grout—ANSI A118.6 or better or ISO CG1 or better. When glass tile is used, specify grout designated by tile and grout manufacturers.
- Epoxy grout, when used—ANSI A118.3 or ISO RG.
- Cementitious bond coat—portland cement paste on a mortar bed that is still workable. For a cured mortar bed, follow recommendations below to select appropriate bonding mortar:
  - When a waterproof membrane is not used—ANSI A118.1 or better or ISO C1 or better.
  - When a waterproof membrane is used—ANSI A118.4 or better or ISO C2S1 or better unless ANSI A118.1 or ISO C1 is recommended by membrane manufacturer.
  - When porcelain tile is used—ANSI A118.4 or better or ISO C2 or better.
  - When glass tile is used, specify mortar designated by tile and mortar manufacturers. Bond coat color will impact the final appearance of translucent glass tile. Specifier shall confirm bond coat color is acceptable.
- Epoxy bond coat, when used—ANSI A118.3 or ISO R1 or better.
  - When glass tile is used, specify epoxy bond coat designated by tile and bond coat manufacturers. Bond coat color will impact the final appearance of translucent glass tile. Specifier shall confirm bond coat color is acceptable.
- Waterproof membrane, when used—ANSI A118.10.
- Mortar bed, reinforcing, and cleavage membrane—ANSI A108.1A.

Materials for Green/Sustainable Design

- See "Green Building Standards and Green Product Selection Guide" and consult manufacturers and suppliers for product sustainability and contribution to green building design.
- Consider specifying tile and installation materials that meet ANSI A138.1, the *American National Standard Specifications for Sustainable Ceramic Tiles, Glass Tiles, and Tile Installation Materials.*

Preparation by Other Trades

- Floor systems, including the framing system and subfloor panels, over which tile will be installed shall be in conformance with the IRC for residential applications, the IBC for commercial applications, or applicable building codes. Maximum allowable floor member live load and concentrated load deflection shall not exceed l/360. See also "Substrate Requirements."
- When concentrated loads (scissor lifts, pallet jacks, automobiles, forklifts, etc.) will be used on an above-ground tiled floor, the engineer and/or specifier shall specify a substrate to accommodate the concentrated loads. Owner/specifier is responsible for protecting the tilework from damage, including allowing sufficient time for installed materials to cure properly.
- Face grain of plywood shall run perpendicular to joists.
- Subfloor—minimum 19⁄32" tongue and groove exterior glue plywood with ⅛" gap between sheets or 1" nominal boards.
- Depressing subfloor between joists on ledger strips permissible in residential use.
- Maximum allowable variation in the installation substrate (plywood)—¼" in 10' from the required plane.

Movement Joint (architect must specify type of joint and show location and details on drawings)

- Movement joints—mandatory according to EJ171.
- When glass tile is used, adhere to more frequent placement recommendations within the ranges listed in EJ171.

Installation Specifications

- Tile—ANSI A108.1A,.1B, or .1C. A108.1B required if membrane or epoxy grout or bond coat to be used.
- Glass tile—ANSI A108.14, .15, .16, or manufacturer's directions.
- Cementitious grout—ANSI A108.10.
- Epoxy mortar/grout—ANSI A108.6.
- Waterproof membrane—ANSI A108.13.
- Movement Joints—EJ171 and ASTM C1193.

Notes

- For hydronic heat application, see RH141.
- When glass tile is used, see "Glass Tile Selection and Installation Guide," and consult manufacturer for recommendations and requirements.
- For thinner mortar bed, see F145.

F145-21

- **Joists max. 16" o.c./Plywood Subfloor**
- **Mortar Bed**
- **Ceramic Tile, Glass Tile**

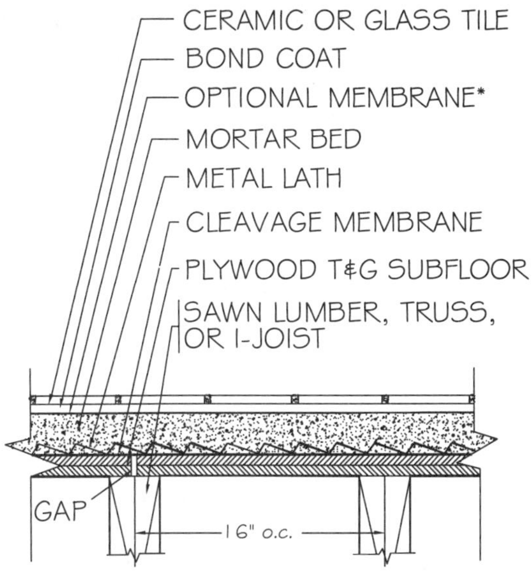

*USE OF A MEMBRANE IS OPTIONAL. SEE MEMBRANE OPTIONS.

Recommended Uses

- For wood substrates where leveling, flattening, or contouring of finish floor height is required, such as when finishing to adjacent floors or where slopes are required.
- For areas where floor flatness is critical, such as when tiles with any edge longer than 15" are specified or where the finished tile assembly is required to achieve in-plane transitions (as may be necessary where accessibility is a concern).

Service Rating

- Light commercial.
- When glass tile is used, service rating may be lower.

Environmental Exposure Classifications

- Res1, 2; Com1, 2.
- May be suitable for increased water exposure. See Membrane Options.
- For installations that may be exposed to staining, specify tile and grout suitable for exposure. Consult product manufacturers; see also "Product Selection Guides."
- For installations that may be exposed to mild chemical attack, specify epoxy grout and tile suitable for exposure. For greater resistance to chemical exposure, also specify an epoxy bonding material. Consult product manufacturers; see also "Product Selection Guides."

Typical Weight of Tile Installation

- 15 pounds/square foot with ¾" mortar bed. Add 3 pounds/square foot for each additional ¼" of mortar bed.
- Does not include weight of substrate. See "Appendix B" for assumptions, included materials, and their individual weights.

Limitations

- Maximum joist spacing 16" on center.

Membrane Options

- A waterproof membrane (A118.10) may be specified to prevent moisture intrusion and protect adjacent walls and building materials. Base flashing should be used for maximum effectiveness.
- Specifier shall indicate if complete waterproofing is required, including if/how membrane connects to drain assembly, if base flashing is required, and treatment at other termination points.
- Check with membrane manufacturers for suitability for applicable conditions, as not all membranes are suitable for steam, high temperature, and/or chemical exposure, exterior use, use over above-ground structural slabs, use over pourable underlayments, use with radiant heat, or use over concrete with excessive moisture vapor transmission and/or alkalinity. Membrane may also affect service rating.
- When glass tile is used, consult glass tile manufacturer for membrane options and recommendations.

Requirements

- Expanded metal lath—minimum 2.5 pounds/square yard nailed or stapled to subfloor over a cleavage membrane.
- Mortar bed thickness—¾" minimum to 1½" maximum.

Materials

- **Multiple options exist for membranes, mortars, grouts, and other materials and *must be clearly specified* to be included. If not specifically indicated, optional materials are not included and mortar/grout choice defaults to minimum performance specification indicated. Consider each system component and intended use to determine minimum requirements and to specify options.**
- Ceramic tile—ANSI A137.1.
- Glass tile, when used—ANSI A137.2; see also "Glass Tile Selection and Installation Guide," and consult tile manufacturer for service rating and environmental exposure classification recommendations. Not all glass tiles are suitable.
- Cementitious grout—ANSI A118.6 or better or ISO CG1 or better. When glass tile is used, specify grout designated by tile and grout manufacturers.
- Epoxy grout, when used—ANSI A118.3 or ISO RG.

- Cementitious bond coat—portland cement paste on a mortar bed that is still workable. For a cured mortar bed, follow recommendations below to select appropriate bonding mortar:
  - When a waterproof membrane is not used—ANSI A118.1 or better or ISO C1 or better.
  - When a waterproof membrane is used—ANSI A118.4 or better or ISO C2S1 or better unless ANSI A118.1 or ISO C1 is recommended by membrane manufacturer.
  - When porcelain tile is used—ANSI A118.4 or better or ISO C2 or better.
  - When glass tile is used, specify mortar designated by tile and mortar manufacturers. Bond coat color will impact the final appearance of translucent glass tile. Specifier shall confirm bond coat color is acceptable.
- Epoxy bond coat, when used—ANSI A118.3 or ISO R1 or better.
  - When glass tile is used, specify epoxy bond coat designated by tile and bond coat manufacturers. Bond coat color will impact the final appearance of translucent glass tile. Specifier shall confirm bond coat color is acceptable.
- Waterproof membrane, when used—ANSI A118.10.
- Mortar bed, metal lath, and cleavage membrane—ANSI A108.1A.

Materials for Green/Sustainable Design

- See "Green Building Standards and Green Product Selection Guide" and consult manufacturers and suppliers for product sustainability and contribution to green building design.
- Consider specifying tile and installation materials that meet ANSI A138.1, the *American National Standard Specifications for Sustainable Ceramic Tiles, Glass Tiles, and Tile Installation Materials.*

Preparation by Other Trades

- Floor systems, including the framing system and subfloor panels, over which tile will be installed shall be in conformance with the IRC for residential applications, the IBC for commercial applications, or applicable building codes. Maximum allowable floor member live load and concentrated load deflection shall not exceed l/360. See also "Substrate Requirements."
- When concentrated loads (scissor lifts, pallet jacks, automobiles, forklifts, etc.) will be used on an above-ground tiled floor, the engineer and/or specifier shall specify a substrate to accommodate the concentrated loads. Owner/specifier is responsible for protecting the tilework from damage, including allowing sufficient time for installed materials to cure properly.
- Face grain of plywood shall run perpendicular to joists.
- Subfloor—minimum 23⁄32" tongue and groove exterior glue plywood with 1⁄8" gap between sheets.
- Maximum allowable variation in the installation substrate (plywood)—1⁄4" in 10' from the required plane.

Movement Joint (architect must specify type of joint and show location and details on drawings)

- Movement joints—mandatory according to EJ171.
- When glass tile is used, adhere to more frequent placement recommendations within the ranges listed in EJ171.

Installation Specifications

- Tile—ANSI A108.1A, .1B, or .1C. A108.1B required if membrane or epoxy grout or bond coat to be used.
- Glass tile—ANSI A108.14, .15, .16, or manufacturer's directions.
- Cementitious grout—ANSI A108.10.
- Epoxy mortar/grout—ANSI A108.6.
- Waterproof membrane—ANSI A108.13.
- Movement Joints—EJ171 and ASTM C1193.

Notes

- For thicker mortar bed, see F141.
- When glass tile is used, see "Glass Tile Selection and Installation Guide," and consult manufacturer for recommendations and requirements.

F142-21

- **Joists max. 16" o.c./Plywood Subfloor**
- **Plywood Underlayment**
- **Organic Adhesive**
- **Ceramic Tile**

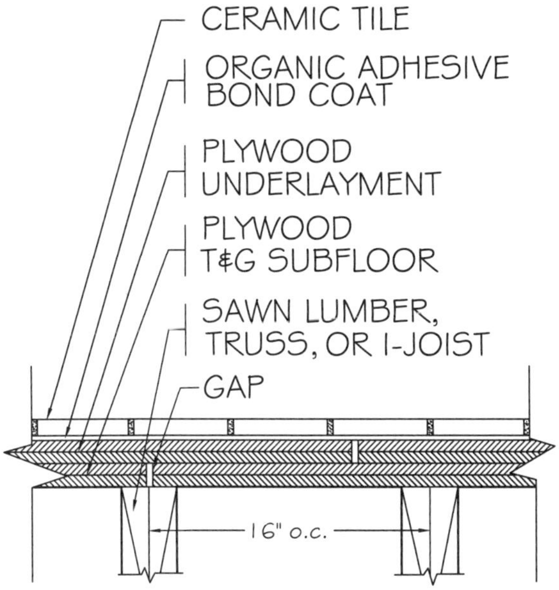

Recommended Uses

- For wood substrates in residential dry areas that will not be subjected to high impact, wheel loads, or point loads.

Service Rating

- Residential.

Environmental Exposure Classifications

- Res1.
- For installations that may be exposed to staining, specify tile and grout suitable for exposure. Consult product manufacturers; see also "Product Selection Guides."
- For installations that may be exposed to mild chemical attack, specify epoxy grout and tile suitable for exposure. Consult product manufacturers; see also "Product Selection Guides."

Typical Weight of Tile Installation

- 7 pounds/square foot.
- Does not include weight of substrate. See "Appendix B" for assumptions, included materials, and their individual weights.

Limitations

- Maximum joist spacing 16" on center.
- Will not withstand high impact, wheel loads, or point loads.
- Not for areas exposed to temperatures exceeding 140°F.
- Maximum tile size 8" × 8" unless organic adhesive manufacturer allows larger tile size. See Notes.

Requirements

- Plywood underlayment—clean and free of dirt, dust, paint, and oily film.
- Gaps between plywood underlayment sheets to be treated per setting material manufacturer's recommendations.

Materials

- **Multiple options exist for membranes, mortars, grouts, and other materials and *must be clearly specified* to be included. If not specifically indicated, optional materials are not included and mortar/grout choice defaults to minimum performance specification indicated. Consider each system component and intended use to determine minimum requirements and to specify options.**
- Ceramic tile—ANSI A137.1.
- Cementitious grout—ANSI A118.6 or better or ISO CG1 or better.
- Epoxy grout, when used—ANSI A118.3 or ISO RG.
- Organic adhesive bond coat—ANSI A136.1 Type I or ISO D2.

Materials for Green/Sustainable Design

- See "Green Building Standards and Green Product Selection Guide" and consult manufacturers and suppliers for product sustainability and contribution to green building design.
- Consider specifying tile and installation materials that meet ANSI A138.1, the *American National Standard Specifications for Sustainable Ceramic Tiles, Glass Tiles, and Tile Installation Materials.*

Preparation by Other Trades

- Floor systems, including the framing system and subfloor panels, over which tile will be installed shall be in conformance with the IRC for residential applications, the IBC for commercial applications, or applicable building codes. Maximum allowable floor member live load and concentrated load deflection shall not exceed l/360. See also "Substrate Requirements."
- When concentrated loads (scissor lifts, pallet jacks, automobiles, forklifts, etc.) will be used on an aboveground tiled floor, the engineer and/or specifier shall specify a substrate to accommodate the concentrated loads. Owner/specifier is responsible for protecting the tilework from damage, including allowing sufficient time for installed materials to cure properly.
- Face grain of plywood subfloor and underlayment shall run perpendicular to joists.
- Subfloor—minimum $^{19}/_{32}$" tongue and groove exterior glue plywood with ⅛" gap between sheets or 1" nominal boards.
- Underlayment—$^{19}/_{32}$" minimum exterior glue plywood with ⅛" gap between sheets.

- Maximum allowable variation in the tile substrate—1⁄16" in 3'. Adjacent edges of plywood sheets not to exceed 1⁄32" difference in height.
- Offset end and edge joints of the underlayment panels by at least 2" from the joints of subfloor panels; they should not coincide with framing below.

Movement Joint (architect must specify type of joint and show location and details on drawings)

- Joints—mandatory according to EJ171.

Installation Specifications

- Tile—ANSI A108.4.
- Cementitious grout—ANSI A108.10.
- Movement Joints—EJ171 and ASTM C1193.

Notes

- As the tile size increases, there is less tolerance for variation in the subfloor from the required plane. Organic adhesive thickness must be thin and uniform; therefore, subfloor flattening may be required.
- Underlayment fasteners should not penetrate joists below.
- Do not use organic adhesives to bond tile to a membrane or other impervious substrate.
- Extended dry time may be required for larger tiles and tiles with low water absorption.

F143-21

- **Joists max. 16" o.c./Plywood Subfloor**
- **Plywood Underlayment**
- **Epoxy Mortar**
- **Ceramic Tile**
- **Epoxy Grout**

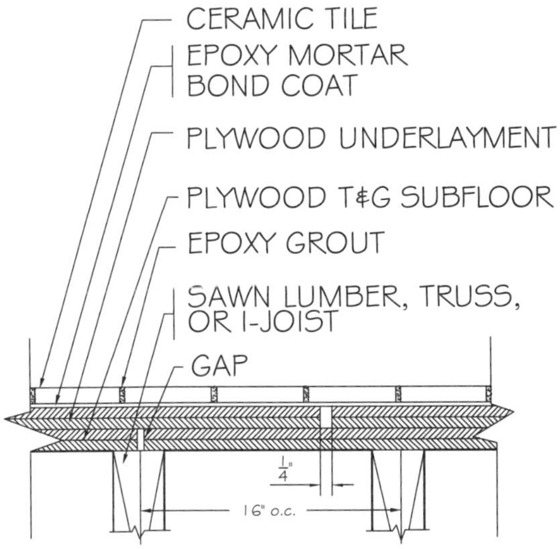

Recommended Uses

- For wood substrates where increased resistance to traffic and/or loading, water, chemicals, and/or aggressive cleaning practices are desired.

Service Rating

- Residential with minimum 15⁄32"-thick plywood underlayment.
- Light commercial with minimum 19⁄32"-thick plywood underlayment.

Environmental Exposure Classifications

- Res1, 2; Com1, 2.

Typical Weight of Tile Installation

- 7 pounds/square foot.
- Does not include weight of substrate. See "Appendix B" for assumptions, included materials, and their individual weights.

Limitations

- Maximum joist spacing 16" on center.
- Not for severe chemical exposure or extreme heat.
- Extreme heat or improper steam cleaning will soften epoxy grouts and wash them out of joints. High temperature resistant formulas are available.

Membrane Options

- A waterproof membrane (A118.10) may be specified to prevent moisture intrusion and protect adjacent walls and building materials. Base flashing should be used for maximum effectiveness.
- Specifier shall indicate if complete waterproofing is required, including if/how membrane connects to drain assembly, if base flashing is required, and treatment at other termination points.
- Check with membrane manufacturers for suitability for applicable conditions, as not all membranes are suitable for steam, high temperature, and/or chemical exposure, exterior use, use over above-ground structural slabs, use over pourable underlayments, use with radiant heat, or use over concrete with excessive moisture vapor transmission and/or alkalinity. Membrane may also affect service rating.

Requirements

- Plywood underlayment—clean and free of dirt, dust, paint, and oily film.
- Gaps between plywood underlayment sheets to be treated per setting material manufacturer's recommendations.

Materials

- **Multiple options exist for membranes, mortars, grouts, and other materials and *must be clearly specified* to be included. If not specifically indicated, optional materials are not included and mortar/grout choice defaults to minimum performance specification indicated. Consider each system component and intended use to determine minimum requirements and to specify options.**
- Ceramic tile—ANSI A137.1.
- Epoxy grout—ANSI A118.3 or ISO RG. For tilework exposed to prolonged high temperatures, use high-temperature, chemical-resistant epoxy grout certified by manufacturer for intended use.
- Epoxy mortar—ANSI A118.3 or ISO R1 or better. For tilework exposed to prolonged high temperatures, use high-temperature, chemical-resistant epoxy mortar certified by manufacturer for intended use.
- Waterproof membrane, when used—ANSI A118.10.

Materials for Green/Sustainable Design

- See "Green Building Standards and Green Product Selection Guide" and consult manufacturers and suppliers for product sustainability and contribution to green building design.
- Consider specifying tile and installation materials that meet ANSI A138.1, the *American National Standard Specifications for Sustainable Ceramic Tiles, Glass Tiles, and Tile Installation Materials.*

Preparation by Other Trades

- Floor systems, including the framing system and subfloor panels, over which tile will be installed shall be in conformance with the IRC for residential applications, the IBC for commercial applications, or applicable building codes. Maximum allowable floor member live load and concentrated load deflection shall not exceed l/360. See also "Substrate Requirements."

- When concentrated loads (scissor lifts, pallet jacks, automobiles, forklifts, etc.) will be used on an above-ground tiled floor, the engineer and/or specifier shall specify a substrate to accommodate the concentrated loads. Owner/specifier is responsible for protecting the tilework from damage, including allowing sufficient time for installed materials to cure properly.
- Face grain of plywood subfloor and underlayment shall run perpendicular to joists.
- Subfloor—minimum $^{19}/_{32}$" tongue and groove plywood with $^{1}/_{8}$" gap between sheets or 1" nominal boards.
- Underlayment—minimum $^{15}/_{32}$" exterior glue plywood with $^{1}/_{4}$" gap between sheets. See also Service Rating.
- Maximum allowable variation in the tile substrate—for tiles with all edges shorter than 15", maximum allowable variation is $^{1}/_{4}$" in 10' from the required plane, with no more than $^{1}/_{16}$" variation in 12" when measured from the high points in the surface. For tiles with at least one edge 15" in length, maximum allowable variation is $^{1}/_{8}$" in 10' from the required plane, with no more than $^{1}/_{16}$" variation in 24" when measured from the high points in the surface. Adjacent edges of plywood sheets not to exceed $^{1}/_{32}$" difference in height.

Movement Joint (architect must specify type of joint and show location and details on drawings)

- Movement joints—mandatory according to EJ171.

Installation Specifications

- Epoxy mortar/grout—ANSI A108.6.
- Movement Joints—EJ171 and ASTM C1193.

Notes

- Specify tile suitable for expected chemical exposure.
- As tile size increases, there is less tolerance for variation in the subfloor from the required plane. Epoxy mortar thickness must be thin and uniform; therefore, subfloor flattening may be required.
- Underlayment fasteners should not penetrate joists below.

F150-21

- **Joists max. 16" o.c./Plywood Subfloor**
- **Plywood Underlayment**
- **Ceramic Tile**

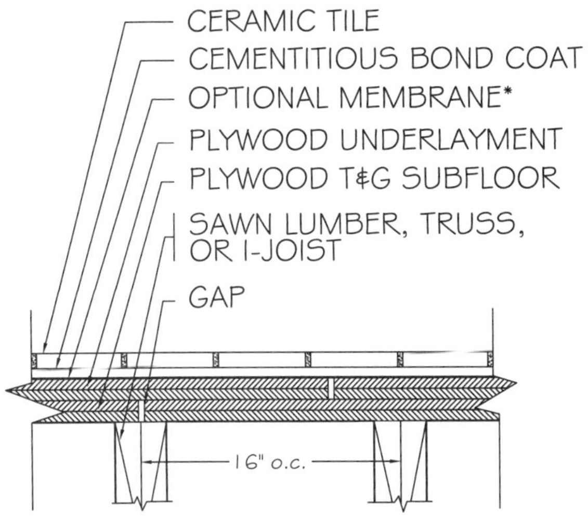

*USE OF A MEMBRANE IS OPTIONAL. SEE MEMBRANE OPTIONS.

Recommended Uses

- For wood substrates in dry areas where increased resistance to traffic and loading is desired.

Service Rating

- Residential with minimum $^{15}/_{32}$"-thick plywood underlayment.
- Light commercial with minimum $^{19}/_{32}$"-thick plywood underlayment.

Environmental Exposure Classifications

- Res1; Com1.
- May be suitable for increased water exposure. See Membrane Options.
- For installations that may be exposed to staining, specify tile and grout suitable for exposure. Consult product manufacturers; see also "Product Selection Guides."
- For installations that may be exposed to mild chemical attack, specify epoxy grout and tile suitable for exposure. Consult product manufacturers; see also "Product Selection Guides."

Typical Weight of Tile Installation

- 7 pounds/square foot.
- Does not include weight of substrate. See "Appendix B" for assumptions, included materials, and their individual weights.

Limitations

- Maximum joist spacing 16" on center.

Membrane Options

- A waterproof membrane (A118.10) may be specified to prevent moisture intrusion and protect adjacent walls and building materials. Base flashing should be used for maximum effectiveness.
- Specifier shall indicate if complete waterproofing is required, including if/how membrane connects to drain assembly, if base flashing is required, and treatment at other termination points.
- Check with membrane manufacturers for suitability for applicable conditions, as not all membranes are suitable for steam, high temperature, and/or chemical exposure, exterior use, use over above-ground structural slabs, use over pourable underlayments, use with radiant heat, or use over concrete with excessive moisture vapor transmission and/or alkalinity. Membrane may also affect service rating.

Requirements

- Plywood underlayment—clean and free of dirt, dust, paint, and oily film.
- Gaps between plywood underlayment sheets to be treated per setting material manufacturer's recommendations.

Materials

- **Multiple options exist for membranes, mortars, grouts, and other materials and *must be clearly specified* to be included. If not specifically indicated, optional materials are not included and mortar/grout choice defaults to minimum performance specification indicated. Consider each system component and intended use to determine minimum requirements and to specify options.**
- Ceramic tile—ANSI A137.1.
- Cementitious grout—ANSI A118.6 or better or ISO CG1 or better.
- Epoxy grout, when used—ANSI A118.3 or ISO RG.
- Cementitious bond coat:
  - When a waterproof membrane is not used—ANSI A118.11 or better or ISO C2S1P1 or better.
  - When a waterproof membrane is used—ANSI A118.4 or better or ISO C2S1 or better, unless ANSI A118.1 or ISO C1 is recommended by membrane manufacturer (mortar for bonding tile to membrane).
- When a waterproof membrane is used, follow membrane manufacturer's directions for bonding membrane to plywood underlayment.
- Waterproof membrane, when used—ANSI A118.10.

Materials for Green/Sustainable Design

- See "Green Building Standards and Green Product Selection Guide" and consult manufacturers and suppliers for product sustainability and contribution to green building design.
- Consider specifying tile and installation materials that meet ANSI A138.1, the *American National Standard*

Specifications for Sustainable Ceramic Tiles, Glass Tiles, and Tile Installation Materials.

Preparation by Other Trades

- Floor systems, including the framing system and subfloor panels, over which tile will be installed shall be in conformance with the IRC for residential applications, the IBC for commercial applications, or applicable building codes. Maximum allowable floor member live load and concentrated load deflection shall not exceed l/360. See also "Substrate Requirements."
- When concentrated loads (scissor lifts, pallet jacks, automobiles, forklifts, etc.) will be used on an above-ground tiled floor, the engineer and/or specifier shall specify a substrate to accommodate the concentrated loads. Owner/specifier is responsible for protecting the tilework from damage, including allowing sufficient time for installed materials to cure properly.
- Face grain of plywood subfloor and underlayment shall run perpendicular to joists.
- Subfloor—minimum 19⁄32" exterior-glue tongue and groove plywood with 1⁄8" gap between sheets.
- Underlayment—minimum 15⁄32" exterior glue plywood with 1⁄8" gap between sheets. See also Service Rating.
- Maximum allowable variation in the tile substrate—for tiles with all edges shorter than 15", maximum allowable variation is 1⁄4" in 10' from the required plane, with no more than 1⁄16" variation in 12" when measured from the high points in the surface. For tiles with at least one edge 15" in length, maximum allowable variation is 1⁄8" in 10' from the required plane, with no more than 1⁄16" variation in 24" when measured from the high points in the surface. Adjacent edges of plywood sheets not to exceed 1⁄32" difference in height.

Movement Joint (architect must specify type of joint and show location and details on drawings)

- Movement joints—mandatory in accordance with EJ171.

Installation Specifications

- Tile—ANSI A108.12.
- Cementitious grout—ANSI A108.10.
- Epoxy grout—ANSI A108.6.
- Waterproof membrane—ANSI A108.13.
- Movement Joints—EJ171 and ASTM C1193.

Notes

- Underlayment fasteners should not penetrate joists below.

F160-21

- **Joists max. 24" o.c./Plywood Subfloor**
- **Exterior Glue/Laminated Wood Panel (EG/LWP) Underlayment**
- **Ceramic Tile**

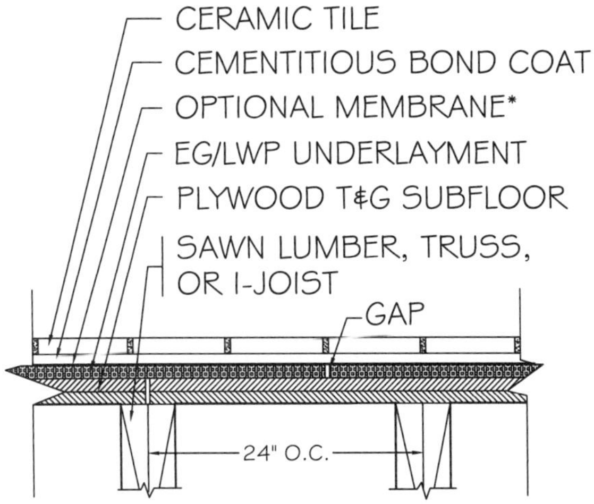

*USE OF A MEMBRANE IS OPTIONAL. SEE MEMBRANE OPTIONS.

Recommended Uses

- For wood substrates in dry areas where joists are spaced 24" on center or less.

Service Rating

- Light commercial.

Environmental Exposure Classifications

- Res1; Com1.
- May be suitable for increased water exposure. See Membrane Options.
- For installations that may be exposed to staining, specify tile and grout suitable for exposure. Consult product manufacturers; see also "Product Selection Guides."
- For installations that may be exposed to mild chemical attack, specify epoxy grout and tile suitable for exposure. Consult product manufacturers; see also "Product Selection Guides."

Typical Weight of Tile Installation

- 7 pounds/square foot.
- Does not include weight of substrate. See "Appendix B" for assumptions, included materials, and their individual weights.

Limitations

- Maximum joist spacing 24" on center.
- 8" × 8" and larger tile only.

Membrane Options

- A waterproof membrane (A118.10) may be specified to prevent moisture intrusion and protect adjacent walls and building materials. Base flashing should be used for maximum effectiveness.
- Specifier shall indicate if complete waterproofing is required, including if/how membrane connects to drain assembly, if base flashing is required, and treatment at other termination points.
- Check with membrane manufacturers for suitability for applicable conditions, as not all membranes are suitable for steam, high temperature, and/or chemical exposure, exterior use, use over above-ground structural slabs, use over pourable underlayments, use with radiant heat, or use over concrete with excessive moisture vapor transmission and/or alkalinity. Membrane may also affect service rating.

Requirements

- Plywood underlayment—clean and free of dirt, dust, paint, and oily film.
- Gaps between plywood underlayment sheets to be treated per setting material manufacturer's recommendations.
- Fasten underlayment with screws every 6" in the field and every 4" along the perimeter. Do not staple.

Materials

- **Multiple options exist for membranes, mortars, grouts, and other materials and *must be clearly specified* to be included. If not specifically indicated, optional materials are not included and mortar/grout choice defaults to minimum performance specification indicated. Consider each system component and intended use to determine minimum requirements and to specify options.**
- Ceramic tile—ANSI A137.1.
- Cementitious grout—ANSI A118.6 or better or ISO CG1 or better.
- Epoxy grout, when used—ANSI A118.3 or ISO RG.
- Cementitious bond coat:
  - When a waterproof membrane is not used—ANSI A118.11 or better or ISO C2S1P1 or better.
  - When a waterproof membrane is used—ANSI A118.4 or better or ISO C2S1 or better, unless ANSI A118.1 or ISO C1 is recommended by membrane manufacturer (mortar for bonding tile to membrane).
- When a waterproof membrane is used, follow membrane manufacturer's directions for bonding membrane to EG/LWP underlayment.
- Waterproof membrane, when used—ANSI A118.10.

Materials for Green/Sustainable Design

- See "Green Building Standards and Green Product Selection Guide" and consult manufacturers and suppliers for product sustainability and contribution to green building design.
- Consider specifying tile and installation materials that meet ANSI A138.1, the *American National Standard*

Specifications for Sustainable Ceramic Tiles, Glass Tiles, and Tile Installation Materials.

Preparation by Other Trades

- Floor systems, including the framing system and subfloor panels, over which tile will be installed shall be in conformance with the IRC for residential applications, the IBC for commercial applications, or applicable building codes. Maximum allowable floor member live load and concentrated load deflection shall not exceed l/360. See also "Substrate Requirements."
- When concentrated loads (scissor lifts, pallet jacks, automobiles, forklifts, etc.) will be used on an aboveground tiled floor, the engineer and/or specifier shall specify a substrate to accommodate the concentrated loads. Owner/specifier is responsible for protecting the tilework from damage, including allowing sufficient time for installed materials to cure properly.
- Face grain of plywood subfloor and EG/LWP underlayment shall run perpendicular to joists.
- Subfloor—minimum 23⁄32" tongue and groove exterior glue plywood with 1⁄8" gap between sheets.
- Underlayment—minimum 3⁄8" exterior-glue laminated birch wood panel with a minimum of 7 plies.
- Maximum allowable variation in the tile substrate—for tiles with all edges shorter than 15", maximum allowable variation is 1⁄4" in 10' from the required plane, with no more than 1⁄16" variation in 12" when measured from the high points in the surface. For tiles with at least one edge 15" in length, maximum allowable variation is 1⁄8" in 10' from the required plane, with no more than 1⁄16" variation in 24" when measured from the high points in the surface. Adjacent edges of plywood sheets not to exceed 1⁄32" difference in height.

Movement Joint (architect must specify type of joint and show location and details on drawings)

- Movement joints—mandatory in accordance with EJ171.

Installation Specifications

- Tile—ANSI A108.12 or .5.
- Cementitious grout—ANSI A108.10.
- Epoxy grout—ANSI A108.6.
- Waterproof membrane—ANSI A108.13.
- Movement Joints—EJ171 and ASTM C1193.

Notes

- Underlayment fasteners should not penetrate joists below.

F144-21

- **Joists max. 16" o.c./Plywood Subfloor**
- **Cement Backer Board or Fiber-Cement Backer Board**
- **Ceramic Tile, Glass Tile**

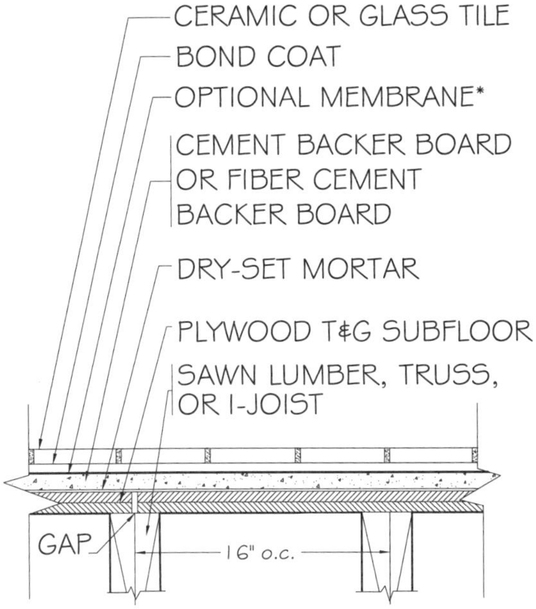

*USE OF A MEMBRANE IS OPTIONAL. SEE MEMBRANE OPTIONS.

Recommended Uses

- For wood substrates where water-resistance is desired.

Service Rating

- Residential with minimum $^{19}/_{32}$"-thick subfloor and minimum ¼"-thick cement backer board.
- Light commercial with minimum $^{19}/_{32}$"-thick subfloor and minimum $^{7}/_{16}$"-thick cement backer board or ¼"-thick fiber cement backer board.
- Light commercial with minimum $^{23}/_{32}$"-thick subfloor and minimum ¼"-thick cement backer board.
- When glass tile is used, service rating may be lower.

Environmental Exposure Classifications

- Res1, 2; Com1, 2.
- May be suitable for increased water exposure. See Membrane Options.
- May be suitable for exterior applications in areas not subject to freeze/thaw cycling when appropriate precautions are taken, including expansion joint placement, proper slope, waterproofing, and material selection.
- For installations that may be exposed to staining, specify tile and grout suitable for exposure. Consult product manufacturers; see also "Product Selection Guides."
- For installations that may be exposed to mild chemical attack, specify epoxy grout and tile suitable for exposure. For greater resistance to chemical exposure, also specify an epoxy bonding material. Consult product manufacturers; see also "Product Selection Guides."

Typical Weight of Tile Installation

- 8 pounds/square foot if ¼" cement backer board or ¼" fiber-cement backer board used.
- 9 pounds/square foot if $^{7}/_{16}$" fiber-cement backer board used.
- 10 pounds/square foot if ½" cement backer board used.
- Does not include weight of substrate. See "Appendix B" for assumptions, included materials, and their individual weights.

Limitations

- Maximum joist spacing 16" on center.

Membrane Options

- A waterproof membrane (A118.10) may be specified to prevent moisture intrusion and protect adjacent walls and building materials. Base flashing should be used for maximum effectiveness.
- Specifier shall indicate if complete waterproofing is required, including if/how membrane connects to drain assembly, if base flashing is required, and treatment at other termination points.
- Check with membrane manufacturers for suitability for applicable conditions, as not all membranes are suitable for steam, high temperature, and/or chemical exposure, exterior use, use over above-ground structural slabs, use over pourable underlayments, use with radiant heat, or use over concrete with excessive moisture vapor transmission and/or alkalinity. Membrane may also affect service rating.
- When glass tile is used, consult glass tile manufacturer for membrane options and recommendations.

Requirements

- Surface of units—clean and free of dirt, dust, paint, and oily film.

Preparation by Backer Board Installers

- Use a sufficient amount of portland cement mortar under the backer board to establish a supporting plane and eliminate voids.
- Fasten backer units with corrosion-resistant fasteners per manufacturer's directions.
- Leave ⅛" perimeter movement gap and ⅛" gap between sheets. Fill joints between sheets solid and tape with portland cement mortar and 2" alkali-resistant glass fiber mesh tape. Do not fill perimeter joint.
- Maximum allowable variation in the tile substrate—for tiles with all edges shorter than 15," maximum allowable variation is ¼" in 10' from the required plane, with no more than $^{1}/_{16}$" variation in 12" when measured from the high points in the surface. For tiles with at least one edge 15" in length, maximum allowable variation is ⅛" in 10' from the required plane, with no more than $^{1}/_{16}$" variation in 24" when measured from the high points in the surface.

- Stagger backer board end and edge joints so as not to coincide with joints in subfloor. Stagger joints in adjacent rows so four corners do not come together within the same plane. Space panel ends and edges in accordance with manufacturer's recommendations.

Materials

- **Multiple options exist for membranes, mortars, grouts, and other materials and *must be clearly specified* to be included. If not specifically indicated, optional materials are not included and mortar/grout choice defaults to minimum performance specification indicated. Consider each system component and intended use to determine minimum requirements and to specify options.**
- Ceramic tile—ANSI A137.1.
- Glass tile, when used—ANSI A137.2; see also "Glass Tile Selection and Installation Guide," and consult tile manufacturer for service rating and environmental exposure classification recommendations. Not all glass tiles are suitable.
- Cementitious grout—ANSI A118.6 or better or ISO CG1 or better. When glass tile is used, specify grout designated by tile and grout manufacturers.
- Epoxy grout, when used—ANSI A118.3 or ISO RG.
- Cementitious bond coat:
  - When a waterproof membrane is not used—ANSI A118.1 or better or ISO C1 or better.
  - When a waterproof membrane is used—ANSI A118.4 or better or ISO C2S1 or better unless ANSI A118.1 or ISO C1 is recommended by membrane manufacturer.
  - When porcelain tile is used—ANSI A118.4 or better or ISO C2 or better.
  - When glass tile is used, specify mortar designated by tile and mortar manufacturers. Bond coat color will impact the final appearance of translucent glass tile. Specifier shall confirm bond coat color is acceptable.
- Epoxy bond coat, when used—ANSI A118.3 or ISO R1 or better.
  - When glass tile is used, specify epoxy bond coat designated by tile and bond coat manufacturers. Bond coat color will impact the final appearance of translucent glass tile. Specifier shall confirm bond coat color is acceptable.
- Cement backer board, when used—ANSI A118.9 or ASTM C1325. See Service Rating for minimum thickness.
- Fiber cement backer board, when used—ASTM C1288. See Service Rating for minimum thickness.
- Mortar under backer board—ANSI A118.1 or better or ISO C1 or better.
- Waterproof membrane, when used—ANSI A118.10.

Materials for Green/Sustainable Design

- See "Green Building Standards and Green Product Selection Guide" and consult manufacturers and suppliers for product sustainability and contribution to green building design.
- Consider specifying tile and installation materials that meet ANSI A138.1, the *American National Standard Specifications for Sustainable Ceramic Tiles, Glass Tiles, and Tile Installation Materials.*

Preparation by Other Trades

- Floor systems, including the framing system and subfloor panels, over which tile will be installed shall be in conformance with the IRC for residential applications, the IBC for commercial applications, or applicable building codes. Maximum allowable floor member live load and concentrated load deflection shall not exceed l/360. See also "Substrate Requirements."
- When concentrated loads (scissor lifts, pallet jacks, automobiles, forklifts, etc.) will be used on an above-ground tiled floor, the engineer and/or specifier shall specify a substrate to accommodate the concentrated loads. Owner/specifier is responsible for protecting the tilework from damage, including allowing sufficient time for installed materials to cure properly.
- Face grain of plywood shall run perpendicular to joists.
- Subfloor—minimum $^{19}/_{32}$" exterior-glue tongue and groove plywood with $^{1}/_{8}$" gap between sheets. See also Service Rating.
- Maximum variation in the plywood surface—$^{1}/_{4}$" in 10' and $^{1}/_{16}$" in 1' from the required plane. Adjacent edges of plywood sheets—maximum $^{1}/_{32}$" above or below each other.

Movement Joint (architect must specify type of joint and show location and details on drawings)

- Movement joints—mandatory in accordance with EJ171.
- When glass tile is used, adhere to more frequent placement recommendations within the ranges listed in EJ171.

Installation Specifications

- Tile—ANSI A108.5.
- Glass tile—ANSI A108.15, .16, or manufacturer's directions.
- Cementitious grout—ANSI A108.10.
- Epoxy mortar/grout—ANSI A108.6.
- Backer board—ANSI A108.11 or manufacturer's directions.
- Waterproof membrane—ANSI A108.13.
- Movement Joints—EJ171 and ASTM C1193.

Notes

- When glass tile is used, see "Glass Tile Selection and Installation Guide," and consult manufacturer for recommendations and requirements.

F146-21

- **Joists max. 16" o.c./Plywood Subfloor**
- **Coated Glass Mat Water-Resistant Gypsum Backer Board**
- **Ceramic Tile, Glass Tile**

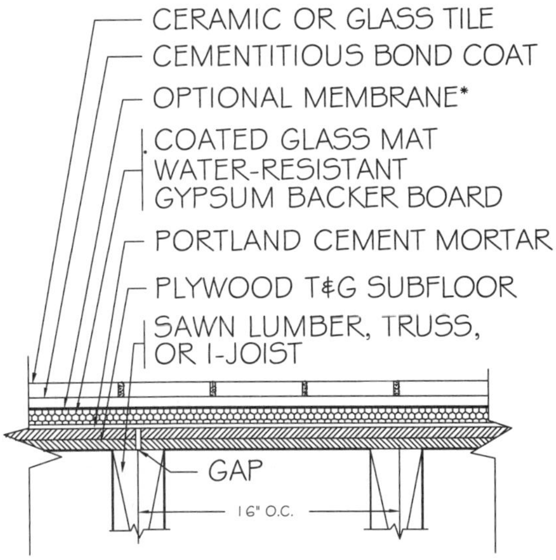

*USE OF A MEMBRANE IS OPTIONAL. SEE MEMBRANE OPTIONS.

Recommended Uses

- For wood substrates where water-resistance is desired.

Service Rating

- Light commercial.
- When glass tile is used, service rating may be lower.

Environmental Exposure Classifications

- Res1, 2; Com1, 2.
- May be suitable for increased water exposure. See Membrane Options.
- For installations that may be exposed to staining, specify tile and grout suitable for exposure. Consult product manufacturers; see also "Product Selection Guides."
- For installations that may be exposed to mild chemical attack, specify epoxy grout and tile suitable for exposure. Consult product manufacturers; see also "Product Selection Guides."

Typical Weight of Tile Installation

- 8 pounds/square foot.
- Does not include weight of substrate. See "Appendix B" for assumptions, included materials, and their individual weights.

Limitations

- Maximum joist spacing 16" on center.

Membrane Options

- A waterproof membrane (A118.10) may be specified to prevent moisture intrusion through seams, corners, fasteners, and other penetrations to protect adjacent walls and building materials. Base flashing should be used for maximum effectiveness. Follow backer board and membrane manufacturer's waterproofing requirements.
- Specifier shall indicate if complete waterproofing is required, including if/how membrane connects to drain assembly, if base flashing is required, and treatment at other termination points.
- Check with membrane manufacturers for suitability for applicable conditions, as not all membranes are suitable for steam, high temperature, and/or chemical exposure, exterior use, use over above-ground structural slabs, use over pourable underlayments, use with radiant heat, or use over concrete with excessive moisture vapor transmission and/or alkalinity. Membrane may also affect service rating.
- When glass tile is used, consult glass tile manufacturer for membrane options and recommendations.
- Some liquid-applied waterproof membranes will require extended cure time. Consult membrane manufacturer.

Requirements

- Caulk or seal penetrations, corners, and abutments to dissimilar materials with flexible sealant.
- Surface of units—clean and free of dirt, dust, paint, and oily film.
- Set tile on positive side of moisture barrier coating.
- When glass tile is used, consult glass tile manufacturer for tile suitability over non-absorptive surface.

Preparation by Backer Board Installers

- Maximum allowable variation in the tile substrate—for tiles with all edges shorter than 15," maximum allowable variation is ¼" in 10' from the required plane, with no more than $^1/_{16}$" variation in 12" when measured from the high points in the surface. For tiles with at least one edge 15" in length, maximum allowable variation is ⅛" in 10' from the required plane, with no more than $^1/_{16}$" variation in 24" when measured from the high points in the surface.
- Fit ends and edges closely to eliminate gaps between panels, including in corners.
- Tape seams and corners with 2" alkali-resistant glass fiber mesh tape and cementitious bonding material, and skim coat over fastener heads, unless waterproofing has been applied.
- Use a sufficient amount of portland cement mortar under the backer board to establish a supporting plane and eliminate voids.
- Fasten backer units with corrosion-resistant fasteners per manufacturer's directions.
- Stagger backer board end and edge joints so as not to coincide with joints in subfloor. Stagger joints in adjacent rows so four corners do not come together within the same plane. Space panel ends and edges in accordance with manufacturer's recommendations.

Materials

- **Multiple options exist for membranes, mortars, grouts, and other materials and *must be clearly specified* to be included. If not specifically indicated, optional materials are not included and mortar/grout choice defaults to minimum performance specification indicated. Consider each system component and intended use to determine minimum requirements and to specify options.**
- Ceramic tile—ANSI A137.1.
- Glass tile, when used—ANSI A137.2; see also "Glass Tile Selection and Installation Guide," and consult tile manufacturer for service rating and environmental exposure classification recommendations. Not all glass tiles are suitable.
- Cementitious grout—ANSI A118.6 or better or ISO CG1 or better. When glass tile is used, specify grout designated by tile and grout manufacturers.
- Epoxy grout, when used—ANSI A118.3 or ISO RG.
- Cementitious bond coat:
  - When a waterproof membrane is not used—ANSI A118.4 or better or ISO C2S1 or better.
  - When a waterproof membrane is used—ANSI A118.4 or better or ISO C2S1 or better unless ANSI A118.1 or ISO C1 is recommended by membrane manufacturer.
  - When glass tile is used, specify mortar designated by tile and mortar manufacturers. Bond coat color will impact the final appearance of translucent glass tile. Specifier shall confirm bond coat color is acceptable.
- Coated glass mat water-resistant gypsum backer board—ASTM C1178.
- Mortar under backer board—ANSI A118.1 or better or ISO C1 or better.
- Waterproof membrane, when used—A118.10.
- Fasteners meeting ASTM F2329 required in wet areas.
- 2" alkali-resistant glass fiber mesh tape.
- Flexible sealant—ASTM C920.

Materials for Green/Sustainable Design

- See "Green Building Standards and Green Product Selection Guide" and consult manufacturers and suppliers for product sustainability and contribution to green building design.
- Consider specifying tile and installation materials that meet ANSI A138.1, the *American National Standard Specifications for Sustainable Ceramic Tiles, Glass Tiles, and Tile Installation Materials.*

Preparation by Other Trades

- Floor systems, including the framing system and subfloor panels, over which tile will be installed shall be in conformance with the IRC for residential applications, the IBC for commercial applications, or applicable building codes. Maximum allowable floor member live load and concentrated load deflection shall not exceed l/360. See also "Substrate Requirements."
- When concentrated loads (scissor lifts, pallet jacks, automobiles, forklifts, etc.) will be used on an above-ground tiled floor, the engineer and/or specifier shall specify a substrate to accommodate the concentrated loads. Owner/specifier is responsible for protecting the tilework from damage, including allowing sufficient time for installed materials to cure properly.
- Face grain of plywood shall run perpendicular to joists.
- Subfloor—minimum $^{19}/_{32}$" exterior-glue tongue and groove plywood with $^{1}/_{8}$" gap between sheets.
- Maximum variation in the plywood surface—$^{1}/_{4}$" in 10' and $^{1}/_{16}$" in 1' from the required plane. Adjacent edges of plywood sheets—maximum $^{1}/_{32}$" above or below each other.

Movement Joint (architect must specify type of joint and show location and details on drawings)

- Movement joints—mandatory according to EJ171.
- When glass tile is used, adhere to more frequent placement recommendations within the ranges listed in EJ171.

Installation Specifications

- Tile—ANSI A108.5.
- Glass tile—manufacturer's directions.
- Cementitious grout—ANSI A108.10.
- Epoxy grout—ANSI A108.6.
- Coated glass mat water-resistant gypsum backer board—manufacturer's directions.
- Waterproof membrane—ANSI A108.13.
- Movement Joints—EJ171 and ASTM C1193.

Notes

- When glass tile is used, see "Glass Tile Selection and Installation Guide," and consult manufacturer for recommendations and requirements.

F170-21

- **Joists max. 16" o.c./Plywood Subfloor**
- **Fiber-Reinforced Water-Resistant Gypsum Backer Board**
- **Ceramic Tile, Glass Tile**

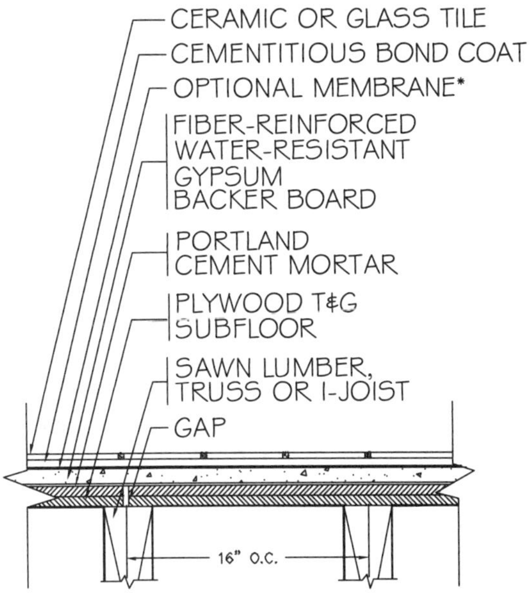

*USE OF A MEMBRANE IS OPTIONAL. SEE MEMBRANE OPTIONS.

Recommended Uses

- For wood substrates where water-resistance is desired.

Service Rating

- Light commercial.
- When glass tile is used, service rating may be lower.

Environmental Exposure Classifications

- Res1, 2; Com1, 2.
- May be suitable for increased water exposure. See Membrane Options.
- For installations that may be exposed to staining, specify tile and grout suitable for exposure. Consult product manufacturers; see also "Product Selection Guides."
- For installations that may be exposed to mild chemical attack, specify epoxy grout and tile suitable for exposure. Consult product manufacturers; see also "Product Selection Guides."

Typical Weight of Tile Installation

- 8 pounds/square foot if ¼" or ⅜" fiber-reinforced water-resistant gypsum backer board used.
- 9 pounds/square foot if ½" fiber-reinforced water-resistant gypsum backer board used.
- Does not include weight of substrate. See "Appendix B" for assumptions, included materials, and their individual weights.

Limitations

- Maximum joist spacing 16" on center.

Membrane Options

- A waterproof membrane (A118.10) may be specified to prevent moisture intrusion and protect adjacent walls and building materials. Base flashing should be used for maximum effectiveness.
- Specifier shall indicate if complete waterproofing is required, including if/how membrane connects to drain assembly, if base flashing is required, and treatment at other termination points.
- Check with membrane manufacturers for suitability for applicable conditions, as not all membranes are suitable for steam, high temperature, and/or chemical exposure, exterior use, use over above-ground structural slabs, use over pourable underlayments, use with radiant heat, or use over concrete with excessive moisture vapor transmission and/or alkalinity. Membrane may also affect service rating.
- When glass tile is used, consult glass tile manufacturer for membrane options and recommendations.

Requirements

- Caulk or seal penetrations and abutments to dissimilar materials.
- Surface of units—clean and free of dirt, dust, paint, and oily film.

Preparation by Backer Board Installers

- Use a sufficient amount of portland cement mortar under the backer board to establish a supporting plane and eliminate voids.
- Fasten backer units with corrosion-resistant fasteners per manufacturer's directions.
- Leave ⅛" perimeter movement gap and ⅛" gap between sheets. Fill joints between sheets solid and tape with portland cement mortar and 2" alkali-resistant glass fiber mesh tape. Do not fill perimeter joint.
- Maximum allowable variation in the tile substrate—for tiles with all edges shorter than 15", maximum allowable variation is ¼" in 10' from the required plane, with no more than ¹⁄₁₆" variation in 12" when measured from the high points in the surface. For tiles with at least one edge 15" in length, maximum allowable variation is ⅛" in 10' from the required plane, with no more than ¹⁄₁₆" variation in 24" when measured from the high points in the surface.
- Stagger backer board end and edge joints so as not to coincide with joints in subfloor. Stagger joints in adjacent rows so four corners do not come together within the same plane. Space panel ends and edges in accordance with manufacturer's recommendations.

Materials

- **Multiple options exist for membranes, mortars, grouts, and other materials and *must be clearly specified* to be included. If not specifically indicated, optional materials are not included and mortar/grout choice defaults to minimum performance specification indicated. Consider each system component and**

intended use to determine minimum requirements and to specify options.

- Ceramic tile—ANSI A137.1.
- Glass tile, when used—ANSI A137.2; see also "Glass Tile Selection and Installation Guide," and consult tile manufacturer for service rating and environmental exposure classification recommendations. Not all glass tiles are suitable.
- Cementitious grout—ANSI A118.6 or better or ISO CG1 or better. When glass tile is used, specify grout designated by tile and grout manufacturers.
- Epoxy grout, when used—ANSI A118.3 or ISO RG.
- Cementitious bond coat:
  - When a waterproof membrane is not used—ANSI A118.4 or better or ISO C2 or better.
  - When a waterproof membrane is used—ANSI A118.4 or better or ISO C2S1 or better unless ANSI A118.1 or ISO C1 is recommended by membrane manufacturer.
  - When glass tile is used, specify mortar designated by tile and mortar manufacturers. Bond coat color will impact the final appearance of translucent glass tile. Specifier shall confirm bond coat color is acceptable.
- Fiber-reinforced water-resistant gypsum backer board—ASTM C1278 (Paragraph 6.1).
- Mortar under backer board—ANSI A118.1 or better or ISO C1 or better.
- Waterproof membrane, when used—A118.10.
- Fasteners meeting ASTM F2329 required in wet areas.
- 2" alkali-resistant glass fiber mesh tape.
- Flexible sealant—ASTM C920.

Materials for Green/Sustainable Design

- See "Green Building Standards and Green Product Selection Guide" and consult manufacturers and suppliers for product sustainability and contribution to green building design.
- Consider specifying tile and installation materials that meet ANSI A138.1, the *American National Standard Specifications for Sustainable Ceramic Tiles, Glass Tiles, and Tile Installation Materials.*

Preparation by Other Trades

- Floor systems, including the framing system and subfloor panels, over which tile will be installed shall be in conformance with the IRC for residential applications, the IBC for commercial applications, or applicable building codes. Maximum allowable floor member live load and concentrated load deflection shall not exceed l/360. See also "Substrate Requirements."
- When concentrated loads (scissor lifts, pallet jacks, automobiles, forklifts, etc.) will be used on an above-ground tiled floor, the engineer and/or specifier shall specify a substrate to accommodate the concentrated loads. Owner/specifier is responsible for protecting the tilework from damage, including allowing sufficient time for installed materials to cure properly.
- Face grain of plywood shall run perpendicular to joists.
- Subfloor—minimum 19⁄32" exterior glue tongue and groove plywood with 1⁄8" gap between sheets.
- Maximum variation in plywood surface—1⁄4" in 10' and 1⁄16" in 1' from the required plane. Adjacent edges of plywood sheets—maximum 1⁄32" above or below each other.

Movement Joint (architect must specify type of joint and show location and details on drawings)

- Movement joints—mandatory in accordance with EJ171.
- When glass tile is used, adhere to more frequent placement recommendations within the ranges listed in EJ171.

Installation Specifications

- Tile—ANSI A108.5.
- Glass tile—manufacturer's directions.
- Cementitious grout—ANSI A108.10.
- Epoxy grout—ANSI A108.6.
- Fiber-reinforced water-resistant gypsum backer board—manufacturer's directions.
- Waterproof membrane—ANSI A108.13.
- Movement Joints—EJ171 and ASTM C1193.

Notes

- When glass tile is used, see "Glass Tile Selection and Installation Guide," and consult manufacturer for recommendations and requirements.

F175-21

- **Joists max. 16" o.c./Plywood Subfloor**
- **Cementitious-Coated Extruded Foam Backer Board**
- **Ceramic Tile**

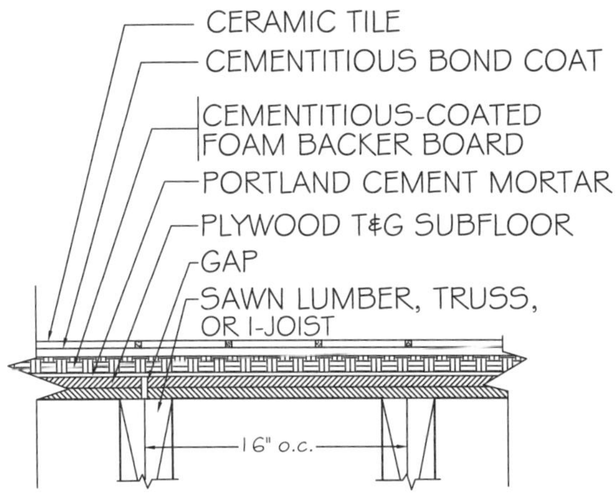

Recommended Uses

- For wood substrates where water-resistance is desired.

Service Rating

- Light commercial.

Environmental Exposure Classifications

- Res1, 2; Com1, 2.
- May be suitable for increased water exposure when backer board manufacturer's waterproofing instructions are followed.
- For installations that may be exposed to staining, specify tile and grout suitable for exposure. Consult product manufacturers; see also "Product Selection Guides."
- For installations that may be exposed to mild chemical attack, specify epoxy grout and tile suitable for exposure. Consult product manufacturers; see also "Product Selection Guides."

Typical Weight of Tile Installation

- 7 pounds/square foot.
- Does not include weight of substrate. See "Appendix B" for assumptions, included materials, and their individual weights.

Limitations

- Maximum joist spacing 16" on center.
- 8" × 8" and larger tile only.

Requirements

- Surface of units—clean and free of dirt, dust, paint, and oily film.
- Specifier shall indicate if complete waterproofing is required, including if/how to connect to drain assembly, if base flashing is required, and treatment at other termination points.

Preparation by Backer Board Installers

- Fasten backer board with board manufacturer's recommended fasteners.
- Use a sufficient amount of portland cement mortar under the backer board to establish a supporting plane and eliminate voids.
- Maximum allowable variation in the tile substrate—for tiles with all edges shorter than 15", maximum allowable variation is ¼" in 10' from the required plane, with no more than ¹⁄₁₆" variation in 12" when measured from the high points in the surface. For tiles with at least one edge 15" in length, maximum allowable variation is ⅛" in 10' from the required plane, with no more than ¹⁄₁₆" variation in 24" when measured from the high points in the surface.
- Fit ends and edges tightly together, including in corners. Apply a bead of manufacturer-recommended sealant in panel joints as panels are being installed, and over fastener heads, per backer board manufacturer's instructions.
- Joints may be taped with 2" alkali-resistant glass fiber mesh tape and cementitious bonding material in dry areas.
- Stagger backer board end and edge joints so as not to coincide with joints in subfloor. Stagger joints in adjacent rows so four corners do not come together within the same plane. Space panel ends and edges in accordance with manufacturer's recommendations.

Materials

- **Multiple options exist for membranes, mortars, grouts, and other materials and *must be clearly specified* to be included. If not specifically indicated, optional materials are not included and mortar/grout choice defaults to minimum performance specification indicated. Consider each system component and intended use to determine minimum requirements and to specify options.**
- Ceramic tile—ANSI A137.1.
- Cementitious grout—ANSI A118.6 or better or ISO CG1 or better.
- Epoxy grout, when used—ANSI A118.3 or ISO RG.
- Cementitious bond coat:
  - ANSI A118.1 or better or ISO C1 or better.
  - When porcelain tile is used—ANSI A118.4 or better or ISO C2 or better.
- Cementitious-coated extruded foam backer board—ASTM C578.
- Mortar under backer board—ANSI A118.1 or better or ISO C1 or better.
- Fasteners—noncorrosive and nonoxidizing.
- 2" alkali-resistant glass fiber mesh tape, when used.
- Flexible sealant—must be recommended by backer board manufacturer.

Materials for Green/Sustainable Design

- See "Green Building Standards and Green Product Selection Guide" and consult manufacturers and suppliers for product sustainability and contribution to green building design.
- Consider specifying tile and installation materials that meet ANSI A138.1, the *American National Standard Specifications for Sustainable Ceramic Tiles, Glass Tiles, and Tile Installation Materials.*

Preparation by Other Trades

- Floor systems, including the framing system and subfloor panels, over which tile will be installed shall be in conformance with the IRC for residential applications, the IBC for commercial applications, or applicable building codes. Maximum allowable floor member live load and concentrated load deflection shall not exceed l/360. See also "Substrate Requirements."
- When concentrated loads (scissor lifts, pallet jacks, automobiles, forklifts, etc.) will be used on an aboveground tiled floor, the engineer and/or specifier shall specify a substrate to accommodate the concentrated loads. Owner/specifier is responsible for protecting the tilework from damage, including allowing sufficient time for installed materials to cure properly.
- Face grain of plywood shall run perpendicular to joists.
- Subfloor—minimum 19⁄32" exterior-glue tongue and groove plywood with 1⁄8" gap between sheets.
- Maximum variation in plywood surface—1⁄4" in 10' and 1⁄16" in 1' from the required plane. Adjacent edges of plywood sheets—maximum 1⁄32" above or below each other.

Movement Joint (architect must specify type of joint and show location and details on drawings)

- Movement joints—mandatory in accordance with EJ171.

Installation Specifications

- Tile—ANSI A108.5.
- Cementitious grout—ANSI A108.10.
- Epoxy grout—ANSI A108.6.
- Cementitious-coated extruded foam backer board—manufacturer's directions.
- Movement Joints—EJ171 and ASTM C1193.

F147-21

- **Joists max. 24" o.c./Plywood Subfloor**
- **Plywood Underlayment**
- **Uncoupling Membrane**
- **Ceramic Tile**

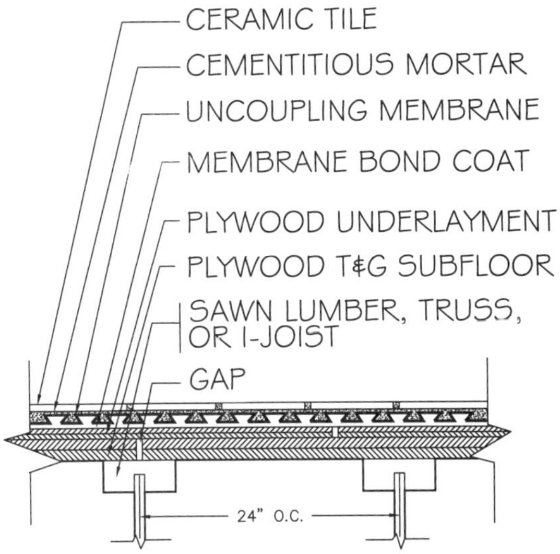

Recommended Uses

- For wood substrates where joists are spaced 24" on center or less and water resistance and uncoupling are desired.

Service Rating

- Residential.

Environmental Exposure Classifications

- Res1, 2.
- May be suitable for increased water exposure when membrane manufacturer's waterproofing instructions are followed. See Notes.
- For installations that may be exposed to staining, specify tile and grout suitable for exposure. Consult product manufacturers; see also "Product Selection Guides."
- For installations that may be exposed to mild chemical attack, specify epoxy grout and tile suitable for exposure. Consult product manufacturers; see also "Product Selection Guides."

Typical Weight of Tile Installation

- 7 pounds/square foot.
- Does not include weight of substrate. See "Appendix B" for assumptions, included materials, and their individual weights.

Limitations

- Maximum joist spacing 24" on center.
- 4" × 4" and larger tile only.

Requirements

- Plywood underlayment—clean and free of dirt, dust, paint, and oily film.
- Specifier shall indicate if complete waterproofing is required, including if/how membrane connects to drain assembly, if base flashing is required, and treatment at other termination points.

Materials

- **Multiple options exist for membranes, mortars, grouts, and other materials and *must be clearly specified* to be included. If not specifically indicated, optional materials are not included and mortar/grout choice defaults to minimum performance specification indicated. Consider each system component and intended use to determine minimum requirements and to specify options.**
- Ceramic tile—ANSI A137.1.
- Cementitious grout—ANSI A118.6 or better or ISO CG1 or better.
- Epoxy grout, when used—ANSI A118.3 or ISO RG.
- Cementitious mortar—as recommended by membrane manufacturer.
- Uncoupling membrane—recommended by manufacturer; must achieve 50 psi or greater shear bond strength in 7 days per the test method in ANSI A118.12 Section 5.1.3.

Materials for Green/Sustainable Design

- See "Green Building Standards and Green Product Selection Guide" and consult manufacturers and suppliers for product sustainability and contribution to green building design.
- Consider specifying tile and installation materials that meet ANSI A138.1, the *American National Standard Specifications for Sustainable Ceramic Tiles, Glass Tiles, and Tile Installation Materials.*

Preparation by Other Trades

- Floor systems, including the framing system and subfloor panels, over which tile will be installed shall be in conformance with the IRC for residential applications, the IBC for commercial applications, or applicable building codes. Maximum allowable floor member live load and concentrated load deflection shall not exceed l/360. See also "Substrate Requirements."
- When concentrated loads (scissor lifts, pallet jacks, automobiles, forklifts, etc.) will be used on an above-ground tiled floor, the engineer and/or specifier shall specify a substrate to accommodate the concentrated loads. Owner/specifier is responsible for protecting the tilework from damage, including allowing sufficient time for installed materials to cure properly.

- Face grain of plywood subfloor and underlayment shall run perpendicular to joists.
- Trusses or I-joists with minimum 2¼" top flange (1½" top flange permissible with 8" × 8" and larger tile); cross-bracing recommended.
- Subfloor—minimum 23⁄32" tongue and groove exterior glue plywood with ⅛" gap between sheets.
- Underlayment—minimum ⅜" exterior glue plywood with ⅛" gap between sheets.
- Maximum allowable variation in the installation substrate (plywood underlayment)—for tiles with all edges shorter than 15", maximum allowable variation is ¼" in 10' from the required plane, with no more than 1⁄16" variation in 12" when measured from the high points in the surface. For tiles with at least one edge 15" in length, maximum allowable variation is ⅛" in 10' from the required plane, with no more than 1⁄16" variation in 24" when measured from the high points in the surface.
- Vapor barrier on crawl space floors according to building code.

Movement Joint (architect must specify type of joint and show location and details on drawings)

- Movement joints—mandatory in accordance with EJ171.

Installation Specifications

- Tile—ANSI A108.5.
- Cementitious grout—ANSI A108.10.
- Epoxy grout—ANSI A108.6.
- Uncoupling membrane—manufacturer's directions.
- Movement Joints—EJ171 and ASTM C1193.

Notes

- Underlayment fasteners should not penetrate joists below.
- Seal seams between membrane sheets per membrane manufacturer's instructions to prevent moisture intrusion and protect adjacent walls and building materials. Base flashing should be used for maximum effectiveness.

F148-21

- **Joists max. 19.2" o.c./Plywood Subfloor**
- **Uncoupling Membrane**
- **Ceramic Tile**

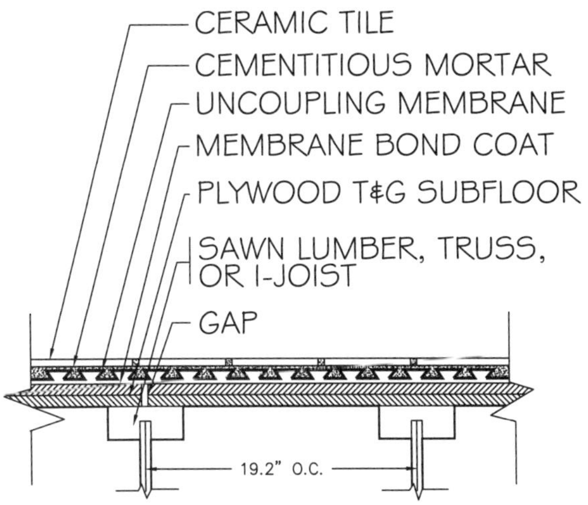

Recommended Uses

- For wood substrates where joists are spaced 19.2" on center or less and water resistance and uncoupling are desired.

Service Rating

- Residential.

Environmental Exposure Classifications

- Res1, 2.
- May be suitable for increased water exposure when membrane manufacturer's waterproofing instructions are followed. See Notes.
- For installations that may be exposed to staining, specify tile and grout suitable for exposure. Consult product manufacturers; see also "Product Selection Guides."
- For installations that may be exposed to mild chemical attack, specify epoxy grout and tile suitable for exposure. Consult product manufacturers; see also "Product Selection Guides."

Typical Weight of Tile Installation

- 6 pounds/square foot.
- Does not include weight of substrate. See "Appendix B" for assumptions, included materials, and their individual weights.

Limitations

- Maximum joist spacing 19.2" on center.
- 3" × 3" and larger tile only.

Requirements

- Plywood subfloor—clean and free of dirt, dust, paint, and oily film.
- Specifier shall indicate if complete waterproofing is required, including if/how membrane connects to drain assembly, if base flashing is required, and treatment at other termination points.

Materials

- **Multiple options exist for membranes, mortars, grouts, and other materials and *must be clearly specified* to be included. If not specifically indicated, optional materials are not included and mortar/grout choice defaults to minimum performance specification indicated. Consider each system component and intended use to determine minimum requirements and to specify options.**
- Ceramic tile—ANSI A137.1.
- Cementitious grout—ANSI A118.6 or better or ISO CG1 or better.
- Epoxy grout, when used—ANSI A118.3 or ISO RG.
- Cementitious mortar—as recommended by membrane manufacturer.
- Uncoupling membrane—recommended by manufacturer; must achieve 50 psi or greater shear bond strength in 7 days per the test method in ANSI A118.12 Section 5.1.3.

Materials for Green/Sustainable Design

- See "Green Building Standards and Green Product Selection Guide" and consult manufacturers and suppliers for product sustainability and contribution to green building design.
- Consider specifying tile and installation materials that meet ANSI A138.1, the *American National Standard Specifications for Sustainable Ceramic Tiles, Glass Tiles, and Tile Installation Materials.*

Preparation by Other Trades

- Floor systems, including the framing system and subfloor panels, over which tile will be installed shall be in conformance with the IRC for residential applications, the IBC for commercial applications, or applicable building codes. Maximum allowable floor member live load and concentrated load deflection shall not exceed l/360. See also "Substrate Requirements."
- When concentrated loads (scissor lifts, pallet jacks, automobiles, forklifts, etc.) will be used on an above-ground tiled floor, the engineer and/or specifier shall specify a substrate to accommodate the concentrated loads. Owner/specifier is responsible for protecting the tilework from damage, including allowing sufficient time for installed materials to cure properly.
- Face grain of plywood shall run perpendicular to joists.
- Subfloor—minimum $^{23}/_{32}$" tongue and groove exterior glue plywood with $1/8$" gap between sheets.
- Maximum allowable variation in the installation substrate (plywood underlayment)—for tiles with all edges

shorter than 15", maximum allowable variation is ¼" in 10' from the required plane, with no more than ¹⁄₁₆" variation in 12" when measured from the high points in the surface. For tiles with at least one edge 15" in length, maximum allowable variation is ⅛" in 10' from the required plane, with no more than ¹⁄₁₆" variation in 24" when measured from the high points in the surface.

- Vapor barrier on crawl space floors according to building code.

Movement Joint (architect must specify type of joint and show location and details on drawings)

- Movement joints—mandatory in accordance with EJ171.

Installation Specifications

- Tile—ANSI A108.5.
- Cementitious grout—ANSI A108.10.
- Epoxy grout—ANSI A108.6.
- Uncoupling membrane—manufacturer's directions.
- Movement Joints—EJ171 and ASTM C1193.

Notes

- Seal seams between membrane sheets per membrane manufacturer's instructions to prevent moisture intrusion and protect adjacent walls and building materials. Base flashing should be used for maximum effectiveness.

F149-21

- **Joists max. 24" o.c./Plywood Subfloor**
- **Plywood Underlayment**
- **Ceramic Tile**

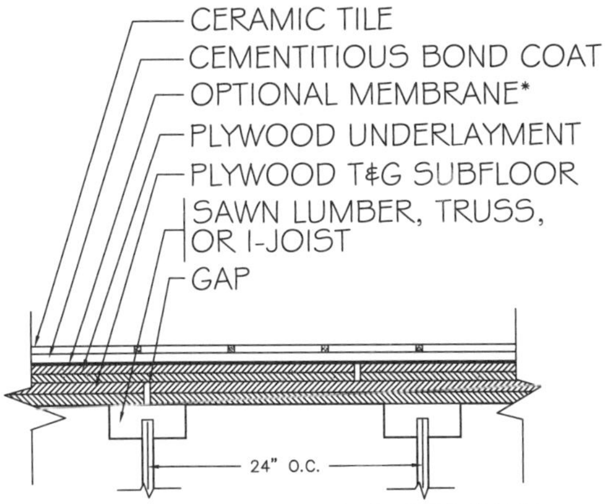

*USE OF A MEMBRANE IS OPTIONAL. SEE MEMBRANE OPTIONS.

Recommended Uses

- For wood substrates in dry areas where joist spacing is 24" on center or less.

Service Rating

- Residential.

Environmental Exposure Classifications

- Res1.
- May be suitable for increased water exposure. See Membrane Options.
- For installations that may be exposed to staining, specify tile and grout suitable for exposure. Consult product manufacturers; see also "Product Selection Guides."
- For installations that may be exposed to mild chemical attack, specify epoxy grout and tile suitable for exposure. Consult product manufacturers; see also "Product Selection Guides."

Typical Weight of Tile Installation

- 7 pounds/square foot.
- Does not include weight of substrate. See "Appendix B" for assumptions, included materials, and their individual weights.

Limitations

- Maximum joist spacing 24" on center.
- 8" × 8" and larger tile only.

Membrane Options

- A waterproof membrane (A118.10) may be specified to prevent moisture intrusion and protect adjacent walls and building materials. Base flashing should be used for maximum effectiveness.
- Specifier shall indicate if complete waterproofing is required, including if/how membrane connects to drain assembly, if base flashing is required, and treatment at other termination points.
- Check with membrane manufacturers for suitability for applicable conditions, as not all membranes are suitable for steam, high temperature, and/or chemical exposure, exterior use, use over above-ground structural slabs, use over pourable underlayments, use with radiant heat, or use over concrete with excessive moisture vapor transmission and/or alkalinity. Membrane may also affect service rating.

Requirements

- Plywood underlayment—clean and free of dirt, dust, paint, and oily film.
- Gaps between plywood underlayment sheets to be treated per setting material manufacturer's recommendations.

Materials

- **Multiple options exist for membranes, mortars, grouts, and other materials and *must be clearly specified* to be included. If not specifically indicated, optional materials are not included and mortar/grout choice defaults to minimum performance specification indicated. Consider each system component and intended use to determine minimum requirements and to specify options.**
- Ceramic tile—ANSI A137.1.
- Cementitious grout—ANSI A118.6 or better or ISO CG1 or better.
- Epoxy grout, when used—ANSI A118.3 or ISO RG.
- Cementitious bond coat:
  - When a waterproof membrane is not used—ANSI A118.11 or better or ISO C2S1P2 or better.
  - When a waterproof membrane is used—ANSI A118.4 or better or ISO C2S1 or better, unless ANSI A118.1 or ISO C1 is recommended by membrane manufacturer (mortar for bonding tile to membrane).
  - When a waterproof membrane is used, follow membrane manufacturer's directions for bonding membrane to plywood underlayment.
- Waterproof membrane, when used—ANSI A118.10.

Materials for Green/Sustainable Design

- See "Green Building Standards and Green Product Selection Guide" and consult manufacturers and suppliers for product sustainability and contribution to green building design.
- Consider specifying tile and installation materials that meet ANSI A138.1, the *American National Standard Specifications for Sustainable Ceramic Tiles, Glass Tiles, and Tile Installation Materials.*

Preparation by Other Trades

- Floor systems, including the framing system and subfloor panels, over which tile will be installed shall be in conformance with the IRC for residential applications, the IBC for commercial applications, or applicable building codes. Maximum allowable floor member live load and concentrated load deflection shall not exceed l/360. See also "Substrate Requirements."
- When concentrated loads (scissor lifts, pallet jacks, automobiles, forklifts, etc.) will be used on an aboveground tiled floor, the engineer and/or specifier shall specify a substrate to accommodate the concentrated loads. Owner/specifier is responsible for protecting the tilework from damage, including allowing sufficient time for installed materials to cure properly.
- Face grain of plywood subfloor and underlayment shall run perpendicular to joists.
- Trusses or I-joists with minimum 1½" top flange or sawn lumber joists; cross bracing recommended.
- Subfloor—minimum 23⁄32" exterior-glue tongue and groove plywood with ⅛" gap between sheets.
- Underlayment—minimum 19⁄32" exterior glue plywood with ⅛" gap between sheets.
- Maximum allowable variation in the tile substrate—for tiles with all edges shorter than 15", maximum allowable variation is ¼" in 10' from the required plane, with no more than 1⁄16" variation in 12" when measured from the high points in the surface. For tiles with at least one edge 15" in length, maximum allowable variation is ⅛" in 10' from the required plane, with no more than 1⁄16" variation in 24" when measured from the high points in the surface. Adjacent edges of plywood sheets not to exceed 1⁄32" difference in height.

Movement Joint (architect must specify type of joint and show location and details on drawings)

- Movement joints—mandatory in accordance with EJ171.

Installation Specifications

- Tile—ANSI A108.12 or .5.
- Cementitious grout—ANSI A108.10.
- Epoxy grout—ANSI A108.6.
- Waterproof membrane—ANSI A108.13.
- Movement Joints—EJ171 and ASTM C1193.

Notes

- Underlayment fasteners should not penetrate joists below.

F151-21

- **Joists max. 24" o.c./Plywood Subfloor**
- **Coated Glass Mat Water-Resistant Gypsum Backer Board**
- **Ceramic Tile**

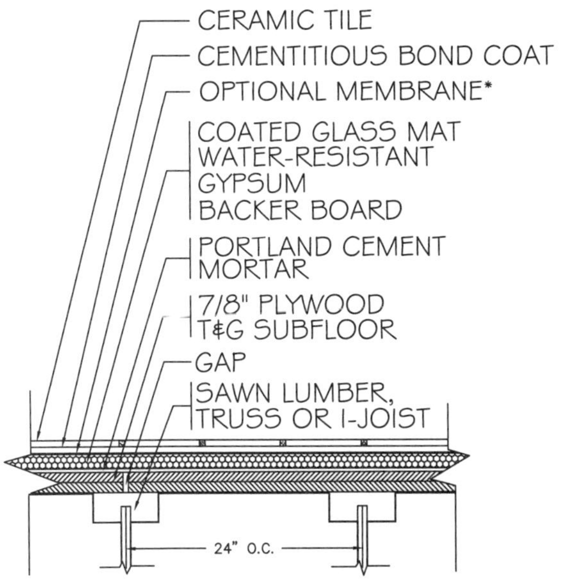

*USE OF A MEMBRANE IS OPTIONAL. SEE MEMBRANE OPTIONS.

Recommended Uses

- For wood substrates where joists are 24" on center or less and water resistance is desired.

Service Rating

- Residential.

Environmental Exposure Classifications

- Res1, 2.
- May be suitable for increased water exposure. See Membrane Options.
- For installations that may be exposed to staining, specify tile and grout suitable for exposure. Consult product manufacturers; see also "Product Selection Guides."
- For installations that may be exposed to mild chemical attack, specify epoxy grout and tile suitable for exposure. Consult product manufacturers; see also "Product Selection Guides."

Typical Weight of Tile Installation

- 8 pounds/square foot.
- Does not include weight of substrate. See "Appendix B" for assumptions, included materials, and their individual weights.

Limitations

- Maximum joist spacing 24" on center.
- 8" × 8" and larger tile only.

Membrane Options

- A waterproof membrane (A118.10) may be specified to prevent moisture intrusion through seams, corners, fasteners, and other penetrations to protect adjacent walls and building materials. Base flashing should be used for maximum effectiveness. Follow backer board and membrane manufacturer's waterproofing requirements.
- Specifier shall indicate if complete waterproofing is required, including if/how membrane connects to drain assembly, if base flashing is required, and treatment at other termination points.
- Check with membrane manufacturers for suitability for applicable conditions, as not all membranes are suitable for steam, high temperature, and/or chemical exposure, exterior use, use over above-ground structural slabs, use over pourable underlayments, use with radiant heat, or use over concrete with excessive moisture vapor transmission and/or alkalinity. Membrane may also affect service rating.
- Some liquid-applied waterproof membranes will require extended cure time. Consult membrane manufacturer.

Requirements

- Caulk or seal penetrations, corners, and abutments to dissimilar materials with flexible sealant.
- Surfaces of units—clean and free of dirt, dust, paint, and oily film.
- Set tiles on positive side of moisture barrier coating.

Preparation by Backer Board Installers

- Maximum allowable variation in the tile substrate—for tiles with all edges shorter than 15", maximum allowable variation is ¼" in 10' from the required plane, with no more than ¹⁄₁₆" variation in 12" when measured from the high points in the surface. For tiles with at least one edge 15" in length, maximum allowable variation is ⅛" in 10' from the required plane, with no more than ¹⁄₁₆" variation in 24" when measured from the high points in the surface.
- Fit ends and edges closely to eliminate gaps between panels, including in corners.
- Tape seams and corners with 2" alkali-resistant glass fiber mesh tape and cementitious bonding material, and skim coat over fastener heads, unless waterproofing has been applied.
- Use a sufficient amount of portland cement mortar under the backer board to establish a supporting plane and eliminate voids.
- Fasten backer units with corrosion-resistant fasteners per manufacturer's directions.
- Stagger backer board end and edge joints so as not to coincide with joints in subfloor. Stagger joints in adjacent rows so four corners do not come together within the

same plane. Space panel ends and edges in accordance with manufacturer's recommendations.

Materials

- **Multiple options exist for membranes, mortars, grouts, and other materials and *must be clearly specified* to be included. If not specifically indicated, optional materials are not included and mortar/grout choice defaults to minimum performance specification indicated. Consider each system component and intended use to determine minimum requirements and to specify options.**
- Ceramic tile—ANSI A137.1.
- Cementitious grout—ANSI A118.6 or better or ISO CG1 or better.
- Epoxy grout, when used—ANSI A118.3 or ISO RG.
- Cementitious bond coat:
  - When a waterproof membrane is not used—ANSI A118.4 or better or ISO C2S1 or better.
  - When a waterproof membrane is used—ANSI A118.4 or better or ISO C2S1 or better unless ANSI A118.1 or ISO C1 is recommended by membrane manufacturer.
- Coated glass mat water-resistant gypsum backer board—ASTM C1178.
- Mortar under backer board—ANSI A118.1 or better or ISO C1 or better.
- Waterproof membrane, when used—A118.10.
- Fasteners meeting ASTM F2329 required in wet areas.
- 2" alkali-resistant glass fiber mesh tape.
- Flexible sealant—ASTM C920.

Materials for Green/Sustainable Design

- See "Green Building Standards and Green Product Selection Guide" and consult manufacturers and suppliers for product sustainability and contribution to green building design.
- Consider specifying tile and installation materials that meet ANSI A138.1, the *American National Standard Specifications for Sustainable Ceramic Tiles, Glass Tiles, and Tile Installation Materials.*

Preparation by Other Trades

- Floor systems, including the framing system and subfloor panels, over which tile will be installed shall be in conformance with the IRC for residential applications, the IBC for commercial applications, or applicable building codes. Maximum allowable floor member live load and concentrated load deflection shall not exceed l/360. See also "Substrate Requirements."
- When concentrated loads (scissor lifts, pallet jacks, automobiles, forklifts, etc.) will be used on an above-ground tiled floor, the engineer and/or specifier shall specify a substrate to accommodate the concentrated loads. Owner/specifier is responsible for protecting the tilework from damage, including allowing sufficient time for installed materials to cure properly.
- Face grain of plywood shall run perpendicular to joists.
- Subfloor—minimum ⅞" exterior-glue tongue and groove plywood with ⅛" gap between sheets.
- Maximum variation in the plywood surface—¼" in 10' and ¹⁄₁₆" in 1' from the required plane. Adjacent edges of plywood sheets—maximum ¹⁄₃₂" above or below each other.
- Install tongue and groove plywood according to manufacturer's installation instructions for tiled floors.

Movement Joint (architect must specify type of joint and show location and details on drawings)

- Movement joints—mandatory in accordance with EJ171.

Installation Specifications

- Tile—ANSI A108.5.
- Cementitious grout—ANSI A108.10.
- Epoxy grout—ANSI A108.6.
- Waterproof membrane—ANSI A108.13.
- Coated glass mat water-resistant gypsum backer board—manufacturer's directions.
- Movement Joints—EJ171 and ASTM C1193.

F152-21

- **Joists max. 24" o.c./Plywood Subfloor**
- **Plywood Underlayment**
- **Bonded Membrane**
- **Ceramic Tile**

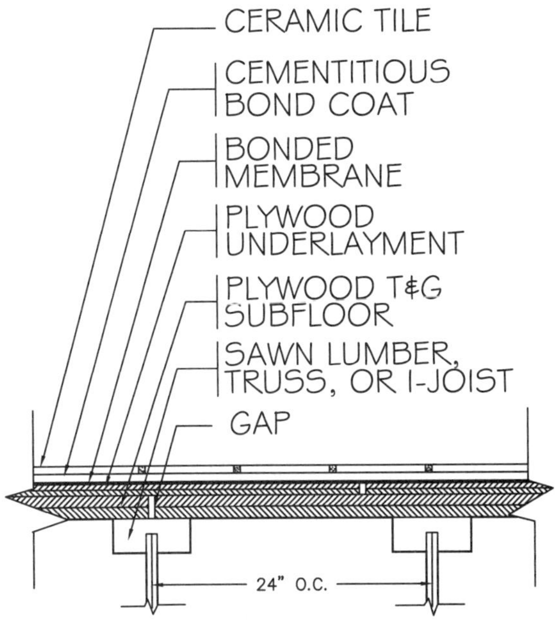

Recommended Uses

- For wood substrates where joists are spaced 24" on center or less and a water-resistant or waterproof floor is desired.

Service Rating

- Residential. Consult membrane manufacturer to verify.

Environmental Exposure Classifications

- Res1, 2.
- May be suitable for increased water exposure when membrane meets ANSI A118.10 and manufacturer's waterproofing instructions are followed. See Notes.
- For installations that may be exposed to staining, specify tile and grout suitable for exposure. Consult product manufacturers; see also "Product Selection Guides."
- For installations that may be exposed to mild chemical attack, specify epoxy grout and tile suitable for exposure. Consult product manufacturers; see also "Product Selection Guides."

Typical Weight of Tile Installation

- 7 pounds/square foot.
- Does not include weight of substrate. See "Appendix B" for assumptions, included materials, and their individual weights.

Limitations

- Maximum joist spacing 24" on center.
- 4" × 4" and larger tile only.

Requirements

- Plywood underlayment—clean and free of dirt, dust, paint, and oily film.
- Consult membrane manufacturer to verify membrane meets residential service rating when tested per ASTM C627 with joists spaced 24" on center and with actual tile selected.
- Specifier shall indicate if complete waterproofing is required, including if/how membrane connects to drain assembly, if base flashing is required, and treatment at other termination points.
- Seal seams between membrane sheets per membrane manufacturer's instructions to prevent moisture intrusion and protect adjacent walls and building materials. Base flashing should be used for maximum effectiveness.

Materials

- **Multiple options exist for membranes, mortars, grouts, and other materials and *must be clearly specified* to be included. If not specifically indicated, optional materials are not included and mortar/grout choice defaults to minimum performance specification indicated. Consider each system component and intended use to determine minimum requirements and to specify options.**
- Ceramic tile—ANSI A137.1.
- Cementitious grout—ANSI A118.6 or better or ISO CG1 or better.
- Epoxy grout, when used—ANSI A118.3 or ISO RG.
- Cementitious bond coat (for bonding tile to membrane)—ANSI A118.4 or better or ISO C2S1 or better.
- Membrane—recommended by manufacturer.

Materials for Green/Sustainable Design

- See "Green Building Standards and Green Product Selection Guide" and consult manufacturers and suppliers for product sustainability and contribution to green building design.
- Consider specifying tile and installation materials that meet ANSI A138.1, the *American National Standard Specifications for Sustainable Ceramic Tiles, Glass Tiles, and Tile Installation Materials.*

Preparation by Other Trades

- Floor systems, including the framing system and subfloor panels, over which tile will be installed shall be in conformance with the IRC for residential applications, the IBC for commercial applications, or applicable building codes. Maximum allowable floor member live load and concentrated load deflection shall not exceed l/360. See also "Substrate Requirements."
- When concentrated loads (scissor lifts, pallet jacks, automobiles, forklifts, etc.) will be used on an above-ground tiled floor, the engineer and/or specifier shall

specify a substrate to accommodate the concentrated loads. Owner/specifier is responsible for protecting the tilework from damage, including allowing sufficient time for installed materials to cure properly.

- Face grain of plywood subfloor and underlayment shall run perpendicular to joists.
- Trusses or I-joists with minimum 2¼" top flange (1½" top flange permissible with 8" × 8" and larger tile); cross-bracing recommended.
- Subfloor—minimum 23⁄32" tongue and groove exterior glue plywood with ⅛" gap between sheets.
- Underlayment—minimum ⅜" exterior glue plywood with ⅛" gap between sheets.
- Maximum allowable variation in the tile substrate—for tiles with all edges shorter than 15", maximum allowable variation is ¼" in 10' from the required plane, with no more than 1⁄16" variation in 12" when measured from the high points in the surface. For tiles with at least one edge 15" in length, maximum allowable variation is ⅛" in 10' from the required plane, with no more than 1⁄16" variation in 24" when measured from the high points in the surface. Adjacent edges of plywood sheets not to exceed 1⁄32" difference in height.

Movement Joint (architect must specify type of joint and show location and details on drawings)

- Movement joints—mandatory in accordance with EJ171.

Installation Specifications

- Tile—ANSI A108.5.
- Cementitious grout—ANSI A108.10.
- Epoxy grout—ANSI A108.6.
- Membrane—manufacturer's directions.
- Movement Joints—EJ171 and ASTM C1193.

Notes

- Underlayment fasteners should not penetrate joists below.
- Not all membranes are recommended for this application. Consult membrane manufacturer.

F155-21

- **Joists max. 24" o.c./Plywood or OSB Subfloor**
- **Plywood Underlayment**
- **Ceramic Tile**

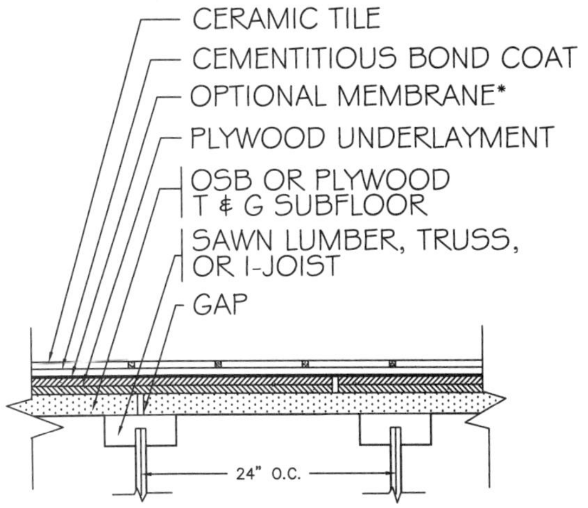

*USE OF A MEMBRANE IS OPTIONAL. SEE MEMBRANE OPTIONS.

Recommended Uses

- For wood substrates in dry areas where joist spacing is 24" on center or less.

Service Rating

- Residential.

Environmental Exposure Classifications

- Res1.
- May be suitable for increased water exposure. See Membrane Options.
- For installations that may be exposed to staining, specify tile and grout suitable for exposure. Consult product manufacturers; see also "Product Selection Guides."
- For installations that may be exposed to mild chemical attack, specify epoxy grout and tile suitable for exposure. Consult product manufacturers; see also "Product Selection Guides."

Typical Weight of Tile Installation

- 7 pounds/square foot.
- Does not include weight of substrate. See "Appendix B" for assumptions, included materials, and their individual weights.

Limitations

- Maximum joist spacing 24" on center.
- 8" × 8" and larger tile only.

Membrane Options

- A waterproof membrane (A118.10) may be specified to prevent moisture intrusion and protect adjacent walls and building materials. Base flashing should be used for maximum effectiveness.
- Specifier shall indicate if complete waterproofing is required, including if/how membrane connects to drain assembly, if base flashing is required, and treatment at other termination points.
- Check with membrane manufacturers for suitability for applicable conditions, as not all membranes are suitable for steam, high temperature, and/or chemical exposure, exterior use, use over above-ground structural slabs, use over pourable underlayments, use with radiant heat, or use over concrete with excessive moisture vapor transmission and/or alkalinity. Membrane may also affect service rating.

Requirements

- Plywood underlayment—clean and free of dirt, dust, paint, and oily film.
- Gaps between underlayment sheets to be treated per setting material manufacturer's recommendations.

Materials

- **Multiple options exist for membranes, mortars, grouts, and other materials and *must be clearly specified* to be included. If not specifically indicated, optional materials are not included and mortar/grout choice defaults to minimum performance specification indicated. Consider each system component and intended use to determine minimum requirements and to specify options.**
- Ceramic tile—ANSI A137.1.
- Cementitious grout—ANSI A118.6 or better or ISO CG1 or better.
- Epoxy grout, when used—ANSI A118.3 or ISO RG.
- Cementitious bond coat:
  - When a waterproof membrane is not used—ANSI A118.11 or better or ISO C2S1P2 or better.
  - When a waterproof membrane is used—ANSI A118.4 or better or ISO C2S1 or better, unless ANSI A118.1 or ISO C1 is recommended by membrane manufacturer (mortar for bonding tile to membrane).
- When a waterproof membrane is used, follow membrane manufacturer's directions for bonding membrane to plywood underlayment.
- Waterproof membrane, when used—ANSI A118.10.

Materials for Green/Sustainable Design

- See "Green Building Standards and Green Product Selection Guide" and consult manufacturers and suppliers for product sustainability and contribution to green building design.
- Consider specifying tile and installation materials that meet ANSI A138.1, the *American National Standard Specifications for Sustainable Ceramic Tiles, Glass Tiles, and Tile Installation Materials.*

Preparation by Other Trades

- Floor systems, including the framing system and subfloor panels, over which tile will be installed shall be in conformance with the IRC for residential applications, the IBC for commercial applications, or applicable building codes. Maximum allowable floor member live load and concentrated load deflection shall not exceed l/360. See also "Substrate Requirements."
- When concentrated loads (scissor lifts, pallet jacks, automobiles, forklifts, etc.) will be used on an aboveground tiled floor, the engineer and/or specifier shall specify a substrate to accommodate the concentrated loads. Owner/specifier is responsible for protecting the tilework from damage, including allowing sufficient time for installed materials to cure properly.
- Strength axis of OSB subfloor and face grain of plywood underlayment shall run perpendicular to joists.
- Trusses or I-joists with minimum 1½" top flange or sawn lumber joists; cross bracing recommended.
- Subfloor—minimum 23⁄32" tongue and groove OSB or plywood with ⅛" gap between sheets.
- Underlayment—minimum 19⁄32" exterior glue plywood with ⅛" gap between sheets.
- Maximum allowable variation in the tile substrate—for tiles with all edges shorter than 15", maximum allowable variation is ¼" in 10' from the required plane, with no more than 1⁄16" variation in 12" when measured from the high points in the surface. For tiles with at least one edge 15" in length, maximum allowable variation is ⅛" in 10' from the required plane, with no more than 1⁄16" variation in 24" when measured from the high points in the surface. Adjacent edges of plywood sheets not to exceed 1⁄32" difference in height.

Movement Joint (architect must specify type of joint and show location and details on drawings)

- Movement joints—mandatory in accordance with EJ171.

Installation Specifications

- Tile—ANSI A108.12 or .5.
- Cementitious grout—ANSI A108.10.
- Epoxy grout—ANSI A108.6.
- Waterproof membrane—ANSI A108.13.
- Movement Joints—EJ171 and ASTM C1193.

Notes

- Underlayment fasteners should not penetrate joists below.

F180-21

- **Joists max. 16" o.c./Plywood Subfloor**
- **Poured Gypsum Underlayment**
- **Bonded Membrane**
- **Ceramic Tile, Glass Tile**

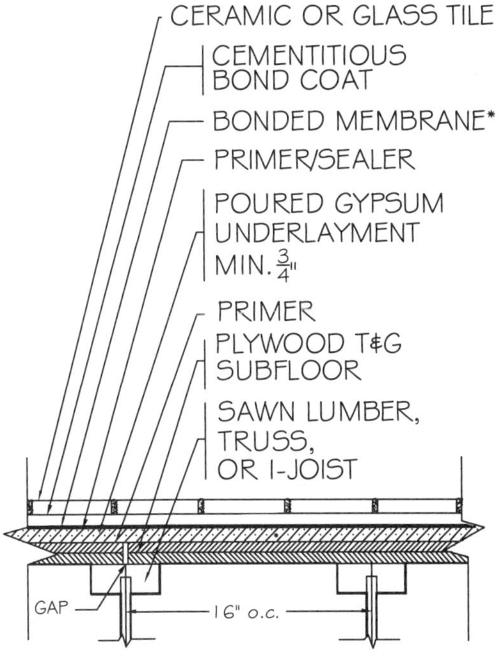

*USE OF A MEMBRANE IS REQUIRED. SEE MEMBRANE OPTIONS.

Recommended Uses

- For wood substrates with a poured gypsum underlayment.

Service Rating

- Light commercial.
- When glass tile is used, service rating may be lower.

Environmental Exposure Classifications

- Res1; Com1.
- May be suitable for limited water exposure areas (Res2; Com2) with waterproof membrane. See Membrane Options.
- For installations that may be exposed to staining, specify tile and grout suitable for exposure. Consult product manufacturers; see also "Product Selection Guides."
- For installations that may be exposed to mild chemical attack, specify epoxy grout and tile suitable for exposure. Consult product manufacturers; see also "Product Selection Guides."

Typical Weight of Tile Installation

- 12 pounds/square foot with ¾" poured gypsum. Add 2¼ pounds/square foot for each additional ¼" of poured gypsum.
- Does not include weight of substrate. See "Appendix B" for assumptions, included materials, and their individual weights.

Limitations

- Maximum joist spacing 16" on center.

Membrane Options

- Requires use of crack isolation (A118.12) and/or waterproof (A118.10) and/or uncoupling membrane.
- Membrane must meet ANSI A118.10 for limited water exposure areas (Res2; Com2). Consult underlayment manufacturer for requirements.
- Specifier shall indicate if complete waterproofing is required, including if/how membrane connects to drain assembly, if base flashing is required, and treatment at other termination points.
- Check with membrane manufacturers for suitability for applicable conditions, as not all membranes are suitable for steam, high temperature, and/or chemical exposure, exterior use, use over above-ground structural slabs, use over pourable underlayments, use with radiant heat, or use over concrete with excessive moisture vapor transmission and/or alkalinity. Membrane may also affect service rating.
- When glass tile is used, consult glass tile manufacturer for membrane options and recommendations.

Requirements

- Proper curing/drying of underlayments prior to application of tile is critical for proper performance. Consult the underlayment manufacturer for specific instructions.

Materials

- **Multiple options exist for membranes, mortars, grouts, and other materials and *must be clearly specified* to be included. If not specifically indicated, optional materials are not included and mortar/grout choice defaults to minimum performance specification indicated. Consider each system component and intended use to determine minimum requirements and to specify options.**
- Ceramic tile—ANSI A137.1.
- Glass tile, when used—ANSI A137.2; see also "Glass Tile Selection and Installation Guide," and consult tile manufacturer for service rating and environmental exposure classification recommendations. Not all glass tiles are suitable.
- Cementitious grout—ANSI A118.6 or better or ISO CG1 or better. When glass tile is used, specify grout designated by tile and grout manufacturers.
- Epoxy grout, when used—ANSI A118.3 or ISO RG.
- Cementitious bond coat:
  - When a crack isolation membrane is used—ANSI A118.4 or better or ISO C2S1 or better.

- When a waterproof membrane is used—ANSI A118.4 or better or ISO C2S1 or better unless ANSI A118.1 or ISO C1 is recommended by membrane manufacturer.
- When an uncoupling membrane is used, check with membrane manufacturer.
- When glass tile is used, specify mortar designated by tile and mortar manufacturers. Bond coat color will impact the final appearance of translucent glass tile. Specifier shall confirm bond coat color is acceptable.

- Crack isolation membrane, when used—ANSI A118.12.
- Waterproof membrane, when used—ANSI A118.10.
- Uncoupling membrane, when used—recommended by uncoupling membrane manufacturer; must achieve 50 psi or greater shear bond strength in 7 days per the test method in ANSI A118.12 Section 5.1.3.

Materials for Green/Sustainable Design

- See "Green Building Standards and Green Product Selection Guide" and consult manufacturers and suppliers for product sustainability and contribution to green building design.
- Consider specifying tile and installation materials that meet ANSI A138.1, the *American National Standard Specifications for Sustainable Ceramic Tiles, Glass Tiles, and Tile Installation Materials.*

Preparation by Other Trades

- Floor systems, including the framing system and subfloor panels, over which tile will be installed shall be in conformance with the IRC for residential applications, the IBC for commercial applications, or applicable building codes. Maximum allowable floor member live load and concentrated load deflection shall not exceed l/360. See also "Substrate Requirements."
- When concentrated loads (scissor lifts, pallet jacks, automobiles, forklifts, etc.) will be used on an above-ground tiled floor, the engineer and/or specifier shall specify a substrate to accommodate the concentrated loads. Owner/specifier is responsible for protecting the tilework from damage, including allowing sufficient time for installed materials to cure properly.
- Poured gypsum underlayment—minimum compressive strength of 2000 psi when tested per ASTM C472 modified and meeting performance requirements of ASTM C627 for the anticipated service level designated by the specifier or intended use.
- Poured gypsum underlayment thickness and application varies, consult the manufacturer for specific recommendations.
- Maximum allowable variation in the tile substrate—for tiles with all edges shorter than 15", maximum allowable variation is ¼" in 10' from the required plane, with no more than ¹⁄₁₆" variation in 12" when measured from the high points in the surface. For tiles with at least one edge 15" in length, maximum allowable variation is ⅛" in 10' from the required plane, with no more than ¹⁄₁₆" variation in 24" when measured from the high points in the surface.
- Face grain of plywood shall run perpendicular to joists.
- Exterior grade plywood subfloor to be clean and free of dirt, dust, paint, and oily film.
- Some manufacturers of poured gypsum underlayments require use of plastic lath. Follow manufacturer's recommendations for installation and use of the lath.
- Surfaces must be prepared and primed according to the underlayment manufacturer's instructions.
- Subfloor—minimum ²³⁄₃₂" tongue and groove exterior glue plywood.
- Poured gypsum underlayment—installed only by a manufacturer-approved applicator in accordance with manufacturer's recommendations.

Movement Joint (architect must specify type of joint and show location and details on drawings)

- Movement joints—mandatory in accordance with EJ171.
- When glass tile is used, adhere to more frequent placement recommendations within the ranges listed in EJ171.

Installation Specifications

- Tile—ANSI A108.5.
- Glass tile—manufacturer's directions.
- Cementitious grout—ANSI A108.10.
- Epoxy grout—ANSI A108.6.
- Crack isolation membrane—ANSI A108.17.
- Waterproof membrane—ANSI A108.13.
- Uncoupling membrane—manufacturer's directions.
- Movement Joints—EJ171 and ASTM C1193.

Notes

- Some manufacturers require special primers prior to the application of the membrane. Follow manufacturer's directions.
- When glass tile is used, see "Glass Tile Selection and Installation Guide," and consult manufacturer for recommendations and requirements.

F185-21

- **Joists max. 19.2" o.c./Plywood Subfloor**
- **Cementitious Self-Leveling Underlayment**
- **Ceramic Tile, Glass Tile**

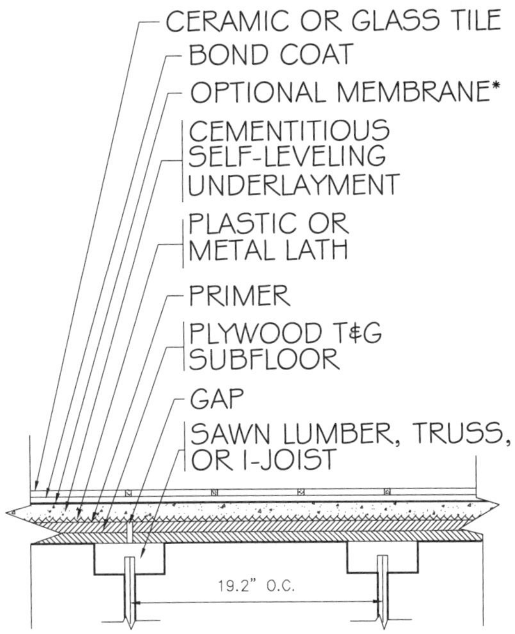

*USE OF A MEMBRANE IS OPTIONAL. SEE MEMBRANE OPTIONS.

Recommended Uses

- For wood substrates where joists are spaced 19.2" on center or less and floor flatness is critical, such as when tiles with an edge longer than 15" are specified or where the finished tile assembly is required to achieve in-plane transitions (as may be necessary where accessibility is a concern).

Service Rating

- Residential.
- When glass tile is used, service rating may be lower.

Environmental Exposure Classifications

- Res1, 2.
- May be suitable for increased water exposure. See Membrane Options.
- For installations that may be exposed to staining, specify tile and grout suitable for exposure. Consult product manufacturers; see also "Product Selection Guides."
- For installations that may be exposed to mild chemical attack, specify epoxy grout and tile suitable for exposure. For greater resistance to chemical exposure, also specify an epoxy bonding material. Consult product manufacturers; see also "Product Selection Guides."

Typical Weight of Tile Installation

- 11 pounds/square foot with ½" cementitious self-leveling underlayment. Add 2¾ pounds/square foot for each additional ¼" of underlayment.
- Does not include weight of substrate. See "Appendix B" for assumptions, included materials, and their individual weights.

Limitations

- Maximum joist spacing 19.2" on center.
- When glass tile is used, maximum joist spacing 16" on center.

Membrane Options

- A waterproof membrane (A118.10) may be required for areas exposed to moisture. Consult underlayment manufacturer for requirements.
- Specifier shall indicate if complete waterproofing is required, including if/how membrane connects to drain assembly, if base flashing is required, and treatment at other termination points.
- A crack isolation membrane may be specified to treat existing in-plane cracks (F125-partial) or for protection against future in-plane cracking (F-125 full). See F125 for more information.
- Check with membrane manufacturers for suitability for applicable conditions, as not all membranes are suitable for steam, high temperature, and/or chemical exposure, exterior use, use over above-ground structural slabs, use over pourable underlayments, use with radiant heat, or use over concrete with excessive moisture vapor transmission and/or alkalinity. Membrane may also affect service rating.
- When glass tile is used, consult glass tile manufacturer for membrane options and recommendations.

Requirements

- Plywood subfloor to be clean and free of dirt, dust, paint, and oily film.
- Surfaces must be prepared and primed according to the underlayment manufacturer's instructions.
- Use lath designed for this purpose, nailed or stapled in accordance with the underlayment manufacturer's instructions.
- Cementitious self-leveling underlayment thickness and application varies, consult the manufacturer for specific recommendations.
- Proper curing/drying of underlayments prior to the application of tile is critical for proper performance. Consult the underlayment manufacturer for specific instructions.

Materials

- **Multiple options exist for membranes, mortars, grouts, and other materials and *must be clearly specified* to be included. If not specifically indicated, optional materials are not included and mortar/grout**

choice defaults to minimum performance specification indicated. Consider each system component and intended use to determine minimum requirements and to specify options.

- Ceramic tile—ANSI A137.1.
- Glass tile, when used—ANSI A137.2; see also "Glass Tile Selection and Installation Guide," and consult tile manufacturer for service rating and environmental exposure classification recommendations. Not all glass tiles are suitable.
- Cementitious grout—ANSI A118.6 or better or ISO CG1 or better. When glass tile is used, specify grout designated by tile and grout manufacturers.
- Epoxy grout, when used—ANSI A118.3 or ISO RG.
- Cementitious bond coat:
  - When a membrane is not used—ANSI A118.4 or better or ISO C2S1 or better.
  - When a crack isolation membrane is used—ANSI A118.4 or better or ISO C2S1 or better.
  - When a waterproof membrane is used—ANSI A118.4 or better or ISO C2S1 or better unless ANSI A118.1 or ISO C1 is recommended by membrane manufacturer.
  - When glass tile is used, specify mortar designated by tile and mortar manufacturers. Bond coat color will impact the final appearance of translucent glass tile. Specifier shall confirm bond coat color is acceptable.
- Epoxy bond coat, when used—ANSI A118.3 or ISO R1 or better.
  - When glass tile is used, specify epoxy bond coat designated by tile and bond coat manufacturers. Bond coat color will impact the final appearance of translucent glass tile. Specifier shall confirm bond coat color is acceptable.
- Crack isolation membrane, when used—ANSI A118.12.
- Waterproof membrane, when used—ANSI A118.10.
- Self-leveling underlayment—meeting performance requirements of ASTM C627 for the anticipated service level designated by the specifier or intended use.

Materials for Green/Sustainable Design

- See "Green Building Standards and Green Product Selection Guide" and consult manufacturers and suppliers for product sustainability and contribution to green building design.
- Consider specifying tile and installation materials that meet ANSI A138.1, the *American National Standard Specifications for Sustainable Ceramic Tiles, Glass Tiles, and Tile Installation Materials.*

Preparation by Other Trades

- Floor systems, including the framing system and subfloor panels, over which tile will be installed shall be in conformance with the IRC for residential applications, the IBC for commercial applications, or applicable building codes. Maximum allowable floor member live load and concentrated load deflection shall not exceed l/360. See also "Substrate Requirements."
- When concentrated loads (scissor lifts, pallet jacks, automobiles, forklifts, etc.) will be used on an above-ground tiled floor, the engineer and/or specifier shall specify a substrate to accommodate the concentrated loads. Owner/specifier is responsible for protecting the tilework from damage, including allowing sufficient time for installed materials to cure properly.
- Face grain of plywood shall run perpendicular to joists.
- Plywood subfloor—minimum 23⁄32" tongue and groove exterior glue plywood with 1⁄8" gap between sheets.
- Maximum allowable variation in the installation substrate (plywood)—1⁄4" in 10' from the required plane.
- Maximum allowable variation in the tile substrate—for tiles with all edges shorter than 15", maximum allowable variation is 1⁄4" in 10' from the required plane, with no more than 1⁄16" variation in 12" when measured from the high points in the surface. For tiles with at least one edge 15" in length, maximum allowable variation is 1⁄8" in 10' from the required plane, with no more than 1⁄16" variation in 24" when measured from the high points in the surface.

Movement Joint (architect must specify type of joint and show location and details on drawings)

- Movement joints—mandatory in accordance with EJ171.
- When glass tile is used, adhere to more frequent placement recommendations within the ranges listed in EJ171.

Installation Specifications

- Tile—ANSI A108.5 or A108.6.
- Glass tile—manufacturer's directions.
- Cementitious grout—ANSI A108.10.
- Epoxy mortar/grout—ANSI A108.6.
- Cementitious self-leveling underlayment—manufacturer's directions.
- Crack isolation membrane—ANSI A108.17.
- Waterproof membrane—ANSI A108.13.
- Movement Joints—EJ171 and ASTM C1193.

Notes

- Some manufacturers require special primers prior to the application of the setting materials or membrane. Follow manufacturer's directions.
- When glass tile is used, see "Glass Tile Selection and Installation Guide," and consult manufacturer for recommendations and requirements.

W201-21

- **Masonry or Concrete**
- **Mortar Bed**
- **Ceramic Tile, Glass Tile**

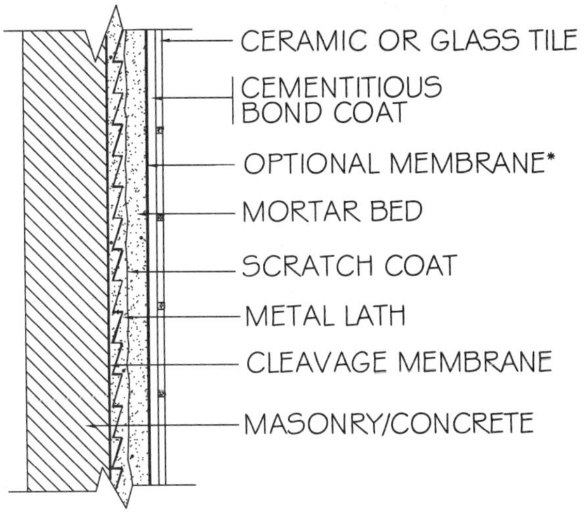

*USE OF A MEMBRANE IS OPTIONAL. SEE MEMBRANE OPTIONS.

Recommended Uses

- For exterior walls of masonry or concrete where flattening or trueing of substrate is required.
- For areas where wall flatness is critical, such as when tiles with any edge longer than 15" are specified.
- For substrates that are cracked, coated, or present bonding issues.
- For isolating the tile installation from the substrate.

Environmental Exposure Classifications

- Res6; Com6.
- For installations that may be exposed to staining, specify tile and grout suitable for exposure. Consult product manufacturers; see also "Product Selection Guides."

Limitations

- Where installation will be subjected to freeze/thaw cycles, degradation can occur over time.

Membrane Options

- Check building code—water-resistive barrier, waterproof membrane, or vapor retarder membrane may be required.
- A waterproof membrane (A118.10) may be specified to prevent moisture intrusion and protect adjacent building materials. Specifier shall indicate if complete waterproofing is required, including base flashing and treatment at other termination points.
- Check with membrane manufacturer for suitability for applicable conditions, as not all membranes are suitable for steam, high-temperature and/or chemical exposure, or exterior use.
- When glass tile is used, consult glass tile manufacturer for membrane options and recommendations.

Requirements

- Cut lath at all movement joints.
- Mortar bed thickness—¾" minimum to 1½" maximum.

Materials

- **Multiple options exist for membranes, mortars, grouts, and other materials and *must be clearly specified* to be included. If not specifically indicated, optional materials are not included and mortar/grout choice defaults to minimum performance specification indicated. Consider each system component and intended use to determine minimum requirements and to specify options.**
- Ceramic tile—ANSI A137.1.
- Glass tile, when used—ANSI A137.2; see also "Glass Tile Selection and Installation Guide," and consult tile manufacturer for environmental exposure classification recommendations. Not all glass tiles are suitable.
- Cementitious grout—ANSI A118.6 or better or ISO CG1 or better. When glass tile is used, specify grout designated by tile and grout manufacturers.
- Cementitious bond coat—portland cement paste on a mortar bed that is still workable. For a cured mortar bed, follow recommendations below to select appropriate bonding mortar:
  - When a waterproof membrane is not used—ANSI A118.15 or better or ISO C2S1 or better.
  - When a waterproof membrane is used—ANSI A118.4 or better or ISO C2S1 or better, unless ANSI A118.1 or ISO C1 is recommended by membrane manufacturer.
  - When glass tile is used, specify mortar designated by tile and mortar manufacturers. Bond coat color will impact the final appearance of translucent glass tile. Specifier shall confirm bond coat color is acceptable.
- Waterproof membrane, when used—ANSI A118.10.
- Mortar bed, metal lath, and cleavage membrane—ANSI A108.1A.

Materials for Green/Sustainable Design

- See "Green Building Standards and Green Product Selection Guide" and consult manufacturers and suppliers for product sustainability and contribution to green building design.
- Consider specifying tile and installation materials that meet ANSI A138.1, the *American National Standard Specifications for Sustainable Ceramic Tiles, Glass Tiles, and Tile Installation Materials.*

Preparation by Other Trades

- Maximum allowable variation in the installation substrate (concrete/masonry)—¼" in 10' from the required plane.

Movement Joint (architect must specify type of joint and show location and details on drawings)

- Movement joints—mandatory according to EJ171.
- When glass tile is used, adhere to more frequent placement recommendations within the ranges listed in EJ171.

Installation Specifications

- Tile—ANSI A108.1A, .1B, or .1C. A108.1B required if membrane to be used.
- Glass tile—ANSI A108.14, .15, .16, or manufacturer's directions.
- Cementitious grout—ANSI A108.10.
- Waterproof membrane—ANSI A108.13.
- Movement Joints—EJ171 and ASTM C1193.

Notes

- For prefabricated exterior panels, see "Additional Products Used in Tile Installations."
- Not all bonding mortars are suitable for exterior use.
- Protection of installation may be required to prevent premature exposure of setting materials to moisture.
- For interior mortar bed applications, see W211, W221, W222, W231/241, B411, B414, SR613, and R612.
- When glass tile is used, see "Glass Tile Selection and Installation Guide," and consult manufacturer for recommendations and requirements.

W202E-21

- **Masonry or Concrete**
- **Ceramic Tile, Glass Tile**

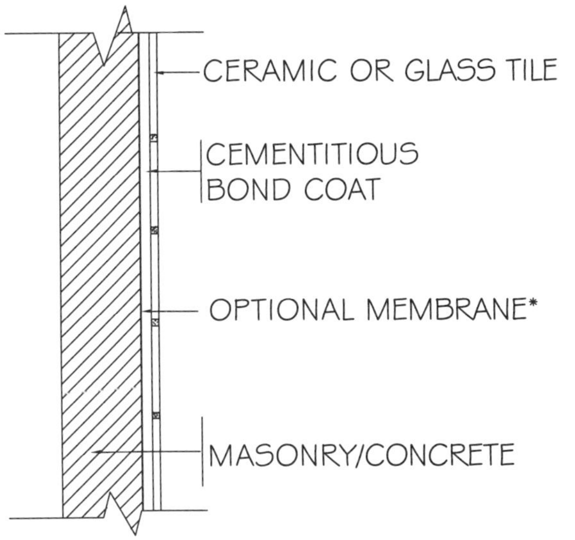

*USE OF A MEMBRANE IS OPTIONAL. SEE MEMBRANE OPTIONS.

Recommended Uses

- For clean, sound exterior walls of masonry or concrete where thin-bed installation of tile is desired.

Environmental Exposure Classifications

- Res6; Com6.
- For installations that may be exposed to staining, specify tile and grout suitable for exposure. Consult product manufacturers; see also "Product Selection Guides."

Limitations

- Where installation will be subjected to freeze/thaw cycles, degradation can occur over time.
- Not for cracked or coated surfaces.

Membrane Options

- Check building code—water-resistive barrier, waterproof membrane, or vapor retarder membrane may be required.
- A waterproof membrane (A118.10) may be specified to prevent moisture intrusion and protect adjacent building materials. Specifier shall indicate if complete waterproofing is required, including base flashing and treatment at other termination points.
- Check with membrane manufacturer for suitability for applicable conditions, as not all membranes are suitable for steam, high-temperature and/or chemical exposure, or exterior use.
- When glass tile is used, consult glass tile manufacturer for membrane options and recommendations.

Requirements

- Masonry or concrete to be well cured, dimensionally stable, and free of cracks, waxy or oily films, and curing compounds.
- Concrete may require bush-hammering or sandblasting to facilitate bonding.

Materials

- **Multiple options exist for membranes, mortars, grouts, and other materials and *must be clearly specified* to be included. If not specifically indicated, optional materials are not included and mortar/grout choice defaults to minimum performance specification indicated. Consider each system component and intended use to determine minimum requirements and to specify options.**
- Ceramic tile—ANSI A137.1.
- Glass tile, when used—ANSI A137.2; see also "Glass Tile Selection and Installation Guide," and consult tile manufacturer for environmental exposure classification recommendations. Not all glass tiles are suitable.
- Cementitious grout—ANSI A118.6 or better or ISO CG1 or better. When glass tile is used, specify grout designated by tile and grout manufacturers.
- Cementitious bond coat:
  - When a waterproof membrane is not used—ANSI A118.15 or better or ISO C2S1 or better.
  - When a waterproof membrane is used—ANSI A118.4 or better or ISO C2S1 or better, unless ANSI A118.1 or ISO C1 is recommended by membrane manufacturer.
  - When glass tile is used, specify mortar designated by tile and mortar manufacturers. Bond coat color will impact the final appearance of translucent glass tile. Specifier shall confirm bond coat color is acceptable.
- Waterproof membrane, when used—ANSI A118.10.

Materials for Green/Sustainable Design

- See "Green Building Standards and Green Product Selection Guide" and consult manufacturers and suppliers for product sustainability and contribution to green building design.
- Consider specifying tile and installation materials that meet ANSI A138.1, the *American National Standard Specifications for Sustainable Ceramic Tiles, Glass Tiles, and Tile Installation Materials.*

Preparation by Other Trades

- Maximum allowable variation in the tile substrate—for tiles with all edges shorter than 15", maximum allowable variation is ¼" in 10' from the required plane, with no more than ¹⁄₁₆" variation in 12" when measured from

the high points in the surface. For tiles with at least one edge 15" in length, maximum allowable variation is ⅛" in 10' from the required plane, with no more than $\frac{1}{16}$" variation in 24" when measured from the high points in the surface.

Movement Joint (architect must specify type of joint and show location and details on drawings)

- Movement joints mandatory—according to EJ171.
- When glass tile is used, adhere to more frequent placement recommendations within the ranges listed in EJ171.

Installation Specifications

- Tile—ANSI A108.5.
- Glass tile—manufacturer's directions.
- Cementitious grout—ANSI A108.10.
- Waterproof membrane—ANSI A108.13.
- Movement Joints—EJ171 and ASTM C1193.

Notes

- For prefabricated exterior panels, see "Additional Products Used in Tile Installations."
- Not all bonding mortars are suitable for exterior use.
- Protection of installation may be required to prevent premature exposure of setting materials to moisture.
- For interior thin-bed applications over masonry or concrete, see W202I and W223.
- When glass tile is used, see "Glass Tile Selection and Installation Guide," and consult manufacturer for recommendations and requirements.

W244E-21

- **Wood or Metal Studs**
- **Cement Backer Board**
- **Ceramic Tile, Glass Tile**

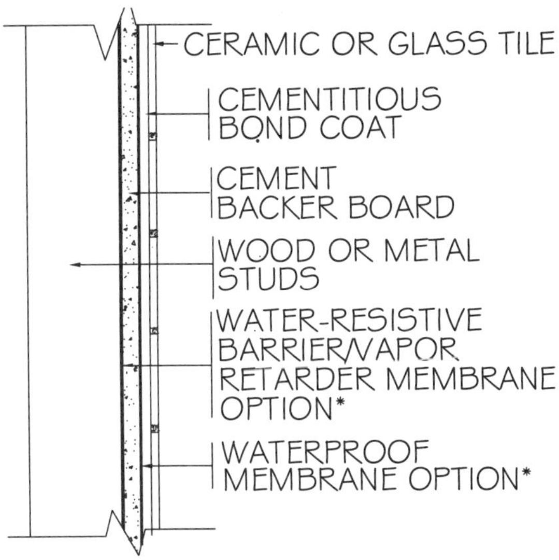

*USE OF A MEMBRANE IS REQUIRED. SEE MEMBRANE OPTIONS.

Recommended Uses

- For exterior walls over wood or metal studs.

Environmental Exposure Classifications

- Res6; Com6.
- For installations that may be exposed to staining, specify tile and grout suitable for exposure. Consult product manufacturers; see also "Product Selection Guides."

Limitations

- Where installation will be subjected to freeze/thaw cycles, degradation can occur over time.
- Maximum stud spacing 16" on center.

Membrane Options

- Check building code—water-resistive barrier, waterproof membrane, or vapor retarder membrane may be required.
- A waterproof membrane (A118.10) may be specified to prevent moisture intrusion and protect adjacent building materials. Specifier shall indicate if complete waterproofing is required, including base flashing and treatment at other termination points.
- If waterproof membrane is applied over cement board, membrane over framing members may not be required.
- Check with membrane manufacturer for suitability for applicable conditions, as not all membranes are suitable for steam, high-temperature and/or chemical exposure, or exterior use.
- When glass tile is used, consult glass tile manufacturer for membrane options and recommendations.

Requirements

- Wood studs—dry and well-braced, minimum depth 3½".
- Metal studs—well-braced; 20 gauge (0.033") or heavier; minimum depth 3½" for residential applications or 3⅝" for commercial applications.
- Water-resistive barrier or waterproof membrane required, per building code. See Membrane Options.

Preparation by Backer Board Installers

- Maximum allowable variation in the tile substrate—for tiles with all edges shorter than 15", maximum allowable variation is ¼" in 10' from the required plane, with no more than 1/16" variation in 12" when measured from the high points in the surface. For tiles with at least one edge 15" in length, maximum allowable variation is ⅛" in 10' from the required plane, with no more than 1/16" variation in 24" when measured from the high points in the surface.
- Horizontal joints—⅛" spacing filled solid and taped with latex-portland cement mortar and 4" alkali-resistant glass fiber mesh tape.
- Vertical joints—fill any space and tape with latex-portland cement mortar and 4" alkali-resistant glass fiber mesh tape.
- Corners—leave space between backer units. Tape joints using skim coat of latex-portland cement mortar, but do not fill.
- Center backer board end or edge joints on framing and stagger joints in adjacent rows so four corners do not come together within the same plane. Space panel ends and edges in accordance with manufacturer's recommendations.

Materials

- **Multiple options exist for membranes, mortars, grouts, and other materials and *must be clearly specified* to be included. If not specifically indicated, optional materials are not included and mortar/grout choice defaults to minimum performance specification indicated. Consider each system component and intended use to determine minimum requirements and to specify options.**
- Ceramic tile—ANSI A137.1.
- Glass tile, when used—ANSI A137.2; see also "Glass Tile Selection and Installation Guide," and consult tile manufacturer for environmental exposure classification recommendations. Not all glass tiles are suitable.
- Cementitious grout—ANSI A118.6 or better or ISO CG1 or better. When glass tile is used, specify grout designated by tile and grout manufacturers.
- Cementitious bond coat:
  - When a waterproof membrane is not used on face of cement backer board—ANSI A118.15 or better or ISO C2S1 or better.

- When a waterproof membrane is used on face of cement backer board—ANSI A118.4 or better or ISO C2S1 or better, unless ANSI A118.1 or ISO C1 is recommended by membrane manufacturer.
- When glass tile is used, specify mortar designated by tile and mortar manufacturers. Bond coat color will impact the final appearance of translucent glass tile. Specifier shall confirm bond coat color is acceptable.

- Cement backer board—ANSI A118.9 or ASTM C1325 (Type A).
- Fasteners—noncorrosive and nonoxidizing.
- Hot-dipped fasteners meeting ASTM F2329 required in wet areas.
- 4" alkali-resistant glass fiber mesh tape.
- Water-resistive barrier/vapor retarder membrane, when used—per building code.
- Waterproof membrane, when used—ANSI A118.10.
- Metal studs—ASTM C645.

Materials for Green/Sustainable Design

- See "Green Building Standards and Green Product Selection Guide" and consult manufacturers and suppliers for product sustainability and contribution to green building design.
- Consider specifying tile and installation materials that meet ANSI A138.1, the *American National Standard Specifications for Sustainable Ceramic Tiles, Glass Tiles, and Tile Installation Materials.*

Preparation by Other Trades

- Wall framing shall meet the general framing requirements of ANSI A108.11-4.0–4.3.

Movement Joint (architect must specify type of joint and show location and details on drawings)

- Movement joints—mandatory according to EJ171.
- When glass tile is used, adhere to more frequent placement recommendations within the ranges listed in EJ171.

Installation Specifications

- Cement backer board—manufacturer's directions.
- Tile—ANSI A108.5.
- Glass tile—ANSI A108.15, .16, or manufacturer's directions.
- Cementitious grout—ANSI A108.10.
- Waterproof membrane—ANSI A108.13.
- Movement Joints—EJ171 and ASTM C1193.

Notes

- For prefabricated exterior panels, see "Additional Products Used in Tile Installations."
- Not all bonding mortars are suitable for exterior use.
- Protection of installation may be required to prevent premature exposure of setting materials to moisture.
- For interior applications, see W244C, B412, B415, and SR614.
- When glass tile is used, see "Glass Tile Selection and Installation Guide," and consult manufacturer for recommendations and requirements.

W211-21

- **Masonry or Concrete**
- **Bonded Mortar Bed**
- **Ceramic Tile, Glass Tile**

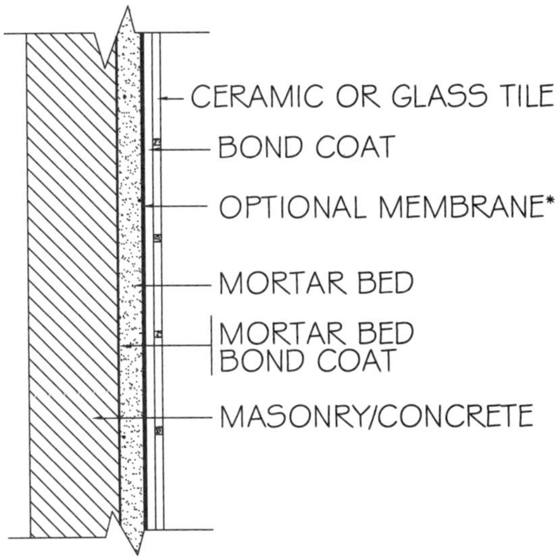

*USE OF A MEMBRANE IS OPTIONAL. SEE MEMBRANE OPTIONS.

Recommended Uses

- For clean, sound interior walls of masonry or concrete where flattening or trueing of substrate is required.
- For areas where wall flatness is critical, such as when tiles with any edge longer than 15" are specified.

Environmental Exposure Classifications

- Res1, 2, 3, 5; Com1, 2, 3, 5.
- May be suitable for exterior applications when appropriate precautions are taken, including flashing, expansion joint placement, waterproofing, and material selection. See Notes.
- For installations that may be exposed to staining, specify tile and grout suitable for exposure. Consult product manufacturers; see also "Product Selection Guides."
- For installations that may be exposed to mild chemical attack, specify epoxy grout and tile suitable for exposure. For greater resistance to chemical exposure, also specify an epoxy bonding material. Consult product manufacturers; see also "Product Selection Guides."

Limitations

- Not for cracked or coated surfaces.
- Exterior—where installation will be subjected to freeze/thaw cycles, degradation can occur over time.

Membrane Options

- A waterproof membrane (A118.10) may be specified to prevent moisture intrusion and protect adjacent building materials. Specifier shall indicate if complete waterproofing is required, including base flashing and treatment at other termination points.
- Check with membrane manufacturer for suitability for applicable conditions, as not all membranes are suitable for steam, high-temperature and/or chemical exposure, or exterior use.
- Exterior—check building code—water-resistive barrier, waterproof membrane, or vapor retarder membrane may be required.
- When glass tile is used, consult glass tile manufacturer for membrane options and recommendations.

Requirements

- Masonry or concrete to be well cured, dimensionally stable, and free of cracks, waxy or oily films, and curing compounds.
- Concrete may require bush-hammering or sandblasting to facilitate bonding.
- Mortar bed thickness—⅜" minimum to ¾" maximum. Scratch coat required if mortar bed will exceed ¾."

Materials

- **Multiple options exist for membranes, mortars, grouts, and other materials and *must be clearly specified* to be included. If not specifically indicated, optional materials are not included and mortar/grout choice defaults to minimum performance specification indicated. Consider each system component and intended use to determine minimum requirements and to specify options.**
- Ceramic tile—ANSI A137.1.
- Glass tile, when used—ANSI A137.2; see also "Glass Tile Selection and Installation Guide," and consult tile manufacturer for environmental exposure classification recommendations. Not all glass tiles are suitable.
- Cementitious grout—ANSI A118.6 or better or ISO CG1 or better. When glass tile is used, specify grout designated by tile and grout manufacturers.
- Epoxy grout, when used—ANSI A118.3 or ISO RG.
- Cementitious bond coat—portland cement paste on a mortar bed that is still workable. For a cured mortar bed, follow recommendations below to select appropriate bonding mortar:
  - When a waterproof membrane is not used—ANSI A118.1 or better or ISO C1 or better.
  - When a waterproof membrane is used—ANSI A118.4 or better or ISO C2S1 or better unless ANSI A118.1 or ISO C1 is recommended by membrane manufacturer.
  - When porcelain tile is used—ANSI A118.4 or better or ISO C2 or better.
  - When glass tile is used, specify mortar designated by tile and mortar manufacturers. Bond coat color will impact the final appearance of translucent glass tile. Specifier shall confirm bond coat color is acceptable.

- Epoxy bond coat, when used—ANSI A118.3 or ISO R1 or better.
  - When glass tile is used, specify epoxy bond coat designated by tile and bond coat manufacturers. Bond coat color will impact the final appearance of translucent glass tile. Specifier shall confirm bond coat color is acceptable.
- Waterproof membrane, when used—ANSI A118.10.
- Mortar bed—ANSI A108.1A.
- Mortar bed bond coat—portland cement slurry.

Materials for Green/Sustainable Design

- See "Green Building Standards and Green Product Selection Guide" and consult manufacturers and suppliers for product sustainability and contribution to green building design.
- Consider specifying tile and installation materials that meet ANSI A138.1, the *American National Standard Specifications for Sustainable Ceramic Tiles, Glass Tiles, and Tile Installation Materials.*

Preparation by Other Trades

- Maximum allowable variation in the installation substrate (concrete/masonry)—¼" in 10' from the required plane.

Movement Joint (architect must specify type of joint and show location and details on drawings)

- Movement joints—mandatory according to EJ171.
- When glass tile is used, adhere to more frequent placement recommendations within the ranges listed in EJ171.

Installation Specifications

- Tile—ANSI A108.1A, .1B, or .1C. A108.1B required if membrane or epoxy bond coat to be used.
- Glass tile—ANSI A108.14, .15, .16, or manufacturer's directions.
- Cementitious grout—ANSI A108.10.
- Epoxy mortar/grout—ANSI A108.6.
- Waterproof membrane—ANSI A108.13.
- Movement Joints—EJ171 and ASTM C1193.

Notes

- Exterior—not all bonding mortars are suitable for exterior use.
- Exterior—protection of installation may be required to prevent premature exposure of setting materials to moisture.
- When glass tile is used, see "Glass Tile Selection and Installation Guide," and consult manufacturer for recommendations and requirements.

W215-21

- **Masonry or Concrete**
- **Spot-Bonding Epoxy**
- **Ceramic Tile**

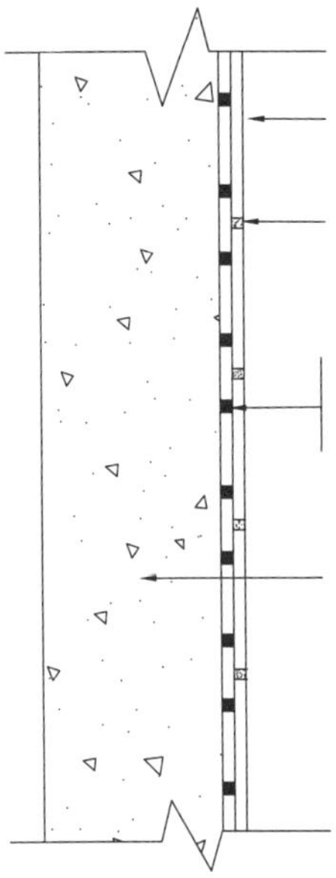

Recommended Uses

- For clean, sound interior walls of masonry or concrete where large format tiles are specified.

Environmental Exposure Classifications

- Res1; Com1.
- For installations that may be exposed to staining, specify tile and grout suitable for exposure. Consult product manufacturers; see also "Product Selection Guides."
- For installations that may be exposed to mild chemical attack, specify epoxy grout and tile suitable for exposure. Consult product manufacturers; see also "Product Selection Guides."

Limitations

- Will not withstand impact.
- Not for cracked or coated surfaces.

Requirements

- Epoxy must be recommended by the manufacturer for spot-bonding.
- Follow epoxy manufacturer's coverage requirements.
- Masonry or concrete to be well cured, dimensionally stable, and free of cracks, waxy or oily films, and curing compounds.
- Concrete may require bush-hammering or sandblasting to facilitate bonding.

Materials

- **Multiple options exist for membranes, mortars, grouts, and other materials and *must be clearly specified* to be included. If not specifically indicated, optional materials are not included and mortar/grout choice defaults to minimum performance specification indicated. Consider each system component and intended use to determine minimum requirements and to specify options.**
- Ceramic tile—ANSI A137.1.
- Cementitious grout—ANSI A118.6 or better or ISO CG1 or better.
- Epoxy grout, when used—ANSI A118.3 or ISO RG.
- Sealant, when used as grout—ASTM C920.
- Spot bonding epoxy adhesive—epoxy recommended by manufacturer for spot bonding.

Materials for Green/Sustainable Design

- See "Green Building Standards and Green Product Selection Guide" and consult manufacturers and suppliers for product sustainability and contribution to green building design.
- Consider specifying tile and installation materials that meet ANSI A138.1, the *American National Standard Specifications for Sustainable Ceramic Tiles, Glass Tiles, and Tile Installation Materials.*

Preparation by Other Trades

- Maximum allowable variation in the tile substrate—¼" in 10' from the required plane.

Movement Joint (architect must specify type of joint and show location and details on drawings)

- Movement joints—mandatory according to EJ171.

Installation Specifications

- Tile—epoxy manufacturer's directions.
- Cementitious grout—ANSI A108.10.
- Epoxy grout—ANSI A108.6.
- Sealant—follow sealant and epoxy manufacturer's instructions for using sealant as grout.
- Movement Joints—EJ171 and ASTM C1193.

W202I-21

- **Masonry or Concrete**
- **Ceramic Tile, Glass Tile**

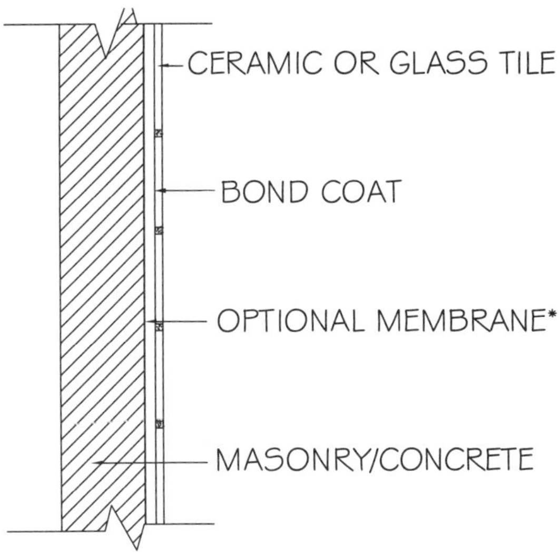

*USE OF A MEMBRANE IS OPTIONAL. SEE MEMBRANE OPTIONS.

Recommended Uses

- For interior walls of masonry or concrete where thin-bed installation of tile is desired.

Environmental Exposure Classifications

- Res1, 2, 3, 5; Com1, 2, 3, 5.
- For installations that may be exposed to staining, specify tile and grout suitable for exposure. Consult product manufacturers; see also "Product Selection Guides."
- For installations that may be exposed to mild chemical attack, specify epoxy grout and tile suitable for exposure. For greater resistance to chemical exposure, also specify an epoxy bonding material. Consult product manufacturers; see also "Product Selection Guides."

Limitations

- Not for cracked or coated surfaces.

Membrane Options

- A waterproof membrane (A118.10) may be specified to prevent moisture intrusion and protect adjacent building materials. Specifier shall indicate if complete waterproofing is required, including base flashing and treatment at other termination points.
- Check with membrane manufacturer for suitability for applicable conditions, as not all membranes are suitable for steam, high-temperature and/or chemical exposure, or exterior use.
- When glass tile is used, consult glass tile manufacturer for membrane options and recommendations.

Requirements

- Masonry or concrete to be well cured, dimensionally stable, and free of cracks, waxy or oily films, and curing compounds.
- Concrete may require bush-hammering or sandblasting to facilitate bonding.

Materials

- **Multiple options exist for membranes, mortars, grouts, and other materials and *must be clearly specified* to be included. If not specifically indicated, optional materials are not included and mortar/grout choice defaults to minimum performance specification indicated. Consider each system component and intended use to determine minimum requirements and to specify options.**
- Ceramic tile—ANSI A137.1.
- Glass tile, when used—ANSI A137.2; see also "Glass Tile Selection and Installation Guide," and consult tile manufacturer for environmental exposure classification recommendations. Not all glass tiles are suitable.
- Cementitious grout—ANSI A118.6 or better or ISO CG1 or better. When glass tile is used, specify grout designated by tile and grout manufacturers.
- Epoxy grout, when used—ANSI A118.3 or ISO RG.
- Cementitious bond coat:
  - When a waterproof membrane is not used—ANSI A118.1 or better or ISO C1 or better.
  - When a waterproof membrane is used—ANSI A118.4 or better or ISO C2S1 or better unless ANSI A118.1 or ISO C1 is recommended by membrane manufacturer.
  - When porcelain tile is used—ANSI A118.4 or better or ISO C2 or better.
  - When glass tile is used, specify mortar designated by tile and mortar manufacturers. Bond coat color will impact the final appearance of translucent glass tile. Specifier shall confirm bond coat color is acceptable.
- Epoxy bond coat, when used—ANSI A118.3 or ISO R1 or better.
  - When glass tile is used, specify epoxy bond coat designated by tile and bond coat manufacturers. Bond coat color will impact the final appearance of translucent glass tile. Specifier shall confirm bond coat color is acceptable.
- Waterproof membrane, when used—ANSI A118.10.

Materials for Green/Sustainable Design

- See "Green Building Standards and Green Product Selection Guide" and consult manufacturers and suppliers for product sustainability and contribution to green building design.

- Consider specifying tile and installation materials that meet ANSI A138.1, the *American National Standard Specifications for Sustainable Ceramic Tiles, Glass Tiles, and Tile Installation Materials.*

Preparation by Other Trades

- Maximum allowable variation in the tile substrate—for tiles with all edges shorter than 15", maximum allowable variation is ¼" in 10' from the required plane, with no more than ¹⁄₁₆" variation in 12" when measured from the high points in the surface. For tiles with at least one edge 15" in length, maximum allowable variation is ⅛" in 10' from the required plane, with no more than ¹⁄₁₆" variation in 24" when measured from the high points in the surface.

Movement Joint (architect must specify type of joint and show location and details on drawings)

- Movement joints mandatory—according to EJ171.
- When glass tile is used, adhere to more frequent placement recommendations within the ranges listed in EJ171.

Installation Specifications

- Tile—ANSI A108.5.
- Glass tile—manufacturer's directions.
- Cementitious grout—ANSI A108.10.
- Epoxy mortar/grout—ANSI A108.6.
- Waterproof membrane—ANSI A108.13.
- Movement Joints—EJ171 and ASTM C1193.

Notes

- For exterior application, see W202E.
- As the tile size increases, there is less tolerance for variation in the substrate from the required plane. Epoxy adhesive thickness must be thin and uniform; therefore, substrate flattening may be required when epoxy bond coat is used.
- When glass tile is used, see "Glass Tile Selection and Installation Guide," and consult manufacturer for recommendations and requirements.

W221-21

- **Solid Backing**
- **Mortar Bed**
- **Ceramic Tile, Glass Tile**

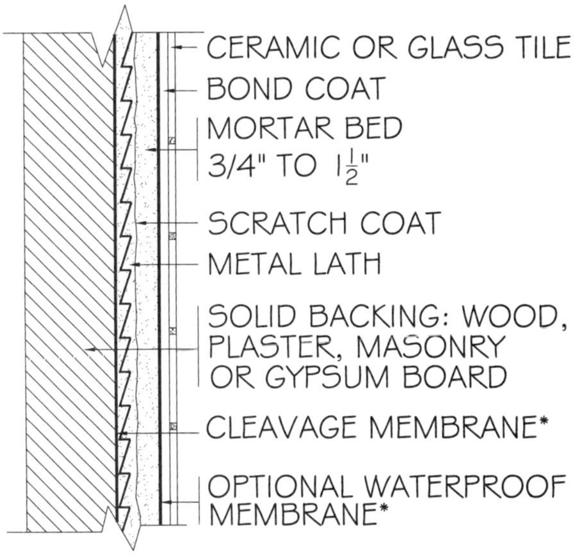

*USE OF A MEMBRANE IS OPTIONAL. SEE MEMBRANE OPTIONS.

Recommended Uses

- For interior walls of masonry, plaster, or other solid backing where flattening or trueing of substrate is required.
- For areas where wall flatness is critical, such as when tiles with any edge longer than 15" are specified.
- For substrates that are cracked, coated, or present bonding issues.

Environmental Exposure Classifications

- With masonry backing—Res1, 2, 3, 5; Com1, 2, 3, 5.
- With gypsum board or plaster backing—Res1, 2; Com1, 2.
- With wood backing—Res1; Com1.
- May be suitable for exterior applications when appropriate precautions are taken, including flashing, expansion joint placement, waterproofing, and material selection. See Notes.
- For installations that may be exposed to staining, specify tile and grout suitable for exposure. Consult product manufacturers; see also "Product Selection Guides."
- For installations that may be exposed to mild chemical attack, specify epoxy grout and tile suitable for exposure. For greater resistance to chemical exposure, also specify an epoxy bonding material. Consult product manufacturers; see also "Product Selection Guides."

Limitations

- When backing is gypsum board, not for areas exposed to temperatures exceeding 125°F.

Membrane Options

- A waterproof membrane (A118.10) may be specified to prevent moisture intrusion and protect adjacent building materials. Specifier shall indicate if complete waterproofing is required, including base flashing and treatment at other termination points.
- If a waterproof membrane (A118.10) is not specified and backing material is moisture sensitive, cleavage membrane must be used behind mortar bed in wet areas.
- Check with membrane manufacturer for suitability for applicable conditions, as not all membranes are suitable for steam, high-temperature and/or chemical exposure, or exterior use.
- Exterior—check building code—water-resistive barrier, waterproof membrane, or vapor retarder membrane may be required.
- When glass tile is used, consult glass tile manufacturer for membrane options and recommendations.

Requirements

- Wood studs—dry and well-braced, minimum depth 3½."
- Metal studs—well-braced; 20 gauge (0.033") or heavier; minimum depth 3½" for residential applications or 3⅝" for commercial applications.
- Solid backing must provide firm anchorage for metal lath.
- Mortar bed thickness—¾" minimum to 1½" maximum.
- Cut lath at all movement joints.

Materials

- **Multiple options exist for membranes, mortars, grouts, and other materials and *must be clearly specified* to be included. If not specifically indicated, optional materials are not included and mortar/grout choice defaults to minimum performance specification indicated. Consider each system component and intended use to determine minimum requirements and to specify options.**
- Ceramic tile—ANSI A137.1.
- Glass tile, when used—ANSI A137.2; see also "Glass Tile Selection and Installation Guide," and consult tile manufacturer for environmental exposure classification recommendations. Not all glass tiles are suitable.
- Cementitious grout—ANSI A118.6 or better or ISO CG1 or better. When glass tile is used, specify grout designated by tile and grout manufacturers.
- Epoxy grout, when used—ANSI A118.3 or ISO RG.
- Cementitious bond coat—portland cement paste on a mortar bed that is still workable. For a cured mortar bed, follow recommendations below to select

appropriate bonding mortar:
  - When a waterproof membrane is not used—ANSI A118.1 or better or ISO C1 or better.
  - When a waterproof membrane is used—ANSI A118.4 or better or ISO C2S1 or better unless ANSI A118.1 or ISO C1 is recommended by membrane manufacturer.
  - When porcelain tile is used—ANSI A118.4 or better or ISO C2 or better.
  - When glass tile is used, specify mortar designated by tile and mortar manufacturers. Bond coat color will impact the final appearance of translucent glass tile. Specifier shall confirm bond coat color is acceptable.
- Epoxy bond coat, when used—ANSI A118.3 or ISO R1 or better.
  - When glass tile is used, specify epoxy bond coat designated by tile and bond coat manufacturers. Bond coat color will impact the final appearance of translucent glass tile. Specifier shall confirm bond coat color is acceptable.
- Waterproof membrane, when used—ANSI A118.10.
- Mortar bed, metal lath, and cleavage membrane—ANSI A108.1A.
- Metal studs—ASTM C645.

Materials for Green/Sustainable Design

- See "Green Building Standards and Green Product Selection Guide" and consult manufacturers and suppliers for product sustainability and contribution to green building design.
- Consider specifying tile and installation materials that meet ANSI A138.1, the *American National Standard Specifications for Sustainable Ceramic Tiles, Glass Tiles, and Tile Installation Materials.*

Preparation by Other Trades

- Maximum allowable variation in the installation substrate (solid backing)—¼" in 10' from the required plane.

Movement Joint (architect must specify type of joint and show location and details on drawings)

- Movement joints—mandatory according to EJ171.
- When glass tile is used, adhere to more frequent placement recommendations within the ranges listed in EJ171.

Installation Specifications

- Tile—ANSI A108.1A, .1B, or .1C. A108.1B required if waterproof membrane or epoxy bond coat to be used.
- Glass tile—ANSI A108.14, .15, .16, or manufacturer's directions.
- Cementitious grout—ANSI A108.10.
- Epoxy mortar/grout—ANSI A108.6.
- Waterproof membrane—ANSI A108.13.
- Movement Joints—EJ171 and ASTM C1193.

Notes

- If lath cannot be attached directly to backing, use furring strips. See W231/W241.
- For thinner mortar bed, see W222.
- Exterior—not all bonding mortars are suitable for exterior use.
- Exterior—protection of installation may be required to prevent premature exposure of setting materials to moisture.
- When glass tile is used, see "Glass Tile Selection and Installation Guide," and consult manufacturer for recommendations and requirements.

W222-21

- **Solid Backing**
- **Mortar Bed (One Coat Method)**
- **Ceramic Tile**

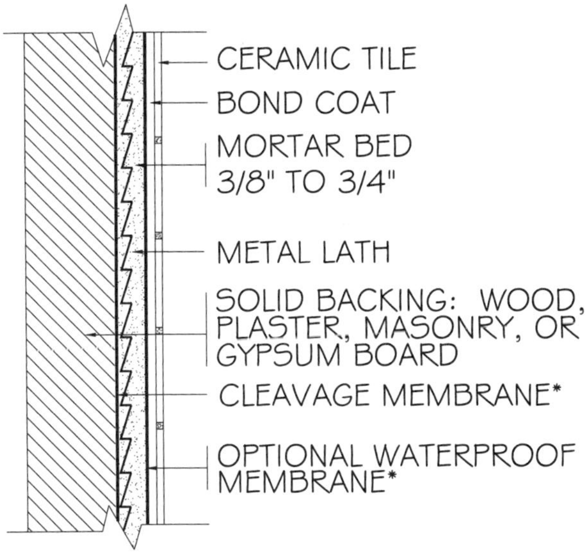

*USE OF A MEMBRANE IS OPTIONAL. SEE MEMBRANE OPTIONS.

Recommended Uses

- For interior walls of masonry, plaster, or other solid backing where flattening or trueing of substrate is required.
- For areas where wall flatness is critical, such as when tiles with any edge longer than 15" are specified.
- For substrates that are cracked, coated, or present bonding issues.

Environmental Exposure Classifications

- With masonry backing—Res1, 2, 5; Com1, 2, 5.
- With gypsum board or plaster backing—Res1, 2; Com1, 2.
- With wood backing—Res1; Com1.
- For Res3 and Com3, see B440 and B441.
- May be suitable for exterior applications when appropriate precautions are taken, including flashing, expansion joint placement, waterproofing, and material selection. See Notes.
- For installations that may be exposed to staining, specify tile and grout suitable for exposure. Consult product manufacturers; see also "Product Selection Guides."
- For installations that may be exposed to mild chemical attack, specify epoxy grout and tile suitable for exposure. For greater resistance to chemical exposure, also specify an epoxy bonding material. Consult product manufacturers; see also "Product Selection Guides."

Limitations

- When backing is gypsum board, not for areas exposed to temperatures exceeding 125°F.

Membrane Options

- A waterproof membrane (A118.10) may be specified to prevent moisture intrusion and protect adjacent building materials. Specifier shall indicate if complete waterproofing is required, including base flashing and treatment at other termination points.
- If a waterproof membrane (A118.10) is not specified and backing material is moisture sensitive, cleavage membrane must be used behind mortar bed in wet areas.
- Check with membrane manufacturer for suitability for applicable conditions, as not all membranes are suitable for steam, high-temperature and/or chemical exposure, or exterior use.
- Exterior—check building code—water-resistive barrier, waterproof membrane, or vapor retarder membrane may be required.

Requirements

- Wood studs—dry and well-braced, minimum depth 3½".
- Metal studs—well-braced; 20 gauge (0.033") or heavier; minimum depth 3½" for residential applications or 3⅝" for commercial applications.
- Solid backing must provide firm anchorage for metal lath.
- Mortar bed thickness—⅜" minimum to ¾" maximum.
- Cut lath at all movement joints.

Materials

- **Multiple options exist for membranes, mortars, grouts, and other materials and *must be clearly specified* to be included. If not specifically indicated, optional materials are not included and mortar/grout choice defaults to minimum performance specification indicated. Consider each system component and intended use to determine minimum requirements and to specify options.**
- Ceramic tile—ANSI A137.1.
- Cementitious grout—ANSI A118.6 or better or ISO CG1 or better.
- Epoxy grout, when used—ANSI A118.3 or ISO RG.
- Cementitious bond coat—portland cement paste on a mortar bed that is still workable. For a cured mortar bed, follow recommendations below to select appropriate bonding mortar:
  - When a waterproof membrane is not used—ANSI A118.1 or better or ISO C1 or better.
  - When a waterproof membrane is used—ANSI A118.4 or better or ISO C2S1 or better unless ANSI A118.1 or ISO C1 is recommended by membrane manufacturer.

  - When porcelain tile is used—ANSI A118.4 or better or ISO C2 or better.
- Epoxy bond coat, when used—ANSI A118.3 or ISO R1 or better.
- Waterproof membrane, when used—ANSI A118.10.
- Mortar bed, metal lath, and cleavage membrane—ANSI A108.1A.
- Metal studs—ASTM C645.

Materials for Green/Sustainable Design

- See "Green Building Standards and Green Product Selection Guide" and consult manufacturers and suppliers for product sustainability and contribution to green building design.
- Consider specifying tile and installation materials that meet ANSI A138.1, the *American National Standard Specifications for Sustainable Ceramic Tiles, Glass Tiles, and Tile Installation Materials.*

Preparation by Other Trades

- Maximum allowable variation in the installation substrate (solid backing)—¼" in 10' from the required plane.

Movement Joint (architect must specify type of joint and show location and details on drawings)

- Movement joints—mandatory according to EJ171.

Installation Specifications

- Tile—ANSI A108.1A, .1B, or .1C. A108.1B required if waterproof membrane or epoxy bond coat to be used.
- Cementitious grout—ANSI A108.10.
- Epoxy mortar/grout—ANSI A108.6.
- Waterproof membrane—ANSI A108.13.
- Movement Joints—EJ171 and ASTM C1193.

Notes

- If lath cannot be attached directly to backing, use furring strips. See W231/W241.
- For thicker mortar bed, see W221.
- For mortar bed directly over wood or metal studs, see W231/W241.
- Exterior—not all bonding mortars are suitable for exterior use.
- Exterior—protection of installation may be required to prevent premature exposure of setting materials to moisture.

W223-21

- **Concrete, Masonry, or Plaster**
- **Organic Adhesive**
- **Ceramic Tile**

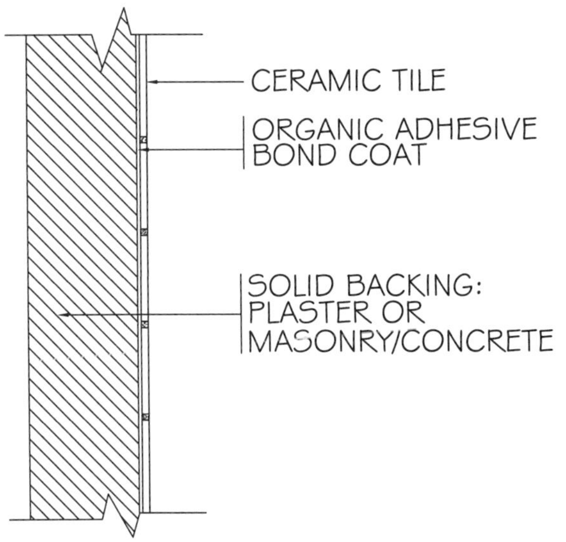

Recommended Uses

- For interior walls of concrete, masonry, plaster, or other backers.

Environmental Exposure Classifications

- Res1, 2; Com1, 2.
- For installations that may be exposed to staining, specify tile and grout suitable for exposure. Consult product manufacturers; see also "Product Selection Guides."
- For installations that may be exposed to mild chemical attack, specify epoxy grout and tile suitable for exposure. Consult product manufacturers; see also "Product Selection Guides."

Limitations

- Maximum tile size 8" × 8" unless organic adhesive manufacturer allows larger tile size. See Notes.
- Not for areas exposed to temperatures exceeding 140°F.

Requirements

- Masonry or concrete—well cured, dimensionally stable, and free of cracks, waxy or oily films, and curing compounds.
- Masonry or concrete—may require bush-hammering or sandblasting to facilitate bonding.
- Wood studs—dry and well-braced, minimum depth 3½."
- Metal studs—well-braced; 20 gauge (0.033") or heavier; minimum depth 3½" for residential applications or 3⅝" for commercial applications.

Materials

- **Multiple options exist for membranes, mortars, grouts, and other materials and *must be clearly specified* to be included. If not specifically indicated, optional materials are not included and mortar/grout choice defaults to minimum performance specification indicated. Consider each system component and intended use to determine minimum requirements and to specify options.**
- Ceramic tile—ANSI A137.1.
- Cementitious grout—ANSI A118.6 or better or ISO CG1 or better.
- Epoxy grout, when used—ANSI A118.3 or ISO RG.
- Organic adhesive bond coat—ANSI A136.1 (Type I or II) or ISO D1 or better.
  - ISO D1 and A136.1 Type II are suitable for dry areas only (Res1; Com1).
- Metal studs—ASTM C645.

Materials for Green/Sustainable Design

- See "Green Building Standards and Green Product Selection Guide" and consult manufacturers and suppliers for product sustainability and contribution to green building design.
- Consider specifying tile and installation materials that meet ANSI A138.1, the *American National Standard Specifications for Sustainable Ceramic Tiles, Glass Tiles, and Tile Installation Materials.*

Preparation by Other Trades

- Maximum allowable variation in the tile substrate—1⁄16" in 3' with no abrupt irregularities greater than 1⁄32."

Movement Joint (architect must specify type of joint and show location and details on drawings)

- Movement joints—mandatory according to EJ171.

Installation Specifications

- Tile—ANSI A108.4.
- Cementitious grout—ANSI A108.10.
- Epoxy grout—ANSI A108.6.
- Movement Joints—EJ171 and ASTM C1193.

Notes

- As the tile size increases, there is less tolerance for variation in the substrate from the required plane. Organic adhesive thickness must be thin and uniform; therefore, substrate flattening may be required when organic adhesive is used.
- For bonding tile to gypsum board with organic adhesive, see W242.
- For bonding tile to cement backer board with organic adhesive, see W244C.
- For bonding tile to fiber-cement backer board with organic adhesive, see W244.

- For bonding tile to fiber-reinforced water-resistant gypsum backer board with organic adhesive, see W247.
- Do not use organic adhesives to bond tile to a membrane or other impervious substrate.
- Extended dry time may be required for larger tiles and tiles with low water absorption when organic adhesive is used.

W231/W241-21

- **Wood or Metal Studs**
- **Mortar Bed**
- **Ceramic Tile, Glass Tile**

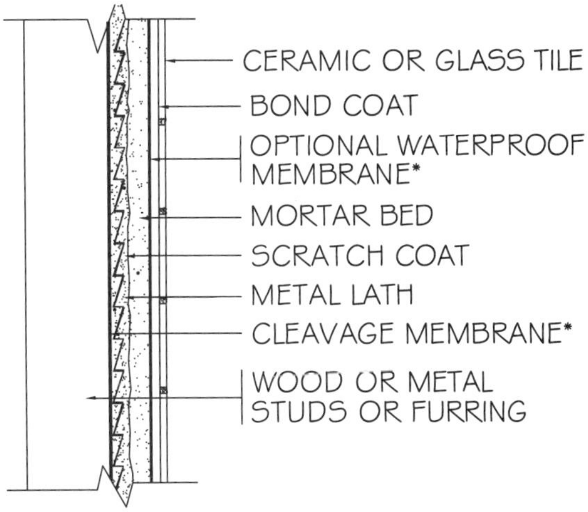

*USE OF A MEMBRANE IS OPTIONAL. SEE MEMBRANE OPTIONS.

Recommended Uses

- For interior walls framed of wood or metal where flattening or trueing of walls is required.
- For areas where wall flatness is critical, such as when tiles with any edge longer than 15" are specified.

Environmental Exposure Classifications

- Res 1, 2, 5; Com1, 2, 5.
- For Res3 and Com3, see B411 and B414.
- For Res4 and Com4, see SR614.
- May be suitable for exterior applications when appropriate precautions are taken, including flashing, expansion joint placement, waterproofing, and material selection. See Notes.
- For installations that may be exposed to staining, specify tile and grout suitable for exposure. Consult product manufacturers; see also "Product Selection Guides."
- For installations that may be exposed to mild chemical attack, specify epoxy grout and tile suitable for exposure. For greater resistance to chemical exposure, also specify an epoxy bonding material. Consult product manufacturers; see also "Product Selection Guides."

Limitations

- Maximum mortar bed thickness—1" for metal studs, 1½" for wood studs.
- Maximum stud spacing 16" on center.

Membrane Options

- A waterproof membrane (A118.10) may be specified to prevent moisture intrusion and protect adjacent building materials. Specifier shall indicate if complete waterproofing is required, including base flashing and treatment at other termination points.
- If a waterproof membrane is applied over the mortar bed, membrane over framing members may not be required.
- Check with membrane manufacturer for suitability for applicable conditions, as not all membranes are suitable for steam, high-temperature and/or chemical exposure, or exterior use.
- Exterior—check building code—water-resistive barrier, waterproof membrane, or vapor retarder membrane may be required.
- When glass tile is used, consult glass tile manufacturer for membrane options and recommendations.

Requirements

- Wood studs—dry and well-braced, minimum depth 3½."
- Metal studs—well-braced; 20 gauge (0.033") or heavier; minimum depth 3½" for residential applications or 3⅝" for commercial applications.
- Mortar bed thickness—¾" minimum to 1½" maximum.

Materials

- **Multiple options exist for membranes, mortars, grouts, and other materials and *must be clearly specified* to be included. If not specifically indicated, optional materials are not included and mortar/grout choice defaults to minimum performance specification indicated. Consider each system component and intended use to determine minimum requirements and to specify options.**
- Ceramic tile—ANSI A137.1.
- Glass tile, when used—ANSI A137.2; see also "Glass Tile Selection and Installation Guide," and consult tile manufacturer for environmental exposure classification recommendations. Not all glass tiles are suitable.
- Cementitious grout—ANSI A118.6 or better or ISO CG1 or better. When glass tile is used, specify grout designated by tile and grout manufacturers.
- Epoxy grout, when used—ANSI A118.3 or ISO RG.
- Cementitious bond coat—portland cement paste on a mortar bed that is still workable. For a cured mortar bed, follow recommendations below to select appropriate bonding mortar:
  - When a waterproof membrane is not used—ANSI A118.1 or better or ISO C1 or better.
  - When a waterproof membrane is used—ANSI A118.4 or better or ISO C2S1 or better unless ANSI A118.1 or ISO C1 is recommended by membrane manufacturer.
  - When porcelain tile is used—ANSI A118.4 or better or ISO C2 or better.
  - When glass tile is used, specify mortar designated by tile and mortar manufacturers. Bond coat color will

impact the final appearance of translucent glass tile. Specifier shall confirm bond coat color is acceptable.

- Epoxy bond coat, when used—ANSI A118.3 or ISO R1 or better.
  - When glass tile is used, specify epoxy bond coat designated by tile and bond coat manufacturers. Bond coat color will impact the final appearance of translucent glass tile. Specifier shall confirm bond coat color is acceptable.
- Water-resistive barrier/vapor retarder membrane, when used—ANSI A108.02-3.8.
- Waterproof membrane, when used—ANSI A118.10.
- Mortar bed, metal lath, and cleavage membrane—ANSI A108.1A.
- Metal studs—ASTM C645.

Materials for Green/Sustainable Design

- See "Green Building Standards and Green Product Selection Guide" and consult manufacturers and suppliers for product sustainability and contribution to green building design.
- Consider specifying tile and installation materials that meet ANSI A138.1, the *American National Standard Specifications for Sustainable Ceramic Tiles, Glass Tiles, and Tile Installation Materials.*

Preparation by Other Trades

- Wall framing shall meet the general framing requirements of ANSI A108.11-4.0-4.3.

Movement Joint (architect must specify type of joint and show location and details on drawings)

- Movement joints—mandatory according to EJ171.
- When glass tile is used, adhere to more frequent placement recommendations within the ranges listed in EJ171.

Installation Specifications

- Tile—ANSI A108.1A, .1B, or .1C. A108.1B required if waterproof membrane or epoxy bond coat to be used.
- Glass tile—ANSI A108.14, .15, .16, or manufacturer's directions.
- Cementitious grout—ANSI A108.10.
- Epoxy mortar/grout—ANSI A108.6.
- Waterproof membrane—ANSI A108.13.
- Movement Joints—EJ171 and ASTM C1193.

Notes

- Exterior—not all bonding mortars are suitable for exterior use.
- Exterior—protection of installation may be required to prevent premature exposure of setting materials to moisture.
- When glass tile is used, see "Glass Tile Selection and Installation Guide," and consult manufacturer for recommendations and requirements.

W242-21

- **Wood or Metal Studs**
- **Gypsum Board**
- **Organic Adhesive**
- **Ceramic Tile**

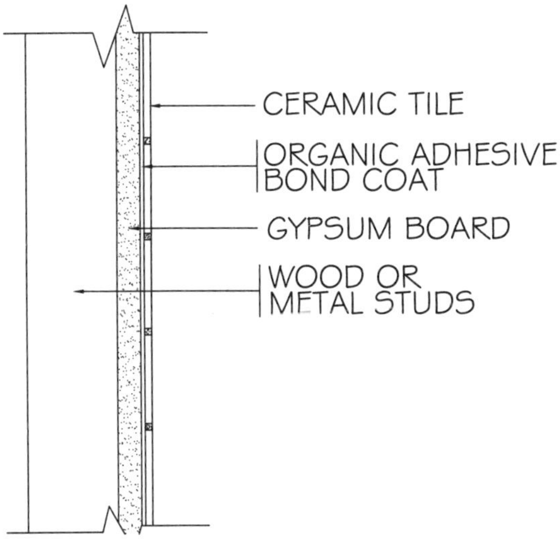

Recommended Uses

- For interior walls in dry areas where gypsum board is the tile backer.
- For fire-resistant, sound-insulated, ceramic tiled walls. See RW800 for fire and sound ratings. (Fire-resistance and sound-insulation ratings calculated on partitions before tiling.)

Environmental Exposure Classifications

- Res1; Com1.
- For installations that may be exposed to staining, specify tile and grout suitable for exposure. Consult product manufacturers; see also "Product Selection Guides."
- For installations that may be exposed to mild chemical attack, specify epoxy grout and tile suitable for exposure. Consult product manufacturers; see also "Product Selection Guides."

Limitations

- Not for areas exposed to temperatures exceeding 125°F.
- Maximum stud spacing 16" on center.
- Maximum tile size 8" × 8" unless organic adhesive manufacturer allows larger tile size. See Notes.

Requirements

- Wood studs—dry and well-braced, minimum depth 3½".
- Metal studs—well-braced; 20 gauge (0.033") or heavier; minimum depth 3½" for residential applications or 3⅝" for commercial applications.

Materials

- **Multiple options exist for membranes, mortars, grouts, and other materials and *must be clearly specified* to be included. If not specifically indicated, optional materials are not included and mortar/grout choice defaults to minimum performance specification indicated. Consider each system component and intended use to determine minimum requirements and to specify options.**
- Ceramic tile—ANSI A137.1.
- Cementitious grout—ANSI A118.6 or better or ISO CG1 or better.
- Epoxy grout, when used—ANSI A118.3 or ISO RG.
- Organic adhesive bond coat—ANSI A136.1 (Type I or II) or ISO D1 or better.
  - ANSI A136.1 Type II and ISO D1 are suitable for dry areas only (Res1; Com1).
- Gypsum board—ASTM C1396/C1396M. Minimum ½" thick for single layer applications.
- Metal studs—ASTM C645.

Materials for Green/Sustainable Design

- See "Green Building Standards and Green Product Selection Guide" and consult manufacturers and suppliers for product sustainability and contribution to green building design.
- Consider specifying tile and installation materials that meet ANSI A138.1, the *American National Standard Specifications for Sustainable Ceramic Tiles, Glass Tiles, and Tile Installation Materials.*

Preparation by Other Trades

- Gypsum board, single or double layer, to be installed per GA-216.
- Gypsum board face layer joints—treated with tape and joint compound, bedding coat only (no finish coats). Nail heads, one coat only.
- Maximum allowable variation in the tile substrate—1/16" in 3' with no abrupt irregularities greater than 1/32".
- Corners, door jambs, etc., must be plumb within ⅛" in 8'.

Movement Joint (architect must specify type of joint and show location and details on drawings)

- Movement joints—mandatory according to EJ171.

Installation Specifications

- Gypsum board—GA-216.
- Tile—ANSI A108.4.
- Cementitious grout—ANSI A108.10.
- Epoxy grout—ANSI A108.6.
- Movement Joints—EJ171 and ASTM C1193.

Notes

- As the tile size increases, there is less tolerance for variation in the substrate from the required plane. Organic adhesive thickness must be thin and uniform; therefore, substrate flattening may be required when organic adhesive is used.
- Do not use organic adhesives to bond tile to a membrane or other impervious substrates.
- Dry time may be required for larger tiles and tiles with low water absorption when organic adhesive is used.

W243-21

- **Wood or Metal Studs**
- **Gypsum Board**
- **Cementitious Bond Coat**
- **Ceramic Tile, Glass Tile**

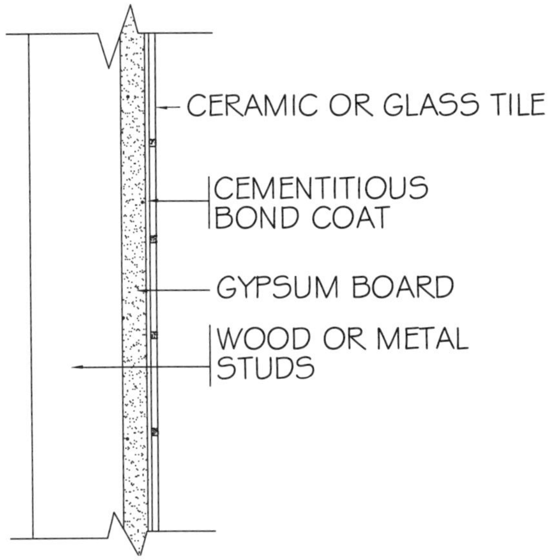

Recommended Uses

- For interior walls in dry areas where gypsum board is the tile backer.
- For fire-resistant, sound-insulated, ceramic tiled walls. See RW800 for fire and sound ratings. (Fire-resistance and sound-insulation ratings calculated on partitions before tiling.)

Environmental Exposure Classifications

- Res1; Com1.
- For installations that may be exposed to staining, specify tile and grout suitable for exposure. Consult product manufacturers; see also "Product Selection Guides."
- For installations that may be exposed to mild chemical attack, specify epoxy grout and tile suitable for exposure. Consult product manufacturers; see also "Product Selection Guides."

Limitations

- Not for areas exposed to temperatures exceeding 125°F.
- Maximum stud spacing 16" on center.

Requirements

- Wood studs—dry and well-braced, minimum depth 3½."
- Metal studs—well-braced; 20 gauge (0.033") or heavier; minimum depth 3½" for residential applications or 3⅝" for commercial applications.

Materials

- **Multiple options exist for membranes, mortars, grouts, and other materials and *must be clearly specified* to be included. If not specifically indicated, optional materials are not included and mortar/grout choice defaults to minimum performance specification indicated. Consider each system component and intended use to determine minimum requirements and to specify options.**
- Ceramic tile—ANSI A137.1.
- Glass tile, when used—ANSI A137.2; see also "Glass Tile Selection and Installation Guide," and consult tile manufacturer for environmental exposure classification recommendations. Not all glass tiles are suitable.
- Cementitious grout—ANSI A118.6 or better or ISO CG1 or better. When glass tile is used, specify grout designated by tile and grout manufacturers.
- Epoxy grout, when used—ANSI A118.3 or ISO RG.
- Cementitious bond coat:
  - ANSI A118.1 or better or ISO C1 or better.
  - When porcelain tile is used—ANSI A118.4 or better or ISO C2 or better.
  - When glass tile is used, specify mortar designated by tile and mortar manufacturers. Bond coat color will impact the final appearance of translucent glass tile. Specifier shall confirm bond coat color is acceptable.
- Gypsum board—ASTM C1396/C1396M. Minimum ½" thick for single layer applications.
- Metal studs—ASTM C645.

Materials for Green/Sustainable Design

- See "Green Building Standards and Green Product Selection Guide" and consult manufacturers and suppliers for product sustainability and contribution to green building design.
- Consider specifying tile and installation materials that meet ANSI A138.1, the *American National Standard Specifications for Sustainable Ceramic Tiles, Glass Tiles, and Tile Installation Materials.*

Preparation by Other Trades

- Gypsum board, single or double layer, to be installed per GA-216.
- Gypsum board face layer joints—treated with tape and joint compound, bedding coat only (no finish coats). Nail heads, one coat only.
- Maximum allowable variation in the tile substrate—for tiles with all edges shorter than 15," maximum allowable variation is ¼" in 10' from the required plane, with no more than ¹⁄₁₆" variation in 12" when measured from the high points in the surface. For tiles with at least one edge 15" in length, maximum allowable variation is ⅛" in 10' from the required plane, with no more than ¹⁄₁₆" variation in 24" when measured from the high points in the surface.

Movement Joint (architect must specify type of joint and show location and details on drawings)

- Movement joints—mandatory according to EJ171.

- When glass tile is used, adhere to more frequent placement recommendations within the ranges listed in EJ171.

Installation Specifications

- Gypsum board—GA-216.
- Tile—ANSI A108.5.
- Glass tile—manufacturer's directions.
- Cementitious grout—ANSI A108.10.
- Epoxy grout—ANSI A108.6.
- Movement Joints—EJ171 and ASTM C1193.

Notes

- When glass tile is used, see "Glass Tile Selection and Installation Guide," and consult manufacturer for recommendations and requirements.

W244C-21

- **Wood or Metal Studs**
- **Cement Backer Board**
- **Ceramic Tile, Glass Tile**

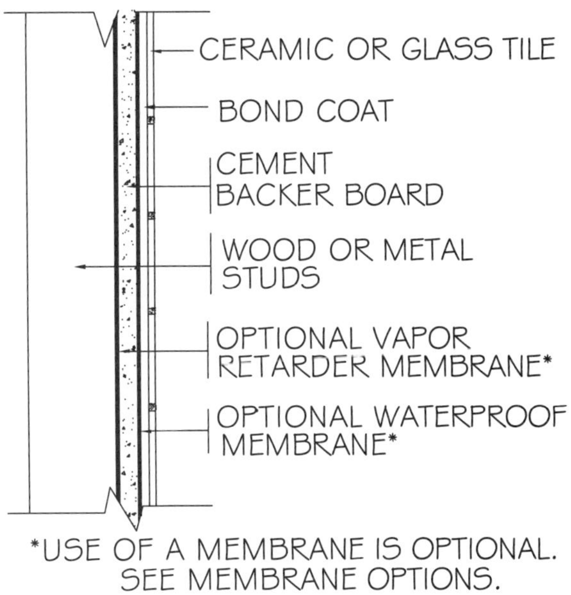

Recommended Uses

- For interior walls over wood or metal studs.

Environmental Exposure Classifications

- With cementitious or epoxy bond coat—Res1, 2, 5; Com1, 2, 5.
- With organic adhesive bond coat—Res1, 2; Com1, 2.
- For Res3 and Com3, see B412 and B415.
- For Res4 and Com4, see SR614.
- For Res6 and Com6, see W244E.
- For installations that may be exposed to staining, specify tile and grout suitable for exposure. Consult product manufacturers; see also "Product Selection Guides."
- For installations that may be exposed to mild chemical attack, specify epoxy grout and tile suitable for exposure. For greater resistance to chemical exposure, also specify an epoxy bonding material. Consult product manufacturers; see also "Product Selection Guides."

Limitations

- Maximum stud spacing 16" on center.
- When organic adhesive is used—not for areas exposed to temperatures exceeding 140°F.
- When organic adhesive is used—maximum tile size 8" × 8" unless organic adhesive manufacturer allows larger tile size. See Notes.

Membrane Options

- A waterproof membrane (A118.10) or vapor retarder membrane (A108.02-3.8) may be specified to prevent moisture intrusion and protect adjacent building materials. Specifier shall indicate if complete waterproofing is required, including base flashing and treatment at other termination points.
- Check with membrane manufacturer for suitability for applicable conditions, as not all membranes are suitable for steam, high-temperature and/or chemical exposure, or exterior use.
- When glass tile is used, consult glass tile manufacturer for membrane options and recommendations.

Requirements

- Wood studs—dry and well-braced, minimum depth 3½."
- Metal studs—well-braced; 20 gauge (0.033") or heavier; minimum depth 3½" for residential applications or 3⅝" for commercial applications.
- Surface of units—clean and free of dirt, dust, paint, and oily film.

Preparation by Backer Board Installers

- Maximum allowable variation in the tile substrate—for tiles with all edges shorter than 15," maximum allowable variation is ¼" in 10' from the required plane, with no more than 1⁄16" variation in 12" when measured from the high points in the surface. For tiles with at least one edge 15" in length, maximum allowable variation is ⅛" in 10' from the required plane, with no more than 1⁄16" variation in 24" when measured from the high points in the surface.
- Horizontal joints—⅛" spacing filled solid and taped with latex-portland cement mortar and 2" alkali-resistant glass fiber mesh tape.
- Vertical joints—fill any space and tape with latex-portland cement mortar and 2" alkali-resistant glass fiber mesh tape.
- Corners—leave space between backer units. Tape joints using skim coat of latex-portland cement mortar, but do not fill.
- Center backer board end or edge joints on framing and stagger joints in adjacent rows so four corners do not come together within the same plane. Space panel ends and edges in accordance with manufacturer's recommendations.

Materials

- **Multiple options exist for membranes, mortars, grouts, and other materials and *must be clearly specified* to be included. If not specifically indicated, optional materials are not included and mortar/grout choice defaults to minimum performance specification indicated. Consider each system component and intended use to determine minimum requirements and to specify options.**
- Ceramic tile—ANSI A137.1.
- Glass tile, when used—ANSI A137.2; see also "Glass

Tile Selection and Installation Guide," and consult tile manufacturer for environmental exposure classification recommendations. Not all glass tiles are suitable.

- Cementitious grout—ANSI A118.6 or better or ISO CG1 or better. When glass tile is used, specify grout designated by tile and grout manufacturers.
- Epoxy grout, when used—ANSI A118.3 or ISO RG.
- Cementitious bond coat:
  - When a waterproof membrane is not used—ANSI A118.1 or better or ISO C1 or better.
  - When a waterproof membrane is used—ANSI A118.4 or better or ISO C2S1 or better unless ANSI A118.1 or ISO C1 is recommended by membrane manufacturer.
  - When porcelain tile is used—ANSI A118.4 or better or ISO C2 or better.
  - When glass tile is used, specify mortar designated by tile and mortar manufacturers. Bond coat color will impact the final appearance of translucent glass tile. Specifier shall confirm bond coat color is acceptable.
- Epoxy bond coat, when used—ANSI A118.3 or ISO R1 or better.
  - When glass tile is used, specify epoxy bond coat designated by tile and bond coat manufacturers. Bond coat color will impact the final appearance of translucent glass tile. Specifier shall confirm bond coat color is acceptable.
- Organic adhesive bond coat, when used—ANSI A136.1 (Type I or II) or ISO D1 or better.
  - ISO D1 and ANSI A136.1 Type II are suitable for dry areas only (Res1; Com1)
- Vapor retarder membrane, when used—ANSI A108.02-3.8.
- Waterproof membrane, when used—ANSI A118.10.
- Cement backer board—ANSI A118.9 or ASTM C1325 (Type B).
- Fasteners—noncorrosive and nonoxidizing.
- 2" alkali-resistant glass fiber mesh tape.
- Metal studs—ASTM C645.

Materials for Green/Sustainable Design

- See "Green Building Standards and Green Product Selection Guide" and consult manufacturers and suppliers for product sustainability and contribution to green building design.
- Consider specifying tile and installation materials that meet ANSI A138.1, the *American National Standard Specifications for Sustainable Ceramic Tiles, Glass Tiles, and Tile Installation Materials.*

Preparation by Other Trades

- Wall framing shall meet the general framing requirements of ANSI A108.11-4.0–4.3.

Movement Joint (architect must specify type of joint and show location and details on drawings)

- Movement joints—mandatory according to EJ171.
- When glass tile is used, adhere to more frequent placement recommendations within the ranges listed in EJ171.

Installation Specifications

- Cement backer board—ANSI A108.11.
- Tile—ANSI A108.4 or A108.5.
- Glass tile—ANSI A108.15, .16, or manufacturer's directions.
- Cementitious grout—ANSI A108.10.
- Epoxy mortar/grout—ANSI A108.6.
- Waterproof membrane—ANSI A108.13.
- Movement Joints—EJ171 and ASTM C1193.

Notes

- As the tile size increases, there is less tolerance for variation in the substrate from the required plane. Organic adhesive or epoxy bond coat thickness must be thin and uniform; therefore, substrate flattening may be required when organic adhesive or epoxy bond coat is used.
- Do not use organic adhesives to bond tile to a membrane or other impervious substrate.
- Extended dry time may be required for larger tiles and tiles with low water absorption when organic adhesive is used.
- When glass tile is used, see "Glass Tile Selection and Installation Guide," and consult manufacturer for recommendations and requirements.

W244F-21

- **Wood or Metal Studs**
- **Fiber-Cement Backer Board**
- **Ceramic Tile, Glass Tile**

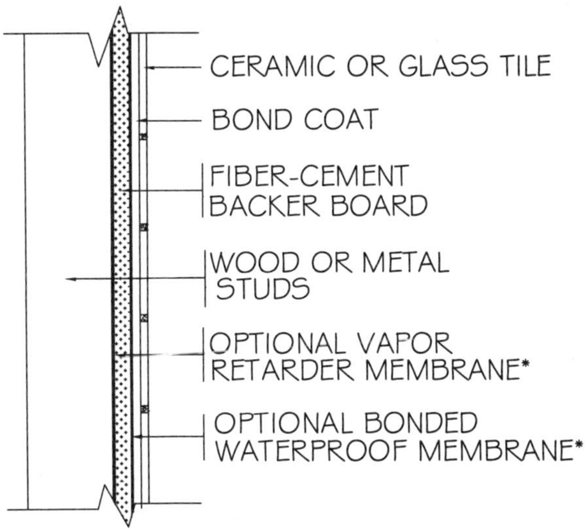

*USE OF A MEMBRANE IS OPTIONAL. SEE MEMBRANE OPTIONS.

Recommended Uses

- For interior walls over wood or metal studs.

Environmental Exposure Classifications

- With cementitious or epoxy bond coat—Res1, 2, 5; Com1, 2, 5.
- With organic adhesive bond coat—Res1, 2; Com1, 2.
- For Res3 and Com3, see B412 and B415.
- For installations that may be exposed to staining, specify tile and grout suitable for exposure. Consult product manufacturers; see also "Product Selection Guides."
- For installations that may be exposed to mild chemical attack, specify epoxy grout and tile suitable for exposure. For greater resistance to chemical exposure, also specify an epoxy bonding material. Consult product manufacturers; see also "Product Selection Guides."

Limitations

- Maximum stud spacing 16" on center.
- When organic adhesive is used—not for areas exposed to temperatures exceeding 140°F.
- When organic adhesive is used—maximum tile size 8" x 8" unless organic adhesive manufacturer allows larger tile size. See Notes.

Membrane Options

- A waterproof membrane (A118.10) or vapor retarder membrane (A108.02-3.8) may be specified to prevent moisture intrusion and protect adjacent building materials. Specifier shall indicate if complete waterproofing is required, including base flashing and treatment at other termination points.
- Check with membrane manufacturer for suitability for applicable conditions, as not all membranes are suitable for steam, high-temperature and/or chemical exposure, or exterior use.
- When glass tile is used, consult glass tile manufacturer for membrane options and recommendations.

Requirements

- Wood studs—dry and well-braced, minimum depth 3½."
- Metal studs—well-braced; 20 gauge (0.033") or heavier; minimum depth 3½" for residential applications or 3⅝" for commercial applications.
- Surface of units—clean and free of dirt, dust, paint, and oily film.

Preparation by Backer Board Installers

- Maximum allowable variation in the tile substrate—for tiles with all edges shorter than 15," maximum allowable variation is ¼" in 10' from the required plane, with no more than $\frac{1}{16}$" variation in 12" when measured from the high points in the surface. For tiles with at least one edge 15" in length, maximum allowable variation is ⅛" in 10' from the required plane, with no more than $\frac{1}{16}$" variation in 24" when measured from the high points in the surface.
- Horizontal joints—⅛" spacing filled solid and taped with latex-portland cement mortar and 2" alkali-resistant glass fiber mesh tape.
- Vertical joints—fill any space and tape with latex-portland cement mortar and 2" alkali-resistant glass fiber mesh tape.
- Corners—leave space between backer units. Tape joints using skim coat of latex-portland cement mortar, but do not fill.
- Center backer board end or edge joints on framing and stagger joints in adjacent rows so four corners do not come together within the same plane. Space panel ends and edges in accordance with manufacturer's recommendations.

Materials

- **Multiple options exist for membranes, mortars, grouts, and other materials and *must be clearly specified* to be included. If not specifically indicated, optional materials are not included and mortar/grout choice defaults to minimum performance specification indicated. Consider each system component and intended use to determine minimum requirements and to specify options.**
- Ceramic tile—ANSI A137.1.
- Glass tile, when used—ANSI A137.2; see also "Glass Tile Selection and Installation Guide," and consult tile

manufacturer for environmental exposure classification recommendations. Not all glass tiles are suitable.

- Cementitious grout—ANSI A118.6 or better or ISO CG1 or better. When glass tile is used, specify grout designated by tile and grout manufacturers.
- Epoxy grout, when used—ANSI A118.3 or ISO RG.
- Cementitious bond coat:
  - When a waterproof membrane is not used—ANSI A118.1 or better or ISO C1 or better.
  - When a waterproof membrane is used—ANSI A118.4 or better or ISO C2S1 or better unless ANSI A118.1 or ISO C1 is recommended by membrane manufacturer.
  - When porcelain tile is used—ANSI A118.4 or better or ISO C2 or better.
  - When glass tile is used, specify mortar designated by tile and mortar manufacturers. Bond coat color will impact the final appearance of translucent glass tile. Specifier shall confirm bond coat color is acceptable.
- Epoxy bond coat, when used—ANSI A118.3 or ISO R1 or better.
  - When glass tile is used, specify epoxy bond coat designated by tile and bond coat manufacturers. Bond coat color will impact the final appearance of translucent glass tile. Specifier shall confirm bond coat color is acceptable.
- Organic adhesive bond coat, when used—ANSI A136.1 (Type I or II) or ISO D1 or better.
  - ISO D1 and ANSI A136.1 Type II are suitable for dry areas only (Res1; Com1).
- Waterproof membrane, when used—ANSI A118.10.
- Vapor retarder membrane, when used—ANSI A108.02-3.8.
- Fiber cement backer board—ASTM C1288.
- Fasteners—noncorrosive and nonoxidizing.
- 2" alkali-resistant glass fiber mesh tape.
- Metal studs—ASTM C645.

Materials for Green/Sustainable Design

- See "Green Building Standards and Green Product Selection Guide" and consult manufacturers and suppliers for product sustainability and contribution to green building design.
- Consider specifying tile and installation materials that meet ANSI A138.1, the *American National Standard Specifications for Sustainable Ceramic Tiles, Glass Tiles, and Tile Installation Materials.*

Preparation by Other Trades

- Wall framing shall meet the general framing requirements of ANSI A108.11-4.0–4.3.

Movement Joint (architect must specify type of joint and show location and details on drawings)

- Movement joints—mandatory according to EJ171.
- When glass tile is used, adhere to more frequent placement recommendations within the ranges listed in EJ171.

Installation Specifications

- Fiber cement backer board—manufacturer's directions.
- Tile—ANSI A108.4 or A108.5.
- Glass tile—manufacturer's directions.
- Cementitious grout—ANSI A108.10.
- Epoxy mortar/grout—ANSI A108.6.
- Waterproof membrane—ANSI A108.13.
- Movement Joints—EJ171 and ASTM C1193.

Notes

- As the tile size increases, there is less tolerance for variation in the substrate from the required plane. Organic adhesive or epoxy bond coat thickness must be thin and uniform; therefore, substrate flattening may be required when organic adhesive or epoxy bond coat is used.
- Do not use organic adhesives to bond tile to a membrane or other impervious substrate.
- Extended dry time may be required for larger tiles and tiles with low water absorption when organic adhesive is used.
- When glass tile is used, see "Glass Tile Selection and Installation Guide," and consult manufacturer for recommendations and requirements.

W245-21

- **Wood or Metal Studs**
- **Coated Glass Mat Water-Resistant Gypsum Backer Board**
- **Ceramic Tile, Glass Tile**

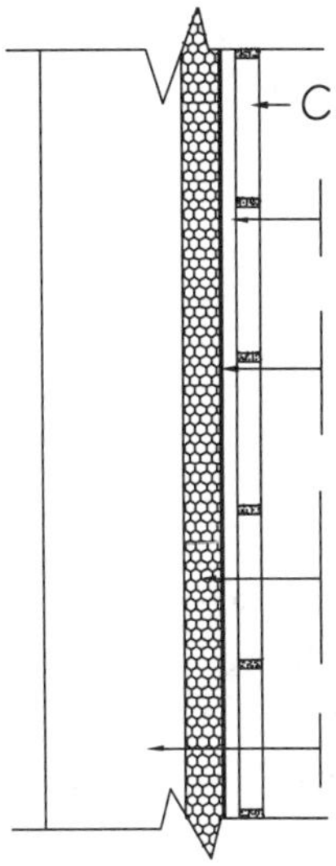

*USE OF A MEMBRANE IS OPTIONAL. SEE MEMBRANE OPTIONS.

Recommended Uses

- For interior walls over wood or metal studs.

Environmental Exposure Classifications

- Res1, 2; Com1, 2.
- For Res3 and Com3, see B419 and B420.
- For installations that may be exposed to staining, specify tile and grout suitable for exposure. Consult product manufacturers; see also "Product Selection Guides."
- For installations that may be exposed to mild chemical attack, specify epoxy grout and tile suitable for exposure. Consult product manufacturers; see also "Product Selection Guides."

Limitations

- Maximum stud spacing 16" on center.
- Not for areas exposed to temperatures exceeding 125°F.

Membrane Options

- A waterproof membrane (A118.10) may be specified to prevent moisture intrusion through seams, corners, fasteners, and other penetrations and to protect adjacent building materials. Follow backer board and membrane manufacturer's waterproofing requirements. Specifier shall indicate if complete waterproofing is required, including base flashing and treatment at other termination points.
- Check with membrane manufacturer for suitability for applicable conditions, as not all membranes are suitable for steam, high-temperature and/or chemical exposure, or exterior use.
- Some liquid-applied waterproof membranes will require extended cure time. Consult membrane manufacturer.

Requirements

- Wood studs—dry and well-braced, minimum depth 3½."
- Metal studs—well-braced; 20 gauge (0.033") or heavier; minimum depth 3½" for residential applications or 3⅝" for commercial applications.
- Caulk or seal penetrations, corners, and abutments to dissimilar materials with flexible sealant.
- When glass tile is used, consult glass tile manufacturer for tile suitability over non-absorptive surface.

Preparation by Backer Board Installers

- Maximum allowable variation in the tile substrate—for tiles with all edges shorter than 15," maximum allowable variation is ¼" in 10' from the required plane, with no more than 1/16" variation in 12" when measured from the high points in the surface. For tiles with at least one edge 15" in length, maximum allowable variation is ⅛" in 10' from the required plane, with no more than 1/16" variation in 24" when measured from the high points in the surface.
- Fit ends and edges closely to eliminate gaps between panels, including in corners.
- Tape seams and corners with 2" alkali-resistant glass fiber mesh tape and cementitious bonding material, and skim coat over fastener heads, unless waterproofing has been applied.
- Center backer board end or edge joints on framing and stagger joints in adjacent rows so four corners do not come together within the same plane. Space panel ends and edges in accordance with manufacturer's recommendations.

Materials

- **Multiple options exist for membranes, mortars, grouts, and other materials and *must be clearly specified* to be included. If not specifically indicated, optional materials are not included and mortar/grout choice defaults to minimum performance specification indicated. Consider each system component and intended use to determine minimum requirements and to specify options.**
- Ceramic tile—ANSI A137.1.
- Glass tile, when used—ANSI A137.2; see also "Glass Tile Selection and Installation Guide," and consult tile manufacturer for environmental exposure classification recommendations. Not all glass tiles are suitable.
- Cementitious grout—ANSI A118.6 or better or ISO CG1 or better. When glass tile is used, specify grout designated by tile and grout manufacturers.
- Epoxy grout, when used—ANSI A118.3 or ISO RG.
- Cementitious bond coat:
  - When a waterproof membrane is not used—ANSI A118.4 or better or ISO C2S1 or better.

    - When a waterproof membrane is used—ANSI A118.4 or better or ISO C2S1 or better unless ANSI A118.1 or ISO C1 is recommended by membrane manufacturer.
    - When glass tile is used, specify mortar designated by tile and mortar manufacturers. Bond coat color will impact the final appearance of translucent glass tile. Specifier shall confirm bond coat color is acceptable.
- Waterproof membrane, when used—ANSI A118.10.
- Coated glass mat, water-resistant gypsum backer board—ASTM C1178.
- Fasteners—noncorrosive and nonoxidizing.
- 2" alkali-resistant glass fiber mesh tape.
- Flexible sealant—ASTM C920.
- Metal studs—ASTM C645.

Materials for Green/Sustainable Design

- See "Green Building Standards and Green Product Selection Guide" and consult manufacturers and suppliers for product sustainability and contribution to green building design.
- Consider specifying tile and installation materials that meet ANSI A138.1, the *American National Standard Specifications for Sustainable Ceramic Tiles, Glass Tiles, and Tile Installation Materials.*

Preparation by Other Trades

- Wall framing shall meet the general framing requirements of ANSI A108.11-4.0–4.3.

Movement Joint (architect must specify type of joint and show location and details on drawings)

- Movement joints—mandatory according to EJ171.
- When glass tile is used, adhere to more frequent placement recommendations within the ranges listed in EJ171.

Installation Specifications

- Coated glass mat water-resistant gypsum backer board—manufacturer's directions.
- Tile—ANSI A108.5.
- Glass tile—manufacturer's directions.
- Cementitious grout—ANSI A108.10.
- Epoxy grout—ANSI A108.6.
- Waterproof membrane—ANSI A108.13.
- Movement Joints—EJ171 and ASTM C1193.

Notes

- When glass tile is used, see "Glass Tile Selection and Installation Guide," and consult manufacturer for recommendations and requirements.

W246-21

- **Wood or Metal Studs**
- **Cementitious-Coated Extruded Foam Backer Board**
- **Ceramic Tile, Glass Tile**

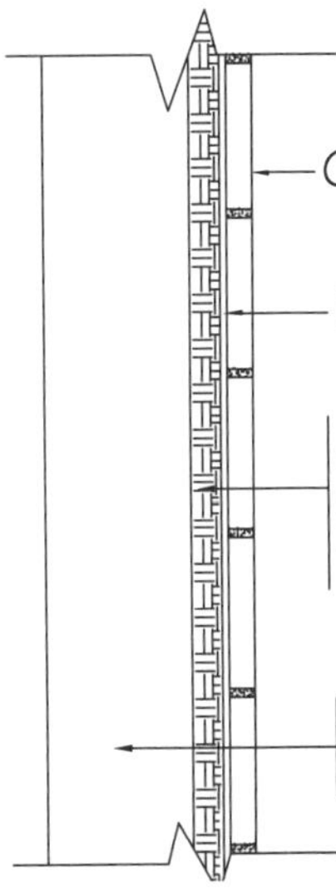

Recommended Uses

- For interior walls over wood or metal studs.

Environmental Exposure Classifications

- Res1, 2; Com1, 2.
- For Res3 and Com3, see B425 and B426.
- For installations that may be exposed to staining, specify tile and grout suitable for exposure. Consult product manufacturers; see also "Product Selection Guides."
- For installations that may be exposed to mild chemical attack, specify epoxy grout and tile suitable for exposure. For greater resistance to chemical exposure, also specify an epoxy bonding material. Consult product manufacturers; see also "Product Selection Guides."

Limitations

- Maximum stud spacing 16" on center.

Requirements

- When glass tile is used, consult glass tile manufacturer for recommendations over sealant (when used).
- Wood studs—dry and well-braced, minimum depth 3½."
- Metal studs—well-braced; 20 gauge (0.033") or heavier; minimum depth 3½" for residential applications or 3⅝" for commercial applications.

Preparation by Backer Board Installers

- Fasten backer board to studs with board manufacturer's recommended fasteners.
- Maximum allowable variation in the tile substrate—for tiles with all edges shorter than 15," maximum allowable variation is ¼" in 10' from the required plane, with no more than ¹⁄₁₆" variation in 12" when measured from the high points in the surface. For tiles with at least one edge 15" in length, maximum allowable variation is ⅛" in 10' from the required plane, with no more than ¹⁄₁₆" variation in 24" when measured from the high points in the surface.
- Fit ends and edges tightly together, including in corners. Apply a bead of manufacturer-recommended sealant in panel joints as panels are being installed, and over fastener heads, per backer board manufacturer's instructions.
- Joints may be taped with 2" alkali-resistant glass fiber mesh tape and cementitious bonding material in dry areas.
- Center backer board end or edge joints on framing and stagger joints in adjacent rows so four corners do not come together within the same plane. Space panel ends and edges in accordance with manufacturer's recommendations.

Materials

- **Multiple options exist for membranes, mortars, grouts, and other materials and *must be clearly specified* to be included. If not specifically indicated, optional materials are not included and mortar/grout choice defaults to minimum performance specification indicated. Consider each system component and intended use to determine minimum requirements and to specify options.**
- Ceramic tile—ANSI A137.1.
- Glass tile, when used—ANSI A137.2; see also "Glass Tile Selection and Installation Guide," and consult tile manufacturer for environmental exposure classification recommendations. Not all glass tiles are suitable.
- Cementitious grout—ANSI A118.6 or better or ISO CG1 or better. When glass tile is used, specify grout designated by tile and grout manufacturers.
- Epoxy grout, when used—ANSI A118.3 or ISO RG.
- Cementitious bond coat:
  - ANSI A118.1 or better or ISO C1 or better.
  - When porcelain tile is used—ANSI A118.4 or better or ISO C2 or better.
  - When glass tile is used, specify mortar designated by tile and mortar manufacturers. Bond coat color will impact the final appearance of translucent glass tile. Specifier shall confirm bond coat color is acceptable.
- Epoxy bond coat, when used—ANSI A118.3 or ISO R1 or better.
  - When glass tile is used, specify epoxy bond coat designated by tile and bond coat manufacturers. Bond coat color will impact the final appearance of translucent glass tile. Specifier shall confirm bond coat color is acceptable.

- Cementitious-coated extruded foam backer board—ASTM C578.
- Fasteners—noncorrosive and nonoxidizing.
- 2" alkali-resistant glass fiber mesh tape, when used.
- Flexible sealant—must be recommended by backer board manufacturer.
- Metal studs—ASTM C645.

Materials for Green/Sustainable Design

- See "Green Building Standards and Green Product Selection Guide" and consult manufacturers and suppliers for product sustainability and contribution to green building design.
- Consider specifying tile and installation materials that meet ANSI A138.1, the *American National Standard Specifications for Sustainable Ceramic Tiles, Glass Tiles, and Tile Installation Materials.*

Preparation by Other Trades

- Wall framing shall meet the general framing requirements of ANSI A108.11-4.0–4.3.

Movement Joint (architect must specify type of joint and show location and details on drawings)

- Movement joints—mandatory according to EJ171.
- When glass tile is used, adhere to more frequent placement recommendations within the ranges listed in EJ171.

Installation Specifications

- Cementitious-coated extruded foam backer board—manufacturer's directions.
- Tile—ANSI A108.5.
- Glass tile—manufacturer's directions.
- Cementitious grout—ANSI A108.10.
- Epoxy mortar/grout—ANSI A108.6.
- Movement Joints—EJ171 and ASTM C1193.

Notes

- As the tile size increases, there is less tolerance for variation in the substrate from the required plane. Epoxy adhesive thickness must be thin and uniform; therefore, substrate flattening may be required when epoxy bond coat is used.
- When glass tile is used, see "Glass Tile Selection and Installation Guide," and consult manufacturer for recommendations and requirements.

W247-21

- **Wood or Metal Studs**
- **Fiber-Reinforced Water-Resistant Gypsum Backer Board**
- **Ceramic Tile, Glass Tile**

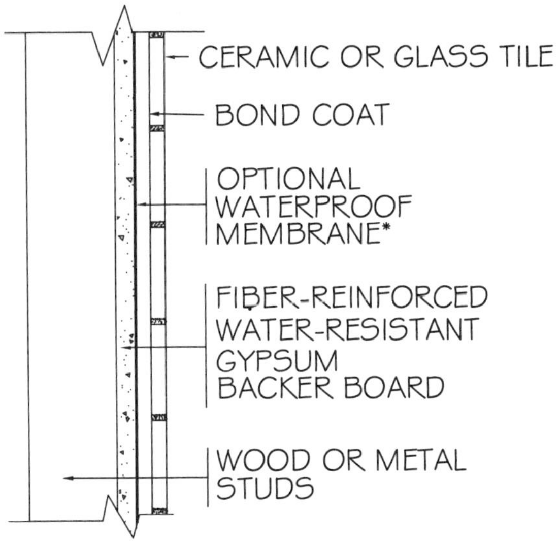

*USE OF A MEMBRANE IS OPTIONAL. SEE MEMBRANE OPTIONS.

Recommended Uses

- For interior walls over wood or metal studs.

Environmental Exposure Classifications

- Res1, 2; Com1, 2.
- For Res3, see B430 and B431.
- For installations that may be exposed to staining, specify tile and grout suitable for exposure. Consult product manufacturers; see also "Product Selection Guides."
- For installations that may be exposed to mild chemical attack, specify epoxy grout and tile suitable for exposure. Consult product manufacturers; see also "Product Selection Guides."

Limitations

- Maximum stud spacing 16" on center.
- Not for areas exposed to temperatures exceeding 125°F.
- When organic adhesive is used—maximum tile size 8" × 8" unless organic adhesive manufacturer allows larger tile size. See Notes.

Membrane Options

- A waterproof membrane (A118.10) may be specified to prevent moisture intrusion and protect adjacent building materials. Specifier shall indicate if complete waterproofing is required, including base flashing and treatment at other termination points.
- Check with membrane manufacturer for suitability for applicable conditions, as not all membranes are suitable for steam, high-temperature and/or chemical exposure, or exterior use.
- When glass tile is used, consult glass tile manufacturer for membrane options and recommendations.

Requirements

- Wood studs—dry and well-braced, minimum depth 3½."
- Metal studs—well-braced; 20 gauge (0.033") or heavier; minimum depth 3½" for residential applications or 3⅝" for commercial applications.
- Caulk or seal penetrations and abutments to dissimilar materials.

Preparation by Backer Board Installers

- Maximum allowable variation in the tile substrate—for tiles with all edges shorter than 15," maximum allowable variation is ¼" in 10' from the required plane, with no more than ⅟₁₆" variation in 12" when measured from the high points in the surface. For tiles with at least one edge 15" in length, maximum allowable variation is ⅛" in 10' from the required plane, with no more than ⅟₁₆" variation in 24" when measured from the high points in the surface.
- Horizontal joints—⅛" spacing filled solid and taped with latex-portland cement mortar and 2" alkali-resistant glass fiber mesh tape.
- Vertical joints—fill any space and tape with latex-portland cement mortar and 2" alkali-resistant glass fiber mesh tape.
- Corners—leave space between backer units. Tape joints using skim coat of latex-portland cement mortar, but do not fill.
- Center backer board end or edge joints on framing and stagger joints in adjacent rows so four corners do not come together within the same plane. Space panel ends and edges in accordance with manufacturer's recommendations.

Materials

- **Multiple options exist for membranes, mortars, grouts, and other materials and *must be clearly specified* to be included. If not specifically indicated, optional materials are not included and mortar/grout choice defaults to minimum performance specification indicated. Consider each system component and intended use to determine minimum requirements and to specify options.**
- Ceramic tile—ANSI A137.1.
- Glass tile, when used—ANSI A137.2; see also "Glass Tile Selection and Installation Guide," and consult tile manufacturer for environmental exposure classification recommendations. Not all glass tiles are suitable.
- Cementitious grout—ANSI A118.6 or better or ISO

CG1 or better. When glass tile is used, specify grout designated by tile and grout manufacturers.
- Epoxy grout, when used—ANSI A118.3 or ISO RG.
- Cementitious bond coat:
  - When a waterproof membrane is not used—ANSI A118.4 or better or ISO C2 or better.
  - When a waterproof membrane is used—ANSI A118.4 or better or ISO C2S1 or better unless ANSI A118.1 or ISO C1 is recommended by membrane manufacturer.
  - When glass tile is used, specify mortar designated by tile and mortar manufacturers. Bond coat color will impact the final appearance of translucent glass tile. Specifier shall confirm bond coat color is acceptable.
- Organic adhesive bond coat, when used—ANSI A136.1 (Type I or II) or ISO D1 or better.
  - ISO D1 and ANSI A136.1 Type II are suitable for dry areas only (Res1; Com1).
- Waterproof membrane, when used—ANSI A118.10.
- Fiber-reinforced water-resistant gypsum backer board—ASTM C1278 (Paragraph 6.1).
- Fasteners—noncorrosive and nonoxidizing.
- 2" alkali-resistant glass fiber mesh tape.
- Metal studs—ASTM C645.
- Flexible sealant—ASTM C920.

Materials for Green/Sustainable Design

- See "Green Building Standards and Green Product Selection Guide" and consult manufacturers and suppliers for product sustainability and contribution to green building design.
- Consider specifying tile and installation materials that meet ANSI A138.1, the *American National Standard Specifications for Sustainable Ceramic Tiles, Glass Tiles, and Tile Installation Materials.*

Preparation by Other Trades

- Wall framing shall meet the general framing requirements of ANSI A108.11-4.0–4.3.

Movement Joint (architect must specify type of joint and show location and details on drawings)

- Movement joints—mandatory in accordance with EJ171.
- When glass tile is used, adhere to more frequent placement recommendations within the ranges listed in EJ171.

Installation Specifications

- Fiber-reinforced water-resistant gypsum backer board—manufacturer's directions.
- Tile—ANSI A108.4 or A108.5.
- Glass tile—manufacturer's directions.
- Cementitious grout—ANSI A108.10.
- Epoxy grout—ANSI A108.6.
- Waterproof membrane—ANSI A108.13.
- Movement Joints—EJ171 and ASTM C1193.

Notes

- As the tile size increases, there is less tolerance for variation in the substrate from the required plane. Organic adhesive thickness must be thin and uniform; therefore, substrate flattening may be required when organic adhesive is used.
- Do not use organic adhesives to bond tile to a membrane or other impervious substrate.
- Extended dry time may be required for larger tiles and tiles with low water absorption when organic adhesive is used.
- When glass tile is used, see "Glass Tile Selection and Installation Guide," and consult manufacturer for recommendations and requirements.

W248-21

- **Wood or Metal Studs**
- **Glass Mat Water-Resistant Gypsum Backer Board**
- **Ceramic Tile, Glass Tile**

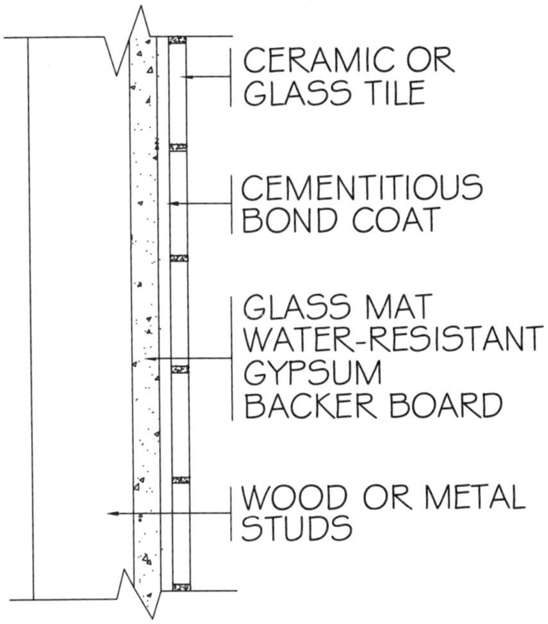

Recommended Uses

- For interior walls over wood or metal studs.

Environmental Exposure Classifications

- Res1; Com1.
- For installations that may be exposed to staining, specify tile and grout suitable for exposure. Consult product manufacturers; see also "Product Selection Guides."
- For installations that may be exposed to mild chemical attack, specify epoxy grout and tile suitable for exposure. Consult product manufacturers; see also "Product Selection Guides."

Limitations

- Maximum stud spacing 16" on center.
- Not for areas exposed to temperatures exceeding 125°F.

Requirements

- Wood studs—dry and well-braced, minimum depth 3½."
- Metal studs—well-braced; 20 gauge (0.033") or heavier; minimum depth 3½" for residential applications or 3⅝" for commercial applications.
- Caulk or seal penetrations, corners, and abutments to dissimilar materials with flexible sealant.

Preparation by Backer Board Installers

- Maximum allowable variation in the tile substrate—for tiles with all edges shorter than 15," maximum allowable variation is ¼" in 10' from the required plane, with no more than 1/16" variation in 12" when measured from the high points in the surface. For tiles with at least one edge 15" in length, maximum allowable variation is ⅛" in 10' from the required plane, with no more than 1/16" variation in 24" when measured from the high points in the surface.
- Fit ends and edges closely to eliminate gaps between panels, including in corners.
- Tape seams and corners with 2" alkali-resistant glass fiber mesh tape and skim coat with cementitious bonding material.
- Center backer board end or edge joints on framing and stagger joints in adjacent rows so four corners do not come together within the same plane. Space panel ends and edges in accordance with manufacturer's recommendations.

Materials

- **Multiple options exist for membranes, mortars, grouts, and other materials and *must be clearly specified* to be included. If not specifically indicated, optional materials are not included and mortar/grout choice defaults to minimum performance specification indicated. Consider each system component and intended use to determine minimum requirements and to specify options.**
- Ceramic tile—ANSI A137.1.
- Glass tile, when used—ANSI A137.2; see also "Glass Tile Selection and Installation Guide," and consult tile manufacturer for environmental exposure classification recommendations. Not all glass tiles are suitable.
- Cementitious grout—ANSI A118.6 or better or ISO CG1 or better. When glass tile is used, specify grout designated by tile and grout manufacturers.
- Epoxy grout, when used—ANSI A118.3 or ISO RG.
- Cementitious bond coat:
  - ANSI A118.4 or better or ISO C2 or better.
  - When glass tile is used, specify mortar designated by tile and mortar manufacturers. Bond coat color will impact the final appearance of translucent glass tile. Specifier shall confirm bond coat color is acceptable.
- Glass mat, water-resistant gypsum backer board—ASTM C1658.
- Fasteners—noncorrosive and nonoxidizing.
- 2" alkali-resistant glass fiber mesh tape.
- Flexible sealant—ASTM C920.
- Metal studs—ASTM C645.

Materials for Green/Sustainable Design

- See "Green Building Standards and Green Product Selection Guide" and consult manufacturers and suppliers for product sustainability and contribution to green building design.

- Consider specifying tile and installation materials that meet ANSI A138.1, the *American National Standard Specifications for Sustainable Ceramic Tiles, Glass Tiles, and Tile Installation Materials.*

Preparation by Other Trades

- Wall framing shall meet the general framing requirements of ANSI A108.11-4.0–4.3.

Movement Joint (architect must specify type of joint and show location and details on drawings)

- Movement joints—mandatory according to EJ171.
- When glass tile is used, adhere to more frequent placement recommendations within the ranges listed in EJ171.

Installation Specifications

- Glass mat water-resistant gypsum backer board—manufacturer's directions.
- Tile—ANSI A108.5.
- Cementitious grout—ANSI A108.10.
- Epoxy grout—ANSI A108.6.
- Movement Joints—EJ171 and ASTM C1193.
- Glass tile—manufacturer's directions.

Notes

- When glass tile is used, see "Glass Tile Selection and Installation Guide," and consult manufacturer for recommendations and requirements.

W260-21

- **Wood or Metal Studs**
- **Cement Backer Board**
- **Spot-Bonding Epoxy**
- **Ceramic Tile**

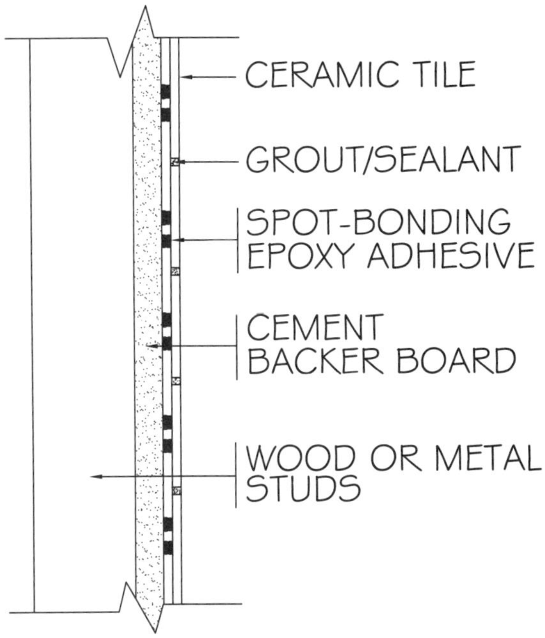

Recommended Uses

- For interior walls where cement backer board is the tile backer and large format tiles are specified.

Environmental Exposure Classifications

- Res1; Com1.
- For installations that may be exposed to staining, specify tile and grout suitable for exposure. Consult product manufacturers; see also "Product Selection Guides."
- Installations that may be exposed to mild chemical attack, specify epoxy grout and tile suitable for exposure. Consult product manufacturers; see also "Product Selection Guides."

Limitations

- Will not withstand impact.
- Maximum stud spacing 16" on center.

Requirements

- Epoxy must be recommended by the manufacturer for spot-bonding.
- Follow epoxy manufacturer's coverage requirements.
- Wood studs—dry and well-braced, minimum depth 3½".
- Metal studs—well-braced; 20 gauge (0.033") or heavier; minimum depth 3½" for residential applications or 3⅝" for commercial applications.
- Surface of units—clean and free of dirt, dust, paint, and oily film.

Preparation by Backer Board Installers

- Maximum allowable variation in the tile substrate—¼" in 10' from the required plane.
- Horizontal joints—⅛" spacing filled solid and taped with latex-portland cement mortar and 2" alkali-resistant glass fiber mesh tape.
- Vertical joints—fill any space and tape with latex-portland cement mortar and 2" alkali-resistant glass fiber mesh tape.
- Corners—leave space between backer units. Tape joints using skim coat of latex-portland cement mortar, but do not fill.
- Center backer board end or edge joints on framing and stagger joints in adjacent rows so four corners do not come together within the same plane. Space panel ends and edges in accordance with manufacturer's recommendations.

Materials

- **Multiple options exist for membranes, mortars, grouts, and other materials and *must be clearly specified* to be included. If not specifically indicated, optional materials are not included and mortar/grout choice defaults to minimum performance specification indicated. Consider each system component and intended use to determine minimum requirements and to specify options.**
- Ceramic tile—ANSI A137.1.
- Cementitious grout—ANSI A118.6 or better or ISO CG1 or better.
- Epoxy grout, when used—ANSI A118.3 or ISO RG.
- Sealant, when used as grout—ASTM C920.
- Spot bonding epoxy adhesive—epoxy recommended by manufacturer for spot bonding.
- Cement backer board—ANSI A118.9 or ASTM C1325 (Type B).
- Fasteners—noncorrosive and nonoxidizing.
- 2" alkali-resistant glass fiber mesh tape.
- Cementitious mortar (for taping joints)—ANSI A118.4 or better or ISO C2 or better.
- Metal studs—ASTM C645.

Materials for Green/Sustainable Design

- See "Green Building Standards and Green Product Selection Guide" and consult manufacturers and suppliers for product sustainability and contribution to green building design.
- Consider specifying tile and installation materials that meet ANSI A138.1, the *American National Standard Specifications for Sustainable Ceramic Tiles, Glass Tiles, and Tile Installation Materials.*

Preparation by Other Trades

- Wall framing shall meet the general framing requirements of ANSI A108.11-4.0–4.3.

Movement Joint (architect must specify type of joint and show location and details on drawings)

- Movement joints—mandatory according to EJ171.

Installation Specifications

- Cement backer board—ANSI A108.11.
- Tile—epoxy manufacturer's directions.
- Cementitious grout—ANSI A108.10.
- Epoxy grout—ANSI A108.6.
- Sealant—sealant and epoxy manufacturer's directions for using sealant as grout.
- Movement Joints—EJ171 and ASTM C1193.

B413-21

- **Bathtub Walls (No Showerhead)**
- **Gypsum Board or Glass Mat Gypsum Board**
- **Ceramic Tile, Glass Tile**

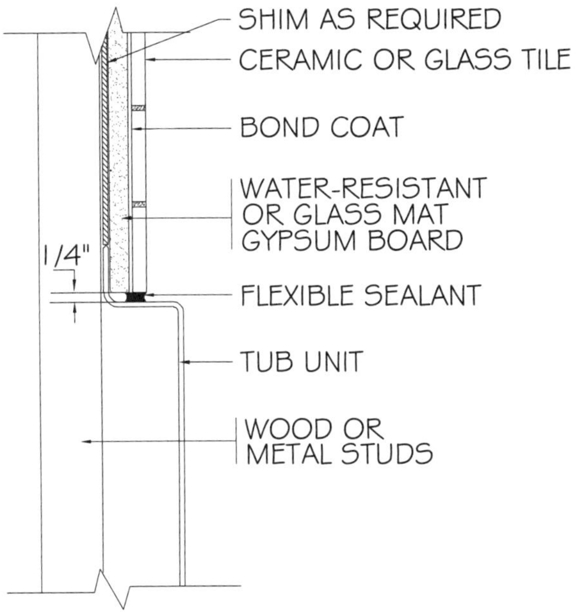

Recommended Uses

- For tub enclosures without a shower head where gypsum board or glass mat water-resistant gypsum backer board is the tile backer.
- For fire-resistant, sound-insulated, ceramic tiled walls. See RW800 for fire and sound ratings. (Fire-resistance and sound-insulation ratings calculated on partitions before tiling.)

Environmental Exposure Classifications

- Res1; Com1.
- For installations that may be exposed to staining, specify tile and grout suitable for exposure. Consult product manufacturers; see also "Product Selection Guides."
- For installations that may be exposed to mild chemical attack, specify epoxy grout and tile suitable for exposure. Consult product manufacturers; see also "Product Selection Guides."

Limitations

- Not for areas exposed to temperatures exceeding 125°F.
- Maximum stud spacing 16" on center.
- When organic adhesive is used, maximum tile size 8" × 8" unless organic adhesive manufacturer allows larger tile size. See Notes.

Requirements

- Wood studs—dry and well-braced, minimum depth 3½".
- Metal studs—well-braced; 20 gauge (0.033") or heavier; minimum depth 3½" for residential applications or 3⅝" for commercial applications.

Materials

- **Multiple options exist for membranes, mortars, grouts, and other materials and *must be clearly specified* to be included. If not specifically indicated, optional materials are not included and mortar/grout choice defaults to minimum performance specification indicated. Consider each system component and intended use to determine minimum requirements and to specify options.**
- Ceramic tile—ANSI A137.1.
- Glass tile, when used—ANSI A137.2; see also "Glass Tile Selection and Installation Guide," and consult tile manufacturer for environmental exposure classification recommendations. Not all glass tiles are suitable.
- Cementitious grout—ANSI A118.6 or better or ISO CG1 or better. When glass tile is used, specify grout designated by tile and grout manufacturers.
- Epoxy grout, when used—ANSI A118.3 or ISO RG.
- Cementitious bond coat:
  - ANSI A118.1 or better or ISO C1 or better.
  - When porcelain tile is used—ANSI A118.4 or better or ISO C2 or better.
  - When glass tile is used, specify mortar designated by tile and mortar manufacturers. Bond coat color will impact the final appearance of translucent glass tile. Specifier shall confirm bond coat color is acceptable.
- Organic adhesive bond coat, when used—ANSI A136.1 (Type I or II) or ISO D1 or better.
  - ANSI A136.1 Type II and ISO D1 are suitable for dry areas only (Res1; Com1).
- Gypsum board, when used—ASTM C1396/C1396M. Minimum ½" thick for single layer applications.
- Glass mat water-resistant gypsum backer board, when used—ASTM C1658.
- Metal studs—ASTM C645.
- Flexible mildew-resistant sealant—ASTM C920.

Materials for Green/Sustainable Design

- See "Green Building Standards and Green Product Selection Guide" and consult manufacturers and suppliers for product sustainability and contribution to green building design.
- Consider specifying tile and installation materials that meet ANSI A138.1, the *American National Standard Specifications for Sustainable Ceramic Tiles, Glass Tiles, and Tile Installation Materials.*

Preparation by Other Trades

- Bathtub—install level and support with metal hangers or on end grain wood blocks.
- Fireproofing behind tub when required.
- Maximum allowable variation in the tile substrate—for tiles with all edges shorter than 15", maximum allowable

variation is ¼" in 10' from the required plane, with no more than ⅟₁₆" variation in 12" when measured from the high points in the surface. For tiles with at least one edge 15" in length, maximum allowable variation is ⅛" in 10' from the required plane, with no more than ⅟₁₆" variation in 24" when measured from the high points in the surface.

- All openings cut in backing board for plumbing and all cut joints between adjoining pieces—seal with adhesive or other materials recommended by manufacturer of backing board.
- Gypsum board face layer joints—treated with tape and joint compound, bedding coat only (no finish coats). Nail heads, one coat only.
- Apply backing board horizontally with the factory paperbound edge spaced a minimum of ¼" above the lip of the tub.

Movement Joint (architect must specify type of joint and show location and details on drawings)

- Movement joints—mandatory according to EJ171.
- When glass tile is used, adhere to more frequent placement recommendations within the ranges listed in EJ171.

Installation Specifications

- Gypsum board—GA-216.
- Glass mat water-resistant gypsum backer board—manufacturer's directions.
- Tile—ANSI A108.4 or A108.5
- Glass Tile—manufacturer's directions.
- Cementitious grout—ANSI A108.10.
- Epoxy grout—ANSI A108.6.
- Movement Joints—EJ171 and ASTM C1193.

Notes

- As the tile size increases, there is less tolerance for variation in the substrate from the required plane. Organic adhesive thickness must be thin and uniform; therefore, substrate flattening may be required when organic adhesive is used.
- Do not use organic adhesives to bond tile to a membrane or other impervious substrate.
- Extended dry time may be required for larger tiles and tiles with low water absorption when organic adhesive is used.
- When glass tile is used, see "Glass Tile Selection and Installation Guide," and consult manufacturer for recommendations and requirements.

B411-21

- **Bathtub or Shower with Prefabricated Receptor**
- **Mortar Bed**
- **Ceramic Tile, Glass Tile**

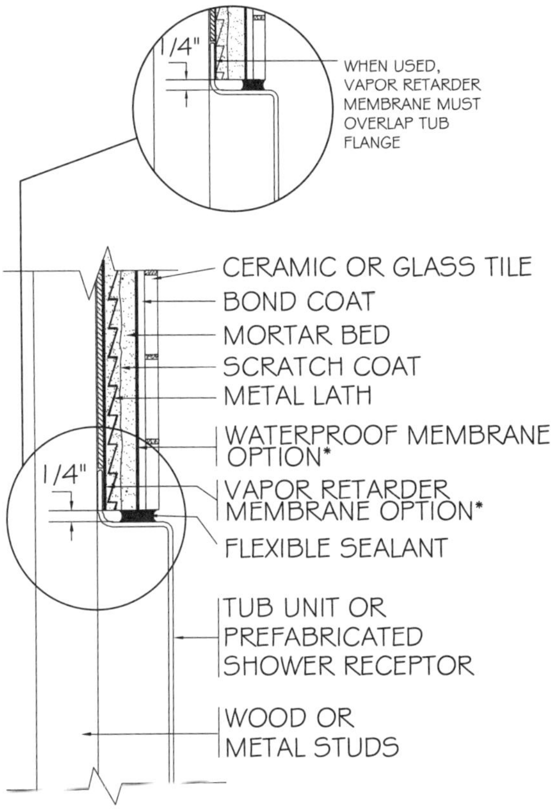

*USE OF A MEMBRANE IS REQUIRED. SEE MEMBRANE OPTIONS.

SHOWER RECEPTORS, CURBS, SEATS, ETC. MUST BE PROPERLY WATERPROOFED AND INSTALLED TO AVOID WATER DAMAGE TO ADJACENT BUILDING MATERIALS. SEE COMMON SHOWER CONFIGURATIONS SECTION.

Recommended Uses

- For tub enclosures or showers with prefabricated receptors, where flattening or trueing of walls is required.
- For areas where wall flatness is critical, such as when tiles with any edge longer than 15" are specified.
- For showers that do not have a prefabricated receptor, see B414.

Environmental Exposure Classifications

- Res1, 2, 3, 5; Com1, 2, 3, 5.
- For Res4 and Com4, see SR614.
- For installations that may be exposed to staining, specify tile and grout suitable for exposure. Consult product manufacturers; see also "Product Selection Guides."
- For installations that may be exposed to mild chemical attack, specify epoxy grout and tile suitable for exposure. For greater resistance to chemical exposure, also specify an epoxy bonding material. Consult product manufacturers; see also "Product Selection Guides."

Limitations

- Maximum mortar bed thickness—1" for metal studs, 1½" for wood studs.
- Maximum stud spacing 16" on center.

Membrane Options

- A waterproof membrane (A118.10) or vapor retarder membrane (A108.02-3.8) must be specified to prevent moisture intrusion and protect adjacent building materials. Specifier shall indicate if complete waterproofing is required, including treatment at termination points.
- If a waterproof membrane is applied over the mortar bed, membrane over framing members may not be required.
- Check with membrane manufacturer for suitability for applicable conditions, as not all membranes are suitable for steam, high-temperature and/or chemical exposure, or exterior use.
- When glass tile is used, consult glass tile manufacturer for membrane options and recommendations.

Requirements

- Membrane behind mortar bed, when used, must lap over face of flange of tub or prefabricated shower receptor.
- Wood studs—dry and well-braced, minimum depth 3½."
- Metal studs—well-braced; 20 gauge (0.033") or heavier; minimum depth 3½" for residential applications or 3⅝" for commercial applications.
- Mortar bed thickness—¾" minimum to 1" maximum (metal studs) or 1½" maximum (wood studs).

Materials

- **Multiple options exist for membranes, mortars, grouts, and other materials and *must be clearly specified* to be included. If not specifically indicated, optional materials are not included and mortar/grout choice defaults to minimum performance specification indicated. Consider each system component and intended use to determine minimum requirements and to specify options.**
- Ceramic tile—ANSI A137.1.
- Glass tile, when used—ANSI A137.2; see also "Glass Tile Selection and Installation Guide," and consult tile manufacturer for environmental exposure classification recommendations. Not all glass tiles are suitable.
- Cementitious grout—ANSI A118.6 or better or ISO CG1 or better. When glass tile is used, specify grout designated by tile and grout manufacturers.

- Epoxy grout, when used—ANSI A118.3 or ISO RG.
- Cementitious bond coat—portland cement paste on a mortar bed that is still workable. For a cured mortar bed, follow recommendations below to select appropriate bonding mortar:
  - When a waterproof membrane is not used—ANSI A118.1 or better or ISO C1 or better.
  - When a waterproof membrane is used—ANSI A118.4 or better or ISO C2S1 or better unless ANSI A118.1 or ISO C1 is recommended by membrane manufacturer.
  - When porcelain tile is used—ANSI A118.4 or better or ISO C2 or better.
  - When glass tile is used, specify mortar designated by tile and mortar manufacturers. Bond coat color will impact the final appearance of translucent glass tile. Specifier shall confirm bond coat color is acceptable.
- Epoxy bond coat, when used—ANSI A118.3 or ISO R1 or better.
  - When glass tile is used, specify epoxy bond coat designated by tile and bond coat manufacturers. Bond coat color will impact the final appearance of translucent glass tile. Specifier shall confirm bond coat color is acceptable.
- Vapor retarder membrane, when used—ANSI A108.02-3.8.
- Waterproof membrane, when used—ANSI A118.10.
- Mortar bed, metal lath, and cleavage membrane—ANSI A108.1A.
- Metal studs—ASTM C645.
- Flexible mildew-resistant sealant—ASTM C920.

Materials for Green/Sustainable Design

- See "Green Building Standards and Green Product Selection Guide" and consult manufacturers and suppliers for product sustainability and contribution to green building design.
- Consider specifying tile and installation materials that meet ANSI A138.1, the *American National Standard Specifications for Sustainable Ceramic Tiles, Glass Tiles, and Tile Installation Materials.*

Preparation by Other Trades

- Wall framing shall meet the general framing requirements of ANSI A108.11-4.0-4.3.
- Opening for recessed bathtubs—not to exceed ½" more than total length of tub.
- Bathtub—install level and support with metal hangers or on end grain wood blocks.
- Fireproofing behind tub when required.

Movement Joint (architect must specify type of joint and show location and details on drawings)

- Movement joints—mandatory according to EJ171.
- When glass tile is used, adhere to more frequent placement recommendations within the ranges listed in EJ171.

Installation Specifications

- Tile—ANSI A108.1A, .1B, or .1C. A108.1B required if waterproof membrane or epoxy bond coat to be used.
- Glass tile—ANSI A108.14, .15, .16, or manufacturer's directions.
- Cementitious grout—ANSI A108.10.
- Epoxy mortar/grout—ANSI A108.6.
- Waterproof membrane—ANSI A108.13.
- Movement Joints—EJ171 and ASTM C1193.

Notes

- All horizontal surfaces, for example shower seats, sills, curbs, etc., must slope towards drain or other surface sloped toward drain. Where present, waterproofing also must be sloped.
- When glass tile is used, see "Glass Tile Selection and Installation Guide," and consult manufacturer for recommendations and requirements.

B440-21

- **Bathtub or Shower with Prefabricated Receptor**
- **Backer Board**
- **Mortar Bed (One Coat Method)**
- **Ceramic Tile**

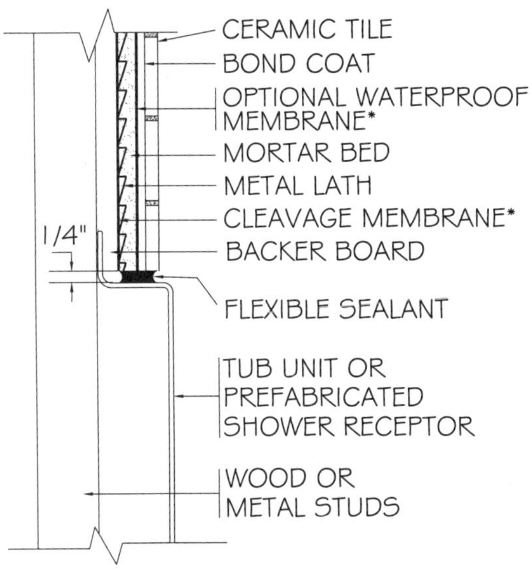

*USE OF A MEMBRANE IS REQUIRED. SEE MEMBRANE OPTIONS.

SHOWER RECEPTORS, CURBS, SEATS, ETC., MUST BE PROPERLY WATERPROOFED AND INSTALLED TO AVOID WATER DAMAGE TO ADJACENT BUILDING MATERIALS. SEE COMMON SHOWER CONFIGURATIONS SECTION.

Recommended Uses

- For tub enclosures or showers with prefabricated receptors, where minor flattening or trueing of walls is required.
- For areas where wall flatness is critical, such as when tiles with any edge longer than 15" are specified.
- For showers that do not have a prefabricated receptor, see B441.

Environmental Exposure Classifications

- With cement backer board or fiber cement backer board—Res1, 2, 3, 5; Com1, 2, 3, 5.
- With coated glass mat water-resistant gypsum backer board or cementitious-coated extruded foam backer board—Res1, 2, 3; Com1, 2, 3.
- With fiber-reinforced water-resistant gypsum backer board—Res 1, 2, 3; Com 1, 2.
- For installations that may be exposed to staining, specify tile and grout suitable for exposure. Consult product manufacturers; see also "Product Selection Guides."
- For installations that may be exposed to mild chemical attack, specify epoxy grout and tile suitable for exposure. For greater resistance to chemical exposure, also specify an epoxy bonding material. Consult product manufacturers; see also "Product Selection Guides."

Limitations

- Maximum mortar bed thickness—¾".
- Maximum stud spacing 16" on center.
- When coated glass mat water-resistant gypsum backer board or fiber-reinforced water-resistant gypsum backer board is used—not for areas exposed to temperatures exceeding 125°F.

Membrane Options

- A cleavage membrane is required. A waterproof membrane (A118.10) may also be specified to prevent moisture intrusion and protect adjacent building materials. Specifier shall indicate if complete waterproofing is required, including treatment at termination points.
- Check with membrane manufacturer for suitability for applicable conditions, as not all membranes are suitable for steam, high-temperature and/or chemical exposure, or exterior use.

Requirements

- Cleavage membrane must lap over face of flange of tub or prefabricated shower receptor.
- Wood studs—dry and well-braced, minimum depth 3½".
- Metal studs—well-braced; 20 gauge (0.033") or heavier; minimum depth 3½" for residential applications or 3⅝" for commercial applications.
- Mortar bed thickness—⅜" minimum to ¾" maximum.
- Backer board must provide firm anchorage for metal lath.
- Cut lath at all movement joints.

Preparation by Backer Board Installers

- Maximum allowable variation in the backer board—⅛" in 10' from the required plane.
- Follow backer board manufacturer's directions for installation, including orientation and spacing of boards, required fasteners, and taping of joints and corners.

Materials

- **Multiple options exist for membranes, mortars, grouts, and other materials and *must be clearly specified* to be included. If not specifically indicated, optional materials are not included and mortar/grout choice defaults to minimum performance specification indicated. Consider each system component and intended use to determine minimum requirements and to specify options.**
- Ceramic tile—ANSI A137.1.
- Cementitious grout—ANSI A118.6 or better or ISO CG1 or better.

- Epoxy grout, when used—ANSI A118.3 or ISO RG.
- Cementitious bond coat—portland cement paste on a mortar bed that is still workable. For a cured mortar bed, follow recommendations below to select appropriate bonding mortar:
    - When a waterproof membrane is not used—ANSI A118.1 or better or ISO C1 or better.
    - When a waterproof membrane is used—ANSI A118.4 or better or ISO C2S1 or better unless ANSI A118.1 or ISO C1 is recommended by membrane manufacturer.
    - When porcelain tile is used—ANSI A118.4 or better or ISO C2 or better.
- Epoxy bond coat, when used—ANSI A118.3 or ISO R1 or better.
- Waterproof membrane, when used—ANSI A118.10.
- Mortar bed, metal lath, and cleavage membrane—ANSI A108.1A.
- Cement backer board, when used—ANSI A118.9 or ASTM C1325 (Type B).
- Fiber cement backer board, when used—ASTM C1288.
- Coated glass mat, water-resistant gypsum backer board, when used—ASTM C1178.
- Cementitious-coated extruded foam backer board, when used—ASTM C578.
- Fiber-reinforced water-resistant gypsum backer board, when used—ASTM C1278 (Paragraph 6.1).
- Fasteners—noncorrosive and nonoxidizing.
- Fasteners meeting ASTM F2329 required in wet areas.
- 2" alkali-resistant glass fiber mesh tape.
- Flexible mildew-resistant sealant—ASTM C920.
- Metal studs—ASTM C645.

Materials for Green/Sustainable Design

- See "Green Building Standards and Green Product Selection Guide" and consult manufacturers and suppliers for product sustainability and contribution to green building design.
- Consider specifying tile and installation materials that meet ANSI A138.1, the *American National Standard Specifications for Sustainable Ceramic Tiles, Glass Tiles, and Tile Installation Materials.*

Preparation by Other Trades

- Wall framing shall meet the general framing requirements of ANSI A108.11-4.0–4.3.
- Opening for recessed bathtubs—not to exceed ½" more than total length of tub.
- Bathtub—install level and support with metal hangers or on end grain wood blocks.
- Fireproofing behind tub when required.
- Fire and sound ratings—extend gypsum board required for ratings down to the floor behind the tub so that construction will be the same as the tested assembly. Backer board may be part of, or installed over, the rated assembly.

Movement Joint (architect must specify type of joint and show location and details on drawings)

- Movement joints—mandatory according to EJ171.

Installation Specifications

- Tile—ANSI A108.1A, .1B, or .1C. A108.1B required if waterproof membrane or epoxy bond coat to be used.
- Cementitious grout—ANSI A108.10.
- Epoxy mortar/grout—ANSI A108.6.
- Waterproof membrane—ANSI A108.13.
- Cement backer board—ANSI A108.11.
- Fiber cement backer board—manufacturer's directions.
- Coated glass mat water-resistant gypsum backer board—manufacturer's directions.
- Cementitious-coated extruded foam backer board—manufacturer's directions.
- Fiber-reinforced water-resistant gypsum backer board—manufacturer's directions.
- Movement Joints—EJ171 and ASTM C1193.

Notes

- All horizontal surfaces, for example shower seats, sills, curbs, etc., must slope towards drain or other surface sloped toward drain. Where present, waterproofing also must be sloped.
- When coated glass mat water-resistant gypsum backer board is used, do not install a vapor barrier behind the backer board.
- If lath cannot be attached directly to backing, use furring strips. See B411.
- For thicker mortar bed, see B411.
- For mortar bed directly over wood or metal studs, see B411.

B412-21

- **Bathtub or Shower with Prefabricated Receptor**
- **Cement Backer Board or Fiber-Cement Backer Board**
- **Ceramic Tile, Glass Tile**

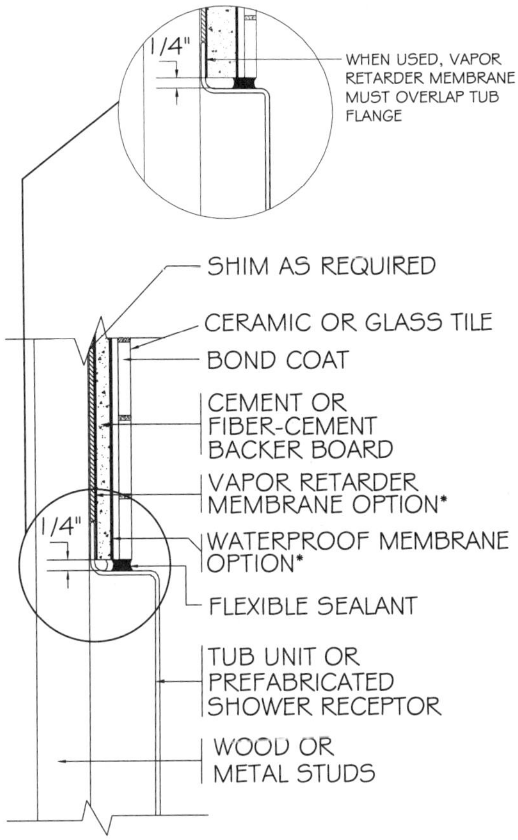

*USE OF A MEMBRANE IS REQUIRED. SEE MEMBRANE OPTIONS.

SHOWER RECEPTORS, CURBS, SEATS, ETC., MUST BE PROPERLY WATERPROOFED AND INSTALLED TO AVOID WATER DAMAGE TO ADJACENT BUILDING MATERIALS. SEE COMMON SHOWER CONFIGURATIONS SECTION.

Recommended Uses

- For tub enclosures or showers with prefabricated receptors.
- For showers that do not have a prefabricated receptor, see B415.

Environmental Exposure Classifications

- With cementitious or epoxy bond coat—Res1, 2, 3, 5; Com1, 2, 3, 5.
- With organic adhesive bond coat—Res1, 2, 3; Com1, 2.
- For Res4 and Com4, see SR614.
- For installations that may be exposed to staining, specify tile and grout suitable for exposure. Consult product manufacturers; see also "Product Selection Guides."
- For installations that may be exposed to mild chemical attack, specify epoxy grout and tile suitable for exposure. For greater resistance to chemical exposure, also specify an epoxy bonding material. Consult product manufacturers; see also "Product Selection Guides."

Limitations

- Maximum stud spacing 16" on center.
- When organic adhesive is used—not for areas exposed to temperatures exceeding 140°F.
- When organic adhesive is used—maximum tile size 8" × 8" unless organic adhesive manufacturer allows larger tile size. See Notes.

Membrane Options

- A waterproof membrane (A118.10) or vapor retarder membrane (A108.02-3.8) must be specified to prevent moisture intrusion and protect adjacent building materials. Specifier shall indicate if complete waterproofing is required, including treatment at termination points.
- Check with membrane manufacturer for suitability for applicable conditions, as not all membranes are suitable for steam, high-temperature and/or chemical exposure, or exterior use.
- When glass tile is used, consult glass tile manufacturer for membrane options and recommendations.

Requirements

- Wood studs—dry and well-braced, minimum depth 3½."
- Metal studs—well-braced; 20 gauge (0.033") or heavier; minimum depth 3½" for residential applications or 3⅝" for commercial applications.
- Surface of units—clean and free of dirt, dust, paint, and oily film.
- Membrane behind backer board, when used, must lap over face of flange of tub or prefabricated shower receptor.

Preparation by Backer Board Installers

- Maximum allowable variation in the tile substrate—for tiles with all edges shorter than 15," maximum allowable variation is ¼" in 10' from the required plane, with no more than ¹⁄₁₆" variation in 12" when measured from the high points in the surface. For tiles with at least one edge 15" in length, maximum allowable variation is ⅛" in 10' from the required plane, with no more than ¹⁄₁₆" variation in 24" when measured from the high points in the surface.
- Horizontal joints—⅛" spacing filled solid and taped with latex-portland cement mortar and 2" alkali-resistant glass fiber mesh tape.
- Vertical joints—fill any space and tape with latex-portland cement mortar and 2" alkali-resistant glass fiber mesh tape.
- Corners—leave space between backer units. Tape joints using skim coat of latex-portland cement mortar and 2" alkali-resistant glass fiber mesh tape, but do not fill.
- Center backer board end or edge joints on framing and stagger joints in adjacent rows so four corners do not come

together within the same plane. Space panel ends and edges in accordance with manufacturer's recommendations.

Materials

- **Multiple options exist for membranes, mortars, grouts, and other materials and *must be clearly specified* to be included. If not specifically indicated, optional materials are not included and mortar/grout choice defaults to minimum performance specification indicated. Consider each system component and intended use to determine minimum requirements and to specify options.**
- Ceramic tile—ANSI A137.1.
- Glass tile, when used—ANSI A137.2; see also "Glass Tile Selection and Installation Guide," and consult tile manufacturer for environmental exposure classification recommendations. Not all glass tiles are suitable.
- Cementitious grout—ANSI A118.6 or better or ISO CG1 or better. When glass tile is used, specify grout designated by tile and grout manufacturers.
- Epoxy grout, when used—ANSI A118.3 or ISO RG.
- Cementitious bond coat:
  - When a waterproof membrane is not used—ANSI A118.1 or better or ISO C1 or better.
  - When a waterproof membrane is used—ANSI A118.4 or better or ISO C2S1 or better unless ANSI A118.1 or ISO C1 is recommended by membrane manufacturer.
  - When porcelain tile is used—ANSI A118.4 or better or ISO C2 or better.
  - When glass tile is used, specify mortar designated by tile and mortar manufacturers. Bond coat color will impact the final appearance of translucent glass tile. Specifier shall confirm bond coat color is acceptable.
- Epoxy bond coat, when used—ANSI A118.3 or ISO R1 or better.
  - When glass tile is used, specify epoxy bond coat designated by tile and bond coat manufacturers. Bond coat color will impact the final appearance of translucent glass tile. Specifier shall confirm bond coat color is acceptable.
- Organic adhesive bond coat, when used—ANSI A136.1 Type I or ISO D2.
- Vapor retarder membrane, when used—ANSI A108.02-3.8.
- Waterproof membrane, when used—ANSI A118.10.
- Cement backer board, when used—ANSI A118.9 or ASTM C1325 (Type B).
- Fiber cement backer board, when used—ASTM C1288.
- Fasteners—noncorrosive and nonoxidizing.
- Fasteners meeting ASTM F2329 required in wet areas.
- 2" alkali-resistant glass fiber mesh tape.
- Flexible mildew-resistant sealant—ASTM C920.
- Metal studs—ASTM C645.

Materials for Green/Sustainable Design

- See "Green Building Standards and Green Product Selection Guide" and consult manufacturers and suppliers for product sustainability and contribution to green building design.
- Consider specifying tile and installation materials that meet ANSI A138.1, the *American National Standard Specifications for Sustainable Ceramic Tiles, Glass Tiles, and Tile Installation Materials.*

Preparation by Other Trades

- Wall framing shall meet the general framing requirements of ANSI A108.11-4.0–4.3.
- Opening for recessed tub—not to exceed ½" more than length of tub.
- Bathtub—install level and support with metal hangers or on wood framing members.
- Fire and sound ratings—extend gypsum board required for ratings down to the floor behind the tub so that construction will be the same as the tested assembly. Backer board may be part of, or installed over, the rated assembly.

Movement Joint (architect must specify type of joint and show location and details on drawings)

- Movement joints—mandatory according to EJ171.
- When glass tile is used, adhere to more frequent placement recommendations within the ranges listed in EJ171.

Installation Specifications

- Cement backer board—ANSI A108.11.
- Fiber cement backer board—manufacturer's directions.
- Tile—ANSI A108.4 or A108.5.
- Glass tile—ANSI A108.15, .16, or manufacturer's directions.
- Cementitious grout—ANSI A108.10.
- Epoxy mortar/grout—ANSI A108.6.
- Waterproof membrane—ANSI A108.13.
- Movement Joints—EJ171 and ASTM C1193.

Notes

- As the tile size increases, there is less tolerance for variation in the substrate from the required plane. Organic adhesive or epoxy bond coat thickness must be thin and uniform; therefore, substrate flattening may be required when organic adhesive or epoxy bond coat is used.
- Do not use organic adhesives to bond tile to a membrane or other impervious substrate.
- Extended dry time may be required for larger tiles and tiles with low water absorption when organic adhesive is used.
- All horizontal surfaces, for example shower seats, sills, curbs, etc., must slope towards drain or other surface sloped toward drain. Where present, waterproofing also must be sloped.
- When glass tile is used, see "Glass Tile Selection and Installation Guide," and consult manufacturer for recommendations and requirements.

B419-21

- **Bathtub or Shower with Prefabricated Receptor**
- **Coated Glass Mat Water-Resistant Gypsum Backer Board**
- **Ceramic Tile, Glass Tile**

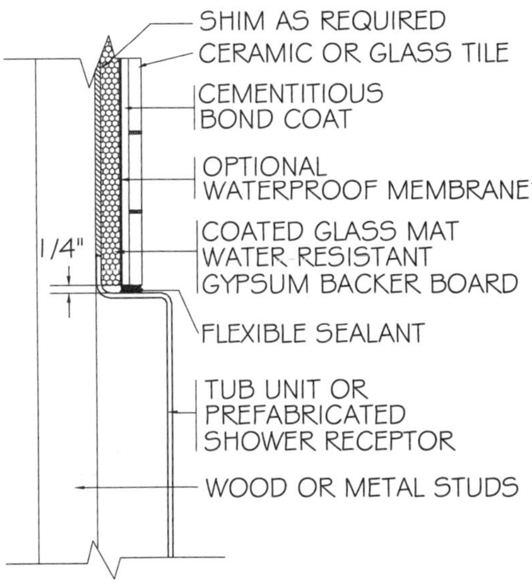

*USE OF A MEMBRANE IS OPTIONAL. SEE MEMBRANE OPTIONS.

SHOWER RECEPTORS, CURBS, SEATS, ETC., MUST BE PROPERLY WATERPROOFED AND INSTALLED TO AVOID WATER DAMAGE TO ADJACENT BUILDING MATERIALS. SEE COMMON SHOWER CONFIGURATIONS SECTION.

Recommended Uses

- For tub enclosures or showers with prefabricated receptors.
- For showers that do not have a prefabricated receptor, see B420.

Environmental Exposure Classifications

- Res1, 2, 3; Com1, 2, 3.
- For installations that may be exposed to staining, specify tile and grout suitable for exposure. Consult product manufacturers; see also "Product Selection Guides."
- For installations that may be exposed to mild chemical attack, specify epoxy grout and tile suitable for exposure. Consult product manufacturers; see also "Product Selection Guides."

Limitations

- Maximum stud spacing 16" on center.
- Not for areas exposed to temperatures exceeding 125°F.

Membrane Options

- A waterproof membrane (A118.10) may be specified to prevent moisture intrusion through seams, corners, fasteners, and other penetrations and to protect adjacent building materials. Follow backer board and membrane manufacturer's waterproofing requirements. Specifier shall indicate if complete waterproofing is required, including treatment at termination points.
- Check with membrane manufacturer for suitability for applicable conditions, as not all membranes are suitable for steam, high-temperature and/or chemical exposure, or exterior use.
- Some liquid-applied waterproof membranes will require extended cure time. Consult membrane manufacturer.

Requirements

- When glass tile is used, consult glass tile manufacturer for tile suitability over non-absorptive surface.
- Wood studs—dry and well-braced, minimum depth 3½".
- Metal studs—well-braced; 20 gauge (0.033") or heavier; minimum depth 3½" for residential applications or 3⅝" for commercial applications.
- Caulk or seal penetrations, corners, and abutments to dissimilar materials with flexible sealant.

Preparation by Backer Board Installers

- Maximum allowable variation in the tile substrate—for tiles with all edges shorter than 15", maximum allowable variation is ¼" in 10' from the required plane, with no more than ¹⁄₁₆" variation in 12" when measured from the high points in the surface. For tiles with at least one edge 15" in length, maximum allowable variation is ⅛" in 10' from the required plane, with no more than ¹⁄₁₆" variation in 24" when measured from the high points in the surface.
- Fit ends and edges closely to eliminate gaps between panels, including in corners.
- Tape seams and corners with 2" alkali resistant glass fiber mesh tape and cementitious bonding material, and skim coat over fastener heads, unless waterproofing has been applied.
- Center backer board end or edge joints on framing and stagger joints in adjacent rows so four corners do not come together within the same plane. Space panel ends and edges in accordance with manufacturer's recommendations.

Materials

- **Multiple options exist for membranes, mortars, grouts, and other materials and *must be clearly specified* to be included. If not specifically indicated, optional materials are not included and mortar/grout choice defaults to minimum performance specification indicated. Consider each system component and intended use to determine minimum requirements and to specify options.**
- Ceramic tile—ANSI A137.1.

- Glass tile, when used—ANSI A137.2; see also "Glass Tile Selection and Installation Guide," and consult tile manufacturer for environmental exposure classification recommendations. Not all glass tiles are suitable.
- Cementitious grout—ANSI A118.6 or better or ISO CG1 or better. When glass tile is used, specify grout designated by tile and grout manufacturers.
- Epoxy grout, when used—ANSI A118.3 or ISO RG.
- Cementitious bond coat:
  - When a waterproof membrane is not used—ANSI A118.4 or better or ISO C2S1 or better.
  - When a waterproof membrane is used—ANSI A118.4 or better or ISO C2S1 or better unless ANSI A118.1 or ISO C1 is recommended by membrane manufacturer.
  - When glass tile is used, specify mortar designated by tile and mortar manufacturers. Bond coat color will impact the final appearance of translucent glass tile. Specifier shall confirm bond coat color is acceptable.
- Waterproof membrane, when used—ANSI A118.10.
- Coated glass mat, water-resistant gypsum backer board—ASTM C1178.
- Fasteners—noncorrosive and nonoxidizing.
- Fasteners meeting ASTM F2329 required in wet areas.
- 2" alkali-resistant glass fiber mesh tape.
- Flexible mildew-resistant sealant—ASTM C920.
- Metal studs—ASTM C645.

Materials for Green/Sustainable Design

- See "Green Building Standards and Green Product Selection Guide" and consult manufacturers and suppliers for product sustainability and contribution to green building design.
- Consider specifying tile and installation materials that meet ANSI A138.1, the *American National Standard Specifications for Sustainable Ceramic Tiles, Glass Tiles, and Tile Installation Materials.*

Preparation by Other Trades

- Wall framing shall meet the general framing requirements of ANSI A108.11-4.0–4.3.
- Opening for recessed tub not to exceed ½" more than length of tub.
- Bathtub—install level and supported with metal hangers or on wood framing members.
- Fire and sound ratings—extend gypsum board required for rating down to the floor behind the tub so that construction will be the same as the tested assembly.

Movement Joint (architect must specify type of joint and show location and details on drawings)

- Movement joints—mandatory according to EJ171.
- When glass tile is used, adhere to more frequent placement recommendations within the ranges listed in EJ171.

Installation Specifications

- Coated glass mat water-resistant gypsum backer board—manufacturer's directions.
- Tile—ANSI A108.5.
- Glass tile—manufacturer's directions.
- Cementitious grout—ANSI A108.10.
- Epoxy grout—ANSI A108.6.
- Waterproof membrane—ANSI A108.13.
- Movement Joints—EJ171 and ASTM C1193.

Notes

- Do not install a vapor barrier behind the tile backer board.
- All horizontal surfaces, for example shower seats, sills, curbs, etc., must slope towards drain or other surface sloped toward drain. Where present, waterproofing also must be sloped.
- When glass tile is used, see "Glass Tile Selection and Installation Guide," and consult manufacturer for recommendations and requirements.

B425-21

- **Bathtub or Shower with Prefabricated Receptor**
- **Cementitious-Coated Extruded Foam Backer Board**
- **Ceramic Tile, Glass Tile**

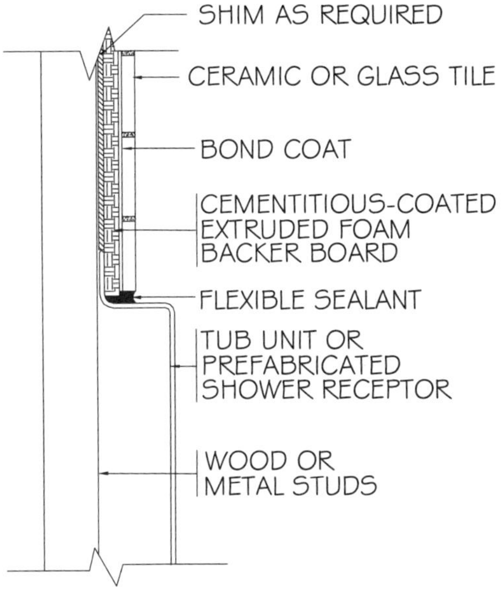

SHOWER RECEPTORS, CURBS, SEATS, ETC., MUST BE PROPERLY WATERPROOFED AND INSTALLED TO AVOID WATER DAMAGE TO ADJACENT BUILDING MATERIALS. SEE COMMON SHOWER CONFIGURATIONS SECTION.

Recommended Uses

- For tub enclosures or showers with prefabricated receptors.
- For showers that do not have a prefabricated receptor, see B426.

Environmental Exposure Classifications

- Res1, 2, 3; Com1, 2, 3.
- For installations that may be exposed to staining, specify tile and grout suitable for exposure. Consult product manufacturers; see also "Product Selection Guides."
- For installations that may be exposed to mild chemical attack, specify epoxy grout and tile suitable for exposure. For greater resistance to chemical exposure, also specify an epoxy bonding material. Consult product manufacturers; see also "Product Selection Guides."

Limitations

- Maximum stud spacing 16" on center.

Requirements

- When glass tile is used, consult glass tile manufacturer for recommendations over sealant (when used).
- Wood studs—dry and well-braced, minimum depth 3½."
- Metal studs—well-braced; 20 gauge (0.033") or heavier; minimum depth 3½" for residential applications or 3⅝" for commercial applications.

Preparation by Backer Board Installers

- Fasten backer board to studs with board manufacturer's recommended fasteners.
- All openings cut in backer board for plumbing and all cut joints between adjoining pieces—seal as recommended by backer board manufacturer.
- Maximum allowable variation in the tile substrate—for tiles with all edges shorter than 15," maximum allowable variation is ¼" in 10' from the required plane, with no more than 1/16" variation in 12" when measured from the high points in the surface. For tiles with at least one edge 15" in length, maximum allowable variation is ⅛" in 10' from the required plane, with no more than 1/16" variation in 24" when measured from the high points in the surface.
- Fit ends and edges tightly together, including in corners. Apply a bead of manufacturer-recommended sealant in panel joints as panels are being installed and over fastener heads per backer board manufacturer's instructions.
- Apply backer board tight to lip of tub and seal to tub with manufacturer-recommended sealant.
- Center backer board end or edge joints on framing and stagger joints in adjacent rows so four corners do not come together within the same plane. Space panel ends and edges in accordance with manufacturer's recommendations.

Materials

- **Multiple options exist for membranes, mortars, grouts, and other materials and *must be clearly specified* to be included. If not specifically indicated, optional materials are not included and mortar/grout choice defaults to minimum performance specification indicated. Consider each system component and intended use to determine minimum requirements and to specify options.**
- Ceramic tile—ANSI A137.1.
- Glass tile, when used—ANSI A137.2; see also "Glass Tile Selection and Installation Guide," and consult tile manufacturer for environmental exposure classification recommendations. Not all glass tiles are suitable.
- Cementitious grout—ANSI A118.6 or better or ISO CG1 or better. When glass tile is used, specify grout designated by tile and grout manufacturers.
- Epoxy grout, when used—ANSI A118.3 or ISO RG.
- Cementitious bond coat:
  - ANSI A118.1 or better or ISO C1 or better.
  - When porcelain tile is used—ANSI A118.4 or better or ISO C2 or better.

    - When glass tile is used, specify mortar designated by tile and mortar manufacturers. Bond coat color will impact the final appearance of translucent glass tile. Specifier shall confirm bond coat color is acceptable.
- Epoxy bond coat, when used—ANSI A118.3 or ISO R1 or better.
    - When glass tile is used, specify epoxy bond coat designated by tile and bond coat manufacturers. Bond coat color will impact the final appearance of translucent glass tile. Specifier shall confirm bond coat color is acceptable.
- Cementitious-coated extruded foam backer board—ASTM C578.
- Fasteners—noncorrosive and nonoxidizing.
- Flexible sealant—must be recommended by backer board manufacturer.
- Metal studs—ASTM C645.

Materials for Green/Sustainable Design

- See "Green Building Standards and Green Product Selection Guide" and consult manufacturers and suppliers for product sustainability and contribution to green building design.
- Consider specifying tile and installation materials that meet ANSI A138.1, the *American National Standard Specifications for Sustainable Ceramic Tiles, Glass Tiles, and Tile Installation Materials.*

Preparation by Other Trades

- Wall framing shall meet the general framing requirements of ANSI A108.11-4.0–4.3.
- Bathtub—install level and support with metal hangers or on wood framing members.
- Opening for recessed tub not to exceed ½" more than length of tub.

Movement Joint (architect must specify type of joint and show location and details on drawings)

- Movement joints—mandatory according to EJ171.
- When glass tile is used, adhere to more frequent placement recommendations within the ranges listed in EJ171.

Installation Specifications

- Cementitious-coated extruded foam backer board—manufacturer's directions.
- Tile—ANSI A108.5.
- Glass tile—manufacturer's directions.
- Cementitious grout—ANSI A108.10.
- Epoxy mortar/grout—ANSI A108.6.
- Movement Joints—EJ171 and ASTM C1193.

Notes

- All horizontal surfaces, for example shower seats, sills, curbs, etc., must slope towards drain or other surface sloped toward drain. Where present, waterproofing also must be sloped.
- As the tile size increases, there is less tolerance for variation in the substrate from the required plane. Epoxy adhesive thickness must be thin and uniform; therefore, substrate flattening may be required when epoxy bond coat is used.
- When glass tile is used, see "Glass Tile Selection and Installation Guide," and consult manufacturer for recommendations and requirements.

B430-21

- **Bathtub or Shower with Prefabricated Receptor**
- **Fiber-Reinforced Water-Resistant Gypsum Backer Board**
- **Ceramic Tile, Glass Tile**

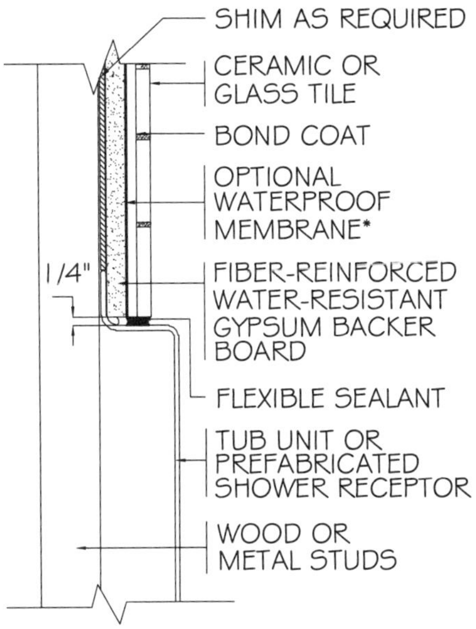

*USE OF A MEMBRANE IS OPTIONAL. SEE MEMBRANE OPTIONS.

SHOWER RECEPTORS, CURBS, SEATS, ETC., MUST BE PROPERLY WATERPROOFED AND INSTALLED TO AVOID WATER DAMAGE TO ADJACENT BUILDING MATERIALS. SEE COMMON SHOWER CONFIGURATIONS SECTION.

Recommended Uses

- For tub enclosures or showers with prefabricated receptors.
- For showers that do not have a prefabricated receptor, see B431.

Environmental Exposure Classifications

- Res1, 2, 3; Com1, 2.
- For installations that may be exposed to staining, specify tile and grout suitable for exposure. Consult product manufacturers; see also "Product Selection Guides."
- For installations that may be exposed to mild chemical attack, specify epoxy grout and tile suitable for exposure. Consult product manufacturers; see also "Product Selection Guides."

Limitations

- Maximum stud spacing 16" on center.
- Not for areas exposed to temperatures exceeding 125°F.
- When organic adhesive is used—maximum tile size 8" × 8" unless organic adhesive manufacturer allows larger tile size. See Notes.

Membrane Options

- A waterproof membrane (A118.10) may be specified to prevent moisture intrusion and protect adjacent building materials. Specifier shall indicate if complete waterproofing is required, including treatment at termination points.
- Check with membrane manufacturer for suitability for applicable conditions, as not all membranes are suitable for steam, high-temperature and/or chemical exposure, or exterior use.
- When glass tile is used, consult glass tile manufacturer for membrane options and recommendations.

Requirements

- Wood studs—dry and well-braced, minimum depth 3½."
- Metal studs—well-braced; 20 gauge (0.033") or heavier; minimum depth 3½" for residential applications or 3⅝" for commercial applications.
- Caulk or seal penetrations and abutments to dissimilar materials.

Preparation by Backer Board Installers

- Maximum allowable variation in the tile substrate—for tiles with all edges shorter than 15," maximum allowable variation is ¼" in 10' from the required plane, with no more than ¹⁄₁₆" variation in 12" when measured from the high points in the surface. For tiles with at least one edge 15" in length, maximum allowable variation is ⅛" in 10' from the required plane, with no more than ¹⁄₁₆" variation in 24" when measured from the high points in the surface.
- Horizontal joints—⅛" spacing filled solid and taped with latex-portland cement mortar and 2" alkali-resistant glass fiber mesh tape.
- Vertical joints—fill any space and tape with latex-portland cement mortar and 2" alkali-resistant glass fiber mesh tape.
- Corners—leave space between backer units. Tape joints using skim coat of latex-portland cement mortar, but do not fill.
- Center backer board end or edge joints on framing and stagger joints in adjacent rows so four corners do not come together within the same plane. Space panel ends and edges in accordance with manufacturer's recommendations.

Materials

- **Multiple options exist for membranes, mortars, grouts, and other materials and *must be clearly specified* to be included. If not specifically indicated, optional materials are not included and mortar/grout choice defaults to minimum performance specification indicated. Consider each system component and**

intended use to determine minimum requirements and to specify options.

- Ceramic tile—ANSI A137.1.
- Glass tile, when used—ANSI A137.2; see also "Glass Tile Selection and Installation Guide," and consult tile manufacturer for environmental exposure classification recommendations. Not all glass tiles are suitable.
- Cementitious grout—ANSI A118.6 or better or ISO CG1 or better. When glass tile is used, specify grout designated by tile and grout manufacturers.
- Epoxy grout, when used—ANSI A118.3 or ISO RG.
- Cementitious bond coat:
  - When a waterproof membrane is not used—ANSI A118.4 or better or ISO C2 or better.
  - When a waterproof membrane is used—ANSI A118.4 or better or ISO C2S1 or better unless ANSI A118.1 or ISO C1 is recommended by membrane manufacturer.
  - When glass tile is used, specify mortar designated by tile and mortar manufacturers. Bond coat color will impact the final appearance of translucent glass tile. Specifier shall confirm bond coat color is acceptable.
- Organic adhesive bond coat, when used—ANSI A136.1 Type I or ISO D2.
- Waterproof membrane, when used—ANSI A118.10.
- Fiber-reinforced water-resistant gypsum backer board—ASTM C1278 (Paragraph 6.1).
- Fasteners—noncorrosive and nonoxidizing.
- 2" alkali-resistant glass fiber mesh tape.
- Flexible mildew-resistant sealant—ASTM C920.
- Metal studs—ASTM C645.

Materials for Green/Sustainable Design

- See "Green Building Standards and Green Product Selection Guide" and consult manufacturers and suppliers for product sustainability and contribution to green building design.
- Consider specifying tile and installation materials that meet ANSI A138.1, the *American National Standard Specifications for Sustainable Ceramic Tiles, Glass Tiles, and Tile Installation Materials.*

Preparation by Other Trades

- Wall framing shall meet the general framing requirements of ANSI A108.11-4.0–4.3.
- Opening for recessed tub not to exceed ½" more than the tub length.
- Bathtub—install level and supported with metal hangers or on wood framing members.
- Fire and sound ratings—extend gypsum board required for rating down to the floor behind the tub so that construction will be the same as the tested assembly.

Movement Joint (architect must specify type of joint and show location and details on drawings)

- Movement joints—mandatory in accordance with EJ171.
- When glass tile is used, adhere to more frequent placement recommendations within the ranges listed in EJ171.

Installation Specifications

- Fiber-reinforced water-resistant gypsum backer board—manufacturer's directions.
- Tile—ANSI A108.4 or A108.5.
- Glass tile—manufacturer's directions.
- Cementitious grout—ANSI A108.10.
- Epoxy grout—ANSI A108.6.
- Waterproof membrane—ANSI A108.13.
- Movement Joints—EJ171 and ASTM C1193.

Notes

- As the tile size increases, there is less tolerance for variation in the substrate from the required plane. Organic adhesive thickness must be thin and uniform; therefore, substrate flattening may be required when organic adhesive is used.
- Do not use organic adhesives to bond tile to a membrane or other impervious substrate.
- Extended dry time may be required for larger tiles and tiles with low water absorption when organic adhesive is used.
- All horizontal surfaces, for example shower seats, sills, curbs, etc., must slope towards drain or other surface sloped toward drain. Where present, waterproofing also must be sloped.
- When glass tile is used, see "Glass Tile Selection and Installation Guide," and consult manufacturer for recommendations and requirements.

B414-21

- **Wood or Metal Studs**
- **Mortar Bed Walls**
- **Mortar Bed Floor**
- **Ceramic Tile, Glass Tile**

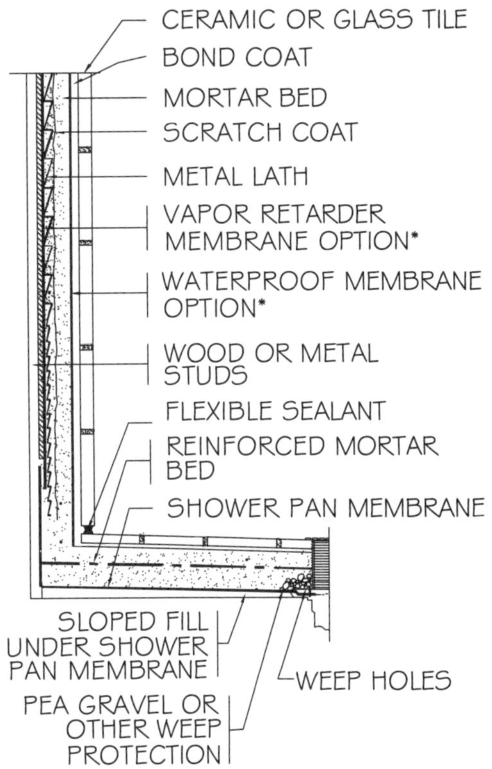

*USE OF A MEMBRANE ON WALLS IS REQUIRED. SEE MEMBRANE OPTIONS.

SHOWER RECEPTORS, CURBS, SEATS, ETC., MUST BE PROPERLY WATERPROOFED AND INSTALLED TO AVOID WATER DAMAGE TO ADJACENT BUILDING MATERIALS. SEE COMMON SHOWER CONFIGURATIONS SECTION.

WALL MEMBRANE LAPS OVER SHOWER PAN.

Recommended Uses

- For showers that do not have prefabricated receptors.
- For areas where wall flatness is critical, such as when tiles with any edge longer than 15" are specified.

Environmental Exposure Classifications

- Res1, 2, 3, 5; Com1, 2, 3, 5.
- For Res4 and Com4, see SR614.
- For installations that may be exposed to staining, specify tile and grout suitable for exposure. Consult product manufacturers; see also "Product Selection Guides."
- For installations that may be exposed to mild chemical attack, specify epoxy grout and tile suitable for exposure. For greater resistance to chemical exposure, also specify an epoxy bonding material. Consult product manufacturers; see also "Product Selection Guides."

Limitations

- Maximum mortar bed thickness (walls)—1" for metal studs, 1½" for wood studs.
- Maximum stud spacing 16" on center.

Membrane Options

- A waterproof membrane (A118.10) or vapor retarder membrane (A108.02-3.8) must be specified for walls to prevent moisture intrusion and protect adjacent building materials. Specifier shall indicate if complete waterproofing of walls is required, including treatment at termination points.
- If a waterproof membrane is applied over the mortar bed walls, membrane over framing members may not be required.
- Check with membrane manufacturer for suitability for applicable conditions, as not all membranes are suitable for steam, high-temperature and/or chemical exposure, or exterior use.
- When glass tile is used, consult glass tile manufacturer for membrane options and recommendations.

Requirements

- Wood studs—dry and well-braced, minimum depth 3½".
- Metal studs—well-braced; 20 gauge (0.033") or heavier; minimum depth 3½" for residential applications or 3⅝" for commercial applications.
- Mortar bed thickness—¾" minimum to 1" maximum (metal studs) or 1½" maximum (wood studs)
- Membrane behind mortar bed, when used, must lap over shower pan membrane.
- Slope shower pan membrane ¼" per foot to weep holes in drain.
- Turn shower pan membrane up walls a minimum of 3" above shower curb (6" above floor in showers without curbs).
- Surround drain with pea gravel or other weep protection to prevent mortar from blocking weep holes.

Materials

- **Multiple options exist for membranes, mortars, grouts, and other materials and *must be clearly specified* to be included. If not specifically indicated, optional materials are not included and mortar/grout choice defaults to minimum performance specification indicated. Consider each system component and intended use to determine minimum requirements and to specify options.**
- Ceramic tile—ANSI A137.1.
- Glass tile, when used—ANSI A137.2; see also "Glass Tile Selection and Installation Guide," and consult tile manufacturer for environmental exposure classification

recommendations. Not all glass tiles are suitable.

- Cementitious grout—ANSI A118.6 or better or ISO CG1 or better. When glass tile is used, specify grout designated by tile and grout manufacturers.
- Epoxy grout, when used—ANSI A118.3 or ISO RG.
- Cementitious bond coat—portland cement paste on a mortar bed that is still workable. For a cured mortar bed, follow recommendations below to select appropriate bonding mortar:
  - When a waterproof membrane is not used—ANSI A118.1 or better or ISO C1 or better.
  - When a waterproof membrane is used—ANSI A118.4 or better or ISO C2S1 or better unless ANSI A118.1 or ISO C1 is recommended by membrane manufacturer.
  - When porcelain tile is used—ANSI A118.4 or better or ISO C2 or better.
  - When glass tile is used, specify mortar designated by tile and mortar manufacturers. Bond coat color will impact the final appearance of translucent glass tile. Specifier shall confirm bond coat color is acceptable.
- Epoxy bond coat, when used—ANSI A118.3 or ISO R1 or better.
  - When glass tile is used, specify epoxy bond coat designated by tile and bond coat manufacturers. Bond coat color will impact the final appearance of translucent glass tile. Specifier shall confirm bond coat color is acceptable.
- Vapor retarder membrane, when used—ANSI A108.02-3.8.
- Waterproof membrane, when used—ANSI A118.10.
- Mortar bed, metal lath, and cleavage membrane—ANSI A108.1A.
- Flexible mildew-resistant sealant—ASTM C920.
- Shower pan membrane—local building code.
- Metal studs—ASTM C645.

Materials for Green/Sustainable Design

- See "Green Building Standards and Green Product Selection Guide" and consult manufacturers and suppliers for product sustainability and contribution to green building design.
- Consider specifying tile and installation materials that meet ANSI A138.1, the *American National Standard Specifications for Sustainable Ceramic Tiles, Glass Tiles, and Tile Installation Materials.*

Preparation by Other Trades

- Wall framing shall meet the general framing requirements of ANSI A108.11-4.0–4.3.
- Apply blocking between the studs to support the shower pan membrane.

Movement Joint (architect must specify type of joint and show location and details on drawings)

- Movement joints—mandatory according to EJ171.
- When glass tile is used, adhere to more frequent placement recommendations within the ranges listed in EJ171.

Installation Specifications

- Shower pan membrane—ANSI A108.01-3.6.
- Tile—ANSI A108.1A, .1B, or .1C. A108.1B required if waterproof membrane or epoxy bond coat to be used.
- Glass tile—ANSI A108.14, .15, .16, or manufacturer's directions.
- Cementitious grout—ANSI A108.10.
- Epoxy mortar/grout—ANSI A108.6.
- Waterproof membrane—ANSI A108.13.
- Movement Joints—EJ171 and ASTM C1193.

Notes

- Test shower pan membrane and drainage fitting for leaks before commencing tilework.
- Materials adversely affected by moisture in areas immediately adjacent to showers, tubs, and roman tubs should be properly protected.
- A sloped portland cement mortar fill or approved preformed slope may be used under shower pan membrane when subfloor is not sloped to drain.
- All horizontal surfaces, for example shower seats, sills, curbs, etc., must slope towards drain or other surface sloped toward drain. Where present, waterproofing also must be sloped.
- When glass tile is used, see "Glass Tile Selection and Installation Guide," and consult manufacturer for recommendations and requirements.
- For curbless shower receptor, see B421C and B422C.

B441-21

- **Wood or Metal Studs**
- **Backer Board**
- **Mortar Bed Walls (One Coat Method)**
- **Mortar Bed Floor**
- **Ceramic Tile**

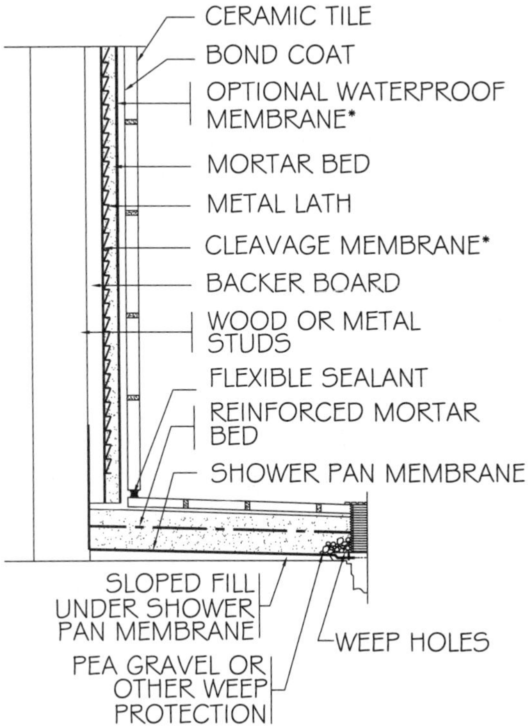

*USE OF A MEMBRANE ON WALLS IS REQUIRED. SEE MEMBRANE OPTIONS.

SHOWER RECEPTORS, CURBS, SEATS, ETC., MUST BE PROPERLY WATERPROOFED AND INSTALLED TO AVOID WATER DAMAGE TO ADJACENT BUILDING MATERIALS. SEE COMMON SHOWER CONFIGURATIONS SECTION.

CLEAVAGE MEMBRANE LAPS OVER SHOWER PAN.

Recommended Uses

- For showers that do not have prefabricated receptors, where minor flattening or trueing of walls is required.
- For areas where wall flatness is critical, such as when tiles with any edge longer than 15" are specified.

Environmental Exposure Classifications

- With cement backer board or fiber cement backer board—Res1, 2, 3, 5; Com1, 2, 3, 5.
- With coated glass mat water-resistant gypsum backer board or cementitious-coated extruded foam backer board—Res1, 2, 3; Com1, 2, 3.
- With fiber-reinforced water-resistant gypsum backer board—Res 1, 2, 3; Com 1, 2.
- For installations that may be exposed to staining, specify tile and grout suitable for exposure. Consult product manufacturers; see also "Product Selection Guides."
- For installations that may be exposed to mild chemical attack, specify epoxy grout and tile suitable for exposure. For greater resistance to chemical exposure, also specify an epoxy bonding material. Consult product manufacturers; see also "Product Selection Guides."

Limitations

- Maximum mortar bed thickness (walls)—¾."
- Maximum stud spacing 16" on center.
- When coated glass mat water-resistant gypsum backer board or fiber-reinforced water-resistant gypsum backer board is used—not for areas exposed to temperatures exceeding 125°F.

Membrane Options

- A cleavage membrane is required. A waterproof membrane (A118.10) may also be specified to prevent moisture intrusion and protect adjacent building materials. Specifier shall indicate if complete waterproofing is required, including treatment at termination points.
- Check with membrane manufacturer for suitability for applicable conditions, as not all membranes are suitable for steam, high-temperature and/or chemical exposure, or exterior use.

Requirements

- Cleavage membrane must lap over shower pan membrane.
- Wood studs—dry and well-braced, minimum depth 3½."
- Metal studs—well-braced; 20 gauge (0.033") or heavier; minimum depth 3½" for residential applications or 3⅝" for commercial applications.
- Mortar bed thickness—⅜" minimum to ¾" maximum.
- Backer board must provide firm anchorage for metal lath.
- Cut lath at all movement joints.
- Slope shower pan membrane ¼" per foot to weep holes in drain.
- Turn shower pan membrane up walls a minimum of 3" above shower curb (6" above floor in showers without curbs).
- Surround drain with pea gravel or other weep protection to prevent mortar from blocking weep holes.

Preparation by Backer Board Installers

- Maximum allowable variation in the backer board—⅛" in 10' from the required plane.
- Follow backer board manufacturer's directions for installation, including orientation and spacing of boards, required fasteners, taping of joints and corners, and special instructions for wet areas.

Materials

- **Multiple options exist for membranes, mortars, grouts, and other materials and *must be clearly specified* to be included. If not specifically indicated,**

optional materials are not included and mortar/grout choice defaults to minimum performance specification indicated. Consider each system component and intended use to determine minimum requirements and to specify options.

- Ceramic tile—ANSI A137.1.
- Cementitious grout—ANSI A118.6 or better or ISO CG1 or better.
- Epoxy grout, when used—ANSI A118.3 or ISO RG.
- Cementitious bond coat—portland cement paste on a mortar bed that is still workable. For a cured mortar bed, follow recommendations below to select appropriate bonding mortar:
  - When a waterproof membrane is not used—ANSI A118.1 or better or ISO C1 or better.
  - When a waterproof membrane is used—ANSI A118.4 or better or ISO C2S1 or better unless ANSI A118.1 or ISO C1 is recommended by membrane manufacturer.
  - When porcelain tile is used—ANSI A118.4 or better or ISO C2 or better.
- Epoxy bond coat, when used—ANSI A118.3 or ISO R1 or better.
- Waterproof membrane, when used—ANSI A118.10.
- Mortar bed, metal lath, and cleavage membrane—ANSI A108.1A.
- Cement backer board, when used—ANSI A118.9 or ASTM C1325 (Type B).
- Fiber cement backer board, when used—ASTM C1288.
- Coated glass mat, water-resistant gypsum backer board, when used—ASTM C1178.
- Cementitious-coated extruded foam backer board, when used—ASTM C578.
- Fiber-reinforced water-resistant gypsum backer board, when used—ASTM C1278 (Paragraph 6.1).
- Fasteners—noncorrosive and nonoxidizing.
- Fasteners meeting ASTM F2329 required in wet areas.
- 2" alkali-resistant glass fiber mesh tape.
- Flexible mildew-resistant sealant—ASTM C920.
- Shower pan membrane—local building code.
- Metal studs—ASTM C645.

Materials for Green/Sustainable Design

- See "Green Building Standards and Green Product Selection Guide" and consult manufacturers and suppliers for product sustainability and contribution to green building design.
- Consider specifying tile and installation materials that meet ANSI A138.1, the *American National Standard Specifications for Sustainable Ceramic Tiles, Glass Tiles, and Tile Installation Materials.*

Preparation by Other Trades

- Wall framing shall meet the general framing requirements of ANSI A108.11-4.0–4.3.
- Apply blocking between the studs to support the shower pan membrane.

Movement Joint (architect must specify type of joint and show location and details on drawings)

- Movement joints—mandatory according to EJ171.

Installation Specifications

- Tile—ANSI A108.1A, .1B, or .1C. A108.1B required if waterproof membrane or epoxy bond coat to be used.
- Cementitious grout—ANSI A108.10.
- Epoxy mortar/grout—ANSI A108.6.
- Waterproof membrane—ANSI A108.13.
- Cement backer board—ANSI A108.11.
- Fiber cement backer board—manufacturer's directions.
- Coated glass mat water-resistant gypsum backer board—manufacturer's directions.
- Cementitious-coated extruded foam backer board—manufacturer's directions.
- Fiber-reinforced water-resistant gypsum backer board—manufacturer's directions.
- Movement Joints—EJ171 and ASTM C1193.
- Shower pan membrane—ANSI A108.01-3.6.

Notes

- All horizontal surfaces, for example shower seats, sills, curbs, etc., must slope towards drain or other surface sloped toward drain. Where present, waterproofing also must be sloped.
- When coated glass mat water-resistant gypsum backer board is used, do not install a vapor barrier behind the backer board.
- If lath cannot be attached directly to backing, use furring strips. See B414.
- Test shower pan membrane and drainage fitting for leaks before commencing tilework.
- Materials adversely affected by moisture in areas immediately adjacent to showers, tubs, and roman tubs should be properly protected.
- A sloped portland cement mortar fill or approved preformed slope may be used under shower pan membrane when subfloor is not sloped to drain.
- For thicker mortar bed, see B414.
- For mortar bed directly over wood or metal studs, see B414.
- For curbless shower receptor, see B421C and B422C.

B415-21

- **Wood or Metal Studs**
- **Cement Backer Board or Fiber-Cement Backer Board Walls**
- **Mortar Bed Floor**
- **Ceramic Tile, Glass Tile**

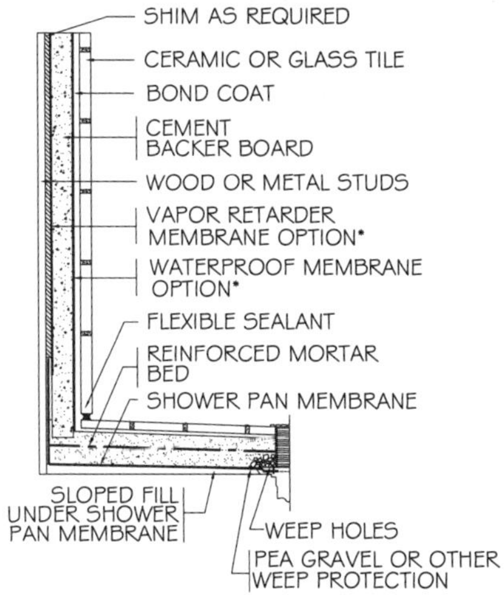

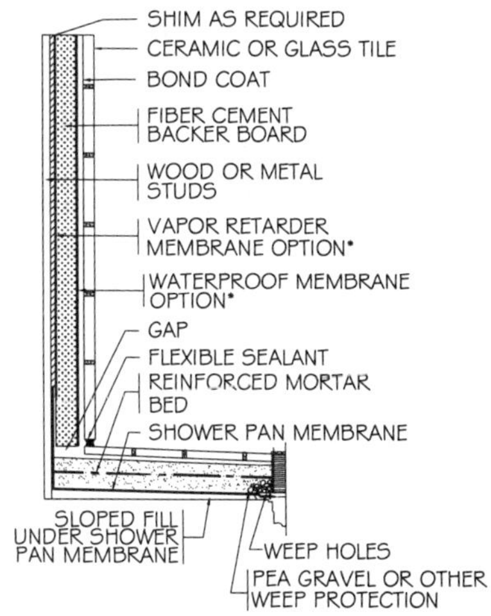

*USE OF A MEMBRANE ON WALLS IS REQUIRED. SEE MEMBRANE OPTIONS. SHOWER RECEPTORS, CURBS, SEATS, ETC., MUST BE PROPERLY WATERPROOFED AND INSTALLED TO AVOID WATER DAMAGE TO ADJACENT BUILDING MATERIALS. SEE COMMON SHOWER CONFIGURATIONS SECTION.

Recommended Uses

- For showers that do not have prefabricated receptors.

Environmental Exposure Classifications

- With cementitious or epoxy bond coat—Res1, 2, 3, 5; Com1, 2, 3, 5.
- With organic adhesive bond coat—Res1, 2, 3; Com1, 2.
- For Res4 and Com4, see SR614.
- For installations that may be exposed to staining, specify tile and grout suitable for exposure. Consult product manufacturers; see also "Product Selection Guides."
- For installations that may be exposed to mild chemical attack, specify epoxy grout and tile suitable for exposure. For greater resistance to chemical exposure, also specify an epoxy bonding material. Consult product manufacturers; see also "Product Selection Guides."

Limitations

- Maximum stud spacing 16" on center.
- When organic adhesive is used—not for areas exposed to temperatures exceeding 140°F.
- When organic adhesive is used—maximum tile size 8" × 8" unless organic adhesive manufacturer allows larger tile size. See Notes.
- Organic adhesive may be used on walls only; do not use on floor or curb.

Membrane Options

- A waterproof membrane (A118.10) or vapor retarder membrane (A108.02-3.8) must be specified for walls to prevent moisture intrusion and protect adjacent building materials. Specifier shall indicate if complete waterproofing of walls is required, including treatment at termination points.
- Check with membrane manufacturer for suitability for applicable conditions, as not all membranes are suitable for steam, high-temperature and/or chemical exposure, or exterior use.
- When glass tile is used, consult glass tile manufacturer for membrane options and recommendations.

Requirements

- Wood studs—dry and well-braced, minimum depth 3½."
- Metal studs—well-braced; 20 gauge (0.033") or heavier; minimum depth 3½" for residential applications or 3⅝" for commercial applications.
- Surface of units—clean and free of dirt, dust, paint, and oily film.
- Membrane behind backer board, when used, must lap over shower pan membrane.
- Slope shower pan membrane ¼" per foot to weep holes in drain.
- Turn shower pan membrane up walls a minimum of 3" above shower curb (6" above floor in showers without curbs).
- Fur out studs above shower pan membrane or notch-out studs behind the shower pan membrane so folds/corners of shower pan membrane do not cause backer board to bow inward, or use Alternate Receptor Base Method.
- Surround drain with pea gravel or other weep protection to prevent mortar from blocking weep holes.

Preparation by Backer Board Installers

- Maximum allowable variation in the tile substrate—for tiles with all edges shorter than 15," maximum allowable variation is ¼" in 10' from the required plane, with no more than 1/16" variation in 12" when measured from the high points in the surface. For tiles with at least one edge 15" in length, maximum allowable variation is ⅛" in 10' from the required plane, with no more than 1/16" variation in 24" when measured from the high points in the surface.
- Horizontal joints—⅛" spacing filled solid and taped with latex-portland cement mortar and 2" alkali-resistant glass fiber mesh tape.

- Vertical joints—fill any space and tape with latex-portland cement mortar and 2" alkali-resistant glass fiber mesh tape.
- Corners—leave space between backer units. Tape joints using skim coat of latex-portland cement mortar, but do not fill.
- Center backer board end or edge joints on framing and stagger joints in adjacent rows so four corners do not come together within the same plane. Space panel ends and edges in accordance with manufacturer's recommendations.

Materials

- **Multiple options exist for membranes, mortars, grouts, and other materials and *must be clearly specified* to be included. If not specifically indicated, optional materials are not included and mortar/grout choice defaults to minimum performance specification indicated. Consider each system component and intended use to determine minimum requirements and to specify options.**
- Ceramic tile—ANSI A137.1.
- Glass tile, when used—ANSI A137.2; see also "Glass Tile Selection and Installation Guide," and consult tile manufacturer for environmental exposure classification recommendations. Not all glass tiles are suitable.
- Cementitious grout—ANSI A118.6 or better or ISO CG1 or better. When glass tile is used, specify grout designated by tile and grout manufacturers.
- Epoxy grout, when used—ANSI A118.3 or ISO RG.
- Cementitious bond coat:
  - When a waterproof membrane is not used—ANSI A118.1 or better or ISO C1 or better.
  - When a waterproof membrane is used—ANSI A118.4 or better or ISO C2S1 or better unless ANSI A118.1 or ISO C1 is recommended by membrane manufacturer.
  - When porcelain tile is used—ANSI A118.4 or better or ISO C2 or better.
  - When glass tile is used, specify mortar designated by tile and mortar manufacturers. Bond coat color will impact the final appearance of translucent glass tile. Specifier shall confirm bond coat color is acceptable.
- Epoxy bond coat, when used—ANSI A118.3 or ISO R1 or better.
  - When glass tile is used, specify epoxy bond coat designated by tile and bond coat manufacturers. Bond coat color will impact the final appearance of translucent glass tile. Specifier shall confirm bond coat color is acceptable.
- Organic adhesive bond coat, when used (walls only)—ANSI A136.1 Type I or ISO D2.
- Vapor retarder membrane, when used—ANSI A108.02-3.8.
- Waterproof membrane, when used—ANSI A118.10.
- Cement backer board, when used—ANSI A118.9 or ASTM C1325 (Type B).
- Fiber-cement backer board, when used—ASTM C1288.
- Fasteners—noncorrosive and nonoxidizing.
- Fasteners meeting ASTM F2329 required in wet areas.
- 2" alkali-resistant glass fiber mesh tape.
- Flexible mildew-resistant sealant—ASTM C920.
- Metal studs—ASTM C645.
- Shower pan membrane—local building code.

Materials for Green/Sustainable Design

- See "Green Building Standards and Green Product Selection Guide" and consult manufacturers and suppliers for product sustainability and contribution to green building design.
- Consider specifying tile and installation materials that meet ANSI A138.1, the *American National Standard Specifications for Sustainable Ceramic Tiles, Glass Tiles, and Tile Installation Materials.*

Preparation by Other Trades

- Wall framing shall meet the general framing requirements of ANSI A108.11-4.0–4.3.
- Apply blocking between the studs to support the shower pan membrane.

Movement Joint (architect must specify type of joint and show location and details on drawings)

- Movement joints—mandatory according to EJ171.
- When glass tile is used, adhere to more frequent placement recommendations within the ranges listed in EJ171.

Installation Specifications

- Shower pan membrane—ANSI A108.01-3.6.
- Fiber-cement backer board—manufacturer's directions.
- Cement backer board—ANSI A108.11.
- Tile—ANSI A108.4 or A108.5.
- Glass tile—ANSI A108.15, .16, or manufacturer's directions.
- Cementitious grout—ANSI A108.10.
- Epoxy mortar/grout—ANSI A108.6.
- Waterproof membrane—ANSI A108.13.
- Movement Joints—EJ171 and ASTM C1193.

Notes

- Test shower pan membrane and drainage fitting for leaks before commencing tilework.
- As the tile size increases, there is less tolerance for variation in the substrate from the required plane. Organic adhesive or epoxy bond coat thickness must be thin and uniform; therefore, substrate flattening may be required when organic adhesive or epoxy bond coat is used.
- Do not use organic adhesives to bond tile to a membrane or other impervious substrate.
- Extended dry time may be required for larger tiles and tiles with low water absorption when organic adhesive is used.
- Materials adversely affected by moisture in areas immediately adjacent to showers, tubs, and roman tubs should be properly protected.
- A sloped portland cement mortar fill or approved preformed slope may be used under shower pan membrane when subfloor is not sloped to drain.
- All horizontal surfaces, for example shower seats, sills, curbs, etc., must slope towards drain or other surface sloped toward drain. Where present, waterproofing also must be sloped.
- When glass tile is used, see "Glass Tile Selection and Installation Guide," and consult manufacturer for recommendations and requirements.
- For curbless shower receptor, see B421C and B422C.

B420-21

- **Wood or Metal Studs**
- **Coated Glass Mat Water-Resistant Gypsum Backer Board Walls**
- **Mortar Bed Floor**
- **Ceramic Tile, Glass Tile**

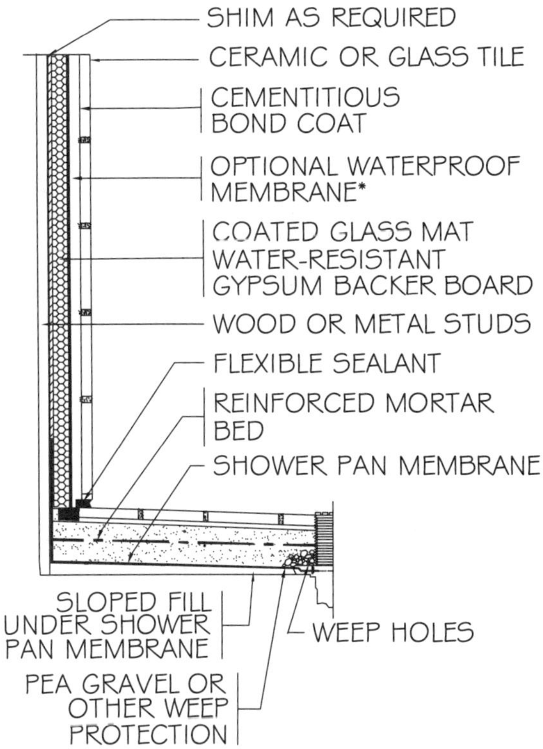

*USE OF A MEMBRANE ON WALLS IS OPTIONAL.
SEE MEMBRANE OPTIONS.

SHOWER RECEPTORS, CURBS, SEATS, ETC., MUST BE PROPERLY WATERPROOFED AND INSTALLED TO AVOID WATER DAMAGE TO ADJACENT BUILDING MATERIALS. SEE COMMON SHOWER CONFIGURATIONS SECTION.

Recommended Uses

- For showers that do not have prefabricated receptors.

Environmental Exposure Classifications

- Res1, 2, 3; Com1, 2, 3.
- For installations that may be exposed to staining, specify tile and grout suitable for exposure. Consult product manufacturers; see also "Product Selection Guides."
- For installations that may be exposed to mild chemical attack, specify epoxy grout and tile suitable for exposure. Consult product manufacturers; see also "Product Selection Guides."

Limitations

- Maximum stud spacing 16" on center.
- Not for areas exposed to temperatures exceeding 125°F.

Membrane Options

- A waterproof membrane (A118.10) may be specified for walls to prevent moisture intrusion through seams, corners, fasteners, and other penetrations and to protect adjacent building materials. Follow backer board and membrane manufacturer's waterproofing requirements. Specifier shall indicate if complete waterproofing of walls is required, including treatment at termination points.
- Check with membrane manufacturer for suitability for applicable conditions, as not all membranes are suitable for steam, high-temperature and/or chemical exposure, or exterior use.
- Some liquid-applied waterproof membranes will require extended cure time. Consult membrane manufacturer.

Requirements

- When glass tile is used, consult glass tile manufacturer for tile suitability over non-absorptive surface.
- Wood studs—dry and well-braced, minimum depth 3½".
- Metal studs—well-braced; 20 gauge (0.033") or heavier; minimum depth 3½" for residential applications or 3⅝" for commercial applications.
- Caulk or seal penetrations, corners, and abutments to dissimilar materials with flexible sealant.
- Slope shower pan membrane ¼" per foot to weep holes in drain.
- Turn shower pan membrane up walls a minimum of 3" above shower curb (6" above floor in showers without curbs).
- Fur out studs above shower pan membrane or notch-out studs behind the shower pan membrane so folds/corners of shower pan membrane do not cause backer board to bow inward, or use Alternate Receptor Base Method.
- Surround drain with pea gravel or other weep protection to prevent mortar from blocking weep holes.

Preparation by Backer Board Installers

- Maximum allowable variation in the tile substrate—for tiles with all edges shorter than 15," maximum allowable variation is ¼" in 10' from the required plane, with no more than ¹⁄₁₆" variation in 12" when measured from the high points in the surface. For tiles with at least one edge 15" in length, maximum allowable variation is ⅛" in 10' from the required plane, with no more than ¹⁄₁₆" variation in 24" when measured from the high points in the surface.
- Fit ends and edges closely to eliminate gaps between panels, including in corners.

- Tape seams and corners with 2" alkali-resistant glass fiber mesh tape and cementitious bonding material, and skim coat over fastener heads, unless waterproofing has been applied.
- Center backer board end or edge joints on framing and stagger joints in adjacent rows so four corners do not come together within the same plane. Space panel ends and edges in accordance with manufacturer's recommendations.

Materials

- **Multiple options exist for membranes, mortars, grouts, and other materials and *must be clearly specified* to be included. If not specifically indicated, optional materials are not included and mortar/grout choice defaults to minimum performance specification indicated. Consider each system component and intended use to determine minimum requirements and to specify options.**
- Ceramic tile—ANSI A137.1.
- Glass tile, when used—ANSI A137.2; see also "Glass Tile Selection and Installation Guide," and consult tile manufacturer for environmental exposure classification recommendations. Not all glass tiles are suitable.
- Cementitious grout—ANSI A118.6 or better or ISO CG1 or better. When glass tile is used, specify grout designated by tile and grout manufacturers.
- Epoxy grout, when used—ANSI A118.3 or ISO RG.
- Cementitious bond coat:
    - When a waterproof membrane is not used—ANSI A118.4 or better or ISO C2S1 or better.
    - When a waterproof membrane is used—ANSI A118.4 or better or ISO C2S1 or better unless ANSI A118.1 or ISO C1 is recommended by membrane manufacturer.
    - When glass tile is used, specify mortar designated by tile and mortar manufacturers. Bond coat color will impact the final appearance of translucent glass tile. Specifier shall confirm bond coat color is acceptable.
- Waterproof membrane, when used—ANSI A118.10.
- Coated glass mat, water-resistant gypsum backer board—ASTM C1178.
- Fasteners—noncorrosive and nonoxidizing.
- Fasteners meeting ASTM F2329 required in wet areas.
- 2" alkali-resistant glass fiber mesh tape.
- Flexible mildew-resistant sealant—ASTM C920.
- Metal studs—ASTM C645.
- Shower pan membrane—local building code.

Materials for Green/Sustainable Design

- See "Green Building Standards and Green Product Selection Guide" and consult manufacturers and suppliers for product sustainability and contribution to green building design.
- Consider specifying tile and installation materials that meet ANSI A138.1, the *American National Standard Specifications for Sustainable Ceramic Tiles, Glass Tiles, and Tile Installation Materials.*

Preparation by Other Trades

- Wall framing shall meet the general framing requirements of ANSI A108.11-4.0–4.3.
- Apply blocking between the studs to support the shower pan.

Movement Joint (architect must specify type of joint and show location and details on drawings)

- Movement joints—mandatory according to EJ171.
- When glass tile is used, adhere to more frequent placement recommendations within the ranges listed in EJ171.

Installation Specifications

- Shower pan membrane—ANSI A108.01-3.6.
- Coated glass mat water-resistant gypsum backer board—manufacturer's directions.
- Tile—A108.5.
- Glass tile—manufacturer's directions.
- Cementitious grout—ANSI A108.10.
- Epoxy grout—ANSI A108.6.
- Waterproof membrane—ANSI A108.13.
- Movement Joints—EJ171 and ASTM C1193.

Notes

- Do not install a vapor barrier behind the tile backer board.
- Test shower pan membrane and drainage fitting for leaks before commencing tilework.
- Materials adversely affected by moisture in areas immediately adjacent to showers, tubs, and roman tubs should be properly protected.
- A sloped portland cement mortar fill or approved preformed slope may be used under shower pan when subfloor is not sloped to drain.
- All horizontal surfaces, for example shower seats, sills, curbs, etc., must slope towards drain or other surface sloped toward drain. Where present, waterproofing also must be sloped.
- When glass tile is used, see "Glass Tile Selection and Installation Guide," and consult manufacturer for recommendations and requirements.
- For curbless shower receptor, see B421C and B422C.

B426-21

- **Wood or Metal Studs**
- **Cementitious-Coated Extruded Foam Backer Board Walls**
- **Mortar Bed Floor**
- **Ceramic Tile, Glass Tile**

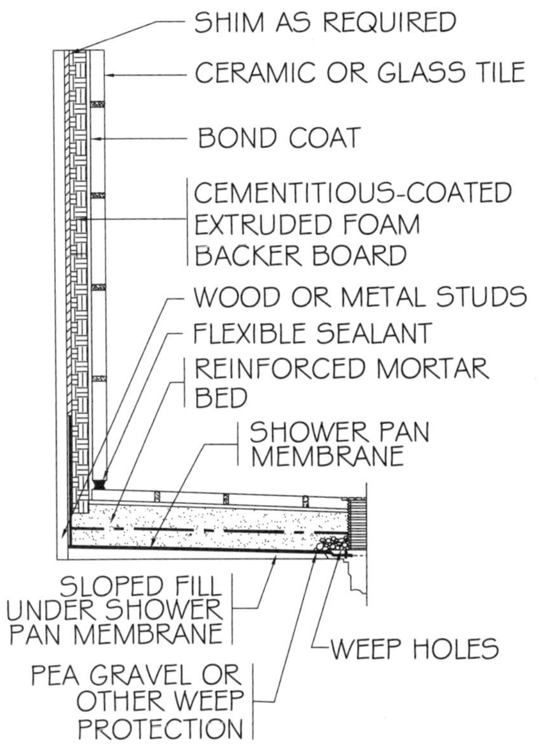

SHOWER RECEPTORS, CURBS, SEATS, ETC., MUST BE PROPERLY WATERPROOFED AND INSTALLED TO AVOID WATER DAMAGE TO ADJACENT BUILDING MATERIALS. SEE COMMON SHOWER CONFIGURATIONS SECTION.

Recommended Uses

- For showers that do not have prefabricated receptors.

Environmental Exposure Classifications

- Res1, 2, 3; Com1, 2, 3.
- For installations that may be exposed to staining, specify tile and grout suitable for exposure. Consult product manufacturers; see also "Product Selection Guides."
- For installations that may be exposed to mild chemical attack, specify epoxy grout and tile suitable for exposure. For greater resistance to chemical exposure, also specify an epoxy bonding material. Consult product manufacturers; see also "Product Selection Guides."

Limitations

- Maximum stud spacing 16" on center.

Requirements

- When glass tile is used, consult glass tile manufacturer for recommendations over sealant (when used).
- Wood studs—dry and well-braced, minimum depth 3½."
- Metal studs—well-braced; 20 gauge (0.033") or heavier; minimum depth 3½" for residential applications or 3⅝" for commercial applications.
- Slope shower pan membrane ¼" per foot to weep holes in drain.
- Turn shower pan membrane up walls a minimum of 3" above shower curb (6" above floor in showers without curbs).
- Fur out studs above shower pan membrane or notch-out studs behind the shower pan membrane so folds/corners of shower pan membrane do not cause backer board to bow inward, or use Alternate Receptor Base Method.
- Surround drain with pea gravel or other weep protection to prevent mortar from blocking weep holes.

Preparation by Backer Board Installers

- Fasten backer board to studs with board manufacturer's recommended fasteners.
- All openings cut in backer board for plumbing and all cut joints between adjoining pieces—seal as recommended by backer board manufacturer.
- Maximum allowable variation in the tile substrate—for tiles with all edges shorter than 15", maximum allowable variation is ¼" in 10' from the required plane, with no more than ¹⁄₁₆" variation in 12" when measured from the high points in the surface. For tiles with at least one edge 15" in length, maximum allowable variation is ⅛" in 10' from the required plane, with no more than ¹⁄₁₆" variation in 24" when measured from the high points in the surface.
- Fit ends and edges tightly together, including in corners. Apply a bead of manufacturer-recommended sealant in panel joints as panels are being installed and over fastener heads per backer board manufacturer's instructions.
- Center backer board end or edge joints on framing and stagger joints in adjacent rows so four corners do not come together within the same plane. Space panel ends and edges in accordance with manufacturer's recommendations.

Materials

- **Multiple options exist for membranes, mortars, grouts, and other materials and *must be clearly specified* to be included. If not specifically indicated, optional materials are not included and mortar/grout choice defaults to minimum performance specification indicated. Consider each system component and intended use to determine minimum requirements and to specify options.**

- Ceramic tile—ANSI A137.1.
- Glass tile, when used—ANSI A137.2; see also "Glass Tile Selection and Installation Guide," and consult tile manufacturer for environmental exposure classification recommendations. Not all glass tiles are suitable.
- Cementitious grout—ANSI A118.6 or better or ISO CG1 or better. When glass tile is used, specify grout designated by tile and grout manufacturers.
- Epoxy grout, when used—ANSI A118.3 or ISO RG.
- Cementitious bond coat:
  - ANSI A118.1 or better or ISO C1 or better.
  - When porcelain tile is used—ANSI A118.4 or better or ISO C2 or better.
  - When glass tile is used, specify mortar designated by tile and mortar manufacturers. Bond coat color will impact the final appearance of translucent glass tile. Specifier shall confirm bond coat color is acceptable.
- Epoxy bond coat, when used—ANSI A118.3 or ISO R1 or better.
  - When glass tile is used, specify epoxy bond coat designated by tile and bond coat manufacturers. Bond coat color will impact the final appearance of translucent glass tile. Specifier shall confirm bond coat color is acceptable.
- Cementitious-coated extruded foam backer board—ASTM C578.
- Fasteners—noncorrosive and nonoxidizing.
- Flexible sealant—must be recommended by backer board manufacturer.
- Shower pan membrane—local building code.
- Metal studs—ASTM C645.

Materials for Green/Sustainable Design

- See "Green Building Standards and Green Product Selection Guide" and consult manufacturers and suppliers for product sustainability and contribution to green building design.
- Consider specifying tile and installation materials that meet ANSI A138.1, the *American National Standard Specifications for Sustainable Ceramic Tiles, Glass Tiles, and Tile Installation Materials.*

Preparation by Other Trades

- Apply blocking between studs to support the shower pan membrane.
- Wall framing shall meet the general framing requirements of ANSI A108.11-4.0–4.3.

Movement Joint (architect must specify type of joint and show location and details on drawings)

- Movement joints—mandatory according to EJ171.
- When glass tile is used, adhere to more frequent placement recommendations within the ranges listed in EJ171.

Installation Specifications

- Shower pan membrane—ANSI A108.01-3.6.
- Cementitious-coated extruded foam backer board—manufacturer's directions.
- Tile—ANSI A108.5.
- Glass tile—manufacturer's directions.
- Cementitious grout—ANSI A108.10.
- Epoxy mortar/grout—ANSI A108.6.
- Movement Joints—EJ171 and ASTM C1193.

Notes

- Test shower pan membrane and drainage fitting for leaks before commencing tilework.
- Materials adversely affected by moisture in areas immediately adjacent to showers, tubs, and roman tubs should be properly protected.
- A sloped portland cement mortar fill or approved preformed slope may be used under shower pan membrane when subfloor is not sloped to drain.
- All horizontal surfaces, for example shower seats, sills, curbs, etc., must slope towards drain or other surface sloped toward drain. Where present, waterproofing also must be sloped.
- As the tile size increases, there is less tolerance for variation in the substrate from the required plane. Epoxy adhesive thickness must be thin and uniform; therefore, substrate flattening may be required when epoxy bond coat is used.
- When glass tile is used, see "Glass Tile Selection and Installation Guide," and consult manufacturer for recommendations and requirements.
- For curbless shower receptor, see B421C and B422C.

B431-21

- **Wood or Metal Studs**
- **Fiber-Reinforced Water-Resistant Gypsum Backer Board Walls**
- **Mortar Bed Floor**
- **Ceramic Tile, Glass Tile**

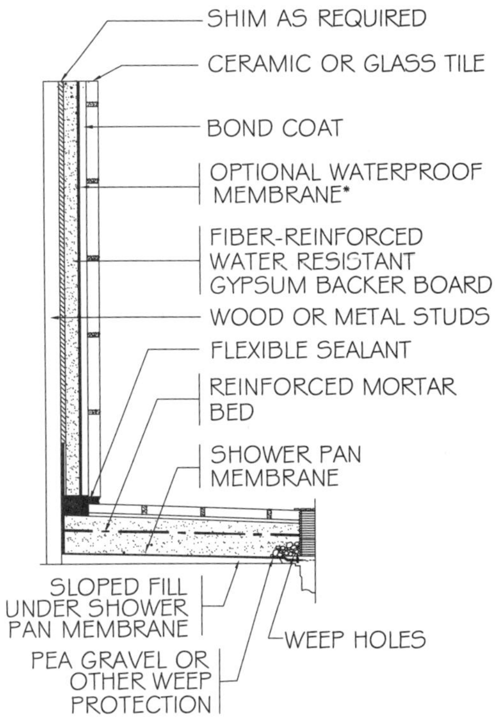

*USE OF A MEMBRANE ON WALLS IS OPTIONAL. SEE MEMBRANE OPTIONS.

SHOWER RECEPTORS, CURBS, SEATS, ETC., MUST BE PROPERLY WATERPROOFED AND INSTALLED TO AVOID WATER DAMAGE TO ADJACENT BUILDING MATERIALS. SEE COMMON SHOWER CONFIGURATIONS SECTION.

Recommended Uses

- For showers that do not have prefabricated receptors.

Environmental Exposure Classifications

- Res1, 2, 3; Com1, 2.
- For installations that may be exposed to staining, specify tile and grout suitable for exposure. Consult product manufacturers; see also "Product Selection Guides."
- For installations that may be exposed to mild chemical attack, specify epoxy grout and tile suitable for exposure. Consult product manufacturers; see also "Product Selection Guides."

Limitations

- Maximum stud spacing 16" on center.
- Not for areas exposed to temperatures exceeding 125°F.
- When organic adhesive is used—maximum tile size 8" × 8" unless organic adhesive manufacturer allows larger tile size. See Notes.
- Organic adhesive may be used on walls only; do not use on floor or curb.

Membrane Options

- A waterproof membrane (A118.10) may be specified for walls to prevent moisture intrusion and protect adjacent building materials. Specifier shall indicate if complete waterproofing of walls is required, including treatment at termination points.
- Check with membrane manufacturer for suitability for applicable conditions, as not all membranes are suitable for steam, high-temperature and/or chemical exposure, or exterior use.
- When glass tile is used, consult glass tile manufacturer for membrane options and recommendations.

Requirements

- Wood studs—dry and well-braced, minimum depth $3\frac{1}{2}$".
- Metal studs—well-braced; 20 gauge (0.033") or heavier; minimum depth $3\frac{1}{2}$" for residential applications or $3\frac{5}{8}$" for commercial applications.
- Slope shower pan membrane $\frac{1}{4}$" per foot to weep holes in drain.
- Turn shower pan membrane up walls a minimum of 3" above shower curb (6" above floor in showers without curbs).
- Fur out studs above shower pan membrane or notch-out studs behind the shower pan membrane so folds/corners of shower pan membrane do not cause backer board to bow inward, or use Alternate Receptor Base Method.
- Surround drain with pea gravel or other weep protection to prevent mortar from blocking weep holes.
- Caulk or seal penetrations and abutments to dissimilar materials.

Preparation by Backer Board Installers

- Maximum allowable variation in the tile substrate—for tiles with all edges shorter than 15", maximum allowable variation is $\frac{1}{4}$" in 10' from the required plane, with no more than $\frac{1}{16}$" variation in 12" when measured from the high points in the surface. For tiles with at least one edge 15" in length, maximum allowable variation is $\frac{1}{8}$" in 10' from the required plane, with no more than $\frac{1}{16}$" variation in 24" when measured from the high points in the surface.
- Horizontal joints—$\frac{1}{8}$" spacing filled solid and taped with latex-portland cement mortar and 2" alkali-resistant glass fiber mesh tape.
- Vertical joints—fill any space and tape with latex-portland cement mortar and 2" alkali-resistant glass fiber mesh tape.
- Corners—leave space between backer units. Tape joints using skim coat of latex-portland cement mortar, but do not fill.
- Center backer board end or edge joints on framing and

stagger joints in adjacent rows so four corners do not come together within the same plane. Space panel ends and edges in accordance with manufacturer's recommendations.

Materials

- **Multiple options exist for membranes, mortars, grouts, and other materials and *must be clearly specified* to be included. If not specifically indicated, optional materials are not included and mortar/grout choice defaults to minimum performance specification indicated. Consider each system component and intended use to determine minimum requirements and to specify options.**
- Ceramic tile—ANSI A137.1.
- Glass tile, when used—ANSI A137.2; see also "Glass Tile Selection and Installation Guide," and consult tile manufacturer for environmental exposure classification recommendations. Not all glass tiles are suitable.
- Cementitious grout—ANSI A118.6 or better or ISO CG1 or better. When glass tile is used, specify grout designated by tile and grout manufacturers.
- Epoxy grout, when used—ANSI A118.3 or ISO RG.
- Cementitious bond coat:
  - When a waterproof membrane is not used—ANSI A118.4 or better or ISO C2 or better.
  - When a waterproof membrane is used—ANSI A118.4 or better or ISO C2S1 or better unless ANSI A118.1 or ISO C1 is recommended by membrane manufacturer.
  - When glass tile is used, specify mortar designated by tile and mortar manufacturers. Bond coat color will impact the final appearance of translucent glass tile. Specifier shall confirm bond coat color is acceptable.
- Organic adhesive bond coat, when used (walls only)—ANSI A136.1 Type I or ISO D2.
- Waterproof membrane, when used—ANSI A118.10.
- Fiber-reinforced water-resistant gypsum backer board—ASTM C1278 (Paragraph 6.1).
- Fasteners—noncorrosive and nonoxidizing.
- 2" alkali-resistant glass fiber mesh tape.
- Flexible mildew-resistant sealant—ASTM C920.
- Shower pan membrane—local building code.
- Metal studs—ASTM C645.

Materials for Green/Sustainable Design

- See "Green Building Standards and Green Product Selection Guide" and consult manufacturers and suppliers for product sustainability and contribution to green building design.
- Consider specifying tile and installation materials that meet ANSI A138.1, the *American National Standard Specifications for Sustainable Ceramic Tiles, Glass Tiles, and Tile Installation Materials.*

Preparation by Other Trades

- Apply blocking between the studs to support the shower pan membrane.
- Wall framing shall meet the general framing requirements of ANSI A108.11-4.0–4.3.

Movement Joint (architect must specify type of joint and show location and details on drawings)

- Movement joints—mandatory in accordance with EJ171.
- When glass tile is used, adhere to more frequent placement recommendations within the ranges listed in EJ171.

Installation Specifications

- Shower pan membrane—ANSI A108.01-3.6.
- Fiber-reinforced water-resistant gypsum backer board—manufacturer's directions.
- Tile—ANSI A108.4 or A108.5.
- Glass tile—manufacturer's directions.
- Cementitious grout—ANSI A108.10.
- Epoxy grout—ANSI A108.6.
- Waterproof membrane—ANSI A108.13.
- Movement Joints—EJ171 and ASTM C1193.

Notes

- As the tile size increases, there is less tolerance for variation in the substrate from the required plane. Organic adhesive thickness must be thin and uniform; therefore, substrate flattening may be required when organic adhesive is used.
- Do not use organic adhesives to bond tile to a membrane or other impervious substrate.
- Extended dry time may be required for larger tiles and tiles with low water absorption when organic adhesive is used.
- Test shower pan membrane and drainage fitting for leaks before commencing tilework.
- Materials adversely affected by moisture in areas immediately adjacent to showers, tubs, and roman tubs should be properly protected.
- A sloped portland cement mortar fill or approved preformed slope may be used under shower pan membrane when subfloor is not sloped to drain.
- All horizontal surfaces, for example shower seats, sills, curbs, etc., must slope towards drain or other surface sloped toward drain. Where present, waterproofing also must be sloped.
- When glass tile is used, see "Glass Tile Selection and Installation Guide," and consult manufacturer for recommendations and requirements.
- For curbless shower receptor, see B421C and B422C.

B421-21

- **Solid Backing**
- **Bonded Waterproof Membrane**
- **Ceramic Tile, Glass Tile**

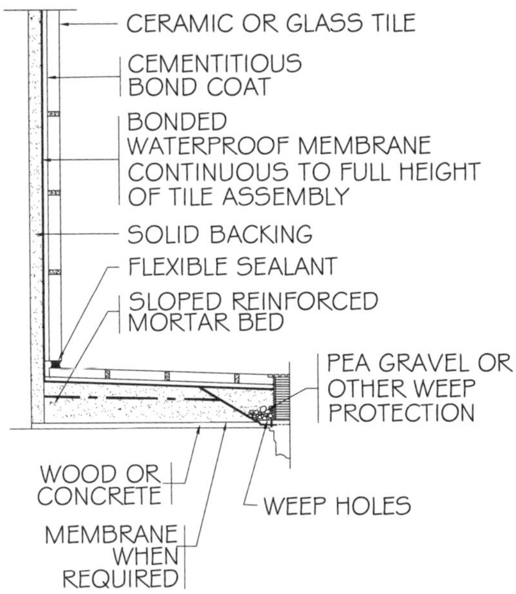

SHOWER RECEPTORS, CURBS, SEATS, ETC., MUST BE PROPERLY WATERPROOFED AND INSTALLED TO AVOID WATER DAMAGE TO ADJACENT BUILDING MATERIALS. SEE COMMON SHOWER CONFIGURATIONS SECTION.

Recommended Uses

- For showers that do not have a prefabricated receptor, where topical waterproofing of floors and walls is desired.

Environmental Exposure Classifications

- Res1, 2, 3; Com1, 2
- May be suitable for Com3, Res5, and Com5 as determined by membrane and backing material manufacturers.
- For Res4 and Com4, see SR613 and SR614.
- For installations that may be exposed to staining, specify tile and grout suitable for exposure. Consult product manufacturers; see also "Product Selection Guides."
- For installations that may be exposed to mild chemical attack, specify epoxy grout and tile suitable for exposure. Consult product manufacturers; see also "Product Selection Guides."

Limitations

- Refer to appropriate wall method for applicable limitations based on type of backing used—W201, W202E, W202I, W211, W221, W222, W231/W241, W243, W244C, W244E, W244F, W245, W246, and W247.

Requirements

- When glass tile is used, consult glass tile manufacturer for tile suitability over non-absorptive surface.
- Bonded waterproof membrane must be continuous, including at changes in plane. Follow membrane manufacturer's requirements for corners, seaming, and overlap.
- Waterproof membrane must extend to full height of tile assembly.
- Surround drain with pea gravel or other weep protection to prevent mortar from blocking weep holes.
- Refer to appropriate wall method for applicable requirements based on type of backing used.
- Slope mortar bed ¼" per foot toward drain and follow membrane manufacturer's instructions for connecting membrane to drain.

Materials

- **Multiple options exist for membranes, mortars, grouts, and other materials and *must be clearly specified* to be included. If not specifically indicated, optional materials are not included and mortar/grout choice defaults to minimum performance specification indicated. Consider each system component and intended use to determine minimum requirements and to specify options.**
- Ceramic tile—ANSI A137.1.
- Glass tile, when used—ANSI A137.2; see also "Glass Tile Selection and Installation Guide," and consult tile manufacturer for environmental exposure classification recommendations. Not all glass tiles are suitable.
- Cementitious grout—ANSI A118.6 or better or ISO CG1 or better. When glass tile is used, specify grout designated by tile and grout manufacturers.
- Epoxy grout, when used—ANSI A118.3 or ISO RG.
- Cementitious bond coat:
  - ANSI A118.4 or better or ISO C2S1 or better.
  - When glass tile is used, specify mortar designated by tile and mortar manufacturers. Bond coat color will impact the final appearance of translucent glass tile. Specifier shall confirm bond coat color is acceptable.
- Waterproof membrane—ANSI A118.10 and recommended by manufacturer of membrane for use over backing type in intended application.
- Flexible mildew-resistant sealant—ASTM C920.

Materials for Green/Sustainable Design

- See "Green Building Standards and Green Product Selection Guide" and consult manufacturers and suppliers for product sustainability and contribution to green building design.
- Consider specifying tile and installation materials that meet ANSI A138.1, the *American National Standard Specifications for Sustainable Ceramic Tiles, Glass Tiles, and Tile Installation Materials.*

Preparation by Other Trades/Backer Board Installers

- Refer to appropriate wall method for applicable preparations by other trades and backer board installers.
- Maximum allowable variation in the tile substrate—for tiles with all edges shorter than 15", maximum allowable variation is ¼" in 10' from the required plane, with no more than ¹⁄₁₆" variation in 12" when measured from the high points in the surface. For tiles with at least one edge 15" in length, maximum allowable variation is ⅛" in 10' from the required plane, with no more than ¹⁄₁₆" variation in 24" when measured from the high points in the surface.
- Center backer board end or edge joints on framing and stagger joints in adjacent rows so four corners do not come together within the same plane. Space panel ends and edges in accordance with manufacturer's recommendations.

Movement Joint (architect must specify type of joint and show location and details on drawings)

- Movement joints—mandatory according to EJ171.
- When glass tile is used, adhere to more frequent placement recommendations within the ranges listed in EJ171.

Installation Specifications

- Waterproof membrane—ANSI A108.13.
- Tile—ANSI A108.5.
- Glass tile—manufacturer's directions.
- Cementitious grout—ANSI A108.10.
- Epoxy grout—ANSI A108.6.
- Movement Joints—EJ171 and ASTM C1193.

Notes

- Test shower pan membrane/waterproof membrane and drainage fitting for leaks before commencing tilework.
- Materials adversely affected by moisture in areas immediately adjacent to showers, tubs, and roman tubs should be properly protected.
- All horizontal surfaces, for example shower seats, sills, curbs, etc., must slope towards drain or other surface sloped toward drain. Where present, waterproofing also must be sloped.
- When glass tile is used, see "Glass Tile Selection and Installation Guide," and consult manufacturer for recommendations and requirements.
- For curbless shower receptor, see B421C and B422C.

B422-21

- **Solid Backing**
- **Bonded Waterproof Membrane**
- **Integrated Bonding Flange**
- **Ceramic Tile, Glass Tile**

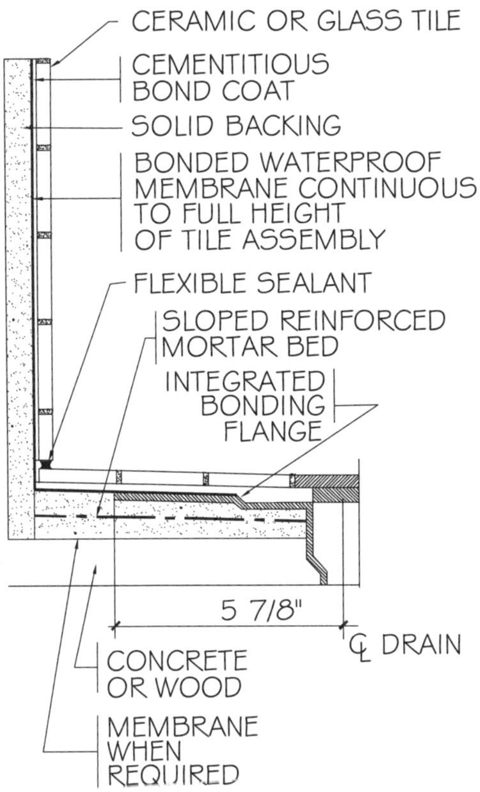

SHOWER RECEPTORS, CURBS, SEATS, ETC., MUST BE PROPERLY WATERPROOFED AND INSTALLED TO AVOID WATER DAMAGE TO ADJACENT BUILDING MATERIALS. SEE COMMON SHOWER CONFIGURATIONS SECTION.

Recommended Uses

- For showers that do not have a prefabricated receptor, where topical waterproofing of floors and walls is desired.

Environmental Exposure Classifications

- Res1, 2, 3; Com1, 2.
- May be suitable for Com3, Res5, and Com5 as determined by membrane and backing material manufacturers.
- For Res4 and Com4, see SR613 and SR614.
- For installations that may be exposed to staining, specify tile and grout suitable for exposure. Consult product manufacturers; see also "Product Selection Guides."
- For installations that may be exposed to mild chemical attack, specify epoxy grout and tile suitable for exposure. Consult product manufacturers; see also "Product Selection Guides."

Limitations

- Refer to appropriate wall method for applicable limitations based on type of backing used—W201, W202E, W202I, W211, W221, W222, W231/W241, W243, W244C, W244E, W244F, W245, W246, and W247.

Requirements

- When glass tile is used, consult glass tile manufacturer for tile suitability over non-absorptive surface.
- Bonded waterproof membrane must be continuous, including at changes in plane. Follow membrane manufacturer's requirements for corners, seaming, and overlap.
- Waterproof membrane must extend to full height of tile assembly.
- Refer to appropriate wall method for applicable requirements based on type of backing used.
- Slope mortar bed ¼" per foot toward drain and follow membrane and integrated bonding flange manufacturers' instructions for connecting membrane to integrated bonding flange.

Materials

- **Multiple options exist for membranes, mortars, grouts, and other materials and *must be clearly specified* to be included. If not specifically indicated, optional materials are not included and mortar/grout choice defaults to minimum performance specification indicated. Consider each system component and intended use to determine minimum requirements and to specify options.**
- Ceramic tile—ANSI A137.1.
- Glass tile, when used—ANSI A137.2; see also "Glass Tile Selection and Installation Guide," and consult tile manufacturer for environmental exposure classification recommendations. Not all glass tiles are suitable.
- Cementitious grout—ANSI A118.6 or better or ISO CG1 or better. When glass tile is used, specify grout designated by tile and grout manufacturers.
- Epoxy grout, when used—ANSI A118.3 or ISO RG.
- Cementitious bond coat:
  - ANSI A118.4 or better or ISO C2S1 or better unless ANSI A118.1 or ISO C1 is recommended by membrane manufacturer.
  - When glass tile is used, specify mortar designated by tile and mortar manufacturers. Bond coat color will impact the final appearance of translucent glass tile. Specifier shall confirm bond coat color is acceptable.
- Waterproof membrane—ANSI A118.10 and recommended by manufacturer of membrane for use over backing type in intended application.
- Flexible mildew-resistant sealant—ASTM C920.

Materials for Green/Sustainable Design

- See "Green Building Standards and Green Product Selection Guide" and consult manufacturers and suppliers for product sustainability and contribution to green building design.
- Consider specifying tile and installation materials that meet ANSI A138.1, the *American National Standard Specifications for Sustainable Ceramic Tiles, Glass Tiles, and Tile Installation Materials.*

Preparation by Other Trades/Backer Board Installers

- Refer to appropriate wall method for applicable preparations by other trades and backer board installers.
- Maximum allowable variation in the tile substrate—for tiles with all edges shorter than 15," maximum allowable variation is ¼" in 10' from the required plane, with no more than $^1/_{16}$" variation in 12" when measured from the high points in the surface. For tiles with at least one edge 15" in length, maximum allowable variation is ⅛" in 10' from the required plane, with no more than $^1/_{16}$" variation in 24" when measured from the high points in the surface.
- Center backer board end or edge joints on framing and stagger joints in adjacent rows so four corners do not come together within the same plane. Space panel ends and edges in accordance with manufacturer's recommendations.

Movement Joint (architect must specify type of joint and show location and details on drawings)

- Movement joints—mandatory according to EJ171.
- When glass tile is used, adhere to more frequent placement recommendations within the ranges listed in EJ171.

Installation Specifications

- Waterproof membrane—ANSI A108.13.
- Tile—ANSI 108.5.
- Glass tile—manufacturer's directions.
- Cementitious grout—ANSI A108.10.
- Epoxy grout—ANSI A108.6.
- Movement Joints—EJ171 and ASTM C1193.

Notes

- Test shower pan membrane/waterproof membrane and drainage fitting for leaks before commencing tilework.
- Materials adversely affected by moisture in areas immediately adjacent to showers, tubs, and roman tubs should be properly protected.
- All horizontal surfaces, for example shower seats, sills, curbs, etc., must slope towards drain or other surface sloped toward drain. Where present, waterproofing also must be sloped.
- When glass tile is used, see "Glass Tile Selection and Installation Guide," and consult manufacturer for recommendations and requirements.
- For curbless shower receptor, see B421C and B422C.

B421C-21

- **Solid Backing (wall)**
- **Concrete Substrate (floor)**
- **Bonded Waterproof Membrane**
- **Ceramic Tile, Glass Tile**

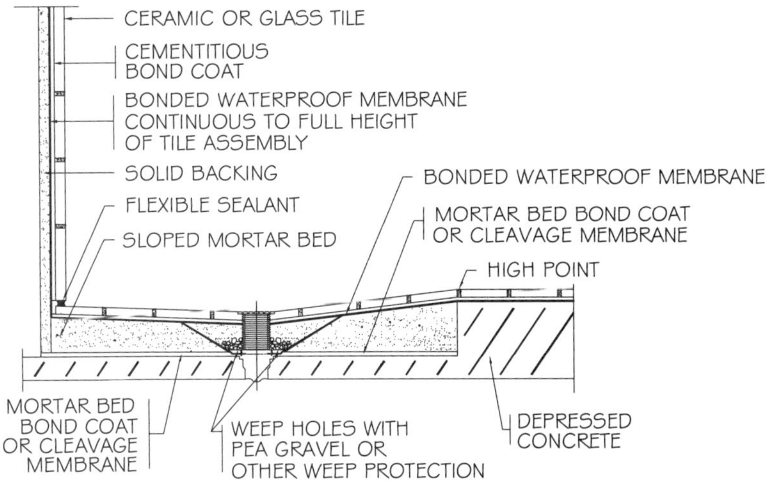

SHOWER RECEPTORS, CURBS, SEATS, ETC., MUST BE PROPERLY WATERPROOFED AND INSTALLED TO AVOID WATER DAMAGE TO ADJACENT BUILDING MATERIALS. SEE COMMON SHOWER CONFIGURATION SECTION.

FOR MORTAR BEDS OVER 65 S.F., WIRE REINFORCING IS REQUIRED.

Recommended Uses

- For construction of a shower without a curb.

Environmental Exposure Classifications

- Res1, 2, 3; Com1, 2
- May be suitable for Com3, Res5, and Com5 as determined by membrane and backing material manufacturers.
- For Res4 and Com4, see SR613 and SR614.
- For installations that may be exposed to staining, specify tile and grout suitable for exposure. Consult product manufacturers; see also "Product Selection Guides."
- For installations that may be exposed to mild chemical attack, specify epoxy grout and tile suitable for exposure. Consult product manufacturers; see also "Product Selection Guides."

Limitations

- Facilitates construction of a curbless shower without adding floor height at restroom entryway, but does not inherently comply with Americans with Disabilities Act (ADA) Standards for Accessible Design. Follow ADA where ADA compliance is required or desired.
- Refer to appropriate wall method for applicable limitations based on type of backing used—W201, W202E, W202I, W211, W221, W222, W231/W241, W243, W244C, W244E, W244F, W245, W246, and W247.

Requirements

- When glass tile is used, consult glass tile manufacturer for tile suitability over non-absorptive surface.
- Bonded waterproof membrane must be continuous, including at changes in plane. Follow membrane manufacturer's requirements for corners, seaming, and overlap.
- Waterproof membrane inside shower area must extend to full height of tile assembly. Also, the floor and wall waterproofing must continue outside the immediate shower area one foot beyond the high point of the floor, but not beyond the tiled area. Additional waterproofing of floor and/or walls outside the shower area may be needed to effectively contain and evacuate shower water and splash water and to protect building materials. When additional waterproofing of floor and/or walls outside the shower area is desired or required, the building design professional must specify all areas to be waterproofed and indicate membrane termination points.
- A secondary drain may be required outside the immediate shower area to facilitate evacuation of shower water and splash water that is not contained in the shower area due to curbless design. When required or desired, building design professional must specify type and location of secondary drain.
- High point of the floor must be outside the shower area, i.e., beyond shower door or curtain, to facilitate evacuation of shower water and splash water that is not contained in the shower area due to curbless design. Location of the high point of the floor is especially critical when secondary drain is not included outside the shower area.
- Surround drain with pea gravel or other weep protection to prevent mortar from blocking weep holes.
- Refer to appropriate wall method for applicable requirements based on type of backing used.

- Slope mortar bed ¼" per foot toward drain and follow membrane manufacturer's instructions for connecting membrane to drain.

Materials

- **Multiple options exist for membranes, mortars, grouts, and other materials and *must be clearly specified* to be included. If not specifically indicated, optional materials are not included and mortar/grout choice defaults to minimum performance specification indicated. Consider each system component and intended use to determine minimum requirements and to specify options.**
- Ceramic tile—ANSI A137.1.
- Glass tile, when used—ANSI A137.2; see also "Glass Tile Selection and Installation Guide," and consult tile manufacturer for environmental exposure classification recommendations. Not all glass tiles are suitable.
- Cementitious grout—ANSI A118.6 or better or ISO CG1 or better. When glass tile is used, specify grout designated by tile and grout manufacturers.
- Epoxy grout, when used—ANSI A118.3 or ISO RG.
- Cementitious bond coat:
  - ANSI A118.4 or better or ISO C2S1 or better unless ANSI A118.1 or ISO C1 is recommended by membrane manufacturer.
  - When glass tile is used, specify mortar designated by tile and mortar manufacturers. Bond coat color will impact the final appearance of translucent glass tile. Specifier shall confirm bond coat color is acceptable.
- Waterproof membrane—ANSI A118.10 and recommended by manufacturer of membrane for use over backing type in intended application.
- Flexible mildew-resistant sealant—ASTM C920.

Materials for Green/Sustainable Design

- See "Green Building Standards and Green Product Selection Guide" and consult manufacturers and suppliers for product sustainability and contribution to green building design.
- Consider specifying tile and installation materials that meet ANSI A138.1, the *American National Standard Specifications for Sustainable Ceramic Tiles, Glass Tiles, and Tile Installation Materials.*

Preparation by Other Trades/Backer Board Installers

- Depressed slab by others. Depressed area must extend beyond shower area to facilitate a finished floor installation that effectively evacuates shower water and splash water, which are often difficult to contain within a curbless shower.
- Refer to appropriate wall method for applicable preparations by other trades and backer board installers.
- Maximum allowable variation in the tile substrate—for tiles with all edges shorter than 15", maximum allowable variation is ¼" in 10' from the required plane, with no more than 1⁄16" variation in 12" when measured from the high points in the surface. For tiles with at least one edge 15" in length, maximum allowable variation is ⅛" in 10' from the required plane, with no more than 1⁄16" variation in 24" when measured from the high points in the surface.
- Center backer board end or edge joints on framing and stagger joints in adjacent rows so four corners do not come together within the same plane. Space panel ends and edges in accordance with manufacturer's recommendations.

Movement Joint (architect must specify type of joint and show location and details on drawings)

- Movement joints—mandatory according to EJ171.
- When glass tile is used, adhere to more frequent placement recommendations within the ranges listed in EJ171.

Installation Specifications

- Waterproof membrane—ANSI A108.13.
- Tile—ANSI A108.5.
- Glass tile—manufacturer's directions.
- Cementitious grout—ANSI A108.10.
- Epoxy grout—ANSI A108.6.
- Movement Joints—EJ171 and ASTM C1193.

Notes

- Test shower pan membrane/waterproof membrane and drainage fitting for leaks before commencing tilework.
- Materials adversely affected by moisture in areas immediately adjacent to showers, tubs, and roman tubs should be properly protected.
- All horizontal surfaces, for example shower seats, sills, curbs, etc., must slope towards drain or other surface sloped toward drain. Where present, waterproofing also must be sloped.
- When glass tile is used, see "Glass Tile Selection and Installation Guide," and consult manufacturer for recommendations and requirements.
- Follow applicable plumbing and building codes.

B422C-21

- **Solid Backing (wall)**
- **Concrete Substrate (floor)**
- **Bonded Waterproof Membrane**
- **Ceramic Tile, Glass Tile**

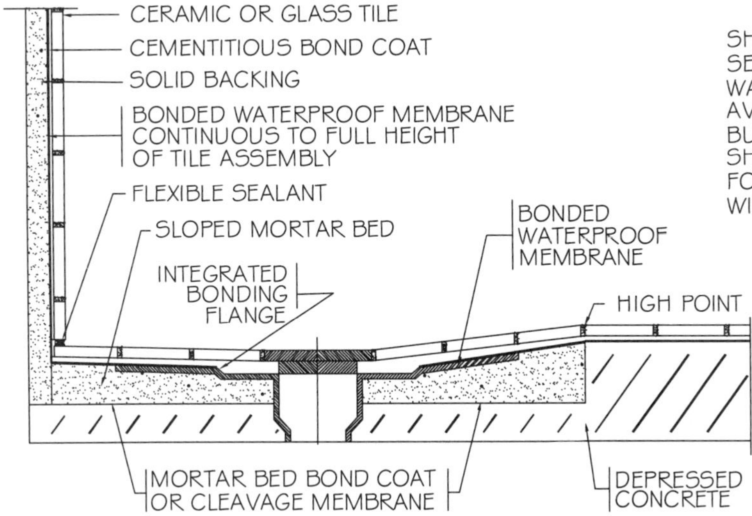

Recommended Uses

- For construction of a shower without a curb.

Environmental Exposure Classifications

- Res1, 2, 3; Com1, 2
- May be suitable for Com3, Res5, and Com5 as determined by membrane and backing material manufacturers.
- For Res4 and Com4, see SR613 and SR614.
- For installations that may be exposed to staining, specify tile and grout suitable for exposure. Consult product manufacturers; see also "Product Selection Guides."
- For installations that may be exposed to mild chemical attack, specify epoxy grout and tile suitable for exposure. Consult product manufacturers; see also "Product Selection Guides."

Limitations

- Facilitates construction of a curbless shower without adding floor height at restroom entryway, but does not inherently comply with Americans with Disabilities Act (ADA) Standards for Accessible Design. Follow ADA where ADA compliance is required or desired.
- Refer to appropriate wall method for applicable limitations based on type of backing used—W201, W202E, W202I, W211, W221, W222, W231/W241, W243, W244C, W244E, W244F, W245, W246, and W247.

Requirements

- When glass tile is used, consult glass tile manufacturer for tile suitability over non-absorptive surface.
- Bonded waterproof membrane must be continuous, including at changes in plane. Follow membrane manufacturer's requirements for corners, seaming, and overlap.
- Waterproof membrane inside shower area must extend to full height of tile assembly. Also, the floor and wall waterproofing must continue outside the immediate shower area one foot beyond the high point of the floor, but not beyond the tiled area. Additional waterproofing of floor and/or walls outside the shower area may be needed to effectively contain and evacuate shower water and splash water and to protect building materials. When additional waterproofing of floor and/or walls outside the shower area is desired or required, the building design professional must specify all areas to be waterproofed and indicate membrane termination points.
- A secondary drain may be required outside the immediate shower area to facilitate evacuation of shower water and splash water that is not contained in the shower area due to curbless design. When required or desired, building design professional must specify type and location of secondary drain.
- High point of the floor must be outside the shower area, i.e., beyond shower door or curtain, to facilitate evacuation

of shower water and splash water that is not contained in the shower area due to curbless design. Location of the high point of the floor is especially critical when secondary drain is not included outside the shower area.

- Refer to appropriate wall method for applicable requirements based on type of backing used.
- Slope mortar bed ¼" per foot toward drain and follow membrane and integrated bonding flange manufacturer's instructions for connecting membrane to integrated bonding flange.

Materials

- **Multiple options exist for membranes, mortars, grouts, and other materials and *must be clearly specified* to be included. If not specifically indicated, optional materials are not included and mortar/grout choice defaults to minimum performance specification indicated. Consider each system component and intended use to determine minimum requirements and to specify options.**
- Ceramic tile—ANSI A137.1.
- Glass tile, when used—ANSI A137.2; see also "Glass Tile Selection and Installation Guide," and consult tile manufacturer for environmental exposure classification recommendations. Not all glass tiles are suitable.
- Cementitious grout—ANSI A118.6 or better or ISO CG1 or better. When glass tile is used, specify grout designated by tile and grout manufacturers.
- Epoxy grout, when used—ANSI A118.3 or ISO RG.
- Cementitious bond coat:
  - ANSI A118.4 or better or ISO C2S1 or better unless ANSI A118.1 or ISO C1 is recommended by membrane manufacturer.
  - When glass tile is used, specify mortar designated by tile and mortar manufacturers. Bond coat color will impact the final appearance of translucent glass tile. Specifier shall confirm bond coat color is acceptable.
- Waterproof membrane—ANSI A118.10 and recommended by manufacturer of membrane for use over backing type in intended application.
- Flexible mildew-resistant sealant—ASTM C920.

Materials for Green/Sustainable Design

- See "Green Building Standards and Green Product Selection Guide" and consult manufacturers and suppliers for product sustainability and contribution to green building design.
- Consider specifying tile and installation materials that meet ANSI A138.1, the *American National Standard Specifications for Sustainable Ceramic Tiles, Glass Tiles, and Tile Installation Materials.*

Preparation by Other Trades/Backer Board Installers

- Depressed slab by others. Depressed area must extend beyond shower area to facilitate a finished floor installation that effectively evacuates shower water and splash water, which are often difficult to contain within a curbless shower.
- Refer to appropriate wall method for applicable preparations by other trades and backer board installers.
- Maximum allowable variation in the tile substrate—for tiles with all edges shorter than 15," maximum allowable variation is ¼" in 10' from the required plane, with no more than ¹⁄₁₆" variation in 12" when measured from the high points in the surface. For tiles with at least one edge 15" in length, maximum allowable variation is ⅛" in 10' from the required plane, with no more than ¹⁄₁₆" variation in 24" when measured from the high points in the surface.
- Center backer board end or edge joints on framing and stagger joints in adjacent rows so four corners do not come together within the same plane. Space panel ends and edges in accordance with manufacturer's recommendations.

Movement Joint (architect must specify type of joint and show location and details on drawings)

- Movement joints—mandatory according to EJ171.
- When glass tile is used, adhere to more frequent placement recommendations within the ranges listed in EJ171.

Installation Specifications

- Waterproof membrane—ANSI A108.13.
- Tile—ANSI A108.5.
- Glass tile—manufacturer's directions.
- Cementitious grout—ANSI A108.10.
- Epoxy grout—ANSI A108.6.
- Movement Joints—EJ171 and ASTM C1193.

Notes

- Test shower pan membrane/waterproof membrane and drainage fitting for leaks before commencing tilework.
- Materials adversely affected by moisture in areas immediately adjacent to showers, tubs, and roman tubs should be properly protected.
- All horizontal surfaces, for example shower seats, sills, curbs, etc., must slope towards drain or other surface sloped toward drain. Where present, waterproofing also must be sloped.
- When glass tile is used, see "Glass Tile Selection and Installation Guide," and consult manufacturer for recommendations and requirements.
- Follow applicable plumbing and building codes.

Mortar Curb*

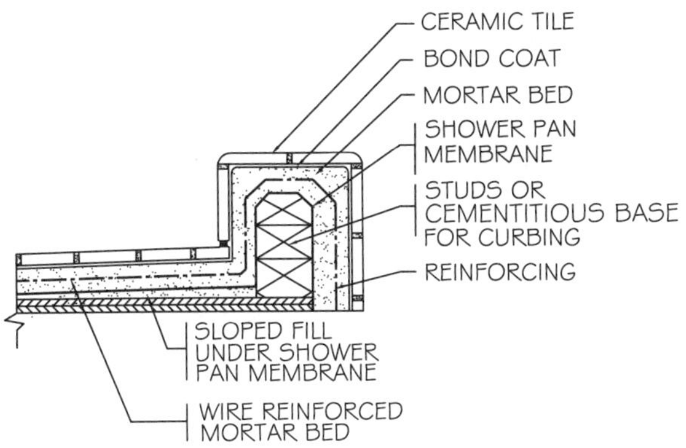

Preformed Curb*

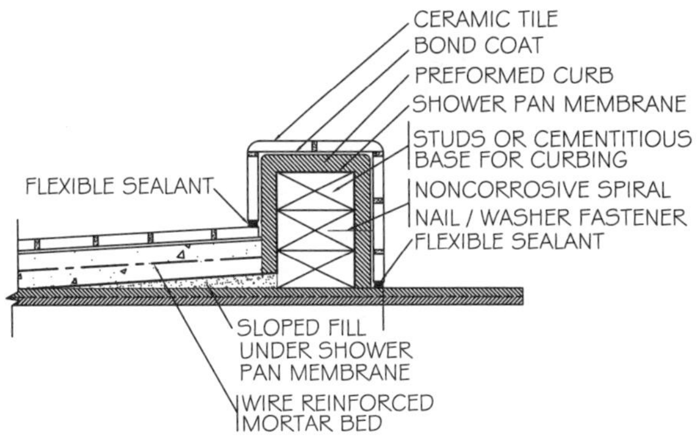

*NOTE: Construct curb such that membrane on top of curb (shower pan membrane or bonded waterproof membrane) is sloped toward drain.

Solid Curb*

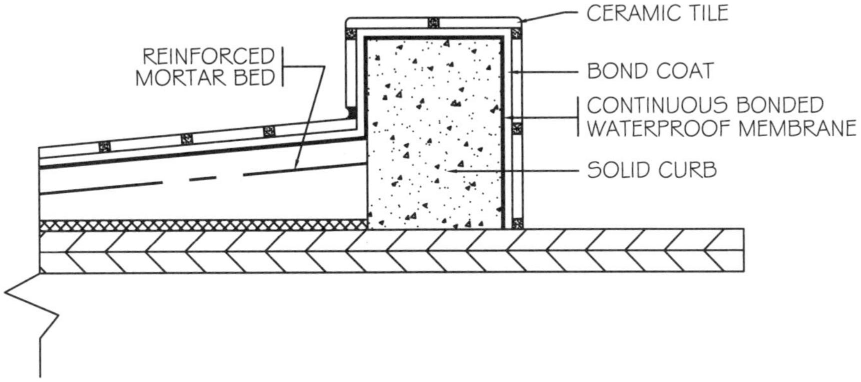

Alternate Receptor Base

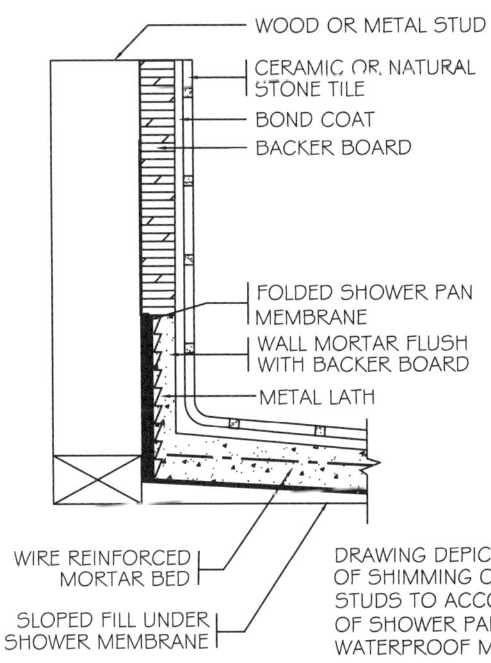

DRAWING DEPICTS USE OF WALL MORTAR IN LIEU OF SHIMMING OUT BACKER BOARD OR NOTCHING STUDS TO ACCOMMODATE THICKNESS OF FOLDS OF SHOWER PAN MEMBRANE. VAPOR RETARDER OR WATERPROOF MEMBRANE AS REQUIRED (NOT SHOWN).

Shower Pan Membrane at Curb/Jamb

1. CONSTRUCT CURB SUCH THAT WATERPROOF MEMBRANE ON TOP SIDE OF CURB IS SLOPED TOWARD THE DRAIN
2. FASTENERS - MIN. 3" ABOVE CURB
3. SHOWER PAN MEMBRANE IS SLOPED MIN. 1/4" PER FOOT TO DRAIN, FOLDED UP WALLS MIN. 3" ABOVE CURB, AND EXTENDS TO FLOOR OUTSIDE OF SHOWER.

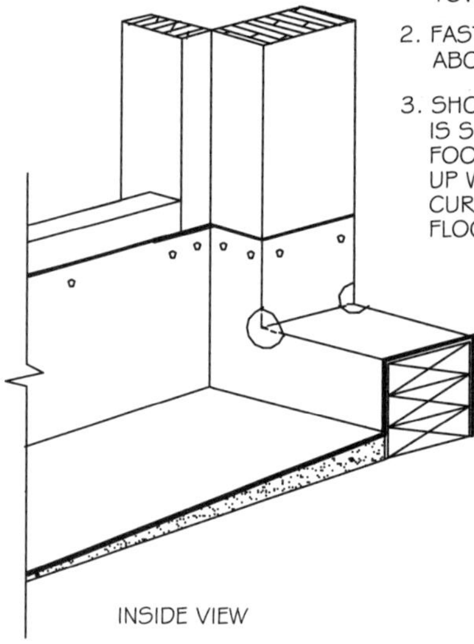

Backer Board Installation Over Shower Pan Membrane

1. SPACE, TAPE, SEAL, ETC., BACKER BOARD PER MANUFACTURER'S REQUIREMENTS, INCLUDING FOR BOTTOM EDGE OF BOARD.
2. BECAUSE OF VARYING REQUIREMENTS, VAPOR RETARDER BEHIND BACKER BOARD NOT SHOWN. INCLUDE AND INSTALL VAPOR RETARDER WHEN AND AS REQUIRED BY BUILDING CODE AND BACKER BOARD MANUFACTURER.
3. FASTENERS - MIN. 3" ABOVE CURB
4. SHOWER PAN MEMBRANE IS SLOPED 1/4" PER FOOT TO DRAIN, FOLDED UP WALLS MIN. 3" ABOVE CURB, AND EXTENDS TO FLOOR OUTSIDE OF SHOWER.
5. WIRE MESH FOR CURB - PER ANSI
6. CONSTRUCT CURB SUCH THAT WATERPROOF MEMBRANE ON TOP SIDE OF CURB IS SLOPED TOWARD THE DRAIN.

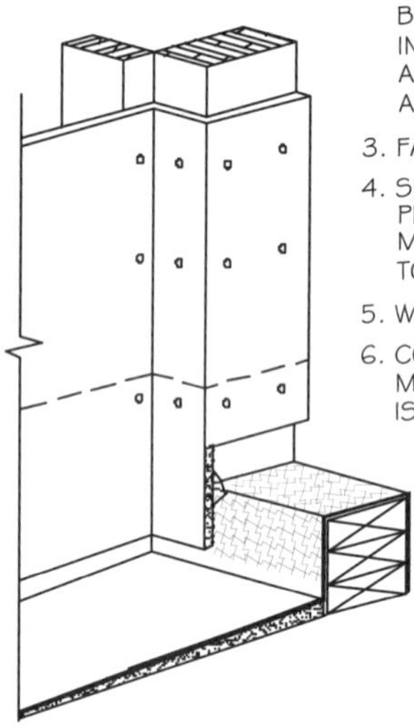

Typical Shower Seat for Backer Board Showers

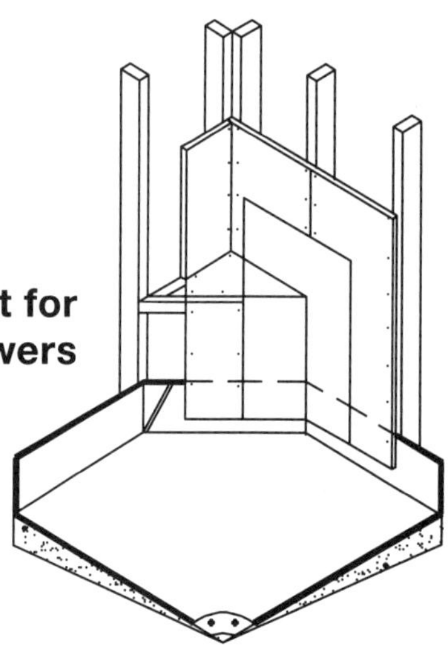

1. CONSTRUCT SHOWER SEAT SUCH THAT BONDED WATERPROOF MEMBRANE ON BACKER BOARD ON TOP SIDE OF SHOWER SEAT IS SLOPED MIN. 1/4" PER FOOT TO DRAIN, AND EXTEND BONDED WATERPROOF MEMBRANE BEYOND SEAT/WALL INTERSECTION, AND BELOW TOP EDGE OF SHOWER PAN MEMBRANE.
2. SHOWER PAN MEMBRANE IS SLOPED MIN. 1/4" PER FOOT TO DRAIN AND FOLDED UP WALLS.
3. SPACE, TAPE, SEAL, ETC., BACKER BOARD PER MANUFACTURER'S REQUIREMENTS, INCLUDING FOR BOTTOM EDGE OF BOARD.
4. BECAUSE OF VARYING REQUIREMENTS, VAPOR RETARDER BEHIND BACKER BOARD NOT SHOWN. INCLUDE AND INSTALL VAPOR RETARDER WHEN AND AS REQUIRED BY BUILDING CODE AND BACKER BOARD MANUFACTURER.

TR418-21

- **Mortar Bed Walls**
- **Mortar Bed Floor**

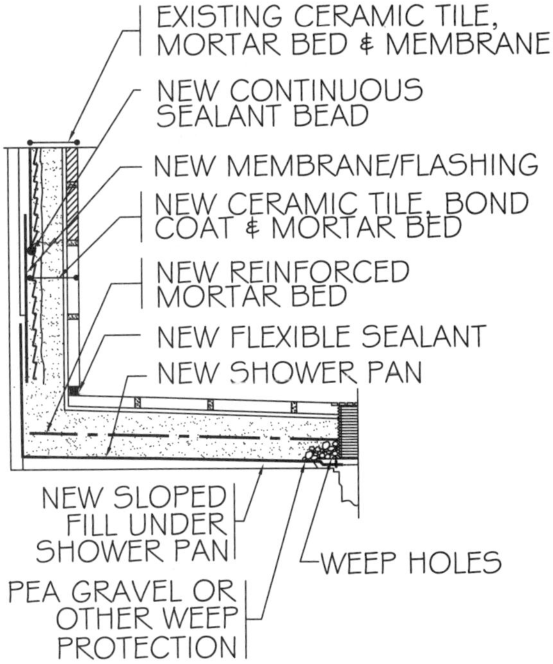

Recommended Uses

- For use where old shower pan has failed.

Requirements

- Remove existing shower receptor, shower pan, and wall tile, as required, to install new shower pan.
- Replace any damaged wall and floor substrate materials.
- New shower pan—slope ¼" per foot to weep holes in drain.
- New shower pan to turn up wall a minimum of 3" above curb (6" above floor in showers without curbs).
- New wall membrane/flashing (ANSI A108.02-3.8) placed behind existing wall membrane, out over new shower pan, and fastened to studs.
- Place continuous bead of sealant on existing wall mortar with new wall mortar brought up tight against it.

Materials for Green/Sustainable Design

- See "Green Building Standards and Green Product Selection Guide" and consult manufacturers and suppliers for product sustainability and contribution to green building design.
- Consider specifying tile and installation materials that meet ANSI A138.1, the *American National Standard Specifications for Sustainable Ceramic Tiles, Glass Tiles, and Tile Installation Materials.*

Notes

- See B414 for complete specifications.

TR420-21

- **Backer Board Walls**
- **Mortar Bed Floor**

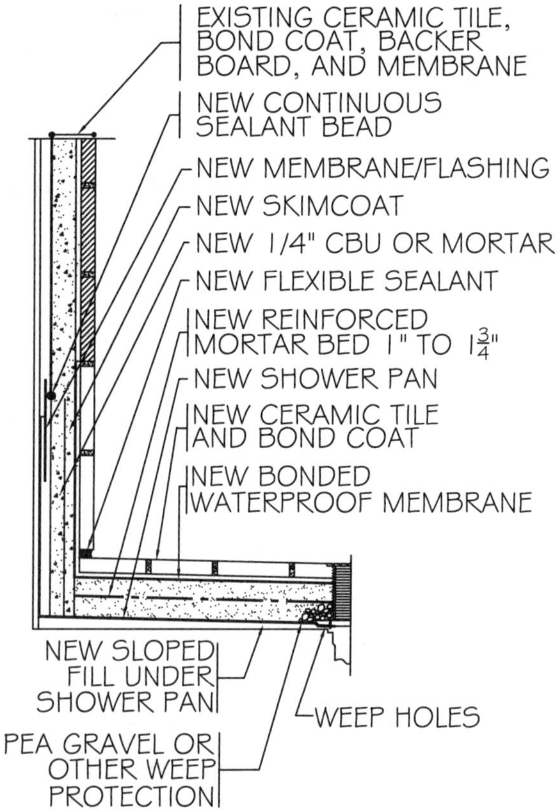

Recommended Uses

- For use where old shower pan has failed.

Requirements

- Remove existing shower receptor, shower pan, and wall tile, as required, to install new shower pan.
- Replace any damaged wall and floor substrate materials.
- New shower pan—slope ¼" per foot to weep holes in drain.
- New shower pan to turn up wall a minimum of 3" above curb (6" above floor in showers without curbs).
- New wall membrane/flashing (ANSI A108.02-3.8) placed behind existing wall membrane, out over new shower pan, and fastened to studs.
- Place continuous bead of sealant on existing wall substrate material with new wall substrate material brought up tight against it.

Materials for Green/Sustainable Design

- See "Green Building Standards and Green Product Selection Guide" and consult manufacturers and suppliers for product sustainability and contribution to green building design.
- Consider specifying tile and installation materials that meet ANSI A138.1, the *American National Standard Specifications for Sustainable Ceramic Tiles, Glass Tiles, and Tile Installation Materials*

Notes

- See B415, B420, B426, or B431 for complete specifications based on wall backing type.

C311-21

• Mortar Bed
• Ceramic Tile, Glass Tile

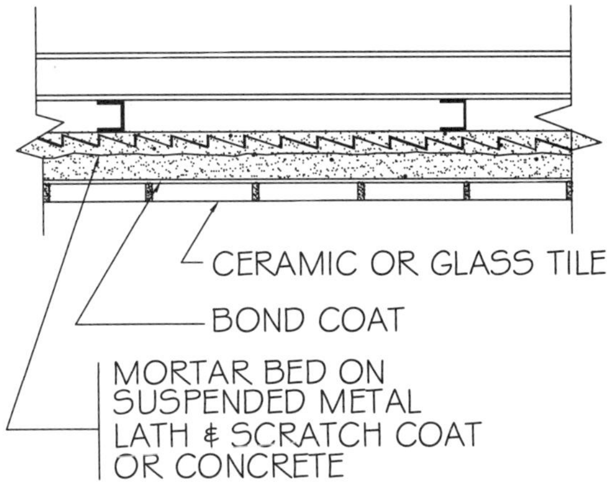

Recommended Uses

- For ceilings where trueing is required.
- For areas where substrate flatness is critical, such as when tiles with any edge longer than 15" are specified.

Environmental Exposure Classifications

- Res1, 2, 3, 5; Com1, 2, 3, 5.
- For Res4 and Com4, see SR613.

Limitations

- Maximum joist spacing 16" on center.

Requirements

- Suspended ceiling systems must be capable of supporting weight of mortar bed and tile.
- Concrete to be well cured, dimensionally stable, and free of cracks, waxy, or oily films.

Materials

- **Multiple options exist for membranes, mortars, grouts, and other materials and *must be clearly specified* to be included. If not specifically indicated, optional materials are not included and mortar/grout choice defaults to minimum performance specification indicated. Consider each system component and intended use to determine minimum requirements and to specify options.**
- Ceramic tile—ANSI A137.1.
- Glass tile, when used—ANSI A137.2; see also "Glass Tile Selection and Installation Guide," and consult tile manufacturer for environmental exposure classification recommendations. Not all glass tiles are suitable.
- Cementitious grout—ANSI A118.6 or better or ISO CG1 or better. When glass tile is used, specify grout designated by tile and grout manufacturers.
- Epoxy grout, when used—ANSI A118.3 or ISO RG.
- Cementitious bond coat—portland cement paste on a mortar bed that is still workable. For a cured mortar bed, follow recommendations below to select appropriate bonding mortar:
  - ANSI A118.1 or better or ISO C1 or better.
  - When porcelain tile is used—ANSI A118.4 or better or ISO C2 or better.
  - When glass tile is used, specify mortar designated by tile and mortar manufacturers. Bond coat color will impact the final appearance of translucent glass tile. Specifier shall confirm bond coat color is acceptable.
- Mortar bed, metal lath, and cleavage membrane—ANSI A108.1A.

Materials for Green/Sustainable Design

- See "Green Building Standards and Green Product Selection Guide" and consult manufacturers and suppliers for product sustainability and contribution to green building design.
- Consider specifying tile and installation materials that meet ANSI A138.1, the *American National Standard Specifications for Sustainable Ceramic Tiles, Glass Tiles, and Tile Installation Materials.*

Preparation by Other Trades

- Maximum allowable variation in the installation substrate—¼" in 10' from the required plane.

Movement Joint (architect must specify type of joint and show location and details on drawings)

- Movement joints—mandatory according to EJ171.
- When glass tile is used, adhere to more frequent placement recommendations within the ranges listed in EJ171.

Installation Specifications

- Tile—ANSI A108.1A, .1B, or .1C.
- Glass tile—ANSI A108.14, .15, .16, or manufacturer's directions.
- Cementitious grout—ANSI A108.10.
- Epoxy grout—ANSI A108.6.
- Movement Joints—EJ171 and ASTM C1193.

Notes

- When glass tile is used, see "Glass Tile Selection and Installation Guide," and consult manufacturer for recommendations and requirements.

C312-21

• Gypsum Board
• Ceramic Tile, Glass Tile

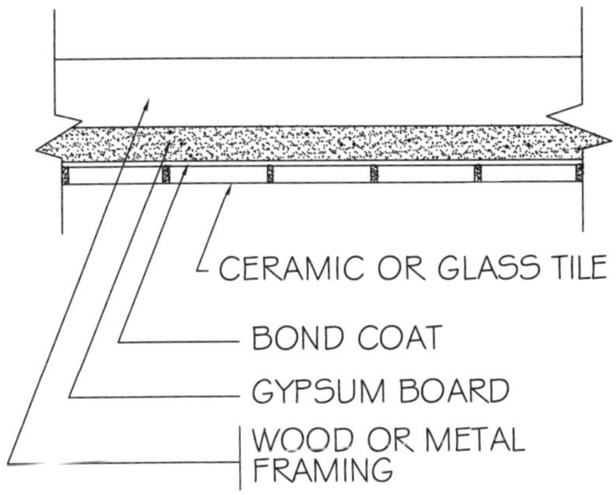

Recommended Uses

- For ceilings in dry areas where gypsum board is the tile backer.

Environmental Exposure Classifications

- Res1; Com1.

Limitations

- The weight of overlaid unsupported insulation shall not exceed 1.3 psf for ½"-thick gypsum board with framing spacing 24" o.c., or 2.2 psf for ½"-thick gypsum board with framing spacing 16" o.c. or ⅝"-thick gypsum board with framing spacing 24" o.c.
- Not for areas exposed to temperatures exceeding 125°F.
- When organic adhesive is used—maximum tile size 8" × 8" unless organic adhesive manufacturer allows larger tile size. See Notes.
- Ceiling framing, furring, or blocking shall not exceed 12" o.c. for ½"-thick water-resistant gypsum backing board and 16" o.c. for ⅝"-thick water-resistant gypsum backing board.

Requirements

- Surface—free of coatings, oil, and wax.
- Ceiling framing must be capable of supporting the weight of backing and tile.
- Minimum thickness of gypsum board—½."

Materials

- **Multiple options exist for membranes, mortars, grouts, and other materials and *must be clearly specified* to be included. If not specifically indicated, optional materials are not included and mortar/grout choice defaults to minimum performance specification indicated. Consider each system component and intended use to determine minimum requirements and to specify options.**
- Ceramic tile—ANSI A137.1.
- Glass tile, when used—ANSI A137.2; see also "Glass Tile Selection and Installation Guide," and consult tile manufacturer for environmental exposure classification recommendations. Not all glass tiles are suitable.
- Cementitious grout—ANSI A118.6 or better or ISO CG1 or better. When glass tile is used, specify grout designated by tile and grout manufacturers.
- Epoxy grout, when used—ANSI A118.3 or ISO RG.
- Cementitious bond coat:
  - ANSI A118.1 or better or ISO C1 or better.
  - When porcelain tile is used—ANSI A118.4 or better or ISO C2 or better.
  - When glass tile is used, specify mortar designated by tile and mortar manufacturers. Bond coat color will impact the final appearance of translucent glass tile. Specifier shall confirm bond coat color is acceptable.
- Organic adhesive bond coat, when used—ANSI A136.1 (Type I or II) or ISO D1 or better.

Materials for Green/Sustainable Design

- See "Green Building Standards and Green Product Selection Guide" and consult manufacturers and suppliers for product sustainability and contribution to green building design.
- Consider specifying tile and installation materials that meet ANSI A138.1, the *American National Standard Specifications for Sustainable Ceramic Tiles, Glass Tiles, and Tile Installation Materials.*

Preparation by Other Trades

- Gypsum board face layer joints—treated with tape and joint compound, bedding coat only (no finish coats). Nail heads, one coat only.
- Maximum allowable variation in the tile substrate—for tiles with all edges shorter than 15," maximum allowable variation is ¼" in 10' from the required plane, with no more than 1/16" variation in 12" when measured from the high points in the surface. For tiles with at least one edge 15" in length, maximum allowable variation is ⅛" in 10' from the required plane, with no more than 1/16" variation in 24" when measured from the high points in the surface.

Movement Joint (architect must specify type of joint and show location and details on drawings)

- Movement joints—mandatory according to EJ171.
- When glass tile is used, adhere to more frequent placement recommendations within the ranges listed in EJ171.

Installation Specifications

- Gypsum board—GA-216.
- Tile—ANSI A108.4 or A108.5.
- Glass tile—manufacturer's directions.

- Cementitious grout—ANSI A108.10.
- Epoxy grout—ANSI A108.6.
- Movement Joints—EJ171 and ASTM C1193.

Notes

- As the tile size increases, there is less tolerance for variation in the substrate from the required plane. Organic adhesive thickness must be thin and uniform; therefore, substrate flattening may be required when organic adhesive is used.
- Do not use organic adhesives to bond tile to a membrane or other impervious substrates.
- Extended dry time may be required for larger tiles and tiles with low water absorption when organic adhesive is used.
- When glass tile is used, see "Glass Tile Selection and Installation Guide," and consult manufacturer for recommendations and requirements.

C315-21

- **Backer Board**
- **Ceramic Tile, Glass Tile**

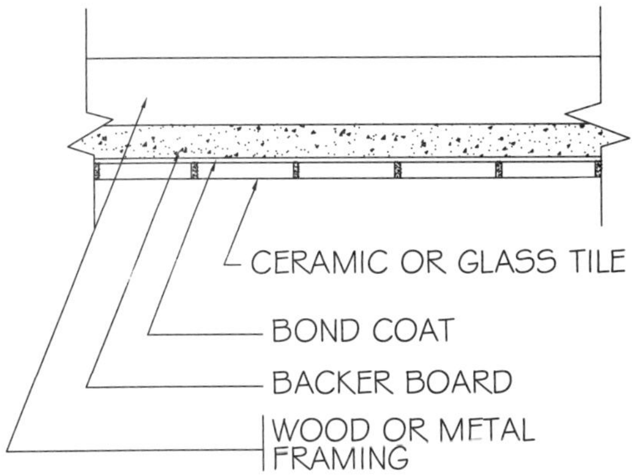

Recommended Uses

- For ceilings in dry or wet areas.

Environmental Exposure Classifications

- With cementitious bond coat—Res1, 2, 3; Com1, 2.
- With organic adhesive—Res1, 2; Com1, 2.
- For Res4, see SR614.

Limitations

- The weight of the tile shall not exceed 5 psf for ½"-thick backer board with framing spacing 24" o.c., or 6 psf for ½"-thick backer board with framing spacing 16" o.c.
- When organic adhesive is used—not for areas exposed to temperatures exceeding 140°F.
- When organic adhesive is used—maximum tile size 8" × 8" unless organic adhesive manufacturer allows larger tile size. See Notes.
- When backer board is coated glass mat gypsum backer board or fiber-reinforced gypsum backer board, not for areas exposed to temperatures exceeding 125°F.

Requirements

- Surface—free of coatings, oil, and wax.
- Ceiling framing must be capable of supporting the weight of backer board and tile.
- Minimum thickness of backer board—½".
- Fasten backer board as recommended by the manufacturer.
- For glass tile over coated glass mat water-resistant gypsum backer board, consult glass tile manufacturer for tile suitability over non-absorptive surface.

Preparation by Backer Board Installers

- Maximum allowable variation in the tile substrate—for tiles with all edges shorter than 15," maximum allowable variation is ¼" in 10' from the required plane, with no more than ¹⁄₁₆" variation in 12" when measured from the high points in the surface. For tiles with at least one edge 15" in length, maximum allowable variation is ⅛" in 10' from the required plane, with no more than ¹⁄₁₆" variation in 24" when measured from the high points in the surface.
- Follow backer board manufacturer's directions for installation, including orientation and spacing of boards, required fasteners, and taping of joints and corners.

Materials

- **Multiple options exist for membranes, mortars, grouts, and other materials and *must be clearly specified* to be included. If not specifically indicated, optional materials are not included and mortar/grout choice defaults to minimum performance specification indicated. Consider each system component and intended use to determine minimum requirements and to specify options.**
- Ceramic tile—ANSI A137.1.
- Glass tile, when used—ANSI A137.2; see also "Glass Tile Selection and Installation Guide," and consult tile manufacturer for environmental exposure classification recommendations. Not all glass tiles are suitable.
- Cementitious grout—ANSI A118.6 or better or ISO CG1 or better. When glass tile is used, specify grout designated by tile and grout manufacturers.
- Epoxy grout, when used—ANSI A118.3 or ISO RG.
- Cementitious bond coat:
  - When cement backer board, fiber-cement backer board, or cementitious-coated extruded foam backer board is used—ANSI A118.1 or better or ISO C1 or better.
  - When coated glass mat water-resistant gypsum or fiber-reinforced water-resistant gypsum backer board is used—ANSI A118.4 or better or ISO C2S1 or better.
  - When porcelain tile is used—ANSI A118.4 or better or ISO C2 or better.
  - When glass tile is used, specify mortar designated by tile and mortar manufacturers. Bond coat color will impact the final appearance of translucent glass tile. Specifier shall confirm bond coat color is acceptable.
- Organic adhesive bond coat, when used—ANSI A136.1 (Type I or II) or ISO D1 or better.
- Cement backer board, when used—ANSI A118.9 or ASTM C1325.
- Fiber-cement backer board, when used—ASTM C1288.
- Cementitious-coated extruded foam backer board, when used—ASTM C578.
- Coated glass mat water-resistant gypsum backer board, when used—ASTM C1178.
- Fiber-reinforced water-resistant gypsum backer board, when used—ASTM C1278 (Paragraph 6.1).

- Metal studs—ASTM C645.

Materials for Green/Sustainable Design

- See "Green Building Standards and Green Product Selection Guide" and consult manufacturers and suppliers for product sustainability and contribution to green building design.
- Consider specifying tile and installation materials that meet ANSI A138.1, the *American National Standard Specifications for Sustainable Ceramic Tiles, Glass Tiles, and Tile Installation Materials.*

Preparation by Other Trades

- Framing shall meet the general framing requirements of ANSI A108.11-4.0-4.3.

Movement Joint (architect must specify type of joint and show location and details on drawings)

- Movement joints—mandatory according to EJ171.
- When glass tile is used, adhere to more frequent placement recommendations within the ranges listed in EJ171.

Installation Specifications

- Backer board—ANSI A108.11 or manufacturer's directions.
- Tile—ANSI A108.4 or A108.5.
- Glass tile—ANSI A108.15, .16, or manufacturer's directions.
- Cementitious grout—ANSI A108.10.
- Epoxy grout—ANSI A108.6.
- Movement Joints—EJ171 and ASTM C1193.

Notes

- As the tile size increases, there is less tolerance for variation in the substrate from the required plane. Organic adhesive thickness must be thin and uniform; therefore, substrate flattening may be required when organic adhesive is used.
- Do not use organic adhesives to bond tile to a membrane or other impervious substrates.
- Extended dry time may be required for larger tiles and tiles with low water absorption when organic adhesive is used.
- When glass tile is used, see "Glass Tile Selection and Installation Guide," and consult manufacturer for recommendations and requirements.

C511-21

- **Plywood Base**
- **Mortar Bed**
- **Ceramic Tile, Glass Tile**

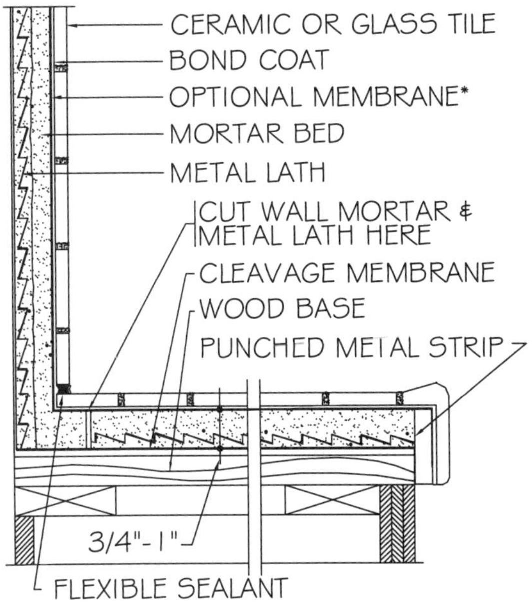

*USE OF A MEMBRANE IS OPTIONAL. SEE MEMBRANE OPTIONS.

Recommended Uses

- On countertops, drainboards, lavatory tops, etc.
- Preferred method where sink or lavatory is to be recessed.

Environmental Exposure Classifications

- Res1, 2; Com1, 2.
- For installations that may be exposed to staining, specify tile and grout suitable for exposure. Consult product manufacturers; see also "Product Selection Guides."
- For installations that may be exposed to mild chemical attack, specify epoxy grout and tile suitable for exposure. Consult product manufacturers; see also "Product Selection Guides."

Membrane Options

- A waterproof membrane (A118.10) may be specified to prevent moisture intrusion. Specifier shall indicate if complete waterproofing is required, including treatment at termination points.
- Check with membrane manufacturers for suitability for applicable conditions, as not all membranes are suitable for steam, high temperature, and/or chemical exposure.
- When glass tile is used, consult glass tile manufacturer for membrane options and recommendations.

Requirements

- Set the bottom edge of the countertop trim the proper distance above the finish floor material to allow clearance for dishwashers, compactors, etc.
- Protect wood substrate with cleavage membrane (ANSI A108.02-3.8).
- Cut lath off at corner as shown.

Materials

- **Multiple options exist for membranes, mortars, grouts, and other materials and *must be clearly specified* to be included. If not specifically indicated, optional materials are not included and mortar/grout choice defaults to minimum performance specification indicated. Consider each system component and intended use to determine minimum requirements and to specify options.**
- Ceramic tile—ANSI A137.1.
- Glass tile, when used—ANSI A137.2; see also "Glass Tile Selection and Installation Guide," and consult tile manufacturer for environmental exposure classification recommendations. Not all glass tiles are suitable.
- Cementitious grout—ANSI A118.6 or better or ISO CG1 or better. When glass tile is used, specify grout designated by tile and grout manufacturers.
- Epoxy grout, when used—ANSI A118.3 or ISO RG.
- Cementitious bond coat—portland cement paste on a mortar bed that is still workable. For a cured mortar bed, follow recommendations below to select appropriate bonding mortar:
  - ANSI A118.1 or better or ISO C1 or better.
  - When porcelain tile is used—ANSI A118.4 or better or ISO C2 or better.
  - When glass tile is used, specify mortar designated by tile and mortar manufacturers. Bond coat color will impact the final appearance of translucent glass tile. Specifier shall confirm bond coat color is acceptable.
- Mortar bed, lath, and cleavage membrane—ANSI A108.1A.
- Flexible sealant—ASTM C920 or proper sealant for food preparation areas. Follow sealant manufacturer's instructions.

Materials for Green/Sustainable Design

- See "Green Building Standards and Green Product Selection Guide" and consult manufacturers and suppliers for product sustainability and contribution to green building design.
- Consider specifying tile and installation materials that meet ANSI A138.1, the *American National Standard Specifications for Sustainable Ceramic Tiles, Glass Tiles, and Tile Installation Materials.*

Preparation by Other Trades

- Wood base—1" × 6" boards with ¼" gap between boards or ¾" exterior glue plywood with dot and dash saw cuts

6" to 8" on center through the length of the plywood board to prevent warping.

- Where overhangs or cantilever counters are used, adequate support must be provided to prevent movement.

Installation Specifications

- Tile—ANSI A108.1A, .1B, or .1C. A108.1B required if membrane to be used.
- Glass tile—ANSI A108.14, .15, .16, or manufacturer's directions.
- Cementitious grout—ANSI A108.10.
- Epoxy grout—ANSI A108.6.
- Waterproof membrane—ANSI A108.13.
- Movement Joints—EJ171 and ASTM C1193.

Notes

- Use tile recommended by manufacturer for countertop applications.
- When glass tile is used, see "Glass Tile Selection and Installation Guide," and consult manufacturer for recommendations and requirements.
- A punched metal strip attached to the front edge of the cabinet is used in some geographical areas as a screed and support for the countertop trim. It is filled with wall mortar.

C512-21

- **Plywood Base**
- **Plywood Underlayment**
- **Ceramic Tile**

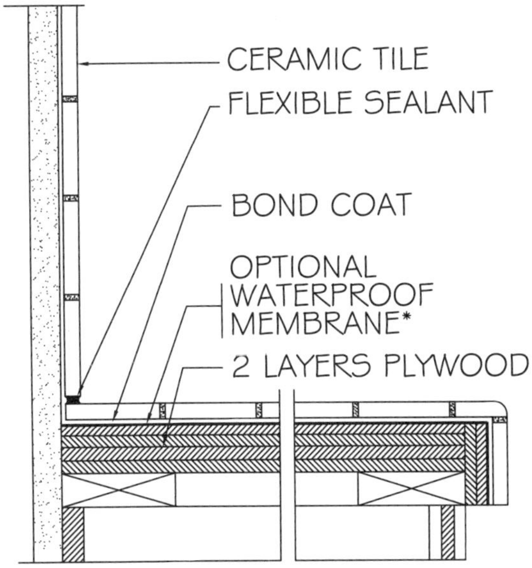

*USE OF A MEMBRANE IS OPTIONAL. SEE MEMBRANE OPTIONS.

Recommended Uses

- On countertops where thin-bed installation is desired.

Environmental Exposure Classifications

- Res1, 2; Com1, 2.
- For installations that may be exposed to staining, specify tile and grout suitable for exposure. Consult product manufacturers; see also "Product Selection Guides."
- For installations that may be exposed to mild chemical attack, specify epoxy grout and tile suitable for exposure. Consult product manufacturers; see also "Product Selection Guides."

Membrane Options

- A waterproof membrane (A118.10) may be specified to prevent moisture intrusion. Specifier shall indicate if complete waterproofing is required, including treatment at termination points.
- Check with membrane manufacturers for suitability for applicable conditions, as not all membranes are suitable for steam, high temperature, and/or chemical exposure.
- Protect plywood from exposure to water and high humidity.
- Use a waterproof membrane (ANSI A118.10) to seal edge and bottom of plywood overhang at face of counter and all other plywood surfaces and edges that will be exposed to water or moisture, especially around automatic dishwashers.

Requirements

- Double plywood layer-clean and free of dirt, dust, and oily film.
- Gaps between plywood sheets to be treated per setting material manufacturer's recommendations.
- Set the bottom edge of the countertop trim the proper distance above the finish floor material to allow clearance for dishwashers, compactors, etc.
- When tile is set with epoxy, completely fill gap between sheets of plywood with epoxy.

Materials

- **Multiple options exist for membranes, mortars, grouts, and other materials and *must be clearly specified* to be included. If not specifically indicated, optional materials are not included and mortar/grout choice defaults to minimum performance specification indicated. Consider each system component and intended use to determine minimum requirements and to specify options.**
- Ceramic tile—ANSI A137.1.
- Cementitious grout—ANSI A118.6 or better or ISO CG1 or better.
- Epoxy grout, when used—ANSI A118.3 or ISO RG.
- Cementitious bond coat:
  - When a waterproof membrane is not used—ANSI A118.11 or better or ISO C2S1P1 or better.
  - When a waterproof membrane is used—ANSI A118.4 or better or ISO C2S1 or better, unless ANSI A118.1 or ISO C1 is recommended by membrane manufacturer (mortar for bonding tile to membrane).
- When a waterproof membrane is used, follow membrane manufacturer's directions for bonding membrane to plywood underlayment.
- Epoxy bond coat, when used—ANSI A118.3 or ISO R1 or better.
- Flexible sealant—ASTM C920 or proper sealant for food preparation areas. Follow sealant manufacturer's instructions.
- Waterproof membrane, when used—ANSI A118.10.

Materials for Green/Sustainable Design

- See "Green Building Standards and Green Product Selection Guide" and consult manufacturers and suppliers for product sustainability and contribution to green building design.
- Consider specifying tile and installation materials that meet ANSI A138.1, the *American National Standard Specifications for Sustainable Ceramic Tiles, Glass Tiles, and Tile Installation Materials.*

Preparation by Other Trades

- When tile is set with epoxy, leave $\frac{1}{4}$" gap between sheets of plywood. Apply batten to underside of sheets to cover gap.
- Wood base—1" × 6" boards with $\frac{1}{4}$" gap between boards or $\frac{3}{4}$" exterior glue plywood with dot and dash saw cuts 6" to 8" on center through the length of the plywood board to prevent warping.
- Where overhangs or cantilever counters are used, adequate support must be provided to prevent movement.
- Bottom layer—$\frac{19}{32}$" exterior glue plywood.
- Underlayment—$\frac{19}{32}$" exterior glue plywood.

Installation Specifications

- Tile—ANSI A108.12.
- Cementitious grout—ANSI A108.10.
- Epoxy mortar/grout—ANSI A108.6.
- Movement Joints—EJ171 and ASTM C1193.

Notes

- Use tile recommended by manufacturer for countertop applications.

C513-21

- **Plywood Base**
- **Backer Board**
- **Ceramic Tile, Glass Tile**

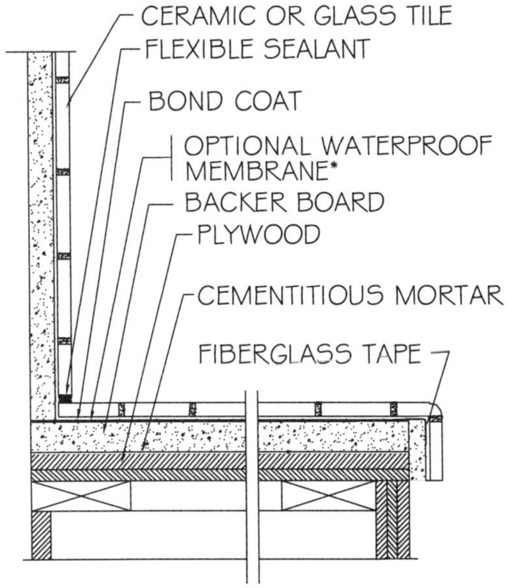

*USE OF A MEMBRANE IS OPTIONAL. SEE MEMBRANE OPTIONS.

Recommended Uses

- Preferred thin-set mortar method where self-rimming sinks and lavatories are desired on countertops, drainboards, lavatory tops, etc.

Environmental Exposure Classifications

- Res1, 2; Com1, 2.
- For installations that may be exposed to staining, specify tile and grout suitable for exposure. Consult product manufacturers; see also "Product Selection Guides."
- For installations that may be exposed to mild chemical attack, specify epoxy grout and tile suitable for exposure. Consult product manufacturers; see also "Product Selection Guides."

Membrane Options

- A waterproof membrane (A118.10) may be specified to prevent moisture intrusion. Specifier shall indicate if complete waterproofing is required, including treatment at termination points.
- Check with membrane manufacturers for suitability for applicable conditions, as not all membranes are suitable for steam, high temperature, and/or chemical exposure.
- When glass tile is used, consult glass tile manufacturer for membrane options and recommendations.
- When coated glass mat water-resistant gypsum backer board is used—some liquid-applied waterproof membranes will require extended cure time. Consult membrane manufacturer.

Requirements

- Set the bottom edge of countertop trim the proper distance above the finish floor material to allow clearance for dishwashers and compactors.
- For glass tile over coated glass mat water-resistant gypsum backer board, consult glass tile manufacturer for tile suitability over non-absorptive surface.

Preparation by Backer Board Installers

- Maximum allowable variation in the tile substrate—for tiles with all edges shorter than 15," maximum allowable variation is ¼" in 10' from the required plane, with no more than 1/16" variation in 12" when measured from the high points in the surface. For tiles with at least one edge 15" in length, maximum allowable variation is ⅛" in 10' from the required plane, with no more than 1/16" variation in 24" when measured from the high points in the surface.
- Follow backer board manufacturer's directions for installation, including orientation and spacing of boards, required fasteners, and taping of joints and corners.

Materials

- **Multiple options exist for membranes, mortars, grouts, and other materials and *must be clearly specified* to be included. If not specifically indicated, optional materials are not included and mortar/grout choice defaults to minimum performance specification indicated. Consider each system component and intended use to determine minimum requirements and to specify options.**
- Ceramic tile—ANSI A137.1.
- Glass tile, when used—ANSI A137.2; see also "Glass Tile Selection and Installation Guide," and consult tile manufacturer for environmental exposure classification recommendations. Not all glass tiles are suitable.
- Cementitious grout—ANSI A118.6 or better or ISO CG1 or better. When glass tile is used, specify grout designated by tile and grout manufacturers.
- Epoxy grout, when used—ANSI A118.3 or ISO RG.
- Cementitious bond coat:
  - When cement backer board, fiber cement backer board, or cementitious-coated extruded foam backer board is used—ANSI A118.1 or better or ISO C1 or better.
  - When coated glass mat water-resistant gypsum or fiber-reinforced water-resistant gypsum backer board is used—ANSI A118.4 or better or ISO C2S1 or better.
  - When porcelain tile is used—ANSI A118.4 or better or ISO C2 or better.
  - When glass tile is used, specify mortar designated by tile and mortar manufacturers. Bond coat color will impact the final appearance of translucent glass tile. Specifier shall confirm bond coat color is acceptable.

- Flexible sealant—ASTM C920 or proper sealant for food preparation areas. Follow sealant manufacturer's instructions.
- Nails—1¼" galvanized roofing type preferably screw shank, or other corrosion-resistant fasteners, applied 6" on center.
- Waterproof membrane, when used—ANSI A118.10.

Materials for Green/Sustainable Design

- See "Green Building Standards and Green Product Selection Guide" and consult manufacturers and suppliers for product sustainability and contribution to green building design.
- Consider specifying tile and installation materials that meet ANSI A138.1, the *American National Standard Specifications for Sustainable Ceramic Tiles, Glass Tiles, and Tile Installation Materials.*

Preparation by Other Trades

- Provide support on overhangs or cantilever counters to prevent movement.
- Maximum allowable variation in plywood surface—⅛" in 10' and $\frac{1}{16}$" in 1' from the required plane.
- Install plywood base flat and level.
- $\frac{23}{32}$" exterior glue plywood base.

Installation Specifications

- Backer board—ANSI A108.11 or manufacturer's directions.
- Tile—ANSI A108.5.
- Glass tile—ANSI A108.15, .16, or manufacturer's directions.
- Cementitious grout—ANSI A108.10.
- Epoxy grout—ANSI A108.6.
- Waterproof membrane—ANSI A108.13.
- Movement Joints—EJ171 and ASTM C1193.

Notes

- Use tile recommended by manufacturer for countertop applications.
- When glass tile is used, see "Glass Tile Selection and Installation Guide," and consult manufacturer for recommendations and requirements.

SR613-21

- **Masonry or Concrete**
- **Mortar Bed Walls**
- **Mortar Bed Floor**
- **Ceramic Tile**

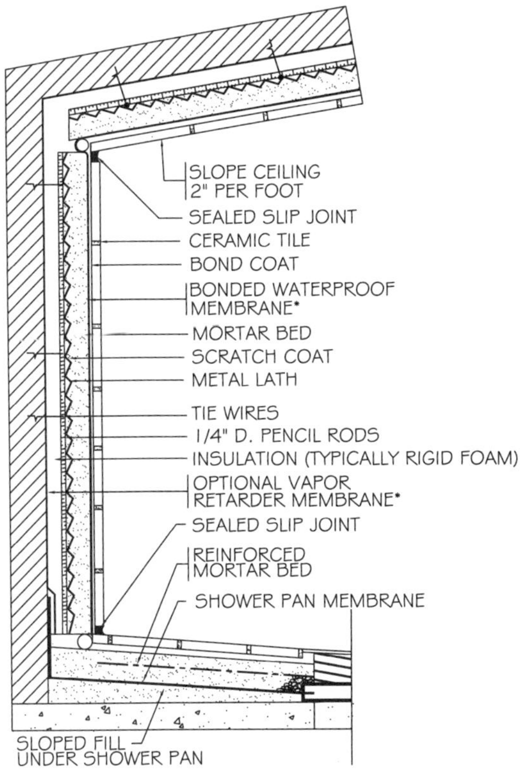

SHOWER RECEPTORS, CURBS, SEATS, ETC., MUST BE PROPERLY DETAILED AND INSTALLED TO AVOID WATER DAMAGE TO ADJACENT BUIILDING MATERIALS. SEE COMMON SHOWER CONFIGURATIONS SECTION.

*FOR CONTINUOUS USE STEAM SHOWERS:
IF BONDED WATERPROOF MEMBRANE HAS A WATER VAPOR PERMEANCE GREATER THAN 0.5 PERMS, A SECONDARY VAPOR RETARDER MEMBRANE IS REQUIRED. SEE REQUIREMENTS. WHEN USED, VAPOR RETARDER MEMBRANE MUST OVERLAP INTO SHOWER PAN MEMBRANE.

CONFIGURATION OF BASE
WITH INTEGRATED BONDING FLANGE

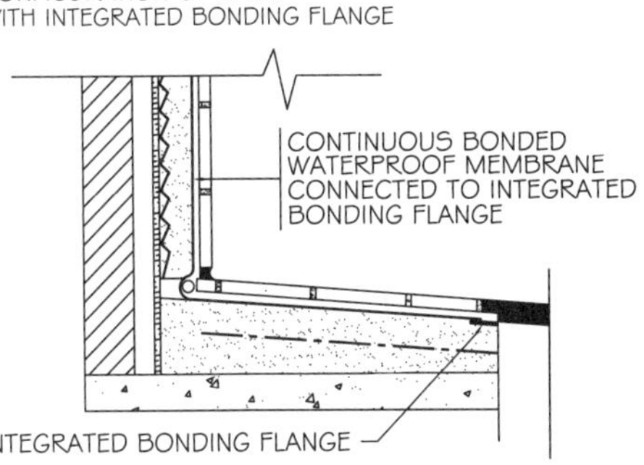

Recommended Uses

- For steam showers and steam rooms where the substrate is concrete or masonry.

Environmental Exposure Classifications

- Res4; Com4.
- For installations that may be exposed to staining, specify tile and grout suitable for exposure. Consult product manufacturers; see also Notes and "Product Selection Guides."

Limitations

- Duration of use as a steam room or steam shower as determined by membrane manufacturer; lower membrane water vapor permeance decreases water vapor transmission.

Requirements

- Steam rooms are highly specialized applications. Design and installation are critical to avoid damage to adjoining materials from vapor migration. Design criteria must include consideration of necessary insulation and temperature and humidity differential.
- Use a tile contractor knowledgeable in steam room installation and experienced with the materials specified.
- Bonded waterproof membrane (sheet, liquid, and trowel-on) must be continuous and must adequately limit vapor transmission into adjacent spaces and building materials, according to intended duration of use as a steam room or steam shower.
- Steam showers designed for continuous use applications should specify a low perm waterproof membrane (a waterproof membrane meeting ANSI A118.10 and with a water vapor permeance rating of 0.5 perms or less when tested per ASTM E96 Procedure E, tested at 90% relative humidity). When a waterproof membrane with a water vapor permeance rating greater than 0.5 perms is specified, a vapor retarder behind the wall assembly is required, and vapor retarder must have a water vapor permeance rating of 0.1 perm or less when tested per ASTM E96 Procedure A, tested at 50% relative humidity. Consult waterproof membrane manufacturer for water vapor permeance rating and vapor retarder requirements.
- Specifier shall indicate how waterproofing and vapor retarding is to be achieved, including details for membrane penetrations such as penetrations for plumbing, lighting fixtures, tie wires, etc. Specifier shall also indicate where and how to waterproof curbs and jambs and where membrane terminates. Area outside steam shower door is a wet area and should be treated accordingly.
- Check with membrane manufacturer for suitability for applicable conditions, as not all membranes are suitable for steam, high temperature and/or chemical exposure.
- Some waterproof membrane manufacturers require use of a vapor retarder membrane in addition to the waterproof membrane. Consult membrane manufacturer for requirements. When used, vapor retarder membrane must weather lap itself and lap into the shower pan membrane.

- If vapor retarder membrane is required, integrated bonding flange cannot be used.
- Follow waterproof membrane manufacturers' directions for interface between drain and membrane(s).
- Design professional to specify adequate insulation on walls and ceilings to reduce condensation. Consult insulation manufacturer for application suitability.
- Seal all membrane penetrations with appropriate sealant according to membrane manufacturer's requirements.
- Specify slip joints at changes in plane, such as where walls and ceiling meet.
- Slope ceiling minimum 2" per foot to avoid condensate dripping onto occupants; sloping ceiling from center can minimize rundown on walls.
- Masonry/concrete must provide firm anchorage for metal lath.
- Attach four equally spaced tie wires through the insulation. Attach ¼" diameter steel pencil rods vertically over the insulation. Attach metal lath to pencil rods on walls and ceilings.
- Mortar bed thickness—¾" minimum to 1½" maximum.
- Cut lath at all slip joints.
- Slope shower pan membrane minimum ¼" per foot to weep holes in drain.
- Turn shower pan membrane up walls a minimum of 3" above shower curb (6" above floor in showers without curbs).
- Surround drain with pea gravel or other weep protection to prevent mortar from blocking weep holes. (Does not apply if integrated bonding flange is used.)

Materials

- **Multiple options exist for membranes, mortars, grouts, and other materials and *must be clearly specified* to be included. If not specifically indicated, optional materials are not included and mortar/grout choice defaults to minimum performance specification indicated. Consider each system component and intended use to determine minimum requirements and to specify options.**
- Ceramic tile—A137.1 and recommended by manufacturer for use in steam rooms.
- Cementitious grout—ANSI A118.6 or better or ISO CG1 or better.
- Epoxy grout, when used—ANSI A118.3 or ISO RG.
- Cementitious bond coat:
  - ANSI A118.15 or better or ISO C2S1 or better, unless ANSI A118.1 or ISO C1 is recommended by membrane manufacturer.
- Waterproof membrane—ANSI A118.10 and recommended by membrane manufacturer for use in specific application. For continuous use applications, see Requirements for water vapor permeance.
- Shower pan membrane—ANSI A118.10, ASTM D4068, or D4551 and meeting applicable building code.
- Mortar bed and reinforcing—ANSI A108.1A
- Vapor retarder membrane, when used—recommended by manufacturer for use in specific application.
- Flexible mildew-resistant sealant—ASTM C920 and recommended by manufacturer for use in steam showers.

Materials for Green/Sustainable Design

- See "Green Building Standards and Green Product Selection Guide" and consult manufacturers and suppliers for product sustainability and contribution to green building design.
- Consider specifying tile and installation materials that meet ANSI A138.1, the *American National Standard Specifications for Sustainable Ceramic Tiles, Glass Tiles, and Tile Installation Materials.*

Preparation by Other Trades

- Maximum allowable variation in the installation substrate (concrete/masonry)—¼" in 10' from the required plane.

Movement Joint (architect must show type of joint and show location and details on drawings)

- Movement joints—mandatory according to EJ171.
- Slip joints at changes in plane, such as where walls and ceiling meet.

Installation Specifications

- Shower pan membrane—ANSI A108.01-3.6.
- Mortar bed and reinforcing—ANSI A108.1A, .1B, and .1C.
- Tile—ANSI A108.5.
- Cementitious grout—ANSI A108.10.
- Epoxy grout—ANSI A108.6.
- Waterproof membrane and slip joint—ANSI A108.13 and membrane manufacturer's slip joint directions.
- Movement Joints—EJ171 and ASTM C1193.

Notes

- Use of softened water in steam showers and steam rooms helps reduce grout and tile staining due to iron and/or hard water. Such stains may require harsh chemicals for removal. Select products suitable for water type and maintenance practices that will be used.
- Standard grouts will need to be periodically maintained over the life of the steam shower.
- Steam unit design must take into consideration the affect of moisture vapor transmission (MVT) on opposite side of steam unit walls. MVT can cause efflorescence and can affect paints and other adhered finishes.

SR614-21

- **Wood or Metal Studs**
- **Mortar Bed or Cement Backer Board Walls**
- **Mortar Bed Floor**
- **Ceramic Tile**

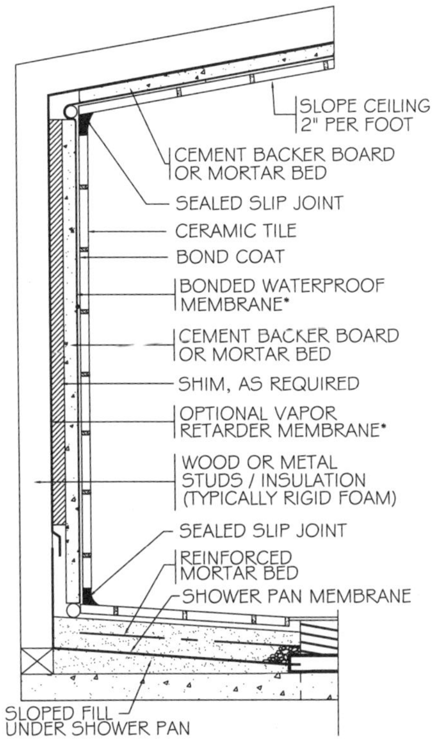

SHOWER RECEPTORS, CURBS, SEATS, ETC., MUST BE PROPERLY DETAILED AND INSTALLED TO AVOID WATER DAMAGE TO ADJACENT BUILDING MATERIALS. SEE COMMON SHOWER CONFIGURATIONS SECTION.

*FOR CONTINUOUS USE STEAM SHOWERS: IF BONDED WATERPROOF MEMBRANE HAS A WATER VAPOR PERMEANCE GREATER THAN 0.5 PERMS, A SECONDARY VAPOR RETARDER MEMBRANE IS REQUIRED. SEE REQUIREMENTS. WHEN USED, VAPOR RETARDER MEMBRANE MUST OVERLAP INTO SHOWER PAN MEMBRANE.

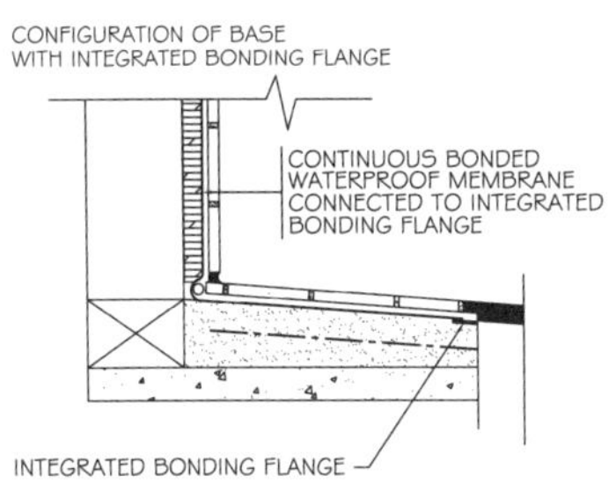

Recommended Uses

- For steam showers and steam rooms framed with wood or metal studs. Specify mortar bed walls option when wall flatness is critical, such as when tiles with any edge longer than 15" are specified.

Environmental Exposure Classifications

- Res4.
- For installations that may be exposed to staining, specify tile and grout suitable for exposure. Consult product manufacturers; see also Notes and "Product Selection Guides."

Limitations

- Duration of use as a steam shower or steam room as determined by membrane and cement backer board (when used) manufacturers; lower membrane water vapor permeance decreases water vapor transmission.
- Maximum stud spacing 16" on center.

Requirements

- Steam rooms are highly specialized applications. Design and installation are critical to avoid damage to adjoining materials from vapor migration. Design criteria must include consideration of necessary insulation and temperature and humidity differential.
- Use a tile contractor knowledgeable in steam applications and experienced with the materials specified.
- Bonded waterproof membrane (sheet, liquid, and trowel-on) must be continuous and must adequately limit vapor transmission into adjacent spaces and building materials, according to intended duration of use as a steam shower.
- Steam showers designed for continuous use applications should specify a low perm waterproof membrane (a waterproof membrane meeting ANSI A118.10 and with a water vapor permeance rating of 0.5 perms or less when tested per ASTM E96 Procedure E, tested at 90% relative humidity). When a waterproof membrane with a water vapor permeance rating greater than 0.5 perms is specified, a vapor retarder behind the wall assembly is required, and vapor retarder must have a water vapor permeance rating of 0.1 perm or less when tested per ASTM E96 Procedure A, tested at 50% relative humidity. Consult waterproof membrane manufacturer for water vapor permeance rating and vapor retarder requirements.
- Specifier shall indicate how waterproofing and vapor retarding is to be achieved, including details for membrane penetrations such as penetrations for plumbing, lighting fixtures, fasteners, etc. Specifier shall also indicate where and how to waterproof curbs and jambs and where membrane terminates. Area outside steam shower door is a wet area and should be treated accordingly.
- Check with membrane manufacturer for suitability for applicable conditions, as not all membranes are suitable for steam, high temperature and/or chemical exposure.
- Some waterproof membrane manufacturers require use of a vapor retarder membrane in addition to the waterproof membrane. Consult membrane manufacturer for requirements. When used, vapor retarder membrane must weather lap itself and lap into the shower pan membrane.
- If vapor retarder membrane is required, integrated bonding flange cannot be used.
- Follow waterproof membrane manufacturers' directions for interface between drain and membrane(s).
- Design professional to specify adequate insulation on walls and ceilings to reduce condensation. Consult insulation manufacturer for application suitability.
- Seal all membrane penetrations with appropriate sealant according to membrane manufacturer's requirements.
- Specify slip joints at change in plane, such as where walls and ceiling meet.

- Slope ceiling minimum 2" per foot to avoid condensate dripping onto occupants; sloping ceiling from center can minimize rundown on walls.
- When mortar bed walls, studs must provide firm anchorage for metal lath.
- When mortar bed walls, mortar bed wall thickness—¾" minimum to 1 ½" maximum.
- When mortar bed walls, cut lath at all slip joints.
- Slope shower pan membrane minimum ¼" per foot to weep holes in drain.
- Turn shower pan membrane up walls a minimum of 3" above shower curb (6" above floor in showers without curbs).
- Surround drain with pea gravel or other weep protection to prevent mortar from blocking weep holes. (Does not apply if integrated bonding flange is used.)
- Wood studs, when used—dry and well braced, minimum depth 3½".
- Metals studs, when used—well-braced; 20 gauge (0.033") or heavier; minimum depth 3½" for residential applications or 3⅝" for commercial applications.

Preparation by Backer Board Installers

- Maximum allowable variation in the tile substrate—for tiles with all edges shorter than 15", maximum allowable variation is ¼" in 10' from the required plane, with no more than ¹⁄₁₆" variation in 12" when measured from the high points in the surface. For tiles with at least one edge 15" in length, maximum allowable variation is ¼" in 10' from the required plane, with no more than ¹⁄₁₆" variation in 24" when measured from the high points in the surface.
- Horizontal joints—⅛" spacing filled solid and taped with latex-portland cement mortar and 2" alkali-resistant glass fiber mesh tape.
- Vertical joints—fill any space and tape with latex-portland cement mortar and 2" alkali-resistant glass fiber mesh tape.
- Corners—leave space between backer units. Tape joints using skim coat of latex-portland cement mortar, but do not fill.
- Center backer board end or edge joints on framing and stagger joints in adjacent rows so four corners do not come together within the same plane. Space panel ends and edges in accordance with manufacturer's recommendations.

Materials

- **Multiple options exist for membranes, mortars, grouts, and other materials and *must be clearly specified* to be included. If not specifically indicated, optional materials are not included and mortar/grout choice defaults to minimum performance specification indicated. Consider each system component and intended use to determine minimum requirements and to specify options.**
- Ceramic tile—ANSI A137.1 and recommended by manufacturer for use in steam showers.
- Cementitious grout—ANSI A118.6 or better or ISO CG1 or better.
- Epoxy grout, when used—ANSI A118.3 or ISO RG.
- Cementitious bond coat:
  - ANSI A118.15 or better or ISO C2S1 or better, unless ANSI A118.1 or ISO C1 is recommended by membrane manufacturer.
- Cement backer board, when used—ANSI A118.9 or ASTM C1325 (Type A).
- Fasteners—noncorrosive and nonoxidizing.
- Hot dipped fasteners meeting ASTM F2329 required in wet areas.
- 2" alkali-resistant glass fiber mesh tape.
- Waterproof membrane—ANSI A118.10 and recommended by membrane manufacturer for use in specific application. For continuous use applications, see Requirements for water vapor permeance.
- Vapor retarder membrane, when used—recommended by manufacturer for use in specific application.
- Metal studs, when used—ASTM C645.
- Shower pan membrane—ANSI A118.10, ASTM D4068, or D4551 and meeting applicable building code.
- Mortar bed and reinforcing—ANSI A108.1A.
- Flexible mildew-resistant sealant—ASTM C920 and recommended by manufacturer for use in steam showers.

Materials for Green/Sustainable Design

- See "Green Building Standards and Green Product Selection Guide" and consult manufacturers and suppliers for product sustainability and contribution to green building design.
- Consider specifying tile and installation materials that meet ANSI A138.1, the *American National Standard Specifications for Sustainable Ceramic Tiles, Glass Tiles, and Tile Installation Materials.*

Preparation by Other Trades

- Wall framing shall meet the general framing requirements of ANSI A108.11-4.0-4.3.

Movement Joint (architect must show type of joint and show location and details on drawings)

- Movement joints—mandatory according to EJ171.
- Slip joints at changes in plane, such as where walls and ceiling meet.

Installation Specifications

- Shower pan membrane—ANSI A108.01-3.6.
- Mortar bed and reinforcing—ANSI A108.1A, .1B, and .1C.
- Tile—ANSI A108.5.
- Cementitious grout—A108.10.
- Epoxy grout—A108.6.
- Waterproof membrane and slip joint—ANSI A108.13 and membrane manufacturer's slip joint directions.
- Movement Joints—EJ171 and ASTM C1193.

Notes

- Use of softened water in steam showers and steam rooms helps reduce grout and tile staining due to iron and/or hard water. Such stains may require harsh chemicals for removal. Select products suitable for water type and maintenance practices that will be used.
- Standard grouts will need to be periodically maintained over the life of the steam shower.
- Steam unit design must take into consideration the affect of moisture vapor transmission (MVT) on opposite side of steam unit walls. MVT can cause efflorescence and can affect paints and other adhered finishes.

P601MB-21

- **Watertight Concrete, Shotcrete, or CMU Pool Tank**
- **Bonded Mortar Bed**
- **Ceramic Tile, Glass Tile**

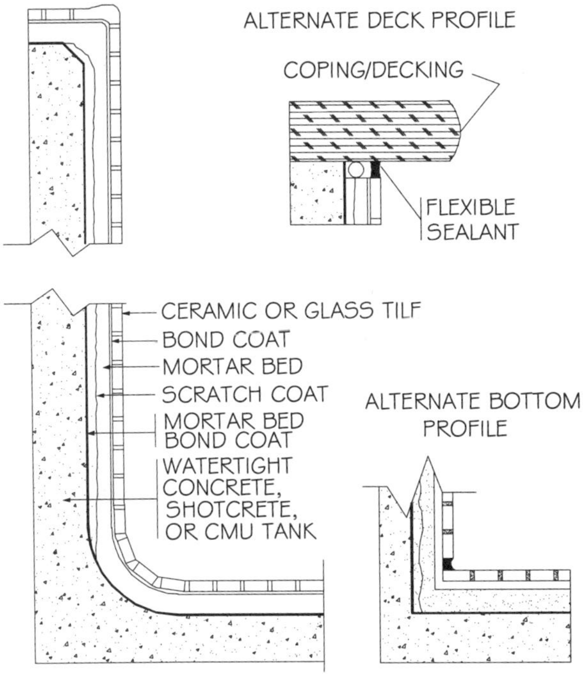

Recommended Uses

- For interior and exterior pools and water features with watertight tanks constructed of reinforced concrete, shotcrete, CMU, etc.
- For commercial, public, and competition pools.

Environmental Exposure Classifications

- Res3, 4, 6, 7; Com3, 4, 6, 7.

Requirements

- Pool tank must be watertight.
- Cement mortar bed nominal thickness—walls ¾", floors 1¼". Walls exceeding ¾" require a cured scratch coat.
- Cure cement mortar beds a minimum of 7 days prior to application of tile.
- Tile bond coat coverage—minimum 95% contact with tile and substrate.
- Pool tank surface must be free of grease, oil, wax, curing compounds, or other bond inhibiting coatings; pressure wash/mechanically scarify if necessary.
- Specify deck profile required, based on pool type/use.

Materials

- **Multiple options exist for membranes, mortars, grouts, and other materials and *must be clearly specified* to be included. If not specifically indicated, optional materials are not included and mortar/grout choice defaults to minimum performance specification indicated. Consider each system component and intended use to determine minimum requirements and to specify options.**
- Ceramic tile—ANSI A137.1 and recommended in writing by manufacturer for use in pools and water features. Tile not covered by A137.1 should be recommended in writing by the manufacturer for use in pools and water features.
- Glass tile, when used—ANSI A137.2 and recommended in writing by manufacturer for use in pools and water features. Tile not covered by A137.2 should be recommended in writing by the manufacturer for use in pools and water features; see also "Glass Tile Selection and Installation Guide," and consult tile manufacturer for environmental exposure classification recommendations. Not all glass tiles are suitable.
- Cementitious grout—ANSI A118.6 or better or ISO CG1 or better. Blue, green, and red grouts may not be suitable for submerged applications; consult manufacturer for application suitability. When glass tile is used, specify grout designated by tile and grout manufacturers.
- Epoxy grout, when used—ANSI A118.3 or ISO RG. Epoxy grout may be affected by UV exposure; consult manufacturer for application suitability.
- Cementitious bond coat—portland cement paste on a mortar bed that is still workable. For a cured mortar bed, follow recommendations below to select appropriate bonding mortar:
  - ANSI A118.15 or better or ISO C2S1 or better.
  - When glass tile is used, specify mortar designated by tile and mortar manufacturers. Bond coat color will impact the final appearance of translucent glass tile. Specifier shall confirm bond coat color is acceptable.
- Epoxy bond coat, when used—ANSI A118.3 or ISO R2 or better.
  - When glass tile is used, specify epoxy bond coat designated by tile and bond coat manufacturers. Bond coat color will impact the final appearance of translucent glass tile. Specifier shall confirm bond coat color is acceptable.
- Portland cement—ASTM C150 Type I.
- Sand—ASTM C144.
- Lime—ASTM C206 Type S or ASTM C207 Type S.
- Water—potable.
- Mortar bed (pool bottom)—1 part portland cement, 4 parts damp sand by volume.
- Scratch coat and mortar bed (pool walls and gutter)—1 part portland cement, ½ part lime, and 4 parts dry sand or 5 parts damp sand; or 1 part portland cement, 3 parts dry sand or 4 parts damp sand.

- Metal lath, when used (for gooseneck gutter)—3.4 pounds galvanized diamond mesh tied to reinforcing rods.
- Flexible Sealant—ASTM C920 and suitable for pool applications. Consult manufacturer.

Materials for Green/Sustainable Design

- See "Green Building Standards and Green Product Selection Guide" and consult manufacturers and suppliers for product sustainability and contribution to green building design.
- Consider specifying tile and installation materials that meet ANSI A138.1, the *American National Standard Specifications for Sustainable Ceramic Tiles, Glass Tiles, and Tile Installation Materials.*

Preparation by Other Trades

- When applicable, based on the soil report, pool tank must be engineered and constructed to support the tile installation and meet the requirements of the ANSI/APSP Pool and Spa Standards 1–5 and applicable building codes.
- Pool tank to be finished with medium-rough bush-hammer finish or with aggregate exposed.
- Deviations from dimensions, contours, or slopes in concrete, shotcrete, or CMU tank must not exceed ½" or encroach on the required minimum thickness of the mortar bed and tilework in order to provide exact dimensional requirements in length and width and specified tolerances.
- For pools and water features completely covered with tile, a water fill test (performed by pool contractor) is required.
- Defects in the tank (i.e., active water leaks, cracks, unsound concrete, shotcrete, CMU) must be repaired prior to the commencement of tilework.

Movement Joint (architect must specify type of joint and show location and details on drawings)

- Directly over any joints in the concrete tank (EJ171A or EJ171E).
- In tilework on 8' to 12' centers (EJ171F).
- Perimeter joint between the tile assembly and coping/decking (EJ171I).
- Perimeter joint at all changes in plane and at all restraining abutments (EJ171G).
- When glass tile is used, adhere to more frequent placement recommendations within the ranges listed in EJ171.

Installation Specifications

- Tile—ANSI A108.1A, .1B, or .1C (with cementitious bond coat) or A108.6 (with epoxy bond coat).
- Glass tile—ANSI A108.14, .15, .16, or manufacturer's directions.
- Cementitious grout—ANSI A108.10.
- Epoxy grout—ANSI A108.6.
- Movement Joints—EJ171 and ASTM C1193.

Notes

- Some tile, glazes, backings, and mounting systems are not suitable for exterior or submerged applications.
- Not all epoxy may be used in exterior applications. Consult manufacturer for epoxy bond coat and epoxy grout suitability.
- Not all bonding mortars are suitable for exterior or submerged use.
- Mortar bed performance may be improved by the use of modified mortar; consult manufacturer for recommendations.
- When glass tile is used, see "Glass Tile Selection and Installation Guide," and consult manufacturer for recommendations and requirements.
- Pool water chemistry and balance will impact the appearance and performance/serviceability of the tile and installation materials. It should be monitored/maintained by a pool maintenance professional. Refer to ANSI APSP-5 for residential or APSP-11 for water quality in public pools and spas for pool start-up and maintenance guidelines.
- Consult setting material and grout manufacturers for minimum cure times prior to submersion. Proper cure time prior to submersion is necessary.
- Pool may need to be protected from direct sunlight, excessive heat, wind, precipitation, and freezing during substrate preparation, installation, and curing.

P601TB-21

- **Watertight Concrete, Shotcrete, or CMU Pool Tank**
- **Ceramic Tile, Glass Tile**

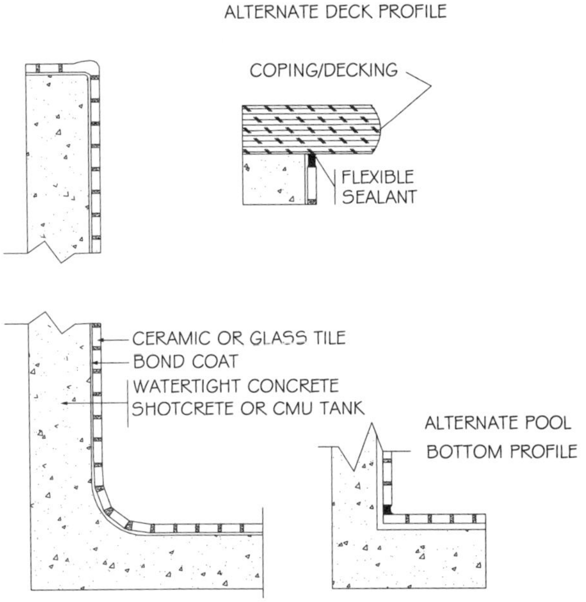

Recommended Uses

- For interior and exterior pools and water features with watertight tanks constructed of reinforced concrete, shotcrete, or CMU, etc., where thin-bed installation of tile is desired.

Environmental Exposure Classifications

- Res3, 4, 6, 7; Com3, 4, 6, 7.

Limitations

- Not for competition pools or other pools or water features with strict dimensional criteria; specify P601MB for such applications.
- Shotcrete tanks typically do not meet required substrate tolerances for thin-bed tile installation.

Requirements

- Pool tank must be watertight.
- Pool tank must meet ⅛" in 10' and ¹⁄₁₆" in 1' finish tolerance and have a concrete surface profile (CSP) of 3 or better. When these tolerances cannot be met, specify P601MB.
- Tile bond coat coverage—minimum 95% contact with tile and substrate.
- Pool tank surface must be free of grease, oil, wax, curing compounds, or other bond inhibiting coatings; pressure wash/mechanically scarify if necessary.
- Specify deck profile required, based on pool type/use.

Materials

- **Multiple options exist for membranes, mortars, grouts, and other materials and *must be clearly specified* to be included. If not specifically indicated, optional materials are not included and mortar/grout choice defaults to minimum performance specification indicated. Consider each system component and intended use to determine minimum requirements and to specify options.**
- Ceramic tile—ANSI A137.1 and recommended in writing by manufacturer for use in pools and water features. Tile not covered by A137.1 should be recommended in writing by the manufacturer for use in pools and water features.
- Glass tile, when used—ANSI A137.2 and recommended in writing by manufacturer for use in pools and water features. Tile not covered by A137.2 should be recommended in writing by the manufacturer for use in pools and water features; see also "Glass Tile Selection and Installation Guide," and consult tile manufacturer for environmental exposure classification recommendations. Not all glass tiles are suitable.
- Cementitious grout—ANSI A118.6 or better or ISO CG1 or better. Blue, green, and red grouts may not be suitable for submerged applications; consult manufacturer for application suitability. When glass tile is used, specify grout designated by tile and grout manufacturers.
- Epoxy grout, when used—ANSI A118.3 or ISO RG. Epoxy grout may be affected by UV exposure; consult manufacturer for application suitability.
- Cementitious bond coat:
  - ANSI A118.15 or better or ISO C2S1 or better.
  - When glass tile is used, specify mortar designated by tile and mortar manufacturers. Bond coat color will impact the final appearance of translucent glass tile. Specifier shall confirm bond coat color is acceptable.
- Epoxy bond coat, when used—ANSI A118.3 or ISO R2 or better.
  - When glass tile is used, specify epoxy bond coat designated by tile and bond coat manufacturers. Bond coat color will impact the final appearance of translucent glass tile. Specifier shall confirm bond coat color is acceptable.
- Flexible Sealant—ASTM C920 and suitable for pool applications. Consult manufacturer.

Materials for Green/Sustainable Design

- See "Green Building Standards and Green Product Selection Guide" and consult manufacturers and suppliers for product sustainability and contribution to green building design.

- Consider specifying tile and installation materials that meet ANSI A138.1, the *American National Standard Specifications for Sustainable Ceramic Tiles, Glass Tiles, and Tile Installation Materials.*

Preparation by Other Trades

- When applicable, based on the soil report, concrete tank must be engineered and constructed to support the tile installation and meet the requirements of the ANSI/APSP Pool and Spa Standards 1–5 and applicable building codes.
- Maximum allowable substrate variation—for tiles with all edges shorter than 15", maximum allowable variation is ⅛" in 10' from the required plane, with no more than ¹⁄₁₆" in 1' from the required plane, when measured from the high points in the surface. For tiles with at least one edge 15" in length, maximum allowable variation is ⅛" in 10' from the required plane, with no more than ¹⁄₁₆" variation in 24" when measured from the high points in the surface.
- For pools and water features completely covered with tile, a water fill test (performed by pool contractor) is required.
- Defects in the tank (i.e., active water leaks, cracks, unsound concrete, shotcrete, CMU) must be repaired prior to the commencement of tilework.

Movement Joint (architect must specify type of joint and show location and details on drawings)

- Directly over any joints in the concrete tank (EJ171A or EJ171E).
- In tilework on 8' to 12' centers (EJ171F).
- Perimeter joint between the tile assembly and coping/decking (EJ171I).
- Perimeter joint at all changes in plane and at all restraining abutments (EJ171G).
- When glass tile is used, adhere to more frequent placement recommendations within the ranges listed in EJ171.

Installation Specifications

- Tile—ANSI A108.5 (with cementitious bond coat) or A108.6 (with epoxy bond coat).
- Glass tile—manufacturer's directions.
- Cementitious grout—ANSI A108.10.
- Epoxy grout—ANSI A108.6.
- Movement Joints—EJ171 and ASTM C1193.

Notes

- Some tile, glazes, backings, and mounting systems are not suitable for exterior or submerged applications.
- Not all epoxy may be used in exterior applications. Consult manufacturer for epoxy bond coat and epoxy grout suitability.
- Not all bonding mortars are suitable for exterior or submerged use.
- When glass tile is used, see "Glass Tile Selection and Installation Guide," and consult manufacturer for recommendations and requirements.
- Pool water chemistry and balance will impact the appearance and performance/serviceability of the tile and installation materials. It should be monitored/maintained by a pool maintenance professional. Refer to ANSI APSP-5 for residential or APSP-11 for water quality in public pools and spas for pool start-up and maintenance guidelines.
- Consult setting material and grout manufacturers for minimum cure times prior to submersion. Proper cure time prior to submersion is necessary.
- Pool may need to be protected from direct sunlight, excessive heat, wind, precipitation, and freezing during substrate preparation, installation, and curing.

P602-21

- **Concrete, Shotcrete, or CMU Pool Tank**
- **Cementitious Waterproofing**
- **Bonded Mortar Bed**
- **Ceramic Tile, Glass Tile**

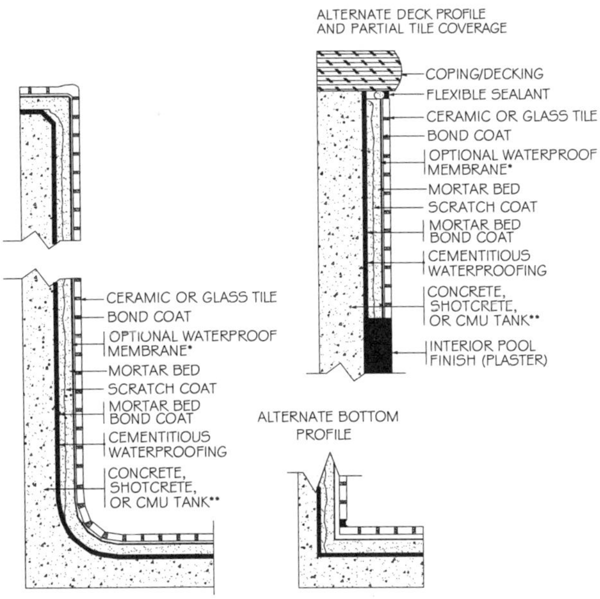

*USE OF AN A118.10 WATERPROOF MEMBRANE IS OPTIONAL. SEE MEMBRANE OPTIONS.

*MAY BE TREATED WITH A PENETRATING COLLOIDAL SILICATE PRIOR TO APPLICATION OF CEMENTITIOUS WATERPROOFING. SEE PREPARATION BY OTHER TRADES.

Recommended Uses

- For interior and exterior pools and water features with tanks constructed of reinforced concrete, shotcrete—wet mix, shotcrete—dry mix, CMU, etc., that are not watertight.

Environmental Exposure Classifications

- Res3, 4, 6, 7; Com3, 4, 6, 7.

Requirements

- Pool tank and water feature construction must be reinforced concrete, shotcrete, CMU, etc.
- Cement mortar bed nominal thickness—walls ¾", floors 1¼". Walls exceeding ¾" require a cured scratch coat.
- Cure cement mortar beds a minimum of 7 days prior to application of tile or bonded waterproof membrane (A118.10).
- Tile bond coat coverage—minimum 95% contact with tile and substrate.
- Pool tank surface must be free of grease, oil, wax, curing compounds, or other bond inhibiting coatings; pressure wash/mechanically scarify if necessary.
- Specify deck profile required, based on pool type/use.

Materials

- **Multiple options exist for membranes, mortars, grouts, and other materials and *must be clearly specified* to be included. If not specifically indicated, optional materials are not included and mortar/grout choice defaults to minimum performance specification indicated. Consider each system component and intended use to determine minimum requirements and to specify options.**
- Ceramic tile—ANSI A137.1 and recommended in writing by manufacturer for use in pools and water features. Tile not covered by A137.1 should be recommended in writing by the manufacturer for use in pools and water features.
- Glass tile, when used—ANSI A137.2 and recommended in writing by manufacturer for use in pools and water features. Tile not covered by A137.2 should be recommended in writing by the manufacturer for use in pools and water features; see also "Glass Tile Selection and Installation Guide," and consult tile manufacturer for environmental exposure classification recommendations. Not all glass tiles are suitable.
- Cementitious grout—ANSI A118.6 or better or ISO CG1 or better. Blue, green, and red grouts may not be suitable for submerged applications; consult manufacturer for application suitability. When glass tile is used, specify grout designated by tile and grout manufacturers.
- Epoxy grout, when used—ANSI A118.3 or ISO RG. Epoxy grout may be affected by UV exposure; consult manufacturer for application suitability.
- Cementitious bond coat:
  - When a waterproof membrane is not used on the face of the mortar bed—ANSI A118.15 or better or ISO C2S1 or better.
  - When a waterproof membrane is used on the face of the mortar bed—ANSI A118.15 or better or ISO C2S1 or better unless ANSI A118.1 or ISO C1 is recommended by membrane manufacturer.
  - When glass tile is used, specify mortar designated by tile and mortar manufacturers. Bond coat color will impact the final appearance of translucent glass tile. Specifier shall confirm bond coat color is acceptable.
- Epoxy bond coat, when used—ANSI A118.3 or ISO R2 or better.
  - When glass tile is used, specify epoxy bond coat designated by tile and bond coat manufacturers. Bond coat color will impact the final appearance of translucent glass tile. Specifier shall confirm bond coat color is acceptable.
- Waterproof membrane on top of the mortar bed, when used—ANSI A118.10.
- Walls—cement mortar bed/scratch coat 1 part portland cement (ASTM C150) to 4 parts damp sand (ASTM C144) by volume.
- Floors—cement mortar bed 1 part portland cement (ASTM C150) to 4 to 5 parts damp sand (ASTM C144) by volume mixed with water to a consistency and workability to allow maximum compaction during tamping (ANSI A108.1A-2.2.2).
- Mortar bed bond coat—latex-portland cement mortar meeting ANSI A118.4 or better or ISO C2S1 or better.
- Cementitious waterproofing—recommended by

manufacturer for use in positive and negative hydrostatic pressure applications and swimming pools. Must be able to support the weight of the bonded mortar bed and tile assembly; consult manufacturer for application suitability.

- Flexible Sealant—ASTM C920 and suitable for pool applications. Consult manufacturer.

Materials for Green/Sustainable Design

- See "Green Building Standards and Green Product Selection Guide" and consult manufacturers and suppliers for product sustainability and contribution to green building design.
- Consider specifying tile and installation materials that meet ANSI A138.1, the *American National Standard Specifications for Sustainable Ceramic Tiles, Glass Tiles, and Tile Installation Materials.*

Membrane Options

- A waterproof membrane (A118.10) may be specified for use over the mortar bed, which may be beneficial in freeze-thaw climates or when additional resistance to positive hydrostatic pressure is required.
- Confirm membrane suitability for specified application with manufacturer. Not all membranes are suitable for exterior use, submersion, high-temperature, chemical exposure, or partial coverage applications such as waterlines.
- When glass tile is used, consult glass tile manufacturer for membrane options and recommendations.

Preparation by Other Trades

- Deviations from dimensions, contours, or slopes in concrete, shotcrete, or CMU tank must not exceed ½" or encroach on the required minimum thickness of the mortar bed and tilework in order to provide exact dimensional requirements in length and width and specified tolerances.
- When applicable, based on the soil report, concrete tank must be engineered and constructed to support the tile installation and meet the requirements of the ANSI/APSP Pool and Spa Standards 1–5 and applicable building codes.
- Concrete tank to be finished per cementitious waterproofing manufacturer's requirements.
- Concrete tank to be made watertight with cementitious waterproofing.
- A penetrating colloidal silicate may be applied to the pool tank prior to application of cementitious waterproofing to densify the tank and reduce the migration of water soluble mineral compounds (efflorescence). Consult penetrating colloidal silicate manufacturer for recommendations. When used, confirm compatibility with cementitious waterproofing manufacturer.
- For pools and water features completely covered with tile, a water fill test (performed by the waterproofing contractor) after the cementitious waterproofing has been applied and cured per manufacturer's directions is required.
- Defects in the tank (i.e., active water leaks, cracks, unsound concrete, shotcrete, CMU, etc.) must be repaired prior to the commencement of tilework.

Movement Joint (architect must show type of joint and show location and details on drawings)

- Directly over any joints in the concrete tank (EJ171A or EJ171E).
- In tilework on 8' to 12' centers (EJ171F).
- Perimeter joint between the tile assembly and coping/decking (EJ171I).
- Perimeter joint at all changes in plane and at all restraining abutments (EJ171G).
- When glass tile is used, adhere to more frequent placement recommendations within the ranges listed in EJ171.

Installation Specifications

- Tile—ANSI A108.5 (with cementitious bond coat) or A108.6 (with epoxy bond coat).
- Glass tile—ANSI A108.14, .15, .16, or manufacturer's directions.
- Cementitious grout—ANSI A108.10.
- Epoxy grout—ANSI A108.6.
- Waterproof membrane—ANSI A108.13.
- Cementitious waterproofing—manufacturer's directions.
- Movement Joints—EJ171 and ASTM C1193.
- Penetrating Colloidal Silicate—manufacturer's directions.

Notes

- Some tile, glazes, backings, and mounting systems are not suitable for exterior or submerged applications.
- Not all epoxy may be used in exterior applications. Consult manufacturer for epoxy bond coat and epoxy grout suitability.
- Not all bonding mortars are suitable for exterior or submerged use.
- In freeze-thaw climates, additional waterproofing consideration may be required to prevent cement mortar bed moisture intrusion.
- Mortar bed performance may be improved by the use of modified mortar; consult manufacturer for recommendations.
- When glass tile is used, see "Glass Tile Selection and Installation Guide," and consult manufacturer for recommendations and requirements.
- Pool water chemistry and balance will impact the appearance and performance/serviceability of the tile and installation materials. It should be monitored/maintained by a pool maintenance professional. Refer to ANSI APSP-5 for residential or APSP-11 for water quality in public pools and spas for pool start-up and maintenance guidelines.
- Consult setting material and grout manufacturers for minimum cure times prior to submersion. Proper cure time prior to submersion is necessary.
- Pool may need to be protected from direct sunlight, excessive heat, wind, precipitation, and freezing during substrate preparation, installation, and curing.

B417-21

- **Concrete Tank**
- **Mortar Bed**

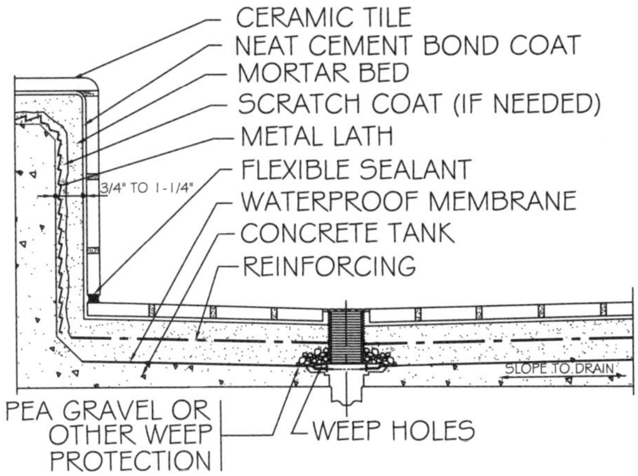

- **Wood Form**
- **Mortar Bed**

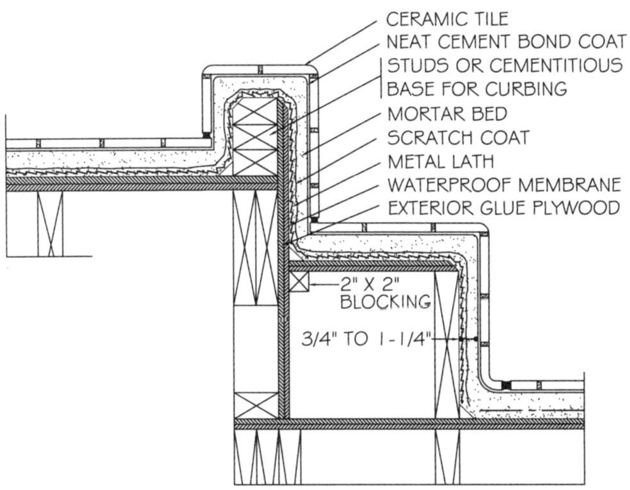

- **Concrete Curb**
- **Mortar Bed**

- **Wood Curb**
- **Mortar Bed**

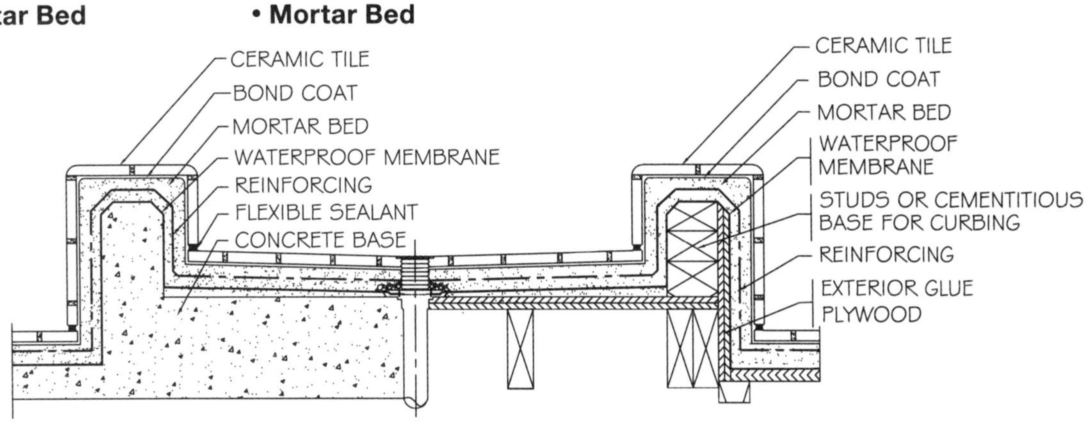

Requirements

- Waterproof membrane—slope membrane ¼" per foot to weep holes in drain.
- Wood framing, when used, should be pressure treated and designed to resist deflection and movement.

Preparation by Other Trades

- Test tank, membrane, and drainage fittings for leaks before starting tilework.

Materials for Green/Sustainable Design

- See "Green Building Standards and Green Product Selection Guide" and consult manufacturers and suppliers for product sustainability and contribution to green building design.
- Consider specifying tile and installation materials that meet ANSI A138.1, the *American National Standard Specifications for Sustainable Ceramic Tiles, Glass Tiles, and Tile Installation Materials.*

Installation Methods

- Attach metal lath only above water line.
- Floor—follow F121.
- Walls—follow W221.

Installation Specifications

- Tile—ANSI A108.1A.
- Grout—ANSI A108.10.
- Movement Joints—EJ171 and ASTM C1193.

NOTE: This detail reflects both concrete and wood substrates. Specify one or the other.

S151-21

• Concrete Stairs

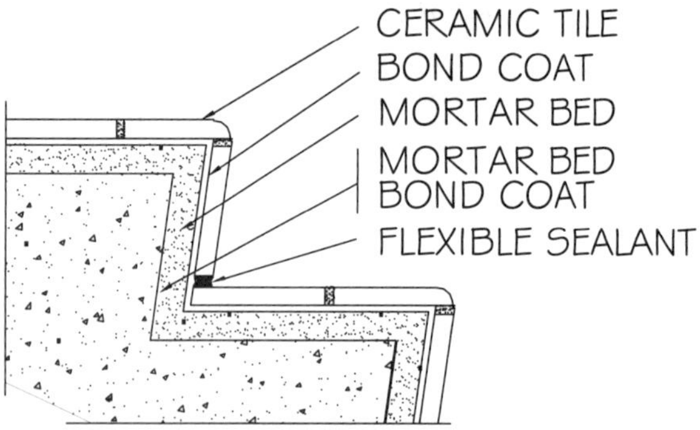

- Concrete to be finished with medium-rough bush-hammer finish and be free of cracks, waxy or oily films, and curing compounds.

• Metal Stairs

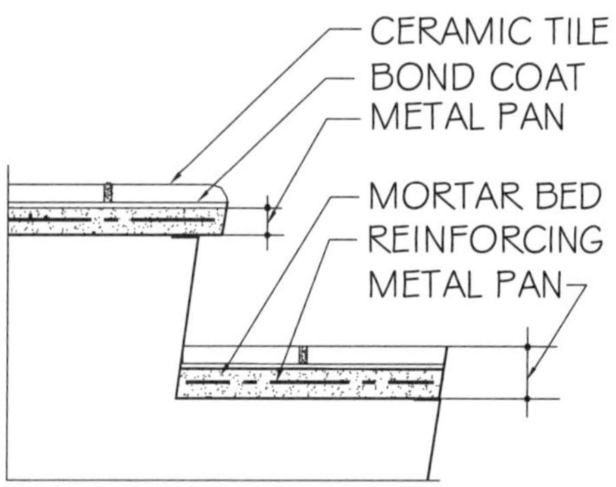

- Reinforcing mesh mandatory. Attach to metal by tack welds or other means.
- Metal stair riser may be tiled. Cut mortar bed and reinforcing at juncture of tread and riser.

• Wood Stairs

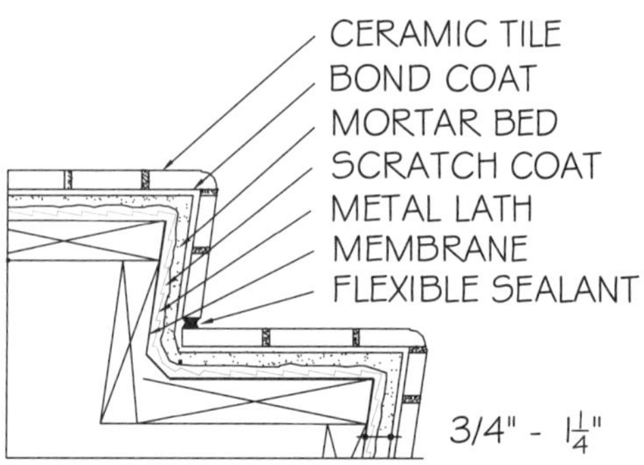

- Waterproof membrane for exterior and wet area installations; cleavage membrane for interiors.

Design Considerations—all stairs

- Use cove tile at junction of riser and tread for easy maintenance. Quarry or paver tile cove can be set horizontally or vertically to facilitate layout.
- Finished step nosings are available in specially shaped quarry and paver tile pieces.
- Use full radius ceramic mosaic bullnose tile for nosings.
- Slip-resistant tile should be specified on stair treads.

Materials for Green/Sustainable Design

- See "Green Building Standards and Green Product Selection Guide" and consult manufacturers and suppliers for product sustainability and contribution to green building design.
- Consider specifying tile and installation materials that meet ANSI A138.1, the *American National Standard Specifications for Sustainable Ceramic Tiles, Glass Tiles, and Tile Installation Materials.*

Installation Specifications

- Metal stairs—See F111. Membrane not required.
- Concrete stairs—See F112; risers see W211.
- Wood stairs—See F141; risers see W231.
- Movement Joints—EJ171 and ASTM C1193.

R612-21

- **Mortar Bed Walls**
- **Mortar Bed Floor**
- **Ceramic Tile**

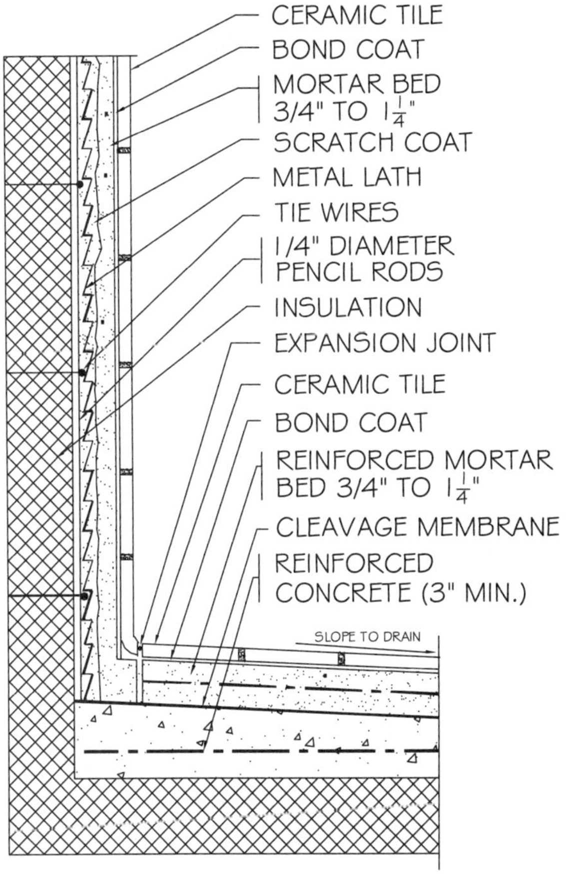

Requirements

- Use tile designated as frost proof by the manufacturer in all rooms subject to freezing temperatures.
- Provide reinforced concrete slab over floor insulation to resist wheel load of hand and mechanical trucks. (3" minimum; 4" or more for forklift trucks.)

Material and Tile Installation

- Attach four equally spaced tie wires to the supporting members and through the insulation. Attach ¼" D steel pencil rods vertically over the insulation.
- Attach metal lath to pencil rods and follow W222, or install on a mortar bed meeting the requirements of W241.
- Floors and movement joints may be installed in accordance to F112, F113, or F131.

Materials for Green/Sustainable Design

- See "Green Building Standards and Green Product Selection Guide" and consult manufacturers and suppliers for product sustainability and contribution to green building design.
- Consider specifying tile and installation materials that meet ANSI A138.1, the *American National Standard Specifications for Sustainable Ceramic Tiles, Glass Tiles, and Tile Installation Materials.*

TR711-21

• Tile Over Other Surfacing Materials/Walls and Floors

Ceramic tile may be considered as a surfacing material over existing wall finishes such as paint, cold glazes (sprayed on plastic), plastic laminates, and steel plate, or existing floor surfacing such as epoxy coatings, paint, vinyl or asphalt tile, seamless flooring,* exposed concrete† and steel plate. Ideally, existing finishes should be completely removed so that the tilework can be placed on the substructure following *Handbook* methods. However, because this is not always practical, the following is intended as a general guide for renovation with ceramic tile. In all cases, consult the setting material manufacturer before starting the work. Consideration should be given to covering the existing surface with a more suitable base. For example: badly cracked or irregular walls should be overlaid with firmly attached gypsum board or backer board to provide a sound tile-setting base.

TR711-21

• with Organic Adhesive, Interiors Only

Suitable Backings

- Smooth walls of all types including plaster, gypsum board, cement backer board, and masonry.
- Smooth floors of all types including wood, concrete, and terrazzo in residences or areas of equivalent residential performance requirements. Deflection not to exceed l/360.
- New gypsum board nailed and/or adhesively applied over existing walls.
- Plastic laminate countertops and walls.

Requirements

- Backing surface must be sound, clean, and dry.
- Abrupt irregularities, such as trowel marks, ridges, and grains, shall be less than $\frac{1}{32}$" above adjacent area.

Preparation

- Roughen surfaces which are glossy, painted, or which have loose surface material by sanding or scarifying.*
- Surface material must be removed if not compatible with adhesive.
- Use primer when recommended by the adhesive manufacturer for particular backings.
- Clean thoroughly to remove all oil, dirt, and dust.
- Apply underlayment as needed according to manufacturer's directions.

Materials for Green/Sustainable Design

- See "Green Building Standards and Green Product Selection Guide" and consult manufacturers and suppliers for product sustainability and contribution to green building design.
- Consider specifying tile and installation materials that meet ANSI A138.1, the *American National Standard Specifications for Sustainable Ceramic Tiles, Glass Tiles, and Tile Installation Materials.*

Installation Specifications

- Tile—ANSI A108.4.
- Grout—ANSI A108.10.
- Movement Joints—EJ171 and ASTM C1193.

---

*WARNING: Do not sand existing resilient flooring. Certain older resilient floor coverings, including sheet vinyl floor covering and vinyl tile, may contain asbestos fibers that are not readily identifiable. Inhalation of asbestos dust may cause serious bodily harm. Smoking greatly increases any such risk of serious bodily harm.

CAUTION: Mechanical or chemical abrasion of tile can release fine particles which could cause harm if inhaled or ingested. Mineral analysis of the tile and glaze should be performed before performing any operation. Appropriate safety equipment should be worn at all times.

†Special installation precautions are necessary when installing tile using a thin-bed method over old concrete floors in bakeries, kitchens, and meat-processing areas. Fats and greases penetrate into concrete floors and cannot be completely neutralized. Note preparation sections included in renovation methods.

TR711-21

• with Cementitious Bond Coat, Interior and Exterior

Suitable Backings

- Prepared portland cement plaster, concrete, concrete masonry, structural clay tile, or brick.
- Backer board applied over existing walls or floors.
- New gypsum board applied over properly furred existing wall in dry areas. Use water-resistant gypsum backer board in wet areas.

Requirements

- The backing surface must be sound, clean, and dry.

Preparation

- Roughen concrete or masonry walls and floors which are glossy, painted, or effloresced, or which have loose surface material. This should be accomplished by sandblasting, chipping, or scarifying.
- Clean thoroughly to remove all sealers, coatings, oil, dirt, and dust to expose masonry surface.

Materials for Green/Sustainable Design

- See "Green Building Standards and Green Product Selection Guide" and consult manufacturers and suppliers for product sustainability and contribution to green building design.
- Consider specifying tile and installation materials that meet ANSI A138.1, the *American National Standard Specifications for Sustainable Ceramic Tiles, Glass Tiles, and Tile Installation Materials.*

Installation Specifications

- Tile—ANSI A108.5.
- Grout—ANSI A108.10.
- Movement Joints—EJ171 and ASTM C1193.

TR711-21

• with Tile-Setting Epoxy Mortar, Epoxy Adhesive, or Modified Epoxy Emulsion Mortar

Suitable Backings

- Generally all sound wall and floor finishes.
- Especially valuable for setting tile floors over non-masonry surfaces where moderate performance level is required.
- Suitable for speedy installation where downtime must be kept to a minimum.

Requirements

- Backing surface must be sound, clean, and dry.

Preparation

- Roughen surfaces which are glossy, painted, or effloresced, or which have loose surface material, by sanding or scarifying.*
- Clean thoroughly to remove all waxes, oil, dirt, and dust.
- With epoxy adhesives, use primer when recommended by the manufacturer as proper for the particular backing.

Materials for Green/Sustainable Design

- See "Green Building Standards and Green Product Selection Guide" and consult manufacturers and suppliers for product sustainability and contribution to green building design.
- Consider specifying tile and installation materials that meet ANSI A138.1, the *American National Standard Specifications for Sustainable Ceramic Tiles, Glass Tiles, and Tile Installation Materials.*

Installation Specifications

- Tile-setting epoxy mortar and epoxy adhesive—ANSI A108.6.
- Modified epoxy emulsion mortar—ANSI A108.9.
- Epoxy formulations vary with respect to chemical resistance and use on vertical surfaces. Consult manufacturer's specifications.
- Grout—ANSI A108.10.
- Movement Joints—EJ171 and ASTM C1193.

*WARNING: Do not sand existing resilient flooring. Certain older resilient floor coverings, including sheet vinyl floor covering and vinyl tile, may contain asbestos fibers that are not readily identifiable. Inhalation of asbestos dust may cause serious bodily harm. Smoking greatly increases any such risk of serious bodily harm.

CAUTION: Mechanical or chemical abrasion of tile can release fine particles which could cause harm if inhaled or ingested. Mineral analysis of the tile and glaze should be performed before performing any operation. Appropriate safety equipment should be worn at all times.

TR712-21

• Tile Over Tile Interior Floors

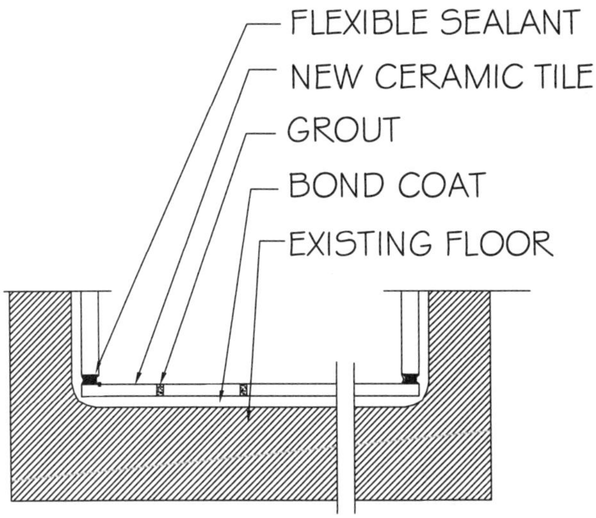

Recommended Uses

- For alteration of ceramic-tiled areas where modernization or a change of design is desired in residences, motels and hotels, restaurants, public rest rooms, malls, etc.
- Also applicable to smooth floors of terrazzo, stone, slate, etc.

Requirements

- Existing installation must be sound, well-bonded, and without structural cracks.
- When possible, remove floor-mounted plumbing and heating fixtures before beginning work.
- Threshold required to adjust between adjacent floors; see TR611.

Preparation

- Remove soap scum, wax, coatings, oil, etc., from existing tile surfaces. Mechanical abrasion with a Carborundum disk followed by a clear water wash is recommended. Other cleaning methods involve use of soapless detergents, commercial tile cleaners, and, in special cases, solvents or acids. Solvents and acids should be used with care and only when necessary because of their hazardous nature. (Appropriate safety equipment should be worn at all times.)
- Installation must be thoroughly rinsed and dry before setting the new tile.

Materials, Grouting, Movement Joints, and Installation Specifications

- Epoxy mortar installation—ANSI A108.6.
- Dry-set mortar or latex-portland cement mortar installation—ANSI A108.5.
- Organic adhesive installation—ANSI A108.4 and follow manufacturer's directions.
- Require current certification from adhesive manufacturer that adhesive conforms with ANSI A136.1.
- Grout—ANSI A108.10.
- Movement Joints—EJ171 and ASTM C1193.

Materials for Green/Sustainable Design

- See "Green Building Standards and Green Product Selection Guide" and consult manufacturers and suppliers for product sustainability and contribution to green building design.
- Consider specifying tile and installation materials that meet ANSI A138.1, the *American National Standard Specifications for Sustainable Ceramic Tiles, Glass Tiles, and Tile Installation Materials.*

Notes

- Use the "Floor Tiling Installation Guide" to select adequate installation method.
- If existing installation is not structurally sound, F111 and F141 may be applicable.

TR713-21

• Tile Over Tile Interior Walls

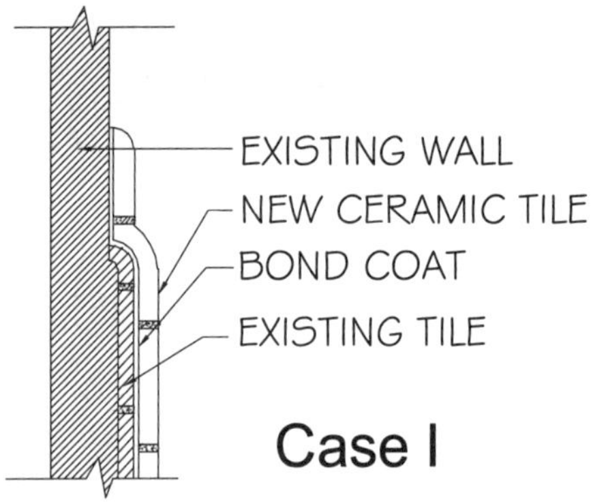

Case I

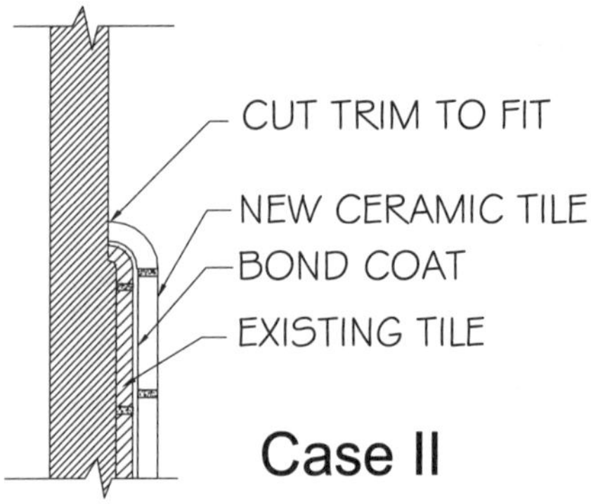

Case II

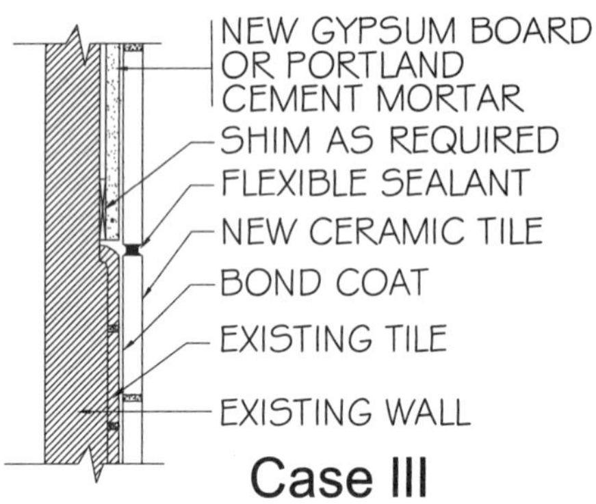

Case III

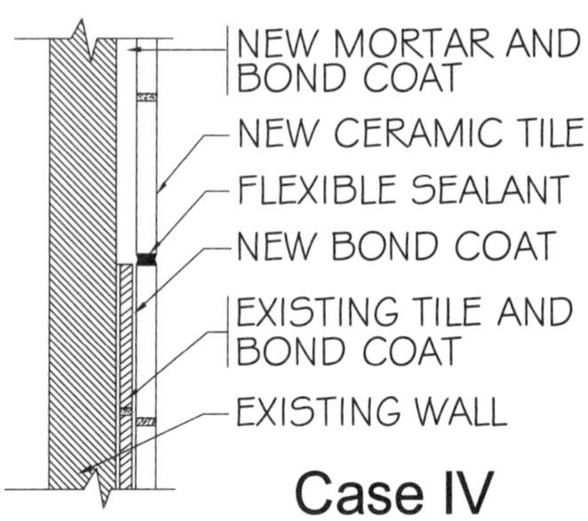

Case IV

Recommended Uses

- For alteration of ceramic-tiled areas where modernization or a change of design is desired in residences, motels and hotels, malls, restaurants, public rest rooms, etc.
- Also applicable to smooth walls of marble, stone, slate, etc.

Requirements

- Existing installation must be sound, well-bonded, and without major structural cracks.

Materials, Grouting, Movement Joints, and Installation Specifications

- For organic adhesive installation, see W223.
- For dry-set or latex-portland cement mortar installation, see W202.
- For epoxy adhesive installation, refer to manufacturer's literature.
- Movement Joints—EJ171 and ASTM C1193.

Materials for Green/Sustainable Design

- See "Green Building Standards and Green Product Selection Guide" and consult manufacturers and suppliers for product sustainability and contribution to green building design.
- Consider specifying tile and installation materials that meet ANSI A138.1, the *American National Standard Specifications for Sustainable Ceramic Tiles, Glass Tiles, and Tile Installation Materials.*

Preparation

- Remove soap scum, wax, coatings, oil, etc., from existing tile surfaces. Mechanical abrasion with a Carborundum disk followed by a clear water wash is recommended. Other cleaning methods involve use of soapless detergents, commercial tile cleaners, and, in special cases, solvents or acids. Solvents and acids should be used with care and only when necessary because of their hazardous nature.
- Installation must be thoroughly rinsed and dry before setting the new tile.
- CASE I—prepare wall above tile to receive trim tile as shown.
- CASE II—cut trim tile to fit over existing trim.
- CASES III and IV—apply new gypsum board above existing wainscot tile to prepare for full wall tiling.

Notes

- Use portland cement mortar or backer board in tub enclosures and shower stalls.
- In wet areas, the application of water-resistant gypsum backer board over any base which creates a vapor barrier, such as old tile or paint, will lead to failure unless such barrier is vented.
- If installation is not structurally sound, W221 and W222 may be applicable.

WS610-21

• Window Stool

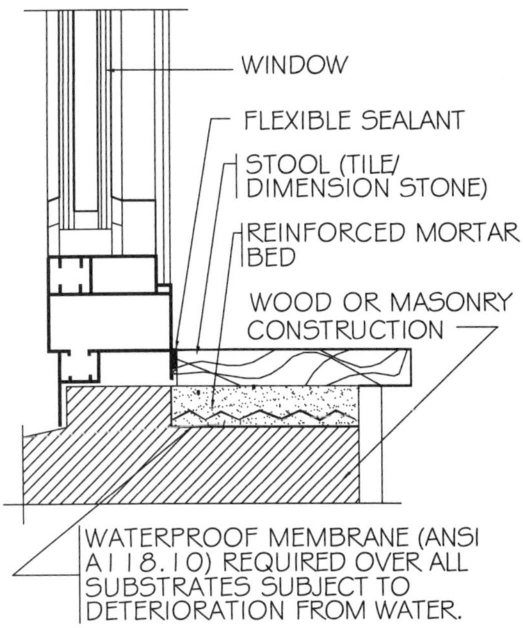

Materials for Green/Sustainable Design

- See "Green Building Standards and Green Product Selection Guide" and consult manufacturers and suppliers for product sustainability and contribution to green building design.
- Consider specifying tile and installation materials that meet ANSI A138.1, the *American National Standard Specifications for Sustainable Ceramic Tiles, Glass Tiles, and Tile Installation Materials.*

TR611-21

• Threshold

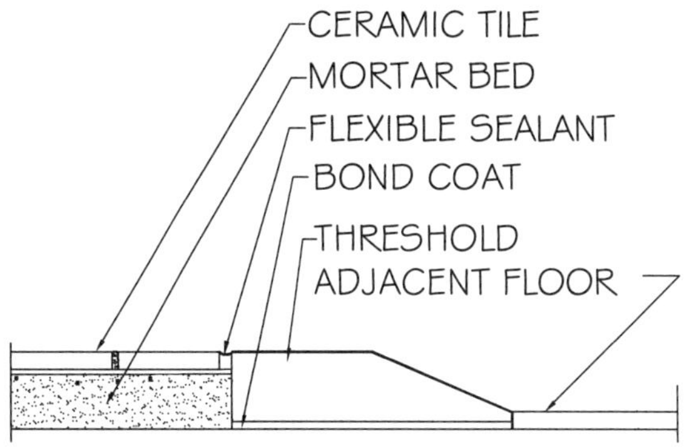

Materials for Green/Sustainable Design

- See "Green Building Standards and Green Product Selection Guide" and consult manufacturers and suppliers for product sustainability and contribution to green building design.
- Consider specifying tile and installation materials that meet ANSI A138.1, the *American National Standard Specifications for Sustainable Ceramic Tiles, Glass Tiles, and Tile Installation Materials.*

Notes

- Thresholds adjust levels between adjacent floors. Commercial and residential thresholds and window stools are available in tile, marble, stone, slate, etc., and can be made in virtually any size and shape to fit special conditions.
- Use 95% coverage of bonding material between threshold and floor or window stool and substrate.

RW800-21

• **Fire and Sound Ratings for Wall Assemblies**

WALL METHOD	RW 811	RW 812	RW 813	RW 814	RW 815	RW 816	RW 817	RW 818	RW 819	RW 820	RW 821	RW 822	RW 823	RW 824	RW 825
Fire Resistance Rating ASTM E119 (hours)	2	2	1	1	1	1	2	2	1	1	2	2	1	-	2
Field Sound Transmission Class (FSTC) ASTM E336, E413					51	53	50	52	61	60					
Sound Transmission Class (STC) ASTM E90, E413														48 / 52(14)	

Recommended Uses

- Wood stud load bearing—RW813, RW814.
- Wood stud load bearing chase wall—RW817, RW818.
- Metal stud non load bearing—3⅝" stud width: RW811, RW812, RW815, RW816, RW823, and RW825.
- Metal stud non load bearing—2½" stud width: RW824.
- Metal stud non load bearing chase wall—RW819, RW820, RW821, and RW822.

Limitations

- Stud spacing not to exceed 16" o.c.
- 3/32" minimum bond coat thickness.
- Cement backer board—maximum screw or nail spacing, 8" o.c.
- Gypsum board—maximum screw or nail spacing, 8" o.c. at edges and 12" o.c. in field for face layer; 24" o.c. for base layer.

Design Requirements (Test Reports)

- Underwriters Laboratory Inc. (ULI)
- United States Gypsum (USG)
- Warnock Hersey International (WHI)
- National Gypsum (NG)
- RW811, RW812—ULI 443 and 443 alt.
- RW813, RW814—ULI U329 and U329 alt.
- RW815, RW816—ULI 442 and 442 alt. (Fire Resistance). USG 840321 and 840313 (Sound Transmission).
- RW817, RW818—WHI 495-0505 and 495-0508 (Fire Resistance). USG 840523 and 840516 (Sound Transmission).
- RW819, RW820—ULI 445 and 445 alt. (Fire Resistance). USG 840524 and 840515 (Sound Transmission).
- RW821, RW822—ULI 444 and 444 alt.
- RW823—WHI-694-0189.
- RW824—NG C3086 and C3087.
- RW825—WHI-694-0400.11.

Preparation by Other Trades

- Gypsum board face layer joints must be taped and finished with minimum of two coats of joint compound. Finish fastener heads.
- Offset studs on each side of chase walls.

Preparation by Backer Board Installers

- Closely fit horizontal and vertical joints and corners without touching, leaving a maximum space of ⅛".
- Embed 2"-wide glass fiber mesh tape in a skim coat of the setting material over joints and corners.
- Center backer board end or edge joints on framing and stagger joints in adjacent rows so four corners do not come together within the same plane. Space panel ends and edges in accordance with manufacturer's recommendations.

Tile Installation

- RW823—any method over solid backing.
- RW824—see W222.
- RW825—see W244C.
- All others—see W223, W244C, or W244F.

Materials (Refer to Details)

- (1) Glazed ceramic wall tile—ANSI A137.1.
- (2) Cementitious bond coat—latex-portland cement mortar ANSI A118.4 or organic adhesive ANSI A136.1,Type 1.
- (3) Cementitious bond coat—dry-set mortar ANSI A118.1.
- (4) $\frac{1}{2}$"-thick cement backer board—certified by manufacturer as suitable for intended use.
- (5) $\frac{7}{16}$"-thick cement backer board—certified by manufacturer as suitable for intended use.
- (6) Metal studs—25 gauge, $3\frac{5}{8}$" × $1\frac{1}{4}$" ASTM C645. Floor and ceiling tracks, $3\frac{5}{8}$" × $1\frac{1}{4}$."
- (7) Metal studs—25 gauge, $2\frac{1}{2}$" ASTM C645. Floor and ceiling tracks, $2\frac{1}{2}$" × $1\frac{1}{4}$."
- (8) Metal studs—20 gauge, $3\frac{5}{8}$" ASTM C645. Floor and ceiling tracks, $3\frac{5}{8}$" × $1\frac{1}{4}$."
- (9) Wood studs—$3\frac{1}{2}$" × $1\frac{1}{2}$". Floor and ceiling plates $3\frac{1}{2}$" × $1\frac{1}{2}$".
- (10) Mineral fiber insulation—3" minimum thickness.
- (11) Mineral fiber insulation—$3\frac{1}{2}$" thick.
- (12) Mineral fiber insulation—$3\frac{5}{8}$" × 6" blanket, 2.5 pounds/cubic foot.
- (13) Mineral fiber insulation—$1\frac{1}{2}$" minimum thickness.
- (14) Mineral fiber insulation—2" thick. Required for STC rating of 53.
- (15) Gypsum board—$\frac{5}{8}$" thick, Type X, ASTM C1396/C1396M.
- (16) Cross brace—$\frac{1}{2}$" × 12" × chase width gypsum panel spaced 48" o.c. horizontally and vertically.
- (17) 4 mil. polyethylene film (when required as moisture barrier).
- (18) Metal lath—self-furring, galvanized or painted, expanded metal lath, 3.4 pounds/square yard.
- (19) Mortar bed—1 part portland cement, 1 part lime, 6 parts sand.
- (20) $\frac{13}{32}$"-thick fiber-cement backer board—certified by the manufacturer as suitable for the intended use.
- Standard tile grout.
- 2" glass fiber mesh tape.
- Screws—$1\frac{1}{4}$" Type S; $1\frac{5}{8}$" screw for double layer gypsum board.
- Nails—$1\frac{1}{2}$" galvanized roofing for cement backer board; $1\frac{3}{8}$" ring shank for gypsum wall board; $1\frac{7}{8}$" nail for double layer gypsum board.

Materials for Green/Sustainable Design

- See "Green Building Standards and Green Product Selection Guide" and consult manufacturers and suppliers for product sustainability and contribution to green building design.
- Consider specifying tile and installation materials that meet ANSI A138.1, the *American National Standard Specifications for Sustainable Ceramic Tiles, Glass Tiles, and Tile Installation Materials.*

RW811-21 **RW812-21**

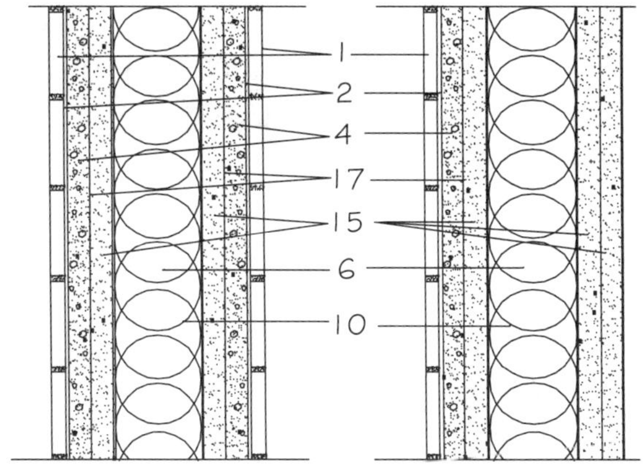

RW813-21 **RW814-21**

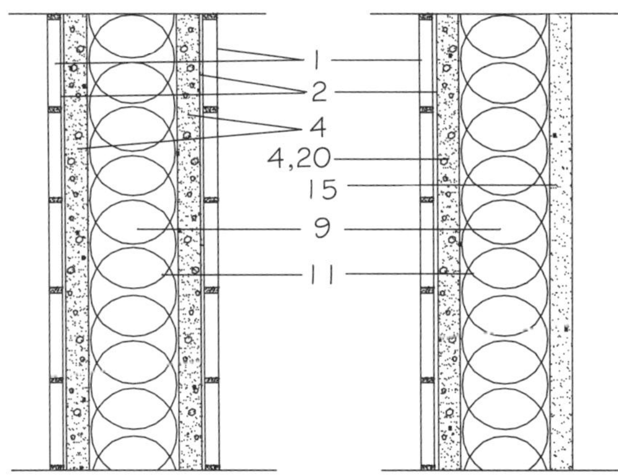

RW815-21 **RW816-21**

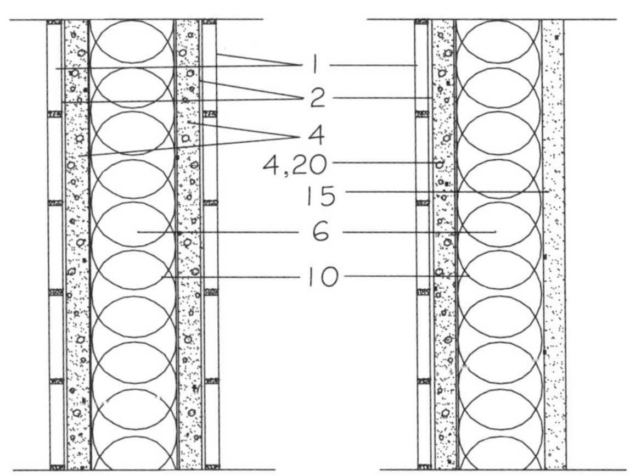

RW817-21 **RW818-21**

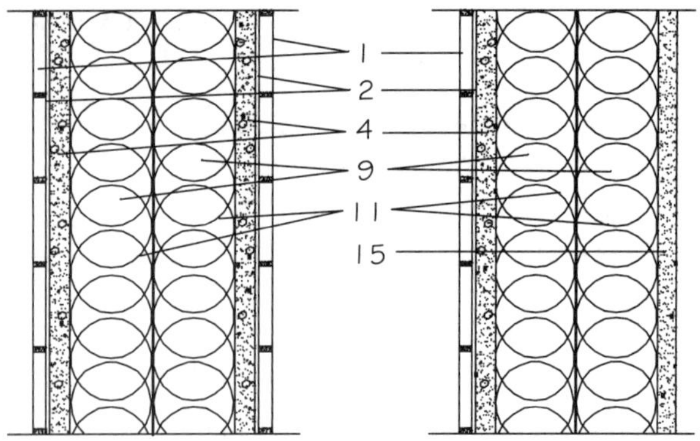

RW823-21

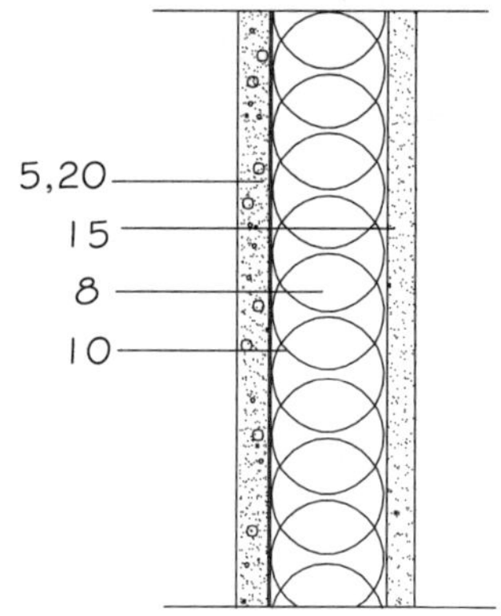

RW819-21 **RW820-21**

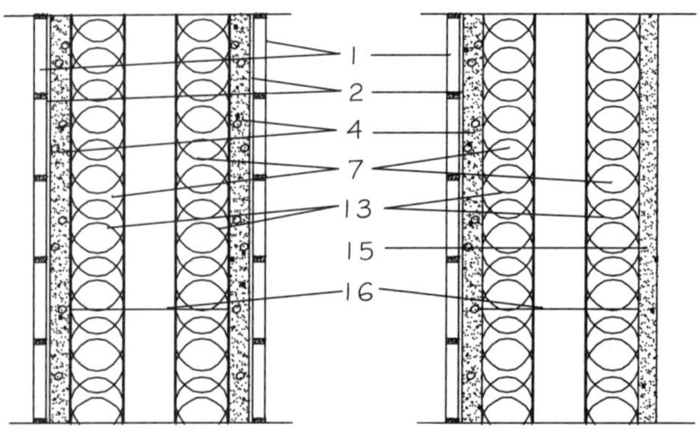

RW824-21

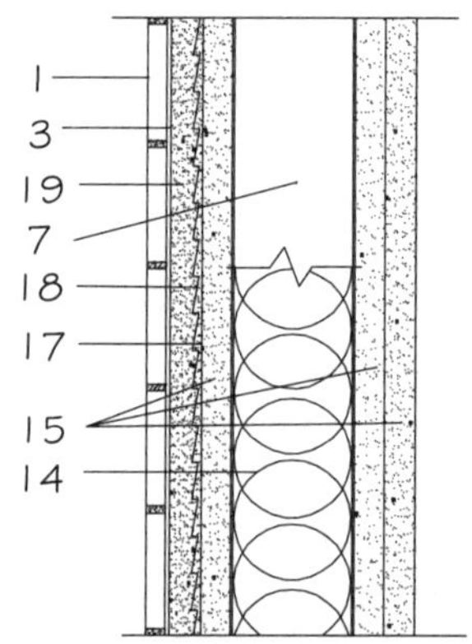

RW821-21 **RW822-21**

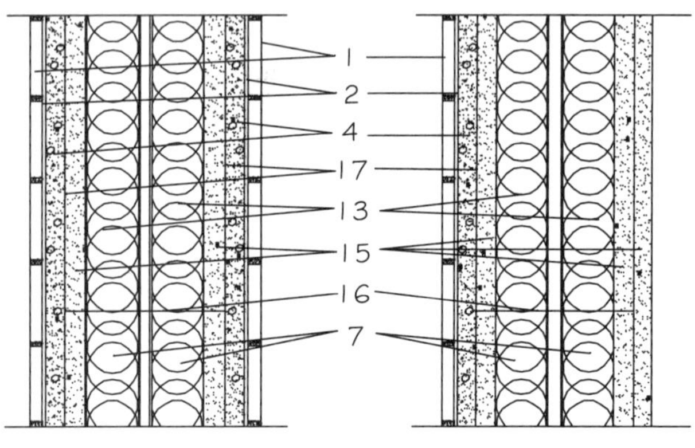

RW825-21

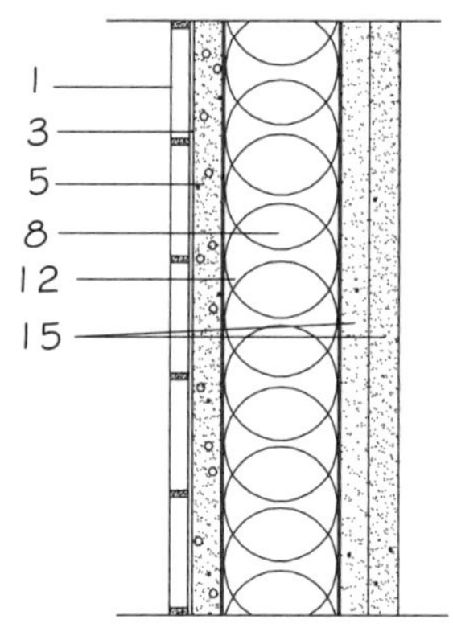

NOTES

- Fire side—both sides of all methods are classified as the fire side except RW824, which is sound-rated only.
- See Materials on page 294 for key.

What is Sound Control?

There are two types of ratings used for sound control: STC (Sound Transmission Class), which is airborne sound, such as speaking, music, etc., and IIC (Impact Insulation Class), which is sound from people walking, moving chairs, dropping objects, etc. Flanking is the transmission of sound through joints, penetrations, or structural components in the assembly. Presenting the greatest challenge for the STC component of sound control is high-frequency sound energy, such as that which is produced by a blow-horn or a whistle. The greatest challenge for sound control in regard to IIC is low-frequency sound energy, such as that which occurs when a basketball is dropped on the floor.

STC Ratings

The term STC refers to the single figure of evaluation used to quantify the transmission of airborne sound through building elements, such as walls or floor systems. These types of sounds would be the equivalent of voices, radio, or television in the context of a multi-family building. The term STC translates to Sound Transmission Class and is measured and stated in accordance with ASTM Standard number C634 and tested via the test methodology of ASTM Test Methods E90, E336, and E596. When done in an accredited test laboratory, these values are stated as an exact number to the right of the initials STC. When done in the field, using ASTM Method E336, the values are designated by the initials FSTC to the left of the numeric value. STC values are in a large part influenced by the solid mass of the structure, but are also dependent on isolation and resilience within the structure.

IIC Ratings

The term IIC refers to the statistical measurement standards used to quantify the transmission of impact sound energy through a floor/ceiling assembly system. These types of sounds would be the equivalent of foot traffic, dropped articles, or furniture moving in the context of a multi-family building. The term IIC translates to Impact Isolation Class and is measured and stated in accordance with ASTM Standard numbers C634 and E989 and tested via the test methodology of ASTM Test Method E492. In addition, a new test protocol for concrete subfloors has been introduced under ASTM E2179. When done in an accredited test laboratory, these values are stated as an exact number to the right of the initials IIC. When done in the field, using ASTM Method E1007, the values are designated by the initials FIIC to the left of the numeric value.

IIC values are not heavily influenced by the presence of solid mass in the structure. IIC values are usually dependent on the presence of a resilient material somewhere in the assembly to isolate and absorb the sonic energy created by impacts.

STC and IIC Ratings and Building Codes

In multi-family construction, in most jurisdictions, there are minimum IIC and STC values that the floor/ceiling assembly must achieve in order to meet the building code standards. Most common are the ICC/BOCA U.B.C. Uniform Building Code and International Building Code (IBC), which call for a minimum 50 IIC and 50 STC value. The higher the IIC or STC, the better the sound attenuation, with 50 considered the minimum for multi-family dwellings. Some states, municipalities, and counties have different building code standards, but the U.B.C. and I.B.C. codes are by far the most common. Consult your local Building Department for the exact code standards applicable in your area.

In addition to the building code standards, some condominium developers and condo homeowners associations have their own minimum standards written into their Covenants, Conditions, and Restrictions (CC&R's), which are often more stringent than the building code in that given jurisdiction. It is wise to consult your condo association in regard to their CC&R standards for the required IIC and STC values before installing hard surface flooring materials in your unit or project.

Challenges

A number of different materials are promoted and used for sound control in floors. Each material is only one component of a complete system in which each component is an essential part of the total assembly. Elimination of any component of the assembly can seriously deplete the sound rating desired. Most of the data available to the market relates to tests of floor/ceiling systems that are comprised of concrete slabs with gypsum wallboard hung on resilient furring channels (suspended or sound-rated ceiling system) with a layer of mineral wool or fiberglass insulation in the cavity. Unfortunately, most of the construction detail in the field does not have this type of ceiling treatment.

In addition, in 2003 a new test protocol for concrete subfloors has been introduced under ASTM E2179. When done in an accredited test laboratory, these values are stated as an exact number to the right of the initials IIC and actually give a separate IIC contribution value number of the product assembly by itself. This number provides an excellent way to evaluate different products and to determine if additional key components (e.g., a suspended or sound-rated ceiling assembly) are needed. Products that have only Field Test reports (FIIC) and no laboratory testing should be considered with caution. Field tests are project specific; comparing different field tests done in different buildings is not an accurate way to make a proper assessment of product performance.

Many materials that are promoted for sound control in hard surface flooring may not be suitable for direct bond tile and stone installations as they are too compressible, and do not provide proper structural support for these installations. In many cases, it is necessary to use additional reinforcing materials, such as mortar beds, poured in place overlays, backer boards, epoxy mortars and grouts, thicker tiles, and other methods to provide a structurally sound installation. It is recommended that all products/systems to be used for sound control be subjected to ASTM C627, commonly referred to as the Robinson Floor Test, and that such products/systems meet a minimum "Residential Rating." All components of the test assembly should be divulged to determine that the product in question is able to achieve the performance rating desired, for the given installation design and the same or similar materials should be used for the sound testing. In particular the size, thickness and type of the tile; the type of mortar and type of grout should be consistent.

ANSI A118.13

In 2010 a new standard was created for "Bonded Sound Reduction Membranes for Thin-set Ceramic Tile Installation." This standard addresses relatively thin bonded membranes for substrates that are approved by the manufacturer and are covered by ANSI Specifications. This standard was created specifically for ceramic tile but can be used for Dimension Stone. It should be recognized that stone tile can be relatively fragile and may be susceptible to damage when placed on sound reduction materials, which are almost always resilient or compressible by nature.

This standard applies to trowel applied, liquid, and flexible sheet membranes used for sound attenuation that are bonded to the substrate.

Key aspects of this standard:

- Fungus and microorganism resistance
- Shear strength of membrane bonded to ceramic tile with cement mortar
- ASTM C627 Robinson Floor Test minimum "Residential Rating"
- ASTM E2179 "Sound Transmission Reduction testing" Δ IIC rating of 10 or greater

Concrete Slab Subfloors

Concrete slabs come in a variety of thicknesses and compositions (e.g., hollow core, post-tensioned, prestressed). The most commonly cited are 6" and 8" concrete slabs with or without a suspended ceiling assembly. Test results for these two thicknesses vary when comparing field tests versus laboratory tests. The field tests result in a much higher range of values than those conducted in the labs. Table 1 demonstrates the wide range of numbers, particularly those within the field reports. If a suspended ceiling assembly is not possible, the most effective method to establish an improved IIC rating is to install a "floating" floor system. This can be done by utilizing a layer of some type of acoustically rated resilient material, followed by a layer of lightweight concrete, mortar bed, or gypsum concrete (typically 3/4" or more, depending on material), onto which the tile or stone is installed.

Wood Joist Floors

For aesthetics and design purposes, wood frame construction will normally have a gypsum wallboard ceiling assembly. To achieve a solid base of sound control, the use of resilient metal channels supporting the gypsum wallboard and sound-absorbing insulating batts in the cavity is recommended. An assembly of this nature with a single layer subfloor has an approximate IIC rating of 45. Adhering a tile or stone surface will actually lower the IIC to around 40. The reason for this is that the hard surface increases noise levels at the higher frequencies, thereby reducing the overall IIC rating. If the subfloor thickness is doubled and the ceiling wallboard panels are doubled as well, the net effect will increase the STC rating, as well as the IIC rating.

Key Points to Remember

- It is not good practice to select materials or systems based solely on field tests. Lab tests are a more accurate model for predicting performance in a range of different construction types; field tests are accurate only for the site where the tests were performed.

- If field tests are being conducted, require both before and after installation testing.
- Testing labs should be NVLAP certified and meet the criteria established in ASTM guidelines E548, E597, and in ASTM Standard E717.
- Sound abatement materials are often quite resilient, so a Robinson Floor Test (ASTM C627) is strongly recommended.
- The size, thickness, and type of tile as well as the type of mortar and grout used in the Robinson Floor Test should be the same/similar as that shown for the sound control testing conducted in the lab or field.
- If a suspended sound rated ceiling is not being used, then a relatively thick assembly on top of the slab may be required (e.g. full mortar bed) to achieve a 50 IIC

Table 1

Concrete Slab Thickness	No Sound Rated Ceiling	Sound Rated Suspended Ceiling*	In Lab IIC**	In Field IIC
6"	X		26 to 30	24 to 32
8"	X		28 to 32	25 to 35
6"		X	45 to 52	33 to 48
8"		X	48 to 55	38 to 52

*Suspended Sound Rated ceiling composed of: 7" plenum, 3" of insulation, resilient channels, ⅝" Type X gypsum wallboard panels.

**Tests were conducted in several different labs. Hence, the range of values for each slab thickness shows the variance between labs, not a variance in the test results within a single lab.

2021 | NATURAL STONE TILE INSTALLATION METHODS

F101-21 STONE

- **On-Ground Concrete**
- **Bonded Mortar Bed**
- **Natural Stone Tile**

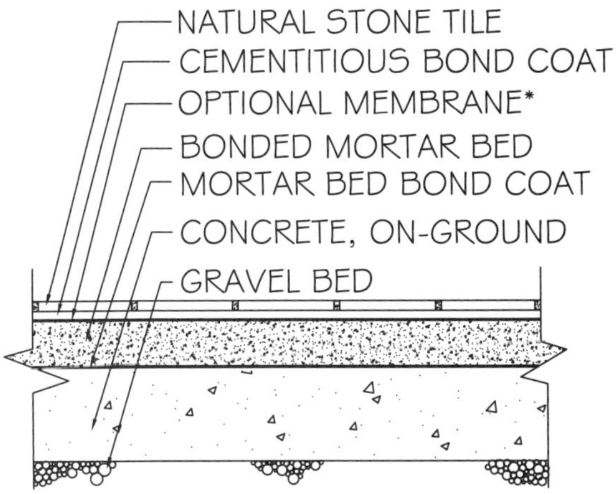

*USE OF A MEMBRANE IS OPTIONAL. SEE MEMBRANE OPTIONS.

Recommended Uses

- For exterior slab-on-ground construction where no bending stresses occur and positive drainage below slab is provided, for example, floors, decks, or patios.
- For areas where leveling, flattening, or contouring of finish floor height is required.
- For areas where floor flatness is critical, such as when tiles with any edge longer than 15" are specified or where the finished tile assembly is required to achieve in-plane transitions (as may be necessary where accessibility is a concern).
- For areas where ungauged stone will be used.

Service Rating

- Heavy, when stone with adequate compressive strength, flexural strength, and resistance to abrasion is used.

Environmental Exposure Classifications

- Res6; Com6.
- For installations that may be exposed to staining and/or chemical attack, specify tile and grout suitable for exposure. Consult product manufacturers; see also "Product Selection Guides."

Typical Weight of Tile Installation

- 17 pounds/square foot with ¾" mortar bed. Add 3 pounds/square foot for each additional ¼" of mortar bed.
- Does not include weight of substrate. See "Appendix B" for assumptions, included materials, and their individual weights.

Limitations

- Not recommended for use with some stones on post-tensioned concrete unless tile installation is 100 square feet or less. Consult stone supplier and installation materials manufacturers.
- Where the installation will be subjected to freeze-thaw cycles, snow and ice accumulation, and/or snow melting chemicals, degradation can occur over time.

Membrane Options

- A waterproof membrane (A118.10) may be specified to prevent moisture intrusion and reduce the need for below-slab drainage. Specifier shall indicate if complete waterproofing is required, including if/how membrane connects to drain assembly, if base flashing is required, and treatment at other termination points.
- Check with membrane manufacturer for suitability for applicable conditions, as not all membranes are suitable for steam, high temperature, and/or chemical exposure, exterior use, use over above-ground structural slabs, use over pourable underlayments, use with radiant heat, or use over concrete with excessive moisture vapor transmission and/or alkalinity. Membrane may also affect service rating.

Requirements

- Slab to be well cured, dimensionally stable, and free of cracks, surface laitance, waxy or oily films, curing compounds, and other surface bond-inhibiting contaminants.
- When bond coat is portland cement paste, cover completed tilework and allow to cure.
- Seal edges of mortar bed with a trowelable membrane/sealant when water intrusion is expected.
- Mortar beds in excess of 2" thick shall be detailed by the architect.
- Mortar bed thickness—¾" minimum to 2" maximum.
- Minimum grout joint width—1/16."

Materials

- **Multiple options exist for membranes, mortars, grouts, and other materials and *must be clearly specified* to be included. If not specifically indicated, optional materials are not included and mortar/grout choice defaults to minimum performance specification indicated. Consider each system component and intended use to determine minimum requirements and to specify options.**
- Natural stone tile—see "Natural Stone Tile Selection and Installation Guide," and consult supplier for application suitability.
- Cementitious grout—ANSI A118.6 or better or ISO CG1 or better. Consider unsanded grout for soft or semisoft stones such as limestone, travertine, marble, onyx, or similar.

- Cementitious bond coat—portland cement paste on a mortar bed that is still workable. For a cured mortar bed, follow recommendations below to select appropriate bonding mortar:
  - When a waterproof membrane is not used—ANSI A118.4 or better or ISO C2 or better.
  - When a waterproof membrane is used—ANSI A118.4 or better or ISO C2S1 or better unless ANSI A118.1 or ISO C1 is recommended by membrane manufacturer.
  - Use white for light-colored marble, limestone, and other stones where staining or darkening from grey setting materials is possible.
- Waterproof membrane, when used—ANSI A118.10.
- Mortar bed—ANSI A108.1A.
- Mortar bed bond coat—portland cement slurry.

Materials for Green/Sustainable Design

- See "Green Building Standards and Green Product Selection Guide" and consult manufacturers and suppliers for product sustainability and contribution to green building design.
- Consider specifying installation materials that meet ANSI A138.1, the *American National Standard Specifications for Sustainable Ceramic Tiles, Glass Tiles, and Tile Installation Materials.*

Preparation by Other Trades

- Gravel bed or other means of drainage below slab.
- Slope slab for complete drainage.
- Slab to have steel trowel and fine broom finish with no curing compounds used. When used, mechanical scarifying is necessary.
- Maximum allowable variation in the installation substrate (concrete)—¼" in 10' from the required plane.

Movement Joint (architect must specify type of joint and show location and details on drawings)

- Movement joints—mandatory according to EJ171.

Installation Specifications

- Tile—ANSI A108.1A, .1B, or .1C. Ungauged stone may require installation per ANSI A108.1A. Material with minimal variation may be allowed per A108.1B with LHT mortar. A108.1B required if membrane to be used.
- Cementitious grout—ANSI A108.10.
- Waterproof membrane—ANSI A108.13.
- Movement Joints—EJ171 and ASTM C1193.

Notes

- Not all bonding mortars are suitable for exterior use.
- Protection of installation may be required to prevent premature exposure of setting materials to moisture.
- If a waterproof membrane is not used, positive drainage below slab must be provided.
- When unsanded grout is used, grout joint width must be 1⁄16" to ⅛" wide (nominal).

F102-21
STONE

- **On-Ground Concrete**
- **Natural Stone Tile**

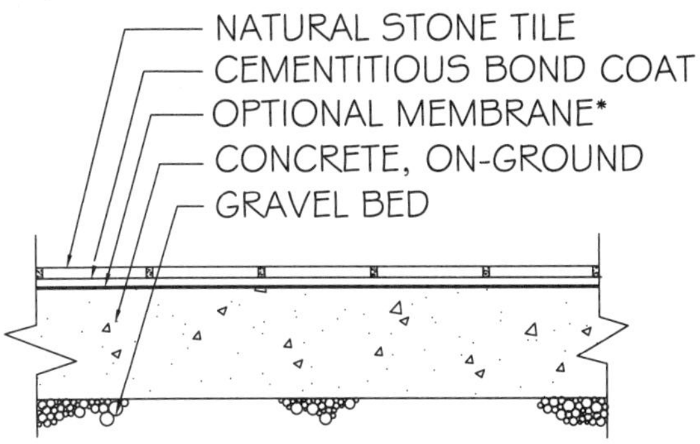

*USE OF A MEMBRANE IS OPTIONAL. SEE MEMBRANE OPTIONS.

Recommended Uses

- For exterior slab-on-ground construction where no bending stresses occur, positive drainage below the slab is provided, and thin-bed installation of tile is desired, for example, floors, decks, or patios.

Service Rating

- Heavy, when stone with adequate compressive strength, flexural strength, and resistance to abrasion is used.

Environmental Exposure Classifications

- Res6; Com6.
- For installations that may be exposed to staining and/or chemical attack, specify tile and grout suitable for exposure. Consult product manufacturers; see also "Product Selection Guides."

Typical Weight of Tile Installation

- 7 pounds/square foot.
- Does not include weight of substrate. See "Appendix B" for assumptions, included materials, and their individual weights.

Limitations

- Not recommended for use with some stones on post-tensioned concrete unless tile installation is 100 square feet or less. Consult stone supplier and installation materials manufacturers.
- Where the installation will be subjected to freeze-thaw cycles, snow and ice accumulation, and/or snow melting chemicals, degradation can occur over time.

Membrane Options

- A waterproof membrane (A118.10) may be specified to prevent moisture intrusion and reduce the need for below-slab drainage. Specifier shall indicate if complete waterproofing is required, including if/how membrane connects to drain assembly, if base flashing is required, and treatment at other termination points.
- A crack isolation membrane (A118.12) may be specified to treat existing in-plane cracks (F125-partial) or for protection against future in-plane cracking (F125-full). See F125 Stone for more information.
- Check with membrane manufacturer for suitability for applicable conditions, as not all membranes are suitable for steam, high temperature, and/or chemical exposure, exterior use, use over above-ground structural slabs, use over pourable underlayments, use with radiant heat, or use over concrete with excessive moisture vapor transmission and/or alkalinity. Membrane may also affect service rating.

Requirements

- Slab to be well cured, dimensionally stable, and free of cracks, surface laitance, waxy or oily films, curing compounds, and other surface bond-inhibiting contaminants.
- Minimum grout joint width—1/16".

Materials

- **Multiple options exist for membranes, mortars, grouts, and other materials and *must be clearly specified* to be included. If not specifically indicated, optional materials are not included and mortar/grout choice defaults to minimum performance specification indicated. Consider each system component and intended use to determine minimum requirements and to specify options.**
- Natural stone tile—see "Natural Stone Tile Selection and Installation Guide," and consult supplier for application suitability.
- Cementitious grout—ANSI A118.6 or better or ISO CG1 or better. Consider unsanded grout for soft or semisoft stones such as limestone, travertine, marble, onyx, or similar.
- Cementitious bond coat:
    - When a membrane is not used—ANSI A118.4 or better or ISO C2 or better.
    - When a crack isolation membrane is used—ANSI A118.4 or better or ISO C2S1 or better.
    - When a waterproof membrane is used—ANSI A118.4 or better or ISO C2S1 or better unless ANSI A118.1 or ISO C1 is recommended by membrane manufacturer.
    - Use white for light-colored marble, limestone, and other stones where staining or darkening from grey setting materials is possible.
- Waterproof membrane, when used—ANSI A118.10.
- Crack isolation membrane, when used—ANSI A118.12.

Materials for Green/Sustainable Design

- See "Green Building Standards and Green Product Selection Guide" and consult manufacturers and suppliers for product sustainability and contribution to green building design.
- Consider specifying installation materials that meet ANSI A138.1, the *American National Standard Specifications for Sustainable Ceramic Tiles, Glass Tiles, and Tile Installation Materials.*

Preparation by Other Trades

- Gravel bed or other means of drainage below slab.
- Slope slab for complete drainage.
- Slab to have steel trowel and fine broom finish with no curing compounds used. When used, mechanical scarifying is necessary.
- Maximum allowable variation in the tile substrate—⅛" in 10' from the required plane when measured from the high points in the surface.

Movement Joint (architect must specify type of joint and show location and details on drawings)

- Movement joints—mandatory according to EJ171.

Installation Specifications

- Tile—ANSI A108.5.
- Cementitious grout—ANSI A108.10.
- Waterproof membrane—ANSI A108.13.
- Crack isolation membrane—ANSI A108.17.
- Movement Joints—EJ171 and ASTM C1193.

Notes

- Not all bonding mortars are suitable for exterior use.
- Protection of installation may be required to prevent premature exposure of setting materials to moisture.
- If a waterproof membrane is not used, positive drainage below slab must be provided.
- When unsanded grout is used, grout joint width must be 1/16" to ⅛" wide (nominal).

F103-21
STONE

- **Roof Deck or Balcony**
- **Drainage Layer**
- **Mortar Bed**
- **Natural Stone Tile**

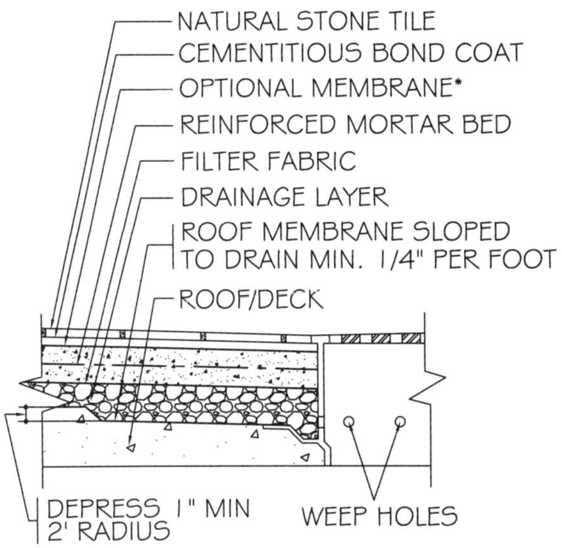

*USE OF A MEMBRANE IS OPTIONAL. SEE MEMBRANE OPTIONS.

Recommended Uses

- For exterior roofs or decks of concrete, steel, or wood where a waterproof roof membrane is used and sloped minimum ¼" per foot to a drain.
- For areas where ungauged stone will be used.

Service Rating

- Heavy, when stone with adequate compressive strength, flexural strength, and resistance to abrasion is used.

Environmental Exposure Classifications

- Res6; Com6.
- For installations that may be exposed to staining and/or chemical attack, specify tile and grout suitable for exposure. Consult product manufacturers; see also "Product Selection Guides."

Typical Weight of Tile Installation

- 33 pounds/square foot with 1¼" mortar bed. Add 3 pounds/square foot for each additional ¼" of mortar bed.
- Does not include weight of substrate. See "Appendix B" for assumptions, included materials, and their individual weights.

Limitations

- Where the installation will be subjected to freeze-thaw cycles, snow and ice accumulation, and/or snow melting chemicals, degradation can occur over time.
- Requires additional consideration by design professional to accommodate movement and/or deflection. Setting materials with improved bond strength and deformability are required.

Membrane Options

- A waterproof membrane (A118.10) may be specified to prevent moisture intrusion. Specifier shall indicate if complete waterproofing is required, including if/how membrane connects to drain assembly, if base flashing is required, and treatment at other termination points.
- Check with membrane manufacturer for suitability for applicable conditions, as not all membranes are suitable for steam, high temperature, and/or chemical exposure, exterior use, use over above-ground structural slabs, use over pourable underlayments, use with radiant heat, or use over concrete with excessive moisture vapor transmission and/or alkalinity. Membrane may also affect service rating.

Requirements

- Mortar beds in excess of 2" thick shall be detailed by the architect.
- Reinforcing mesh mandatory.
- Protect weepholes to prevent blockage.
- When bond coat is portland cement paste, cover completed tilework and allow to cure.
- Mortar bed thickness—1¼" minimum to 2" maximum.
- Minimum grout joint width—1⁄16."
- Above-ground installations are inherently more susceptible to vibration and deflection. Grout and mortar manufacturers to warrant product suitability.

Materials

- **Multiple options exist for membranes, mortars, grouts, and other materials and *must be clearly specified* to be included. If not specifically indicated, optional materials are not included and mortar/grout choice defaults to minimum performance specification indicated. Consider each system component and intended use to determine minimum requirements and to specify options.**
- Natural stone tile—see "Natural Stone Tile Selection and Installation Guide," and consult supplier for application suitability.
- Cementitious grout—ANSI A118.6 or better or ISO CG1 or better. Consider unsanded grout for soft or semisoft stones such as limestone, travertine, marble, onyx, or similar.
- Cementitious bond coat—portland cement paste on a mortar bed that is still workable. For a cured mortar bed, follow recommendations below to select appropriate bonding mortar:
  - Must be recommended by manufacturer for above-ground use.

- When a waterproof membrane is not used—ANSI A118.15 or better or ISO C2S1 or better.
- When a waterproof membrane is used—ANSI A118.15 or better or ISO C2S1 or better unless ANSI A118.1 or ISO C1 is recommended by membrane manufacturer.
- Use white for light-colored marble, limestone, and other stones where staining or darkening from grey setting materials is possible.

- Waterproof membrane, when used—ANSI A118.10.
- Mortar bed and reinforcing—ANSI A108.1A.
- Drainage layer—minimum 1"-thick bed of crushed stone (½" maximum size).
- Filter fabric—burlap or closely woven cheesecloth.

Materials for Green/Sustainable Design

- See "Green Building Standards and Green Product Selection Guide" and consult manufacturers and suppliers for product sustainability and contribution to green building design.
- Consider specifying installation materials that meet ANSI A138.1, the *American National Standard Specifications for Sustainable Ceramic Tiles, Glass Tiles, and Tile Installation Materials.*

Preparation by Other Trades

- Floor systems over which tile will be installed shall be in conformance with the IRC for residential applications, the IBC for commercial applications, or applicable building codes. Maximum allowable substrate deflection under live load not to exceed l/360. See also "Substrate Requirements."
- When concentrated loads (scissor lifts, pallet jacks, automobiles, forklifts, etc.) will be used on an above-ground tiled floor, the engineer and/or specifier shall specify a substrate to accommodate the concentrated loads. Owner/specifier is responsible for protecting the tilework from damage, including allowing sufficient time for installed materials to cure properly.
- Roof drains and membrane by other trades—provide complete drainage at roof membrane level by use of weep holes as shown or other methods. Flat deck with poor or no drainage will not perform well.
- Maximum allowable variation in the roof/deck—¼" in 10' from the required plane.

Movement Joint (architect must specify type of joint and show location and details on drawings)

- Movement joints—mandatory according to EJ171. For above-ground installations, additional movement joints are required.
- Movement joints should not go through the gravel bed; they should extend only to the bottom of the setting bed.

Installation Specifications

- Tile—ANSI A108.1A, .1B, or .1C. Ungauged stone may require installation per ANSI A108.1A. Material with minimal variation may be allowed per A108.1B with LHT mortar. A108.1B required if membrane to be used.
- Cementitious grout—ANSI A108.10.
- Waterproof membrane—ANSI A108.13.
- Movement Joints—EJ171 and ASTM C1193.

Notes

- Not all bonding mortars are suitable for exterior use.
- Not all mortars, grouts, and membranes are suitable for above-ground use. Check manufacturer recommendations.
- Protection of installation may be required to prevent premature exposure of setting materials to moisture.
- Some substrate materials used in wet areas are subject to deterioration from moisture. See ANSI A108.01-2.4.
- When unsanded grout is used, grout joint width must be $\frac{1}{16}$" to ⅛" wide (nominal).

F103B-21
STONE

- **Roof Deck or Balcony**
- **Drainage Mat**
- **Mortar Bed**
- **Natural Stone Tile**

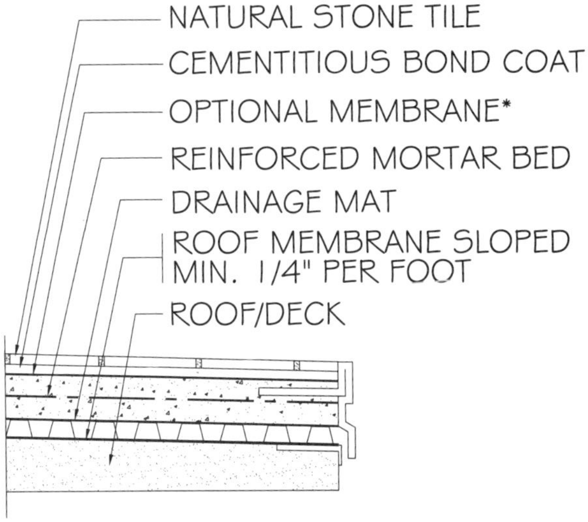

*USE OF A MEMBRANE IS OPTIONAL. SEE MEMBRANE OPTIONS.

Recommended Uses

- For exterior roofs or decks of concrete, steel, or wood where a waterproof roof membrane is used and sloped minimum ¼" per foot.
- For areas where ungauged stone will be used.

Service Rating

- Consult drainage mat manufacturer.

Environmental Exposure Classifications

- Res6; Com6.
- For installations that may be exposed to staining and/or chemical attack, specify tile and grout suitable for exposure. Consult product manufacturers; see also "Product Selection Guides."

Typical Weight of Tile Installation

- 23 pounds/square foot with 1¼" mortar bed. Add 3 pounds/square foot for each additional ¼" of mortar bed.
- Does not include weight of substrate. See "Appendix B" for assumptions, included materials, and their individual weights.

Limitations

- Where the installation will be subjected to freeze-thaw cycles, snow and ice accumulation, and/or snow melting chemicals, degradation can occur over time.
- Requires additional consideration by design professional to accommodate movement and/or deflection. Setting materials with improved bond strength and deformability are required.

Membrane Options

- A waterproof membrane (A118.10) may be specified to prevent moisture intrusion. Specifier shall indicate if complete waterproofing is required, including if/how membrane connects to drain assembly, if base flashing is required, and treatment at other termination points.
- Check with membrane manufacturer for suitability for applicable conditions, as not all membranes are suitable for steam, high temperature, and/or chemical exposure, exterior use, use over above-ground structural slabs, use over pourable underlayments, use with radiant heat, or use over concrete with excessive moisture vapor transmission and/or alkalinity. Membrane may also affect service rating.

Requirements

- Mortar beds in excess of 2" thick shall be detailed by the architect.
- Reinforcing mesh mandatory.
- When bond coat is portland cement paste, cover completed tilework and allow to cure.
- Mortar bed thickness—1¼" minimum to 2" maximum.
- Minimum grout joint width—¹⁄₁₆".
- Above-ground installations are inherently more susceptible to vibration and deflection. Grout and mortar manufacturers to warrant product suitability.

Materials

- **Multiple options exist for membranes, mortars, grouts, and other materials and *must be clearly specified* to be included. If not specifically indicated, optional materials are not included and mortar/grout choice defaults to minimum performance specification indicated. Consider each system component and intended use to determine minimum requirements and to specify options.**
- Natural stone tile—see "Natural Stone Tile Selection and Installation Guide," and consult supplier for application suitability.
- Cementitious grout—ANSI A118.6 or better or ISO CG1 or better. Consider unsanded grout for soft or semisoft stones such as limestone, travertine, marble, onyx, or similar.
- Cementitious bond coat—portland cement paste on a mortar bed that is still workable; For a cured mortar bed, follow recommendations below to select appropriate bonding mortar:
  - Must be recommended by manufacturer for above-ground use.
  - When a waterproof membrane is not used—ANSI A118.15 or better or ISO C2S1 or better.
  - When a waterproof membrane is used—ANSI

A118.15 or better or ISO C2S1 or better unless ANSI A118.1 or ISO C1 is recommended by membrane manufacturer.

- Use white for light-colored marble, limestone, and other stones where staining or darkening from grey setting materials is possible.

- Waterproof membrane, when used—ANSI A118.10.
- Mortar bed and reinforcing—ANSI A108.1A.
- Drainage mat—recommended by manufacturer for intended use or application.

Materials for Green/Sustainable Design

- See "Green Building Standards and Green Product Selection Guide" and consult manufacturers and suppliers for product sustainability and contribution to green building design.
- Consider specifying installation materials that meet ANSI A138.1, the *American National Standard Specifications for Sustainable Ceramic Tiles, Glass Tiles, and Tile Installation Materials.*

Preparation by Other Trades

- Floor systems over which tile will be installed shall be in conformance with the IRC for residential applications, the IBC for commercial applications, or applicable building codes. Maximum allowable substrate deflection under live load not to exceed l/360. See also "Substrate Requirements."
- When concentrated loads (scissor lifts, pallet jacks, automobiles, forklifts, etc.) will be used on an above-ground tiled floor, the engineer and/or specifier shall specify a substrate to accommodate the concentrated loads. Owner/specifier is responsible for protecting the tilework from damage, including allowing sufficient time for installed materials to cure properly.
- Roof membrane must be sloped and provide complete protection from moisture. Flat deck with poor or no drainage will not perform well.
- Maximum allowable variation in the roof/deck—¼" in 10' from the required plane.

Movement Joint (architect must specify type of joint and show location and details on drawings)

- Movement joints—mandatory according to EJ171. For above-ground installations, additional movement joints are required.

Installation Specifications

- Tile—ANSI A108.1A, .1B, or .1C. Ungauged stone may require installation per ANSI A108.1A. Material with minimal variation may be allowed per A108.1B with LHT mortar. A108.1B required if membrane to be used.
- Cementitious grout—ANSI A108.10.
- Waterproof membrane—ANSI A108.13.
- Drainage mat—manufacturer's directions.
- Movement Joints—EJ171 and ASTM C1193.

Notes

- Not all bonding mortars are suitable for exterior use.
- Not all mortars, grouts, and membranes are suitable for above-ground use. Check manufacturer recommendations.
- Protection of installation may be required to prevent premature exposure of setting materials to moisture.
- Some substrate materials used in wet areas are subject to deterioration from moisture. See ANSI A108.01-2.4.
- When unsanded grout is used, grout joint width must be ¹⁄₁₆" to ⅛" wide (nominal).

F104-21
STONE

- **Roof Deck or Balcony**
- **Natural Stone Tile**

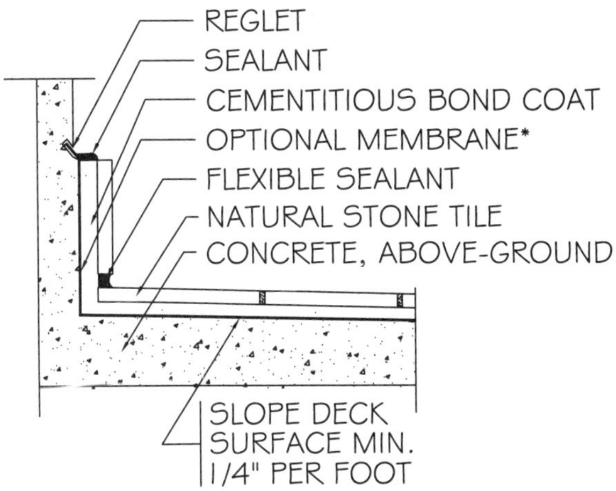

*USE OF A MEMBRANE IS OPTIONAL. SEE MEMBRANE OPTIONS.

Recommended Uses

- For exterior balconies or decks of concrete over unoccupied space where waterproof roof membrane is not used, where slab is sloped minimum ¼" per foot, and where thin-bed installation of tile is desired.

Service Rating

- Heavy, when stone with adequate compressive strength, flexural strength, and resistance to abrasion is used.

Environmental Exposure Classifications

- Res6; Com6.
- For installations that may be exposed to staining and/or chemical attack, specify tile and grout suitable for exposure. Consult product manufacturers; see also "Product Selection Guides."

Typical Weight of Tile Installation

- 7 pounds/square foot.
- Does not include weight of substrate. See "Appendix B" for assumptions, included materials, and their individual weights.

Limitations

- Not recommended for use with some stones on post-tensioned concrete unless tile installation is 100 square feet or less. Consult stone supplier and installation materials manufacturers.
- Where the installation will be subjected to freeze-thaw cycles, snow and ice accumulation, and/or snow melting chemicals, degradation can occur over time.
- Requires additional consideration by design professional to accommodate movement and/or deflection. Setting materials with improved bond strength and deformability are required.

Membrane Options

- A waterproof membrane (A118.10) may be specified to prevent moisture intrusion. Specifier shall indicate if complete waterproofing is required, including if/how membrane connects to drain assembly, if base flashing is required, and treatment at other termination points.
- A crack isolation membrane (A118.12) may be specified to treat existing in-plane cracks (F125-partial) or for protection against future in-plane cracking (F125-full). See F125 Stone for more information.
- Check with membrane manufacturer for suitability for applicable conditions, as not all membranes are suitable for steam, high temperature, and/or chemical exposure, exterior use, use over above-ground structural slabs, use over pourable underlayments, use with radiant heat, or use over concrete with excessive moisture vapor transmission and/or alkalinity. Membrane may also affect service rating.

Requirements

- Slab to be well cured, dimensionally stable, and free of cracks, surface laitance, waxy or oily films, curing compounds, and other surface bond-inhibiting contaminants.
- Minimum grout joint width—1⁄16".
- Above-ground installations are inherently more susceptible to vibration and deflection. Grout and mortar manufacturers to warrant product suitability.

Materials

- **Multiple options exist for membranes, mortars, grouts, and other materials and *must be clearly specified* to be included. If not specifically indicated, optional materials are not included and mortar/grout choice defaults to minimum performance specification indicated. Consider each system component and intended use to determine minimum requirements and to specify options.**
- Natural stone tile—see "Natural Stone Tile Selection and Installation Guide," and consult supplier for application suitability.
- Cementitious grout—ANSI A118.6 or better or ISO CG1 or better. Consider unsanded grout for soft or semisoft stones such as limestone, travertine, marble, onyx, or similar.
- Cementitious bond coat:
  - Must be recommended by manufacturer for above-ground use.
  - When a membrane is not used—ANSI A118.15 or better or ISO C2S1 or better.
  - When a crack isolation membrane is used—ANSI

A118.15 or better or ISO C2S1 or better.

- When a waterproof membrane is used—ANSI A118.15 or better or ISO C2S1 or better unless ANSI A118.1 or ISO C1 is recommended by membrane manufacturer.
- Use white for light-colored marble, limestone, and other stones where staining or darkening from grey setting materials is possible.

- Waterproof membrane, when used—ANSI A118.10.
- Crack isolation membrane, when used—ANSI A118.12.

Materials for Green/Sustainable Design

- See "Green Building Standards and Green Product Selection Guide" and consult manufacturers and suppliers for product sustainability and contribution to green building design.
- Consider specifying installation materials that meet ANSI A138.1, the *American National Standard Specifications for Sustainable Ceramic Tiles, Glass Tiles, and Tile Installation Materials.*

Preparation by Other Trades

- Floor systems over which tile will be installed shall be in conformance with the IRC for residential applications, the IBC for commercial applications, or applicable building codes. Maximum allowable substrate deflection under live load not to exceed l/360. See also "Substrate Requirements."
- When concentrated loads (scissor lifts, pallet jacks, automobiles, forklifts, etc.) will be used on an above-ground tiled floor, the engineer and/or specifier shall specify a substrate to accommodate the concentrated loads. Owner/specifier is responsible for protecting the tilework from damage, including allowing sufficient time for installed materials to cure properly.
- Maximum allowable variation in the tile substrate—⅛" in 10' from the required plane when measured from the high points in the surface.
- Slope slab minimum ¼" per foot. Flat deck with poor or no drainage will not perform well.
- Slab to have steel trowel finish with no curing compounds used. When used, mechanical scarification is necessary.

Movement Joint (architect must specify type of joint and show location and details on drawings)

- Movement joints—mandatory in accordance with EJ171. For above-ground installations, additional movement joints are required.

Installation Specifications

- Tile—ANSI A108.5.
- Cementitious grout—ANSI A108.10.
- Waterproof membrane—ANSI A108.13.
- Crack isolation membrane—ANSI A108.17.
- Movement Joints—EJ171 and ASTM C1193.

Notes

- Not all bonding mortars are suitable for exterior use.
- Not all mortars, grouts, and membranes are suitable for above-ground use. Check manufacturer recommendations.
- Protection of installation may be required to prevent premature exposure of setting materials to moisture.
- When unsanded grout is used, grout joint width must be $\frac{1}{16}$" to ⅛" wide (nominal).

F105-21
STONE

- **Roof Deck or Balcony**
- **Structural Ribbed Self-Supporting Boards (SRSB) Filled with Portland Cement Mortar**
- **Natural Stone Tile**

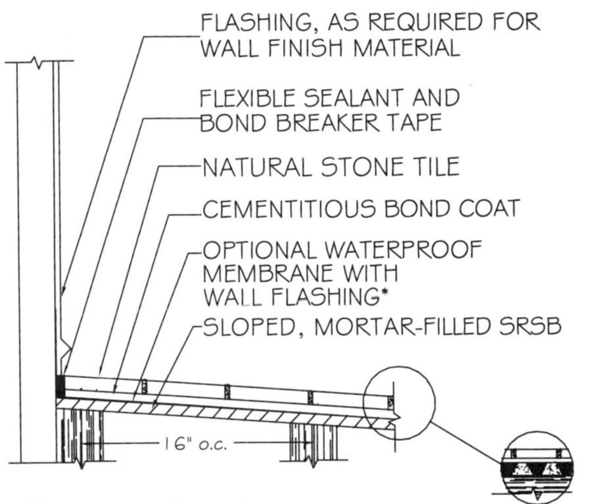

*Waterproof membrane meeting ANSI A118.10 required when subject to freeze/thaw cycling. See Membrane Options.

Front View

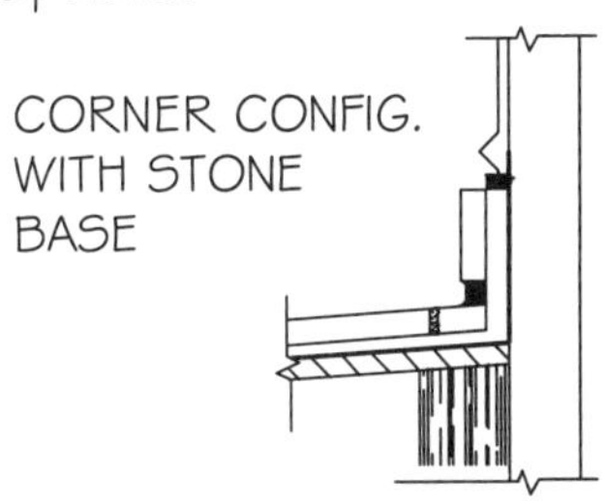

Recommended Uses

- For residential balconies and decks over unoccupied space, constructed of a joist system sloped min. ¼" per foot.

Service Rating

- Light, when stone with adequate compressive strength, flexural strength, and resistance to abrasion is used.

Environmental Classifications

- Res6; Com6.
- For installations that may be exposed to staining, specify tile and grout suitable for exposure. Consult product manufacturers; see also "Product Selection Guides."

Typical Weight of Tile Installation

- 7 pounds/square foot.
- Does not include weight of substrate (mortar-filled SRSB). See "Appendix B" for assumptions, included materials, and their individual weights.

Limitations

- 12" × 12" and larger tile only.
- Maximum joist spacing 12" on center.
- Where the installation will be subjected to freeze-thaw cycles, snow and ice accumulation, and/or snow melting chemicals, degradation can occur over time.
- Requires additional consideration by design professional to accommodate movement and/or deflection. Setting materials with improved bond strength and deformability are required.

Membrane Options

- A waterproof membrane (A118.10) is required for installations that will be exposed to freeze/thaw cycling. When used, membrane flashing is required.
- Specifier shall indicate if complete waterproofing is required, including if/how membrane connects to drain assembly, if base flashing is required, and treatment at other termination points.
- Check with membrane manufacturers for suitability for applicable conditions, as not all membranes are suitable for steam, high temperature, and/or chemical exposure, exterior use, use over above-ground structural slabs, use over pourable underlayments, use with radiant heat, or use over concrete with excessive moisture vapor transmission and/or alkalinity. Membrane may also affect service rating.

Requirements

- Minimum grout joint width 1/16".
- Above-ground installations are inherently more susceptible to vibration and deflection. Grout and mortar manufacturers to warrant product suitability.
- Mortar bed thickness—screed fill channels of SRSB boards flush to top of ribs and allow to cure 24 hours.

Preparation by SRSB Installers

- The ends of SRSB must be fully supported. Cantilevering is not permitted.
- Do not extend movement joints through SRSB boards.
- Fasteners—1⅝" pan head screws, non-corrosive and non-oxidizing.

Materials

- **Multiple options exist for membranes, mortars, grouts, and other materials and *must be clearly specified* to be included. If not specifically indicated, optional materials are not included and mortar/grout choice defaults to minimum performance specification indicated. Consider each system component and intended use to determine minimum requirements and to specify options.**
- Natural stone tile—see "Natural Stone Tile Selection and Installation Guide," and consult supplier for application suitability.

- Cementitious grout—ANSI A118.7 or better or ISO CG2 or better. Consider unsanded grout for soft or semisoft stones such as limestone, travertine, marble, onyx, or similar.
- Cementitious bond coat:
  - Must be recommended by manufacturer for above-ground use.
  - When a membrane is not used—ANSI A118.15 or better or ISO C2S1 or better.
  - When a waterproof membrane is used—ANSI A118.15 or better or ISO C2S1 or better unless ANSI A118.1 or ISO C1 is recommended by membrane manufacturer.
  - Use white for light-colored marble, limestone, and other stones where staining or darkening from grey setting materials is possible.
- Waterproof membrane, when used—ANSI A118.10.
- Waterproof membrane flashing, when used—per applicable building codes.
- Mortar bed—ANSI A108.1A.
- Structural ribbed self-supporting boards—recommended by manufacturer for intended use or application and meeting the following requirements:
  - Water absorption < 3% when tested per ASTM D570.
  - Passes 300 cycles when tested per ASTM C67, Section 9.
  - Flexural strength ≥ 2,600 psf when tested per ASTM D790.
  - Direct screw withdrawal resistance (using SRSB manufacturer's recommended fastener) ≥ 250 psf when tested per ASTM D1037, Section 16.
  - Passes Robinson Floor Test (ASTM C627) cycles 1-10 when tested with standard 12" × 12" ANSI A137.1 porcelain tile, ANSI A118.6 grout, and ANSI A118.4 mortar; and joists spaced 16" o.c.
  - Passes Robinson Floor Test (ASTM C627) cycles 1-10 when tested with standard 12" × 12" ANSI A137.1 porcelain tile, ANSI A118.6 grout, and ANSI A118.4 mortar; joists spaced 16" o.c.; and modified to accommodate a 14' suspended floor assembly.
  - Receives flame spread and smoke development indices of 0-0-0-0 when the assembly is exposed to fire, when tested per ASTM E84.

Materials for Green/Sustainable Design

- See "Green Building Standards and Green Product Selection Guide" and consult manufacturers and suppliers for product sustainability and contribution to green building design.
- Consider specifying tile and installation materials that meet ANSI A138.1, the *American National Standard Specifications for Sustainable Ceramic Tiles, Glass Tiles, and Tile Installation Materials.*

Preparation by Other Trades

- For decks with 8 foot and longer joists, cross bracing may be needed between joists to prevent racking. Consult SRSB manufacturer.
- Floor systems over which tile will be installed shall be in conformance with the IRC for residential applications, the IBC for commercial applications, or applicable building codes. Maximum allowable substrate deflection under live load not to exceed l/720. See also "Substrate Requirements."
- When concentrated loads (scissor lifts, pallet jacks, automobiles, forklifts, etc.) will be used on an above-ground tiled floor, the engineer and/or specifier shall specify a substrate to accommodate the concentrated loads. Owner/specifier is responsible for protecting the tilework from damage, including allowing sufficient time for installed materials to cure properly.
- Maximum allowable variation in the joist system—¼" in 10' and ¹⁄₁₆" in 1' from the required plane.
- Slope joist system min. ¼" per foot. Flat deck with poor or no drainage will not perform well.

Movement Joint (architect must specify type of joint and show location and details on drawing)

- Movement joints—mandatory according to EJ171. For above-ground installations additional movement joints are required.

Installation Specifications

- Tile—ANSI A108.5
- Cementitious grout—ANSI A108.10.
- Waterproof membrane—ANSI A108.13.
- Mortar bed—ANSI A108.1B
- SRSB—manufacturer's directions
- Movement Joints—EJ171 and ASTM C1193.

Notes

- State and local building codes for exterior deck construction, membrane requirements, and membrane flashing requirements vary; check applicable codes for requirements.
- SRSB are structural panels. Specifier to ensure compliance with building codes and expected load. Consult with SRSB manufacturer for design criteria.
- Not all bonding mortars are suitable for exterior use.
- Not all mortars, grouts, and membranes are suitable for above-ground use. Check manufacturer recommendations.
- Protection of installation may be required to prevent premature exposure of setting materials to moisture.
- Some substrate materials used in wet areas are subject to deterioration from moisture. See ANSI A108.01-2.4.
- When unsanded grout is used, grout joint width must be ¹⁄₁₆" to ⅛" wide (nominal).

F111-21
STONE

- **On-Ground or Above-Ground Concrete**
- **Unbonded Mortar Bed**
- **Natural Stone Tile**

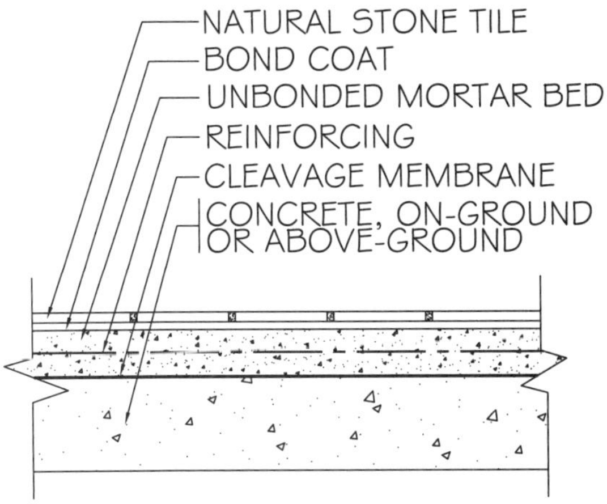

Recommended Uses

- For above-ground structural slabs and for slab-on-ground construction.
- For isolating the tile installation from the concrete substrate.
- For areas where leveling, flattening, or contouring of finish floor height is required.
- For areas where floor flatness is critical, such as when tiles with any edge longer than 15" are specified or where the finished tile assembly is required to achieve in-plane transitions (as may be necessary where accessibility is a concern).
- For concrete substrates that are cracked or present bonding issues.
- For areas with in-slab hydronic tubing.
- For areas where ungauged stone will be used.

Service Rating

- Heavy, when stone with adequate compressive strength, flexural strength, and resistance to abrasion is used.

Environmental Exposure Classifications

- Res1, 2, 3, 5; Com1, 2, 3, 5.
- May be suitable for exterior applications in areas not subject to freeze/thaw cycling when appropriate precautions are taken, including expansion joint placement, proper slope, waterproofing, and material selection.
- For installations that may be exposed to staining and/or chemical attack, specify tile and grout suitable for exposure. Consult product manufacturers; see also "Product Selection Guides."

Typical Weight of Tile Installation

- 23 pounds/square foot with 1¼" mortar bed. Add 3 pounds/square foot for each additional ¼" of mortar bed.
- Does not include weight of substrate. See "Appendix B" for assumptions, included materials, and their individual weights.

Limitations

- Above-ground structural slabs and other floors subject to movement and/or deflection—requires additional consideration by design professional to accommodate movement and/or deflection. Setting materials with improved bond strength and deformability are required.

Requirements

- Reinforcing mesh mandatory.
- Mortar bed thickness—1¼" minimum to 2" maximum.
- Mortar beds in excess of 2" thick shall be detailed by the architect.
- Minimum grout joint width—1/16".
- Above-ground—above-ground installations are inherently more susceptible to vibration and deflection. Grout and mortar manufacturers to warrant product suitability.

Materials

- **Multiple options exist for membranes, mortars, grouts, and other materials and *must be clearly specified* to be included. If not specifically indicated, optional materials are not included and mortar/grout choice defaults to minimum performance specification indicated. Consider each system component and intended use to determine minimum requirements and to specify options.**
- Natural stone tile—see "Natural Stone Tile Selection and Installation Guide," and consult supplier for application suitability.
- Cementitious grout—ANSI A118.6 or better or ISO CG1 or better. Consider unsanded grout for soft or semisoft stones such as limestone, travertine, marble, onyx, or similar.
- Cementitious bond coat (on-ground)—portland cement paste on a mortar bed that is still workable. For a cured mortar bed, follow recommendations below to select appropriate bonding mortar:
  - ANSI A118.1 or better or ISO C1 or better.
  - Use white for light-colored marble, limestone, and other stones where staining or darkening from grey setting materials is possible.
- Cementitious bond coat (above ground)—portland cement paste on a mortar bed that is still workable. For a cured mortar bed, follow recommendations below to select appropriate bonding mortar:
  - Must be recommended by manufacturer for above-ground use.
  - ANSI A118.15 or better or ISO C2S1 or better.
  - Use white for light-colored marble, limestone, and other stones where staining or darkening from grey setting materials is possible.

- Epoxy bond coat, when used—ANSI A118.3 or ISO R1 or better. For above-ground use, must also be recommended by manufacturer. See Water Sensitivity and Fiberglass Mesh Reinforced Stone in the "Natural Stone Tile Selection and Installation Guide."
- Mortar bed, reinforcing, and cleavage membrane—ANSI A108.1A.

Materials for Green/Sustainable Design

- See "Green Building Standards and Green Product Selection Guide" and consult manufacturers and suppliers for product sustainability and contribution to green building design.
- Consider specifying installation materials that meet ANSI A138.1, the *American National Standard Specifications for Sustainable Ceramic Tiles, Glass Tiles, and Tile Installation Materials.*

Preparation by Other Trades

- Floor systems over which tile will be installed shall be in conformance with the IRC for residential applications, the IBC for commercial applications, or applicable building codes. Maximum allowable substrate deflection under live load not to exceed l/360. See also "Substrate Requirements."
- When concentrated loads (scissor lifts, pallet jacks, automobiles, forklifts, etc.) will be used on an above-ground tiled floor, the engineer and/or specifier shall specify a substrate to accommodate the concentrated loads. Owner/specifier is responsible for protecting the tilework from damage, including allowing sufficient time for installed materials to cure properly.
- Where slab is depressed to accommodate mortar bed thickness, slab depression to be accurate with steel trowel finish.
- Slope, when required, to be in subfloor.
- Maximum allowable variation in the installation substrate (concrete)—¼" in 10' from the required plane.

Movement Joint (architect must specify type of joint and show location and details on drawings)

- Movement joints—mandatory according to EJ171. For above-ground installations, additional movement joints are required.

Installation Specifications

- Tile—ANSI A108.1A, .1B, or .1C. Ungauged stone may require installation per ANSI A108.1A. Material with minimal variation may be allowed per A108.1B with LHT mortar. A108.1B required if epoxy bond coat to be used.
- Cementitious grout—ANSI A108.10.
- Movement Joints—EJ171 and ASTM C1193.

Notes

- For waterproof application when using a mortar bed, see F121 Stone or F112 Stone.
- For hydronic tubing on top of the slab when using a mortar bed, see RH117 Stone.
- Above-ground—not all mortars and grouts are suitable for above-ground use. Check manufacturer recommendations.
- When unsanded grout is used, grout joint width must be 1⁄16" to ⅛" wide (nominal).

F112-21
STONE

- **On-Ground or Above-Ground Concrete**
- **Bonded Mortar Bed**
- **Natural Stone Tile**

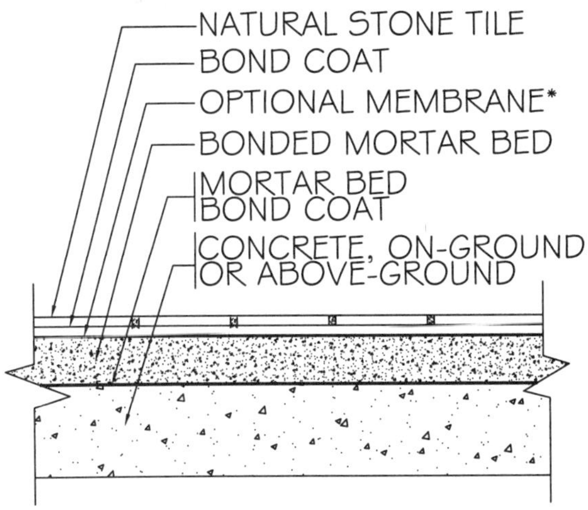

*USE OF A MEMBRANE IS OPTIONAL.
SEE MEMBRANE OPTIONS.

Recommended Uses

- For slab-on-ground construction where no bending stresses occur and for above-ground structural slabs where tile installation is 100 square feet or less.
- For areas where leveling, flattening, or contouring of finish floor height is required.
- For areas where floor flatness is critical, such as when tiles with any edge longer than 15" are specified or where the finished tile assembly is required to achieve in-plane transitions (as may be necessary where accessibility is a concern).
- For areas with in-slab hydronic tubing.
- For areas where ungauged stone will be used.

Service Rating

- Heavy, when stone with adequate compressive strength, flexural strength, and resistance to abrasion is used.

Environmental Exposure Classifications

- Res1, 2, 3, 5; Com1, 2, 3, 5.
- May be suitable for exterior applications in areas not subject to freeze/thaw cycling when appropriate precautions are taken, including expansion joint placement, proper slope, waterproofing, and material selection.
- For installations that may be exposed to staining and/or chemical attack, specify tile and grout suitable for exposure. Consult product manufacturers; see also "Product Selection Guides."

Typical Weight of Tile Installation

- 17 pounds/square foot with ¾" mortar bed. Add 3 pounds/square foot for each additional ¼" of mortar bed.
- Does not include weight of substrate. See "Appendix B" for assumptions, included materials, and their individual weights.

Limitations

- Not recommended for use with some stones on post-tensioned concrete unless tile installation is 100 square feet or less. Consult stone supplier and installation materials manufacturers.
- Not for use over above-ground structural slabs or other floors subject to movement and/or deflection except where tile installation is 100 square feet or less. See Notes.
- Above-ground structural slabs and other floors subject to movement and/or deflection—requires additional consideration by design professional to accommodate movement and/or deflection. Setting materials with improved bond strength and deformability are required.

Membrane Options

- A waterproof membrane (A118.10) may be specified to prevent moisture intrusion. Specifier shall indicate if complete waterproofing is required, including if/how membrane connects to drain assembly, if base flashing is required, and treatment at other termination points.
- Check with membrane manufacturers for suitability for applicable conditions, as not all membranes are suitable for steam, high temperature, and/or chemical exposure, exterior use, use over above-ground structural slabs, use over pourable underlayments, use with radiant heat, or use over concrete with excessive moisture vapor transmission and/or alkalinity. Membrane may also affect service rating.

Requirements

- Mortar bed thickness—¾" minimum to 2" maximum.
- Slab to be well cured, dimensionally stable, and free of cracks, surface laitance, waxy or oily films, curing compounds, and other surface bond-inhibiting contaminants.
- Mortar beds in excess of 2" thick shall be detailed by architect.
- Minimum grout joint width—1/16."
- Above-ground—above-ground installations are inherently more susceptible to vibration and deflection. Grout and mortar manufacturers to warrant product suitability.

Materials

- **Multiple options exist for membranes, mortars, grouts, and other materials and *must be clearly specified* to be included. If not specifically indicated, optional materials are not included and mortar/grout choice defaults to minimum performance specification indicated. Consider each system component and intended use to determine minimum requirements and to specify options.**

- Natural stone tile—see "Natural Stone Tile Selection and Installation Guide," and consult supplier for application suitability.
- Cementitious grout—ANSI A118.6 or better or ISO CG1 or better. Consider unsanded grout for soft or semisoft stones such as limestone, travertine, marble, onyx, or similar.
- Cementitious bond coat (on-ground)—portland cement paste on a mortar bed that is still workable. For a cured mortar bed, follow recommendations below to select appropriate bonding mortar:
  - When a waterproof membrane is not used—ANSI A118.1 or better or ISO C1 or better.
  - When a waterproof membrane is used—ANSI A118.4 or better or ISO C2S1 or better unless ANSI A118.1 or ISO C1 is recommended by membrane manufacturer.
  - Use white for light-colored marble, limestone, and other stones where staining or darkening from grey setting materials is possible.
- Cementitious bond coat (above ground)—portland cement paste on a mortar bed that is still workable. For a cured mortar bed, follow recommendations below to select appropriate bonding mortar:
  - Must be recommended by manufacturer for above-ground use.
  - When a waterproof membrane is not used—ANSI A118.15 or better or ISO C2S1 or better.
  - When a waterproof membrane is used—ANSI A118.4 or better or ISO C2S1 or better unless ANSI A118.1 or ISO C1 is recommended by membrane manufacturer.
  - Use white for light-colored marble, limestone, and other stones where staining or darkening from grey setting materials is possible.
- Epoxy bond coat, when used—ANSI A118.3 or ISO R1 or better. For above-ground use, must also be recommended by manufacturer. See Water Sensitivity and Fiberglass Mesh Reinforced Stone in the "Natural Stone Tile Selection and Installation Guide."
- Waterproof membrane, when used—ANSI A118.10.
- Mortar bed—ANSI A108.1A.
- Mortar bed bond coat—portland cement slurry.

Materials for Green/Sustainable Design

- See "Green Building Standards and Green Product Selection Guide" and consult manufacturers and suppliers for product sustainability and contribution to green building design.
- Consider specifying installation materials that meet ANSI A138.1, the *American National Standard Specifications for Sustainable Ceramic Tiles, Glass Tiles, and Tile Installation Materials.*

Preparation by Other Trades

- Floor systems over which tile will be installed shall be in conformance with the IRC for residential applications, the IBC for commercial applications, or applicable building codes. Maximum allowable substrate deflection under live load not to exceed l/360. See also "Substrate Requirements."
- When concentrated loads (scissor lifts, pallet jacks, automobiles, forklifts, etc.) will be used on an above-ground tiled floor, the engineer and/or specifier shall specify a substrate to accommodate the concentrated loads. Owner/specifier is responsible for protecting the tilework from damage, including allowing sufficient time for installed materials to cure properly.
- Slab to have steel trowel and fine broom finish with no curing compounds used. When used, mechanical scarifying is necessary.
- Slope, when required, to be in subfloor.
- Maximum allowable variation in the installation substrate (concrete)—¼" in 10' from the required plane.

Movement Joint (architect must specify type of joint and show location and details on drawings)

- Movement joints—mandatory according to EJ171. For above-ground installations, additional movement joints are required.

Installation Specifications

- Tile—ANSI A108.1A, .1B, or .1C. Ungauged stone may require installation per ANSI A108.1A. Material with minimal variation may be allowed per A108.1B with LHT mortar. A108.1B required if membrane or epoxy bond coat to be used.
- Cementitious grout—ANSI A108.10.
- Waterproof membrane—ANSI A108.13.
- Movement Joints—EJ171 and ASTM C1193.

Notes

- F111 is preferred over above-ground structural slabs and other floors subject to movement and/or deflection.
- Above-ground—not all mortars, grouts and membranes are suitable for above-ground use. Check manufacturer recommendations.
- For hydronic tubing on top of the slab when using a mortar bed, see RH117 Stone.
- For waterproofing below mortar bed (unbonded), see F121 Stone.
- When unsanded grout is used, grout joint width must be 1/16" to ⅛" wide (nominal).

F113-21 STONE

- **On-Ground Concrete**
- **Natural Stone Tile**

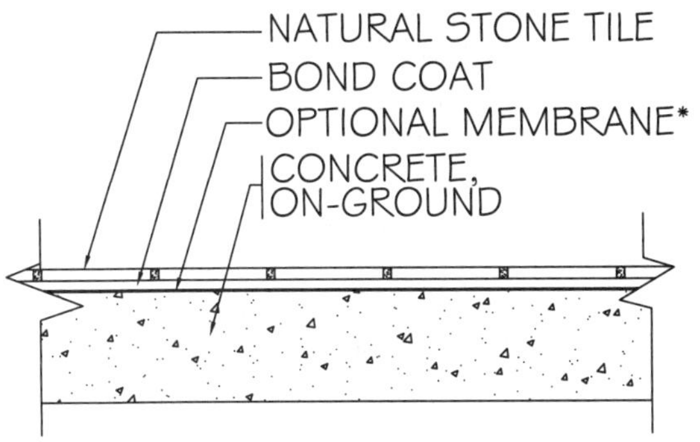

Recommended Uses

- For slab-on-ground construction where no bending stresses occur and thin-bed installation of tile is desired.

Service Rating

- Heavy, when stone with adequate compressive strength, flexural strength, and resistance to abrasion is used.

Environmental Exposure Classifications

- Res1, 2, 3, 5; Com1, 2, 3, 5.
- May be suitable for exterior applications in areas not subject to freeze/thaw cycling when appropriate precautions are taken, including expansion joint placement, proper slope, waterproofing, and material selection.
- For installations that may be exposed to staining and/or chemical attack, specify tile and grout suitable for exposure. Consult product manufacturers; see also "Product Selection Guides."

Typical Weight of Tile Installation

- 7 pounds/square foot.
- Does not include weight of substrate. See "Appendix B" for assumptions, included materials, and their individual weights.

Limitations

- Not recommended for use with some stones on post-tensioned concrete unless tile installation is 100 square feet or less. Consult stone supplier and installation materials manufacturers.
- Not for use over above-ground structural slabs and other floors subject to movement and/or deflection. See F113A Stone.

Membrane Options

- A crack isolation membrane may be specified to treat existing in-plane cracks (F125-partial) or for protection against future in-plane cracking (F-125 full). See F125 Stone for more information.
- An uncoupling membrane may be specified to accommodate subfloor movement. Check with membrane manufacturer for recommendations and requirements.
- Check with membrane manufacturers for suitability for applicable conditions, as not all membranes are suitable for steam, high temperature, and/or chemical exposure, exterior use, use over above-ground structural slabs, use over pourable underlayments, use with radiant heat, or use over concrete with excessive moisture vapor transmission and/or alkalinity. Membrane may also affect service rating.

Requirements

- Slab to be well cured, dimensionally stable, and free of cracks, surface laitance, waxy or oily films, curing compounds, and other surface bond-inhibiting contaminants.
- Minimum grout joint width—1⁄16".

Materials

- **Multiple options exist for membranes, mortars, grouts, and other materials and *must be clearly specified* to be included. If not specifically indicated, optional materials are not included and mortar/grout choice defaults to minimum performance specification indicated. Consider each system component and intended use to determine minimum requirements and to specify options.**
- Natural stone tile—see "Natural Stone Tile Selection and Installation Guide," and consult supplier for application suitability.
- Cementitious grout—ANSI A118.6 or better or ISO CG1 or better. Consider unsanded grout for soft or semisoft stones such as limestone, travertine, marble, onyx, or similar.
- Cementitious bond coat:
  - When a membrane is not used—ANSI A118.1 or better or ISO C1 or better.
  - When a crack isolation membrane is used—ANSI A118.4 or better or ISO C2S1 or better.
  - When an uncoupling membrane is used, check with membrane manufacturer.
  - Use white for light-colored marble, limestone, and other stones where staining or darkening from grey setting materials is possible.
- Epoxy bond coat, when used—ANSI A118.3 or ISO R1 or better. See Water Sensitivity and Fiberglass Mesh Reinforced Stone in the "Natural Stone Tile Selection and Installation Guide."
- Crack isolation membrane, when used—ANSI A118.12.
- Uncoupling membrane, when used—recommended by manufacturer; must achieve 50 psi or greater shear

bond strength in 7 days per the test method in ANSI A118.12 Section 5.1.3.

Materials for Green/Sustainable Design

- See "Green Building Standards and Green Product Selection Guide" and consult manufacturers and suppliers for product sustainability and contribution to green building design.
- Consider specifying installation materials that meet ANSI A138.1, the *American National Standard Specifications for Sustainable Ceramic Tiles, Glass Tiles, and Tile Installation Materials.*

Preparation by Other Trades

- Slab to have steel trowel and fine broom finish free of curing compounds. When used, mechanical scarifying is necessary.
- Slope, when required, to be in subfloor.
- Maximum allowable variation in the tile substrate—⅛" in 10' from the required plane when measured from the high points in the surface.

Movement Joint (architect must specify type of joint and show location and details on drawings)

- Movement joints—mandatory according to EJ171.

Installation Specifications

- Tile—ANSI A108.5 or A108.6.
- Cementitious grout—ANSI A108.10.
- Crack isolation membrane—ANSI A108.17.
- Uncoupling membrane—manufacturer's directions.
- Movement Joints—EJ171 and ASTM C1193.

Notes

- For waterproof application where thin-bed installation is desired, see F122 Stone.
- For in-slab hydronic tubing, see RH110 Stone.
- As the tile size increases, there is less tolerance for variation in the substrate from the required plane. Epoxy bond coat thickness must be thin and uniform; therefore, substrate flattening may be required when epoxy bond coat is used.
- When unsanded grout is used, grout joint width must be $\frac{1}{16}$" to ⅛" wide (nominal).

F113A-21
STONE

- **Above-Ground Concrete**
- **Natural Stone Tile**

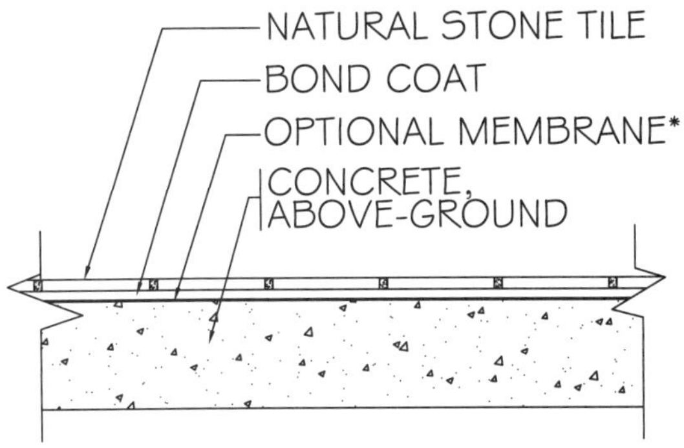

*USE OF A MEMBRANE IS OPTIONAL. SEE MEMBRANE OPTIONS.

Recommended Uses

- For above-ground structural slabs where thin-bed installation of tile is desired.

Service Rating

- Heavy, when stone with adequate compressive strength, flexural strength, and resistance to abrasion is used.

Environmental Exposure Classifications

- Res1, 2, 3, 5; Com1, 2, 3, 5.
- May be suitable for exterior applications in areas not subject to freeze/thaw cycling when appropriate precautions are taken, including expansion joint placement, proper slope, waterproofing and material selection.
- For installations that may be exposed to staining and/or chemical attack, specify tile and grout suitable for exposure. Consult product manufacturers; see also "Product Selection Guides."

Typical Weight of Tile Installation

- 7 pounds/square foot.
- Does not include weight of substrate. See "Appendix B" for assumptions, included materials, and their individual weights.

Limitations

- Not recommended for use with some stones on post-tensioned concrete unless tile installation is 100 square feet or less. Consult stone supplier and installation materials manufacturers.
- Requires additional consideration by design professional to accommodate movement and/or deflection. Setting materials with improved bond strength and deformability are required.

Membrane Options

- A crack isolation membrane (A118.12) may be specified to treat existing in-plane cracks (F125-partial) or for protection against future in-plane cracking (F125-full). See F125 Stone for more information.
- An uncoupling membrane may be specified to accommodate subfloor movement. Check with membrane manufacturer for recommendations and requirements.
- Check with membrane manufacturer for suitability for applicable conditions, as not all membranes are suitable for steam, high temperature, and/or chemical exposure, exterior use, use over above-ground structural slabs, use over pourable underlayments, use with radiant heat, or use over concrete with excessive moisture vapor transmission and/or alkalinity. Membrane may also affect service rating.

Requirements

- Slab to be well cured, dimensionally stable, and free of cracks, surface laitance, waxy or oily films, curing compounds, and other surface bond-inhibiting contaminants.
- Minimum grout joint width—1/16".
- Above-ground installations are inherently more susceptible to vibration and deflection. Grout and mortar manufacturers to warrant product suitability.

Materials

- **Multiple options exist for membranes, mortars, grouts, and other materials and *must be clearly specified* to be included. If not specifically indicated, optional materials are not included and mortar/grout choice defaults to minimum performance specification indicated. Consider each system component and intended use to determine minimum requirements and to specify options.**
- Natural stone tile—see "Natural Stone Tile Selection and Installation Guide," and consult supplier for application suitability.
- Cementitious grout—ANSI A118.6 or better or ISO CG1 or better. Consider unsanded grout for soft or semisoft stones such as limestone, travertine, marble, onyx, or similar.
- Cementitious bond coat:
  - Must be recommended by manufacturer for above-ground use.
  - When a membrane is not used—ANSI A118.15 or better or ISO C2S1 or better.
  - When a crack isolation membrane is used—ANSI A118.4 or better or ISO C2S1 or better.
  - When an uncoupling membrane is used, check with membrane manufacturer.
  - Use white for light-colored marble, limestone, and other stones where staining or darkening from grey setting materials is possible.
- Epoxy bond coat, when used—ANSI A118.3 or ISO R1 or better. Must also be recommended by

manufacturer for above-ground use. See Water Sensitivity and Fiberglass Mesh Reinforced Stone in the "Natural Stone Tile Selection and Installation Guide."

- Crack isolation membrane, when used—A118.12.
- Uncoupling membrane, when used—recommended by manufacturer; must achieve 50 psi or greater shear bond strength in 7 days per the test method in ANSI A118.12 Section 5.1.3.

Materials for Green/Sustainable Design

- See "Green Building Standards and Green Product Selection Guide" and consult manufacturers and suppliers for product sustainability and contribution to green building design.
- Consider specifying installation materials that meet ANSI A138.1, the *American National Standard Specifications for Sustainable Ceramic Tiles, Glass Tiles, and Tile Installation Materials.*

Preparation by Other Trades

- Floor systems over which tile will be installed shall be in conformance with the IRC for residential applications, the IBC for commercial applications, or applicable building codes. Maximum allowable substrate deflection under live load not to exceed l/360. See also "Substrate Requirements."
- When concentrated loads (scissor lifts, pallet jacks, automobiles, forklifts, etc.) will be used on an above-ground tiled floor, the engineer and/or specifier shall specify a substrate to accommodate the concentrated loads. Owner/specifier is responsible for protecting the tilework from damage, including allowing sufficient time for installed materials to cure properly.
- Slab to have steel trowel and fine broom finish free of curing compounds. When used, mechanical scarifying is necessary.
- Slope, when required, to be in subfloor.
- Maximum allowable variation in the tile substrate—⅛" in 10' from the required plane when measured from the high points in the surface.

Movement Joint (architect must specify type of joint and show location and details on drawings)

- Movement joints—mandatory according to EJ171. For above-ground installations, additional movement joints are required.

Installation Specifications

- Tile—ANSI A108.5 or A108.6.
- Cementitious grout—ANSI A108.10.
- Crack isolation membrane—A108.17.
- Uncoupling membrane—manufacturer's directions.
- Movement Joints—EJ171 and ASTM C1193.

Notes

- F111 is preferred over above-ground structural slabs and other floors subject to movement and/or deflection.
- Not all mortars, grouts, and membranes are suitable for above-ground installations. Check manufacturer recommendations.
- For waterproof application where thin-bed installation is desired, see F122A Stone.
- For in-slab hydronic tubing, see RH110A Stone.
- As the tile size increases, there is less tolerance for variation in the substrate from the required plane. Epoxy bond coat thickness must be thin and uniform; therefore, substrate flattening may be required when epoxy bond coat is used.
- When unsanded grout is used, grout joint width must be 1⁄16" to ⅛" wide (nominal).

F200-21
STONE

- **On-Ground Concrete**
- **Poured Gypsum Underlayment**
- **Bonded Membrane**
- **Natural Stone Tile**

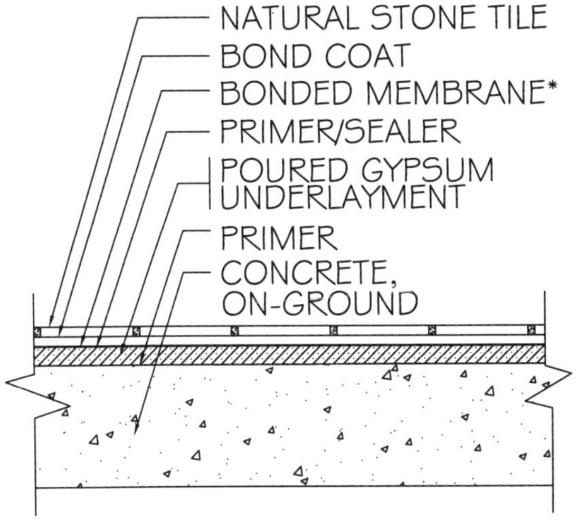

*USE OF A MEMBRANE IS REQUIRED. SEE MEMBRANE OPTIONS.

Recommended Uses

- For slab-on-ground construction where no bending stresses occur and with a poured gypsum underlayment.

Service Rating

- Moderate, when stone with adequate compressive strength, flexural strength, and resistance to abrasion is used.

Environmental Exposure Classifications

- Res1; Com1.
- May be suitable for limited water exposure areas (Res2; Com2). See Membrane Options.
- For installations that may be exposed to staining and/or chemical attack, specify tile and grout suitable for exposure. Consult product manufacturers; see also "Product Selection Guides."

Typical Weight of Tile Installation

- 14 pounds/square foot with ¾" poured gypsum. Add 2¼ pounds/square foot for each additional ¼" of poured gypsum.
- Does not include weight of substrate. See "Appendix B" for assumptions, included materials, and their individual weights.

Limitations

- Not recommended for use with some stones on post-tensioned concrete unless tile installation is 100 square feet or less. Consult stone supplier and installation materials manufacturers.
- Not for below-ground installation.
- Not for use over above-ground structural slabs and other floors subject to movement and/or deflection. See F200A Stone.
- Not for use where moisture vapor transmission rate exceeds gypsum underlayment manufacturer's limitations.

Membrane Options

- Requires use of crack isolation (A118.12) and/or waterproof (A118.10) and/or uncoupling membrane.
- Membrane must meet ANSI A118.10 for limited water exposure areas (Res2, Com2). Consult underlayment manufacturer for requirements.
- Specifier shall indicate if complete waterproofing is required, including if/how membrane connects to drain assembly, if base flashing is required, and treatment at other termination points.
- Check with membrane manufacturers for suitability for applicable conditions, as not all membranes are suitable for steam, high temperature, and/or chemical exposure, exterior use, use over above-ground structural slabs, use over pourable underlayments, use with radiant heat, or use over concrete with excessive moisture vapor transmission and/or alkalinity. Membrane may also affect service rating.

Requirements

- Proper curing/drying of gypsum underlayments prior to application of tile is critical for proper performance. Consult the underlayment manufacturer for specific instructions.
- Minimum grout joint width—1⁄16."

Materials

- **Multiple options exist for membranes, mortars, grouts, and other materials and *must be clearly specified* to be included. If not specifically indicated, optional materials are not included and mortar/grout choice defaults to minimum performance specification indicated. Consider each system component and intended use to determine minimum requirements and to specify options.**
- Natural stone tile—see "Natural Stone Tile Selection and Installation Guide," and consult supplier for application suitability.
- Cementitious grout—ANSI A118.6 or better or ISO CG1 or better. Consider unsanded grout for soft or semisoft stones such as limestone, travertine, marble, onyx, or similar.
- Cementitious bond coat:
  - When a crack isolation membrane is used—ANSI A118.4 or better or ISO C2S1 or better.
  - When a waterproof membrane is used—ANSI A118.4 or better or ISO C2S1 or better unless ANSI A118.1 or

ISO C1 is recommended by membrane manufacturer.

- When an uncoupling membrane is used, check with membrane manufacturer.
- Use white for light-colored marble, limestone, and other stones where staining or darkening from grey setting materials is possible.

- Epoxy bond coat, when used—ANSI A118.3 or ISO R1 or better. See Water Sensitivity and Fiberglass Mesh Reinforced Stone in the "Natural Stone Tile Selection and Installation Guide."
- Crack isolation membrane, when used—ANSI A118.12.
- Waterproof membrane, when used—ANSI A118.10.
- Uncoupling membrane, when used—recommended by uncoupling membrane manufacturer; must achieve 50 psi or greater shear bond strength in 7 days per the test method in ANSI A118.12 Section 5.1.3.

Materials for Green/Sustainable Design

- See "Green Building Standards and Green Product Selection Guide" and consult manufacturers and suppliers for product sustainability and contribution to green building design.
- Consider specifying installation materials that meet ANSI A138.1, the *American National Standard Specifications for Sustainable Ceramic Tiles, Glass Tiles, and Tile Installation Materials.*

Preparation by Other Trades

- Poured gypsum underlayment—minimum compressive strength of 2000 psi when tested per ASTM C472 modified and meeting performance requirements of ASTM C627 for the anticipated service level designated by the specifier or intended use.
- Consult the manufacturer for maximum moisture vapor transmission limitations prior to application of underlayment.
- Maximum allowable variation in the tile substrate—⅛" in 10' from the required plane when measured from the high points in the surface.
- Slab to be well cured, dimensionally stable, and free of cracks, surface laitance, waxy or oily films, curing compounds, and other surface bond-inhibiting contaminants.
- Poured gypsum underlayment thickness and application varies, consult the manufacturer for specific recommendations.
- Surfaces must be prepared and primed according to the underlayment manufacturer's instructions.
- Poured gypsum underlayment—installed only by a manufacturer-approved applicator in accordance with manufacturer's recommendations.

Movement Joint (architect must specify type of joint and show location and details on drawings)

- Movement joints—mandatory in accordance with EJ171.

Installation Specifications

- Tile—ANSI A108.5 or A108.6.
- Cementitious grout—ANSI A108.10.
- Crack isolation membrane—ANSI A108.17.
- Waterproof membrane—ANSI A108.13.
- Uncoupling membrane—manufacturer's directions.
- Movement Joints—EJ171 and ASTM C1193.

Notes

- Some manufacturers require special primers prior to the application of the membrane. Follow manufacturer's directions.
- As the tile size increases, there is less tolerance for variation in the substrate from the required plane. Epoxy bond coat thickness must be thin and uniform; therefore, substrate flattening may be required when epoxy bond coat is used.
- When unsanded grout is used, grout joint width must be 1⁄16" to ⅛" wide (nominal).

F200A-21
STONE

- **Above-Ground Concrete**
- **Poured Gypsum Underlayment**
- **Bonded Membrane**
- **Natural Stone Tile**

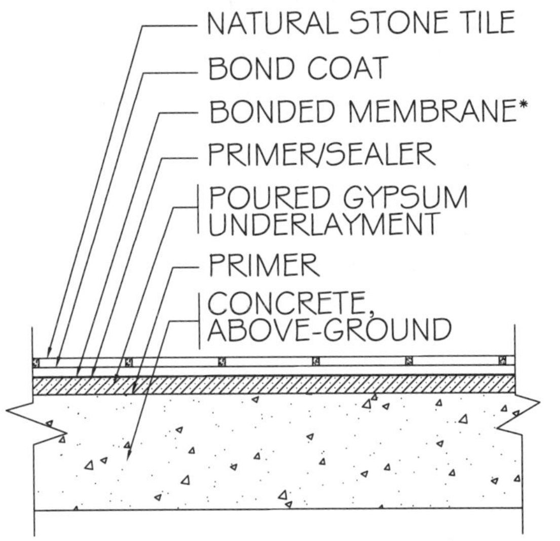

*USE OF A MEMBRANE IS REQUIRED. SEE MEMBRANE OPTIONS.

Recommended Uses

- For above-ground structural slabs with a poured gypsum underlayment.

Service Rating

- Moderate, when stone with adequate compressive strength, flexural strength, and resistance to abrasion is used.

Environmental Exposure Classifications

- Res1; Com1.
- May be suitable for limited water exposure areas (Res2; Com2). See Membrane Options.
- For installations that may be exposed to staining and/or chemical attack, specify tile and grout suitable for exposure. Consult product manufacturers; see also "Product Selection Guides."

Typical Weight of Tile Installation

- 14 pounds/square foot with ¾" poured gypsum. Add 2¼ pounds/square foot for each additional ¼" of poured gypsum.
- Does not include weight of substrate. See "Appendix B" for assumptions, included materials, and their individual weights.

Limitations

- Not recommended for use with some stones on post-tensioned concrete unless tile installation is 100 square feet or less. Consult stone supplier and installation materials manufacturers.
- Not for use where moisture vapor transmission rate exceeds gypsum underlayment manufacturer's limitations.
- Requires additional consideration by design professional to accommodate movement and/or deflection. Setting materials with improved bond strength and deformability are required.

Membrane Options

- Requires use of crack isolation (A118.12) and/or waterproof (A118.10) and/or uncoupling membrane.
- Membrane must meet ANSI A118.10 for limited water exposure areas (Res2, Com2). Consult underlayment manufacturer for requirements.
- Specifier shall indicate if complete waterproofing is required, including if/how membrane connects to drain assembly, if base flashing is required, and treatment at other termination points.
- Check with membrane manufacturers for suitability for applicable conditions, as not all membranes are suitable for steam, high temperature, and/or chemical exposure, exterior use, use over above-ground structural slabs, use over pourable underlayments, use with radiant heat, or use over concrete with excessive moisture vapor transmission and/or alkalinity. Membrane may also affect service rating.

Requirements

- Proper curing/drying of underlayments prior to application of tile is critical for proper performance. Consult the underlayment manufacturer for specific instructions.
- Minimum grout joint width—$\frac{1}{16}$".
- Above-ground installations are inherently more susceptible to vibration and deflection. Grout and mortar manufacturers to warrant product suitability.

Materials

- **Multiple options exist for membranes, mortars, grouts, and other materials and *must be clearly specified* to be included. If not specifically indicated, optional materials are not included and mortar/grout choice defaults to minimum performance specification indicated. Consider each system component and intended use to determine minimum requirements and to specify options.**
- Natural stone tile—see "Natural Stone Tile Selection and Installation Guide," and consult supplier for application suitability.
- Cementitious grout—ANSI A118.6 or better or ISO CG1 or better. Consider unsanded grout for soft or semisoft stones such as limestone, travertine, marble, onyx, or similar.

- Cementitious bond coat:
  - Must be recommended by manufacturer for above-ground use.
  - When a crack isolation membrane is used—ANSI A118.4 or better or ISO C2S1 or better.
  - When a waterproof membrane is used—ANSI A118.4 or better or ISO C2S1 or better unless ANSI A118.1 or ISO C1 is recommended by membrane manufacturer.
  - When an uncoupling membrane is used, check with membrane manufacturer.
  - Use white for light-colored marble, limestone, and other stones where staining or darkening from grey setting materials is possible.
- Epoxy bond coat, when used—ANSI A118.3 or ISO R1 or better. Must also be recommended by manufacturer for above-ground use. See Water Sensitivity and Fiberglass Mesh Reinforced Stone in the "Natural Stone Tile Selection and Installation Guide."
- Crack isolation membrane, when used—ANSI A118.12.
- Waterproof membrane, when used—ANSI A118.10.
- Uncoupling membrane, when used—recommended by uncoupling membrane manufacturer; must achieve 50 psi or greater shear bond strength in 7 days per the test method in ANSI A118.12 Section 5.1.3.

Materials for Green/Sustainable Design

- See "Green Building Standards and Green Product Selection Guide" and consult manufacturers and suppliers for product sustainability and contribution to green building design.
- Consider specifying installation materials that meet ANSI A138.1, the *American National Standard Specifications for Sustainable Ceramic Tiles, Glass Tiles, and Tile Installation Materials.*

Preparation by Other Trades

- Floor systems over which tile will be installed shall be in conformance with the IRC for residential applications, the IBC for commercial applications, or applicable building codes. Maximum allowable substrate deflection under live load not to exceed l/360. See also "Substrate Requirements."
- When concentrated loads (scissor lifts, pallet jacks, automobiles, forklifts, etc.) will be used on an above-ground tiled floor, the engineer and/or specifier shall specify a substrate to accommodate the concentrated loads. Owner/specifier is responsible for protecting the tilework from damage, including allowing sufficient time for installed materials to cure properly.
- Slab to be well cured, dimensionally stable, and free of cracks, surface laitance, waxy or oily films, curing compounds, and other surface bond-inhibiting contaminants.
- Consult the manufacturer for maximum moisture vapor transmission limitations prior to application of underlayment.
- Maximum allowable variation in the tile substrate—⅛" in 10' from the required plane when measured from the high points in the surface.
- Poured gypsum underlayment thickness and application varies, consult the manufacturer for specific recommendations.
- Surfaces must be prepared and primed according to the underlayment manufacturer's instructions.
- Poured gypsum underlayment—minimum compressive strength of 2000 psi when tested per ASTM C472 modified and meeting performance requirements of ASTM C627 for the anticipated service level designated by the specifier or intended use.
- Poured gypsum underlayment—installed only by a manufacturer-approved applicator in accordance with manufacturer's recommendations.

Movement Joint (architect must specify type of joint and show location and details on drawings)

- Movement joints—mandatory according to EJ171. For above-ground installations, additional movement joints are required.

Installation Specifications

- Tile—ANSI A108.5 or A108.6.
- Cementitious grout—ANSI A108.10.
- Crack isolation membrane—ANSI A108.17.
- Waterproof membrane—ANSI A108.13.
- Uncoupling membrane—manufacturer's directions.
- Movement Joints—EJ171 and ASTM C1193.

Notes

- Some manufacturers require special primers prior to the application of the membrane. Follow manufacturer's directions.
- As the tile size increases, there is less tolerance for variation in the substrate from the required plane. Epoxy bond coat thickness must be thin and uniform; therefore, substrate flattening may be required when epoxy bond coat is used.
- Not all mortars, grouts, and membranes are suitable for above-ground use. Check manufacturer recommendations.
- F111 is preferred over above-ground structural slabs and other floors subject to movement and/or deflection.
- When unsanded grout is used, grout joint width must be 1/16" to ⅛" wide (nominal).

F205-21
STONE

- **On-Ground Concrete**
- **Flowable Hydraulic Cement Underlayment/ Cementitious Self-Leveling Underlayment**
- **Natural Stone Tile**

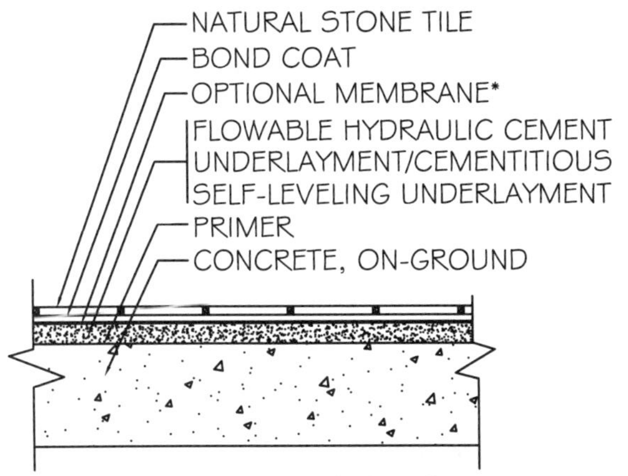

*USE OF A MEMBRANE IS OPTIONAL. SEE MEMBRANE OPTIONS.

Recommended Uses

- For areas where leveling, flattening, or raising the finished floor height is required.
- For slab-on-ground construction where no bending stresses occur and floor flatness is critical, such as when tiles with any edge longer than 15" are specified or where the finished tile assembly is required to achieve in-plane transitions (as may be necessary where accessibility is a concern).

Service Rating

- Heavy, when stone with adequate compressive strength, flexural strength, and resistance to abrasion is used.

Environmental Exposure Classifications

- Res1; Com1.
- May be suitable for increased water exposure. See Membrane Options.
- For installations that may be exposed to staining and/or chemical attack, specify tile and grout suitable for exposure. Consult product manufacturers; see also "Product Selection Guides."

Typical Weight of Tile Installation

- 13 pounds/square foot with ½" flowable hydraulic cement underlayment/cementitious self leveling underlayment. Add 2¾ pounds/square foot for each additional ¼" of underlayment.
- Does not include weight of substrate. See "Appendix B" for assumptions, included materials, and their individual weights.

Limitations

- Not recommended for use with some stones on post-tensioned concrete unless tile installation is 100 square feet or less. Consult stone supplier and installation materials manufacturers.
- Not for use where moisture vapor transmission rate exceeds underlayment manufacturer's limitations.
- Not for use over above-ground structural slabs and other floors subject to movement and/or deflection. See F205A Stone.
- Not for below-ground installation.

Membrane Options

- A waterproof membrane (A118.10) may be required for areas exposed to moisture. Consult underlayment manufacturer for requirements.
- Specifier shall indicate if complete waterproofing is required, including if/how membrane connects to drain assembly, if base flashing is required, and treatment at other termination points.
- A crack isolation membrane may be specified to treat existing in-plane cracks (F125-partial) or for protection against future in-plane cracking (F-125 full). See F125 Stone for more information.
- Check with membrane manufacturers for suitability for applicable conditions, as not all membranes are suitable for steam, high temperature, and/or chemical exposure, exterior use, use over above-ground structural slabs, use over pourable underlayments, use with radiant heat, or use over concrete with excessive moisture vapor transmission and/or alkalinity. Membrane may also affect service rating.

Requirements

- Consult the manufacturer for maximum moisture vapor transmission limitations prior to application of underlayment.
- Slab to be well cured, dimensionally stable, and free of cracks, surface laitance, waxy or oily films, curing compounds, and other surface bond-inhibiting contaminants.
- Flowable hydraulic cement underlayment/cementitious self-leveling underlayment thickness and application varies, consult the manufacturer for specific recommendations.
- When used to flatten the substrate and/or raise the finished tile assembly, the height of the finished tile assembly should be indicated by the specifier/design professional.
  - The field conditions must be evaluated by the installation professional prior to determining the amount of material necessary.
- Surfaces must be prepared and primed according to manufacturer's instructions.

- Proper curing/drying of underlayments prior to application of tile is critical for proper performance. Consult the underlayment manufacturer for specific instructions.
- Minimum grout joint width—1⁄16."

Materials

- **Multiple options exist for membranes, mortars, grouts, and other materials and *must be clearly specified* to be included. If not specifically indicated, optional materials are not included and mortar/grout choice defaults to minimum performance specification indicated. Consider each system component and intended use to determine minimum requirements and to specify options.**
- Natural stone tile—see "Natural Stone Tile Selection and Installation Guide," and consult supplier for application suitability.
- Cementitious grout—ANSI A118.6 or better or ISO CG1 or better. Consider unsanded grout for soft or semisoft stones such as limestone, travertine, marble, onyx, or similar.
- Cementitious bond coat:
  - When a membrane is not used—ANSI A118.4 or better or ISO C2 or better.
  - When a crack isolation membrane is used—ANSI A118.4 or better or ISO C2S1 or better.
  - When a waterproof membrane is used—ANSI A118.4 or better or ISO C2S1 or better unless ANSI A118.1 or ISO C1 is recommended by membrane manufacturer.
  - Use white for light-colored marble, limestone, and other stones where staining or darkening from grey setting materials is possible.
- Epoxy bond coat, when used—ANSI A118.3 or ISO R1 or better. See Water Sensitivity and Fiberglass Mesh Reinforced Stone in the "Natural Stone Tile Selection and Installation Guide."
- Crack isolation membrane, when used—ANSI A118.12.
- Waterproof membrane, when used—ANSI A118.10.
- Flowable hydraulic cement underlayment/cementitious self-leveling underlayment—meeting performance requirements of ASTM C627 for the anticipated service level designated by the specifier or intended use.

Materials for Green/Sustainable Design

- See "Green Building Standards and Green Product Selection Guide" and consult manufacturers and suppliers for product sustainability and contribution to green building design.
- Consider specifying installation materials that meet ANSI A138.1, the *American National Standard Specifications for Sustainable Ceramic Tiles, Glass Tiles, and Tile Installation Materials.*

Preparation by Other Trades

- Slab to have a fine broom finish with no curing compounds used. When used, mechanical scarifying is necessary.
- Maximum allowable variation in the installation substrate (concrete)—1⁄4" in 10' from the required plane.
- Maximum allowable variation in the tile substrate—1⁄8" in 10' from the required plane when measured from the high points in the surface.

Movement Joint (architect must specify type of joint and show location and details on drawings)

- Movement joints—mandatory according to EJ171.

Installation Specifications

- Tile—ANSI A108.5 or A108.6.
- Cementitious grout—ANSI A108.10.
- Flowable hydraulic cement underlayment/ cementitious self-leveling underlayment—manufacturer's directions.
- Crack isolation membrane—ANSI A108.17.
- Waterproof membrane—ANSI A108.13.
- Movement Joints—EJ171 and ASTM C1193.

Notes

- Some manufacturers require special primers prior to the application of the setting materials or membrane. Follow manufacturer's directions.
- When unsanded grout is used, grout joint width must be 1⁄16" to 1⁄8" wide (nominal).

F205A-21
STONE

- **Above-Ground Concrete**
- **Flowable Hydraulic Cement Underlayment/ Cementitious Self-Leveling Underlayment**
- **Natural Stone Tile**

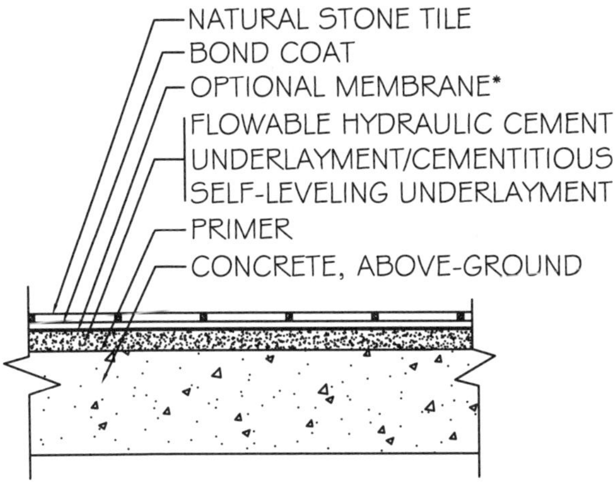

*USE OF A MEMBRANE IS OPTIONAL. SEE MEMBRANE OPTIONS.

Recommended Uses

- For areas where leveling, flattening, or raising the finished floor height is required.
- For slab-on-ground construction where no bending stresses occur and floor flatness is critical, such as when tiles with any edge longer than 15" are specified or where the finished tile assembly is required to achieve in-plane transitions (as may be necessary where accessibility is a concern).

Service Rating

- Heavy, when stone with adequate compressive strength, flexural strength, and resistance to abrasion is used.

Environmental Exposure Classifications

- Res1; Com1.
- May be suitable for increased water exposure. See Membrane Options.
- For installations that may be exposed to staining and/or chemical attack, specify tile and grout suitable for exposure. Consult product manufacturers; see also "Product Selection Guides."

Typical Weight of Tile Installation

- 13 pounds/square foot with ½" flowable hydraulic cement underlayment/cementitious self-leveling underlayment. Add 2¾ pounds/square foot for each additional ¼" of underlayment.
- Does not include weight of substrate. See "Appendix B" for assumptions, included materials, and their individual weights.

Limitations

- Not recommended for use with some stones on post-tensioned concrete unless tile installation is 100 square feet or less. Consult stone supplier and installation materials manufacturers.
- Not for use where moisture vapor transmission rate exceeds underlayment manufacturer's limitations.
- Requires additional consideration by design professional to accommodate movement and/or deflection. Setting materials with improved bond strength and deformability are required.

Membrane Options

- A waterproof membrane (A118.10) may be required for areas exposed to moisture. Consult underlayment manufacturer for requirements.
- Specifier shall indicate if complete waterproofing is required, including if/how membrane connects to drain assembly, if base flashing is required, and treatment at other termination points.
- A crack isolation membrane may be specified to treat existing in-plane cracks (F125-partial) or for protection against future in-plane cracking (F-125 full). See F125 Stone for more information.
- Check with membrane manufacturers for suitability for applicable conditions, as not all membranes are suitable for steam, high temperature, and/or chemical exposure, exterior use, use over above-ground structural slabs, use over pourable underlayments, use with radiant heat, or use over concrete with excessive moisture vapor transmission and/or alkalinity. Membrane may also affect service rating.

Requirements

- Consult the manufacturer for maximum moisture vapor transmission limitations prior to application of underlayment.
- Slab to be well cured, dimensionally stable, and free of cracks, surface laitance, waxy or oily films, curing compounds, and other surface bond-inhibiting contaminants.
- Flowable hydraulic cement underlayment/cementitious self-leveling underlayment thickness and application varies, consult the manufacturer for specific recommendations.
- When used to flatten the substrate and/or raise the finished tile assembly, the height of the finished tile assembly should be indicated by the specifier/design professional.
    - The field conditions must be evaluated by the installation professional prior to determining the amount of material necessary.
- Surfaces must be prepared and primed according to manufacturer's instructions.

- Proper curing/drying of underlayments prior to application of tile is critical for proper performance. Consult the underlayment manufacturer for specific instructions.
- Minimum grout joint width—1⁄16".
- Above-ground installations are inherently more susceptible to vibration and deflection. Grout and mortar manufacturers to warrant product suitability.

Materials

- **Multiple options exist for membranes, mortars, grouts, and other materials and *must be clearly specified* to be included. If not specifically indicated, optional materials are not included and mortar/grout choice defaults to minimum performance specification indicated. Consider each system component and intended use to determine minimum requirements and to specify options.**
- Natural stone tile—see "Natural Stone Tile Selection and Installation Guide," and consult supplier for application suitability.
- Cementitious grout—ANSI A118.6 or better or ISO CG1 or better. Consider unsanded grout for soft or semisoft stones such as limestone, travertine, marble, onyx, or similar.
- Cementitious bond coat:
  - Must be recommended by manufacturer for above-ground use.
  - When a membrane is not used—ANSI A118.15 or better or ISO C2S1 or better.
  - When a crack isolation membrane is used—ANSI A118.4 or better or ISO C2S1 or better.
  - When a waterproof membrane is used—ANSI A118.4 or better or ISO C2S1 or better unless ANSI A118.1 or ISO C1 is recommended by membrane manufacturer.
  - Use white for light-colored marble, limestone, and other stones where staining or darkening from grey setting materials is possible.
- Epoxy bond coat, when used—ANSI A118.3 or ISO R1 or better. Must also be recommended by manufacturer for above-ground use. See Water Sensitivity and Fiberglass Mesh Reinforced Stone in the "Natural Stone Tile Selection and Installation Guide."
- Crack isolation membrane, when used—ANSI A118.12.
- Waterproof membrane, when used—ANSI A118.10.
- Flowable hydraulic cement underlayment/cementitious self-leveling underlayment—meeting performance requirements of ASTM C627 for the anticipated service level designated by the specifier or intended use.

Materials for Green/Sustainable Design

- See "Green Building Standards and Green Product Selection Guide" and consult manufacturers and suppliers for product sustainability and contribution to green building design.
- Consider specifying installation materials that meet ANSI A138.1, the *American National Standard Specifications for Sustainable Ceramic Tiles, Glass Tiles, and Tile Installation Materials.*

Preparation by Other Trades

- Slab to have a fine broom finish with no curing compounds used. When used, mechanical scarifying is necessary.
- Maximum allowable variation in the installation substrate (concrete)—1⁄4" in 10' from the required plane.
- Maximum allowable variation in the tile substrate—1⁄8" in 10' from the required plane when measured from the high points in the surface.
- Floor systems over which tile will be installed shall be in conformance with the IRC for residential applications, the IBC for commercial applications, or applicable building codes. Maximum allowable substrate deflection under live load not to exceed l/360. See also "Substrate Requirements."
- When concentrated loads (scissor lifts, pallet jacks, automobiles, forklifts, etc.) will be used on an above-ground tiled floor, the engineer and/or specifier shall specify a substrate to accommodate the concentrated loads. Owner/specifier is responsible for protecting the tilework from damage, including allowing sufficient time for installed materials to cure properly.

Movement Joint (architect must specify type of joint and show location and details on drawings)

- Movement joints—mandatory according to EJ171. For above-ground installations, additional movement joints are required.

Installation Specifications

- Tile—ANSI A108.5 or A108.6.
- Cementitious grout—ANSI A108.10.
- Flowable hydraulic cement underlayment/cementitious self-leveling underlayment—manufacturer's directions.
- Crack isolation membrane—ANSI A108.17.
- Waterproof membrane—ANSI A108.13.
- Movement Joints—EJ171 and ASTM C1193.

Notes

- F111 is preferred over above-ground structural slabs and other floors subject to movement and/or deflection.
- Some manufacturers require special primers prior to the application of the setting materials or membrane. Follow manufacturer's directions.
- Not all mortars, grouts, and membranes are suitable for above-ground use. Check manufacturer recommendations.
- When unsanded grout is used, grout joint width must be 1⁄16" to 1⁄8" wide (nominal).

F121-21 STONE

- **On-Ground or Above-Ground Concrete**
- **Unbonded Mortar Bed**
- **Waterproof Membrane**
- **Natural Stone Tile**

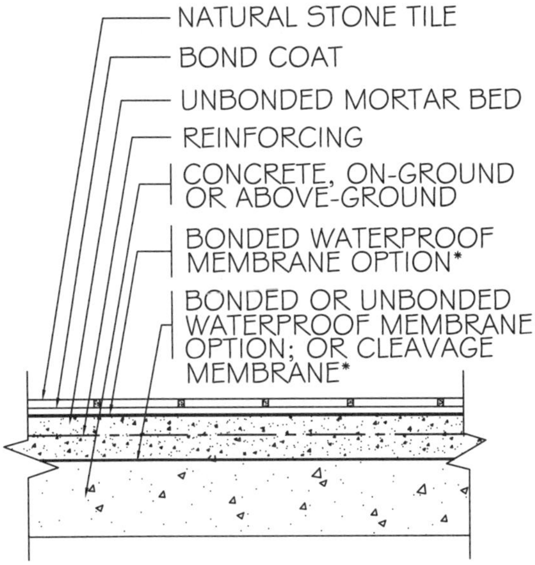

*USE OF A MEMBRANE IS REQUIRED. SEE MEMBRANE OPTIONS.

Recommended Uses

- For above-ground structural slabs and for slab-on-ground construction where a waterproof floor is required.
- For areas where leveling, flattening, or contouring of finish floor height is required and a waterproof floor is required.
- For areas where floor flatness is critical, such as when tiles with any edge longer than 15" are specified or where the finished tile assembly is required to achieve in-plane transitions (as may be necessary where accessibility is a concern), and a waterproof floor is required.
- For concrete substrates that are cracked or present bonding issues and a waterproof floor is required.
- For areas where ungauged stone will be used.

Service Rating

- Heavy, when stone with adequate compressive strength, flexural strength, and resistance to abrasion is used. Consult membrane manufacturer to verify.

Environmental Exposure Classifications

- Res1, 2, 3, 5; Com1, 2, 3, 5.
- May be suitable for exterior applications in areas not subject to freeze/thaw cycling when appropriate precautions are taken, including expansion joint placement, proper slope, waterproofing, and material selection.
- For installations that may be exposed to staining and/or chemical attack, specify tile and grout suitable for exposure. Consult product manufacturers; see also "Product Selection Guides."

Typical Weight of Tile Installation

- 23 pounds/square foot with 1¼" mortar bed. Add 3 pounds/square foot for each additional ¼" of mortar bed.
- Does not include weight of substrate. See "Appendix B" for assumptions, included materials, and their individual weights.

Limitations

- Above-ground structural slabs and other floors subject to movement and/or deflection—requires additional consideration by design professional to accommodate movement and/or deflection. Setting materials with improved bond strength and deformability are required.

Membrane Options

- A waterproof membrane and a cleavage membrane are required.
- If a bonded waterproof membrane (ANSI A118.10) is used, it may be placed above or below the mortar bed. If a shower pan membrane (ASTM D4551 or D4068) is used, it must be placed below the mortar bed.
- The waterproof membrane can also function as the cleavage membrane if loose laid (not bonded to concrete) below the mortar bed. Check with membrane manufacturer if membrane may be loose laid.
- If waterproof membrane is bonded to the top of the mortar bed, a cleavage membrane must be placed below the mortar bed.
- Specifier shall indicate if complete waterproofing is required, including if/how membrane connects to drain assembly, if base flashing is required, and treatment at other termination points.
- Check with membrane manufacturer for suitability for applicable conditions, as not all membranes are suitable for steam, high temperature, and/or chemical exposure, exterior use, use over above-ground structural slabs, use over pourable underlayments, use with radiant heat, or use over concrete with excessive moisture vapor transmission and/or alkalinity. Membrane may also affect service rating.

Requirements

- Mortar bed thickness—1¼" minimum to 2" maximum.
- Mortar beds in excess of 2" thick shall be detailed by the architect.
- Reinforcing mesh mandatory.
- Minimum grout joint width—$\frac{1}{16}$".
- Above-ground—above-ground installations are inherently more susceptible to vibration and deflection. Grout and mortar manufacturers to warrant product suitability.

Materials

- **Multiple options exist for membranes, mortars, grouts, and other materials and *must be clearly specified* to be included. If not specifically indicated, optional materials are not included and mortar/grout**

choice defaults to minimum performance specification indicated. Consider each system component and intended use to determine minimum requirements and to specify options.

- Natural stone tile—see "Natural Stone Tile Selection and Installation Guide," and consult supplier for application suitability.
- Cementitious grout—ANSI A118.6 or better or ISO CG1 or better. Consider unsanded grout for soft or semisoft stones such as limestone, travertine, marble, onyx, or similar.
- Cementitious bond coat (on-ground)—portland cement paste on a mortar bed that is still workable. For a cured mortar bed, follow recommendations below to select appropriate bonding mortar:
  - When waterproof membrane is placed below mortar bed—ANSI A118.1 or better or ISO C1 or better.
  - When waterproof membrane is placed above mortar bed—ANSI A118.4 or better or ISO C2S1 or better unless ANSI A118.1 or ISO C1 is recommended by membrane manufacturer.
  - Use white for light-colored marble, limestone, and other stones where staining or darkening from grey setting materials is possible.
- Cementitious bond coat (above-ground)—portland cement paste on a mortar bed that is still workable. For a cured mortar bed, follow recommendations below to select appropriate bonding mortar:
  - Must be recommended by manufacturer for above-ground use.
  - When waterproof membrane is placed below mortar bed—ANSI A118.15 or better or ISO C2S1 or better.
  - When waterproof membrane is placed above mortar bed—ANSI A118.4 or better or ISO C2S1 or better unless ANSI A118.1 or ISO C1 is recommended by membrane manufacturer.
  - Use white for light-colored marble, limestone, and other stones where staining or darkening from grey setting materials is possible.
- Epoxy bond coat, when used—ANSI A118.3 or ISO R1 or better. For above-ground use, must also be recommended by manufacturer. See Water Sensitivity and Fiberglass Mesh Reinforced Stone in the "Natural Stone Tile Selection and Installation Guide."
- Mortar bed, reinforcing, and cleavage membrane—ANSI A108.1A.
- Waterproof membrane, when used—ANSI A118.10.
- Shower pan membrane, when used—ASTM D4551 or D4068.

Materials for Green/Sustainable Design

- See "Green Building Standards and Green Product Selection Guide" and consult manufacturers and suppliers for product sustainability and contribution to green building design.
- Consider specifying installation materials that meet ANSI A138.1, the *American National Standard Specifications for Sustainable Ceramic Tiles, Glass Tiles, and Tile Installation Materials.*

Preparation by Other Trades

- Maximum allowable variation in the installation substrate (concrete)—¼" in 10' from the required plane.
- Where slab is depressed to accommodate mortar bed thickness, slab depression to be accurate, with steel trowel finish.
- Slope subfloor ¼" per foot to drain.
- Floor systems over which tile will be installed shall be in conformance with the IRC for residential applications, the IBC for commercial applications, or applicable building codes. Maximum allowable substrate deflection under live load not to exceed l/360. See also "Substrate Requirements."
- When concentrated loads (scissor lifts, pallet jacks, automobiles, forklifts, etc.) will be used on an above-ground tiled floor, the engineer and/or specifier shall specify a substrate to accommodate the concentrated loads. Owner/specifier is responsible for protecting the tilework from damage, including allowing sufficient time for installed materials to cure properly.

Movement Joint (architect must specify type of joint and show location and details on drawings)

- Movement joints—mandatory according to EJ171. For above-ground installations, additional movement joints are required.

Installation Specifications

- Tile—ANSI A108.1A, .1B, or .1C. Ungauged stone may require installation per ANSI A108.1A. Material with minimal variation may be allowed per A108.1B with LHT mortar. A108.1B required if membrane to be used on top of the mortar bed and/or epoxy bond coat to be used.
- Cementitious grout—ANSI A108.10.
- Waterproof membrane—ANSI A108.13 or manufacturer's directions. Comply with applicable plumbing and building codes.
- Shower pan membrane—manufacturer's directions. Comply with applicable plumbing and building codes.
- Movement Joints—EJ171 and ASTM C1193.

Notes

- For hydronic tubing on top of the slab when using a mortar bed, see RH117 Stone.
- See shower receptor details in stone methods B414, B421, and B422 for drain connections.
- For a waterproof application, but with a bonded mortar bed, see F112 Stone.
- Above-ground—not all mortars, grouts, and membranes are suitable for above-ground use. Check manufacturer recommendations.
- When unsanded grout is used, grout joint width must be 1/16" to ⅛" wide (nominal).

F122-21
STONE

- **On-Ground Concrete**
- **Waterproof Membrane**
- **Natural Stone Tile**

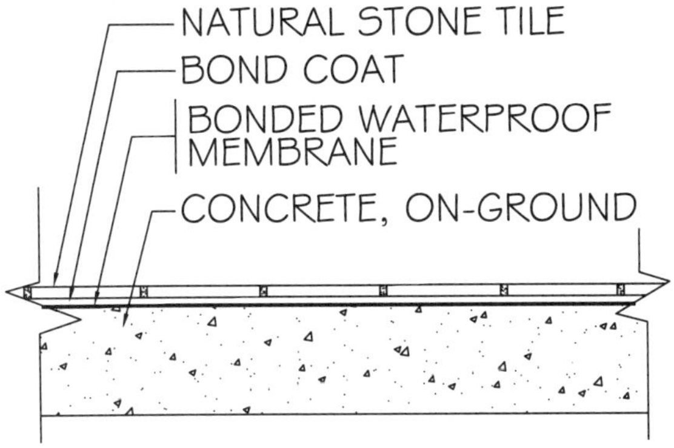

Recommended Uses

- For slab-on-ground construction where no bending stresses occur, where thin-bed installation of tile is desired, and a waterproof floor is required.

Service Rating

- Moderate, when stone with adequate compressive strength, flexural strength, and resistance to abrasion is used. Consult membrane manufacturer to verify.

Environmental Exposure Classifications

- Res1, 2, 3, 5; Com1, 2, 3, 5.
- May be suitable for exterior applications in areas not subject to freeze/thaw cycling when appropriate precautions are taken, including expansion joint placement, proper slope, waterproofing, and material selection.
- For installations that may be exposed to staining and/or chemical attack, specify tile and grout suitable for exposure. Consult product manufacturers; see also "Product Selection Guides."

Typical Weight of Tile Installation

- 8 pounds/square foot.
- Does not include weight of substrate. See "Appendix B" for assumptions, included materials, and their individual weights.

Limitations

- Not recommended for use with some stones on post-tensioned concrete unless tile installation is 100 square feet or less. Consult stone supplier and installation materials manufacturers.
- Not for use over above-ground structural slabs and other floors subject to movement and/or deflection. See F122A Stone.

Requirements

- Slab to be well cured, dimensionally stable, and free of cracks, surface laitance, waxy or oily films, curing compounds, and other surface bond-inhibiting contaminants.
- Check with membrane manufacturers for suitability for applicable conditions, as not all membranes are suitable for steam, high temperature, and/or chemical exposure, exterior use, use over above-ground structural slabs, use over pourable underlayments, use with radiant heat, or use over concrete with excessive moisture vapor transmission and/or alkalinity. Membrane may also affect service rating.
- Specifier shall indicate if complete waterproofing is required, including if/how membrane connects to drain assembly, if base flashing is required, and treatment at other termination points.
- Minimum grout joint width—1⁄16".

Materials

- **Multiple options exist for membranes, mortars, grouts, and other materials and *must be clearly specified* to be included. If not specifically indicated, optional materials are not included and mortar/grout choice defaults to minimum performance specification indicated. Consider each system component and intended use to determine minimum requirements and to specify options.**
- Natural stone tile—see "Natural Stone Tile Selection and Installation Guide," and consult supplier for application suitability.
- Cementitious grout—ANSI A118.6 or better or ISO CG1 or better. Consider unsanded grout for soft or semisoft stones such as limestone, travertine, marble, onyx, or similar.
- Cementitious bond coat:
  - ANSI A118.4 or better or ISO C2S1 or better unless ANSI A118.1 or ISO C1 is recommended by membrane manufacturer.
  - Use white for light-colored marble, limestone, and other stones where staining or darkening from grey setting materials is possible.
- Epoxy bond coat, when used—ANSI A118.3 or ISO R1 or better. See Water Sensitivity and Fiberglass Mesh Reinforced Stone in the "Natural Stone Tile Selection and Installation Guide."
- Waterproof membrane—ANSI A118.10.

Materials for Green/Sustainable Design

- See "Green Building Standards and Green Product Selection Guide" and consult manufacturers and suppliers for product sustainability and contribution to green building design.

- Consider specifying installation materials that meet ANSI A138.1, the *American National Standard Specifications for Sustainable Ceramic Tiles, Glass Tiles, and Tile Installation Materials.*

Preparation by Other Trades

- Slab to have steel trowel finish free of curing compounds. When used, mechanical scarification is necessary.
- Maximum allowable variation in the tile substrate—⅛" in 10' from the required plane when measured from the high points in the surface.
- Slope subfloor ¼" per foot to drain.

Movement Joint (architect must specify type of joint and show location and details on drawings)

- Movement joints—mandatory according to EJ171.

Installation Specifications

- Tile—ANSI A108.5 or A108.6.
- Cementitious grout—ANSI A108.10.
- Waterproof membrane—ANSI A108.13 or manufacturer's directions. Comply with applicable plumbing and building codes.
- Movement Joints—EJ171 and ASTM C1193.

Notes

- See shower receptor details B421 Stone and B422 Stone for drain connections.
- As the tile size increases, there is less tolerance for variation in the substrate from the required plane. Epoxy bond coat thickness must be thin and uniform; therefore, substrate flattening may be required when epoxy bond coat is used.
- When unsanded grout is used, grout joint width must be 1⁄16" to ⅛" wide (nominal).

F122A-21
STONE

- **Above-Ground Concrete**
- **Waterproof Membrane**
- **Natural Stone Tile**

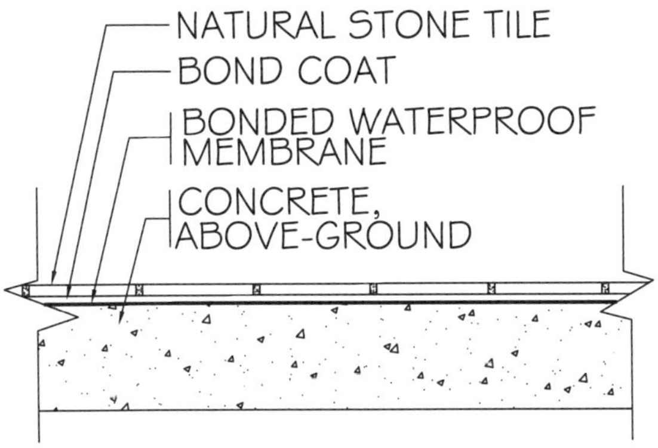

Recommended Uses

- For above-ground structural slabs where thin-bed installation of tile is desired and a waterproof floor is required.

Service Rating

- Moderate, when stone with adequate compressive strength, flexural strength, and resistance to abrasion is used. Consult membrane manufacturer to verify.

Environmental Exposure Classifications

- Res1, 2, 3, 5; Com1, 2, 3, 5.
- May be suitable for exterior applications in areas not subject to freeze/thaw cycling when appropriate precautions are taken, including expansion joint placement, proper slope, waterproofing and material selection.
- For installations that may be exposed to staining and/or chemical attack, specify tile and grout suitable for exposure. Consult product manufacturers; see also "Product Selection Guides."

Typical Weight of Tile Installation

- 8 pounds/square foot.
- Does not include weight of substrate. See "Appendix B" for assumptions, included materials, and their individual weights.

Limitations

- Not recommended for use with some stones on post-tensioned concrete unless tile installation is 100 square feet or less. Consult stone supplier and installation materials manufacturers.
- Requires additional consideration by design professional to accommodate movement and/or deflection. Setting materials with improved bond strength and deformability are required.

Requirements

- Slab to be well cured, dimensionally stable, and free of cracks, surface laitance, waxy or oily films, curing compounds, and other surface bond-inhibiting contaminants.
- Check with membrane manufacturers for suitability for applicable conditions, as not all membranes are suitable for steam, high temperature, and/or chemical exposure, exterior use, use over above-ground structural slabs, use over pourable underlayments, use with radiant heat, or use over concrete with excessive moisture vapor transmission and/or alkalinity. Membrane may also affect service rating.
- Specifier shall indicate if complete waterproofing is required, including if/how membrane connects to drain assembly, if base flashing is required, and treatment at other termination points.
- Minimum grout joint width—1⁄16".
- Above-ground installations are inherently more susceptible to vibration and deflection. Grout and mortar manufacturers to warrant product suitability.

Materials

- **Multiple options exist for membranes, mortars, grouts, and other materials and *must be clearly specified* to be included. If not specifically indicated, optional materials are not included and mortar/grout choice defaults to minimum performance specification indicated. Consider each system component and intended use to determine minimum requirements and to specify options.**
- Natural stone tile—see "Natural Stone Tile Selection and Installation Guide," and consult supplier for application suitability.
- Cementitious grout—ANSI A118.6 or better or ISO CG1 or better. Consider unsanded grout for soft or semisoft stones such as limestone, travertine, marble, onyx, or similar.
- Cementitious bond coat:
  - Must be recommended by manufacturer for above-ground use.
  - ANSI A118.4 or better or ISO C2S1 or better unless ANSI A118.1 or ISO C1 is recommended by membrane manufacturer.
  - Use white for light-colored marble, limestone, and other stones where staining or darkening from grey setting materials is possible.
- Epoxy bond coat, when used—ANSI A118.3 or ISO R1 or better. Must also be recommended by manufacturer for above-ground use. See Water Sensitivity and Fiberglass Mesh Reinforced Stone in the "Natural Stone Tile Selection and Installation Guide."
- Waterproof membrane—ANSI A118.10.

Materials for Green/Sustainable Design

- See "Green Building Standards and Green Product Selection Guide" and consult manufacturers and suppliers for product sustainability and contribution to green building design.
- Consider specifying installation materials that meet ANSI A138.1, the *American National Standard Specifications for Sustainable Ceramic Tiles, Glass Tiles, and Tile Installation Materials.*

Preparation by Other Trades

- Floor systems over which tile will be installed shall be in conformance with the IRC for residential applications, the IBC for commercial applications, or applicable building codes. Maximum allowable substrate deflection under live load not to exceed l/360. See also "Substrate Requirements."
- When concentrated loads (scissor lifts, pallet jacks, automobiles, forklifts, etc.) will be used on an above-ground tiled floor, the engineer and/or specifier shall specify a substrate to accommodate the concentrated loads. Owner/specifier is responsible for protecting the tilework from damage, including allowing sufficient time for installed materials to cure properly.
- Slab to have steel trowel finish free of curing compounds. When used, mechanical scarification is necessary.
- Maximum allowable variation in the tile substrate—⅛" in 10' from the required plane when measured from the high points in the surface.
- Slope subfloor ¼" per foot to drain.

Movement Joint (architect must specify type of joint and show location and details on drawings)

- Movement joints—mandatory according to EJ171. For above-ground installations, additional movement joints are required.

Installation Specifications

- Tile—ANSI A108.5 or A108.6.
- Cementitious grout—ANSI A108.10.
- Waterproof membrane—ANSI A108.13 or manufacturer's directions. Comply with applicable plumbing and building codes.
- Movement Joints—EJ171 and ASTM C1193.

Notes

- F111 is preferred over above-ground structural slabs and other floors subject to movement and/or deflection.
- Not all mortars, grouts, and membranes are suitable for above-ground installations. Check manufacturer recommendations.
- See shower receptor details B421 and B422 for drain connections.
- As the tile size increases, there is less tolerance for variation in the substrate from the required plane. Epoxy bond coat thickness must be thin and uniform; therefore, substrate flattening may be required when epoxy bond coat is used.
- When unsanded grout is used, grout joint width must be 1⁄16" to ⅛" wide (nominal).

F125-Partial-21
STONE

- **On- or Above-Ground Concrete**
- **Crack Isolation Membrane (Partial Coverage)**
- **Natural Stone Tile**

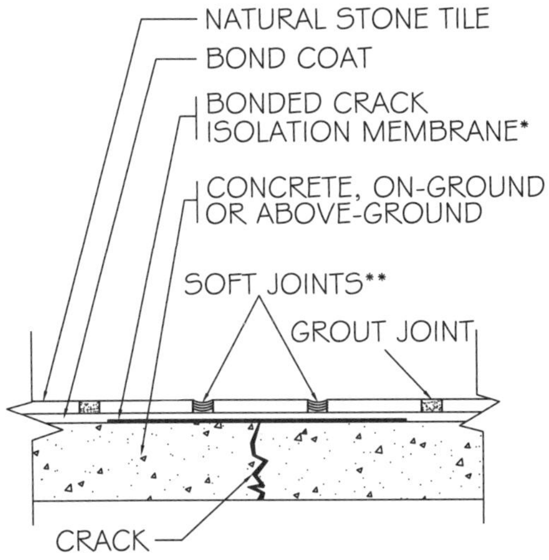

*CONSULT MANUFACTURER FOR MEMBRANE WIDTH

**ONE OR TWO SOFT JOINTS MAY BE REQUIRED (CONSULT MANUFACTURER)

F125-Full-21
STONE

- **On- or Above-Ground Concrete**
- **Crack Isolation Membrane (Full Coverage)**
- **Natural Stone Tile**

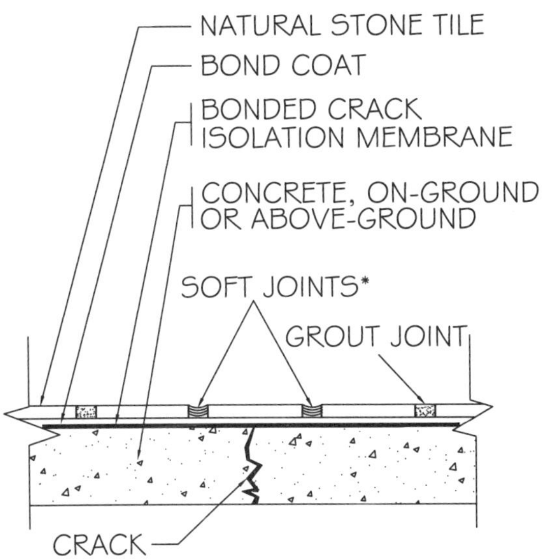

*ONE OR TWO SOFT JOINTS MAY BE REQUIRED (CONSULT MANUFACTURER)

Recommended Uses

- For use in conjunction with other concrete substrate installation methods.
- For slab-on-ground construction and above-ground structural slabs where thin-bed installation of tile is desired and crack isolation is required to treat existing in-plane cracks (F125-partial).
- For slab-on-ground construction and above-ground structural slabs where thin-bed installation of tile is desired and protection against future in-plane cracking is desired, full-coverage is recommended and must be additionally specified (F125-full).

Service Rating

- Different crack isolation membranes will produce different service ratings when tested per ASTM C627. Consult membrane manufacturer to verify service rating.

Environmental Exposure Classifications

- Refer to appropriate installation method.

Typical Weight of Tile Installation

- 8 pounds/square foot.
- Does not include weight of substrate. See "Appendix B" for assumptions, included materials, and their individual weights.

Limitations

- Not recommended for use with some stones on post-tensioned concrete unless tile installation is 100 square feet or less. Consult stone supplier and installation materials manufacturers.
- Above-ground structural slabs and other floors subject to movement and/or deflection—requires additional consideration by design professional to accommodate movement and/or deflection. Setting materials with improved bond strength and deformability are required.
- May not always be 100% effective in preventing reflective cracking.
- For lateral (in-plane) movement only.
- Standard performance membranes can accommodate up to $\frac{1}{16}$" in-plane movement. Consult manufacturer for membrane performance rating.
- High performance membranes can accommodate up to $\frac{1}{8}$" in-plane movement. Consult manufacturer for membrane performance rating.
- Not for cracking associated with deflection of above-ground installations.

Requirements

- For existing cracks, soft joints may be required adjacent to cracks per manufacturer's instructions.
- Specifier to identify areas of coverage (partial or full). Unless full coverage is clearly specified, installing contractor's responsibility is limited to partial coverage.

- Slab to be well cured, dimensionally stable, and free of cracks, surface laitance, waxy or oily films, curing compounds, and other surface bond-inhibiting contaminants.
- Check with membrane manufacturers for suitability for applicable conditions, as not all membranes are suitable for steam, high temperature, and/or chemical exposure, exterior use, use over above-ground structural slabs, use over pourable underlayments, use with radiant heat, or use over concrete with excessive moisture vapor transmission and/or alkalinity. Membrane may also affect service rating.
- Minimum grout joint width—1⁄16".
- Above-ground—above-ground installations are inherently more susceptible to vibration and deflection. Grout and mortar manufacturers to warrant product suitability.

Materials

- **Multiple options exist for membranes, mortars, grouts, and other materials and *must be clearly specified* to be included. If not specifically indicated, optional materials are not included and mortar/grout choice defaults to minimum performance specification indicated. Consider each system component and intended use to determine minimum requirements and to specify options. Unless full coverage of crack isolation membrane is clearly specified, installing contractor's responsibility is limited to partial coverage.**
- Refer to appropriate installation method.
- Sealant—ASTM C920.
- Crack isolation membrane—ANSI A118.12.

Materials for Green/Sustainable Design

- See "Green Building Standards and Green Product Selection Guide" and consult manufacturers and suppliers for product sustainability and contribution to green building design.
- Consider specifying installation materials that meet ANSI A138.1, the *American National Standard Specifications for Sustainable Ceramic Tiles, Glass Tiles, and Tile Installation Materials.*

Preparation by Other Trades

- Refer to appropriate installation method.
- Floor systems over which tile will be installed shall be in conformance with the IRC for residential applications, the IBC for commercial applications, or applicable building codes. Maximum allowable substrate deflection under live load not to exceed l/360. See also "Substrate Requirements."
- When concentrated loads (scissor lifts, pallet jacks, automobiles, forklifts, etc.) will be used on an above-ground tiled floor, the engineer and/or specifier shall specify a substrate to accommodate the concentrated loads. Owner/specifier is responsible for protecting the tilework from damage, including allowing sufficient time for installed materials to cure properly.
- Maximum allowable variation in the tile substrate—1⁄8" in 10' from the required plane when measured from the high points in the surface.
- Slab to have steel trowel and fine broom finish free of curing compounds. When used, mechanical scarifying is necessary.
- Slope, when required, to be in subfloor.

Movement Joint (architect must specify type of joint and show location and details on drawings)

- Movement joints—mandatory according to EJ171. For above-ground installations, additional movement joints are required.

Installation Specifications

- Refer to appropriate installation method.
- Crack isolation membrane—ANSI A108.17.
- Movement Joints—EJ171 and ASTM C1193.

Notes

- Stone tiles with reduced flexural strength, or natural planes of separation within the stone fabric may be vulnerable to breakage either due to elongation strain of the membrane or deflection over softer, compressible membrane materials. Consult membrane manufacturer and stone supplier for suitability. Testing per ANSI A118.12, modified by conducting the test using the actual stone tile can be performed to evaluate its performance under test as compared with the ceramic tile meeting ANSI A137.1 specified by the test.
- When unsanded grout is used, grout joint width must be 1⁄16" to 1⁄8" wide (nominal).

RH110-21
STONE

- **On-Ground Concrete Encapsulating Hydronic Tubing**
- **Natural Stone Tile**

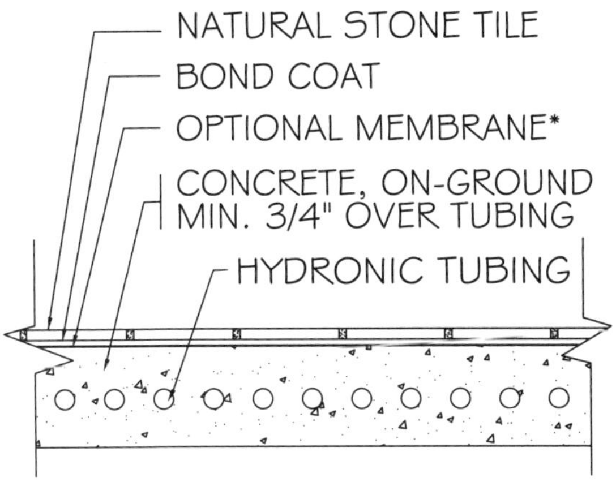

*USE OF A MEMBRANE IS OPTIONAL. SEE MEMBRANE OPTIONS.

Recommended Uses

- For slab-on-ground construction where no bending stresses occur, thin-bed installation of tile is desired, and hydronic tubing is encapsulated within the slab.

Service Rating

- Moderate, when stone with adequate compressive strength, flexural strength, and resistance to abrasion is used.

Environmental Exposure Classifications

- Res1, 2, 3, 5; Com1, 2, 3, 5.
- May be suitable for exterior applications in areas not subject to freeze/thaw cycling when appropriate precautions are taken, including expansion joint placement, proper slope, waterproofing, and material selection.
- For installations that may be exposed to staining and/or chemical attack, specify tile and grout suitable for exposure. Consult product manufacturers; see also "Product Selection Guides."

Typical Weight of Tile Installation

- 7 pounds/square foot.
- Does not include weight of substrate. See "Appendix B" for assumptions, included materials, and their individual weights.

Limitations

- Not recommended for use with some stones on post-tensioned concrete unless tile installation is 100 square feet or less. Consult stone supplier and installation materials manufacturers.
- Not for use on above-ground structural slabs and other floors subject to movement and/or deflection. See RH110A Stone.

Membrane Options

- A crack isolation membrane (A118.12) may be required. Check with heat system and setting material manufacturers.
- A crack isolation membrane may be specified to treat existing in-plane cracks (F125-partial) or for protection against future in-plane cracking (F-125 full). See F125 Stone for more information.
- Crack isolation membrane may also provide waterproofing. Consult membrane manufacturer.
- Check with membrane manufacturers for suitability for applicable conditions, as not all membranes are suitable for steam, high temperature, and/or chemical exposure, exterior use, use over above-ground structural slabs, use over pourable underlayments, use with radiant heat, or use over concrete with excessive moisture vapor transmission and/or alkalinity. Membrane may also affect service rating.

Requirements

- Slab to be well cured, dimensionally stable, and free of cracks, surface laitance, waxy or oily films, curing compounds, and other surface bond-inhibiting contaminants.
- Minimum grout joint width—1/16".

Materials

- **Multiple options exist for membranes, mortars, grouts, and other materials and *must be clearly specified* to be included. If not specifically indicated, optional materials are not included and mortar/grout choice defaults to minimum performance specification indicated. Consider each system component and intended use to determine minimum requirements and to specify options.**
- Natural stone tile—see "Natural Stone Tile Selection and Installation Guide," and consult supplier for application suitability.
- Cementitious grout—ANSI A118.6 or better or ISO CG1 or better. Consider unsanded grout for soft or semisoft stones such as limestone, travertine, marble, onyx, or similar.
- Cementitious bond coat:
  - When a crack isolation membrane is not used—ANSI A118.15 or better or ISO C2S1 or better.
  - When a crack isolation membrane is used—ANSI A118.4 or better or ISO C2S1 or better.
  - Use white for light-colored marble, limestone, and other stones where staining or darkening from grey setting materials is possible.
- Epoxy bond coat, when used—ANSI A118.3 or ISO R1 or better. See Water Sensitivity and Fiberglass Mesh Reinforced Stone in the "Natural Stone Tile Selection and Installation Guide."
- Crack isolation membrane, when used—ANSI A118.12.

Materials for Green/Sustainable Design

- See "Green Building Standards and Green Product Selection Guide" and consult manufacturers and suppliers for product sustainability and contribution to green building design.
- Consider specifying installation materials that meet ANSI A138.1, the *American National Standard Specifications for Sustainable Ceramic Tiles, Glass Tiles, and Tile Installation Materials.*

Preparation by Other Trades

- Slab to have steel trowel and fine broom finish free of curing compounds. When used, mechanical scarifying is necessary.
- Slope, when required, to be in subfloor.
- Maximum allowable variation in the tile substrate—⅛" in 10' from the required plane when measured from the high points in the surface.
- Radiant tubing by others.
- Portland cement concrete poured minimum ¾" over top of hydronic tubing.

Movement Joint (architect must specify type of joint and show location and details on drawings)

- Movement joints—mandatory according to EJ171.

Installation Specifications

- Tile—ANSI A108.5 or A108.6.
- Cementitious grout—ANSI A108.10.
- Crack isolation membrane—ANSI A108.17.
- Movement Joints—EJ171 and ASTM C1193.

Notes

- For waterproof application, see F122 Stone.
- As the tile size increases, there is less tolerance for variation in the substrate from the required plane. Epoxy bond coat thickness must be thin and uniform; therefore, substrate flattening may be required when epoxy bond coat is used.
- When unsanded grout is used, grout joint width must be 1/16" to ⅛" wide (nominal).

RH110A-21 STONE

- **Above-Ground Concrete Encapsulating Hydronic Tubing**
- **Natural Stone Tile**

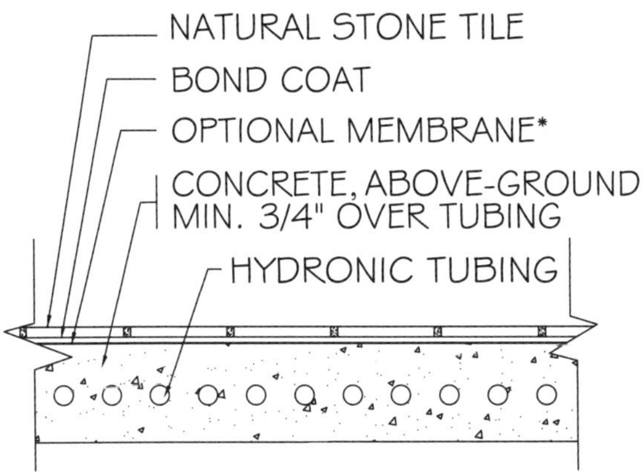

*USE OF A MEMBRANE IS OPTIONAL. SEE MEMBRANE OPTIONS.

Recommended Uses

- For above-ground structural slabs where thin-bed installation of tile is desired and hydronic tubing is encapsulated within the slab.

Service Rating

- Moderate, when stone with adequate compressive strength, flexural strength, and resistance to abrasion is used.

Environmental Exposure Classifications

- Res1, 2, 3, 5; Com1, 2, 3, 5.
- May be suitable for exterior applications in areas not subject to freeze/thaw cycling when appropriate precautions are taken, including expansion joint placement, proper slope, waterproofing, and material selection.
- For installations that may be exposed to staining and/or chemical attack, specify tile and grout suitable for exposure. Consult product manufacturers; see also "Product Selection Guides."

Typical Weight of Tile Installation

- 7 pounds/square foot.
- Does not include weight of substrate. See "Appendix B" for assumptions, included materials, and their individual weights.

Limitations

- Not recommended for use with some stones on post-tensioned concrete unless tile installation is 100 square feet or less. Consult stone supplier and installation materials manufacturers.
- Requires additional consideration by design professional to accommodate movement and/or deflection. Setting materials with improved bond strength and deformability are required.

Membrane Options

- A crack isolation membrane (A118.12) may be required. Check with heat system and setting material manufacturers.
- A crack isolation membrane may be specified to treat existing in-plane cracks (F125-partial) or for protection against future in-plane cracking (F-125 full). See F125 Stone for more information.
- Crack isolation membrane may also provide waterproofing. Consult membrane manufacturer.
- Check with membrane manufacturers for suitability for applicable conditions, as not all membranes are suitable for steam, high temperature, and/or chemical exposure, exterior use, use over above-ground structural slabs, use over pourable underlayments, use with radiant heat, or use over concrete with excessive moisture vapor transmission and/or alkalinity. Membrane may also affect service rating.

Requirements

- Slab to be well cured, dimensionally stable, and free of cracks, surface laitance, waxy or oily films, curing compounds, and other surface bond-inhibiting contaminants.
- Minimum grout joint width—1/16".
- Above-ground installations are inherently more susceptible to vibration and deflection. Grout and mortar manufacturers to warrant product suitability.

Materials

- **Multiple options exist for membranes, mortars, grouts, and other materials and *must be clearly specified* to be included. If not specifically indicated, optional materials are not included and mortar/grout choice defaults to minimum performance specification indicated. Consider each system component and intended use to determine minimum requirements and to specify options.**
- Natural stone tile—see "Natural Stone Tile Selection and Installation Guide," and consult supplier for application suitability.
- Cementitious grout—ANSI A118.6 or better or ISO CG1 or better. Consider unsanded grout for soft or semisoft stones such as limestone, travertine, marble, onyx, or similar.
- Cementitious bond coat:
  - Must be recommended by manufacturer for above-ground use.
  - When a crack isolation membrane is not used—ANSI A118.15 or better or ISO C2S1 or better.

- When a crack isolation membrane is used—ANSI A118.4 or better or ISO C2S1 or better.
- Use white for light-colored marble, limestone, and other stones where staining or darkening from grey setting materials is possible.

- Epoxy bond coat, when used—ANSI A118.3 or ISO R1 or better. Must also be recommended by manufacturer for above-ground use. See Water Sensitivity and Fiberglass Mesh Reinforced Stone in the "Natural Stone Tile Selection and Installation Guide."
- Crack isolation membrane, when used—ANSI A118.12.

Materials for Green/Sustainable Design

- See "Green Building Standards and Green Product Selection Guide" and consult manufacturers and suppliers for product sustainability and contribution to green building design.
- Consider specifying installation materials that meet ANSI A138.1, the *American National Standard Specifications for Sustainable Ceramic Tiles, Glass Tiles, and Tile Installation Materials.*

Preparation by Other Trades

- Floor systems over which tile will be installed shall be in conformance with the IRC for residential applications, the IBC for commercial applications, or applicable building codes. Maximum allowable substrate deflection under live load not to exceed l/360. See also "Substrate Requirements."
- When concentrated loads (scissor lifts, pallet jacks, automobiles, forklifts, etc.) will be used on an above-ground tiled floor, the engineer and/or specifier shall specify a substrate to accommodate the concentrated loads. Owner/specifier is responsible for protecting the tilework from damage, including allowing sufficient time for installed materials to cure properly.
- Slab to have steel trowel and fine broom finish free of curing compounds. When used, mechanical scarifying is necessary.
- Slope, when required, to be in subfloor.
- Maximum allowable variation in the tile substrate—⅛" in 10' from the required plane when measured from the high points in the surface.
- Radiant tubing by others.
- Portland cement concrete poured minimum ¾" over top of hydronic tubing.

Movement Joint (architect must specify type of joint and show location and details on drawings)

- Movement joints—mandatory according to EJ171. For above-ground installations, additional movement joints are required.

Installation Specifications

- Tile—ANSI A108.5 or A108.6.
- Cementitious grout—ANSI A108.10.
- Crack isolation membrane—A108.17.
- Movement Joints—EJ171 and ASTM C1193.

Notes

- F111 is preferred over above-ground structural slabs and other floors subject to movement and/or deflection.
- Not all mortars, grouts, and membranes are suitable for above-ground installations. Check manufacturer recommendations.
- For waterproof application, see F122A Stone.
- As the tile size increases, there is less tolerance for variation in the substrate from the required plane. Epoxy bond coat thickness must be thin and uniform; therefore, substrate flattening may be required when epoxy bond coat is used.
- When unsanded grout is used, grout joint width must be 1/16" to ⅛" wide (nominal).

RH117-21
STONE

- **On-Ground or Above-Ground Concrete**
- **Unbonded Mortar Bed Encapsulating Hydronic Tubing**
- **Natural Stone Tile**

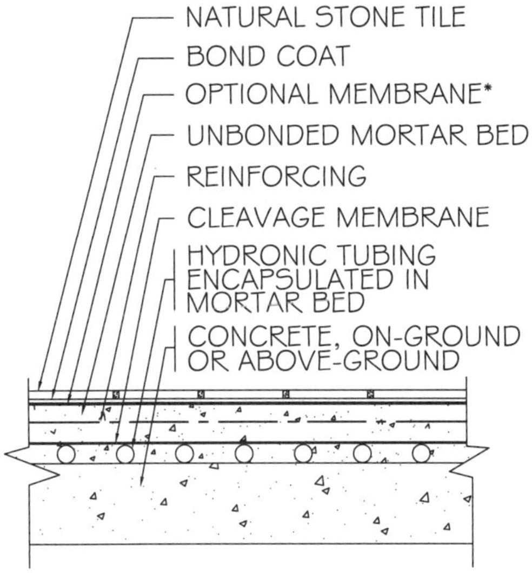

*USE OF A MEMBRANE IS OPTIONAL SEE MEMBRANE OPTIONS

Recommended Uses

- For above-ground structural slabs and for slab-on-ground construction where hydronic tubing will be placed over the slab.
- For areas where leveling, flattening, or contouring of finish floor height is required where hydronic tubing will be placed over the slab.
- For areas where floor flatness is critical, such as when tiles with any edge longer than 15" are specified or where the finished tile assembly is required to achieve in-plane transitions (as may be necessary where accessibility is a concern), and where hydronic tubing will be placed over the slab.
- For concrete substrates that are cracked or present bonding issues and hydronic tubing will be placed over the slab.
- For areas where ungauged stone will be used.

Service Rating

- Heavy, when stone with adequate compressive strength, flexural strength, and resistance to abrasion is used.

Environmental Exposure Classifications

- Res1, 2, 3, 5; Com1, 2, 3, 5.
- May be suitable for exterior applications in areas not subject to freeze/thaw cycling when appropriate precautions are taken, including expansion joint placement, proper slope, waterproofing, and material selection.
- For installations that may be exposed to staining and/or chemical attack, specify tile and grout suitable for exposure. Consult product manufacturers; see also "Product Selection Guides."

Typical Weight of Tile Installation

- 32 pounds/square foot with 1¼" mortar bed over hydronic tubing. Add 3 pounds/square foot for each additional ¼" of mortar bed.
- Does not include weight of substrate. See "Appendix B" for assumptions, included materials, and their individual weights.

Limitations

- Above-ground structural slabs and other floors subject to movement and/or deflection—requires additional consideration by design professional to accommodate movement and/or deflection. Setting materials with improved bond strength and deformability are required.

Membrane Options

- A crack isolation membrane (A118.12) may be required. Check with heat system and setting material manufacturers.
- Crack isolation membrane may also provide waterproofing. Consult membrane manufacturer.
- A waterproof membrane (A118.10) may be specified to prevent moisture intrusion. Specifier shall indicate if complete waterproofing is required, including if/how membrane connects to drain assembly, if base flashing is required, and treatment at other termination points.
- Check with membrane manufacturers for suitability for applicable conditions, as not all membranes are suitable for steam, high temperature, and/or chemical exposure, exterior use, use over above-ground structural slabs, use over pourable underlayments, use with radiant heat, or use over concrete with excessive moisture vapor transmission and/or alkalinity. Membrane may also affect service rating.

Requirements

- Reinforcing mesh mandatory.
- Mortar bed thickness—1¼" minimum to 2" maximum.
- Mortar beds in excess of 2" thick shall be detailed by the architect.
- Screed fill flush to top of radiant tubing before placing cleavage membrane and mortar bed.
- Minimum grout joint width—$\frac{1}{16}$".
- Above-ground—above-ground installations are inherently more susceptible to vibration and deflection. Grout and mortar manufacturers to warrant product suitability.

Materials

- **Multiple options exist for membranes, mortars, grouts, and other materials and *must be clearly specified* to be included. If not specifically indicated, optional materials are not included and mortar/grout**

choice defaults to minimum performance specification indicated. Consider each system component and intended use to determine minimum requirements and to specify options.

- Natural stone tile—see "Natural Stone Tile Selection and Installation Guide," and consult supplier for application suitability.
- Cementitious grout—ANSI A118.6 or better or ISO CG1 or better. Consider unsanded grout for soft or semisoft stones such as limestone, travertine, marble, onyx, or similar.
- Cementitious bond coat (on-ground)—portland cement paste on a mortar bed that is still workable. For a cured mortar bed, follow recommendations below to select appropriate bonding mortar:
  - When a membrane is not used—ANSI A118.15 or better or ISO C2S1 or better.
  - When a crack isolation membrane is used—ANSI A118.4 or better or ISO C2S1 or better.
  - When a waterproof membrane is used—ANSI A118.4 or better or ISO C2S1 or better unless ANSI A118.1 or ISO C1 is recommended by membrane manufacturer.
  - Use white for light-colored marble, limestone, and other stones where staining or darkening from grey setting materials is possible.
- Cementitious bond coat (above-ground)—portland cement paste on a mortar bed that is still workable. For a cured mortar bed, follow recommendations below to select appropriate bonding mortar:
  - Must be recommended by manufacturer for above-ground use.
  - When a membrane is not used—ANSI A118.15 or better or ISO C2S1 or better.
  - When a crack isolation membrane is used—ANSI A118.4 or better or ISO C2S1 or better.
  - When a waterproof membrane is used—ANSI A118.4 or better or ISO C2S1 or better unless ANSI A118.1 or ISO C1 is recommended by membrane manufacturer.
  - Use white for light-colored marble, limestone, and other stones where staining or darkening from grey setting materials is possible.
- Epoxy bond coat, when used—ANSI A118.3 or ISO R1 or better. For above-ground use, must also be recommended by manufacturer. See Water Sensitivity and Fiberglass Mesh Reinforced Stone in the "Natural Stone Tile Selection and Installation Guide."
- Crack isolation membrane, when used—ANSI A118.12.
- Waterproof membrane, when used—ANSI A118.10.
- Mortar bed, reinforcing, and cleavage membrane—ANSI A108.1A.

Materials for Green/Sustainable Design

- See "Green Building Standards and Green Product Selection Guide" and consult manufacturers and suppliers for product sustainability and contribution to green building design.
- Consider specifying installation materials that meet ANSI A138.1, the *American National Standard Specifications for Sustainable Ceramic Tiles, Glass Tiles, and Tile Installation Materials.*

Preparation by Other Trades

- Floor systems over which tile will be installed shall be in conformance with the IRC for residential applications, the IBC for commercial applications, or applicable building codes. Maximum allowable substrate deflection under live load not to exceed l/360. See also "Substrate Requirements."
- When concentrated loads (scissor lifts, pallet jacks, automobiles, forklifts, etc.) will be used on an above-ground tiled floor, the engineer and/or specifier shall specify a substrate to accommodate the concentrated loads. Owner/specifier is responsible for protecting the tilework from damage, including allowing sufficient time for installed materials to cure properly.
- Where slab is depressed to accommodate mortar bed thickness, slab depression to be accurate with steel trowel finish.
- Slope, when required, to be in subfloor.
- Maximum allowable variation in the installation substrate (concrete)—¼" in 10' from the required plane.
- Radiant tubing by others.

Movement Joint (architect must specify type of joint and show location and details on drawings)

- Movement joints—mandatory according to EJ171. For above-ground installations, additional movement joints are required.

Installation Specifications

- Tile—ANSI A108.1A, .1B, or .1C. Ungauged stone may require installation per ANSI A108.1A. Material with minimal variation may be allowed per A108.1B with LHT mortar. A108.1B required if membrane or epoxy bond coat to be used.
- Cementitious grout—ANSI A108.10.
- Crack isolation membrane—ANSI A108.17.
- Waterproof membrane—ANSI A108.13.
- Movement Joints—EJ171 and ASTM C1193.

Notes

- For in-slab hydronic tubing, see F111 Stone.
- Above-ground—not all mortars, grouts, and membranes are suitable for above-ground use. Check manufacturer recommendations.
- When unsanded grout is used, grout joint width must be ¹⁄₁₆" to ⅛" wide (nominal).

RH111-21 STONE

- **On-Ground Concrete**
- **Poured Gypsum Underlayment Encapsulating Hydronic Tubing**
- **Bonded Membrane**
- **Natural Stone Tile**

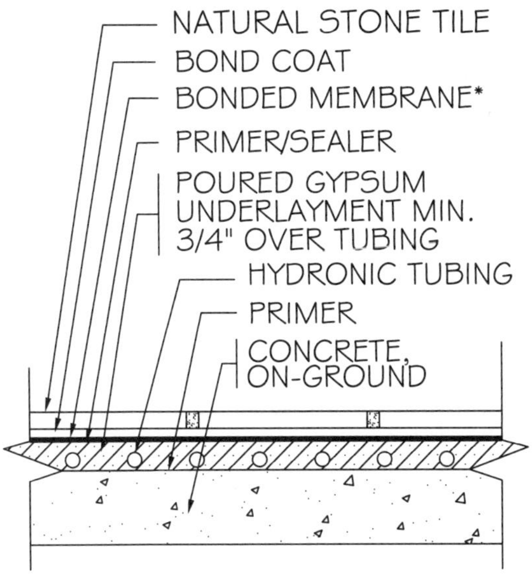

*USE OF A MEMBRANE IS REQUIRED. SEE MEMBRANE OPTIONS.

Recommended Uses

- For slab-on-ground construction where no bending stresses occur and with a poured gypsum underlayment encapsulating hydronic tubing.

Service Rating

- Moderate when stone with adequate compressive strength, flexural strength and resistance to abrasion is used.

Environmental Exposure Classifications

- Res1; Com1.
- May be suitable for limited water exposure areas (Res2; Com2). See Membrane Options.
- For installations that may be exposed to staining and/or chemical attack, specify tile and grout suitable for exposure. Consult product manufacturers; see also "Product Selection Guides."

Typical Weight of Tile Installation

- 21 pounds/square foot with ¾" poured gypsum over hydronic tubing. Add 2¼ pounds/square foot for each additional ¼" of poured gypsum.
- Does not include weight of substrate. See "Appendix B" for assumptions, included materials, and their individual weights.

Limitations

- Not for below-ground installation.
- Not for use over above-ground structural slabs and other floors subject to movement and/or deflection. See RH111A Stone.
- Not for use where moisture vapor transmission rate exceeds gypsum underlayment manufacturer's limitations.
- Not recommended for use with some stones on post tensioned concrete unless tile installation is 100 square feet or less. Consult stone supplier and installation materials manufacturer.

Membrane Options

- Requires use of crack isolation (A118.12) and/or waterproof (A118.10) and/or uncoupling membrane.
- Membrane must meet ANSI A118.10 for limited water exposure areas (Res2; Com2). Consult underlayment manufacturer for requirements.
- Specifier shall indicate if complete waterproofing is required, including if/how membrane connects to drain assembly, if base flashing is required, and treatment at other termination points.
- Check with membrane manufacturer for suitability for applicable conditions, as not all membranes are suitable for steam, high temperature, and/or chemical exposure, exterior use, use over above-ground structural slabs, use over pourable underlayments, use with radiant heat, or use over concrete with excessive moisture vapor transmission and/or alkalinity. Membrane may also affect service rating.

Requirements

- Proper curing/drying of gypsum underlayments prior to application of tile is critical for proper performance. Consult the underlayment manufacturer for specific instructions.
- Minimum grout joint width—$\frac{1}{16}$".

Materials

- **Multiple options exist for membranes, mortars, grouts, and other materials and *must be clearly specified* to be included. If not specifically indicated, optional materials are not included and mortar/grout choice defaults to minimum performance specification indicated. Consider each system component and intended use to determine minimum requirements and to specify options.**
- Natural stone tile—See "Natural Stone Tile Selection and Installation Guide" and consult supplier for application suitability.
- Cementitious grout—ANSI A118.6 or better or ISO CG1 or better. Consider unsanded grout for soft or semi-soft stones such as limestone, travertine, marble, onyx or similar.

- Cementitious bond coat:
  - When a crack isolation membrane is used—ANSI A118.4 or better or ISO C2S1 or better.
  - When a waterproof membrane is used—ANSI A118.4 or better or ISO C2S1 or better, unless ANSI A118.1 or ISO C1 is recommended by membrane manufacturer.
  - When an uncoupling membrane is used, check with membrane manufacturer.
  - Use white for light-colored marble, limestone, and other stones where staining or darkening from grey setting materials is possible.
- Epoxy bond coat, when used—ANSI A118.3 or ISO R1 or better. See Water Sensitivity and Fiberglass Mesh Reinforced Stone in the "Natural Stone Tile Selection and Installation Guide."
- Crack isolation membrane, when used—ANSI A118.12.
- Waterproof membrane, when used—ANSI A118.10.
- Uncoupling membrane, when used—recommended by uncoupling membrane manufacturer; must achieve 50 psi or greater shear bond strength in 7 days per the test method in ANSI A118.12 Section 5.1.3.

Materials for Green/Sustainable Design

- See "Green Building Standards and Green Product Selection Guide" and consult manufacturers and suppliers for product sustainability and contribution to green building design.
- Consider specifying installation materials that meet ANSI A138.1, the *American National Standard Specifications for Sustainable Ceramic Tiles, Glass Tiles, and Tile Installation Materials.*

Preparation by Other Trades

- Consult the manufacturer for maximum moisture vapor transmission limitations prior to application of underlayment.
- Maximum allowable variation in the tile substrate—⅛" in 10' from the required plane when measured from the high points in the surface.
- Slab to be well cured, dimensionally stable, and free of cracks, surface laitance, waxy or oily films, curing compounds, and other surface bond-inhibiting contaminants.
- Poured gypsum underlayment thickness and application varies, consult the manufacturer for specific recommendations when used with floor warming systems. Minimum thickness is ¾" above hydronic tubing.
- Surfaces must be prepared and primed according to the underlayment manufacturer's instructions.
- Radiant tubing by others.
- Poured gypsum underlayment—minimum compressive strength of 2000 psi when tested per ASTM C472 modified and meeting performance requirements of ASTM C627 for the anticipated service level designated by the specifier or intended use.
- Poured gypsum underlayment—installed only by a manufacturer-approved applicator in accordance with manufacturer's recommendations.

Movement Joint (architect must specify type of joint and show location and details on drawings)

- Movement joints—mandatory in accordance with EJ171.

Installation Specifications

- Tile—ANSI A108.5.
- Cementitious grout—ANSI A108.10.
- Epoxy mortar—ANSI A108.6.
- Crack isolation membrane—ANSI A108.17.
- Waterproof membrane—ANSI A108.13.
- Uncoupling membrane—manufacturer's directions.
- Movement Joints—EJ171 and ASTM C1193.

Notes

- Consult underlayment manufacturer for minimum cure/dry time of installation before radiant heat system is activated.
- Some manufacturers require special primers prior to the application of the membrane. Follow manufacturer's directions.
- When unsanded grout is used, grout joint width must be $\frac{1}{16}$" to ⅛" wide (nominal).

RH111A-21
STONE

- **Above-Ground Concrete**
- **Poured Gypsum Underlayment Encapsulating Hydronic Tubing**
- **Bonded Membrane**
- **Natural Stone Tile**

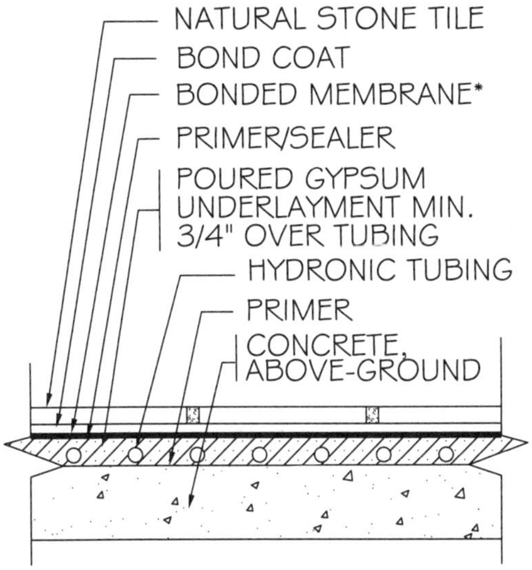

*USE OF A MEMBRANE IS REQUIRED. SEE MEMBRANE OPTIONS.

Recommended Uses

- For above-ground structural slabs with a poured gypsum underlayment encapsulating hydronic tubing.

Service Rating

- Moderate when stone with adequate compressive strength, flexural strength and resistance to abrasion is used.

Environmental Exposure Classifications

- Res1; Com1.
- May be suitable for limited water exposure areas (Res2; Com2). See Membrane Options.
- For installations that may be exposed to staining and/or chemical attack, specify tile and grout suitable for exposure. Consult product manufacturers; see also "Product Selection Guides."

Typical Weight of Tile Installation

- 21 pounds/square foot with ¾" poured gypsum over hydronic tubing. Add 2¼ pounds/square foot for each additional ¼" of poured gypsum.
- Does not include weight of substrate. See "Appendix B" for assumptions, included materials, and their individual weights.

Limitations

- Requires additional consideration by design professional to accommodate movement and/or deflection. Setting materials with improved bond strength and deformability are required.
- Not for use where moisture vapor transmission rate exceeds gypsum underlayment manufacturer's limitations.
- Not recommended for use with some stones on post tensioned concrete unless tile installation is 100 square feet or less. Consult stone supplier and installation materials manufacturer.

Membrane Options

- Requires use of crack isolation (A118.12) and/or waterproof (A118.10) and/or uncoupling membrane.
- Membrane must meet ANSI A118.10 for limited water exposure areas (Res2; Com2). Consult underlayment manufacturer for requirements.
- Specifier shall indicate if complete waterproofing is required, including if/how membrane connects to drain assembly, if base flashing is required, and treatment at other termination points.
- Check with membrane manufacturers for suitability for applicable conditions, as not all membranes are suitable for steam, high temperature, and/or chemical exposure, exterior use, use over above-ground structural slabs, use over pourable underlayments, use with radiant heat, or use over concrete with excessive moisture vapor transmission and/or alkalinity. Membrane may also affect service rating.

Requirements

- Proper curing/drying of underlayment prior to application of tile is critical for proper performance. Consult the underlayment manufacturer for specific instructions.
- Above-ground installations are inherently more susceptible to vibration and deflection. Grout and mortar manufacturers to warrant product suitability.
- Minimum grout joint width—1⁄16."

Materials

- **Multiple options exist for membranes, mortars, grouts, and other materials and *must be clearly specified* to be included. If not specifically indicated, optional materials are not included and mortar/grout choice defaults to minimum performance specification indicated. Consider each system component and intended use to determine minimum requirements and to specify options.**
- Natural stone tile—See "Natural Stone Tile Selection and Installation Guide" and consult supplier for application suitability.
- Cementitious grout—ANSI A118.6 or better or ISO CG1 or better. Consider unsanded grout for soft or semi-soft stones such as limestone, travertine, marble, onyx, or similar.
- Cementitious bond coat:
  - Must be recommended by manufacturer for above-ground use.

  - When a crack isolation membrane is used—ANSI A118.4 or better or ISO C2S1 or better.
  - When a waterproof membrane is used—ANSI A118.4 or better or ISO C2S1 or better, unless ANSI A118.1 or ISO C1 is recommended by membrane manufacturer.
  - When an uncoupling membrane is used, check with membrane manufacturer.
  - Use white for light-colored marble, limestone, and other stones where staining or darkening from grey setting materials is possible.
- Epoxy bond coat, when used—ANSI A118.3 or ISO R1 or better. See Water Sensitivity and Fiberglass Mesh Reinforced Stone in the "Natural Stone Tile Selection and Installation Guide."
- Crack isolation membrane, when used—ANSI A118.12.
- Waterproof membrane, when used—ANSI A118.10.
- Uncoupling membrane, when used—recommended by uncoupling membrane manufacturer; must achieve 50 psi or greater shear bond strength in 7 days per the test method in ANSI A118.12 Section 5.1.3.

Materials for Green/Sustainable Design

- See "Green Building Standards and Green Product Selection Guide" and consult manufacturers and suppliers for product sustainability and contribution to green building design.
- Consider specifying installation materials that meet ANSI A138.1, the *American National Standard Specifications for Sustainable Ceramic Tiles, Glass Tiles, and Tile Installation Materials.*

Preparation by Other Trades

- Floor systems over which tile will be installed shall be in conformance with the IRC for residential applications, the IBC for commercial applications, or applicable building codes. Maximum allowable substrate deflection under live load not to exceed l/360. See also "Substrate Requirements."
- When concentrated loads (scissor lifts, pallet jacks, automobiles, forklifts, etc.) will be used on an above-ground tiled floor, the engineer and/or specifier shall specify a substrate to accommodate the concentrated loads. Owner/specifier is responsible for protecting the tilework from damage, including allowing sufficient time for installed materials to cure properly.
- Slab to be well cured, dimensionally stable, and free of cracks, surface laitance, waxy or oily films, curing compounds, and other surface bond-inhibiting contaminants.
- Consult the manufacturer for maximum moisture vapor transmission limitations prior to application of underlayment.
- Maximum allowable variation in the tile substrate—⅛" in 10' from the required plane when measured from the high points in the surface.
- Radiant tubing by others.
- Poured gypsum underlayment thickness and application varies, consult the manufacturer for specific recommendations when used with floor warming systems. Minimum thickness is ¾" above hydronic tubing.
- Surfaces must be prepared and primed according to the underlayment manufacturer's instructions.
- Poured gypsum underlayment—minimum compressive strength of 2000 psi when tested per ASTM C472 modified and meeting performance requirements of ASTM C627 for the anticipated service level designated by the specifier or intended use.
- Poured gypsum underlayment—installed only by a manufacturer-approved applicator in accordance with manufacturer's recommendations.

Movement Joint (architect must specify type of joint and show location and details on drawings)

- Movement joints—mandatory according to EJ171. For above-ground installations, additional movement joints are required.

Installation Specifications

- Tile—ANSI A108.5.
- Cementitious grout—ANSI A108.10.
- Epoxy mortar—ANSI A108.6.
- Crack isolation membrane—ANSI A108.17.
- Waterproof membrane—ANSI A108.13.
- Uncoupling membrane—manufacturer's directions.
- Movement Joints—EJ171 and ASTM C1193.

Notes

- Consult underlayment manufacturer for minimum cure/dry time of installation before radiant heat system is activated.
- Some manufacturers require special primers prior to the application of the membrane. Follow manufacturer's directions.
- RH117 Stone is preferred over above-ground structural slabs and other floors subject to movement and/or deflection.
- Not all mortars, grouts, and membranes are suitable for above-ground use. Check manufacturer recommendations.
- When unsanded grout is used, grout joint width must be ¹⁄₁₆" to ⅛" wide (nominal).

RH112-21
STONE

- **On-Ground Concrete**
- **Cementitious Self-Leveling Underlayment Encapsulating Hydronic Tubing**
- **Natural Stone Tile**

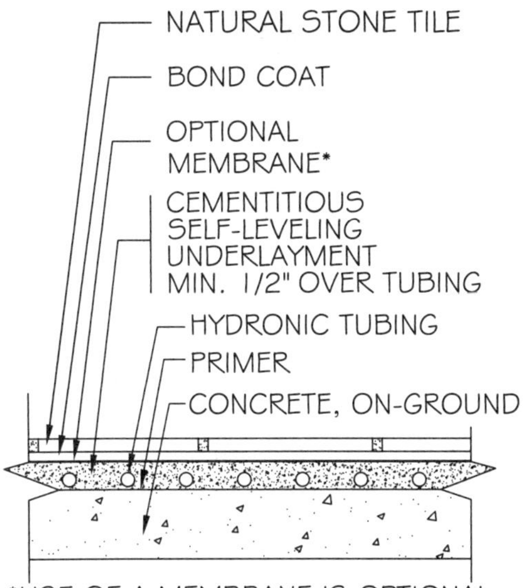

*USE OF A MEMBRANE IS OPTIONAL. SEE MEMBRANE OPTIONS.

Recommended Uses

- For slab-on-ground construction where no bending stresses occur and where hydronic tubing will be placed on the slab.

Service Rating

- Moderate when stone with adequate compressive strength, flexural strength and resistance to abrasion is used.

Environmental Exposure Classifications

- Res1; Com1.
- May be suitable for increased water exposure. See Membrane Options.
- For installations that may be exposed to staining and/or chemical attack, specify tile and grout suitable for exposure. Consult product manufacturers; see also "Product Selection Guides."

Typical Weight of Tile Installation

- 21 pounds/square foot with ½" cementitious self-leveling underlayment over hydronic tubing. Add 2¾ pounds/square foot for each additional ¼" of underlayment.
- Does not include weight of substrate. See "Appendix B" for assumptions, included materials, and their individual weights.

Limitations

- Not for below-ground installations.
- Not for use where moisture vapor transmission rate exceeds underlayment manufacturer's limitations.
- Not for use over above-ground structural slabs and other floors subject to movement and/or deflection. See RH112A Stone.
- Not recommended for use with some stones on post tensioned concrete unless tile installation is 100 square feet or less. Consult stone supplier and installation materials manufacturer.

Membrane Options

- A crack isolation membrane (A118.12) may be required. Check with heat system and setting material manufacturers.
- Crack isolation membrane may also provide waterproofing. Consult membrane manufacturer.
- A waterproof membrane (A118.10) may be required for areas exposed to moisture. Consult underlayment manufacturer for requirements.
- Specifier shall indicate if complete waterproofing is required, including if/how membrane connects to drain assembly, if base flashing is required, and treatment at other termination points.
- A crack isolation membrane may be specified to treat existing in-plane cracks (F125-Partial Stone) or for protection against future in-plane cracking (F-125 Full Stone). See F125 Stone for more information.
- An uncoupling membrane may be specified to accommodate subfloor movement. Check with membrane manufacturer for recommendations and requirements.
- Check with membrane manufacturers for suitability for applicable conditions, as not all membranes are suitable for steam, high temperature, and/or chemical exposure, exterior use, use over above-ground structural slabs, use over pourable underlayments, use with radiant heat, or use over concrete with excessive moisture vapor transmission and/or alkalinity. Membrane may also affect service rating.

Requirements

- Consult the manufacturer for maximum moisture vapor transmission limitations prior to application of underlayment.
- Slab to be well cured, dimensionally stable, and free of cracks, surface laitance, waxy or oily films, curing compounds, and other surface bond-inhibiting contaminants.
- Cementitious self-leveling underlayment thickness and application varies, consult the manufacturer for specific recommendations when used with floor warming systems. Minimum thickness is ½" above hydronic tubing.
- Surfaces must be prepared and primed according to manufacturer's instructions.
- Proper curing/drying of underlayments prior to appli-

cation of tile is critical for proper performance. Consult the underlayment manufacturer for specific instructions.

- Minimum grout joint width—1/16".

Materials

- **Multiple options exist for membranes, mortars, grouts, and other materials and *must be clearly specified* to be included. If not specifically indicated, optional materials are not included and mortar/grout choice defaults to minimum performance specification indicated. Consider each system component and intended use to determine minimum requirements and to specify options.**
- Natural stone tile—See "Natural Stone Tile Selection and Installation Guide" and consult supplier for application suitability.
- Cementitious grout—ANSI A118.6 or better or ISO CG1 or better. Consider unsanded grout for soft or semi-soft stones such as limestone, travertine, marble, onyx, or similar.
- Cementitious bond coat:
  - When a membrane is not used—ANSI A118.15 or better or ISO C2S1 or better.
  - When a crack isolation membrane is used—ANSI A118.4 or better or ISO C2S1 or better.
  - When a waterproof membrane is used—ANSI A118.4 or better or ISO C2S1 or better, unless ANSI A118.1 or ISO C1 is recommended by membrane manufacturer.
  - When an uncoupling membrane is used, check with membrane manufacturer.
  - Use white for light-colored marble, limestone, and other stones where staining or darkening from grey setting materials is possible.
- Epoxy bond coat, when used—ANSI A118.3 or ISO R1 or better. See Water Sensitivity and Fiberglass Mesh Reinforced Stone in the "Natural Stone Tile Selection and Installation Guide."
- Crack isolation membrane, when used—ANSI A118.12.
- Waterproof membrane, when used—ANSI A118.10.
- Uncoupling membrane, when used—recommended by uncoupling membrane manufacturer; must achieve 50 psi or greater shear bond strength in 7 days per the test method in ANSI A118.12 Section 5.1.3.
- Self-leveling underlayment—meeting performance requirements of ASTM C627 for the anticipated service level designated by the specifier or intended use.

Materials for Green/Sustainable Design

- See "Green Building Standards and Green Product Selection Guide" and consult manufacturers and suppliers for product sustainability and contribution to green building design.
- Consider specifying installation materials that meet ANSI A138.1, the *American National Standard Specifications for Sustainable Ceramic Tiles, Glass Tiles, and Tile Installation Materials.*

Preparation by Other Trades

- Slab to have fine broom finish with no curing compounds. When used, mechanical scarification is necessary.
- Maximum allowable variation in the installation substrate (concrete)—1/4" in 10' from the required plane.
- Maximum allowable variation in the tile substrate—1/8" in 10' from the required plane when measured from the high points in the surface.
- Radiant tubing by others.

Movement Joint (architect must specify type of joint and show location and details on drawings)

- Movement joints—mandatory in accordance with EJ171.

Installation Specifications

- Tile—ANSI A108.5.
- Cementitious grout—ANSI A108.10.
- Epoxy mortar—ANSI A108.6.
- Cementitious self-leveling underlayment—manufacturer's directions.
- Crack isolation membrane—ANSI A108.17.
- Waterproof membrane—ANSI A108.13.
- Uncoupling membrane—manufacturer's directions.
- Movement Joints—EJ171 and ASTM C1193.

Notes

- Consult underlayment manufacturer for minimum cure/dry time of installation before radiant heat system is activated.
- Some manufacturers require special primers prior to the application of the setting materials or membrane. Follow manufacturer's directions.
- When unsanded grout is used, grout joint width must be 1/16" to 1/8" wide (nominal).

RH112A-21
STONE

- **Above-Ground Concrete**
- **Cementitious Self-Leveling Underlayment Encapsulating Hydronic Tubing**
- **Natural Stone Tile**

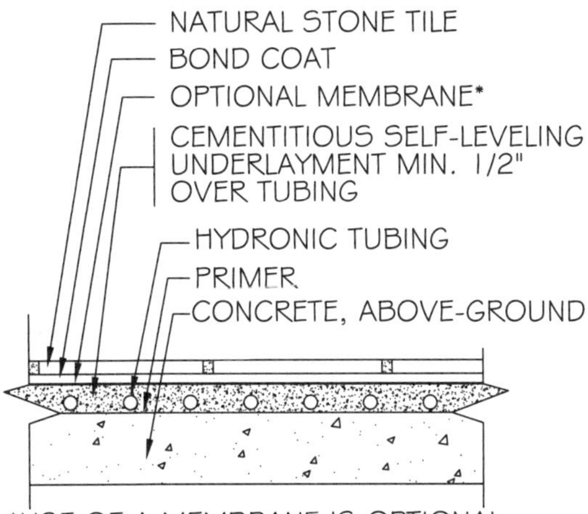

*USE OF A MEMBRANE IS OPTIONAL. SEE MEMBRANE OPTIONS.

Recommended Uses

- For above-ground structural slabs where hydronic tubing will be placed on the slab.

Service Rating

- Moderate when stone with adequate compressive strength, flexural strength and resistance to abrasion is used.

Environmental Exposure Classifications

- Res1; Com1.
- May be suitable for increased water exposure. See Membrane Options.
- For installations that may be exposed to staining and/or chemical attack, specify tile and grout suitable for exposure. Consult product manufacturers; see also "Product Selection Guides."

Typical Weight of Tile Installation

- 21 pounds/square foot with ½" cementitious self-leveling underlayment over hydronic tubing. Add 2¾ pounds/square foot for each additional ¼" of underlayment.
- Does not include weight of substrate. See "Appendix B" for assumptions, included materials, and their individual weights.

Limitations

- Requires additional consideration by design professional to accommodate movement and/or deflection. Setting materials with improved bond strength and deformability are required.
- Not for use where moisture vapor transmission rate exceeds underlayment manufacturer's limitations.
- Not recommended for use with some stones on post tensioned concrete unless tile installation is 100 square feet or less. Consult stone supplier and installation materials manufacturer.

Membrane Options

- A crack isolation membrane (A118.12) may be required. Check with heat system and setting material manufacturers.
- Crack isolation membrane may also provide waterproofing. Consult membrane manufacturer.
- A waterproof membrane (A118.10) may be required for areas exposed to moisture. Consult underlayment manufacturer for requirements.
- Specifier shall indicate if complete waterproofing is required, including if/how membrane connects to drain assembly, if base flashing is required, and treatment at other termination points.
- A crack isolation membrane may be specified to treat existing in-plane cracks (F125-Partial Stone) or for protection against future in-plane cracking (F-125 Full Stone). See F125 Stone for more information.
- An uncoupling membrane may be specified to accommodate subfloor movement. Check with membrane manufacturer for recommendations and requirements.
- Check with membrane manufacturers for suitability for applicable conditions, as not all membranes are suitable for steam, high temperature, and/or chemical exposure, exterior use, use over above-ground structural slabs, use over pourable underlayments, use with radiant heat, or use over concrete with excessive moisture vapor transmission and/or alkalinity. Membrane may also affect service rating.

Requirements

- Slab to be well cured, dimensionally stable, and free of cracks, surface laitance, waxy or oily films, curing compounds, and other surface bond-inhibiting contaminants.
- Cementitious self-leveling underlayment thickness and application varies, consult the manufacturer for specific recommendations when used with floor warming systems. Minimum thickness is ½" above hydronic tubing.
- Surfaces must be prepared and primed according to the manufacturer's instructions.
- Proper curing/drying of underlayment prior to application of tile is critical for proper performance. Consult the underlayment manufacturer for specific instructions.
- Consult the manufacturer for maximum moisture vapor transmission limitations prior to application of underlayment.
- Above-ground installations are inherently more susceptible to vibration and deflection. Grout and mortar manufacturers to warrant product suitability.

- Minimum grout joint width—1⁄16".

Materials

- **Multiple options exist for membranes, mortars, grouts, and other materials and *must be clearly specified* to be included. If not specifically indicated, optional materials are not included and mortar/grout choice defaults to minimum performance specification indicated. Consider each system component and intended use to determine minimum requirements and to specify options.**
- Natural stone tile—See "Natural Stone Tile Selection and Installation Guide" and consult supplier for application suitability.
- Cementitious grout—ANSI A118.6 or better or ISO CG1 or better. Consider unsanded grout for soft or semi-soft stones such as limestone, travertine, marble, onyx, or similar.
- Cementitious bond coat:
  - Must be recommended by manufacturer for above-ground use.
  - When a membrane is not used—ANSI A118.15 or better or ISO C2S1 or better.
  - When a crack isolation membrane is used—ANSI A118.4 or better or ISO C2S1 or better.
  - When a waterproof membrane is used—ANSI A118.4 or better or ISO C2S1 or better, unless ANSI A118.1 or ISO C1 is recommended by membrane manufacturer.
  - When an uncoupling membrane is used, check with membrane manufacturer.
  - Use white for light-colored marble, limestone, and other stones where staining or darkening from grey setting materials is possible.
- Epoxy bond coat, when used—ANSI A118.3 or ISO R1 or better. See Water Sensitivity and Fiberglass Mesh Reinforced Stone in the "Natural Stone Tile Selection and Installation Guide."
- Crack isolation membrane, when used—ANSI A118.12.
- Waterproof membrane, when used—ANSI A118.10.
- Uncoupling membrane, when used—recommended by uncoupling membrane manufacturer; must achieve 50 psi or greater shear bond strength in 7 days per the test method in ANSI A118.12 Section 5.1.3.
- Self-leveling underlayment—meeting performance requirements of ASTM C627 for the anticipated service level designated by the specifier or intended use.

Materials for Green/Sustainable Design

- See "Green Building Standards and Green Product Selection Guide" and consult manufacturers and suppliers for product sustainability and contribution to green building design.
- Consider specifying installation materials that meet ANSI A138.1, the *American National Standard Specifications for Sustainable Ceramic Tiles, Glass Tiles, and Tile Installation Materials.*

Preparation by Other Trades

- Floor systems over which tile will be installed shall be in conformance with the IRC for residential applications, the IBC for commercial applications, or applicable building codes. Maximum allowable substrate deflection under live load not to exceed l/360. See also "Substrate Requirements."
- When concentrated loads (scissor lifts, pallet jacks, automobiles, forklifts, etc.) will be used on an above-ground tiled floor, the engineer and/or specifier shall specify a substrate to accommodate the concentrated loads. Owner/specifier is responsible for protecting the tilework from damage, including allowing sufficient time for installed materials to cure properly.
- Slab to have fine broom finish with no curing compounds used. When used, mechanical scarification is necessary.
- Maximum allowable variation in the installation substrate (concrete)—1⁄4" in 10' from the required plane.
- Maximum allowable variation in the tile substrate—1⁄8" in 10' from the required plane when measured from the high points in the surface.
- Radiant tubing by others.

Movement Joint (architect must specify type of joint and show location and details on drawings)

- Movement joints—mandatory according to EJ171. For above-ground installations, additional movement joints are required.

Installation Specifications

- Tile—ANSI A108.5.
- Cementitious grout—ANSI A108.10.
- Epoxy mortar—ANSI A108.6.
- Cementitious self-leveling underlayment—manufacturer's directions.
- Crack isolation membrane—ANSI A108.17.
- Waterproof membrane—ANSI A108.13.
- Uncoupling membrane—manufacturer's directions.
- Movement Joints—EJ171 and ASTM C1193.

Notes

- RH117 Stone is preferred over above-ground structural slabs and other floors subject to movement and/or deflection.
- Some manufacturers require special primers prior to the application of the setting materials or membrane. Follow manufacturer's directions.
- Consult underlayment manufacturer for minimum cure/dry time of installation before radiant heat system is activated.
- Not all mortars, grouts, and membranes are suitable for above-ground use. Check manufacturer recommendations.
- When unsanded grout is used, grout joint width must be 1⁄16" to 1⁄8" wide (nominal).

RH116-21
STONE

- **On-Ground Concrete**
- **Electric Radiant Heat System Encapsulated in Cementitious Self-Leveling Underlayment**
- **Natural Stone Tile**

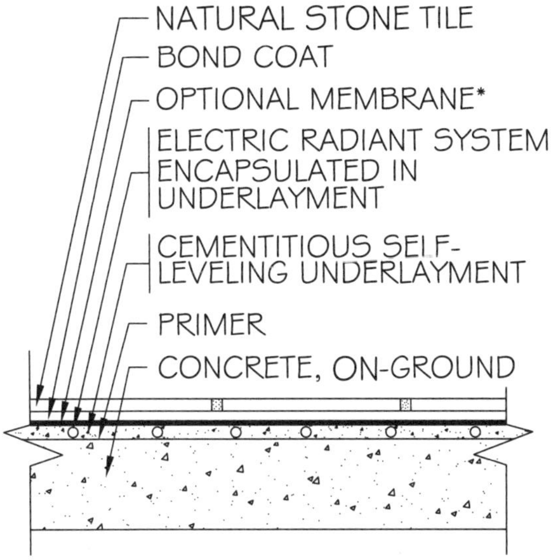

*USE OF A MEMBRANE IS OPTIONAL. SEE MEMBRANE OPTIONS.

Recommended Uses

- For slab-on-ground construction where no bending stresses occur and where floor leveling and electric radiant in-floor heat are desired.

Service Rating

- Moderate when stone with adequate compressive strength, flexural strength and resistance to abrasion is used.

Environmental Exposure Classifications

- Res1; Com1.
- May be suitable for increased water exposure. See Membrane Options.
- For installations that may be exposed to staining and/or chemical attack, specify tile and grout suitable for exposure. Consult product manufacturers; see also "Product Selection Guides."

Typical Weight of Tile Installation

- 13 pounds/square foot with ½" cementitious self-leveling underlayment. Add 2¾ pounds/square foot for each additional ¼" of underlayment.
- Does not include weight of substrate. See "Appendix B" for assumptions, included materials, and their individual weights.

Limitations

- Not for below-ground installations.
- Not for use where moisture vapor transmission rate exceeds underlayment manufacturer's limitations.
- Not for use over above-ground structural slabs and other floors subject to movement and/or deflection. See RH116A Stone.
- Not recommended for use with some stones on post tensioned concrete unless tile installation is 100 square feet or less. Consult stone supplier and installation materials manufacturer.

Membrane Options

- A crack isolation membrane (A118.12) may be required. Check with heat system and setting material manufacturers.
- Crack isolation membrane may also provide waterproofing. Consult membrane manufacturer.
- A waterproof membrane (A118.10) may be required for areas exposed to moisture. Consult underlayment manufacturer for requirements.
- Specifier shall indicate if complete waterproofing is required, including if/how membrane connects to drain assembly, if base flashing is required, and treatment at other termination points.
- A crack isolation membrane may be specified to treat existing in-plane cracks (F125-Partial Stone) or for protection against future in-plane cracking (F-125 Full Stone). See F125 Stone for more information.
- Check with membrane manufacturers for suitability for applicable conditions, as not all membranes are suitable for steam, high temperature, and/or chemical exposure, exterior use, use over above-ground structural slabs, use over pourable underlayments, use with radiant heat, or use over concrete with excessive moisture vapor transmission and/or alkalinity. Membrane may also affect service rating.

Requirements

- Consult the manufacturer for maximum moisture vapor transmission limitations prior to application of underlayment.
- Slab to be well cured, dimensionally stable, and free of cracks, surface laitance, waxy or oily films, curing compounds, and other surface bond-inhibiting contaminants.
- Cementitious self-leveling underlayment thickness and application varies, consult the manufacturer for specific recommendations.
- Surfaces must be prepared and primed according to manufacturer's instructions.
- Proper curing/drying of underlayments prior to application of tile is critical for proper performance. Consult the underlayment manufacturer for specific instructions.

- Electric radiant system should not be installed over building expansion joints.
- Minimum grout joint width—1⁄16".

Materials

- **Multiple options exist for membranes, mortars, grouts, and other materials and *must be clearly specified* to be included. If not specifically indicated, optional materials are not included and mortar/grout choice defaults to minimum performance specification indicated. Consider each system component and intended use to determine minimum requirements and to specify options.**
- Natural stone tile—See "Natural Stone Tile Selection and Installation Guide" and consult supplier for application suitability.
- Cementitious grout—ANSI A118.6 or better or ISO CG1 or better. Consider unsanded grout for soft or semi-soft stones such as limestone, travertine, marble, onyx, or similar.
- Cementitious bond coat:
    - When a membrane is not used—ANSI A118.15 or better or ISO C2S1 or better.
    - When a crack isolation membrane is used—ANSI A118.4 or better or ISO C2S1 or better.
    - When a waterproof membrane is used—ANSI A118.4 or better or ISO C2S1 or better, unless ANSI A118.1 or ISO C1 is recommended by membrane manufacturer.
    - Use white for light-colored marble, limestone, and other stones where staining or darkening from grey setting materials is possible.
- Epoxy bond coat, when used—ANSI A118.3 or ISO R1 or better. See Water Sensitivity and Fiberglass Mesh Reinforced Stone in the "Natural Stone Tile Selection and Installation Guide."
- Crack isolation membrane, when used—ANSI A118.12.
- Waterproof membrane, when used—ANSI A118.10.
- Self-leveling underlayment—meeting performance requirements of ASTM C627 for the anticipated service level designated by the specifier or intended use.

Materials for Green/Sustainable Design

- See "Green Building Standards and Green Product Selection Guide" and consult manufacturers and suppliers for product sustainability and contribution to green building design.
- Consider specifying installation materials that meet ANSI A138.1, the *American National Standard Specifications for Sustainable Ceramic Tiles, Glass Tiles, and Tile Installation Materials.*

Preparation by Other Trades

- Slab to have a fine broom finish with no curing compounds used. When used, mechanical scarifying is necessary.
- Maximum allowable variation in the installation substrate (concrete)—1⁄4" in 10' from the required plane.
- Maximum allowable variation in the tile substrate—1⁄8" in 10' from the required plane when measured from the high points in the surface.
- Electrician or qualified contractor to wire electric radiant system to power source.

Movement Joint (architect must specify type of joint and show location and details on drawings)

- Movement joints—mandatory according to EJ171.

Installation Specifications

- Tile—ANSI A108.5.
- Cementitious grout—ANSI A108.10.
- Epoxy mortar—ANSI A108.6.
- Cementitious self-leveling underlayment—manufacturer's directions.
- Crack isolation membrane—ANSI A108.17.
- Waterproof membrane—ANSI A108.13.
- Electric radiant system—UL (US) 1693; UL (CAN/CSA) C22.2 #217; NEC Article 424.
- Movement Joints—EJ171 and ASTM C1193.

Notes

- Consult underlayment manufacturer for minimum cure/dry time of installation before radiant heat system is activated.
- Some manufacturers require special primers prior to the application of the setting materials or membrane. Follow manufacturer's directions.
- When unsanded grout is used, grout joint width must be 1⁄16" to 1⁄8" wide (nominal).

RH116A-21
STONE

- **Above-Ground Concrete**
- **Electric Radiant Heat System Encapsulated in Cementitious Self-Leveling Underlayment**
- **Natural Stone Tile**

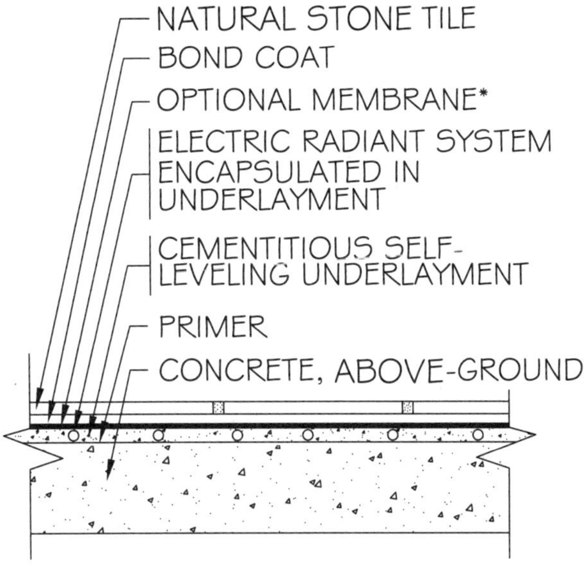

*USE OF A MEMBRANE IS OPTIONAL. SEE MEMBRANE OPTIONS.

Recommended Uses

- For above-ground structural slabs where floor leveling and electric radiant in-floor heat are desired.

Service Rating

- Moderate when stone with adequate compressive strength, flexural strength and resistance to abrasion is used.

Environmental Exposure Classifications

- Res1; Com1.
- May be suitable for increased water exposure. See Membrane Options.
- For installations that may be exposed to staining and/or chemical attack, specify tile and grout suitable for exposure. Consult product manufacturers; see also "Product Selection Guides."

Typical Weight of Tile Installation

- 13 pounds/square foot with ½" cementitious self leveling underlayment. Add 2¾ pounds/square foot for each additional ¼" of underlayment.
- Does not include weight of substrate. See "Appendix B" for assumptions, included materials, and their individual weights.

Limitations

- Requires additional consideration by design professional to accommodate movement and/or deflection. Setting materials with improved bond strength and deformability are required.
- Not for use where moisture vapor transmission rate exceeds underlayment manufacturer's limitations.
- Not recommended for use with some stones on post tensioned concrete unless tile installation is 100 square feet or less. Consult stone supplier and installation materials manufacturer.

Membrane Options

- A crack isolation membrane (A118.12) may be required. Check with heat system and setting material manufacturers.
- Crack isolation membrane may also provide waterproofing. Consult membrane manufacturer.
- A waterproof membrane (A118.10) may be required for areas exposed to moisture. Consult underlayment manufacturer for requirements.
- Specifier shall indicate if complete waterproofing is required, including if/how membrane connects to drain assembly, if base flashing is required, and treatment at other termination points.
- A crack isolation membrane may be specified to treat existing in-plane cracks (F125-Partial Stone) or for protection against future in-plane cracking (F-125 Full Stone). See F125 Stone for more information.
- Check with membrane manufacturers for suitability for applicable conditions, as not all membranes are suitable for steam, high temperature, and/or chemical exposure, exterior use, use over above-ground structural slabs, use over pourable underlayments, use with radiant heat, or use over concrete with excessive moisture vapor transmission and/or alkalinity. Membrane may also affect service rating.

Requirements

- Consult the manufacturer for maximum moisture vapor transmission limitations prior to application of underlayment.
- Slab to be well cured, dimensionally stable, and free of cracks, surface laitance, waxy or oily films, curing compounds, and other surface bond-inhibiting contaminants.
- Cementitious self-leveling underlayment thickness and application varies, consult the manufacturer for specific recommendations.
- Surfaces must be prepared and primed according to the underlayment manufacturer's instructions.
- Proper curing/drying of underlayments prior to application of tile is critical for proper performance. Consult the underlayment manufacturer for specific instructions.
- Electric radiant system should not be installed over building expansion joints.
- Above-ground installations are inherently more susceptible to vibration and deflection. Grout and mortar

manufacturers to warrant product suitability.

- Minimum grout joint width—1⁄16".

Materials

- **Multiple options exist for membranes, mortars, grouts, and other materials and *must be clearly specified* to be included. If not specifically indicated, optional materials are not included and mortar/grout choice defaults to minimum performance specification indicated. Consider each system component and intended use to determine minimum requirements and to specify options.**
- Natural stone tile—See "Natural Stone Tile Selection and Installation Guide" and consult supplier for application suitability.
- Cementitious grout—ANSI A118.6 or better or ISO CG1 or better. Consider unsanded grout for soft or semi-soft stones such as limestone, travertine, marble, onyx, or similar.
- Cementitious bond coat:
  - Must be recommended by manufacturer for above-ground use.
  - When a membrane is not used—ANSI A118.15 or better or ISO C2S1 or better.
  - When a crack isolation membrane is used—ANSI A118.4 or better or ISO C2S1 or better.
  - When a waterproof membrane is used—ANSI A118.4 or better or ISO C2S1 or better, unless ANSI A118.1 or ISO C1 is recommended by membrane manufacturer.
  - Use white for light-colored marble, limestone, and other stones where staining or darkening from grey setting materials is possible.
- Epoxy bond coat, when used—ANSI A118.3 or ISO R1 or better. See Water Sensitivity and Fiberglass Mesh Reinforced Stone in the "Natural Stone Tile Selection and Installation Guide."
- Crack isolation membrane, when used—ANSI A118.12.
- Waterproof membrane, when used—ANSI A118.10.
- Self-leveling underlayment—meeting performance requirements of ASTM C627 for the anticipated service level designated by the specifier or intended use.

Materials for Green/Sustainable Design

- See "Green Building Standards and Green Product Selection Guide" and consult manufacturers and suppliers for product sustainability and contribution to green building design.
- Consider specifying installation materials that meet ANSI A138.1, the *American National Standard Specifications for Sustainable Ceramic Tiles, Glass Tiles, and Tile Installation Materials.*

Preparation by Other Trades

- Floor systems over which tile will be installed shall be in conformance with the IRC for residential applications, the IBC for commercial applications, or applicable building codes. Maximum allowable substrate deflection under live load not to exceed l/360. See also "Substrate Requirements."
- When concentrated loads (scissor lifts, pallet jacks, automobiles, forklifts, etc.) will be used on an above-ground tiled floor, the engineer and/or specifier shall specify a substrate to accommodate the concentrated loads. Owner/specifier is responsible for protecting the tilework from damage, including allowing sufficient time for installed materials to cure properly.
- Slab to have a fine broom finish with no curing compounds used. When used, mechanical scarifying is necessary.
- Maximum allowable variation in the installation substrate (concrete)—1⁄4" in 10' from the required plane.
- Maximum allowable variation in the tile substrate—1⁄8" in 10' from the required plane when measured from the high points in the surface.
- Electrician or qualified contractor to wire electric radiant system to power source.

Movement Joint (architect must specify type of joint and show location and details on drawings)

- Movement joints—mandatory according to EJ171. For above-ground installations, additional movement joints are required.

Installation Specifications

- Tile—ANSI A108.5.
- Cementitious grout—ANSI A108.10.
- Epoxy mortar—ANSI A108.6.
- Cementitious self-leveling underlayment—manufacturer's directions.
- Crack isolation membrane—ANSI A108.17.
- Waterproof membrane—ANSI A108.13.
- Electric radiant system—UL (US) 1693; UL (CAN/CSA) C22.2 #217; NEC Article 424.
- Movement Joints—EJ171 and ASTM C1193.

Notes

- Consult underlayment manufacturer for minimum cure/dry time of installation before radiant heat system is activated.
- RH117 Stone is preferred over above-ground structural slabs and other floors subject to movement and/or deflection.
- Some manufacturers require special primers prior to the application of the setting materials or membrane. Follow manufacturer's directions.
- Not all mortars, grouts, and membranes are suitable for above-ground use. Check manufacturer recommendations.
- When unsanded grout is used, grout joint width must be 1⁄16" to 1⁄8" wide (nominal).

RH141-21
STONE

- **Joists max. 16" o.c./ Plywood Subfloor**
- **Unbonded Mortar Bed Encapsulating Hydronic Tubing**
- **Natural Stone Tile**

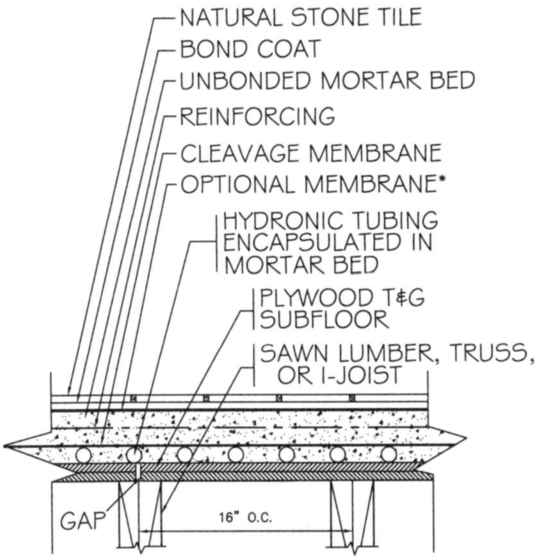

*USE OF A MEMBRANE IS OPTIONAL. SEE MEMBRANE OPTIONS.

Recommended Uses

- For floor leveling or flattening, especially where multiple finish floor heights must be accommodated or slopes are required and where hydronic tubing will be placed.
- For installations that must be at least 1½" thick (nominal, including tile) and where hydronic tubing will be placed.
- For isolating the tile installation from the floor framing where hydronic tubing will be placed.
- For areas where ungauged stone will be used.

Service Rating

- Light commercial, when stone with adequate compressive strength, flexural strength, and resistance to abrasion is used.

Environmental Exposure Classifications

- Res1, 2; Com1, 2.
- May be suitable for increased water exposure. See Membrane Options.
- May be suitable for exterior applications in areas not subject to freeze/thaw cycling when appropriate precautions are taken, including expansion joint placement, proper slope, waterproofing, and material selection.
- For installations that may be exposed to staining and/or chemical attack, specify tile and grout suitable for exposure. Consult product manufacturers; see also "Product Selection Guides."

Typical Weight of Tile Installation

- 32 pounds/square foot with 1¼" mortar bed over hydronic tubing. Add 3 pounds/square foot for each additional ¼" of mortar bed.
- Does not include weight of substrate. See "Appendix B" for assumptions, included materials, and their individual weights.

Limitations

- Maximum joist spacing 16" on center.

Membrane Options

- A crack isolation membrane (A118.12) may be required. Check with heat system and setting material manufacturers.
- Crack isolation membrane may also provide waterproofing. Consult membrane manufacturer.
- A waterproof membrane (A118.10) may be specified to prevent moisture intrusion and protect adjacent walls and building materials. Base flashing should be used for maximum effectiveness.
- Specifier shall indicate if complete waterproofing is required, including if/how membrane connects to drain assembly, if base flashing is required, and treatment at other termination points.
- Check with membrane manufacturers for suitability for applicable conditions, as not all membranes are suitable for steam, high temperature, and/or chemical exposure, exterior use, use over above-ground structural slabs, use over pourable underlayments, use with radiant heat, or use over concrete with excessive moisture vapor transmission and/or alkalinity. Membrane may also affect service rating.

Requirements

- Screed fill flush to top of hydronic tubing before placing a cleavage membrane and reinforced mortar bed.
- Reinforcing mesh mandatory.
- Mortar bed thickness—1¼" minimum to 2" maximum.
- Mortar beds in excess of 2" thick shall be detailed by the architect.
- Minimum grout joint width—¹⁄₁₆".

Materials

- **Multiple options exist for membranes, mortars, grouts, and other materials and *must be clearly specified* to be included. If not specifically indicated, optional materials are not included and mortar/grout choice defaults to minimum performance specification indicated. Consider each system component and intended use to determine minimum requirements and to specify options.**
- Natural stone tile—see "Natural Stone Tile Selection and Installation Guide," and consult supplier for application suitability.

- Cementitious grout—ANSI A118.6 or better or ISO CG1 or better. Consider unsanded grout for soft or semisoft stones such as limestone, travertine, marble, onyx, or similar.
- Cementitious bond coat—portland cement paste on a mortar bed that is still workable. For a cured mortar bed, follow recommendations below to select appropriate bonding mortar:
  - When a membrane is not used on top of the mortar bed—ANSI A118.15 or better or ISO C2S1 or better.
  - When a crack isolation membrane is used on top of the mortar bed—ANSI A118.4 or better or ISO C2S1 or better.
  - When a waterproof membrane is used on top of the mortar bed—ANSI A118.4 or better or ISO C2S1 or better, unless ANSI A118.1 or ISO C1 is recommended by membrane manufacturer.
  - Use white for light-colored marble, limestone, and other stones where staining or darkening from grey setting materials is possible.
- Epoxy bond coat, when used—ANSI A118.3 or ISO R1 or better. See Water Sensitivity and Fiberglass Mesh Reinforced Stone in the "Natural Stone Tile Selection and Installation Guide."
- Mortar bed, reinforcing, and cleavage membrane—ANSI A108.1A.
- Crack isolation membrane, when used—ANSI A118.12.
- Waterproof membrane, when used—ANSI A118.10.

Materials for Green/Sustainable Design

- See "Green Building Standards and Green Product Selection Guide" and consult manufacturers and suppliers for product sustainability and contribution to green building design.
- Consider specifying installation materials that meet ANSI A138.1, the *American National Standard Specifications for Sustainable Ceramic Tiles, Glass Tiles, and Tile Installation Materials.*

Preparation by Other Trades

- Floor systems, including the framing system and subfloor panels, over which tile will be installed shall be in conformance with the IRC for residential applications, the IBC for commercial applications, or applicable building codes and maximum allowable floor member live load and concentrated load deflection shall not exceed l/720. See also "Substrate Requirements."
- When concentrated loads (scissor lifts, pallet jacks, forklifts, etc.) will be used on an above-ground tiled floor, the engineer and/or specifier shall specify a substrate to accommodate the concentrated loads. Owner/specifier is responsible for protecting the tilework from damage, including allowing sufficient time for installed materials to cure properly.
- Face grain of plywood shall run perpendicular to joists.
- Subfloor—minimum 19⁄32" tongue and groove plywood with 1⁄8" gap between sheets or 1" nominal boards.
- Depressing subfloor between joists on ledger strips permissible in residential use.
- Maximum allowable variation in the installation substrate (plywood)—1⁄4" in 10' from the required plane.

Movement Joint (architect must specify type of joint and show location and details on drawings)

- Movement joints—mandatory according to EJ171.

Installation Specifications

- Tile—ANSI A108.1A,.1B, or .1C. Ungauged stone may require installation per ANSI A108.1A. Material with minimal variation may be allowed per A108.1B with LHT mortar. A108.1B required if membrane or epoxy bond coat to be used.
- Cementitious grout—ANSI A108.10.
- Crack isolation membrane—A108.17.
- Waterproof membrane—A108.13.
- Movement Joints—EJ171 and ASTM C1193.

Notes

- When unsanded grout is used, grout joint width must be 1⁄16" to 1⁄8" wide (nominal).

F141-21
STONE

- **Joists max. 16" o.c./ Plywood Subfloor**
- **Unbonded Mortar Bed**
- **Natural Stone Tile**

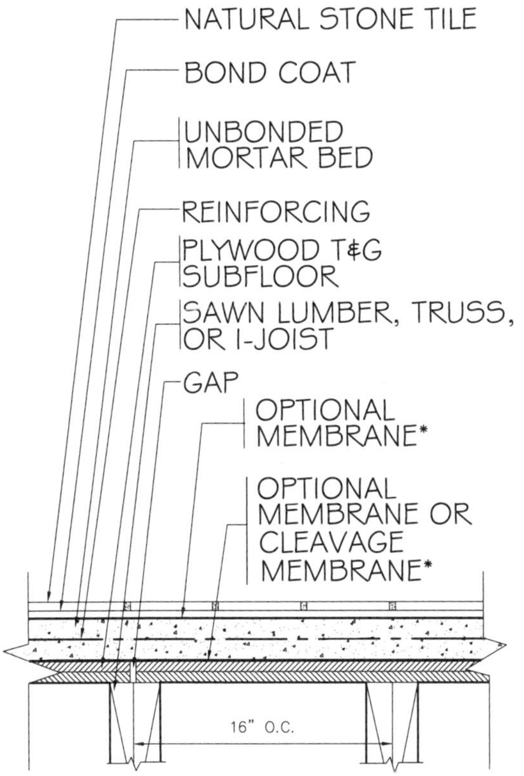

*USE OF A MEMBRANE IS OPTIONAL. SEE MEMBRANE OPTIONS.

Recommended Uses

- For wood substrates where leveling, flattening, or contouring of finish floor height is required, such as when finishing to adjacent floors or where slopes are required.
- For areas where floor flatness is critical, such as when tiles with any edge longer than 15" are specified or where the finished tile assembly is required to achieve in-plane transitions (as may be necessary where accessibility is a concern).
- For areas where ungauged stone will be used.

Service Rating

- Light commercial, when stone with adequate compressive strength, flexural strength, and resistance to abrasion is used.

Environmental Exposure Classifications

- Res1, 2; Com1, 2.
- May be suitable for increased water exposure. See Membrane Options.
- May be suitable for exterior applications in areas not subject to freeze/thaw cycling when appropriate precautions are taken, including expansion joint placement, proper slope, waterproofing, and material selection.
- For installations that may be exposed to staining and/or chemical attack, specify tile and grout suitable for exposure. Consult product manufacturers; see also "Product Selection Guides."

Typical Weight of Tile Installation

- 23 pounds/square foot with 1¼" mortar bed. Add 3 pounds/square foot for each additional ¼" of mortar bed.
- Does not include weight of substrate. See "Appendix B" for assumptions, included materials, and their individual weights.

Limitations

- Maximum joist spacing 16" on center.

Membrane Options

- A waterproof membrane (A118.10) may be specified to prevent moisture intrusion and protect adjacent walls and building materials. Base flashing should be used for maximum effectiveness.
- Specifier shall indicate if complete waterproofing is required, including if/how membrane connects to drain assembly, if base flashing is required, and treatment at other termination points.
- A cleavage membrane is required.
- If a bonded waterproof membrane (ANSI A118.10) is used, it may be placed above or below the mortar bed. If a shower pan membrane (ASTM D4551 or D4068) is used, it must be placed below the mortar bed.
- The waterproof membrane can also function as the cleavage membrane if loose laid (not bonded to plywood) below the mortar bed. Check with membrane manufacturer if membrane may be loose laid.
- If waterproof membrane is bonded to the top of the mortar bed, a cleavage membrane must be placed below the mortar bed.
- Check with membrane manufacturers for suitability for applicable conditions, as not all membranes are suitable for steam, high temperature, and/or chemical exposure, exterior use, use over above-ground structural slabs, use over pourable underlayments, use with radiant heat, or use over concrete with excessive moisture vapor transmission and/or alkalinity. Membrane may also affect service rating.

Requirements

- Reinforcing mesh mandatory.
- Mortar bed thickness—1¼" minimum to 2" maximum.
- Mortar beds in excess of 2" thick shall be detailed by the architect.
- Minimum grout joint width—$\frac{1}{16}$".

Materials

- **Multiple options exist for membranes, mortars, grouts, and other materials and *must be clearly specified* to be included. If not specifically indicated, optional materials are not included and mortar/grout choice defaults to minimum performance specification indicated. Consider each system component and intended use to determine minimum requirements and to specify options.**
- Natural stone tile—see "Natural Stone Tile Selection and Installation Guide," and consult supplier for application suitability.
- Cementitious grout—ANSI A118.6 or better or ISO CG1 or better. Consider unsanded grout for soft or semisoft stones such as limestone, travertine, marble, onyx, or similar.
- Cementitious bond coat—portland cement paste on a mortar bed that is still workable. For a cured mortar bed, follow recommendations below to select appropriate bonding mortar:
  - When a waterproof membrane is not used—ANSI A118.1 or better or ISO C1 or better.
  - When a waterproof membrane is used—ANSI A118.4 or better or ISO C2S1 or better unless ANSI A118.1 or ISO C1 is recommended by membrane manufacturer.
  - Use white for light-colored marble, limestone, and other stones where staining or darkening from grey setting materials is possible.
- Epoxy bond coat, when used—ANSI A118.3 or ISO R1 or better. See Water Sensitivity and Fiberglass Mesh Reinforced Stone in the "Natural Stone Tile Selection and Installation Guide."
- Waterproof membrane, when used—ANSI A118.10.
- Mortar bed, reinforcing, and cleavage membrane—ANSI A108.1A.

Materials for Green/Sustainable Design

- See "Green Building Standards and Green Product Selection Guide" and consult manufacturers and suppliers for product sustainability and contribution to green building design.
- Consider specifying installation materials that meet ANSI A138.1, the *American National Standard Specifications for Sustainable Ceramic Tiles, Glass Tiles, and Tile Installation Materials.*

Preparation by Other Trades

- Floor systems, including the framing system and subfloor panels, over which tile will be installed shall be in conformance with the IRC for residential applications, the IBC for commercial applications, or applicable building codes and maximum allowable floor member live load and concentrated load deflection shall not exceed l/720. See also "Substrate Requirements."
- Face grain of plywood shall run perpendicular to joists.
- Subfloor—minimum 19⁄32" tongue and groove exterior glue plywood with 1⁄8" gap between sheets or 1" nominal boards.
- Depressing subfloor between joists on ledger strips permissible in residential use.
- Maximum allowable variation in the installation substrate (plywood)—1⁄4" in 10' from the required plane.

Movement Joint (architect must specify type of joint and show location and details on drawings)

- Movement joints—mandatory according to EJ171.

Installation Specifications

- Tile—ANSI A108.1A,.1B, or .1C. Ungauged stone may require installation per ANSI A108.1A. Material with minimal variation may be allowed per A108.1B with LHT mortar. A108.1B required if membrane or epoxy bond coat to be used.
- Cementitious grout—ANSI A108.10.
- Waterproof membrane—ANSI A108.13.
- Movement Joints—EJ171 and ASTM C1193.

Notes

- For hydronic heat application, see RH141 Stone.
- When unsanded grout is used, grout joint width must be 1⁄16" to 1⁄8" wide (nominal).

F250-21
STONE

- **Joists max. 16" o.c./Plywood Subfloor**
- **Plywood Underlayment**
- **Backer Board**
- **Natural Stone Tile**

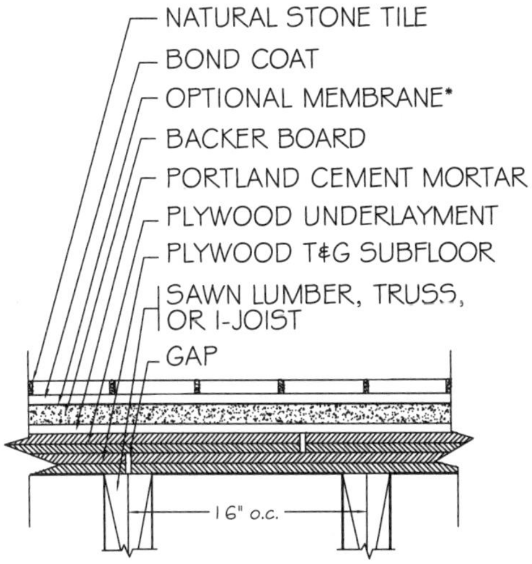

*USE OF A MEMBRANE IS OPTIONAL. SEE MEMBRANE OPTIONS.

Recommended Uses

- For wood substrates where thin-bed installation is desired.

Service Rating

- Residential with minimum 15⁄32"-thick plywood underlayment, when stone with adequate compressive strength, flexural strength, and resistance to abrasion is used.
- Light commercial with minimum 19⁄32"-thick plywood underlayment, when stone with adequate compressive strength, flexural strength, and resistance to abrasion is used.

Environmental Exposure Classifications

- Res1, 2; Com1, 2.
- May be suitable for increased water exposure. See Membrane Options.
- For installations that may be exposed to staining and/or chemical attack, specify tile and grout suitable for exposure. Consult product manufacturers; see also "Product Selection Guides."

Typical Weight of Tile Installation

- 10 pounds/square foot with 1⁄4" or 1⁄2" cementitious coated extruded foam backer board over 15⁄32" or 19⁄32" plywood underlayment.
- 11 pounds/square foot with 1⁄4" or 1⁄2" coated glass mat water-resistant gypsum backer board over 15⁄32" plywood underlayment; or with 1⁄4" board over 19⁄32" plywood underlayment.
- 11 pounds/square foot with 1⁄4" cement backer board over 15⁄32" plywood underlayment.
- 11 pounds/square foot with 1⁄4" or 3⁄8" fiber-reinforced water-resistant gypsum backer board over 15⁄32" plywood underlayment; or with 1⁄4" board over 19⁄32" plywood underlayment.
- 11 pounds/square foot with 1⁄4" fiber-cement backer board over 15⁄32" plywood underlayment.
- 12 pounds/square foot with 1⁄4" or 7⁄16" fiber-cement backer board over 19⁄32" plywood underlayment; or with 7⁄16" board over 15⁄32" plywood underlayment.
- 12 pounds/square foot with 1⁄4" cement backer board over 19⁄32" plywood underlayment.
- 12 pounds/square foot with 1⁄2" coated glass mat water-resistant gypsum backer board over 19⁄32" plywood underlayment.
- 12 pounds/square foot with 3⁄8" fiber-reinforced, water-resistant gypsum backer board over 19⁄32" plywood underlayment; or with 1⁄2" board over 15⁄32" plywood underlayment.
- 13 pounds/square foot with 1⁄2" cement backer board over 15⁄32" plywood underlayment.
- 13 pounds/square foot with 1⁄2" fiber-reinforced, water-resistant gypsum backer board over 19⁄32" plywood underlayment.
- 14 pounds/square foot with 1⁄2" cement backer board over 19⁄32" plywood underlayment.
- Does not include weight of substrate. See "Appendix B" for assumptions, included materials, and their individual weights.

Limitations

- Maximum joist spacing 16" on center.
- 8" × 8" or larger tile required when cementitious coated extruded foam backer board used.

Membrane Options

- Membrane options are available according to type of backer board used. See ceramic tile methods F144, F146, F170, and F175 for options.
- Specifier shall indicate if complete waterproofing is required, including if/how membrane connects to drain assembly, if base flashing is required, and treatment at other termination points.
- Check with membrane manufacturers for suitability for applicable conditions, as not all membranes are suitable for steam, high temperature, and/or chemical exposure, exterior use, use over above-ground structural slabs, use over pourable underlayments, use with radiant heat, or use over concrete with excessive moisture vapor transmission and/or alkalinity. Membrane may also affect service rating.
- When coated glass mat water-resistant gypsum backer board is used—some liquid-applied waterproof membranes will require extended cure time. Consult membrane manufacturer.

Requirements

- See ceramic tile methods F144, F146, F170, or F175 for requirements based on backer board type.
- Minimum grout joint width—1⁄16."

Preparation by Backer Board Installers

- Use a sufficient amount of portland cement mortar under the backer board to establish a supporting plane and eliminate voids.
- Installation of backer board varies according to type of backer board used. See ceramic tile methods F144, F146, F170, and F175.
- Maximum allowable variation in the tile substrate—⅛" in 10' from the required plane when measured from the high points in the surface.
- Stagger backer board end and edge joints so as not to coincide with joints in subfloor. Stagger joints in adjacent rows so four corners do not come together within the same plane. Space panel ends and edges in accordance with manufacturer's recommendations.

Materials

- **Multiple options exist for membranes, mortars, grouts, and other materials and *must be clearly specified* to be included. If not specifically indicated, optional materials are not included and mortar/grout choice defaults to minimum performance specification indicated. Consider each system component and intended use to determine minimum requirements and to specify options.**
- Natural stone tile—see "Natural Stone Tile Selection and Installation Guide," and consult supplier for application suitability.
- Cementitious grout—ANSI A118.6 or better or ISO CG1 or better. Consider unsanded grout for soft or semisoft stones such as limestone, travertine, marble, onyx, or similar.
- Cementitious bond coat:
  - When a waterproof membrane is not used—ANSI A118.4 or better or ISO C2S1 or better.
  - When a waterproof membrane is used—ANSI A118.4 or better or ISO C2S1 or better unless ANSI A118.1 or ISO C1 is recommended by membrane manufacturer.
  - Use white for light-colored marble, limestone, and other stones where staining or darkening from grey setting materials is possible.
- Epoxy bond coat, when used—ANSI A118.3 or ISO R1 or better. See Water Sensitivity and Fiberglass Mesh Reinforced Stone in the "Natural Stone Tile Selection and Installation Guide."
- Backer board—cement backer board meeting ANSI A118.9 or ASTM C1325, fiber-cement backer board meeting ASTM C1288, coated glass mat water-resistant gypsum backer board meeting ASTM C1178, fiber-reinforced water-resistant gypsum backer board meeting ASTM C1278 Section 6, or cementitious-coated extruded foam backer board meeting ASTM C578.
- Mortar under backer board—ANSI A118.1 or better or ISO C1 or better.
- Waterproof membrane, when used—ANSI A118.10.

Materials for Green/Sustainable Design

- See "Green Building Standards and Green Product Selection Guide" and consult manufacturers and suppliers for product sustainability and contribution to green building design.
- Consider specifying installation materials that meet ANSI A138.1, the *American National Standard Specifications for Sustainable Ceramic Tiles, Glass Tiles, and Tile Installation Materials.*

Preparation by Other Trades

- Floor systems, including the framing system and subfloor panels, over which tile will be installed shall be in conformance with the IRC for residential applications, the IBC for commercial applications, or applicable building codes and maximum allowable floor member live load and concentrated load deflection shall not exceed l/720. See also "Substrate Requirements."
- When concentrated loads (scissor lifts, pallet jacks, forklifts, etc.) will be used on an above-ground tiled floor, the engineer and/or specifier shall specify a substrate to accommodate the concentrated loads. Owner/specifier is responsible for protecting the tilework from damage, including allowing sufficient time for installed materials to cure properly.
- Face grain of plywood subfloor and underlayment shall run perpendicular to joists.
- Subfloor—minimum 19⁄32" exterior-glue tongue and groove plywood with ⅛" gap between sheets.
- Underlayment—minimum 15⁄32" exterior glue plywood with ¼" gap between sheets. See also Service Rating.
- Maximum allowable variation in the plywood underlayment—⅛" in 10' from the required plane, when measured from the high points in the surface. Adjacent edges of plywood sheets not to exceed 1⁄32" difference in height.

Movement Joint (architect must specify type of joint and show location and details on drawings)

- Movement joints—mandatory in accordance with EJ171.

Installation Specifications

- Tile—ANSI A108.5 or A108.6.
- Cementitious grout—ANSI A108.10.
- Waterproof membrane—ANSI A108.13.
- Backer board—ANSI A108.11 or manufacturer's directions.
- Movement Joints—EJ171 and ASTM C1193.

Notes

- Underlayment fasteners should not penetrate joists below.
- As the tile size increases, there is less tolerance for variation in the substrate from the required plane. Epoxy bond coat thickness must be thin and uniform; therefore, substrate flattening may be required when epoxy bond coat is used.
- When unsanded grout is used, grout joint width must be 1⁄16" to ⅛" wide (nominal).

W201-21
STONE

- **Masonry or Concrete**
- **Mortar Bed**
- **Natural Stone Tile**

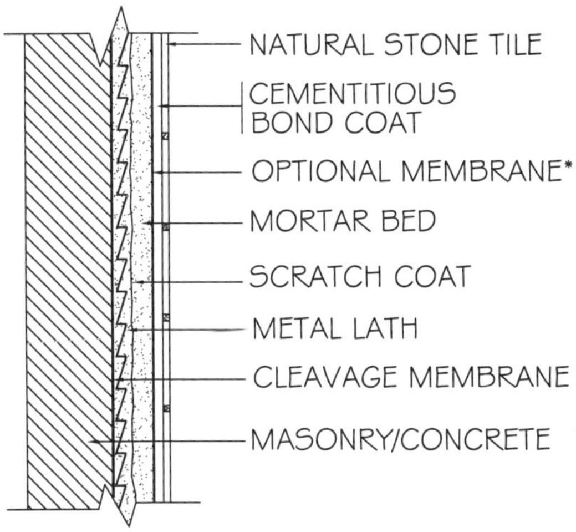

*USE OF A MEMBRANE IS OPTIONAL. SEE MEMBRANE OPTIONS.

Recommended Uses

- For exterior walls of masonry or concrete where flattening or trueing of substrate is required.
- For areas where wall flatness is critical, such as when tiles with any edge longer than 15" are specified.
- For substrates that are cracked, coated, or present bonding issues.
- For isolating the tile installation from the substrate.
- For areas where ungauged stone will be used.

Environmental Exposure Classifications

- Res6; Com6.
- For installations that may be exposed to staining and/or chemical attack, specify tile and grout suitable for exposure. Consult product manufacturers; see also "Product Selection Guides."

Limitations

- Where installation will be subjected to freeze/thaw cycles, degradation can occur over time.

Membrane Options

- Check building code—water-resistive barrier, waterproof membrane, or vapor retarder membrane may be required.
- A waterproof membrane (A118.10) may be specified to prevent moisture intrusion and protect adjacent building materials. Specifier shall indicate if complete waterproofing is required, including base flashing and treatment at other termination points.
- Check with membrane manufacturer for suitability for applicable conditions, as not all membranes are suitable for steam, high-temperature and/or chemical exposure, or exterior use.

Requirements

- Cut lath at all movement joints.
- Mortar bed thickness—¾" minimum to 1½" maximum.
- Minimum grout joint width—1⁄16."

Materials

- **Multiple options exist for membranes, mortars, grouts, and other materials and *must be clearly specified* to be included. If not specifically indicated, optional materials are not included and mortar/grout choice defaults to minimum performance specification indicated. Consider each system component and intended use to determine minimum requirements and to specify options.**
- Natural stone tile—see "Natural Stone Tile Selection and Installation Guide," and consult supplier for application suitability.
- Cementitious grout—ANSI A118.6 or better or ISO CG1 or better. Consider unsanded grout for soft or semisoft stones such as limestone, travertine, marble, onyx, or similar.
- Cementitious bond coat—portland cement paste on a mortar bed that is still workable. For a cured mortar bed, follow recommendations below to select appropriate bonding mortar:
  - When a waterproof membrane is not used—ANSI A118.15 or better or ISO C2S1 or better.
  - When a waterproof membrane is used—ANSI A118.4 or better or ISO C2S1 or better, unless ANSI A118.1 or ISO C1 is recommended by membrane manufacturer.
  - Use white for light-colored marble, limestone, and other stones where staining or darkening from grey setting materials is possible.
- Waterproof membrane, when used—ANSI A118.10.
- Mortar bed, metal lath, and cleavage membrane—ANSI A108.1A.

Materials for Green/Sustainable Design

- See "Green Building Standards and Green Product Selection Guide" and consult manufacturers and suppliers for product sustainability and contribution to green building design.
- Consider specifying installation materials that meet ANSI A138.1, the *American National Standard Specifications for Sustainable Ceramic Tiles, Glass Tiles, and Tile Installation Materials.*

Preparation by Other Trades

- Maximum allowable variation in the installation substrate (concrete/masonry)—¼" in 10' from the required plane.

Movement Joint (architect must specify type of joint and show location and details on drawings)

- Movement joints—mandatory according to EJ171.

Installation Specifications

- Tile—ANSI A108.1A, .1B, or .1C. Ungauged stone may require installation per ANSI A108.1A. Material with minimal variation may be allowed per A108.1B with LHT mortar. A108.1B required if membrane to be used.
- Cementitious grout—ANSI A108.10.
- Waterproof membrane—ANSI A108.13.
- Movement Joints—EJ171 and ASTM C1193.

Notes

- For prefabricated exterior panels, see "Additional Products Used in Tile Installations."
- Not all bonding mortars are suitable for exterior use.
- Protection of installation may be required to prevent premature exposure of setting materials to moisture.
- For interior mortar bed applications, see stone methods W211, W221, W222, W231/W241, B411, and B414.
- When unsanded grout is used, grout joint width must be $\frac{1}{16}$" to $\frac{1}{8}$" wide (nominal).

W202E-21
STONE

- **Masonry or Concrete**
- **Natural Stone Tile**

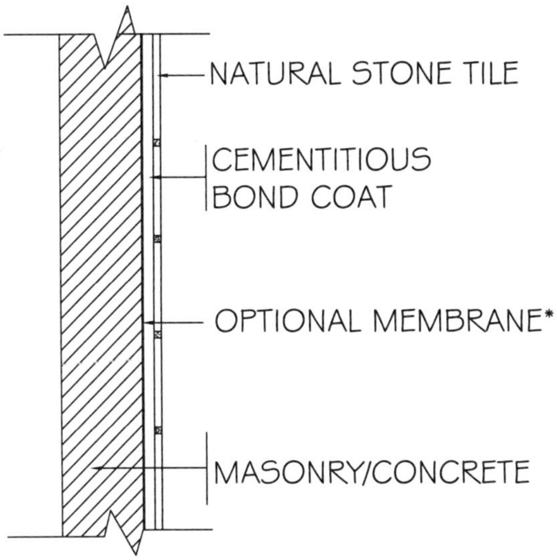

*USE OF A MEMBRANE IS OPTIONAL. SEE MEMBRANE OPTIONS.

Recommended Uses

- For clean, sound exterior walls of masonry or concrete where thin-bed installation of tile is desired.

Environmental Exposure Classifications

- Res6; Com6.
- For installations that may be exposed to staining and/or chemical attack, specify tile and grout suitable for exposure. Consult product manufacturers; see also "Product Selection Guides."

Limitations

- Where installation will be subjected to freeze/thaw cycles, degradation can occur over time.
- Not for cracked or coated surfaces.

Membrane Options

- Check building code—water-resistive barrier, waterproof membrane, or vapor retarder membrane may be required.
- A waterproof membrane (A118.10) may be specified to prevent moisture intrusion and protect adjacent building materials. Specifier shall indicate if complete waterproofing is required, including base flashing and treatment at other termination points.
- Check with membrane manufacturer for suitability for applicable conditions, as not all membranes are suitable for steam, high-temperature and/or chemical exposure, or exterior use.

Requirements

- Masonry or concrete to be well cured, dimensionally stable, and free of cracks, waxy or oily films, and curing compounds.
- Concrete may require bush-hammering or sandblasting to facilitate bonding.
- Minimum grout joint width—1⁄16."

Materials

- **Multiple options exist for membranes, mortars, grouts, and other materials and *must be clearly specified* to be included. If not specifically indicated, optional materials are not included and mortar/grout choice defaults to minimum performance specification indicated. Consider each system component and intended use to determine minimum requirements and to specify options.**
- Natural stone tile—see "Natural Stone Tile Selection and Installation Guide," and consult supplier for application suitability.
- Cementitious grout—ANSI A118.6 or better or ISO CG1 or better. Consider unsanded grout for soft or semisoft stones such as limestone, travertine, marble, onyx, or similar.
- Cementitious bond coat:
  - When a waterproof membrane is not used—ANSI A118.15 or better or ISO C2S1 or better.
  - When a waterproof membrane is used—ANSI A118.4 or better or ISO C2S1 or better, unless ANSI A118.1 or ISO C1 is recommended by membrane manufacturer.
  - Use white for light-colored marble, limestone, and other stones where staining or darkening from grey setting materials is possible.
- Waterproof membrane, when used—ANSI A118.10.

Materials for Green/Sustainable Design

- See "Green Building Standards and Green Product Selection Guide" and consult manufacturers and suppliers for product sustainability and contribution to green building design.
- Consider specifying installation materials that meet ANSI A138.1, the *American National Standard Specifications for Sustainable Ceramic Tiles, Glass Tiles, and Tile Installation Materials.*

Preparation by Other Trades

- Maximum allowable variation in the tile substrate—1⁄8" in 10' from the required plane when measured from the high points in the surface.

Movement Joint (architect must specify type of joint and show location and details on drawings)

- Movement joints mandatory—according to EJ171.

Installation Specifications

- Tile—ANSI A108.5.
- Cementitious grout—ANSI A108.10.
- Waterproof membrane—ANSI A108.13.
- Movement Joints—EJ171 and ASTM C1193.

Notes

- For prefabricated exterior panels, see "Additional Products Used in Tile Installations."
- Not all bonding mortars are suitable for exterior use.
- Protection of installation may be required to prevent premature exposure of setting materials to moisture.
- For interior thin-bed applications over masonry or concrete, see W202I Stone.
- When unsanded grout is used, grout joint width must be 1/16" to 1/8" wide (nominal).

W244E-21
STONE

- **Wood or Metal Studs**
- **Cement Backer Board**
- **Natural Stone Tile**

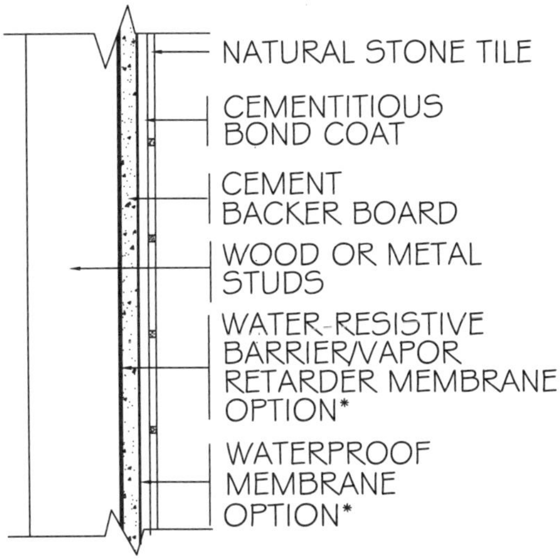

*USE OF A MEMBRANE IS REQUIRED. SEE MEMBRANE OPTIONS.

Recommended Uses

- For exterior walls over wood or metal studs.

Environmental Exposure Classifications

- Res6; Com6.
- For installations that may be exposed to staining and/or chemical attack, specify tile and grout suitable for exposure. Consult product manufacturers; see also "Product Selection Guides."

Limitations

- Where installation will be subjected to freeze/thaw cycles, degradation can occur over time.
- Maximum stud spacing 16" on center.

Membrane Options

- Check building code—water-resistive barrier, waterproof membrane, or vapor retarder membrane may be required.
- A waterproof membrane (A118.10) may be specified to prevent moisture intrusion and protect adjacent building materials. Specifier shall indicate if complete waterproofing is required, including base flashing and treatment at other termination points.
- If waterproof membrane is applied over cement board, membrane over framing members may not be required.
- Check with membrane manufacturer for suitability for applicable conditions, as not all membranes are suitable for steam, high-temperature and/or chemical exposure, or exterior use.

Requirements

- Wood studs—dry and well-braced, minimum depth 3½".
- Metal studs—well-braced; 20 gauge (0.033") or heavier; minimum depth 3½" for residential applications or 3⅝" for commercial applications.
- Water-resistive barrier or waterproof membrane required, per building code. See Membrane Options.
- Minimum grout joint width—1/16".

Preparation by Backer Board Installers

- Maximum allowable variation in the tile substrate—⅛" in 10' from the required plane when measured from the high points in the surface.
- Horizontal joints—⅛" spacing filled solid and taped with latex-portland cement mortar and 4" alkali-resistant glass fiber mesh tape.
- Vertical joints—fill any space and tape with latex-portland cement mortar and 4" alkali-resistant glass fiber mesh tape.
- Corners—leave space between backer units. Tape joints using skim coat of latex-portland cement mortar, but do not fill.
- Center backer board end or edge joints on framing and stagger joints in adjacent rows so four corners do not come together within the same plane. Space panel ends and edges in accordance with manufacturer's recommendations.

Materials

- **Multiple options exist for membranes, mortars, grouts, and other materials and *must be clearly specified* to be included. If not specifically indicated, optional materials are not included and mortar/grout choice defaults to minimum performance specification indicated. Consider each system component and intended use to determine minimum requirements and to specify options.**
- Natural stone tile—see "Natural Stone Tile Selection and Installation Guide," and consult supplier for application suitability.
- Cementitious grout—ANSI A118.6 or better or ISO CG1 or better. Consider unsanded grout for soft or semisoft stones such as limestone, travertine, marble, onyx, or similar.
- Cementitious bond coat:
  - When a waterproof membrane is not used on face of cement backer board—ANSI A118.15 or better or ISO C2S1 or better.
  - When a waterproof membrane is used on face of cement backer board—ANSI A118.4 or better or ISO C2S1 or better, unless ANSI A118.1 or ISO C1 is recommended by membrane manufacturer.

  - Use white for light-colored marble, limestone, and other stones where staining or darkening from grey setting materials is possible.
- Cement backer board—ANSI A118.9 or ASTM C1325 (Type A).
- Fasteners—noncorrosive and nonoxidizing.
- Hot-dipped fasteners meeting ASTM F2329 required in wet areas.
- 4" alkali-resistant glass fiber mesh tape.
- Water-resistive barrier/vapor retarder membrane, when used—per building code.
- Waterproof membrane, when used—ANSI A118.10.
- Metal studs—ASTM C645.

Materials for Green/Sustainable Design

- See "Green Building Standards and Green Product Selection Guide" and consult manufacturers and suppliers for product sustainability and contribution to green building design.
- Consider specifying installation materials that meet ANSI A138.1, the *American National Standard Specifications for Sustainable Ceramic Tiles, Glass Tiles, and Tile Installation Materials.*

Preparation by Other Trades

- Wall framing shall meet the general framing requirements of ANSI A108.11-4.0–4.3.

Movement Joint (architect must specify type of joint and show location and details on drawings)

- Movement joints—mandatory according to EJ171.

Installation Specifications

- Cement backer board—manufacturer's directions.
- Tile—ANSI A108.5.
- Cementitious grout—ANSI A108.10.
- Waterproof membrane—ANSI A108.13.
- Movement Joints—EJ171 and ASTM C1193.

Notes

- For prefabricated exterior panels, see "Additional Products Used in Tile Installations."
- Not all bonding mortars are suitable for exterior use.
- Protection of installation may be required to prevent premature exposure of setting materials to moisture.
- For interior applications, see stone methods W244C, B412, and B415.
- When unsanded grout is used, grout joint width must be 1/16" to 1/8" wide (nominal).

W211-21
STONE

- **Masonry or Concrete**
- **Bonded Mortar Bed**
- **Natural Stone Tile**

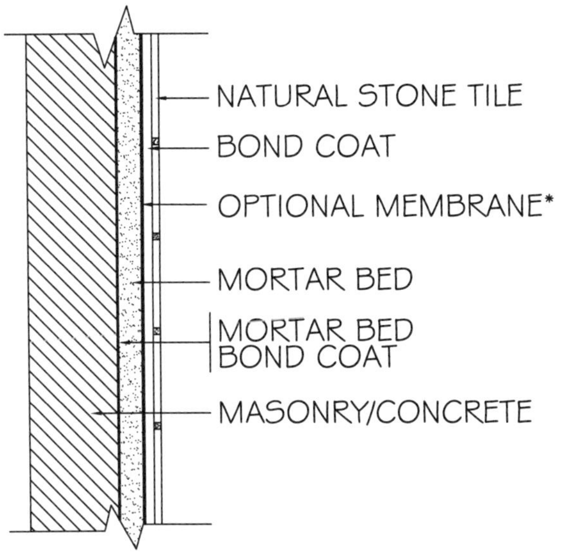

*USE OF A MEMBRANE IS OPTIONAL. SEE MEMBRANE OPTIONS.

Recommended Uses

- For clean, sound interior walls of masonry or concrete where flattening or trueing of substrate is required.
- For areas where wall flatness is critical, such as when tiles with any edge longer than 15" are specified.
- For areas where ungauged stone will be used.

Environmental Exposure Classifications

- Res1, 2, 3, 5; Com1, 2, 3, 5.
- May be suitable for exterior applications when appropriate precautions are taken, including flashing, expansion joint placement, waterproofing, and material selection. See Notes.
- For installations that may be exposed to staining and/or chemical attack, specify tile and grout suitable for exposure. Consult product manufacturers; see also "Product Selection Guides."

Limitations

- Not for cracked or coated surfaces.
- Exterior—where installation will be subjected to freeze/thaw cycles, degradation can occur over time.

Membrane Options

- A waterproof membrane (A118.10) may be specified to prevent moisture intrusion and protect adjacent building materials. Specifier shall indicate if complete waterproofing is required, including base flashing and treatment at other termination points.
- Check with membrane manufacturer for suitability for applicable conditions, as not all membranes are suitable for steam, high-temperature and/or chemical exposure, or exterior use.
- Exterior—check building code—water-resistive barrier, waterproof membrane, or vapor retarder membrane may be required.

Requirements

- Masonry or concrete to be well cured, dimensionally stable, and free of cracks, waxy or oily films, and curing compounds.
- Concrete may require bush-hammering or sandblasting to facilitate bonding.
- Mortar bed thickness—3/8" minimum to 3/4" maximum. Scratch coat required if mortar bed will exceed 3/4."
- Minimum grout joint width—1/16."

Materials

- **Multiple options exist for membranes, mortars, grouts, and other materials and *must be clearly specified* to be included. If not specifically indicated, optional materials are not included and mortar/grout choice defaults to minimum performance specification indicated. Consider each system component and intended use to determine minimum requirements and to specify options.**
- Natural stone tile—see "Natural Stone Tile Selection and Installation Guide," and consult supplier for application suitability.
- Cementitious grout—ANSI A118.6 or better or ISO CG1 or better. Consider unsanded grout for soft or semisoft stones such as limestone, travertine, marble, onyx, or similar.
- Cementitious bond coat—portland cement paste on a mortar bed that is still workable. For a cured mortar bed, follow recommendations below to select appropriate bonding mortar:
  - When a waterproof membrane is not used—ANSI A118.1 or better or ISO C1 or better.
  - When a waterproof membrane is used—ANSI A118.4 or better or ISO C2S1 or better unless ANSI A118.1 or ISO C1 is recommended by membrane manufacturer.
  - Use white for light-colored marble, limestone, and other stones where staining or darkening from grey setting materials is possible.
- Epoxy bond coat, when used—ANSI A118.3 or ISO R1 or better. See Water Sensitivity and Fiberglass Mesh Reinforced Stone in the "Natural Stone Tile Selection and Installation Guide."
- Waterproof membrane, when used—ANSI A118.10.
- Mortar bed—ANSI A108.1A.
- Mortar bed bond coat—portland cement slurry.

Materials for Green/Sustainable Design

- See "Green Building Standards and Green Product Selection Guide" and consult manufacturers and suppliers for product sustainability and contribution to green building design.
- Consider specifying installation materials that meet ANSI A138.1, the *American National Standard Specifications for Sustainable Ceramic Tiles, Glass Tiles, and Tile Installation Materials.*

Preparation by Other Trades

- Maximum allowable variation in the installation substrate (concrete/masonry)—¼" in 10' from the required plane.

Movement Joint (architect must specify type of joint and show location and details on drawings)

- Movement joints—mandatory according to EJ171.

Installation Specifications

- Tile—ANSI A108.1A, .1B, or .1C. Ungauged stone may require installation per ANSI A108.1A. Material with minimal variation may be allowed per A108.1B with LHT mortar. A108.1B required if membrane or epoxy bond coat to be used.
- Cementitious grout—ANSI A108.10.
- Waterproof membrane—ANSI A108.13.
- Movement Joints—EJ171 and ASTM C1193.

Notes

- Exterior—not all bonding mortars are suitable for exterior use.
- Exterior—protection of installation may be required to prevent premature exposure of setting materials to moisture.
- When unsanded grout is used, grout joint width must be $\frac{1}{16}$" to ⅛" wide (nominal).

W215-21
STONE

- **Masonry or Concrete**
- **Spot-Bonding Epoxy**
- **Natural Stone Tile**

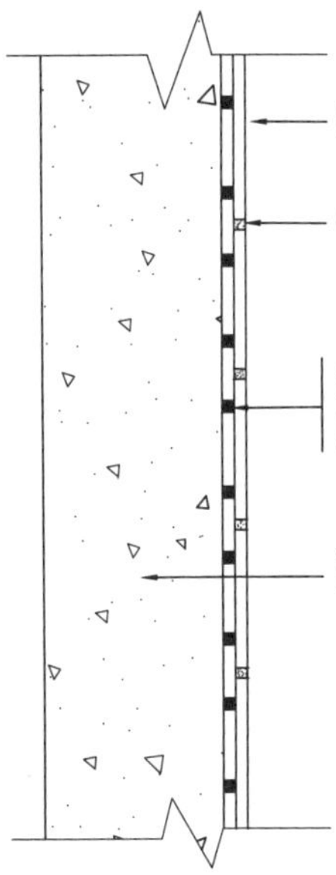

Recommended Uses

- For clean, sound interior walls of masonry or concrete where large format tiles are specified.

Environmental Exposure Classifications

- Res1; Com1.
- For installations that may be exposed to staining and/or chemical attack, specify tile and grout suitable for exposure. Consult product manufacturers; see also "Product Selection Guides."

Limitations

- Will not withstand impact.
- Not for cracked or coated surfaces.

Requirements

- Epoxy must be recommended by the manufacturer for spot-bonding.
- Follow epoxy manufacturer's coverage requirements.
- Masonry or concrete to be well cured, dimensionally stable, and free of cracks, waxy or oily films, and curing compounds.
- Concrete may require bush-hammering or sandblasting to facilitate bonding.
- Minimum grout joint width—1⁄16."

Materials

- **Multiple options exist for membranes, mortars, grouts, and other materials and *must be clearly specified* to be included. If not specifically indicated, optional materials are not included and mortar/grout choice defaults to minimum performance specification indicated. Consider each system component and intended use to determine minimum requirements and to specify options.**
- Natural stone tile—see "Natural Stone Tile Selection and Installation Guide," and consult supplier for application suitability.
- Cementitious grout—ANSI A118.6 or better or ISO CG1 or better. Consider unsanded grout for soft or semisoft stones such as limestone, travertine, marble, onyx, or similar.
- Sealant, when used as grout—ASTM C920.
- Spot bonding epoxy adhesive—epoxy recommended by manufacturer for spot bonding.

Materials for Green/Sustainable Design

- See "Green Building Standards and Green Product Selection Guide" and consult manufacturers and suppliers for product sustainability and contribution to green building design.
- Consider specifying installation materials that meet ANSI A138.1, the *American National Standard Specifications for Sustainable Ceramic Tiles, Glass Tiles, and Tile Installation Materials.*

Preparation by Other Trades

- Maximum allowable variation in the tile substrate—1⁄4" in 10' from the required plane.

Movement Joint (architect must specify type of joint and show location and details on drawings)

- Movement joints—mandatory according to EJ171.

Installation Specifications

- Tile—epoxy manufacturer's directions.
- Cementitious grout—ANSI A108.10.
- Sealant—follow sealant and epoxy manufacturer's directions.
- Movement Joints—EJ171 and ASTM C1193.

Notes

- When unsanded grout is used, grout joint width must be 1⁄16" to 1⁄8" wide (nominal).

W202I-21
STONE

- **Masonry or Concrete**
- **Natural Stone Tile**

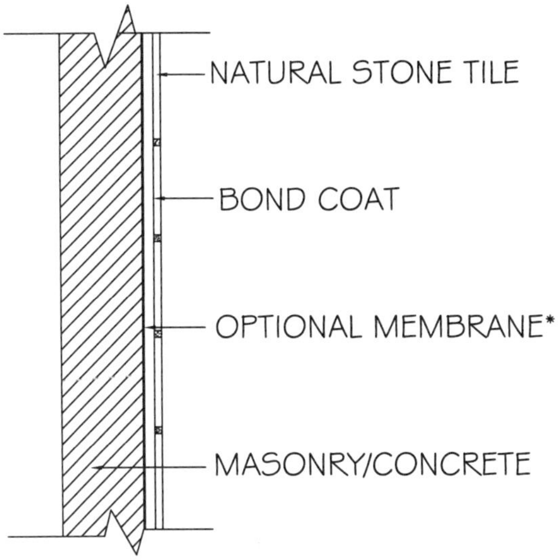

*USE OF A MEMBRANE IS OPTIONAL. SEE MEMBRANE OPTIONS.

Recommended Uses

- For interior walls of masonry or concrete where thin-bed installation of tile is desired.

Environmental Exposure Classifications

- Res1, 2, 3, 5; Com1, 2, 3, 5.
- For Res6 and Com6, see W202E Stone.
- For installations that may be exposed to staining and/or chemical attack, specify tile and grout suitable for exposure. Consult product manufacturers; see also "Product Selection Guides."

Limitations

- Not for cracked or coated surfaces.

Membrane Options

- A waterproof membrane (A118.10) may be specified to prevent moisture intrusion and protect adjacent building materials. Specifier shall indicate if complete waterproofing is required, including base flashing and treatment at other termination points.
- Check with membrane manufacturer for suitability for applicable conditions, as not all membranes are suitable for steam, high-temperature and/or chemical exposure, or exterior use.

Requirements

- Masonry or concrete to be well cured, dimensionally stable, and free of cracks, waxy or oily films, and curing compounds.
- Concrete may require bush-hammering or sandblasting to facilitate bonding.
- Minimum grout joint width—1/16."

Materials

- **Multiple options exist for membranes, mortars, grouts, and other materials and *must be clearly specified* to be included. If not specifically indicated, optional materials are not included and mortar/grout choice defaults to minimum performance specification indicated. Consider each system component and intended use to determine minimum requirements and to specify options.**
- Natural stone tile—see "Natural Stone Tile Selection and Installation Guide," and consult supplier for application suitability.
- Cementitious grout—ANSI A118.6 or better or ISO CG1 or better. Consider unsanded grout for soft or semisoft stones such as limestone, travertine, marble, onyx, or similar.
- Cementitious bond coat:
  - When a waterproof membrane is not used—ANSI A118.1 or better or ISO C1 or better.
  - When a waterproof membrane is used—ANSI A118.4 or better or ISO C2S1 or better unless ANSI A118.1 or ISO C1 is recommended by membrane manufacturer.
  - Use white for light-colored marble, limestone, and other stones where staining or darkening from grey setting materials is possible.
- Epoxy bond coat, when used—ANSI A118.3 or ISO R1 or better. See Water Sensitivity and Fiberglass Mesh Reinforced Stone in the "Natural Stone Tile Selection and Installation Guide."
- Waterproof membrane, when used—ANSI A118.10.

Materials for Green/Sustainable Design

- See "Green Building Standards and Green Product Selection Guide" and consult manufacturers and suppliers for product sustainability and contribution to green building design.
- Consider specifying installation materials that meet ANSI A138.1, the *American National Standard Specifications for Sustainable Ceramic Tiles, Glass Tiles, and Tile Installation Materials.*

Preparation by Other Trades

- Maximum allowable variation in the tile substrate—1/8" in 10' from the required plane when measured from the high points in the surface.

Movement Joint (architect must specify type of joint and show location and details on drawings)

- Movement joints mandatory—according to EJ171.

Installation Specifications

- Tile—ANSI A108.5 or A108.6.
- Cementitious grout—ANSI A108.10.
- Waterproof membrane—ANSI A108.13.
- Movement Joints—EJ171 and ASTM C1193.

Notes

- For exterior application, see W202E Stone.
- As the tile size increases, there is less tolerance for variation in the substrate from the required plane. Epoxy bond coat thickness must be thin and uniform; therefore, substrate flattening may be required when epoxy bond coat is used.
- When unsanded grout is used, grout joint width must be $\frac{1}{16}$" to $\frac{1}{8}$" wide (nominal).

W221-21
STONE

- **Solid Backing**
- **Mortar Bed**
- **Natural Stone Tile**

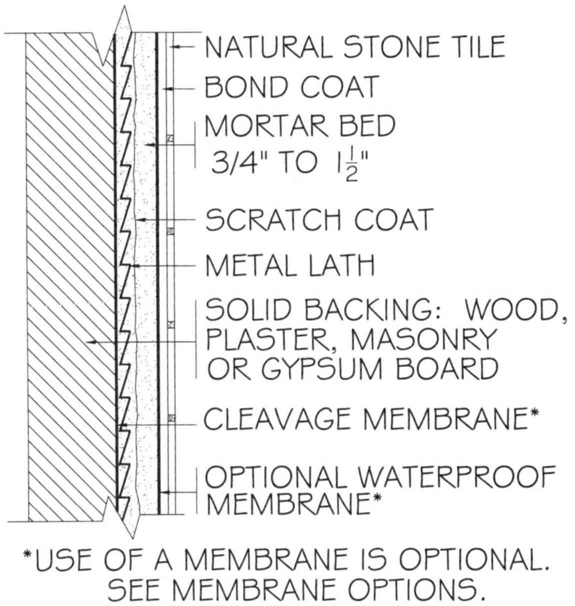

Recommended Uses

- For interior walls of masonry, plaster, or other solid backing where flattening or trueing of substrate is required.
- For areas where wall flatness is critical, such as when tiles with any edge longer than 15" are specified.
- For substrates that are cracked, coated, or present bonding issues.
- For areas where ungauged stone will be used.

Environmental Exposure Classifications

- With masonry backing—Res1, 2, 3, 5; Com1, 2, 3, 5.
- With gypsum board or plaster backing—Res1, 2; Com1, 2.
- With wood backing—Res1; Com1.
- May be suitable for exterior applications when appropriate precautions are taken, including flashing, expansion joint placement, waterproofing, and material selection. See Notes.
- For installations that may be exposed to staining and/or chemical attack, specify tile and grout suitable for exposure. Consult product manufacturers; see also "Product Selection Guides."

Limitations

- When backing is gypsum board, not for areas exposed to temperatures exceeding 125°F.

Membrane Options

- A waterproof membrane (A118.10) may be specified to prevent moisture intrusion and protect adjacent building materials. Specifier shall indicate if complete waterproofing is required, including base flashing and treatment at other termination points.
- If a waterproof membrane (A118.10) is not specified and backing material is moisture sensitive, cleavage membrane must be used behind mortar bed in wet areas.
- Check with membrane manufacturer for suitability for applicable conditions, as not all membranes are suitable for steam, high-temperature and/or chemical exposure, or exterior use.
- Exterior—check building code—water-resistive barrier, waterproof membrane, or vapor retarder membrane may be required.

Requirements

- Wood studs—dry and well-braced, minimum depth 3½."
- Metal studs—well-braced; 20 gauge (0.033") or heavier; minimum depth 3½" for residential applications or 3⅝" for commercial applications.
- Solid backing must provide firm anchorage for metal lath.
- Mortar bed thickness—¾" minimum to 1½" maximum.
- Cut lath at all movement joints.
- Minimum grout joint width—1/16."

Materials

- **Multiple options exist for membranes, mortars, grouts, and other materials and *must be clearly specified* to be included. If not specifically indicated, optional materials are not included and mortar/grout choice defaults to minimum performance specification indicated. Consider each system component and intended use to determine minimum requirements and to specify options.**
- Natural stone tile—see "Natural Stone Tile Selection and Installation Guide," and consult supplier for application suitability.
- Cementitious grout—ANSI A118.6 or better or ISO CG1 or better. Consider unsanded grout for soft or semisoft stones such as limestone, travertine, marble, onyx, or similar.
- Cementitious bond coat—portland cement paste on a mortar bed that is still workable. For a cured mortar bed, follow recommendations below to select appropriate bonding mortar:
  - When a waterproof membrane is not used—ANSI A118.1 or better or ISO C1 or better.
  - When a waterproof membrane is used—ANSI A118.4 or better or ISO C2S1 or better unless ANSI A118.1 or ISO C1 is recommended by membrane manufacturer.

  - Use white for light-colored marble, limestone, and other stones where staining or darkening from grey setting materials is possible.
- Epoxy bond coat, when used—ANSI A118.3 or ISO R1 or better. See Water Sensitivity and Fiberglass Mesh Reinforced Stone in the "Natural Stone Tile Selection and Installation Guide."
- Waterproof membrane, when used—ANSI A118.10.
- Mortar bed, metal lath, and cleavage membrane—ANSI A108.1A.
- Metal studs—ASTM C645.

Materials for Green/Sustainable Design

- See "Green Building Standards and Green Product Selection Guide" and consult manufacturers and suppliers for product sustainability and contribution to green building design.
- Consider specifying installation materials that meet ANSI A138.1, the *American National Standard Specifications for Sustainable Ceramic Tiles, Glass Tiles, and Tile Installation Materials.*

Preparation by Other Trades

- Maximum allowable variation in the installation substrate (solid backing)—¼" in 10' from the required plane.

Movement Joint (architect must specify type of joint and show location and details on drawings)

- Movement joints—mandatory according to EJ171.

Installation Specifications

- Tile—ANSI A108.1A, .1B, or .1C. Ungauged stone may require installation per ANSI A108.1A. Material with minimal variation may be allowed per A108.1B with LHT mortar. A108.1B required if waterproof membrane or epoxy bond coat to be used.
- Cementitious grout—ANSI A108.10.
- Waterproof membrane—ANSI A108.13.
- Movement Joints—EJ171 and ASTM C1193.

Notes

- If lath cannot be attached directly to backing, use furring strips. See W231/W241 Stone.
- For thinner mortar bed, see W222 Stone.
- Exterior—not all bonding mortars are suitable for exterior use.
- Exterior—protection of installation may be required to prevent premature exposure of setting materials to moisture.
- When unsanded grout is used, grout joint width must be ¹⁄₁₆" to ⅛" wide (nominal).

W222-21
STONE

- **Solid Backing**
- **Mortar Bed (One Coat Method)**
- **Natural Stone Tile**

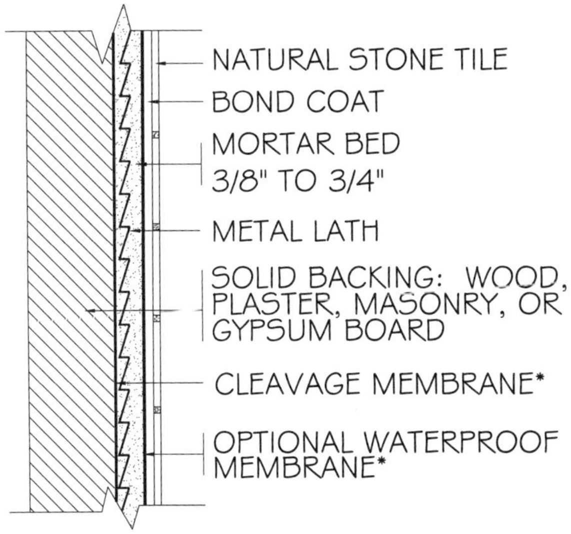

*USE OF A MEMBRANE IS OPTIONAL. SEE MEMBRANE OPTIONS.

Recommended Uses

- For interior walls of masonry, plaster, or other solid backing where flattening or trueing of substrate is required.
- For areas where wall flatness is critical, such as when tiles with any edge longer than 15" are specified.
- For substrates that are cracked, coated, or present bonding issues.
- For areas where ungauged stone will be used.

Environmental Exposure Classifications

- With masonry backing—Res1, 2, 5; Com1, 2, 5.
- With gypsum board or plaster backing—Res1, 2; Com1, 2.
- With wood backing—Res1; Com1.
- For Res3 and Com3, see B440 Stone and B441 Stone.
- May be suitable for exterior applications when appropriate precautions are taken, including flashing, expansion joint placement, waterproofing, and material selection. See Notes.
- For installations that may be exposed to staining and/or chemical attack, specify tile and grout suitable for exposure. Consult product manufacturers; see also "Product Selection Guides."

Limitations

- When backing is gypsum board, not for areas exposed to temperatures exceeding 125°F.

Membrane Options

- A waterproof membrane (A118.10) may be specified to prevent moisture intrusion and protect adjacent building materials. Specifier shall indicate if complete waterproofing is required, including base flashing and treatment at other termination points.
- If a waterproof membrane (A118.10) is not specified and backing material is moisture sensitive, cleavage membrane must be used behind mortar bed in wet areas.
- Check with membrane manufacturer for suitability for applicable conditions, as not all membranes are suitable for steam, high-temperature and/or chemical exposure, or exterior use.
- Exterior—check building code—water-resistive barrier, waterproof membrane, or vapor retarder membrane may be required.

Requirements

- Wood studs—dry and well-braced, minimum depth 3½."
- Metal studs—well-braced; 20 gauge (0.033") or heavier; minimum depth 3½" for residential applications or 3⅝" for commercial applications.
- Solid backing must provide firm anchorage for metal lath.
- Mortar bed thickness—⅜" minimum to ¾" maximum.
- Cut lath at all movement joints.
- Minimum grout joint width—1/16."

Materials

- **Multiple options exist for membranes, mortars, grouts, and other materials and *must be clearly specified* to be included. If not specifically indicated, optional materials are not included and mortar/grout choice defaults to minimum performance specification indicated. Consider each system component and intended use to determine minimum requirements and to specify options.**
- Natural stone tile—see "Natural Stone Tile Selection and Installation Guide," and consult supplier for application suitability.
- Cementitious grout—ANSI A118.6 or better or ISO CG1 or better. Consider unsanded grout for soft or semisoft stones such as limestone, travertine, marble, onyx, or similar.
- Cementitious bond coat—portland cement paste on a mortar bed that is still workable. For a cured mortar bed, follow recommendations below to select appropriate bonding mortar:
  - When a waterproof membrane is not used—ANSI A118.1 or better or ISO C1 or better.
  - When a waterproof membrane is used—ANSI A118.4 or better or ISO C2S1 or better unless ANSI A118.1 or ISO C1 is recommended by membrane manufacturer.

  - Use white for light-colored marble, limestone, and other stones where staining or darkening from grey setting materials is possible.
- Epoxy bond coat, when used—ANSI A118.3 or ISO R1 or better. See Water Sensitivity and Fiberglass Mesh Reinforced Stone in the "Natural Stone Tile Selection and Installation Guide."
- Waterproof membrane, when used—ANSI A118.10.
- Mortar bed, metal lath, and cleavage membrane—ANSI A108.1A.
- Metal studs—ASTM C645.

Materials for Green/Sustainable Design

- See "Green Building Standards and Green Product Selection Guide" and consult manufacturers and suppliers for product sustainability and contribution to green building design.
- Consider specifying installation materials that meet ANSI A138.1, the *American National Standard Specifications for Sustainable Ceramic Tiles, Glass Tiles, and Tile Installation Materials.*

Preparation by Other Trades

- Maximum allowable variation in the installation substrate (solid backing)—¼" in 10' from the required plane.

Movement Joint (architect must specify type of joint and show location and details on drawings)

- Movement joints—mandatory according to EJ171.

Installation Specifications

- Tile—ANSI A108.1A, .1B, or .1C. Ungauged stone may require installation per ANSI A108.1A. Material with minimal variation may be allowed per A108.1B with LHT mortar. A108.1B required if waterproof membrane or epoxy bond coat to be used.
- Cementitious grout—ANSI A108.10.
- Waterproof membrane—ANSI A108.13.
- Movement Joints—EJ171 and ASTM C1193.

Notes

- If lath cannot be attached directly to backing, use furring strips. See W231/W241 Stone.
- For thicker mortar bed, see W221 Stone.
- For mortar bed directly over wood or metal studs, see W231/W241 Stone.
- Exterior—not all bonding mortars are suitable for exterior use.
- Exterior—protection of installation may be required to prevent premature exposure of setting materials to moisture.
- When unsanded grout is used, grout joint width must be $^{1}/_{16}$" to ⅛" wide (nominal).

W231/W241-21 STONE

- **Wood or Metal Studs**
- **Mortar Bed**
- **Natural Stone Tile**

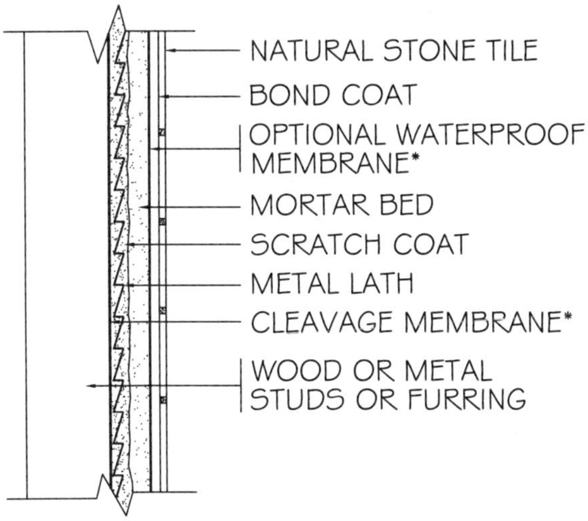

*USE OF A MEMBRANE IS OPTIONAL. SEE MEMBRANE OPTIONS.

Recommended Uses

- For interior walls of wood or metal framing where flattening or trueing of walls is required.
- For areas where wall flatness is critical, such as when tiles with any edge longer than 15" are specified.
- For areas where ungauged stone will be used.

Environmental Exposure Classifications

- Res 1, 2, 5; Com1, 2, 5.
- For Res3 and Com3, see B411 Stone and B414 Stone.
- May be suitable for exterior applications when appropriate precautions are taken, including flashing, expansion joint placement, waterproofing, and material selection. See Notes.
- For installations that may be exposed to staining and/or chemical attack, specify tile and grout suitable for exposure. Consult product manufacturers; see also "Product Selection Guides."

Limitations

- Maximum mortar bed thickness—1" for metal studs, 1½" for wood studs.
- Maximum stud spacing 16" on center.

Membrane Options

- A waterproof membrane (A118.10) may be specified to prevent moisture intrusion and protect adjacent building materials. Specifier shall indicate if complete waterproofing is required, including base flashing and treatment at other termination points.
- If a waterproof membrane is applied over the mortar bed, membrane over framing members may not be required.
- Check with membrane manufacturer for suitability for applicable conditions, as not all membranes are suitable for steam, high-temperature and/or chemical exposure, or exterior use.
- Exterior—check building code—water-resistive barrier, waterproof membrane, or vapor retarder membrane may be required.

Requirements

- Wood studs—dry and well-braced, minimum depth 3½".
- Metal studs—well-braced; 20 gauge (0.033") or heavier; minimum depth 3½" for residential applications or 3⅝" for commercial applications.
- Mortar bed thickness—¾" minimum to 1½" maximum.
- Minimum grout joint width—1/16".

Materials

- **Multiple options exist for membranes, mortars, grouts, and other materials and *must be clearly specified* to be included. If not specifically indicated, optional materials are not included and mortar/grout choice defaults to minimum performance specification indicated. Consider each system component and intended use to determine minimum requirements and to specify options.**
- Natural stone tile—see "Natural Stone Tile Selection and Installation Guide," and consult supplier for application suitability.
- Cementitious grout—ANSI A118.6 or better or ISO CG1 or better. Consider unsanded grout for soft or semisoft stones such as limestone, travertine, marble, onyx, or similar.
- Cementitious bond coat—portland cement paste on a mortar bed that is still workable. For a cured mortar bed, follow recommendations below to select appropriate bonding mortar:
  - When a waterproof membrane is not used—ANSI A118.1 or better or ISO C1 or better.
  - When a waterproof membrane is used—ANSI A118.4 or better or ISO C2S1 or better unless ANSI A118.1 or ISO C1 is recommended by membrane manufacturer.
  - Use white for light-colored marble, limestone, and other stones where staining or darkening from grey setting materials is possible.
- Epoxy bond coat, when used—ANSI A118.3 or ISO R1 or better. See Water Sensitivity and Fiberglass Mesh Reinforced Stone in the "Natural Stone Tile Selection and Installation Guide."
- Water-resistive barrier/vapor retarder membrane, when used—ANSI A108.02-3.8.

- Waterproof membrane, when used—ANSI A118.10.
- Mortar bed, metal lath, and cleavage membrane—ANSI A108.1A.
- Metal studs—ASTM C645.

Materials for Green/Sustainable Design

- See "Green Building Standards and Green Product Selection Guide" and consult manufacturers and suppliers for product sustainability and contribution to green building design.
- Consider specifying installation materials that meet ANSI A138.1, the *American National Standard Specifications for Sustainable Ceramic Tiles, Glass Tiles, and Tile Installation Materials.*

Preparation by Other Trades

- Wall framing shall meet the general framing requirements of ANSI A108.11-4.0-4.3.

Movement Joint (architect must specify type of joint and show location and details on drawings)

- Movement joints—mandatory according to EJ171.

Installation Specifications

- Tile—ANSI A108.1A, .1B, or .1C. Ungauged stone may require installation per ANSI A108.1A. Material with minimal variation may be allowed per A108.1B with LHT mortar. A108.1B required if waterproof membrane or epoxy bond coat to be used.
- Cementitious grout—ANSI A108.10.
- Epoxy mortar/grout—ANSI A108.6.
- Waterproof membrane—ANSI A108.13.
- Movement Joints—EJ171 and ASTM C1193.

Notes

- Exterior—not all bonding mortars are suitable for exterior use.
- Exterior—protection of installation may be required to prevent premature exposure of setting materials to moisture.
- When unsanded grout is used, grout joint width must be $\frac{1}{16}$" to $\frac{1}{8}$" wide (nominal).

W243-21
STONE

- **Wood or Metal Studs**
- **Gypsum Board**
- **Natural Stone Tile**

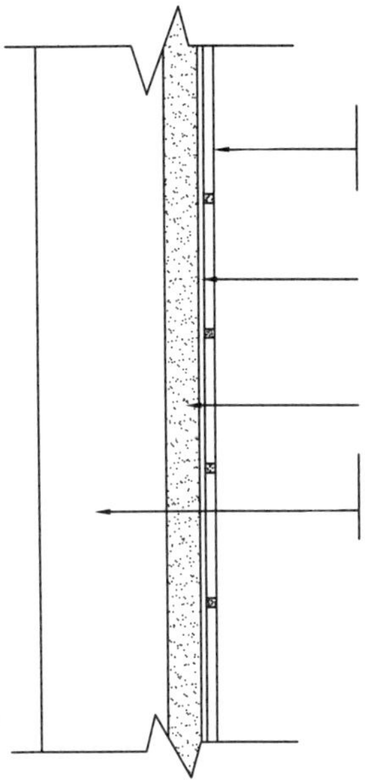

Recommended Uses

- For interior walls in dry areas where gypsum board is the tile backer.
- For fire-resistant, sound-insulated, ceramic tiled walls. See RW800 for fire and sound ratings. (Fire-resistance and sound-insulation ratings calculated on partitions before tiling.)

Environmental Exposure Classifications

- Res1; Com1.
- For installations that may be exposed to staining and/or chemical attack, specify tile and grout suitable for exposure. Consult product manufacturers; see also "Product Selection Guides."

Limitations

- Not for areas exposed to temperatures exceeding 125°F.
- Maximum stud spacing 16" on center.

Requirements

- Wood studs—dry and well-braced, minimum depth 3½."
- Metal studs—well-braced; 20 gauge (0.033") or heavier; minimum depth 3½" for residential applications or 3⅝" for commercial applications.
- Minimum grout joint width—1/16."

Materials

- **Multiple options exist for membranes, mortars, grouts, and other materials and *must be clearly specified* to be included. If not specifically indicated, optional materials are not included and mortar/grout choice defaults to minimum performance specification indicated. Consider each system component and intended use to determine minimum requirements and to specify options.**
- Natural stone tile—see "Natural Stone Tile Selection and Installation Guide," and consult supplier for application suitability.
- Cementitious grout—ANSI A118.6 or better or ISO CG1 or better. Consider unsanded grout for soft or semisoft stones such as limestone, travertine, marble, onyx, or similar.
- Cementitious bond coat:
  - ANSI A118.1 or better or ISO C1 or better.
  - Use white for light-colored marble, limestone, and other stones where staining or darkening from grey setting materials is possible.
- Epoxy bond coat, when used—ANSI A118.3 or ISO R1 or better. See Water Sensitivity and Fiberglass Mesh Reinforced Stone in the "Natural Stone Tile Selection and Installation Guide."
- Gypsum board—ASTM C1396/C1396M. Minimum ½" thick for single layer applications.
- Metal studs—ASTM C645.

Materials for Green/Sustainable Design

- See "Green Building Standards and Green Product Selection Guide" and consult manufacturers and suppliers for product sustainability and contribution to green building design.
- Consider specifying installation materials that meet ANSI A138.1, the *American National Standard Specifications for Sustainable Ceramic Tiles, Glass Tiles, and Tile Installation Materials.*

Preparation by Other Trades

- Gypsum board, single or double layer, to be installed per GA-216.
- Gypsum board face layer joints—treated with tape and joint compound, bedding coat only (no finish coats). Nail heads, one coat only.
- Maximum allowable variation in the tile substrate—⅛" in 10' from the required plane when measured from the high points in the surface.

Movement Joint (architect must specify type of joint and show location and details on drawings)

- Movement joints—mandatory according to EJ171.

Installation Specifications

- Gypsum board—GA-216.
- Tile—ANSI A108.5 or A108.6.
- Cementitious grout—ANSI A108.10.
- Movement Joints—EJ171 and ASTM C1193.

Notes

- As the tile size increases, there is less tolerance for variation in the substrate from the required plane. Epoxy bond coat thickness must be thin and uniform; therefore, substrate flattening may be required when epoxy bond coat is used.
- When unsanded grout is used, grout joint width must be 1⁄16" to 1⁄8" wide (nominal).

W244C-21
STONE

- **Wood or Metal Studs**
- **Cement Backer Board**
- **Natural Stone Tile**

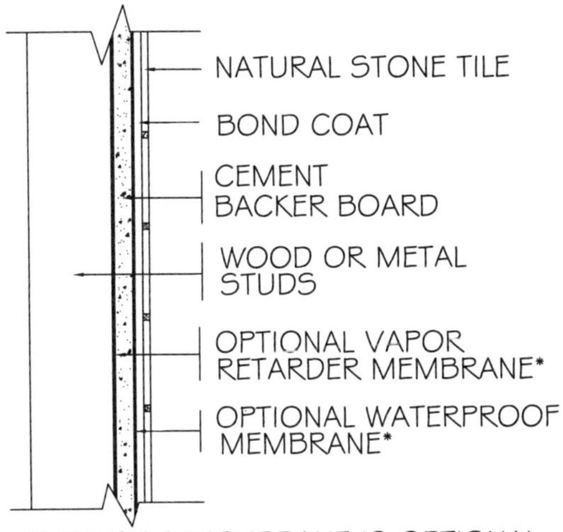

*USE OF A MEMBRANE IS OPTIONAL. SEE MEMBRANE OPTIONS.

Recommended Uses

- For interior walls over wood or metal studs.

Environmental Exposure Classifications

- Res1, 2, 5; Com1, 2, 5.
- For Res3 and Com3, see B412 Stone and B415 Stone.
- For Res6 and Com6, see W244F Stone.
- For installations that may be exposed to staining and/or chemical attack, specify tile and grout suitable for exposure. Consult product manufacturers; see also "Product Selection Guides."

Limitations

- Maximum stud spacing 16" on center.

Membrane Options

- A waterproof membrane (A118.10) or vapor retarder membrane (A108.02-3.8) may be specified to prevent moisture intrusion and protect adjacent building materials. Specifier shall indicate if complete waterproofing is required, including base flashing and treatment at other termination points.
- Check with membrane manufacturer for suitability for applicable conditions, as not all membranes are suitable for steam, high-temperature and/or chemical exposure, or exterior use.

Requirements

- Wood studs—dry and well-braced, minimum depth 3½."
- Metal studs—well-braced; 20 gauge (0.033") or heavier; minimum depth 3½" for residential applications or 3⅝" for commercial applications.
- Surface of units—clean and free of dirt, dust, paint, and oily film.
- Minimum grout joint width—¹⁄₁₆."

Preparation by Backer Board Installers

- Maximum allowable variation in the tile substrate—⅛" in 10' from the required plane when measured from the high points in the surface.
- Horizontal joints—⅛" spacing filled solid and taped with latex-portland cement mortar and 2" alkali-resistant glass fiber mesh tape.
- Vertical joints—fill any space and tape with latex-portland cement mortar and 2" alkali-resistant glass fiber mesh tape.
- Corners—leave space between backer units. Tape joints using skim coat of latex-portland cement mortar, but do not fill.
- Center backer board end or edge joints on framing and stagger joints in adjacent rows so four corners do not come together within the same plane. Space panel ends and edges in accordance with manufacturer's recommendations.

Materials

- **Multiple options exist for membranes, mortars, grouts, and other materials and *must be clearly specified* to be included. If not specifically indicated, optional materials are not included and mortar/grout choice defaults to minimum performance specification indicated. Consider each system component and intended use to determine minimum requirements and to specify options.**
- Natural stone tile—see "Natural Stone Tile Selection and Installation Guide," and consult supplier for application suitability.
- Cementitious grout—ANSI A118.6 or better or ISO CG1 or better. Consider unsanded grout for soft or semisoft stones such as limestone, travertine, marble, onyx, or similar.
- Cementitious bond coat:
  - When a waterproof membrane is not used—ANSI A118.1 or better or ISO C1 or better.
  - When a waterproof membrane is used—ANSI A118.4 or better or ISO C2S1 or better unless ANSI A118.1 or ISO C1 is recommended by membrane manufacturer.
  - Use white for light-colored marble, limestone, and other stones where staining or darkening from grey setting materials is possible.
- Epoxy bond coat, when used—ANSI A118.3 or ISO R1 or better. See Water Sensitivity and Fiberglass Mesh Reinforced Stone in the "Natural Stone Tile Selection and Installation Guide."
- Vapor retarder membrane, when used—ANSI A108.02-3.8.

- Waterproof membrane, when used—ANSI A118.10.
- Cement backer board—ANSI A118.9 or ASTM C1325 (Type B).
- Fasteners—noncorrosive and nonoxidizing.
- Fasteners meeting ASTM F2329 required in wet areas.
- 2" alkali-resistant glass fiber mesh tape.
- Metal studs—ASTM C645.

Materials for Green/Sustainable Design

- See "Green Building Standards and Green Product Selection Guide" and consult manufacturers and suppliers for product sustainability and contribution to green building design.
- Consider specifying installation materials that meet ANSI A138.1, the *American National Standard Specifications for Sustainable Ceramic Tiles, Glass Tiles, and Tile Installation Materials.*

Preparation by Other Trades

- Wall framing shall meet the general framing requirements of ANSI A108.11-4.0–4.3.

Movement Joint (architect must specify type of joint and show location and details on drawings)

- Movement joints—mandatory according to EJ171.

Installation Specifications

- Cement backer board—ANSI A108.11.
- Tile—ANSI A108.5 or A108.6.
- Cementitious grout—ANSI A108.10.
- Waterproof membrane—ANSI A108.13.
- Movement Joints—EJ171 and ASTM C1193.

Notes

- As the tile size increases, there is less tolerance for variation in the substrate from the required plane. Epoxy bond coat thickness must be thin and uniform; therefore, substrate flattening may be required when epoxy bond coat is used.
- When unsanded grout is used, grout joint width must be 1/16" to 1/8" wide (nominal).

W244F-21
STONE

- **Wood or Metal Studs**
- **Fiber-Cement Backer Board**
- **Natural Stone Tile**

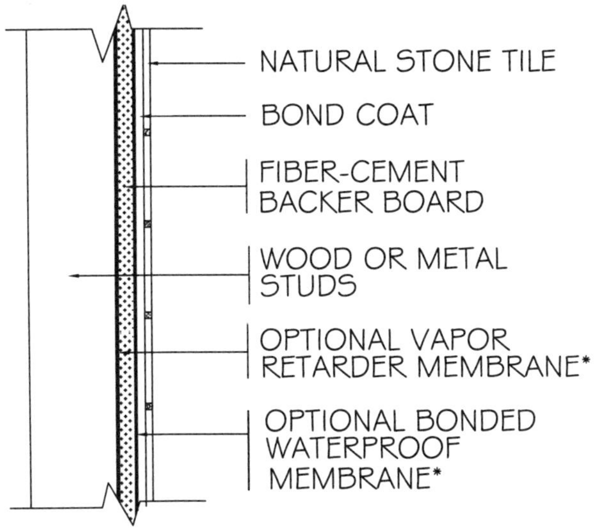

*USE OF A MEMBRANE IS OPTIONAL. SEE MEMBRANE OPTIONS.

Recommended Uses

- For interior walls over wood or metal studs.

Environmental Exposure Classifications

- Res1, 2, 5; Com1, 2, 5.
- For Res3 and Com3, see B412 Stone and B415 Stone.
- For installations that may be exposed to staining and/or chemical attack, specify tile and grout suitable for exposure. Consult product manufacturers; see also "Product Selection Guides."

Limitations

- Maximum stud spacing 16" on center.

Membrane Options

- A waterproof membrane (A118.10) or vapor retarder membrane (A108.02-3.8) may be specified to prevent moisture intrusion and protect adjacent building materials. Specifier shall indicate if complete waterproofing is required, including base flashing and treatment at other termination points.
- Check with membrane manufacturer for suitability for applicable conditions, as not all membranes are suitable for steam, high-temperature and/or chemical exposure, or exterior use.

Requirements

- Wood studs—dry and well-braced, minimum depth 3½."
- Metal studs—well-braced; 20 gauge (0.033") or heavier; minimum depth 3½" for residential applications or 3⅝" for commercial applications.
- Surface of units—clean and free of dirt, dust, paint, and oily film.
- Minimum grout joint width—¹⁄₁₆."

Preparation by Backer Board Installers

- Maximum allowable variation in the tile substrate—⅛" in 10' from the required plane when measured from the high points in the surface.
- Horizontal joints—⅛" spacing filled solid and taped with latex-portland cement mortar and 2" alkali-resistant glass fiber mesh tape.
- Vertical joints—fill any space and tape with latex-portland cement mortar and 2" alkali-resistant glass fiber mesh tape.
- Corners—leave space between backer units. Tape joints using skim coat of latex-portland cement mortar, but do not fill.
- Center backer board end or edge joints on framing and stagger joints in adjacent rows so four corners do not come together within the same plane. Space panel ends and edges in accordance with manufacturer's recommendations.

Materials

- **Multiple options exist for membranes, mortars, grouts, and other materials and *must be clearly specified* to be included. If not specifically indicated, optional materials are not included and mortar/grout choice defaults to minimum performance specification indicated. Consider each system component and intended use to determine minimum requirements and to specify options.**
- Natural stone tile—see "Natural Stone Tile Selection and Installation Guide," and consult supplier for application suitability.
- Cementitious grout—ANSI A118.6 or better or ISO CG1 or better. Consider unsanded grout for soft or semisoft stones such as limestone, travertine, marble, onyx, or similar.
- Cementitious bond coat:
  - When a waterproof membrane is not used—ANSI A118.1 or better or ISO C1 or better.
  - When a waterproof membrane is used—ANSI A118.4 or better or ISO C2S1 or better unless ANSI A118.1 or ISO C1 is recommended by membrane manufacturer.
  - Use white for light-colored marble, limestone, and other stones where staining or darkening from grey setting materials is possible.
- Epoxy bond coat, when used—ANSI A118.3 or ISO R1 or better. See Water Sensitivity and Fiberglass Mesh Reinforced Stone in the "Natural Stone Tile Selection and Installation Guide."
- Vapor retarder membrane, when used—ANSI A108.02-3.8.

- Waterproof membrane, when used—ANSI A118.10.
- Fiber cement backer board—ASTM C1288.
- Fasteners—noncorrosive and nonoxidizing.
- Fasteners meeting ASTM F2329 required in wet areas.
- 2" alkali-resistant glass fiber mesh tape.
- Metal studs—ASTM C645.

Materials for Green/Sustainable Design

- See "Green Building Standards and Green Product Selection Guide" and consult manufacturers and suppliers for product sustainability and contribution to green building design.
- Consider specifying installation materials that meet ANSI A138.1, the *American National Standard Specifications for Sustainable Ceramic Tiles, Glass Tiles, and Tile Installation Materials.*

Preparation by Other Trades

- Wall framing shall meet the general framing requirements of ANSI A108.11-4.0–4.3.

Movement Joint (architect must specify type of joint and show location and details on drawings)

- Movement joints—mandatory according to EJ171.

Installation Specifications

- Fiber cement backer board—manufacturer's directions.
- Tile—ANSI A108.5 or A108.6.
- Cementitious grout—ANSI A108.10.
- Waterproof membrane—ANSI A108.13.
- Movement Joints—EJ171 and ASTM C1193.

Notes

- As the tile size increases, there is less tolerance for variation in the substrate from the required plane. Epoxy bond coat thickness must be thin and uniform; therefore, substrate flattening may be required when epoxy bond coat is used.
- When unsanded grout is used, grout joint width must be 1/16" to 1/8" wide (nominal).

W245-21 STONE

- **Wood or Metal Studs**
- **Coated Glass Mat Water-Resistant Gypsum Backer Board**
- **Natural Stone Tile**

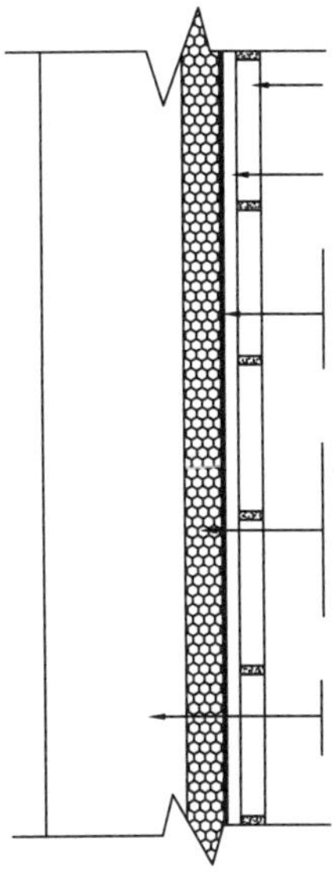

*USE OF A MEMBRANE IS OPTIONAL. SEE MEMBRANE OPTIONS.

Recommended Uses

- For interior walls over wood or metal studs.

Environmental Exposure Classifications

- Res1, 2; Com1, 2.
- For Res3 and Com3, see B419 Stone and B420 Stone.
- For installations that may be exposed to staining and/or chemical attack, specify tile and grout suitable for exposure. Consult product manufacturers; see also "Product Selection Guides."

Limitations

- Maximum stud spacing 16" on center.
- Not for areas exposed to temperatures exceeding 125°F.

Membrane Options

- A waterproof membrane (A118.10) may be specified to prevent moisture intrusion through seams, corners, fasteners, and other penetrations and to protect adjacent building materials. Follow backer board and membrane manufacturer's waterproofing requirements. Specifier shall indicate if complete waterproofing is required, including base flashing and treatment at other termination points.
- Check with membrane manufacturer for suitability for applicable conditions, as not all membranes are suitable for steam, high-temperature and/or chemical exposure, or exterior use.
- Some liquid-applied waterproof membranes will require extended cure time. Consult membrane manufacturer.

Requirements

- Wood studs—dry and well-braced, minimum depth 3½."
- Metal studs—well-braced; 20 gauge (0.033") or heavier; minimum depth 3½" for residential applications or 3⅝" for commercial applications.
- Caulk or seal penetrations, corners, and abutments to dissimilar materials with flexible sealant.
- Minimum grout joint width—1/16."

Preparation by Backer Board Installers

- Maximum allowable variation in the tile substrate—⅛" in 10' from the required plane when measured from the high points in the surface.
- Fit ends and edges closely to eliminate gaps between panels, including in corners.
- Tape seams and corners with 2" alkali-resistant glass fiber mesh tape and cementitious bonding material, and skim coat over fastener heads, unless waterproofing has been applied.
- Center backer board end or edge joints on framing and stagger joints in adjacent rows so four corners do not come together within the same plane. Space panel ends and edges in accordance with manufacturer's recommendations.

Materials

- **Multiple options exist for membranes, mortars, grouts, and other materials and *must be clearly specified* to be included. If not specifically indicated, optional materials are not included and mortar/grout choice defaults to minimum performance specification indicated. Consider each system component and intended use to determine minimum requirements and to specify options.**
- Natural stone tile—see "Natural Stone Tile Selection and Installation Guide," and consult supplier for application suitability.
- Cementitious grout—ANSI A118.6 or better or ISO CG1 or better. Consider unsanded grout for soft or semisoft stones such as limestone, travertine, marble, onyx, or similar.
- Cementitious bond coat:
  - When a waterproof membrane is not used—ANSI A118.4 or better or ISO C2S1 or better.
  - When a waterproof membrane is used—ANSI A118.4 or better or ISO C2S1 or better unless ANSI A118.1 or ISO C1 is recommended by membrane manufacturer.
  - Use white for light-colored marble, limestone, and other stones where staining or darkening from grey setting materials is possible.

- Epoxy bond coat, when used—ANSI A118.3 or ISO R1 or better. See Water Sensitivity and Fiberglass Mesh Reinforced Stone in the "Natural Stone Tile Selection and Installation Guide."
- Waterproof membrane, when used—ANSI A118.10.
- Coated glass mat, water-resistant gypsum backer board—ASTM C1178.
- Fasteners—noncorrosive and nonoxidizing.
- Fasteners meeting ASTM F2329 required in wet areas.
- 2" alkali-resistant glass fiber mesh tape.
- Flexible sealant—ASTM C920.
- Metal studs—ASTM C645.

Materials for Green/Sustainable Design

- See "Green Building Standards and Green Product Selection Guide" and consult manufacturers and suppliers for product sustainability and contribution to green building design.
- Consider specifying installation materials that meet ANSI A138.1, the *American National Standard Specifications for Sustainable Ceramic Tiles, Glass Tiles, and Tile Installation Materials.*

Preparation by Other Trades

- Wall framing shall meet the general framing requirements of ANSI A108.11-4.0–4.3.

Movement Joint (architect must specify type of joint and show location and details on drawings)

- Movement joints—mandatory according to EJ171.

Installation Specifications

- Coated glass mat water-resistant gypsum backer board—manufacturer's directions.
- Tile—ANSI A108.5 or A108.6.
- Cementitious grout—ANSI A108.10.
- Waterproof membrane—ANSI A108.13.
- Movement Joints—EJ171 and ASTM C1193.

Notes

- As the tile size increases, there is less tolerance for variation in the substrate from the required plane. Epoxy bond coat thickness must be thin and uniform; therefore, substrate flattening may be required when epoxy bond coat is used.
- When unsanded grout is used, grout joint width must be $\frac{1}{16}$" to $\frac{1}{8}$" wide (nominal).

W246-21
STONE

- **Wood or Metal Studs**
- **Cementitious-Coated Extruded Foam Backer Board**
- **Natural Stone Tile**

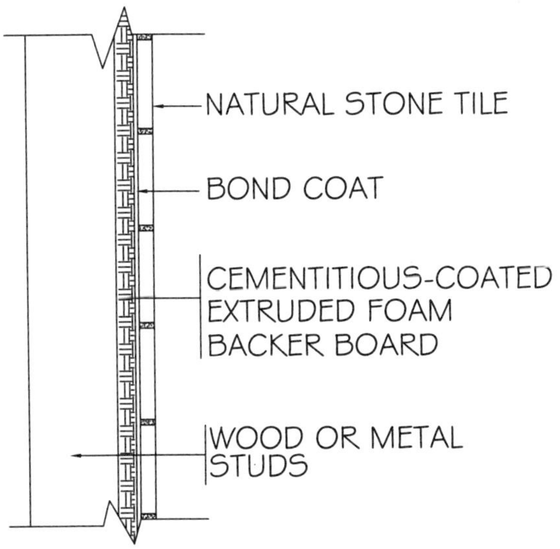

Recommended Uses

- For interior walls over wood or metal studs.

Environmental Exposure Classifications

- Res1, 2; Com1, 2.
- For Res3 and Com3, see B425 Stone and B426 Stone.
- For installations that may be exposed to staining and/or chemical attack, specify tile and grout suitable for exposure. Consult product manufacturers; see also "Product Selection Guides."

Limitations

- Maximum stud spacing 16" on center.

Requirements

- Wood studs—dry and well-braced, minimum depth 3½."
- Metal studs—well-braced; 20 gauge (0.033") or heavier; minimum depth 3½" for residential applications or 3⅝" for commercial applications.
- Minimum grout joint width—1/16."

Preparation by Backer Board Installers

- Fasten backer board to studs with board manufacturer's recommended fasteners.
- Maximum allowable variation in the tile substrate—⅛" in 10' from the required plane when measured from the high points in the surface.
- Fit ends and edges tightly together, including in corners. Apply a bead of manufacturer-recommended sealant in panel joints as panels are being installed, and over fastener heads, per backer board manufacturer's instructions.
- Joints may be taped with 2" alkali-resistant glass fiber mesh tape and cementitious bonding material in dry areas.
- Center backer board end or edge joints on framing and stagger joints in adjacent rows so four corners do not come together within the same plane. Space panel ends and edges in accordance with manufacturer's recommendations.

Materials

- **Multiple options exist for membranes, mortars, grouts, and other materials and *must be clearly specified* to be included. If not specifically indicated, optional materials are not included and mortar/grout choice defaults to minimum performance specification indicated. Consider each system component and intended use to determine minimum requirements and to specify options.**
- Natural stone tile—see "Natural Stone Tile Selection and Installation Guide," and consult supplier for application suitability.
- Cementitious grout—ANSI A118.6 or better or ISO CG1 or better. Consider unsanded grout for soft or semisoft stones such as limestone, travertine, marble, onyx, or similar.
- Cementitious bond coat:
  - ANSI A118.1 or better or ISO C1 or better.
  - Use white for light-colored marble, limestone, and other stones where staining or darkening from grey setting materials is possible.
- Epoxy bond coat, when used—ANSI A118.3 or ISO R1 or better. See Water Sensitivity and Fiberglass Mesh Reinforced Stone in the "Natural Stone Tile Selection and Installation Guide."
- Cementitious-coated extruded foam backer board—ASTM C578.
- Fasteners—noncorrosive and nonoxidizing.
- 2" alkali-resistant glass fiber mesh tape, when used.
- Flexible sealant—must be recommended by backer board manufacturer.
- Metal studs—ASTM C645.

Materials for Green/Sustainable Design

- See "Green Building Standards and Green Product Selection Guide" and consult manufacturers and suppliers for product sustainability and contribution to green building design.
- Consider specifying installation materials that meet ANSI A138.1, the *American National Standard Specifications for Sustainable Ceramic Tiles, Glass Tiles, and Tile Installation Materials.*

Preparation by Other Trades

- Wall framing shall meet the general framing requirements of ANSI A108.11-4.0–4.3.

Movement Joint (architect must specify type of joint and show location and details on drawings)

- Movement joints—mandatory according to EJ171.

Installation Specifications

- Cementitious-coated extruded foam backer board—manufacturer's directions.
- Tile—ANSI A108.5 or A108.6.
- Cementitious grout—ANSI A108.10.
- Movement Joints—EJ171 and ASTM C1193.

Notes

- As the tile size increases, there is less tolerance for variation in the substrate from the required plane. Epoxy bond coat thickness must be thin and uniform; therefore, substrate flattening may be required when epoxy bond coat is used.
- When unsanded grout is used, grout joint width must be 1/16" to 1/8" wide (nominal).

W247-21
STONE

- **Wood or Metal Studs**
- **Fiber-Reinforced Water-Resistant Gypsum Backer Board**
- **Natural Stone Tile**

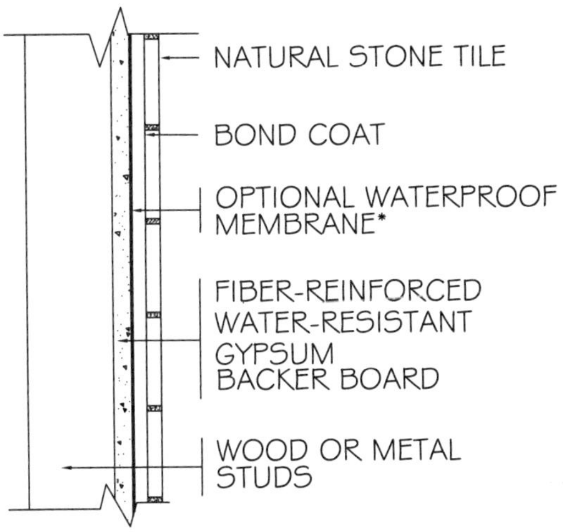

*USE OF A MEMBRANE IS OPTIONAL. SEE MEMBRANE OPTIONS.

Recommended Uses

- For interior walls over wood or metal studs.

Environmental Exposure Classifications

- Res1, 2; Com1, 2.
- For Res3, see B430 Stone and B431 Stone.
- For installations that may be exposed to staining and/or chemical attack, specify tile and grout suitable for exposure. Consult product manufacturers; see also "Product Selection Guides."

Limitations

- Maximum stud spacing 16" on center.
- Not for areas exposed to temperatures exceeding 125°F.

Membrane Options

- A waterproof membrane (A118.10) may be specified to prevent moisture intrusion and protect adjacent building materials. Specifier shall indicate if complete waterproofing is required, including base flashing and treatment at other termination points.
- Check with membrane manufacturer for suitability for applicable conditions, as not all membranes are suitable for steam, high-temperature and/or chemical exposure, or exterior use.

Requirements

- Wood studs—dry and well-braced, minimum depth 3½."
- Metal studs—well-braced; 20 gauge (0.033") or heavier; minimum depth 3½" for residential applications or 3⅝" for commercial applications.
- Caulk or seal penetrations and abutments to dissimilar materials.
- Minimum grout joint width—1/16."

Preparation by Backer Board Installers

- Maximum allowable variation in the tile substrate—⅛" in 10' from the required plane when measured from the high points in the surface.
- Horizontal joints—⅛" spacing filled solid and taped with latex-portland cement mortar and 2" alkali-resistant glass fiber mesh tape.
- Vertical joints—fill any space and tape with latex-portland cement mortar and 2" alkali-resistant glass fiber mesh tape.
- Corners—leave space between backer units. Tape joints using skim coat of latex-portland cement mortar, but do not fill.
- Center backer board end or edge joints on framing and stagger joints in adjacent rows so four corners do not come together within the same plane. Space panel ends and edges in accordance with manufacturer's recommendations.

Materials

- **Multiple options exist for membranes, mortars, grouts, and other materials and *must be clearly specified* to be included. If not specifically indicated, optional materials are not included and mortar/grout choice defaults to minimum performance specification indicated. Consider each system component and intended use to determine minimum requirements and to specify options.**
- Natural stone tile—see "Natural Stone Tile Selection and Installation Guide," and consult supplier for application suitability.
- Cementitious grout—ANSI A118.6 or better or ISO CG1 or better. Consider unsanded grout for soft or semisoft stones such as limestone, travertine, marble, onyx, or similar.
- Cementitious bond coat:
  - When a waterproof membrane is not used—ANSI A118.4 or better or ISO C2 or better.
  - When a waterproof membrane is used—ANSI A118.4 or better or ISO C2S1 or better unless ANSI A118.1 or ISO C1 is recommended by membrane manufacturer.
  - Use white for light-colored marble, limestone, and other stones where staining or darkening from grey setting materials is possible.
- Epoxy bond coat, when used—ANSI A118.3 or ISO R1 or better. See Water Sensitivity and Fiberglass Mesh Reinforced Stone in the "Natural Stone Tile Selection and Installation Guide."
- Waterproof membrane, when used—ANSI A118.10.

- Fiber-reinforced water-resistant gypsum backer board—ASTM C1278 (Paragraph 6.1).
- Fasteners—noncorrosive and nonoxidizing.
- 2" alkali-resistant glass fiber mesh tape.
- Metal studs—ASTM C645.
- Flexible sealant—ASTM C920.

Materials for Green/Sustainable Design

- See "Green Building Standards and Green Product Selection Guide" and consult manufacturers and suppliers for product sustainability and contribution to green building design.
- Consider specifying installation materials that meet ANSI A138.1, the *American National Standard Specifications for Sustainable Ceramic Tiles, Glass Tiles, and Tile Installation Materials.*

Preparation by Other Trades

- Wall framing shall meet the general framing requirements of ANSI A108.11-4.0–4.3.

Movement Joint (architect must specify type of joint and show location and details on drawings)

- Movement joints—mandatory in accordance with EJ171.

Installation Specifications

- Fiber-reinforced water-resistant gypsum backer board—manufacturer's directions.
- Tile—ANSI A108.5 or A108.6.
- Cementitious grout—ANSI A108.10.
- Waterproof membrane—ANSI A108.13.
- Movement Joints—EJ171 and ASTM C1193.

Notes

- As the tile size increases, there is less tolerance for variation in the substrate from the required plane. Epoxy bond coat thickness must be thin and uniform; therefore, substrate flattening may be required when epoxy bond coat is used.
- When unsanded grout is used, grout joint width must be $\frac{1}{16}$" to $\frac{1}{8}$" wide (nominal).

W248-21
STONE

- **Wood or Metal Studs**
- **Glass Mat Water-Resistant Gypsum Backer Board**
- **Natural Stone Tile**

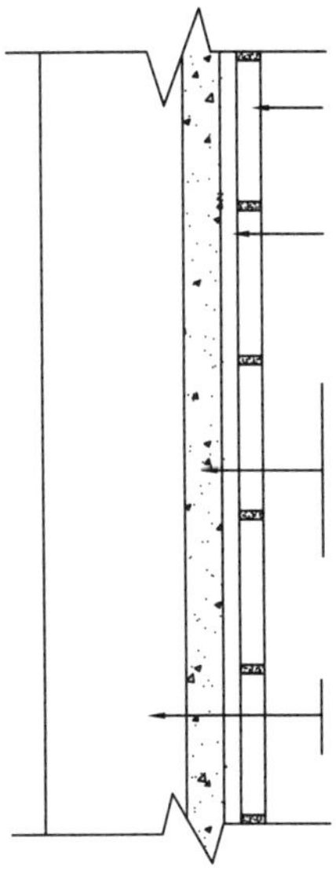

Recommended Uses

- For interior walls over wood or metal studs.

Environmental Exposure Classifications

- Res1; Com1.
- For installations that may be exposed to staining and/or chemical attack, specify tile and grout suitable for exposure. Consult product manufacturers; see also "Product Selection Guides."

Limitations

- Maximum stud spacing 16" on center.
- Not for areas exposed to temperatures exceeding 125°F.

Requirements

- Wood studs—dry and well-braced, minimum depth 3½."
- Metal studs—well-braced; 20 gauge (0.033") or heavier; minimum depth 3½" for residential applications or 3⅝" for commercial applications.
- Caulk or seal penetrations, corners, and abutments to dissimilar materials with flexible sealant.
- Minimum grout joint width—1/16."

Preparation by Backer Board Installers

- Maximum allowable variation in the tile substrate—⅛" in 10' from the required plane when measured from the high points in the surface.
- Fit ends and edges closely to eliminate gaps between panels, including in corners.
- Tape seams and corners with 2" alkali-resistant glass fiber mesh tape and skim coat with cementitious bonding material.
- Center backer board end or edge joints on framing and stagger joints in adjacent rows so four corners do not come together within the same plane. Space panel ends and edges in accordance with manufacturer's recommendations.

Materials

- **Multiple options exist for membranes, mortars, grouts, and other materials and *must be clearly specified* to be included. If not specifically indicated, optional materials are not included and mortar/grout choice defaults to minimum performance specification indicated. Consider each system component and intended use to determine minimum requirements and to specify options.**
- Natural stone tile—see "Natural Stone Tile Selection and Installation Guide," and consult supplier for application suitability.
- Cementitious grout—ANSI A118.6 or better or ISO CG1 or better. Consider unsanded grout for soft or semisoft stones such as limestone, travertine, marble, onyx, or similar.
- Cementitious bond coat:
  - ANSI A118.4 or better or ISO C2 or better.
  - Use white for light-colored marble, limestone, and other stones where staining or darkening from grey setting materials is possible.
- Epoxy bond coat, when used—ANSI A118.3 or ISO R1 or better. See Water Sensitivity and Fiberglass Mesh Reinforced Stone in the "Natural Stone Tile Selection and Installation Guide."
- Glass mat, water-resistant gypsum backer board—ASTM C1658.
- Fasteners—noncorrosive and nonoxidizing.
- 2" alkali-resistant glass fiber mesh tape.
- Flexible sealant—ASTM C920.
- Metal studs—ASTM C645.

Materials for Green/Sustainable Design

- See "Green Building Standards and Green Product Selection Guide" and consult manufacturers and suppliers for product sustainability and contribution to green building design.
- Consider specifying installation materials that meet ANSI A138.1, the *American National Standard Specifications for Sustainable Ceramic Tiles, Glass Tiles, and Tile Installation Materials.*

Preparation by Other Trades

- Wall framing shall meet the general framing requirements of ANSI A108.11-4.0–4.3.

Movement Joint (architect must specify type of joint and show location and details on drawings)

- Movement joints—mandatory according to EJ171.

Installation Specifications

- Glass mat water-resistant gypsum backer board—manufacturer's directions.
- Tile—ANSI A108.5 or A108.6.
- Cementitious grout—ANSI A108.10.
- Movement Joints—EJ171 and ASTM C1193.

Notes

- As the tile size increases, there is less tolerance for variation in the substrate from the required plane. Epoxy bond coat thickness must be thin and uniform; therefore, substrate flattening may be required when epoxy bond coat is used.
- When unsanded grout is used, grout joint width must be $\frac{1}{16}$" to $\frac{1}{8}$" wide (nominal).

W260-21
STONE

- **Wood or Metal Studs**
- **Cement Backer Board**
- **Spot-Bonding Epoxy**
- **Natural Stone Tile**

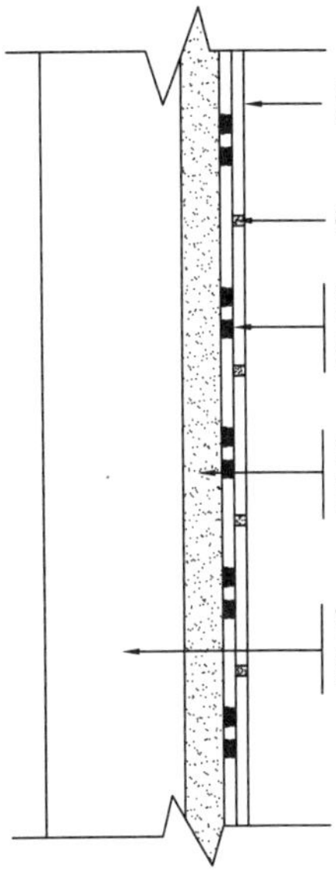

Recommended Uses

- For interior walls where cement backer board is the tile backer and large format tiles are specified.

Environmental Exposure Classifications

- Res1; Com1.
- For installations that may be exposed to staining and/or chemical attack, specify tile and grout suitable for exposure. Consult product manufacturers; see also "Product Selection Guides."

Limitations

- Will not withstand impact.
- Maximum stud spacing 16" on center.

Requirements

- Epoxy must be recommended by the manufacturer for spot-bonding.
- Follow epoxy manufacturer's coverage requirements.
- Wood studs—dry and well-braced, minimum depth 3½."
- Metal studs—well-braced; 20 gauge (0.033") or heavier; minimum depth 3½" for residential applications or 3⅝" for commercial applications.
- Surface of units—clean and free of dirt, dust, paint, and oily film.
- Minimum grout joint width—1/16."

Preparation by Backer Board Installers

- Maximum allowable variation in the tile substrate—¼" in 10' from the required plane.
- Horizontal joints—⅛" spacing filled solid and taped with latex-portland cement mortar and 2" alkali-resistant glass fiber mesh tape.
- Vertical joints—fill any space and tape with latex-portland cement mortar and 2" alkali-resistant glass fiber mesh tape.
- Corners—leave space between backer units. Tape joints using skim coat of latex-portland cement mortar, but do not fill.
- Center backer board end or edge joints on framing and stagger joints in adjacent rows so four corners do not come together within the same plane. Space panel ends and edges in accordance with manufacturer's recommendations.

Materials

- **Multiple options exist for membranes, mortars, grouts, and other materials and *must be clearly specified* to be included. If not specifically indicated, optional materials are not included and mortar/grout choice defaults to minimum performance specification indicated. Consider each system component and intended use to determine minimum requirements and to specify options.**
- Natural stone tile—see "Natural Stone Tile Selection and Installation Guide," and consult supplier for application suitability.
- Cementitious grout—ANSI A118.6 or better or ISO CG1 or better. Consider unsanded grout for soft or semisoft stones such as limestone, travertine, marble, onyx, or similar.
- Sealant, when used as grout—ASTM C920.
- Spot bonding epoxy adhesive—epoxy recommended by manufacturer for spot bonding.
- Cement backer board—ANSI A118.9 or ASTM C1325 (Type B).
- Fasteners—noncorrosive and nonoxidizing.
- 2" alkali-resistant glass fiber mesh tape.
- Cementitious mortar (for taping joints)—ANSI A118.4 or better or ISO C2 or better.
- Metal studs—ASTM C645.

Materials for Green/Sustainable Design

- See "Green Building Standards and Green Product Selection Guide" and consult manufacturers and suppliers for product sustainability and contribution to green building design.
- Consider specifying installation materials that meet ANSI A138.1, the *American National Standard Specifications for Sustainable Ceramic Tiles, Glass Tiles, and Tile Installation Materials.*

Preparation by Other Trades

- Wall framing shall meet the general framing requirements of ANSI A108.11-4.0–4.3.

Movement Joint (architect must specify type of joint and show location and details on drawings)

- Movement joints—mandatory according to EJ171.

Installation Specifications

- Cement backer board—ANSI A108.11.
- Tile—epoxy manufacturer's directions.
- Cementitious grout—ANSI A108.10.
- Sealant—sealant and epoxy manufacturer's instructions for using sealant as grout.
- Movement Joints—EJ171 and ASTM C1193.

Notes

- When unsanded grout is used, grout joint width must be $\frac{1}{16}$" to $\frac{1}{8}$" wide (nominal).

B413-21 STONE

- **Bathtub Walls (No Showerhead) Gypsum Board or Glass Mat Gypsum Board**
- **Natural Stone Tile**

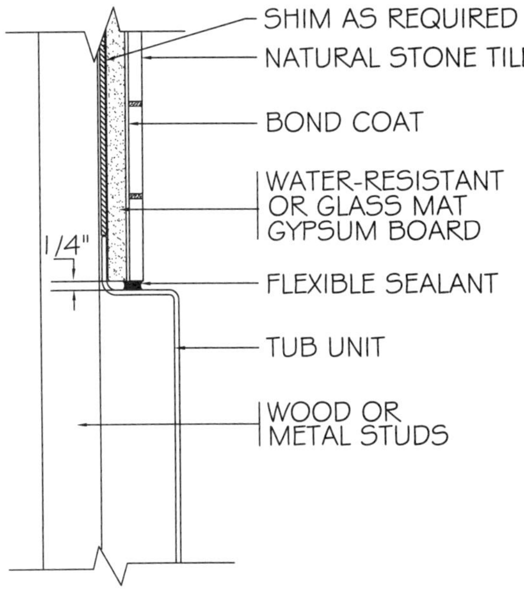

Recommended Uses

- For tub enclosures without a shower head where gypsum board or glass mat water-resistant gypsum backer board is the tile backer.
- For fire-resistant, sound-insulated, ceramic tiled walls. See RW800 for fire and sound ratings. (Fire-resistance and sound-insulation ratings calculated on partitions before tiling.)

Environmental Exposure Classifications

- Res1; Com1.
- For installations that may be exposed to staining and/or chemical attack, specify tile and grout suitable for exposure. Consult product manufacturers; see also "Product Selection Guides."

Limitations

- Not for areas exposed to temperatures exceeding 125°F.
- Maximum stud spacing 16" on center.

Requirements

- Wood studs—dry and well-braced, minimum depth 3½."
- Metal studs—well-braced; 20 gauge (0.033") or heavier; minimum depth 3½" for residential applications or 3⅝" for commercial applications.
- Minimum grout joint width—1/16."

Materials

- **Multiple options exist for membranes, mortars, grouts, and other materials and *must be clearly specified* to be included. If not specifically indicated, optional materials are not included and mortar/grout choice defaults to minimum performance specification indicated. Consider each system component and intended use to determine minimum requirements and to specify options.**
- Natural stone tile—see "Natural Stone Tile Selection and Installation Guide," and consult supplier for application suitability.
- Cementitious grout—ANSI A118.6 or better or ISO CG1 or better. Consider unsanded grout for soft or semisoft stones such as limestone, travertine, marble, onyx, or similar.
- Cementitious bond coat:
  - ANSI A118.1 or better or ISO C1 or better.
  - Use white for light-colored marble, limestone, and other stones where staining or darkening from grey setting materials is possible.
- Epoxy bond coat, when used—ANSI A118.3 or ISO R1 or better. See Water Sensitivity and Fiberglass Mesh Reinforced Stone in the "Natural Stone Tile Selection and Installation Guide."
- Gypsum board, when used—ASTM C1396/C1396M. Minimum ½" thick for single layer applications.
- Glass mat water-resistant gypsum backer board, when used—ASTM C1658.
- Metal studs—ASTM C645.
- Flexible mildew-resistant sealant—ASTM C920.

Materials for Green/Sustainable Design

- See "Green Building Standards and Green Product Selection Guide" and consult manufacturers and suppliers for product sustainability and contribution to green building design.
- Consider specifying installation materials that meet ANSI A138.1, the *American National Standard Specifications for Sustainable Ceramic Tiles, Glass Tiles, and Tile Installation Materials.*

Preparation by Other Trades

- Bathtub—install level and support with metal hangers or on end grain wood blocks.
- Fireproofing behind tub when required.
- Maximum allowable variation in the tile substrate—⅛" in 10' from the required plane when measured from the high points in the surface.
- All openings cut in backing board for plumbing and all cut joints between adjoining pieces—seal with adhesive or other materials recommended by manufacturer of backing board.
- Gypsum board face layer joints—treated with tape and joint compound, bedding coat only (no finish coats). Nail heads, one coat only.
- Apply water-resistant gypsum backing board horizontally with the factory paperbound edge spaced a minimum of ¼" above the lip of the tub.

Movement Joint (architect must specify type of joint and show location and details on drawings)

- Movement joints—mandatory according to EJ171.

Installation Specifications

- Gypsum board—GA-216.
- Glass mat water-resistant gypsum backer board—manufacturer's directions.
- Tile—ANSI A108.5 or A108.6.
- Cementitious grout—ANSI A108.10.
- Movement Joints—EJ171 and ASTM C1193.

Notes

- As the tile size increases, there is less tolerance for variation in the substrate from the required plane. Epoxy bond coat thickness must be thin and uniform; therefore, substrate flattening may be required when epoxy bond coat is used.
- When unsanded grout is used, grout joint width must be $\frac{1}{16}$" to $\frac{1}{8}$" wide (nominal).

B411-21
STONE

- **Bathtub or Shower with Prefabricated Receptor**
- **Mortar Bed**
- **Natural Stone Tile**

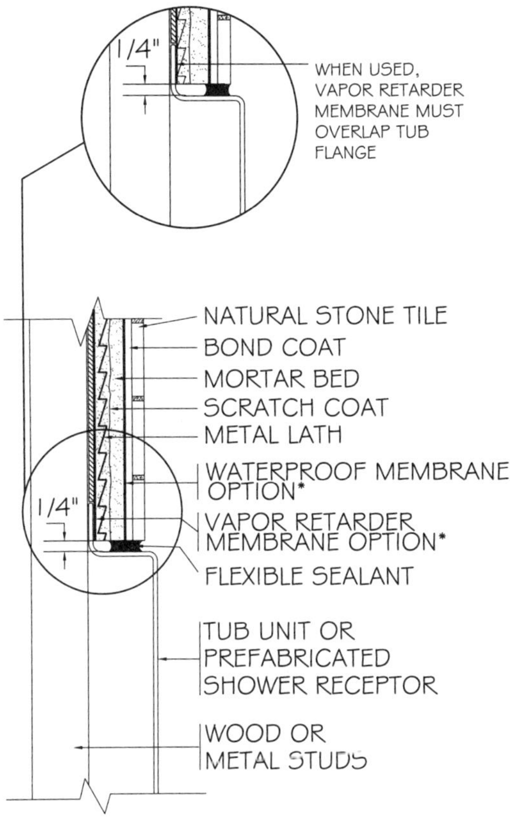

*USE OF A MEMBRANE IS REQUIRED. SEE MEMBRANE OPTIONS.

SHOWER RECEPTORS, CURBS, SEATS, ETC., MUST BE PROPERLY WATERPROOFED AND INSTALLED TO AVOID WATER DAMAGE TO ADJACENT BUILDING MATERIALS. SEE COMMON SHOWER CONFIGURATIONS SECTION.

Recommended Uses

- For tub enclosures or showers with prefabricated receptors, where flattening or trueing of walls is required.
- For areas where wall flatness is critical, such as when tiles with any edge longer than 15" are specified.
- For areas where ungauged stone will be used.
- For showers that do not have a prefabricated receptor, see B414 Stone.

Environmental Exposure Classifications

- Res1, 2, 3, 5; Com1, 2, 3, 5.
- For installations that may be exposed to staining and/or chemical attack, specify tile and grout suitable for exposure. Consult product manufacturers; see also "Product Selection Guides."

Limitations

- Maximum mortar bed thickness—1" for metal studs, 1½" for wood studs.
- Maximum stud spacing 16" on center.

Membrane Options

- A waterproof membrane (A118.10) or vapor retarder membrane (A108.02-3.8) must be specified for walls to prevent moisture intrusion and protect adjacent building materials. Specifier shall indicate if complete waterproofing of walls is required, including treatment at termination points.
- If a waterproof membrane is applied over the mortar bed, membrane over framing members may not be required.
- Check with membrane manufacturer for suitability for applicable conditions, as not all membranes are suitable for steam, high-temperature and/or chemical exposure, when used, or exterior use.

Requirements

- Membrane behind mortar bed, when used, must lap over face of flange of tub or prefabricated shower receptor.
- Wood studs—dry and well-braced, minimum depth 3½."
- Metal studs—well-braced; 20 gauge (0.033") or heavier; minimum depth 3½" for residential applications or 3⅝" for commercial applications.
- Mortar bed thickness—¾" minimum to 1½" maximum.
- Minimum grout joint width—1/16."

Materials

- **Multiple options exist for membranes, mortars, grouts, and other materials and *must be clearly specified* to be included. If not specifically indicated, optional materials are not included and mortar/grout choice defaults to minimum performance specification indicated. Consider each system component and intended use to determine minimum requirements and to specify options.**
- Natural stone tile—see "Natural Stone Tile Selection and Installation Guide," and consult supplier for application suitability.
- Cementitious grout—ANSI A118.6 or better or ISO CG1 or better. Consider unsanded grout for soft or semisoft stones such as limestone, travertine, marble, onyx, or similar.
- Cementitious bond coat—portland cement paste on a mortar bed that is still workable. For a cured mortar bed, follow recommendations below to select appropriate bonding mortar:
  - When a waterproof membrane is not used—ANSI A118.1 or better or ISO C1 or better.
  - When a waterproof membrane is used—ANSI

A118.4 or better or ISO C2S1 or better unless ANSI A118.1 or ISO C1 is recommended by membrane manufacturer.

- Use white for light-colored marble, limestone, and other stones where staining or darkening from grey setting materials is possible.

- Epoxy bond coat, when used—ANSI A118.3 or ISO R1 or better. See Water Sensitivity and Fiberglass Mesh Reinforced Stone in the "Natural Stone Tile Selection and Installation Guide."
- Vapor retarder membrane, when used—ANSI A108.02-3.8.
- Waterproof membrane, when used—ANSI A118.10.
- Mortar bed, metal lath, and cleavage membrane—ANSI A108.1A.
- Metal studs—ASTM C645.
- Flexible mildew-resistant sealant—ASTM C920.

Materials for Green/Sustainable Design

- See "Green Building Standards and Green Product Selection Guide" and consult manufacturers and suppliers for product sustainability and contribution to green building design.
- Consider specifying installation materials that meet ANSI A138.1, the *American National Standard Specifications for Sustainable Ceramic Tiles, Glass Tiles, and Tile Installation Materials.*

Preparation by Other Trades

- Wall framing shall meet the general framing requirements of ANSI A108.11-4.0-4.3.
- Opening for recessed bathtubs—not to exceed ½" more than total length of tub.
- Bathtub—install level and support with metal hangers or on end grain wood blocks.
- Fireproofing behind tub when required.

Movement Joint (architect must specify type of joint and show location and details on drawings)

- Movement joints—mandatory according to EJ171.

Installation Specifications

- Tile—ANSI A108.1A, .1B, or .1C. Ungauged stone may require installation per ANSI A108.1A. Material with minimal variation may be allowed per A108.1B with LHT mortar. A108.1B required if waterproof membrane or epoxy bond coat to be used.
- Epoxy mortar—ANSI A108.6.
- Cementitious grout—ANSI A108.10.
- Waterproof membrane—ANSI A108.13.
- Movement Joints—EJ171 and ASTM C1193.

Notes

- All horizontal surfaces, for example shower seats, sills, curbs, etc., must slope towards drain or other surface sloped toward drain. Where present, waterproofing also must be sloped.
- When unsanded grout is used, grout joint width must be 1⁄16" to ⅛" wide (nominal).

B440-21
STONE

- **Bathtub or Shower with Prefabricated Receptor**
- **Backer Board**
- **Mortar Bed (One Coat Method)**
- **Natural Stone Tile**

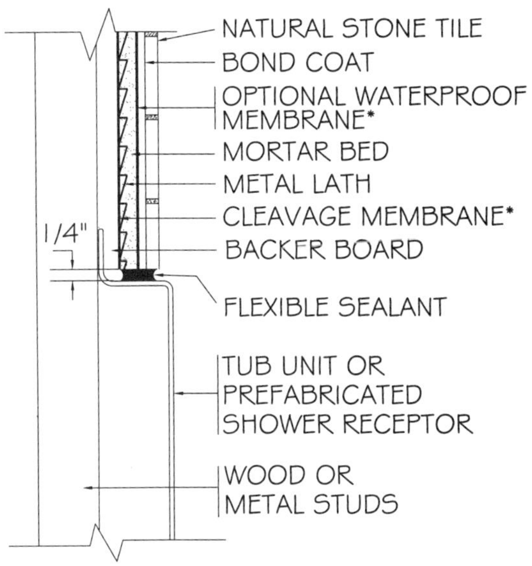

*USE OF A MEMBRANE IS REQUIRED. SEE MEMBRANE OPTIONS.

SHOWER RECEPTORS, CURBS, SEATS, ETC., MUST BE PROPERLY WATERPROOFED AND INSTALLED TO AVOID WATER DAMAGE TO ADJACENT BUILDING MATERIALS. SEE COMMON SHOWER CONFIGURATIONS SECTION.

Recommended Uses

- For tub enclosures or showers with prefabricated receptors, where minor flattening or trueing of walls is required.
- For areas where wall flatness is critical, such as when tiles with any edge longer than 15" are specified.
- For areas where ungauged stone will be used.
- For showers that do not have a prefabricated receptor, see B441 Stone.

Environmental Exposure Classifications

- With cement backer board or fiber cement backer board—Res1, 2, 3, 5; Com1, 2, 3, 5.
- With coated glass mat water-resistant gypsum backer board or cementitious-coated extruded foam backer board—Res1, 2, 3; Com1, 2, 3.
- With fiber-reinforced water-resistant gypsum backer board—Res 1, 2, 3; Com 1, 2.
- For installations that may be exposed to staining and/or chemical attack, specify tile and grout suitable for exposure. Consult product manufacturers; see also "Product Selection Guides."

Limitations

- Maximum mortar bed thickness—¾."
- Maximum stud spacing 16" on center.
- When coated glass mat water-resistant gypsum backer board or fiber-reinforced water-resistant gypsum backer board is used—not for areas exposed to temperatures exceeding 125°F.

Membrane Options

- A cleavage membrane is required. A waterproof membrane (A118.10) may also be specified to prevent moisture intrusion and protect adjacent building materials. Specifier shall indicate if complete waterproofing is required, including treatment at termination points.
- Check with membrane manufacturer for suitability for applicable conditions, as not all membranes are suitable for steam, high-temperature and/or chemical exposure, or exterior use.

Requirements

- Cleavage membrane must lap over face of flange of tub or prefabricated shower receptor.
- Wood studs—dry and well-braced, minimum depth 3½."
- Metal studs—well-braced; 20 gauge (0.033") or heavier; minimum depth 3½" for residential applications or 3⅝" for commercial applications.
- Mortar bed thickness—⅜" minimum to ¾" maximum.
- Backer board must provide firm anchorage for metal lath.
- Cut lath at all movement joints.
- Minimum grout joint width—1/16."

Preparation by Backer Board Installers

- Maximum allowable variation in backer board—⅛" in 10' from the required plane.
- Follow backer board manufacturer's directions for installation, including orientation and spacing of boards, required fasteners, and taping of joints and corners.

Materials

- **Multiple options exist for membranes, mortars, grouts, and other materials and *must be clearly specified* to be included. If not specifically indicated, optional materials are not included and mortar/grout choice defaults to minimum performance specification indicated. Consider each system component and intended use to determine minimum requirements and to specify options.**
- Natural stone tile—see "Natural Stone Tile Selection and Installation Guide," and consult supplier for application suitability.
- Cementitious grout—ANSI A118.6 or better or ISO CG1 or better. Consider unsanded grout for soft or semisoft stones such as limestone, travertine, marble, onyx, or similar.

- Cementitious bond coat—portland cement paste on a mortar bed that is still workable. For a cured mortar bed, follow recommendations below to select appropriate bonding mortar:
  - When a waterproof membrane is not used—ANSI A118.1 or better or ISO C1 or better.
  - When a waterproof membrane is used—ANSI A118.4 or better or ISO C2S1 or better unless ANSI A118.1 or ISO C1 is recommended by membrane manufacturer.
  - Use white for light-colored marble, limestone, and other stones where staining or darkening from grey setting materials is possible.
- Epoxy bond coat, when used—ANSI A118.3 or ISO R1 or better. See Water Sensitivity and Fiberglass Mesh Reinforced Stone in the "Natural Stone Tile Selection and Installation Guide."
- Waterproof membrane, when used—ANSI A118.10.
- Mortar bed, metal lath, and cleavage membrane—ANSI A108.1A.
- Cement backer board, when used—ANSI A118.9 or ASTM C1325 (Type B).
- Fiber cement backer board, when used—ASTM C1288.
- Coated glass mat, water-resistant gypsum backer board, when used—ASTM C1178.
- Cementitious-coated extruded foam backer board, when used—ASTM C578.
- Fiber-reinforced water-resistant gypsum backer board, when used—ASTM C1278 (Paragraph 6.1).
- Fasteners—noncorrosive and nonoxidizing.
- Fasteners meeting ASTM F2329 required in wet areas.
- 2" alkali-resistant glass fiber mesh tape.
- Flexible mildew-resistant sealant—ASTM C920.
- Metal studs—ASTM C645.

Materials for Green/Sustainable Design

- See "Green Building Standards and Green Product Selection Guide" and consult manufacturers and suppliers for product sustainability and contribution to green building design.
- Consider specifying installation materials that meet ANSI A138.1, the *American National Standard Specifications for Sustainable Ceramic Tiles, Glass Tiles, and Tile Installation Materials.*

Preparation by Other Trades

- Wall framing shall meet the general framing requirements of ANSI A108.11-4.0–4.3.
- Opening for recessed bathtubs—not to exceed ½" more than total length of tub.
- Bathtub—install level and support with metal hangers or on end grain wood blocks.
- Fireproofing behind tub when required.

Movement Joint (architect must specify type of joint and show location and details on drawings)

- Movement joints—mandatory according to EJ171.

Installation Specifications

- Tile—ANSI A108.1A, .1B, or .1C. Ungauged stone may require installation per ANSI A108.1A. Material with minimal variation may be allowed per A108.1B with LHT mortar. A108.1B required if waterproof membrane or epoxy bond coat to be used.
- Cementitious grout—ANSI A108.10.
- Epoxy mortar—ANSI A108.6.
- Waterproof membrane—ANSI A108.13.
- Cement backer board—ANSI A108.11.
- Fiber cement backer board—manufacturer's directions.
- Coated glass mat water-resistant gypsum backer board—manufacturer's directions.
- Cementitious-coated extruded foam backer board—manufacturer's directions.
- Fiber-reinforced water-resistant gypsum backer board—manufacturer's directions.
- Movement Joints—EJ171 and ASTM C1193.

Notes

- All horizontal surfaces, for example shower seats, sills, curbs, etc., must slope towards drain or other surface sloped toward drain. Where present, waterproofing also must be sloped.
- When coated glass mat water-resistant gypsum backer board is used, do not install a vapor barrier behind the backer board.
- When unsanded grout is used, grout joint width must be ¹⁄₁₆" to ⅛" wide (nominal).
- If lath cannot be attached directly to backing, use furring strips. See B411 Stone.
- For thicker mortar bed, see B411 Stone.
- For mortar bed directly over wood or metal studs, see B411 Stone.

B412-21
STONE

- **Bathtub or Shower with Prefabricated Receptor**
- **Cement Backer Board or Fiber-Cement Backer Board**
- **Natural Stone Tile**

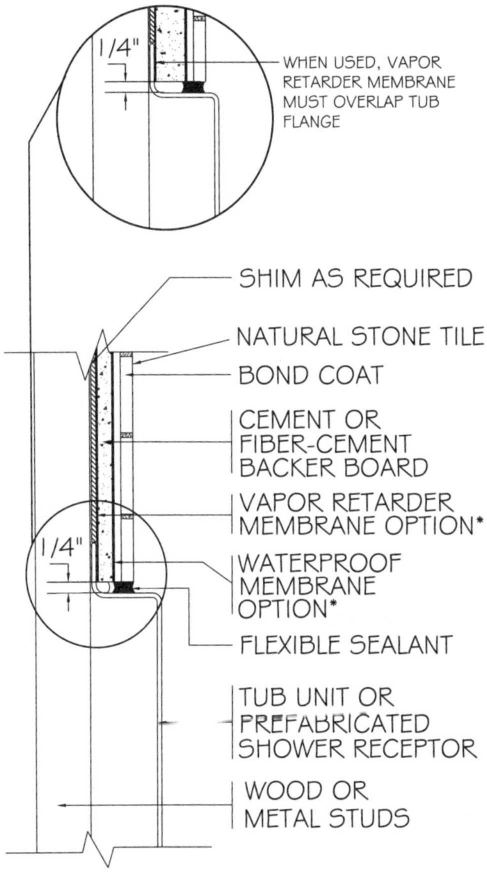

*USE OF A MEMBRANE IS REQUIRED. SEE MEMBRANE OPTIONS.

SHOWER RECEPTORS, CURBS, SEATS, ETC. MUST BE PROPERLY WATERPROOFED AND INSTALLED TO AVOID WATER DAMAGE TO ADJACENT BUILDING MATERIALS. SEE COMMON SHOWER CONFIGURATIONS SECTION.

Recommended Uses

- For tub enclosures or showers with prefabricated receptors.
- For showers that do not have a prefabricated receptor, see B415 Stone.

Environmental Exposure Classifications

- Res1, 2, 3, 5; Com1, 2, 3, 5.
- For installations that may be exposed to staining and/or chemical attack, specify tile and grout suitable for exposure. Consult product manufacturers; see also "Product Selection Guides."

Limitations

- Maximum stud spacing 16" on center.

Membrane Options

- A waterproof membrane (A118.10) or vapor retarder membrane (A108.02-3.8) must be specified for walls to prevent moisture intrusion and protect adjacent building materials. Specifier shall indicate if complete waterproofing of walls is required, including treatment at termination points.
- Check with membrane manufacturer for suitability for applicable conditions, as not all membranes are suitable for steam, high-temperature and/or chemical exposure, or exterior use.

Requirements

- Wood studs—dry and well-braced, minimum depth 3½".
- Metal studs—well-braced; 20 gauge (0.033") or heavier; minimum depth 3½" for residential applications or 3⅝" for commercial applications.
- Surface of units—clean and free of dirt, dust, paint, and oily film.
- Membrane behind backer unit, when used, must lap over face of flange of tub or prefabricated shower receptor.
- Minimum grout joint width—1/16".

Preparation by Backer Board Installers

- Maximum allowable variation in the tile substrate—⅛" in 10' from the required plane when measured from the high points in the surface.
- Horizontal joints—⅛" spacing filled solid and taped with latex-portland cement mortar and 2" alkali-resistant glass fiber mesh tape.
- Vertical joints—fill any space and tape with latex-portland cement mortar and 2" alkali-resistant glass fiber mesh tape.
- Corners—leave space between backer units. Tape joints using skim coat of latex-portland cement mortar and 2" alkali-resistant glass fiber mesh tape, but do not fill.
- Center backer board end or edge joints on framing and stagger joints in adjacent rows so four corners do not come together within the same plane. Space panel ends and edges in accordance with manufacturer's recommendations.

Materials

- **Multiple options exist for membranes, mortars, grouts, and other materials and *must be clearly specified* to be included. If not specifically indicated, optional materials are not included and mortar/grout choice defaults to minimum performance specification indicated. Consider each system component and intended use to determine minimum requirements and to specify options.**

- Natural stone tile—see "Natural Stone Tile Selection and Installation Guide," and consult supplier for application suitability.
- Cementitious grout—ANSI A118.6 or better or ISO CG1 or better. Consider unsanded grout for soft or semisoft stones such as limestone, travertine, marble, onyx, or similar.
- Cementitious bond coat:
  - When a waterproof membrane is not used—ANSI A118.1 or better or ISO C1 or better.
  - When a waterproof membrane is used—ANSI A118.4 or better or ISO C2S1 or better unless ANSI A118.1 or ISO C1 is recommended by membrane manufacturer.
  - Use white for light-colored marble, limestone, and other stones where staining or darkening from grey setting materials is possible.
- Epoxy bond coat, when used—ANSI A118.3 or ISO R1 or better. See Water Sensitivity and Fiberglass Mesh Reinforced Stone in the "Natural Stone Tile Selection and Installation Guide."
- Vapor retarder membrane, when used—ANSI A108.02-3.8.
- Waterproof membrane, when used—ANSI A118.10.
- Cement backer board, when used—ANSI A118.9 or ASTM C1325 (Type B).
- Fiber cement backer board, when used—ASTM C1288.
- Fasteners—noncorrosive and nonoxidizing.
- Fasteners meeting ASTM F2329 required in wet areas.
- 2" alkali-resistant glass fiber mesh tape.
- Flexible mildew-resistant sealant—ASTM C920.
- Metal studs—ASTM C645.

Materials for Green/Sustainable Design

- See "Green Building Standards and Green Product Selection Guide" and consult manufacturers and suppliers for product sustainability and contribution to green building design.
- Consider specifying installation materials that meet ANSI A138.1, the *American National Standard Specifications for Sustainable Ceramic Tiles, Glass Tiles, and Tile Installation Materials.*

Preparation by Other Trades

- Wall framing shall meet the general framing requirements of ANSI A108.11-4.0–4.3.
- Opening for recessed tub—not to exceed ½" more than length of tub.
- Bathtub—install level and support with metal hangers or on wood framing members.
- Fire and sound ratings—extend gypsum board required for ratings down to the floor behind the tub so that construction will be the same as the tested assembly. Backer board may be part of, or installed over, the rated assembly.

Movement Joint (architect must specify type of joint and show location and details on drawings)

- Movement joints—mandatory according to EJ171.

Installation Specifications

- Cement backer board—ANSI A108.11.
- Fiber cement backer board—manufacturer's directions.
- Tile—ANSI A108.5 or A108.6.
- Cementitious grout—ANSI A108.10.
- Waterproof membrane—ANSI A108.13.
- Movement Joints—EJ171 and ASTM C1193.

Notes

- As the tile size increases, there is less tolerance for variation in the substrate from the required plane. Epoxy bond coat thickness must be thin and uniform; therefore, substrate flattening may be required when epoxy bond coat is used.
- All horizontal surfaces, for example shower seats, sills, curbs, etc., must slope towards drain or other surface sloped toward drain. Where present, waterproofing also must be sloped.
- When unsanded grout is used, grout joint width must be $\frac{1}{16}$" to ⅛" wide (nominal).

B419-21
STONE

- **Bathtub or Shower with Prefabricated Receptor**
- **Coated Glass Mat Water-Resistant Gypsum Backer Board**
- **Natural Stone Tile**

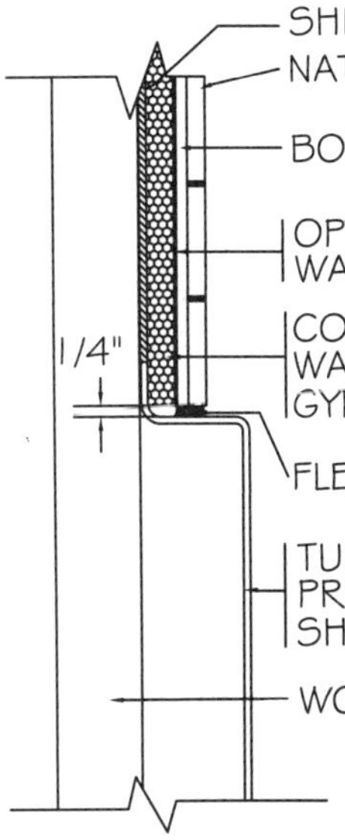

*USE OF A MEMBRANE IS OPTIONAL. SEE MEMBRANE OPTIONS.

SHOWER RECEPTORS, CURBS, SEATS, ETC., MUST BE PROPERLY WATERPROOFED AND INSTALLED TO AVOID WATER DAMAGE TO ADJACENT BUILDING MATERIALS. SEE COMMON SHOWER CONFIGURATIONS SECTION.

Recommended Uses

- For tub enclosures or showers with prefabricated receptors.
- For showers that do not have a prefabricated receptor, see B420 Stone.

Environmental Exposure Classifications

- Res1, 2, 3; Com1, 2, 3.
- For installations that may be exposed to staining and/or chemical attack, specify tile and grout suitable for exposure. Consult product manufacturers; see also "Product Selection Guides."

Limitations

- Maximum stud spacing 16" on center.
- Not for areas exposed to temperatures exceeding 125°F.

Membrane Options

- A waterproof membrane (A118.10) may be specified for walls to prevent moisture intrusion through seams, corners, fasteners, and other penetrations and to protect adjacent building materials. Follow backer board and membrane manufacturer's waterproofing requirements. Specifier shall indicate if complete waterproofing of walls is required, including treatment at termination points.
- Check with membrane manufacturer for suitability for applicable conditions, as not all membranes are suitable for steam, high-temperature and/or chemical exposure, or exterior use.
- Some liquid-applied waterproof membranes will require extended cure time. Consult membrane manufacturer.

Requirements

- Wood studs—dry and well-braced, minimum depth 3½."
- Metal studs—well-braced; 20 gauge (0.033") or heavier; minimum depth 3½" for residential applications or 3⅝" for commercial applications.
- Caulk or seal penetrations, corners, and abutments to dissimilar materials with flexible sealant.
- Minimum grout joint width—1/16."

Preparation by Backer Board Installers

- Maximum allowable variation in the tile substrate—⅛" in 10' from the required plane when measured from the high points in the surface.
- Fit ends and edges closely to eliminate gaps between panels, including in corners.
- Tape seams and corners with 2" alkali-resistant glass fiber mesh tape and cementitious bonding material, and skim coat over fastener heads, unless waterproofing has been applied.
- Center backer board end or edge joints on framing and stagger joints in adjacent rows so four corners do not come together within the same plane. Space panel ends and edges in accordance with manufacturer's recommendations.

Materials

- **Multiple options exist for membranes, mortars, grouts, and other materials and *must be clearly specified* to be included. If not specifically indicated, optional materials are not included and mortar/grout choice defaults to minimum performance specification indicated. Consider each system component and intended use to determine minimum requirements and to specify options.**
- Natural stone tile—see "Natural Stone Tile Selection and Installation Guide," and consult supplier for application suitability.
- Cementitious grout—ANSI A118.6 or better or ISO CG1 or better. Consider unsanded grout for soft or semisoft stones such as limestone, travertine, marble, onyx, or similar.
- Cementitious bond coat:
  - When a waterproof membrane is not used—ANSI A118.4 or better or ISO C2S1 or better.

- When a waterproof membrane is used—ANSI A118.4 or better or ISO C2S1 or better unless ANSI A118.1 or ISO C1 is recommended by membrane manufacturer.
- Use white for light-colored marble, limestone, and other stones where staining or darkening from grey setting materials is possible.

- Epoxy bond coat, when used—ANSI A118.3 or ISO R1 or better. See Water Sensitivity and Fiberglass Mesh Reinforced Stone in the "Natural Stone Tile Selection and Installation Guide."
- Waterproof membrane, when used—ANSI A118.10.
- Coated glass mat, water-resistant gypsum backer board—ASTM C1178.
- Fasteners—noncorrosive and nonoxidizing.
- Fasteners meeting ASTM F2329 required in wet areas.
- 2" alkali-resistant glass fiber mesh tape.
- Flexible mildew-resistant sealant—ASTM C920.
- Metal studs—ASTM C645.

Materials for Green/Sustainable Design

- See "Green Building Standards and Green Product Selection Guide" and consult manufacturers and suppliers for product sustainability and contribution to green building design.
- Consider specifying installation materials that meet ANSI A138.1, the *American National Standard Specifications for Sustainable Ceramic Tiles, Glass Tiles, and Tile Installation Materials.*

Preparation by Other Trades

- Wall framing shall meet the general framing requirements of ANSI A108.11-4.0–4.3.
- Opening for recessed tub not to exceed ½" more than length of tub.
- Bathtub—install level and supported with metal hangers or on wood framing members.
- Fire and sound ratings—extend gypsum board required for rating down to the floor behind the tub so that construction will be the same as the tested assembly.

Movement Joint (architect must specify type of joint and show location and details on drawings)

- Movement joints—mandatory according to EJ171.

Installation Specifications

- Coated glass mat water-resistant gypsum backer board—manufacturer's directions.
- Tile—ANSI A108.5 or A108.6.
- Cementitious grout—ANSI A108.10.
- Waterproof membrane—ANSI A108.13.
- Movement Joints—EJ171 and ASTM C1193.

Notes

- Do not install a vapor barrier behind the tile backer board.
- All horizontal surfaces, for example shower seats, sills, curbs, etc., must slope towards drain or other surface sloped toward drain. Where present, waterproofing also must be sloped.
- As the tile size increases, there is less tolerance for variation in the substrate from the required plane. Epoxy bond coat thickness must be thin and uniform; therefore, substrate flattening may be required when epoxy bond coat is used.
- When unsanded grout is used, grout joint width must be 1⁄16" to ⅛" wide (nominal).

B425-21 STONE

- **Bathtub or Shower with Prefabricated Receptor**
- **Cementitious-Coated Extruded Foam Backer Board**
- **Natural Stone Tile**

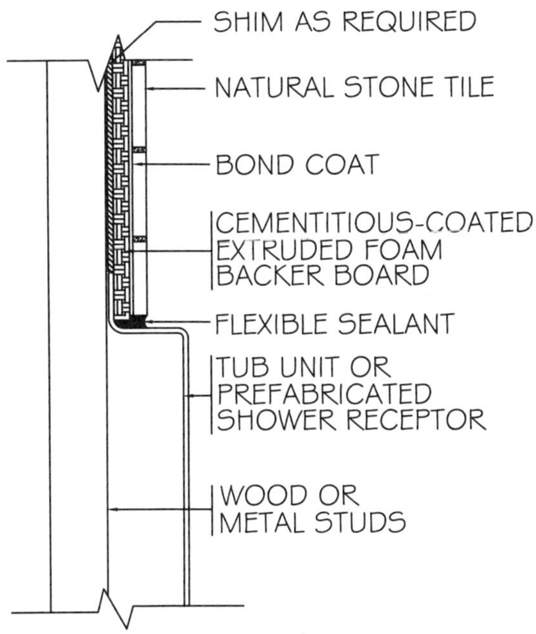

SHOWER RECEPTORS, CURBS, SEATS, ETC., MUST BE PROPERLY WATERPROOFED AND INSTALLED TO AVOID WATER DAMAGE TO ADJACENT BUILDING MATERIALS. SEE COMMON SHOWER CONFIGURATIONS SECTION.

Recommended Uses

- For tub enclosures or showers with prefabricated receptors.
- For showers that do not have a prefabricated receptor, see B426 Stone.

Environmental Exposure Classifications

- Res1, 2, 3; Com1, 2, 3.
- For installations that may be exposed to staining and/or chemical attack, specify tile and grout suitable for exposure. Consult product manufacturers; see also "Product Selection Guides."

Limitations

- Maximum stud spacing 16" on center.

Requirements

- Wood studs—dry and well-braced, minimum depth 3½."
- Metal studs—well-braced; 20 gauge (0.033") or heavier; minimum depth 3½" for residential applications or 3⅝" for commercial applications.
- Minimum grout joint width—1⁄16."

Preparation by Backer Board Installers

- Fasten backer board to studs with board manufacturer's recommended fasteners.
- All openings cut in backer board for plumbing and all cut joints between adjoining pieces—seal with waterproofing as recommended by backer board manufacturer.
- Maximum allowable variation in the tile substrate—⅛" in 10' from the required plane when measured from the high points in the surface.
- Fit ends and edges tightly together, including in corners. Apply a bead of manufacturer-recommended sealant in panel joints as panels are being installed and over fastener heads per backer board manufacturer's instructions.
- Apply backer board tight to lip of tub and seal to tub with manufacturer-recommended sealant.
- Center backer board end or edge joints on framing and stagger joints in adjacent rows so four corners do not come together within the same plane. Space panel ends and edges in accordance with manufacturer's recommendations.

Materials

- **Multiple options exist for membranes, mortars, grouts, and other materials and *must be clearly specified* to be included. If not specifically indicated, optional materials are not included and mortar/grout choice defaults to minimum performance specification indicated. Consider each system component and intended use to determine minimum requirements and to specify options.**
- Natural stone tile—see "Natural Stone Tile Selection and Installation Guide," and consult supplier for application suitability.
- Cementitious grout—ANSI A118.6 or better or ISO CG1 or better. Consider unsanded grout for soft or semisoft stones such as limestone, travertine, marble, onyx, or similar.
- Cementitious bond coat:
    - ANSI A118.1 or better or ISO C1 or better.
    - Use white for light-colored marble, limestone, and other stones where staining or darkening from grey setting materials is possible.
- Epoxy bond coat, when used—ANSI A118.3 or ISO R1 or better. See Water Sensitivity and Fiberglass Mesh Reinforced Stone in the "Natural Stone Tile Selection and Installation Guide."
- Cementitious-coated extruded foam backer board—ASTM C578.
- Fasteners—noncorrosive and nonoxidizing.
- Flexible sealant—must be recommended by backer board manufacturer.
- Metal studs—ASTM C645.

Materials for Green/Sustainable Design

- See "Green Building Standards and Green Product Selection Guide" and consult manufacturers and suppliers for product sustainability and contribution to green building design.
- Consider specifying installation materials that meet ANSI A138.1, the *American National Standard Specifications for Sustainable Ceramic Tiles, Glass Tiles, and Tile Installation Materials.*

Preparation by Other Trades

- Wall framing shall meet the general framing requirements of ANSI A108.11-4.0–4.3.
- Bathtub—install level and support with metal hangers or on wood framing members.
- Opening for recessed tub not to exceed ½" more than length of tub.

Movement Joint (architect must specify type of joint and show location and details on drawings)

- Movement joints—mandatory according to EJ171.

Installation Specifications

- Cementitious-coated extruded foam backer board—manufacturer's directions.
- Tile—ANSI A108.5 or A108.6.
- Cementitious grout—ANSI A108.10.
- Movement Joints—EJ171 and ASTM C1193.

Notes

- All horizontal surfaces, for example shower seats, sills, curbs, etc., must slope towards drain or other surface sloped toward drain. Where present, waterproofing also must be sloped.
- As the tile size increases, there is less tolerance for variation in the substrate from the required plane. Epoxy bond coat thickness must be thin and uniform; therefore, substrate flattening may be required when epoxy bond coat is used.
- When unsanded grout is used, grout joint width must be 1/16" to ⅛" wide (nominal).

B430-21
STONE

- **Bathtub or Shower with Prefabricated Receptor**
- **Fiber-Reinforced Water-Resistant Gypsum Backer Board**
- **Natural Stone Tile**

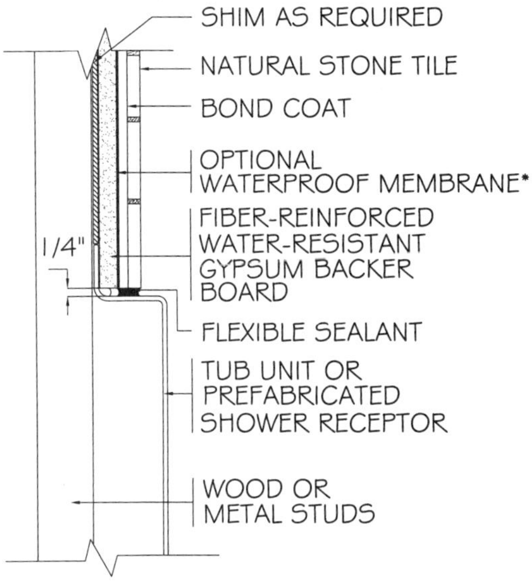

*USE OF A MEMBRANE IS OPTIONAL. SEE MEMBRANE OPTIONS.

SHOWER RECEPTORS, CURBS, SEATS, ETC., MUST BE PROPERLY WATERPROOFED AND INSTALLED TO AVOID WATER DAMAGE TO ADJACENT BUILDING MATERIALS. SEE COMMON SHOWER CONFIGURATIONS SECTION.

Recommended Uses

- For tub enclosures or showers with prefabricated receptors.
- For showers that do not have a prefabricated receptor, see B431 Stone.

Environmental Exposure Classifications

- Res1, 2, 3; Com1, 2.
- For installations that may be exposed to staining and/or chemical attack, specify tile and grout suitable for exposure. Consult product manufacturers; see also "Product Selection Guides."

Limitations

- Maximum stud spacing 16" on center.
- Not for areas exposed to temperatures exceeding 125°F.

Membrane Options

- A waterproof membrane (A118.10) may be specified for walls to prevent moisture intrusion and protect adjacent building materials. Specifier shall indicate if complete waterproofing of walls is required, including treatment at termination points.
- Check with membrane manufacturer for suitability for applicable conditions, as not all membranes are suitable for steam, high-temperature and/or chemical exposure, or exterior use.

Requirements

- Wood studs—dry and well-braced, minimum depth 3½."
- Metal studs—well-braced; 20 gauge (0.033") or heavier; minimum depth 3½" for residential applications or 3⅝" for commercial applications.
- Caulk or seal penetrations and abutments to dissimilar materials.
- Minimum grout joint width—1/16."

Preparation by Backer Board Installers

- Maximum allowable variation in the tile substrate—⅛" in 10' from the required plane when measured from the high points in the surface.
- Horizontal joints—⅛" spacing filled solid and taped with latex-portland cement mortar and 2" alkali-resistant glass fiber mesh tape.
- Vertical joints—fill any space and tape with latex-portland cement mortar and 2" alkali-resistant glass fiber mesh tape.
- Corners—leave space between backer units. Tape joints using skim coat of latex-portland cement mortar, but do not fill.
- Center backer board end or edge joints on framing and stagger joints in adjacent rows so four corners do not come together within the same plane. Space panel ends and edges in accordance with manufacturer's recommendations.

Materials

- **Multiple options exist for membranes, mortars, grouts, and other materials and *must be clearly specified* to be included. If not specifically indicated, optional materials are not included and mortar/grout choice defaults to minimum performance specification indicated. Consider each system component and intended use to determine minimum requirements and to specify options.**
- Natural stone tile—see "Natural Stone Tile Selection and Installation Guide," and consult supplier for application suitability.
- Cementitious grout—ANSI A118.6 or better or ISO CG1 or better. Consider unsanded grout for soft or semisoft stones such as limestone, travertine, marble, onyx, or similar.
- Cementitious bond coat:
  - When a waterproof membrane is not used—ANSI A118.4 or better or ISO C2 or better.
  - When a waterproof membrane is used—ANSI A118.4 or better or ISO C2S1 or better unless ANSI A118.1 or ISO C1 is recommended by membrane manufacturer.

    - Use white for light-colored marble, limestone, and other stones where staining or darkening from grey setting materials is possible.
- Epoxy bond coat, when used—ANSI A118.3 or ISO R1 or better. See Water Sensitivity and Fiberglass Mesh Reinforced Stone in the "Natural Stone Tile Selection and Installation Guide."
- Waterproof membrane, when used—ANSI A118.10.
- Fiber-reinforced water-resistant gypsum backer board—ASTM C1278 (Paragraph 6.1).
- Fasteners—noncorrosive and nonoxidizing.
- 2" alkali-resistant glass fiber mesh tape.
- Flexible mildew-resistant sealant—ASTM C920.
- Metal studs—ASTM C645.

Materials for Green/Sustainable Design

- See "Green Building Standards and Green Product Selection Guide" and consult manufacturers and suppliers for product sustainability and contribution to green building design.
- Consider specifying installation materials that meet ANSI A138.1, the *American National Standard Specifications for Sustainable Ceramic Tiles, Glass Tiles, and Tile Installation Materials.*

Preparation by Other Trades

- Wall framing shall meet the general framing requirements of ANSI A108.11-4.0–4.3.
- Opening for recessed tub not to exceed ½" more than the tub length.
- Bathtub—install level and supported with metal hangers or on wood framing members.
- Fire and sound ratings—extend gypsum board required for rating down to the floor behind the tub so that construction will be the same as the tested assembly.

Movement Joint (architect must specify type of joint and show location and details on drawings)

- Movement joints—mandatory in accordance with EJ171.

Installation Specifications

- Fiber-reinforced water-resistant gypsum backer board—manufacturer's directions.
- Tile—ANSI A108.5 or A108.6.
- Cementitious grout—ANSI A108.10.
- Waterproof membrane—ANSI A108.13.
- Movement Joints—EJ171 and ASTM C1193.

Notes

- As the tile size increases, there is less tolerance for variation in the substrate from the required plane. Epoxy bond coat thickness must be thin and uniform; therefore, substrate flattening may be required when epoxy bond coat is used.
- All horizontal surfaces, for example shower seats, sills, curbs, etc., must slope towards drain or other surface sloped toward drain. Where present, waterproofing also must be sloped.
- When unsanded grout is used, grout joint width must be $^1/_{16}$" to ⅛" wide (nominal).

B414-21
STONE

- **Wood or Metal Studs**
- **Mortar Bed Walls**
- **Mortar Bed Floor**
- **Natural Stone Tile**

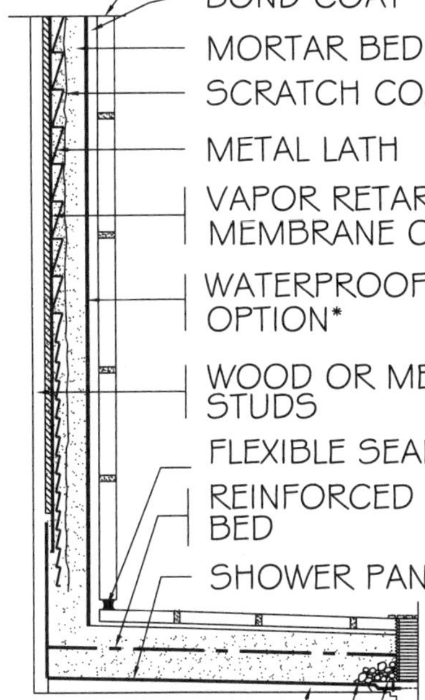

*USE OF A MEMBRANE ON WALLS IS REQUIRED. SEE MEMBRANE OPTIONS.

SHOWER RECEPTORS, CURBS, SEATS, ETC., MUST BE PROPERLY WATERPROOFED AND INSTALLED TO AVOID WATER DAMAGE TO ADJACENT BUILDING MATERIALS. SEE COMMON SHOWER CONFIGURATIONS SECTION.

Recommended Uses

- For showers that do not have prefabricated receptors.
- For areas where wall flatness is critical, such as when tiles with any edge longer than 15" are specified.
- For areas where ungauged stone will be used.

Environmental Exposure Classifications

- Res1, 2, 3, 5; Com1, 2, 3, 5.
- For installations that may be exposed to staining and/or chemical attack, specify tile and grout suitable for exposure. Consult product manufacturers; see also "Product Selection Guides."

Limitations

- Maximum mortar bed thickness (walls)—1" for metal studs, 1½" for wood studs.
- Maximum stud spacing 16" on center.

Membrane Options

- A waterproof membrane (A118.10) or vapor retarder membrane (A108.02-3.8) must be specified for walls to prevent moisture intrusion and protect adjacent building materials. Specifier shall indicate if complete waterproofing of walls is required, including treatment at termination points.
- If a waterproof membrane is applied over the mortar bed walls, membrane over framing members may not be required.
- Check with membrane manufacturer for suitability for applicable conditions, as not all membranes are suitable for steam, high-temperature and/or chemical exposure, when used, or exterior use.

Requirements

- Wood studs—dry and well-braced, minimum depth 3½."
- Metal studs—well-braced; 20 gauge (0.033") or heavier; minimum depth 3½" for residential applications or 3⅝" for commercial applications.
- Mortar bed thickness—¾" minimum to 1" maximum (metal studs) or 1½" maximum (wood studs).
- Membrane behind mortar bed, when used, must lap over shower pan membrane.
- Slope shower pan membrane ¼" per foot to weep holes in drain.
- Turn shower pan membrane up walls a minimum of 3" above shower curb (6" above floor in showers without curbs).
- Surround drain with pea gravel or other weep protection to prevent mortar from blocking weep holes.
- Minimum grout joint width—1/16."

Materials

- **Multiple options exist for membranes, mortars, grouts, and other materials and *must be clearly specified* to be included. If not specifically indicated, optional materials are not included and mortar/grout choice defaults to minimum performance specification indicated. Consider each system component and intended use to determine minimum requirements and to specify options.**
- Natural stone tile—see "Natural Stone Tile Selection and Installation Guide," and consult supplier for application suitability.
- Cementitious grout—ANSI A118.6 or better or ISO CG1

or better. Consider unsanded grout for soft or semisoft stones such as limestone, travertine, marble, onyx, or similar.

- Cementitious bond coat—portland cement paste on a mortar bed that is still workable. For a cured mortar bed, follow recommendations below to select appropriate bonding mortar:
  - When a waterproof membrane is not used—ANSI A118.1 or better or ISO C1 or better.
  - When a waterproof membrane is used—ANSI A118.4 or better or ISO C2S1 or better unless ANSI A118.1 or ISO C1 is recommended by membrane manufacturer.
  - Use white for light-colored marble, limestone, and other stones where staining or darkening from grey setting materials is possible.
- Epoxy bond coat, when used—ANSI A118.3 or ISO R1 or better. See Water Sensitivity and Fiberglass Mesh Reinforced Stone in the "Natural Stone Tile Selection and Installation Guide."
- Vapor retarder membrane, when used—ANSI A108.02-3.8.
- Waterproof membrane, when used—ANSI A118.10.
- Mortar bed, metal lath, and cleavage membrane—ANSI A108.1A.
- Flexible mildew-resistant sealant—ASTM C920.
- Shower pan membrane—local building code.
- Metal studs—ASTM C645.

Materials for Green/Sustainable Design

- See "Green Building Standards and Green Product Selection Guide" and consult manufacturers and suppliers for product sustainability and contribution to green building design.
- Consider specifying installation materials that meet ANSI A138.1, the *American National Standard Specifications for Sustainable Ceramic Tiles, Glass Tiles, and Tile Installation Materials.*

Preparation by Other Trades

- Wall framing shall meet the general framing requirements of ANSI A108.11-4.0-4.3.
- Apply blocking between the studs to support the shower pan membrane.

Movement Joint (architect must specify type of joint and show location and details on drawings)

- Movement joints—mandatory according to EJ171.

Installation Specifications

- Shower pan membrane—ANSI A108.01-3.6.
- Tile—ANSI A108.1A, .1B, or .1C. Ungauged stone may require installation per ANSI A108.1A. Material with minimal variation may be allowed per A108.1B with LHT mortar. A108.1B required if waterproof membrane or epoxy bond coat to be used.
- Cementitious grout—ANSI A108.10.
- Waterproof membrane—ANSI A108.13.
- Movement Joints—EJ171 and ASTM C1193.

Notes

- Test shower pan membrane and drainage fitting for leaks before commencing tilework.
- Materials adversely affected by moisture in areas immediately adjacent to showers, tubs, and roman tubs should be properly protected.
- A sloped portland cement mortar fill or approved preformed slope may be used under shower pan membrane when subfloor is not sloped to drain.
- All horizontal surfaces, for example shower seats, sills, curbs, etc., must slope towards drain or other surface sloped toward drain. Where present, waterproofing also must be sloped.
- When unsanded grout is used, grout joint width must be $\frac{1}{16}$" to $\frac{1}{8}$" wide (nominal).

B441-21
STONE

- **Wood or Metal Studs**
- **Backer Board**
- **Mortar Bed Walls (One Coat Method)**
- **Mortar Bed Floor**
- **Natural Stone Tile**

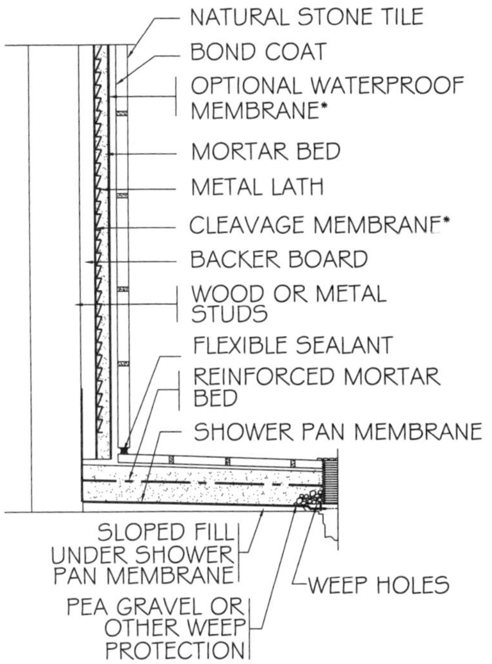

*USE OF A MEMBRANE ON WALLS IS REQUIRED. SEE MEMBRANE OPTIONS.
SHOWER RECEPTORS, CURBS, SEATS, ETC., MUST BE PROPERLY WATERPROOFED AND INSTALLED TO AVOID WATER DAMAGE TO ADJACENT BUILDING MATERIALS.
SEE COMMON SHOWER CONFIGURATIONS SECTION.

CLEAVAGE MEMBRANE LAPS OVER SHOWER PAN.

Recommended Uses

- For showers that do not have prefabricated receptors, where minor flattening or trueing of walls is required.
- For areas where wall flatness is critical, such as when tiles with any edge longer than 15" are specified.
- For areas where ungauged stone will be used.

Environmental Exposure Classifications

- With cement backer board or fiber cement backer board—Res1, 2, 3, 5; Com1, 2, 3, 5.
- With coated glass mat water-resistant gypsum backer board or cementitious-coated extruded foam backer board—Res1, 2, 3; Com1, 2, 3.
- With fiber-reinforced water-resistant gypsum backer board—Res 1, 2, 3; Com 1, 2.
- For installations that may be exposed to staining and/or chemical attack, specify tile and grout suitable for exposure. Consult product manufacturers; see also "Product Selection Guides."

Limitations

- Maximum mortar bed thickness (walls)—$\frac{3}{4}$".
- Maximum stud spacing 16" on center.
- When coated glass mat water-resistant gypsum backer board or fiber-reinforced water-resistant gypsum backer board is used—not for areas exposed to temperatures exceeding 125°F.

Membrane Options

- A cleavage membrane is required. A waterproof membrane (A118.10) may also be specified to prevent moisture intrusion and protect adjacent building materials. Specifier shall indicate if complete waterproofing is required, including treatment at termination points.
- Check with membrane manufacturer for suitability for applicable conditions, as not all membranes are suitable for steam, high-temperature and/or chemical exposure, or exterior use.

Requirements

- Cleavage membrane must lap over shower pan membrane.
- Wood studs—dry and well-braced, minimum depth $3\frac{1}{2}$".
- Metal studs—well-braced; 20 gauge (0.033") or heavier; minimum depth $3\frac{1}{2}$" for residential applications or $3\frac{5}{8}$" for commercial applications.
- Mortar bed thickness—$\frac{3}{8}$" minimum to $\frac{3}{4}$" maximum.
- Backer board must provide firm anchorage for metal lath.
- Cut lath at all movement joints.
- Slope shower pan membrane $\frac{1}{4}$" per foot to weep holes in drain.
- Turn shower pan membrane up walls a minimum of 3" above shower curb (6" above floor in showers without curbs).
- Surround drain with pea gravel or other weep protection to prevent mortar from blocking weep holes.
- Minimum grout joint width—$\frac{1}{16}$".

Preparation by Backer Board Installers

- Maximum allowable variation in the backer board—$\frac{1}{8}$" in 10' from the required plane.
- Follow backer board manufacturer's directions for installation, including orientation and spacing of boards, required fasteners, taping of joints and corners, and special instructions for wet areas.

Materials

- **Multiple options exist for membranes, mortars, grouts, and other materials and *must be clearly specified* to be included. If not specifically indicated, optional materials are not included and mortar/grout choice defaults to minimum performance specification indicated. Consider each system component and intended use to determine minimum requirements and to specify options.**

- Natural stone tile—see "Natural Stone Tile Selection and Installation Guide," and consult supplier for application suitability.
- Cementitious grout—ANSI A118.6 or better or ISO CG1 or better. Consider unsanded grout for soft or semisoft stones such as limestone, travertine, marble, onyx, or similar.
- Cementitious bond coat—portland cement paste on a mortar bed that is still workable. For a cured mortar bed, follow recommendations below to select appropriate bonding mortar:
  - When a waterproof membrane is not used—ANSI A118.1 or better or ISO C1 or better.
  - When a waterproof membrane is used—ANSI A118.4 or better or ISO C2S1 or better unless ANSI A118.1 or ISO C1 is recommended by membrane manufacturer.
  - Use white for light-colored marble, limestone, and other stones where staining or darkening from grey setting materials is possible.
- Epoxy bond coat, when used—ANSI A118.3 or ISO R1 or better. See Water Sensitivity and Fiberglass Mesh Reinforced Stone in the "Natural Stone Tile Selection and Installation Guide."
- Waterproof membrane, when used—ANSI A118.10.
- Mortar bed, metal lath, and cleavage membrane—ANSI A108.1A.
- Cement backer board, when used—ANSI A118.9 or ASTM C1325 (Type B).
- Fiber cement backer board, when used—ASTM C1288.
- Coated glass mat, water-resistant gypsum backer board, when used—ASTM C1178.
- Cementitious-coated extruded foam backer board, when used—ASTM C578.
- Fiber-reinforced water-resistant gypsum backer board, when used—ASTM C1278 (Paragraph 6.1).
- Fasteners—noncorrosive and nonoxidizing.
- Fasteners meeting ASTM F2329 required in wet areas.
- 2" alkali-resistant glass fiber mesh tape.
- Flexible mildew-resistant sealant—ASTM C920.
- Shower pan membrane—local building code.
- Metal studs—ASTM C645.

Materials for Green/Sustainable Design

- See "Green Building Standards and Green Product Selection Guide" and consult manufacturers and suppliers for product sustainability and contribution to green building design.
- Consider specifying installation materials that meet ANSI A138.1, the *American National Standard Specifications for Sustainable Ceramic Tiles, Glass Tiles, and Tile Installation Materials.*

Preparation by Other Trades

- Wall framing shall meet the general framing requirements of ANSI A108.11-4.0–4.3.
- Apply blocking between the studs to support the shower pan membrane.

Movement Joint (architect must specify type of joint and show location and details on drawings)

- Movement joints—mandatory according to EJ171.

Installation Specifications

- Tile—ANSI A108.1A, .1B, or .1C. Ungauged stone may require installation per ANSI A108.1A. Material with minimal variation may be allowed per A108.1B with LHT mortar. A108.1B required if waterproof membrane or epoxy bond coat to be used.
- Cementitious grout—ANSI A108.10.
- Epoxy mortar—ANSI A108.6.
- Waterproof membrane—ANSI A108.13.
- Cement backer board—ANSI A108.11.
- Fiber cement backer board—manufacturer's directions.
- Coated glass mat water-resistant gypsum backer board—manufacturer's directions.
- Cementitious-coated extruded foam backer board—manufacturer's directions.
- Fiber-reinforced water-resistant gypsum backer board—manufacturer's directions.
- Movement Joints—EJ171 and ASTM C1193.
- Shower pan membrane—ANSI A108.01-3.6.

Notes

- All horizontal surfaces, for example shower seats, sills, curbs, etc., must slope towards drain or other surface sloped toward drain. Where present, waterproofing also must be sloped.
- When coated glass mat water-resistant gypsum backer board is used, do not install a vapor barrier behind the backer board.
- When unsanded grout is used, grout joint width must be 1⁄16" to 1⁄8" wide (nominal).
- If lath cannot be attached directly to backing, use furring strips. See B414 Stone.
- Test shower pan membrane and drainage fitting for leaks before commencing tilework.
- Materials adversely affected by moisture in areas immediately adjacent to showers, tubs, and roman tubs should be properly protected.
- A sloped portland cement mortar fill or approved preformed slope may be used under shower pan membrane when subfloor is not sloped to drain.
- For thicker mortar bed, see B414 Stone.
- For mortar bed directly over wood or metal studs, see B414 Stone.

B415-21
STONE

- **Wood or Metal Studs**
- **Cement Backer Board or Fiber Cement Backer Board Walls**
- **Mortar Bed Floor**
- **Natural Stone Tile**

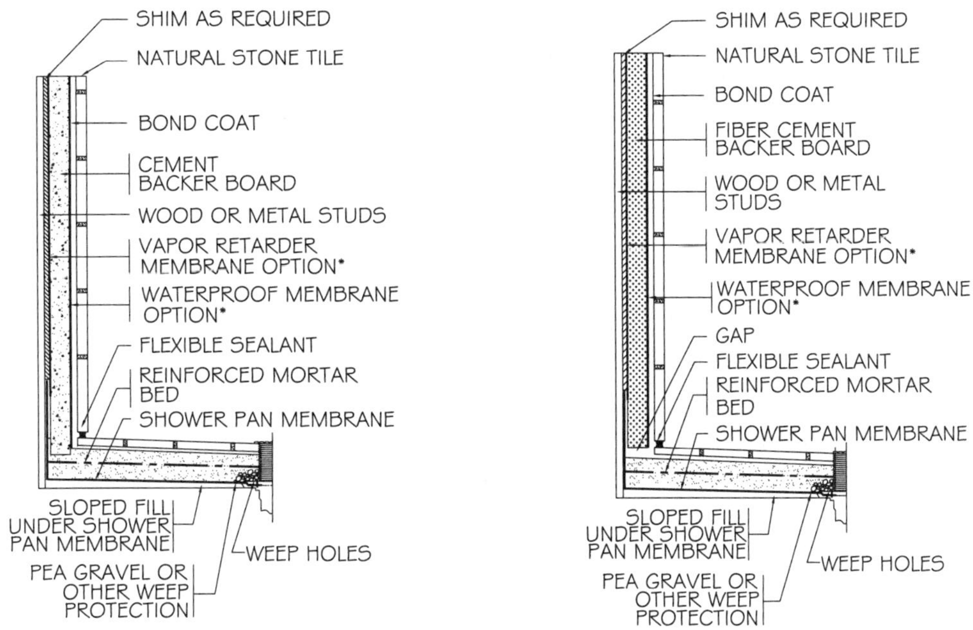

*USE OF A MEMBRANE ON WALLS IS REQUIRED. SEE MEMBRANE OPTIONS. SHOWER RECEPTORS, CURBS, SEATS, ETC., MUST BE PROPERLY WATERPROOFED AND INSTALLED TO AVOID WATER DAMAGE TO ADJACENT BUILDING MATERIALS. SEE COMMON SHOWER CONFIGURATIONS SECTION.

Recommended Uses

- For showers that do not have prefabricated receptors.

Environmental Exposure Classifications

- Res1, 2, 3, 5; Com1, 2, 3, 5.
- For installations that may be exposed to staining and/or chemical attack, specify tile and grout suitable for exposure. Consult product manufacturers; see also "Product Selection Guides."

Limitations

- Maximum stud spacing 16" on center.

Membrane Options

- A waterproof membrane (A118.10) or vapor retarder membrane (A108.02-3.8) must be specified for walls to prevent moisture intrusion and protect adjacent-building materials. Specifier shall indicate if complete waterproofing of walls is required, including treatment at termination points.
- Check with membrane manufacturer for suitability for applicable conditions, as not all membranes are suitable for steam, high-temperature and/or chemical exposure, or exterior use.

Requirements

- Wood studs—dry and well-braced, minimum depth 3½."
- Metal studs—well-braced; 20 gauge (0.033") or heavier; minimum depth 3½" for residential applications or 3⅝" for commercial applications.
- Surface of units—clean and free of dirt, dust, paint, and oily film.
- Membrane behind backer unit, when used, must lap over shower pan membrane.
- Slope shower pan membrane ¼" per foot to weep holes in drain.
- Turn shower pan membrane up walls a minimum of 3" above shower curb (6" above floor in showers without curbs).
- Fur out studs above shower pan membrane or notch-out studs behind the shower pan membrane so folds/corners of shower pan membrane do not cause backer board to bow inward, or use Alternate Receptor Base Method.
- Surround drain with pea gravel or other weep protection to prevent mortar from blocking weep holes.
- Minimum grout joint width—1/16."

Preparation by Backer Board Installers

- Maximum allowable variation in the tile substrate—⅛" in 10' from the required plane when measured from the high points in the surface.
- Horizontal joints—⅛" spacing filled solid and taped with latex-portland cement mortar and 2" alkali-resistant glass fiber mesh tape.
- Vertical joints—fill any space and tape with latex-portland cement mortar and 2" alkali-resistant glass fiber mesh tape.
- Corners—leave space between backer units. Tape joints using skim coat of latex-portland cement mortar, but do not fill.
- Center backer board end or edge joints on framing and stagger joints in adjacent rows so four corners do not come together within the same plane. Space panel ends and edges in accordance with manufacturer's recommendations.

Materials

- **Multiple options exist for membranes, mortars, grouts, and other materials and *must be clearly specified* to be included. If not specifically indicated, optional materials are not included and mortar/grout choice defaults to minimum performance specification indicated. Consider each system component and intended use to determine minimum requirements and to specify options.**
- Natural stone tile—see "Natural Stone Tile Selection and Installation Guide," and consult supplier for application suitability.
- Cementitious grout—ANSI A118.6 or better or ISO CG1 or better. Consider unsanded grout for soft or semisoft stones such as limestone, travertine, marble, onyx, or similar.
- Cementitious bond coat:
  - When a waterproof membrane is not used—ANSI A118.1 or better or ISO C1 or better.
  - When a waterproof membrane is used—ANSI A118.4 or better or ISO C2S1 or better unless ANSI A118.1 or ISO C1 is recommended by membrane manufacturer.
  - Use white for light-colored marble, limestone, and other stones where staining or darkening from grey setting materials is possible.
- Epoxy bond coat, when used—ANSI A118.3 or ISO R1 or better. See Water Sensitivity and Fiberglass Mesh Reinforced Stone in the "Natural Stone Tile Selection and Installation Guide."
- Vapor retarder membrane, when used—ANSI A108.02-3.8.
- Waterproof membrane, when used—ANSI A118.10.
- Cement backer board, when used—ANSI A118.9 or ASTM C1325 (Type B).
- Fiber cement backer board, when used—ASTM C1288.
- Fasteners—noncorrosive and nonoxidizing.
- Fasteners meeting ASTM F2329 required in wet areas.
- 2" alkali-resistant glass fiber mesh tape.
- Flexible mildew-resistant sealant—ASTM C920.
- Metal studs—ASTM C645.
- Shower pan membrane—local building code.

Materials for Green/Sustainable Design

- See "Green Building Standards and Green Product Selection Guide" and consult manufacturers and suppliers for product sustainability and contribution to green building design.
- Consider specifying installation materials that meet ANSI A138.1, the *American National Standard Specifications for Sustainable Ceramic Tiles, Glass Tiles, and Tile Installation Materials.*

Preparation by Other Trades

- Wall framing shall meet the general framing requirements of ANSI A108.11-4.0–4.3.
- Apply blocking between the studs to support the shower pan membrane.

Movement Joint (architect must specify type of joint and show location and details on drawings)

- Movement joints—mandatory according to EJ171.

Installation Specifications

- Shower pan membrane—ANSI A108.01-3.6.
- Fiber cement backer board—manufacturer's directions.
- Cement backer board—ANSI A108.11.
- Tile—ANSI A108.5 or A108.6.
- Cementitious grout—ANSI A108.10.
- Waterproof membrane—ANSI A108.13.
- Movement Joints—EJ171 and ASTM C1193.

Notes

- Test shower pan membrane and drainage fitting for leaks before commencing tilework.
- As the tile size increases, there is less tolerance for variation in the substrate from the required plane. Epoxy bond coat thickness must be thin and uniform; therefore, substrate flattening may be required when epoxy bond coat is used.
- Materials adversely affected by moisture in areas immediately adjacent to showers, tubs, and roman tubs should be properly protected.
- A sloped portland cement mortar fill or approved preformed slope may be used under shower pan membrane when subfloor is not sloped to drain.
- All horizontal surfaces, for example shower seats, sills, curbs, etc., must slope towards drain or other surface sloped toward drain. Where present, waterproofing also must be sloped.
- When unsanded grout is used, grout joint width must be ¹⁄₁₆" to ⅛" wide (nominal).

B420-21
STONE

- **Wood or Metal Studs**
- **Coated Glass Mat Water-Resistant Gypsum Backer Board Walls**
- **Mortar Bed Floor**
- **Natural Stone Tile**

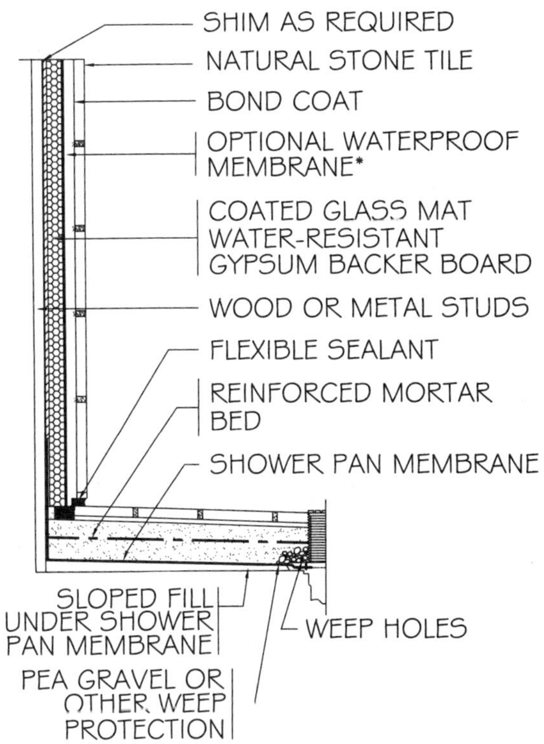

*USE OF A MEMBRANE ON WALLS IS OPTIONAL. SEE MEMBRANE OPTIONS.

SHOWER RECEPTORS, CURBS, SEATS, ETC., MUST BE PROPERLY WATERPROOFED AND INSTALLED TO AVOID WATER DAMAGE TO ADJACENT BUILDING MATERIALS. SEE COMMON SHOWER CONFIGURATIONS SECTION.

Recommended Uses

- For showers that do not have prefabricated receptors.

Environmental Exposure Classifications

- Res1, 2, 3; Com1, 2, 3.
- For installations that may be exposed to staining and/or chemical attack, specify tile and grout suitable for exposure. Consult product manufacturers; see also "Product Selection Guides."

Limitations

- Maximum stud spacing 16" on center.
- Not for areas exposed to temperatures exceeding 125°F.

Membrane Options

- A waterproof membrane (A118.10) may be specified for walls to prevent moisture intrusion through seams, corners, fasteners, and other penetrations and to protect adjacent building materials. Follow backer board and membrane manufacturer's waterproofing requirements. Specifier shall indicate if complete waterproofing of walls is required, including treatment at termination points.
- Check with membrane manufacturer for suitability for applicable conditions, as not all membranes are suitable for steam, high-temperature and/or chemical exposure, or exterior use.
- Some liquid-applied waterproof membranes will require extended cure time. Consult membrane manufacturer.

Requirements

- Wood studs—dry and well-braced, minimum depth 3½."
- Metal studs—well-braced; 20 gauge (0.033") or heavier; minimum depth 3½" for residential applications or 3⅝" for commercial applications.
- Caulk or seal penetrations, corners, and abutments to dissimilar materials with flexible sealant.
- Slope shower pan membrane ¼" per foot to weep holes in drain.
- Turn shower pan membrane up walls a minimum of 3" above shower curb (6" above floor in showers without curbs).
- Fur out studs above shower pan membrane or notch-out studs behind the shower pan membrane so folds/corners of shower pan membrane do not cause backer board to bow inward, or use Alternate Receptor Base Method.
- Surround drain with pea gravel or other weep protection to prevent mortar from blocking weep holes.
- Minimum grout joint width—1/16."

Preparation by Backer Board Installers

- Maximum allowable variation in the tile substrate—⅛" in 10' from the required plane when measured from the high points in the surface.
- Fit ends and edges closely to eliminate gaps between panels, including in corners.
- Tape seams and corners with 2" alkali-resistant glass fiber mesh tape and cementitious bonding material, and skim coat over fastener heads, unless waterproofing has been applied.
- Center backer board end or edge joints on framing and stagger joints in adjacent rows so four corners do not come together within the same plane. Space panel ends and edges in accordance with manufacturer's recommendations.

Materials

- **Multiple options exist for membranes, mortars, grouts, and other materials and *must be clearly specified* to be included. If not specifically indicated,**

optional materials are not included and mortar/grout choice defaults to minimum performance specification indicated. Consider each system component and intended use to determine minimum requirements and to specify options.

- Natural stone tile—see "Natural Stone Tile Selection and Installation Guide," and consult supplier for application suitability.
- Cementitious grout—ANSI A118.6 or better or ISO CG1 or better. Consider unsanded grout for soft or semisoft stones such as limestone, travertine, marble, onyx, or similar.
- Cementitious bond coat:
  - When a waterproof membrane is not used—ANSI A118.4 or better or ISO C2S1 or better.
  - When a waterproof membrane is used—ANSI A118.4 or better or ISO C2S1 or better unless ANSI A118.1 or ISO C1 is recommended by membrane manufacturer.
  - Use white for light-colored marble, limestone, and other stones where staining or darkening from grey setting materials is possible.
- Epoxy bond coat, when used—ANSI A118.3 or ISO R1 or better. See Water Sensitivity and Fiberglass Mesh Reinforced Stone in the "Natural Stone Tile Selection and Installation Guide."
- Waterproof membrane, when used—ANSI A118.10.
- Coated glass mat, water-resistant gypsum backer board—ASTM C1178.
- Fasteners—noncorrosive and nonoxidizing.
- Fasteners meeting ASTM F2329 required in wet areas.
- 2" alkali-resistant glass fiber mesh tape.
- Flexible mildew-resistant sealant—ASTM C920.
- Metal studs—ASTM C645.
- Shower pan membrane—local building code.

Materials for Green/Sustainable Design

- See "Green Building Standards and Green Product Selection Guide" and consult manufacturers and suppliers for product sustainability and contribution to green building design.
- Consider specifying installation materials that meet ANSI A138.1, the *American National Standard Specifications for Sustainable Ceramic Tiles, Glass Tiles, and Tile Installation Materials.*

Preparation by Other Trades

- Wall framing shall meet the general framing requirements of ANSI A108.11-4.0–4.3.
- Apply blocking between the studs to support the shower pan.

Movement Joint (architect must specify type of joint and show location and details on drawings)

- Movement joints—mandatory according to EJ171.

Installation Specifications

- Shower pan membrane—ANSI A108.01-3.6.
- Coated glass mat water-resistant gypsum backer board—manufacturer's directions.
- Tile—A108.5 or A108.6.
- Cementitious grout—ANSI A108.10.
- Waterproof membrane—ANSI A108.13.
- Movement Joints—EJ171 and ASTM C1193.

Notes

- Test shower pan membrane and drainage fitting for leaks before commencing tilework.
- Materials adversely affected by moisture in areas immediately adjacent to showers, tubs, and roman tubs should be properly protected.
- A sloped portland cement mortar fill or approved preformed slope may be used under shower pan when subfloor is not sloped to drain.
- All horizontal surfaces, for example shower seats, sills, curbs, etc., must slope towards drain or other surface sloped toward drain. Where present, waterproofing also must be sloped.
- As the tile size increases, there is less tolerance for variation in the substrate from the required plane. Epoxy bond coat thickness must be thin and uniform; therefore, substrate flattening may be required when epoxy bond coat is used.
- When unsanded grout is used, grout joint width must be 1/16" to 1/8" wide (nominal).

B426-21
STONE

- **Wood or Metal Studs**
- **Cementitious-Coated Extruded Foam Backer Board Walls**
- **Mortar Bed Floor**
- **Natural Stone Tile**

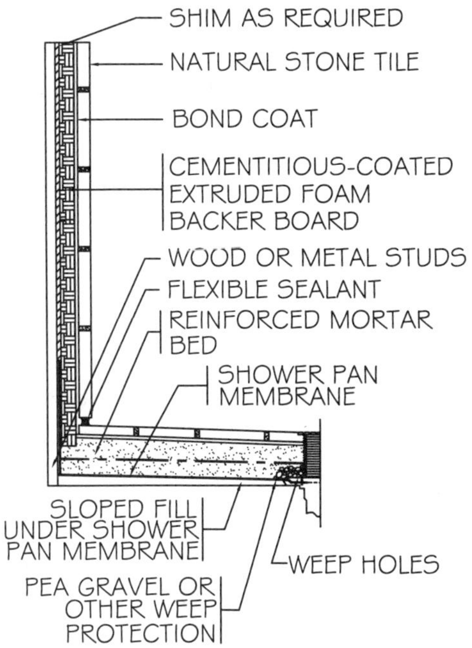

SHOWER RECEPTORS, CURBS, SEATS, ETC., MUST BE PROPERLY WATERPROOFED AND INSTALLED TO AVOID WATER DAMAGE TO ADJACENT BUILDING MATERIALS. SEE COMMON SHOWER CONFIGURATIONS SECTION.

Recommended Uses

- For showers that do not have prefabricated receptors.

Environmental Exposure Classifications

- Res1, 2, 3; Com1, 2, 3.
- For installations that may be exposed to staining and/or chemical attack, specify tile and grout suitable for exposure. Consult product manufacturers; see also "Product Selection Guides."

Limitations

- Maximum stud spacing 16" on center.

Requirements

- Wood studs—dry and well-braced, minimum depth 3½".
- Metal studs—well-braced; 20 gauge (0.033") or heavier; minimum depth 3½" for residential applications or 3⅝" for commercial applications.
- Slope shower pan membrane ¼" per foot to weep holes in drain.
- Turn shower pan membrane up walls a minimum of 3" above shower curb (6" above floor in showers without curbs).
- Fur out studs above shower pan membrane or notch-out studs behind the shower pan membrane so folds/corners of shower pan membrane do not cause backer board to bow inward, or use Alternate Receptor Base Method.
- Surround drain with pea gravel or other weep protection to prevent mortar from blocking weep holes.
- Minimum grout joint width—1/16".

Preparation by Backer Board Installers

- Fasten backer board to studs with board manufacturer's recommended fasteners.
- All openings cut in backer board for plumbing and all cut joints between adjoining pieces—seal with waterproofing as recommended by backer board manufacturer.
- Maximum allowable variation in the tile substrate—⅛" in 10' from the required plane when measured from the high points in the surface.
- Fit ends and edges tightly together, including in corners. Apply a bead of manufacturer-recommended sealant in panel joints as panels are being installed and over fastener heads per backer board manufacturer's instructions.
- Center backer board end or edge joints on framing and stagger joints in adjacent rows so four corners do not come together within the same plane. Space panel ends and edges in accordance with manufacturer's recommendations.

Materials

- **Multiple options exist for membranes, mortars, grouts, and other materials and *must be clearly specified* to be included. If not specifically indicated, optional materials are not included and mortar/grout choice defaults to minimum performance specification indicated. Consider each system component and intended use to determine minimum requirements and to specify options.**
- Natural stone tile—see "Natural Stone Tile Selection and Installation Guide," and consult supplier for application suitability.
- Cementitious grout—ANSI A118.6 or better or ISO CG1 or better. Consider unsanded grout for soft or semisoft stones such as limestone, travertine, marble, onyx, or similar.
- Cementitious bond coat:
  - ANSI A118.1 or better or ISO C1 or better.
  - Use white for light-colored marble, limestone, and other stones where staining or darkening from grey setting materials is possible.
- Epoxy bond coat, when used—ANSI A118.3 or ISO R1 or better. See Water Sensitivity and Fiberglass Mesh Reinforced Stone in the "Natural Stone Tile Selection and Installation Guide."
- Cementitious-coated extruded foam backer board—ASTM C578.

- Fasteners—noncorrosive and nonoxidizing.
- Flexible sealant—must be recommended by backer board manufacturer.
- Shower pan membrane—local building code.
- Metal studs—ASTM C645.

Materials for Green/Sustainable Design

- See "Green Building Standards and Green Product Selection Guide" and consult manufacturers and suppliers for product sustainability and contribution to green building design.
- Consider specifying installation materials that meet ANSI A138.1, the *American National Standard Specifications for Sustainable Ceramic Tiles, Glass Tiles, and Tile Installation Materials.*

Preparation by Other Trades

- Apply blocking between studs to support the shower pan membrane.
- Wall framing shall meet the general framing requirements of ANSI A108.11-4.0-4.3.

Movement Joint (architect must specify type of joint and show location and details on drawings)

- Movement joints—mandatory according to EJ171.

Installation Specifications

- Shower pan membrane—ANSI A108.01-3.6.
- Cementitious-coated extruded foam backer board—manufacturer's directions.
- Tile—ANSI A108.5 or A108.6.
- Cementitious grout—ANSI A108.10.
- Movement Joints—EJ171 and ASTM C1193.

Notes

- Test shower pan membrane and drainage fitting for leaks before commencing tilework.
- Materials adversely affected by moisture in areas immediately adjacent to showers, tubs, and roman tubs should be properly protected.
- A sloped portland cement mortar fill or approved preformed slope may be used under shower pan membrane when subfloor is not sloped to drain.
- All horizontal surfaces, for example shower seats, sills, curbs, etc., must slope towards drain or other surface sloped toward drain. Where present, waterproofing also must be sloped.
- As the tile size increases, there is less tolerance for variation in the substrate from the required plane. Epoxy bond coat thickness must be thin and uniform; therefore, substrate flattening may be required when epoxy bond coat is used.
- When unsanded grout is used, grout joint width must be 1⁄16" to 1⁄8" wide (nominal).

B431-21
STONE

- **Wood or Metal Studs**
- **Fiber-Reinforced Water-Resistant Gypsum Backer Board Walls**
- **Mortar Bed Floor**
- **Natural Stone Tile**

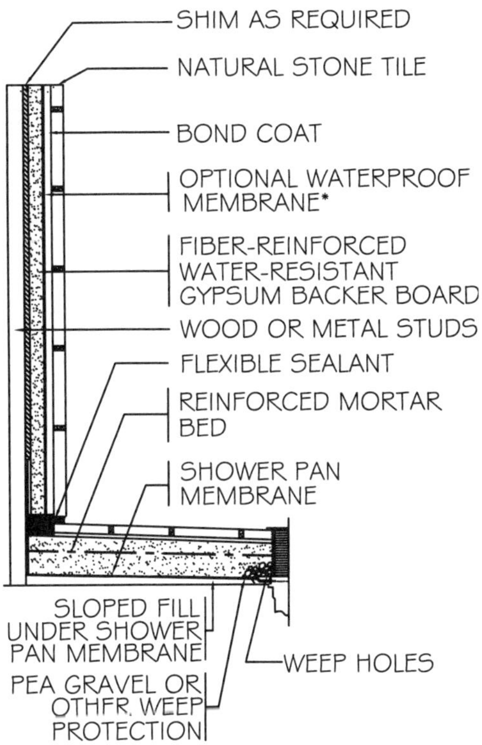

*USE OF A MEMBRANE ON WALLS IS OPTIONAL. SEE MEMBRANE OPTIONS.

SHOWER RECEPTORS, CURBS, SEATS, ETC., MUST BE PROPERLY WATERPROOFED AND INSTALLED TO AVOID WATER DAMAGE TO ADJACENT BUILDING MATERIALS. SEE COMMON SHOWER CONFIGURATIONS SECTION.

Recommended Uses

- For showers that do not have prefabricated receptors.

Environmental Exposure Classifications

- Res1, 2, 3; Com1, 2.
- For installations that may be exposed to staining and/or chemical attack, specify tile and grout suitable for exposure. Consult product manufacturers; see also "Product Selection Guides."

Limitations

- Maximum stud spacing 16" on center.
- Not for areas exposed to temperatures exceeding 125°F.

Membrane Options

- A waterproof membrane (A118.10) may be specified for walls to prevent moisture intrusion and protect adjacent building materials. Specifier shall indicate if complete waterproofing of walls is required, including treatment at termination points.
- Check with membrane manufacturer for suitability for applicable conditions, as not all membranes are suitable for steam, high-temperature and/or chemical exposure, or exterior use.

Requirements

- Wood studs—dry and well-braced, minimum depth 3½."
- Metal studs—well-braced; 20 gauge (0.033") or heavier; minimum depth 3½" for residential applications or 3⅝" for commercial applications.
- Slope shower pan membrane ¼" per foot to weep holes in drain.
- Turn shower pan membrane up walls a minimum of 3" above shower curb (6" above floor in showers without curbs).
- Fur out studs above shower pan membrane or notch-out studs behind the shower pan membrane so folds/corners of shower pan membrane do not cause backer board to bow inward, or use Alternate Receptor Base Method.
- Surround drain with pea gravel or other weep protection to prevent mortar from blocking weep holes.
- Caulk or seal penetrations and abutments to dissimilar materials.
- Minimum grout joint width—¹⁄₁₆."

Preparation by Backer Board Installers

- Maximum allowable variation in the tile substrate—⅛" in 10' from the required plane when measured from the high points in the surface.
- Horizontal joints—⅛" spacing filled solid and taped with latex-portland cement mortar and 2" alkali-resistant glass fiber mesh tape.
- Vertical joints—fill any space and tape with latex-portland cement mortar and 2" alkali-resistant glass fiber mesh tape.
- Corners—leave space between backer units. Tape joints using skim coat of latex-portland cement mortar, but do not fill.
- Center backer board end or edge joints on framing and stagger joints in adjacent rows so four corners do not come together within the same plane. Space panel ends and edges in accordance with manufacturer's recommendations.

Materials

- **Multiple options exist for membranes, mortars, grouts, and other materials and *must be clearly specified* to be included. If not specifically indicated, optional materials are not included and mortar/grout**

choice defaults to minimum performance specification indicated. Consider each system component and intended use to determine minimum requirements and to specify options.

- Natural stone tile—see "Natural Stone Tile Selection and Installation Guide," and consult supplier for application suitability.
- Cementitious grout—ANSI A118.6 or better or ISO CG1 or better. Consider unsanded grout for soft or semisoft stones such as limestone, travertine, marble, onyx, or similar.
- Cementitious bond coat:
  - When a waterproof membrane is not used—ANSI A118.4 or better or ISO C2 or better.
  - When a waterproof membrane is used—ANSI A118.4 or better or ISO C2S1 or better unless ANSI A118.1 or ISO C1 is recommended by membrane manufacturer.
  - Use white for light-colored marble, limestone, and other stones where staining or darkening from grey setting materials is possible.
- Epoxy bond coat, when used—ANSI A118.3 or ISO R1 or better. See Water Sensitivity and Fiberglass Mesh Reinforced Stone in the "Natural Stone Tile Selection and Installation Guide."
- Waterproof membrane, when used—ANSI A118.10.
- Fiber-reinforced water-resistant gypsum backer board—ASTM C1278 (Paragraph 6.1).
- Fasteners—noncorrosive and nonoxidizing.
- 2" alkali-resistant glass fiber mesh tape.
- Flexible mildew-resistant sealant—ASTM C920.
- Shower pan membrane—local building code.
- Metal studs—ASTM C645.

Materials for Green/Sustainable Design

- See "Green Building Standards and Green Product Selection Guide" and consult manufacturers and suppliers for product sustainability and contribution to green building design.
- Consider specifying installation materials that meet ANSI A138.1, the *American National Standard Specifications for Sustainable Ceramic Tiles, Glass Tiles, and Tile Installation Materials.*

Preparation by Other Trades

- Apply blocking between the studs to support the shower pan membrane.
- Wall framing shall meet the general framing requirements of ANSI A108.11-4.0–4.3.

Movement Joint (architect must specify type of joint and show location and details on drawings)

- Movement joints—mandatory in accordance with EJ171.

Installation Specifications

- Shower pan membrane—ANSI A108.01-3.6.
- Fiber-reinforced water-resistant gypsum backer board—manufacturer's directions.
- Tile—ANSI A108.5 or A108.6.
- Cementitious grout—ANSI A108.10.
- Waterproof membrane—ANSI A108.13.
- Movement Joints—EJ171 and ASTM C1193.

Notes

- As the tile size increases, there is less tolerance for variation in the substrate from the required plane. Epoxy bond coat thickness must be thin and uniform; therefore, substrate flattening may be required when epoxy bond coat is used.
- Test shower pan membrane and drainage fitting for leaks before commencing tilework.
- Materials adversely affected by moisture in areas immediately adjacent to showers, tubs, and roman tubs should be properly protected.
- A sloped portland cement mortar fill or approved preformed slope may be used under shower pan membrane when subfloor is not sloped to drain.
- All horizontal surfaces, for example shower seats, sills, curbs, etc., must slope towards drain or other surface sloped toward drain. Where present, waterproofing also must be sloped.
- When unsanded grout is used, grout joint width must be $\frac{1}{16}$" to $\frac{1}{8}$" wide (nominal).

B421-21 STONE

- **Solid Backing**
- **Bonded Waterproof Membrane**
- **Natural Stone Tile**

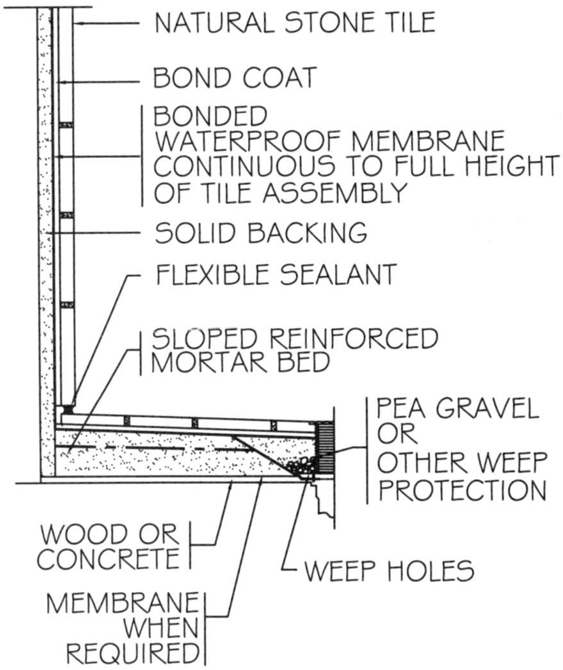

SHOWER RECEPTORS, CURBS, SEATS, ETC., MUST BE PROPERLY WATERPROOFED AND INSTALLED TO AVOID WATER DAMAGE TO ADJACENT BUILDING MATERIALS. SEE COMMON SHOWER CONFIGURATIONS SECTION.

Recommended Uses

- For showers that do not have a prefabricated receptor, where topical waterproofing of floors and walls is desired.

Environmental Exposure Classifications

- Res1, 2, 3; Com1, 2.
- May be suitable for Com3 as determined by membrane and backing material manufacturers.
- For installations that may be exposed to staining and/or chemical attack, specify tile and grout suitable for exposure. Consult product manufacturers; see also "Product Selection Guides."

Limitations

- Refer to appropriate wall method for applicable limitations based on type of backing used—W201, W202E, W202I, W211, W221, W222, W231/W241, W243, W244C. W244E, W244F, W245, W246, and W247.

Requirements

- Bonded waterproof membrane must be continuous, including at changes in plane. Follow membrane manufacturer's requirements for corners, seaming, and overlap.
- Waterproof membrane must extend to full height of tile assembly.
- Surround drain with pea gravel or other weep protection to prevent mortar from blocking weep holes.
- Refer to appropriate wall method for applicable requirements based on type of backing used.
- Slope mortar bed ¼" per foot toward drain and follow membrane manufacturer's instructions for connecting membrane to drain.
- Minimum grout joint width—$\frac{1}{16}$".

Materials

- **Multiple options exist for membranes, mortars, grouts, and other materials and *must be clearly specified* to be included. If not specifically indicated, optional materials are not included and mortar/grout choice defaults to minimum performance specification indicated. Consider each system component and intended use to determine minimum requirements and to specify options.**
- Natural stone tile—see "Natural Stone Tile Selection and Installation Guide," and consult supplier for application suitability.
- Cementitious grout—ANSI A118.6 or better or ISO CG1 or better. Consider unsanded grout for soft or semisoft stones such as limestone, travertine, marble, onyx, or similar.
- Cementitious bond coat:
  - ANSI A118.4 or better or ISO C2S1 or better.
  - Use white for light-colored marble, limestone, and other stones where staining or darkening from grey setting materials is possible.
- Epoxy bond coat, when used—ANSI A118.3 or ISO R1 or better. See Water Sensitivity and Fiberglass Mesh Reinforced Stone in the "Natural Stone Tile Selection and Installation Guide."
- Waterproof membrane—ANSI A118.10 and recommended by manufacturer of membrane for use over backing type in intended application.
- Flexible mildew-resistant sealant—ASTM C920.

Materials for Green/Sustainable Design

- See "Green Building Standards and Green Product Selection Guide" and consult manufacturers and suppliers for product sustainability and contribution to green building design.

- Consider specifying installation materials that meet ANSI A138.1, the *American National Standard Specifications for Sustainable Ceramic Tiles, Glass Tiles, and Tile Installation Materials.*

Preparation by Other Trades/Backer Board Installers

- Refer to appropriate wall method for applicable preparations by other trades and backer board installers.
- Maximum allowable variation in the tile substrate—⅛" in 10' from the required plane when measured from the high points in the surface.
- Center backer board end or edge joints on framing and stagger joints in adjacent rows so four corners do not come together within the same plane. Space panel ends and edges in accordance with manufacturer's recommendations.

Movement Joint (architect must specify type of joint and show location and details on drawings)

- Movement joints—mandatory according to EJ171.

Installation Specifications

- Waterproof membrane—ANSI A108.13.
- Tile—ANSI A108.5 or A108.6.
- Cementitious grout—ANSI A108.10.
- Movement Joints—EJ171 and ASTM C1193.

Notes

- Test shower pan membrane/waterproof membrane and drainage fitting for leaks before commencing tilework.
- Materials adversely affected by moisture in areas immediately adjacent to showers, tubs, and roman tubs should be properly protected.
- All horizontal surfaces, for example shower seats, sills, curbs, etc., must slope towards drain or other surface sloped toward drain. Where present, waterproofing also must be sloped.
- When unsanded grout is used, grout joint width must be $\frac{1}{16}$" to ⅛" wide (nominal).

B422-21
STONE

- **Solid Backing**
- **Bonded Waterproof Membrane**
- **Integrated Bonding Flange**
- **Natural Stone Tile**

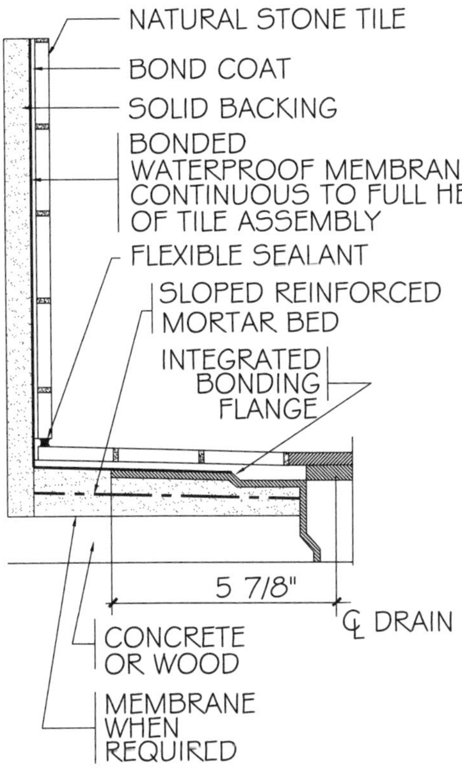

SHOWER RECEPTORS, CURBS, SEATS, ETC., MUST BE PROPERLY WATERPROOFED AND INSTALLED TO AVOID WATER DAMAGE TO ADJACENT BUILDING MATERIALS. SEE COMMON SHOWER CONFIGURATIONS SECTION.

Recommended Uses

- For showers that do not have a prefabricated receptor, where topical waterproofing of floors and walls is desired.

Environmental Exposure Classifications

- Res1, 2, 3; Com1, 2
- May be suitable for Com3 as determined by membrane and backing material manufacturers.
- For installations that may be exposed to staining and/or chemical attack, specify tile and grout suitable for exposure. Consult product manufacturers; see also "Product Selection Guides."

Limitations

- Refer to appropriate wall method for applicable limitations based on type of backing used—W201, W202E, W202I, W211, W221, W222, W231/W241, W243, W244C. W244E, W244F, W245, W246, and W247.

Requirements

- Bonded waterproof membrane must be continuous, including at changes in plane. Follow membrane manufacturer's requirements for corners, seaming, and overlap.
- Waterproof membrane must extend to full height of tile assembly.
- Refer to appropriate wall method for applicable requirements based on type of backing used.
- Slope mortar bed ¼" per foot toward drain and follow membrane and integrated bonding flange manufacturers' instructions for connecting membrane to integrated bonding flange.
- Minimum grout joint width—1⁄16".

Materials

- **Multiple options exist for membranes, mortars, grouts, and other materials and *must be clearly specified* to be included. If not specifically indicated, optional materials are not included and mortar/grout choice defaults to minimum performance specification indicated. Consider each system component and intended use to determine minimum requirements and to specify options.**
- Natural stone tile—see "Natural Stone Tile Selection and Installation Guide," and consult supplier for application suitability.
- Cementitious grout—ANSI A118.6 or better or ISO CG1 or better. Consider unsanded grout for soft or semisoft stones such as limestone, travertine, marble, onyx, or similar.
- Cementitious bond coat:
    - ANSI A118.4 or better or ISO C2S1 or better unless ANSI A118.1 or ISO C1 is recommended by membrane manufacturer.
    - Use white for light-colored marble, limestone, and other stones where staining or darkening from grey setting materials is possible.
- Epoxy bond coat, when used—ANSI A118.3 or ISO R1 or better. See Water Sensitivity and Fiberglass Mesh Reinforced Stone in the "Natural Stone Tile Selection and Installation Guide."
- Waterproof membrane—ANSI A118.10 and recommended by manufacturer of membrane for use over backing type in intended application.
- Flexible mildew-resistant sealant—ASTM C920.

Materials for Green/Sustainable Design

- See "Green Building Standards and Green Product Selection Guide" and consult manufacturers and suppliers for product sustainability and contribution to green building design.
- Consider specifying installation materials that meet ANSI A138.1, the *American National Standard Specifications for Sustainable Ceramic Tiles, Glass Tiles, and Tile Installation Materials.*

Preparation by Other Trades/Backer Board Installers

- Refer to appropriate wall method for applicable preparations by other trades and backer board installers.
- Maximum allowable variation in the tile substrate—⅛" in 10' from the required plane when measured from the high points in the surface.
- Center backer board end or edge joints on framing and stagger joints in adjacent rows so four corners do not come together within the same plane. Space panel ends and edges in accordance with manufacturer's recommendations.

Movement Joint (architect must specify type of joint and show location and details on drawings)

- Movement joints—mandatory according to EJ171.

Installation Specifications

- Waterproof membrane—ANSI A108.13.
- Tile—ANSI A108.5 or A108.6.
- Cementitious grout—ANSI A108.10.
- Movement Joints—EJ171 and ASTM C1193.

Notes

- Test shower pan membrane/waterproof membrane and drainage fitting for leaks before commencing tilework.
- Materials adversely affected by moisture in areas immediately adjacent to showers, tubs, and roman tubs should be properly protected.
- All horizontal surfaces, for example shower seats, sills, curbs, etc., must slope towards drain or other surface sloped toward drain. Where present, waterproofing also must be sloped.
- When unsanded grout is used, grout joint width must be 1⁄16" to ⅛" wide (nominal).

C311-21 STONE

- **Mortar Bed**
- **Natural Stone Tile**

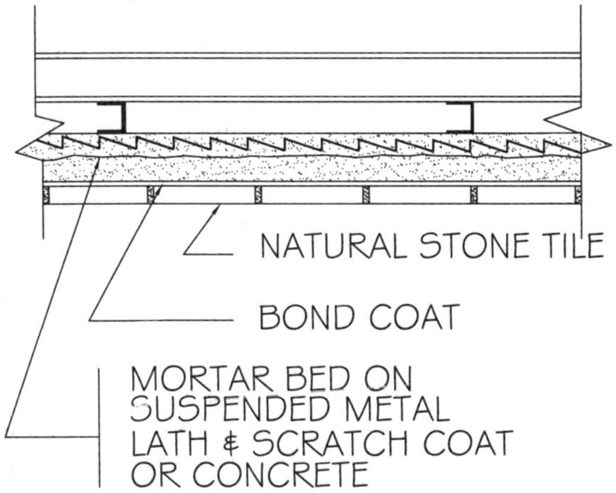

Recommended Uses

- For ceilings where trueing is required.
- For areas where substrate flatness is critical, such as when tiles with any edge longer than 15" are specified.
- For areas where ungauged stone will be used.

Environmental Exposure Classifications

- Res1, 2, 3, 5; Com1, 2, 3, 5.
- For installations that may be exposed to staining and/or chemical attack, specify tile and grout suitable for exposure. Consult product manufacturers; see also "Product Selection Guides."

Limitations

- Maximum joist spacing 16" on center.

Requirements

- Suspended ceiling systems must be capable of supporting weight of mortar bed and tile.
- Concrete to be well cured, dimensionally stable, and free of cracks, waxy, or oily films.
- Minimum grout joint width—1/16".

Materials

- **Multiple options exist for membranes, mortars, grouts, and other materials and *must be clearly specified* to be included. If not specifically indicated, optional materials are not included and mortar/grout choice defaults to minimum performance specification indicated. Consider each system component and intended use to determine minimum requirements and to specify options.**
- Natural stone tile—see "Natural Stone Tile Selection and Installation Guide," and consult supplier for application suitability.
- Cementitious grout—ANSI A118.6 or better or ISO CG1 or better. Consider unsanded grout for soft or semisoft stones such as limestone, travertine, marble, onyx, or similar.
- Cementitious bond coat—portland cement paste on a mortar bed that is still workable. For a cured mortar bed, follow recommendations below to select appropriate bonding mortar:
  - ANSI A118.1 or better or ISO C1 or better.
  - Use white for light-colored marble, limestone, and other stones where staining or darkening from grey setting materials is possible.
- Epoxy bond coat, when used—ANSI A118.3 or ISO R1 or better. See Water Sensitivity and Fiberglass Mesh Reinforced Stone in the "Natural Stone Tile Selection and Installation Guide."
- Mortar bed, metal lath, and cleavage membrane—ANSI A108.1A.

Materials for Green/Sustainable Design

- See "Green Building Standards and Green Product Selection Guide" and consult manufacturers and suppliers for product sustainability and contribution to green building design.
- Consider specifying installation materials that meet ANSI A138.1, the *American National Standard Specifications for Sustainable Ceramic Tiles, Glass Tiles, and Tile Installation Materials.*

Preparation by Other Trades

- Maximum allowable variation in the installation substrate—1/4" in 10' from the required plane.

Movement Joint (architect must specify type of joint and show location and details on drawings)

- Movement joints—mandatory according to EJ171.

Installation Specifications

- Tile—ANSI A108.1A, .1B, or .1C. Ungauged stone may require installation per ANSI A108.1A. Material with minimal variation may be allowed per A108.1B with LHT mortar. A108.1B required if epoxy bond coat to be used.
- Cementitious grout—ANSI A108.10.
- Movement Joints—EJ171 and ASTM C1193.

Notes

- When unsanded grout is used, grout joint width must be 1/16" to 1/8" wide (nominal).

C312-21
STONE

- **Gypsum Board**
- **Natural Stone Tile**

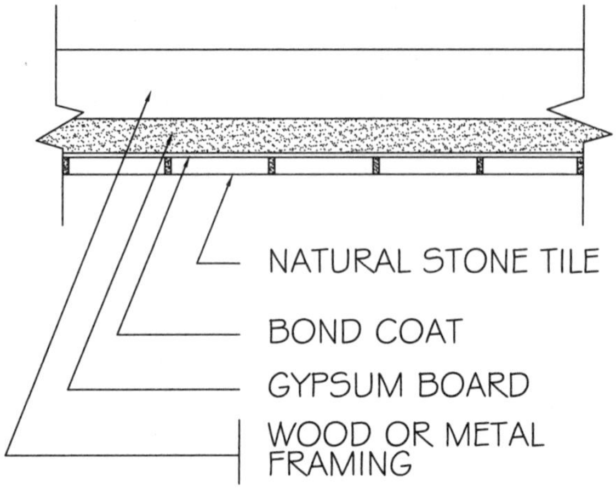

Recommended Uses

- For ceilings in dry areas where gypsum board is the tile backer.

Environmental Exposure Classifications

- Res1; Com1.

Limitations

- The weight of overlaid unsupported insulation shall not exceed 1.3 psf for ½"-thick gypsum board with framing spacing 24" o.c. or 2.2 psf for ½"-thick gypsum board with framing spacing 16" o.c. or ⅝"-thick gypsum board with framing spacing 24" o.c.
- Not for areas exposed to temperatures exceeding 125°F.
- Ceiling framing, furring, or blocking shall not exceed 12" o.c. for ½"-thick water-resistant gypsum backing board and 16" o.c. for ⅝"-thick water-resistant gypsum backing board.

Requirements

- Surface—free of coatings, oil, and wax.
- Ceiling framing must be capable of supporting the weight of backing and tile.
- Minimum thickness of gypsum board—½."
- Minimum grout joint width—1/16."

Materials

- **Multiple options exist for membranes, mortars, grouts, and other materials and *must be clearly specified* to be included. If not specifically indicated, optional materials are not included and mortar/grout choice defaults to minimum performance specification indicated. Consider each system component and intended use to determine minimum requirements and to specify options.**
- Natural stone tile—see "Natural Stone Tile Selection and Installation Guide," and consult supplier for application suitability.
- Cementitious grout—ANSI A118.6 or better or ISO CG1 or better. Consider unsanded grout for soft or semisoft stones such as limestone, travertine, marble, onyx, or similar.
- Cementitious bond coat:
  - ANSI A118.1 or better or ISO C1 or better.
  - Use white for light-colored marble, limestone, and other stones where staining or darkening from grey setting materials is possible.
- Epoxy bond coat, when used—ANSI A118.3 or ISO R1 or better. See Water Sensitivity and Fiberglass Mesh Reinforced Stone in the "Natural Stone Tile Selection and Installation Guide."

Materials for Green/Sustainable Design

- See "Green Building Standards and Green Product Selection Guide" and consult manufacturers and suppliers for product sustainability and contribution to green building design.
- Consider specifying installation materials that meet ANSI A138.1, the *American National Standard Specifications for Sustainable Ceramic Tiles, Glass Tiles, and Tile Installation Materials.*

Preparation by Other Trades

- Gypsum board face layer joints—treated with tape and joint compound, bedding coat only (no finish coats). Nail heads, one coat only.
- Maximum allowable variation in the tile substrate—⅛" in 10' from the required plane when measured from the high points in the surface.

Movement Joint (architect must specify type of joint and show location and details on drawings)

- Movement joints—mandatory according to EJ171.

Installation Specifications

- Gypsum board—GA-216.
- Tile—ANSI A108.5 or A108.6.
- Cementitious grout—ANSI A108.10.
- Movement Joints—EJ171 and ASTM C1193.

Notes

- As the tile size increases, there is less tolerance for variation in the substrate from the required plane. Epoxy bond coat thickness must be thin and uniform; therefore, substrate flattening may be required when epoxy bond coat is used.
- When unsanded grout is used, grout joint width must be 1/16" to ⅛" wide (nominal).

C315-21
STONE

- **Backer Board**
- **Natural Stone Tile**

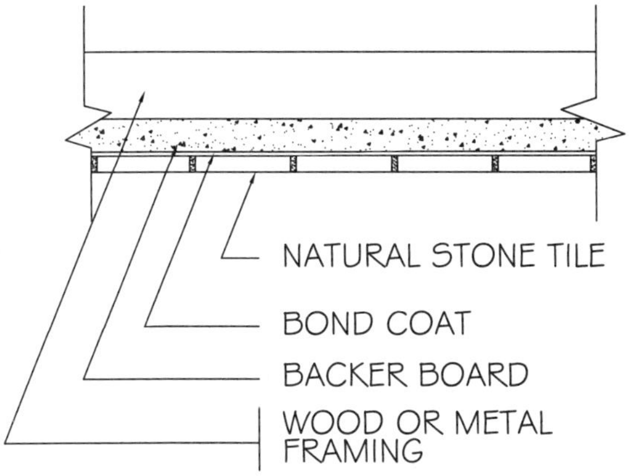

Recommended Uses

- For ceilings in dry or wet areas.

Environmental Exposure Classifications

- Res1, 2, 3; Com1, 2.

Limitations

- The weight of the tile shall not exceed 5 psf for ½"-thick backer board with framing spacing 24" o.c., or 6 psf for ½"-thick backer board with framing spacing 16" o.c.
- When backer board is coated glass mat gypsum backer board or fiber-reinforced gypsum backer board, not for areas exposed to temperatures exceeding 125°F.

Requirements

- Surface—free of coatings, oil, and wax.
- Ceiling framing must be capable of supporting the weight of backer board and tile.
- Minimum thickness of backer board—½".
- Fasten backer board as recommended by the manufacturer.
- Minimum grout joint width—1/16".

Preparation by Backer Board Installers

- Maximum allowable variation in the tile substrate—⅛" in 10' from the required plane when measured from the high points in the surface.
- Follow backer board manufacturer's directions for installation, including orientation and spacing of boards, required fasteners, and taping of joints and corners.

Materials

- **Multiple options exist for membranes, mortars, grouts, and other materials and *must be clearly specified* to be included. If not specifically indicated, optional materials are not included and mortar/grout choice defaults to minimum performance specification indicated. Consider each system component and intended use to determine minimum requirements and to specify options.**
- Natural stone tile—see "Natural Stone Tile Selection and Installation Guide," and consult supplier for application suitability.
- Cementitious grout—ANSI A118.6 or better or ISO CG1 or better. Consider unsanded grout for soft or semisoft stones such as limestone, travertine, marble, onyx, or similar.
- Cementitious bond coat:
  - When cement backer board, fiber cement backer board, or cementitious-coated extruded foam backer board is used—ANSI A118.1 or better or ISO C1 or better.
  - When coated glass mat water-resistant gypsum or fiber-reinforced water-resistant gypsum backer board is used—ANSI A118.4 or better or ISO C2S1 or better.
  - Use white for light-colored marble, limestone, and other stones where staining or darkening from grey setting materials is possible.
- Epoxy bond coat, when used—ANSI A118.3 or ISO R1 or better. See Water Sensitivity and Fiberglass Mesh Reinforced Stone in the "Natural Stone Tile Selection and Installation Guide."
- Cement backer board, when used—ANSI A118.9 or ASTM C1325.
- Fiber cement backer board, when used—ASTM C1288.
- Cementitious-coated extruded foam backer board, when used—ASTM C578.
- Coated glass mat water-resistant gypsum backer board, when used—ASTM C1178.
- Fiber-reinforced water-resistant gypsum backer board, when used—ASTM C1278 (Paragraph 6.1).
- Metal studs—ASTM C645.

Materials for Green/Sustainable Design

- See "Green Building Standards and Green Product Selection Guide" and consult manufacturers and suppliers for product sustainability and contribution to green building design.
- Consider specifying installation materials that meet ANSI A138.1, the *American National Standard Specifications for Sustainable Ceramic Tiles, Glass Tiles, and Tile Installation Materials.*

Preparation by Other Trades

- Framing shall meet the general framing requirements of ANSI A108.11-4.0-4.3.

Movement Joint (architect must specify type of joint and show location and details on drawings)

- Movement joints—mandatory according to EJ171.

Installation Specifications

- Backer board—ANSI A108.11 or manufacturer's directions.
- Tile—ANSI A108.5 or A108.6.
- Cementitious grout—ANSI A108.10.
- Movement Joints—EJ171 and ASTM C1193.

Notes

- As the tile size increases, there is less tolerance for variation in the substrate from the required plane. Epoxy bond coat thickness must be thin and uniform; therefore, substrate flattening may be required when epoxy bond coat is used.
- When unsanded grout is used, grout joint width must be 1/16" to 1/8" wide (nominal).

2021 | MOVEMENT JOINTS

EJ171 MOVEMENT JOINT GUIDELINES FOR CERAMIC, GLASS, AND STONE

Perimeter and field movement joints within a tile installation are essential and required. This section provides general recommendations and guidelines, including the means by which movement joints in substrates are carried through and incorporated into tile installations.

Because of the limitless conditions and structural systems on which tile can be installed, the design professional or engineer shall show the specific locations and details of movement joints on project drawings. Preparation of openings left by the tile contractor and installation of backup strip and sealant should be specified in the Caulking and Sealant section of the job specification.

Location and Frequency of Joints

Interior—maximum of 25' in each direction.

Exterior—8' to 12' in each direction. More frequent joint placement may be required depending on materials and environmental conditions.

Interior tilework exposed to direct sunlight (heat) or moisture—maximum of 12' in each direction.

Above-ground concrete slab substrate—maximum of 12' in each direction.

Perimeter Joints—movement joints are required where tilework abuts restraining surfaces such as perimeter walls, dissimilar floor finishes, curbs, columns, pipes, ceilings, and where changes occur in backing materials, but not at drain strainers.

Change of plane, exterior—movement joints required at all inside and outside corners.

Change of plane, interior—movement joints required at all inside corners.

All expansion, control, construction, cold, saw-cut, isolation, contraction, and seismic joints in the structure should continue through the tilework, including such joints at vertical surfaces.

If a proprietary crack isolation membrane is specified over a saw cut joint to relocate a movement joint, contrary to EJ171, the tile contractor is not responsible for cracking in grout joints or tile over the saw cut, provided the tile, membrane, and other materials were installed correctly. This includes curling and/or deformation of the concrete occurring after installation of the membrane.

Joint Width

The following recommendations do not consider frequency and width for substrates subject to deflection, shrinkage, and expansion. All example calculations only consider the thermal linear expansion of the tile. All other considerations are specifically excluded and must be estimated by the design professional or engineer.

Exterior

Width shall be a minimum ⅜" for joints 8' on center and minimum ½" for joints 12' on center. Minimum widths must be increased 1/16" for each 15°F tile surface temperature change greater than 100°F between summer high and winter low. For example, for a 115°F temperature change, add 1/16" to the minimum joint width; for a 130°F change, add ⅛."

Interior

Perimeter joints at perimeter walls—not less than ¼" and commonly covered by tile trim, cove base, or shoe molding.

Perimeter joints, other than perimeter walls—preferred not less than ¼" but never less than ⅛."

Change of plane—same as grout joint but never less than ⅛."

Dry Interior Generic Movement Joints Not Exposed to Direct Sunlight

Formula for Calculating

To calculate the approximate minimum movement joint width (sealant joint) for dry interior generic movement joints not exposed to direct sunlight considering (with respect to expected movement) only linear thermal expansion of the tile, use the formula:

$$W_M = \frac{(L \times \Delta T \times \alpha)}{\left(\frac{S}{100}\right)}$$

W_M Sealant joint width required for movement

L Total linear dimension between movement joints expressed in inches (in.)

ΔT Change in temperature expressed in degrees Celsius (°C)

α Coefficient of linear thermal expansion expressed in inches per inch per degree Celsius (in./in.°C)

S Sealant movement rating or class per ASTM C920

Note: Calculation does not consider concrete shrinkage from curing and other forms of substrate shrinkage or expansion.

Sample calculation for approximate minimum required movement joint width when:

coefficient of linear thermal expansion of the tile (α):
7×10^{-6} in./in.°C

total linear dimension between movement joints (L):
240" (every 20')

change in temperature (ΔT):
27.8°C (10°C to 37.8°C, or 50°F to 100°F)

sealant movement rating or class (S):
25

$$W_M = \frac{(240\text{in.} \times 27.8°\text{C} \times .000007\text{in./in.°C})}{(25/100)} = 0.1868 \text{ in. (approximately } 3/16\text{")}$$

Using the formula, in the above sample calculation, 3/16" movement joints are needed every 20'.

EJ171 TABLE 1: **Pre-calculated approximate minimum required movement joint widths for dry interior generic movement joints not exposed to direct sunlight, based (with respect to movement) only on linear thermal expansion of the tile and various options for joint frequency, assuming a class 25 sealant and minimum and maximum temperatures of 50°F (10°C) and 100°F (37.8°C), respectively.**

Tile's Coefficient of Linear Thermal Expansion (α)	Frequency of Joints				
	8' (96")	12' (144")	16' (192")	20' (240")	25' (300")
3×10^{-6}	1/8"	1/8"	1/8"	1/8"	1/8"
4×10^{-6}	1/8"	1/8"	1/8"	1/8"	3/16"
5×10^{-6}	1/8"	1/8"	1/8"	3/16"	3/16"
6×10^{-6}	1/8"	1/8"	3/16"	3/16"	1/4"
7×10^{-6}	1/8"	1/8"	3/16"	3/16"	1/4"
8×10^{-6}	1/8"	3/16"	3/16"	1/4"	5/16"
9×10^{-6}	1/8"	3/16"	1/4"	1/4"	5/16"
10×10^{-6}	1/8"	3/16"	1/4"	5/16"	3/8"

In practical application, dry interior movement joints not exposed to direct sunlight shall be no less than as shown in Table 1 and in no case shall the movement joints be less than 1/8" wide.

Consult tile manufacturer for coefficient of linear thermal expansion in ambient temperature range of the installation.

Note: Some glass tiles and some stone tiles have a coefficient of linear thermal expansion greater than 10 and as high as 22 for some stones.

Design professional or engineer should also refer to ASTM C1193 "Standard Guide for the Use of Joint Sealants" and ASTM C1472 "Standard Guide for Calculating Movement and Other Effects When Establishing Sealant Joint Width" to determine the appropriate movement joint width when considering parameters other than linear thermal expansion.

Caution: These guidelines and calculations do not consider concrete shrinkage from curing and other forms of substrate shrinkage or expansion.

Interior Generic Movement Joints Subject to Direct Sunlight or Moisture

Quarry Tile—same as grout joint, but not less than 1/4."

All other tile—preferred 1/4" but never less than 1/8."

Joint Width Over Other Trades' Substrate Joints

Joints in tile and setting materials shall never be less than the width of the saw-cut control joint width.

Joints through tilework directly over structural joints must never be narrower than the structural joint.

Joint Depth

Exterior

For joints 1/2" wide or less—1/4."

For joints greater than 1/2" wide—joint depth shall be determined by a ratio of 2:1 (width/depth).

For joints greater than 1" wide—joint shall not exceed 1/2."

Interior

Preferred not less than 1/4" nor greater than 3/8" (width/depth ratio not applied for static joint)

Materials

Backup strip shall be a flexible and compressible type of closed-cell foam polyethylene, butyl rubber, or open cell and closed cell polyurethane, rounded at surface to contact sealant, as shown in details, and as recommended by sealant manufacturers. It must fit neatly into the joint without compacting and to such a height to allow a sealant depth of 1/2 the width of the joint. Sealant must not bond to the backup material.

Suitable sealants include silicone, urethane, and polysulfide. Sealants are available in both single- and multi-component formulations. Either formulation is generally suitable for movement joints in tilework. Single-component sealants are furnished in prepacked cartridges or other forms requiring no jobsite mixing. Multi-component sealants require jobsite mixing, but cure faster than single-component counterparts, making them advantageous for traffic areas.

Urethane sealants are recommended for exterior vertical tile surfaces and both exterior and interior horizontal tile surfaces, including tiled traffic areas. Because of their abrasion and penetration resistance, urethane sealants are recommended for movement joints in tiled traffic areas.

Silicone sealants may be used on both exterior and interior vertical tile surfaces. Single-component, mildew-resistant silicone sealants are formulated with fungicide for sealing interior joints in ceramic tile showers and around tubs, sinks and plumbing fixtures.

Sealants in traffic areas require a Shore A hardness of 25 or greater.

Use sealants complying with ASTM C920, which designates sealants according to type, grade, class, and uses. The following are suitable for use in tilework:

- Type S—single-component sealant.
- Type M—multi-component sealant.
- Grade P—sealants for joints on horizontal surfaces.
- Grade NS—non-sagging sealants for joints in vertical surfaces.
- Class 25—sealants that can withstand an increase and decrease of ±25% of joint width.
- Use T—sealants for pedestrian and vehicle traffic exposure.
- Use NT—sealants for non-traffic exposures.
- Uses M and G—sealants that will remain adhered to mortar (M) and glass (G).

The performance requirements of some applications such as exterior swimming pools, dairies, food plants, etc., may exceed the minimum requirements of these sealant specifications. Therefore, follow recommendations of experienced manufacturers as to specific sealants suitable in the job environment. In some severe environments, a program for regular maintenance of sealant in joints may be required.

Silicone sealants contain plasticizers that may stain some natural stones. See "Natural Stone Tile Selection and Installation Guide."

Manufactured/preformed joint profiles are available. Consult manufacturer.

Materials for Green/Sustainable Design

See "Green Building Standards and Green Product Selection Guide" and consult manufacturers and suppliers for product sustainability and contribution to green building design.

Consider specifying tile and installation materials that meet ANSI A138.1, the *American National Standard Specifications for Sustainable Ceramic Tiles, Glass Tiles, and Tile Installation Materials.*

Installation

Install movement joints per ASTM C1193 Standard Guide for Use of Joint Sealants.

Tile edges to which the sealant will bond must be clean and dry.

Some sealants require edge priming. Consult manufacturer's specifications. If required, care must be taken to keep primer off tile faces.

To ensure that location of joints in tilework align with existing joints in substrate, joints in tilework should be constructed during installation of mortar beds and/or tile, rather than saw-cutting joints after installation.

Keep movement joint cavities open and free of dirt, debris, grout, mortar, and setting materials.

Set compressible backup strip when mortar is placed or utilize removable wood strip to provide space for backup after mortar has cured.

Install sealant after tilework and grout are dry. Follow sealant manufacturer's recommendations.

Movement Joint Types In Concrete

Construction/cold joint—The surface where two successive placements of concrete meet, across which it may be desirable to achieve bond and through which reinforcement may be continuous. A cold joint becomes a weakened joint that, upon movement, will crack, permitting leakage or buckling and cracking of a tile floor set over the slab. Such joints should be shown on architectural drawings.

Contraction/control joint—Formed, sawed, or tooled groove in a concrete structure to create a weakened plane and regulate the location of cracking resulting from the dimensional change of different parts of the structure. Also referred to as "saw-cut joint."

Expansion joint—(1) A separation provided between adjoining parts of a structure to allow movement where expansion is likely to exceed contraction; (2) a separation between pavement slabs on grade, filled with a compressible filler material; (3) an isolation joint intended to allow independent movement between adjoining parts.

Isolation joint—A separation between adjoining parts of a concrete structure, usually a vertical plane, at a designated location such as to interfere least with

performance of the structure, yet such as to allow relative movement in three directions and avoid formation of cracks elsewhere in the concrete and through which all or part of the bonded reinforcement is interrupted.

Some large slabs on-grade are poured monolithically, then later saw-cut at intervals, providing control/contraction joints to allow for cracking at these weakened points.

Movement Joint Types, Other

Acoustical Joint—A non-drying, non-hardening, rubber-like seal at the perimeter and at all penetrations and retaining surfaces of a floor installation in which a bonded sound reduction membrane is used for sound reduction. The primary function of an acoustical joint is to minimize flanking, the transmission of sound through joints, penetrations, or structural components in the assembly.

Non-Linear Joint—Generic movement joints (also known as field joints) in the tilework not located over an existing construction joint, contraction (control) joint, expansion joint, or isolation joint in the substrate, can be installed in a non-linear configuration (e.g., sawtooth, toothed, zipper), which may be necessary for tile patterns with non-continuous grout joints. The generic movement joint must be properly designed per this section, and the sealant shall meet ASTM C920, or as indicated in the project specification. The intended application must be approved by the sealant manufacturer, and the sealant must be properly installed per the sealant manufacturers' requirements. (See EJ171M).

Wall Tile Movement Joints in Framed Wall Assemblies

Location and frequency of movement joints shall be determined by the design professional or engineer, including the design of the joint itself with framing, sheathing, and backer board considerations. Consult backer board manufacturer for specific recommendations. Design details must be specified before the wall is assembled as movement accommodation cannot be retroactively added.

EJ171C-21

- **Expansion Joint**

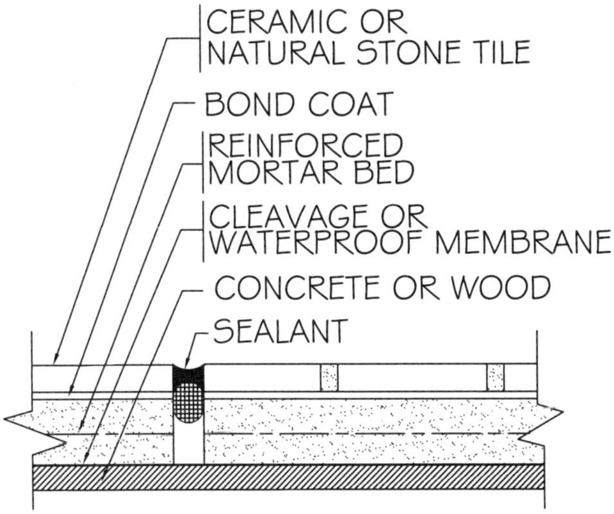

EJ171D-21

- **Isolation/Expansion Joint**

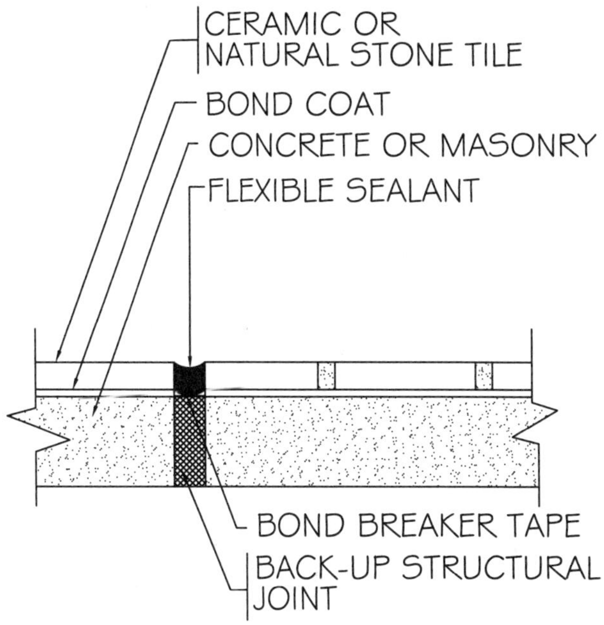

EJ171-21

- **Expansion Joint**

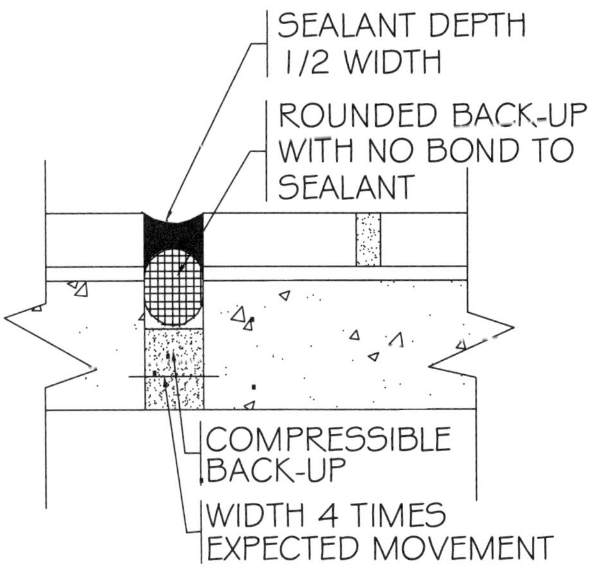

EJ171E-21

- **Expansion Joint, Cement Mortar, Bonded**

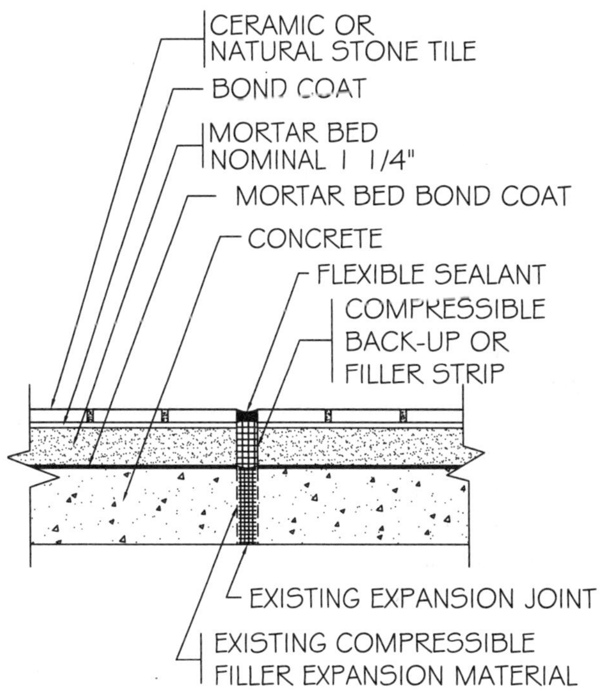

EJ171H-21

• **Expansion Joint, Cement Mortar, Cleavage Membrane**

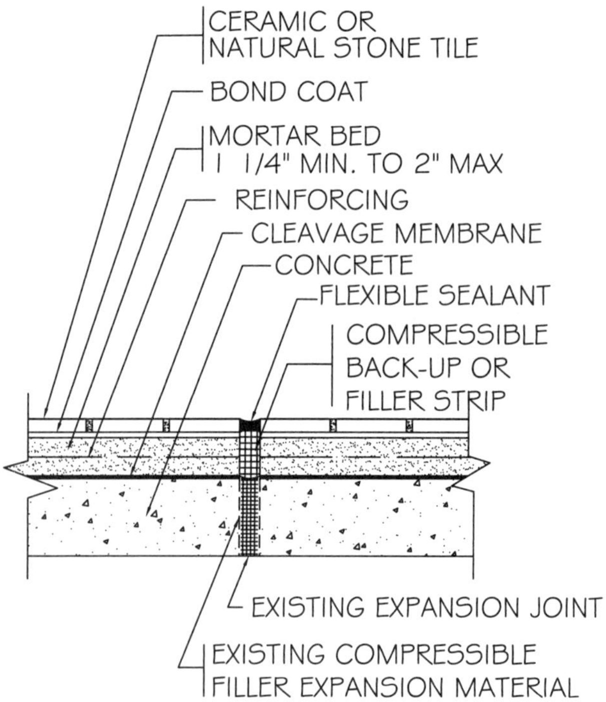

EJ171B-21

• **Contraction Joint**

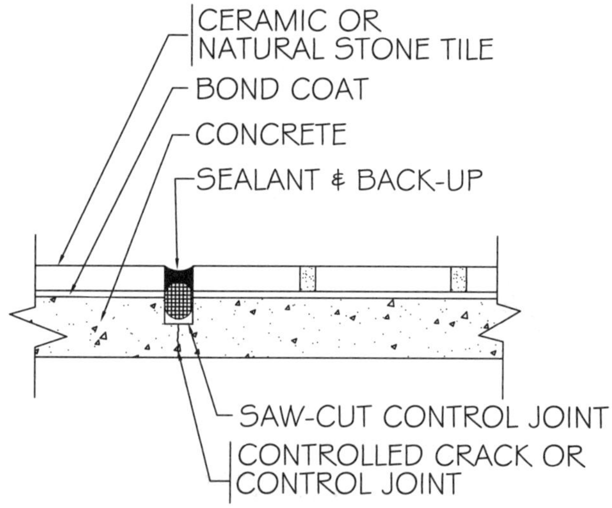

EJ171A-21

• **Construction Joint**

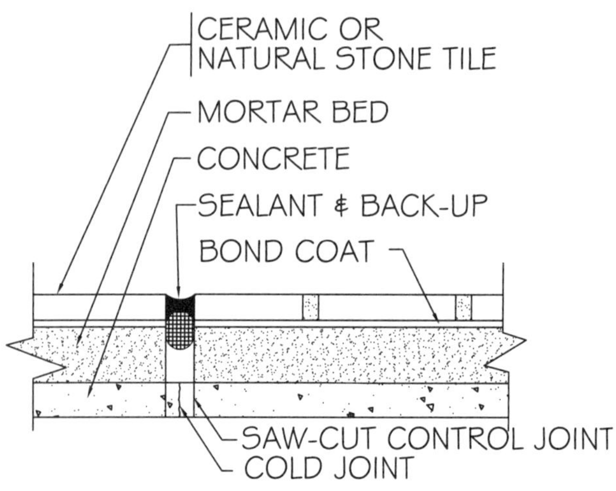

EJ171G-21

• **Perimeter Joint**

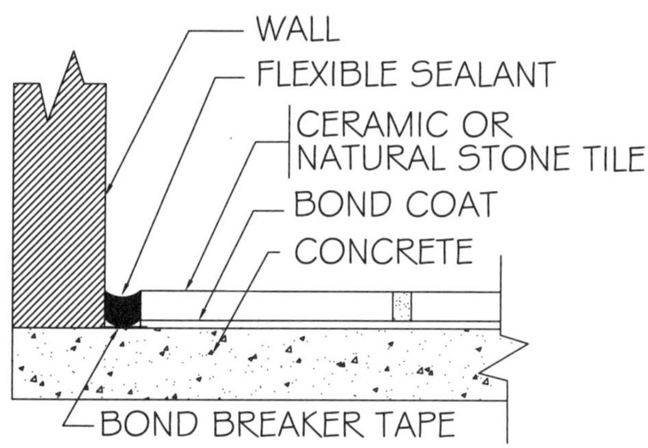

EJ171I-21

• Perimeter Joint

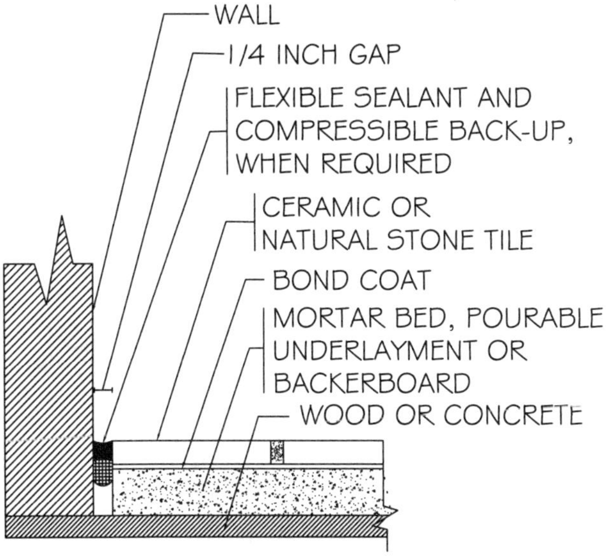

EJ171J-21

• Perimeter Movement Joint

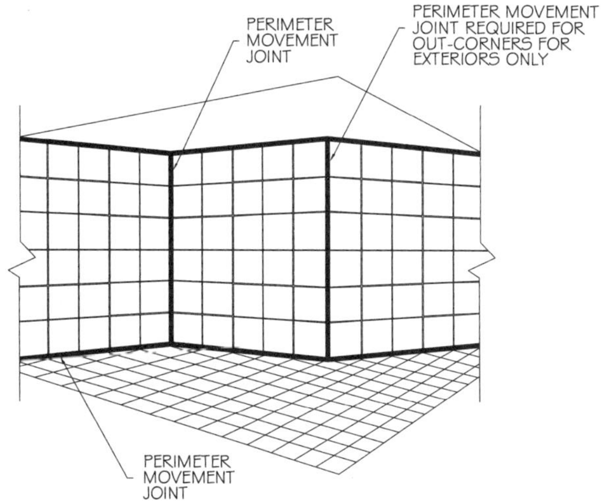

EJ171F-21

• Generic Movement Joint

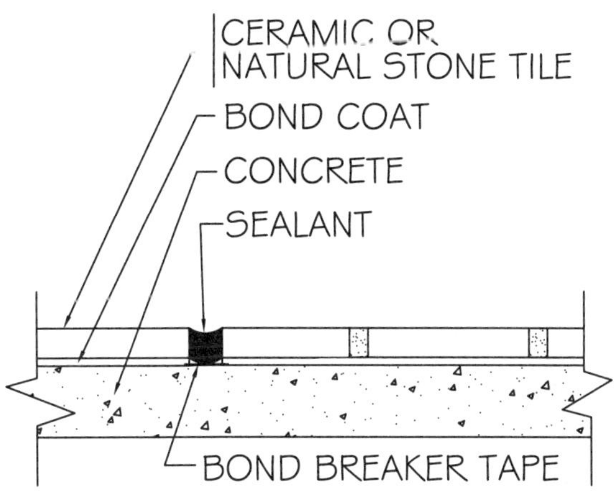

EJ171K-21

• Movement Joint in Tile and Backerboard

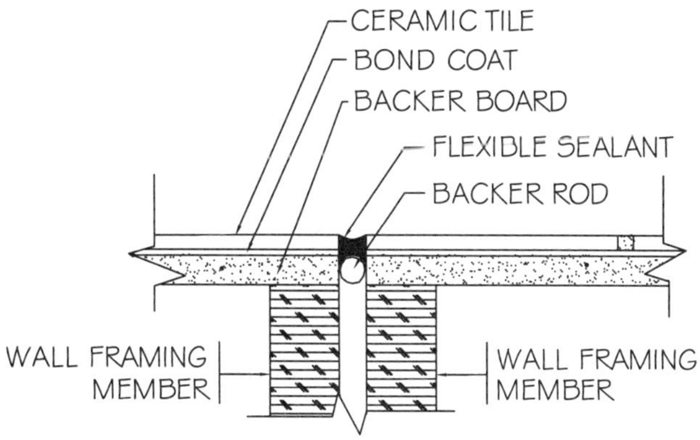

EJ171L-21

• Generic Movement Joint with Backerboard

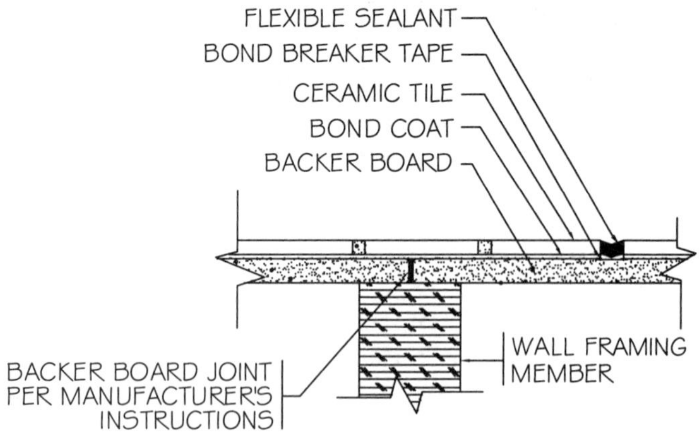

EJ171M-21

• Examples of Non-Linear Movement Joints per ANSI 108.01, Figure 11

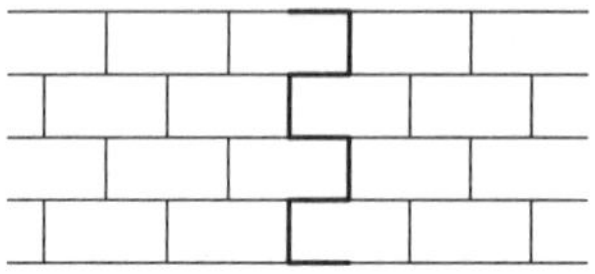

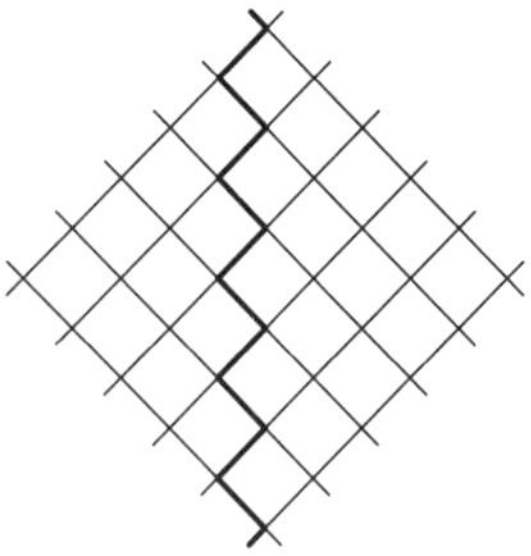

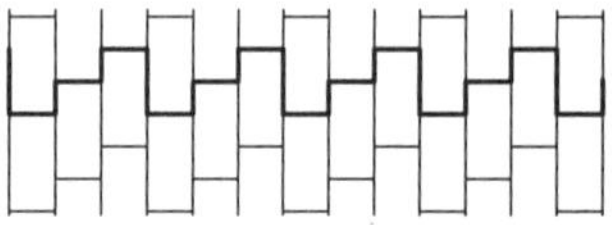

2021 | APPENDICES

APPENDIX A. ISO 13007 STANDARDS FOR ADHESIVES AND GROUTS

ISO installation material standards for mortars, adhesives, and grouts are more fully described in the following standards:

13007-1 Ceramic Tiles—Grouts and adhesives—Part 1: Terms, definitions, and specifications for adhesives
13007-2 Ceramic Tiles—Grouts and adhesives—Part 2: Test methods for adhesives
13007-3 Ceramic Tiles—Grouts and adhesives—Part 3: Terms, definitions and specifications for grouts
13007-4 Ceramic Tiles—Grouts and adhesives—Part 4: Test methods for grouts

Where ANSI installation material standards refer to "mortars," ISO installation material standards refer to such mortars as either adhesives or mortar adhesives.

ISO adhesive and grout classifications are designated using letters to identify chemical make-up and optional properties. Numbers are used to identify performance levels.

ISO Performance Requirements for Adhesives

C–Cementitious Mortars

C1: Normal—Tensile bond strength of ≥ 72.5 psi (≥ 0.5 N/mm^2) when tested per ISO 13007 curing parameters, including water immersion, heat aging, and freeze-thaw cycling. Twenty-eight day tensile adhesion bond strength ≥ 0.5 N/mm^2 (72.5 psi) after 20 minutes open time.

C2: Improved—Tensile bond strength of ≥ 145 psi (≥ 1 N/mm^2) when tested per ISO 13007 curing parameters, including water immersion, heat aging, and freeze-thaw cycling. Twenty-eight-day tensile adhesion bond strength ≥ 0.5 N/mm^2 (72.5 psi) after 20 minutes open time.

Note: A Tensile Bond Test quantifies the amount of vertical force required to pull a tile or stone from a given substrate after a prescribed period of time. This test can be performed both in the lab and at the jobsite.

D–Dispersion Adhesives (Mastics)

D1: Normal—Shear adhesion strength of ≥ 145 psi (≥ 1 N/mm^2) when tested per ISO 13007 curing parameters, including heat aging. Twenty-eight-day tensile adhesion bond strength ≥ 0.5 N/mm^2 (72.5 psi) after 20 minutes open time.

D2: Improved—D1 performance plus additional shear adhesion strength of ≥ 72.5 psi (0.5 N/mm^2) after 21-day air cure and 7-day water immersion; also shear adhesion strength of ≥ 145 psi (1 N/mm^2) at elevated temperatures.

R–Reaction Resin Adhesives (epoxy or polyurethane)

R1: Normal—Shear adhesion strength of ≥ 290 psi (2 N/mm^2) when tested per ISO 13007 curing parameters, including water immersion. Twenty-eight-day tensile adhesion bond strength ≥ 0.5 N/mm^2 (72.5 psi) after 20 minutes open time.

R2: Improved—R1 performance plus additional shear adhesion strength of ≥ 290 psi (2 N/mm^2) after thermal shock testing.

ISO Optional Performance Characteristics for Adhesives

F–Fast-Setting/Fast-Drying

F: Fast-setting (cementitious) with a minimum tensile bond strength of ≥ 72.5 psi (0.5 N/mm^2) achieved within 24 hours. Twenty-eight-day tensile adhesion bond strength ≥ 0.5 N/mm^2 (72.5 psi) after 10 minutes open time.

F: Fast-drying (D2 dispersion adhesive/mastic) with a minimum shear bond strength of ≥ 72.5 psi (0.5 N/mm^2) after 7-day air cure and 7-day water immersion; also shear adhesion strength of ≥ 145 psi (1 N/mm^2) at elevated temperatures.

E–Extended Open Time

E: "Extended open time" extends the open time in which a tile can be laid into the adhesive without loss of minimum bond strength. This would extend the "Open Time" of a C1 or C2, D1 or D2 adhesive from 20 minutes to 30 minutes for a 28-day tensile adhesion bond strength ≥ 0.5 N/mm^2 (72.5 psi).

T–Thixotropic [Non-Slip/Non-Sag]

T: Slip resistant adhesives must not allow a specified test tile sample applied on a vertical surface to slip more than 0.0197" (0.5 mm).

S–Deformability

Transverse deformation is the ability of the mortar to accommodate movement between the tile and the substrate.

S1: Normal—Deformability ≥ 0.1" (2.5 mm) and < 0.2" (5 mm).

S2: Improved—Deformability ≥ 0.2" (5 mm).

P–Plywood Adhesion Bond Strength

P1: Normal—Tensile adhesion bond strength of ≥ 72.5 psi (0.5 N/mm^2) after 28-day cure

P2: Improved—Tensile adhesion bond strength of ≥ 145 psi (1 N/mm^2) after 28-day cure

ISO Performance Requirements for Cementitious Grouts

CG–Cementitious Grouts

CG1: Normal

Abrasion resistance < 0.12 in^3 (2000 mm^3)

Flexural strength under standard conditions > 362.5 psi (2.5 N/mm^2)

Flexural strength after freeze-thaw cycles > 362.5 psi (2.5 N/mm^2)

Compressive strength under standard conditions > 2176 psi (15 N/mm^2)

Compressive strength after freeze-thaw cycles > 2176 psi (15 N/mm^2)

Shrinkage < 3 mm/m

Water absorption after 30 min < 5g

Water absorption after 240 min < 10g

CG2: Improved

Meets CG1 criteria with either improved abrasion resistance, reduced water absorption or both as given below.

ISO Optional Performance Characteristics for Cementitious Grouts

A–Abrasion Resistance

A: High abrasion resistance ≤ 0.06 cubic inch (1000 mm^3).

W–Water Absorption

W: Reduced water absorption after 30 min ≤ 2g; reduced water absorption after 240 min ≤ 5g

F–Fast Setting (CG1 or CG2)

F: Fast-setting with compressive strength in ≤ 24 hours under standard conditions > 2176 psi (15 N/mm^2)

ISO Performance Requirements for Reaction Resin Grouts

RG–Reaction Resin Grouts

Abrasion resistance ≤ 0.015 cubic inch (250 mm^3)

Flexural strength under standard conditions ≥ 4351 psi (30 N/mm^2)

Compressive strength under standard conditions ≥ 6527 psi (45 N/mm^2)

Shrinkage ≤ 1.5 mm/m

Water absorption after 240 min <0.1g

Note: An RG grout will always have higher performance characteristics than the best CG2AW grout in addition to higher chemical resistance to a wider variety of chemicals. Always check with the grout manufacturer for chemical resistance to a specific chemical.

Code			C–Cementitious Mortars
Type	Class	Optional Characteristics	
C	1	—	Cementitious mortar with normal tensile bond strength • Tensile adhesion bond strength of ≥ 0.5 N/mm² (72.5 psi) when tested per 13007 curing parameters • 28-day tensile adhesion bond strength of ≥ 0.5 N/mm² (72.5 psi) after 20 minutes open time
C	2	—	Cementitious mortar with improved tensile bond strength • Tensile adhesion bond strength of ≥ 1 N/mm² (145 psi) when tested per 13007 curing parameters • Baseline "C1" open time characteristics
C	1 or 2	E	Cementitious mortar with extended open time • 28-day tensile adhesion bond strength of ≥ 0.5 N/mm² (72.5 psi) after 30 minutes open time
C	1 or 2	F	Cementitious mortar, fast-setting • 24-hour tensile adhesion bond strength of ≥ 0.5 N/mm² (72.5 psi) • 28-day tensile adhesion bond strength of ≥ 0.5 N/mm² (72.5 psi) after 10 minutes open time
C	1 or 2	T	Cementitious mortar with slip resistance • Will not allow a tile applied on a vertical surface to slip more than 0.5 mm (.0196")
C	1 or 2	S1	Cementitious mortar with normal deformability • Deformability ≥ 2.5 mm (0.1") and < 5 mm (0.2") when tested per 13007
C	1 or 2	S2	Cementitious mortar with improved deformability • Deformability ≥ 5 mm (0.2") when tested per 13007
C	1 or 2	P1	Cementitious mortar with normal adhesion to EGP (exterior glue plywood) • 28-day tensile adhesion bond strength of ≥ 0.5 N/mm² (72.5 psi)
C	1 or 2	P2	Cementitious mortar with improved adhesion to EGP (exterior glue plywood) • 28-day tensile adhesion bond strength of ≥ 1 N/mm² (145 psi) • Note: It is recommended with EGP installations over 16" O.C. that the "S2" code also be specified in addition to the "P2" code.

Code			D–Dispersion (Mastic) Adhesives
Type	Class	Optional Characteristics	
D	1	—	Dispersion (Mastic) adhesive with normal shear bond adhesion strength • Shear adhesion bond strength of ≥ 1 N/mm² (145 psi) when tested per 13007 curing parameters • 28-day tensile adhesion bond strength of ≥ 0.5 N/mm² (72.5 psi) after 20 minutes open time
D	2	—	Dispersion (Mastic) adhesive with improved shear bond adhesion strength • Shear adhesion bond strength of ≥ 0.5 N/mm² (72.5 psi) after 21-day air cure, 7-day water immersion • Shear adhesion bond strength of ≥ 1 N/mm² (145 psi) at elevated temperatures • Baseline "D1" open time and adhesion strength characteristics for all other curing parameters
D	1 or 2	T	Dispersion (Mastic) adhesive with slip resistance • Will not allow a tile applied on a vertical surface to slip more than 0.5 mm (.0196")
D	2	F or A	Dispersion (Mastic) adhesive, fast (accelerated)-drying • Shear adhesion bond strength of ≥ 0.5 N/mm² (72.5 psi) after 7-day air cure, 7-day water immersion • Shear adhesion bond strength of ≥ 1 N/mm² (145 psi) at elevated temperatures
D	1 or 2	E	Dispersion (Mastic) adhesive with extended open time • 28-day tensile adhesion bond strength of ≥ 0.5 N/mm² (72.5 psi) after 30 minutes open time

Code			R–Reaction Resin (Epoxy or Urethane) Adhesives
Type	Class	Optional Characteristics	
R	1	—	Reaction resin (epoxy or urethane) adhesive • Shear adhesion bond strength of ≥ 2 N/mm² (290 psi) when tested per 13007 curing parameters • 28-day tensile adhesion bond strength of ≥ 0.5 N /mm² (72.5 psi) after 20 minutes open time
R	2	—	Reaction resin (epoxy or urethane) adhesive • Shear adhesion bond strength of ≥ 2 N/mm² (290 psi) after high temperature thermal shock • Baseline "R1" open time and adhesion strength characteristics for all other curing parameters
R	1 or 2	T	Reaction resin (epoxy or urethane) adhesive with slip resistance • Will not allow a tile applied on a vertical surface to slip more than 0.5 mm (.0196")

Code			CG–Cementitious Grouts
Type	Class	Optional Characteristics	
CG	1	—	Normal cementitious grout • Abrasion resistance < 2000 mm³ (0.12 in³) • Flexural strength under standard conditions > 2.5 N/mm² (362.5 psi) • Flexural strength after freeze-thaw cycles > 2.5 N/mm² (362.5 psi) • Compressive strength under standard conditions > 15 N/mm² (2176 psi) • Compressive strength after freeze-thaw cycles > 15 N/mm² (2176 psi) • Shrinkage < 3 mm/m • Water absorption after 30 min < 5 g • Water absorption after 240 min < 10 g
CG	2	A	Improved cementitious grout with high abrasion resistance • Baseline "CG1" characteristics • Abrasion resistance < 1000 mm³ (0.06 in³)
CG	2	W	Improved cementitious grout with reduced water absorption • Baseline "CG1" characteristics • 30 minute water absorption < 2 g • 240 minute water absorption < 5 g
CG	1 or 2	F	Fast-setting cementitious grout • Baseline "CG1," "CG2A," or "CG2W" characteristics • 24-hour compressive strength under standard conditions > 15 N/mm² (2176 psi)

Code			RG–Reaction Resin (Epoxy or Urethane) Grouts
Type	Class	Optional Characteristics	
RG	—	—	• Abrasion resistance ≤ 250 mm³ (0.15 in³) • Flexural strength under standard conditions ≥ 30 N/mm² (4351 psi) • Compressive strength under standard conditions ≥ 45 N/mm² (6527 psi) • Shrinkage ≤ 1.5 mm/m • Water absorption after 240 min ≤ 0.1 g

APPENDIX B. ESTIMATED WEIGHTS FOR FLOOR INSTALLATIONS

The estimated weight of each floor installation method is provided to aid the design professional in calculating the contribution of the tile installation to the total dead load. Loads from all fixed objects and materials are dead loads. Dead load is defined in the 2009 International Building Code (IBC) as "the weight of materials of construction incorporated into the building, including but not limited to walls, floors, roofs, ceilings, stairways, built-in partitions, finishes, cladding and other similarly incorporated architectural and structural items, and the weight of fixed service equipment, such as cranes, plumbing stacks and risers, electrical feeders, heating, ventilating and air-conditioning systems and fire sprinkler systems."

The estimated weights are based on typical weights of the materials used in the installation methods. They include materials installed above the substrate, not the substrate itself.

For each flooring installation method, a typical weight per square foot is provided. Where material thickness can vary, the weight per relevant unit of measure is given. For the actual weight of any installation, the actual products used and thickness of application must be considered.

Typical weights for individual components used to calculate the estimated weights are provided in Tables 1–6.

General Assumptions

Tile: The estimated weights assume a 3⁄8"-thick ceramic tile weighing 4 pounds/square foot, except in F134, in which the weight of a 1¼"-thick packing house tile was used.

Stone Tile: The estimated weights assume a weight of 6 pounds/square foot for stone tiles. Stone densities vary from about 120 pounds/cubic foot for high-porosity limestones to 200 pounds/cubic foot for the black, igneous rocks commonly sold as "granite." Additionally, stone tile thicknesses are generally 10, 12, or 15 mm, giving a range of weight/square foot as shown in the table below.

Weight per ft.² of Stone Tiles			
Stone Density (lbs./cu. ft.)	Stone Thickness		
	10 mm 13⁄32 in.	12 mm 15⁄32 in.	15 mm 19⁄32 in.
120	3.9 lbs.	4.7 lbs.	5.9 lbs.
140	4.6 lbs.	5.5 lbs.	6.9 lbs.
160	5.2 lbs.	6.3 lbs.	7.9 lbs.
180	5.9 lbs.	7.1 lbs.	8.9 lbs.
200	6.6 lbs.	7.9 lbs.	9.8 lbs.

Mortar Bed Assumptions

The estimated weights assume a ¾"-thick or 1¼"-thick mortar bed, depending on the minimum thickness required by the method. Where hydronic tubes are encapsulated in the mortar bed, the estimated weight assumes a 2"-thick mortar bed (¾" of mortar to encapsulate the ⅝" O.D. tubing and 1¼" of mortar over that.) Thicker mortar beds can be calculated for by adding 3 pounds/square foot for each additional ¼" of mortar. Estimated weights assume use of the standard materials of a cured mortar bed followed by a thin-bed tile application. Metal lath (2.5 pounds/square yard), where applicable, adds 0.3 pounds/square foot, per manufacturer literature.

Mortar Bed Bond Coat: The weight of a 1⁄16"-thick cementitious bond coat was used for all bonded mortar bed methods.

Chemical-Resistant Membrane: The typical weight of a ¼"-thick chemical-resistant membrane was used in the weight calculation for F134, and it is assumed that the membrane is self-adhering and proportional in weight to a typical crack isolation membrane.

Cleavage Membrane: Cleavage membrane weight was assumed to be insignificant and not factored into the estimated weights.

Cementitious Self-Leveling Underlayment Assumptions

The estimated weights assume a ½"-thick cementitious self-leveling layer, with or without electric radiant systems. For each method with hydronic tubes, the estimated weight assumes a 1¼"-thick cementitious self-leveling layer (¾" of self-leveler to encapsulate ⅝" O.D. tubing and ½" of self-leveler over that). Thicker self-leveling application can be calculated for by adding 2¾ pounds/square foot for each additional ¼" of self-leveler.

Metal/Plastic Lath: For methods in which use of metal or plastic lath is optional, metal lath is included in the estimated weights at 2.5 pounds/square yard, adding 0.3 pounds/square foot. The weight of plastic lath is assumed to be insignificant and is not factored into the estimated weights of installation methods that reference only plastic lath.

Membranes: The typical weight of a crack isolation membrane with a 1⁄16" cementitious bond coat is included in the estimated weights of all methods with hydronic tubing.

Primer: The weights of primers is assumed to be insignificant and is not factored into the estimated weights.

Poured Gypsum Assumptions

The estimated weights assume $\frac{3}{4}$" of poured gypsum, whether over concrete or plywood. For methods with hydronic radiant heating tubes, the estimated weight assumes $1\frac{1}{2}$" of poured gypsum ($\frac{3}{4}$" of poured gypsum to encapsulate $\frac{5}{8}$" O.D. tubing and $\frac{3}{4}$" of poured gypsum over that). Thicker applications of poured gypsum can be calculated for by adding $2\frac{1}{4}$ pounds/square foot for each additional $\frac{1}{4}$" of poured gypsum.

Crack Isolation Membranes: The typical weight of a crack isolation membrane with a $\frac{1}{16}$" cementitious bond coat is included in the estimated weights of all poured gypsum methods.

Primer/Sealer: The weights of primers/sealers atop poured gypsum is assumed to be insignificant and is not factored into the estimated weights.

Backer Board and Wood Underlayment Panel Assumptions

The weight of a $\frac{1}{8}$"-thick cementitious mortar bedding coat is included in the estimated weights of all backer board methods. For wood underlayment panel methods, the weight of a bedding coat is not included, nor is the weight of fasteners, which is assumed to be insignificant.

Cement Backer Board: A weight of 4 pounds/square foot is assumed for $\frac{1}{2}$" cement backer board, and a weight of 2 pounds/square foot is assumed for $\frac{1}{4}$" board, although some manufacturers list weights closer to 3 pounds/square foot and 1.5 pounds/square foot, respectively.

Other Backer Boards: Weights of other types of backer boards, including cementitious-coated extruded foam, coated glass mat water resistant gypsum, fiber-cement, and fiber-reinforced water-resistant gypsum were derived from weights listed in the product literature of various manufacturers.

Membrane Assumptions

A $\frac{1}{16}$"-thick cementitious bond coat is used for all sheet applied, non-self-adhering membrane methods. Partial-coverage crack isolation membrane weights are assumed to be insignificant.

Table 1: Direct Bond Methods

Method	Weight (lbs. per sq. ft.)					
	Mortar Fill	Tile/Stone Bond Coat	Ceramic Tile	Stone	Total Assembly* (Tile)	Total Assembly* (Stone)
F102/F102S		1¼	4	6	5	7
F104/F104S		1¼	4	6	5	7
F105/F105S		1¼	4	6	5	7
F113/F113S		1¼	4	6	5	7
F113A/F113AS		1¼	4	6	5	7
F115		1¼	4		5	
F115A		1¼	4		5	
F116E		1½	4		6	
F116O		¾	4		5	
F131		1½	4		6	
F133		1¼	4		5	
RH110/RH110S		1¼	4	6	5	7
RH110A/RH110AS		1¼	4	6	5	7
RH115	1¾	1¼	4		7	
RH115A	1¾	1¼	4		7	

*Note: For presentation purposes, total assembly weights are rounded to the nearest pound, and individual component weights are rounded to the nearest ¼ pound. However, non-rounded values were used for all calculations.

Table 2: Mortar Bed Methods

Method	Weight (lbs. per sq. ft.)										
	Drainage Layer	Metal Lath	Mortar Bed Bond Coat	Wire	Mortar Bed	Membrane	Tile or Stone Bond Coat	Ceramic Tile	Stone	Total Assembly* (Tile)	Total Assembly* (Stone)
F101/F101S: ¾" Mortar Bed			½		9		1¼	4	6	15	17
F103/F103S: 1¼" Mortar Bed	10½			¼	15		1¼	4	6	31	33
F103B/F103BS: 1¼" Mortar Bed	¼			¼	15		1¼	4	6	21	23
F111/F111S: 1¼" Mortar Bed				¼	15		1¼	4	6	21	23
F112/F112S: ¾" Mortar Bed			½		9		1¼	4	6	15	17
F114: 1¼" Mortar Bed				¼	15		1¼	4		21	
F121/F121S: 1¼" Mortar Bed				¼	15		1¼	4	6	21	23
F132: ¾" Mortar Bed			½		9		1½	4		15	
F134: ¾" Mortar Bed			½		9	1¾	1½	13		26	
RH117/RH117S: 1¼" Mortar Bed				¼	24		1¼	4	6	30	32
RH141/RH141S: 1¼" Mortar Bed				¼	24		1¼	4	6	30	32
F141/F141S: 1¼" Mortar Bed				¼	15		1¼	4	6	21	23
F145: ¾" Mortar Bed		¼			9		1¼	4		15	

*Note: For presentation purposes, total assembly weights are rounded to the nearest pound, and individual component weights are rounded to the nearest ¼ pound. However, non-rounded values were used for all calculations.

Table 3: Cementitious Self-Leveling Underlayment (SLU) Methods

Method	Weight (lbs. per sq. ft.)								
	Metal Lath	SLU	Membrane Bond Coat	Membrane	Tile or Stone Bond Coat	Ceramic Tile	Stone	Total Assembly* (Tile)	Total Assembly* (Stone)
F205/F205S: ½" SLU		5½			1¼	4	6	11	13
F205A/F205AS: ½" SLU		5½			1¼	4	6	11	13
RH112/RH112S: 1¼" SLU		13½	½	¼	1¼	4	6	19	21
RH112A/ RH112AS: 1¼" SLU		13½	½	¼	1¼	4	6	19	21
RH116/ RH116S: ½" SLU		5½			1¼	4	6	11	13
RH116A/RH116AS: ½" SLU		5½			1¼	4	6	11	13
RH123: 1¼" SLU		13½	½	¼	1¼	4		19	
RH140: ½" SLU		5½			1¼	4		11	
F185: ½" SLU	¼	5½			1¼	4		11	

Table 4: Poured Gypsum Methods

Method	Weight (lbs. per sq. ft.)							
	Gypsum	Membrane Bond Coat	Membrane	Tile Bond Coat	Ceramic Tile	Stone	Total Assembly* (Tile)	Total Assembly* (Stone)
F200/F200S: ¾" Gypsum	6½	½	¼	1¼	4	6	12	14
F200A/F200AS: ¾" Gypsum	6½	½	¼	1¼	4	6	12	14
RH111/RH111S: 1½" Gypsum	13	½	¼	1¼	4	6	19	21
RH111A/RH111AS: 1½" Gypsum	13	½	¼	1¼	4	6	19	21
RH122: 1½" Gypsum	13	½	¼	1¼	4		19	
F180: ¾" Gypsum	6½	½	¼	1¼	4		12	

*Note: For presentation purposes, total assembly weights are rounded to the nearest pound, and individual component weights are rounded to the nearest ¼ pound. However, non-rounded values were used for all calculations.

Table 5: Backer Board and Wood Underlayment Panel Methods

Method	Backer Board or Wood Underlayment	Weight (lbs. per sq. ft.)								
		Mortar Under Board	Plywood Underlayment	Backer Board	Radiant System Mortar	Tile Bond Coat	Ceramic Tile	Stone	Total Assembly* (Tile)	Total Assembly* (Stone)
RH130	15/32" Plywood		1½		1¾	1¼	4		7	
RH130	19/32" Plywood		1¾		1¾	1¼	4		7	
RH135	¼" CBU	1		2	1¾	1¼	4		10	
RH135	½" CBU	1		4	1¾	1¼	4		12	
RH135	¼" Cementitious-Coated Foam	1		½	1¾	1¼	4		8	
RH135	½" Cementitious-Coated Foam	1		½	1¾	1¼	4		8	
RH135	¼" Coated-Glass Mat Gypsum	1		1½	1¾	1¼	4		9	
RH135	½" Coated-Glass Mat Gypsum	1		2	1¾	1¼	4		10	
RH135	¼" Fiber-Cement	1		2	1¾	1¼	4		10	
RH135	7/16" Fiber-Cement	1		2½	1¾	1¼	4		10	
RH135	¼" Fiber Gypsum	1		1½	1¾	1¼	4		9	
RH135	⅜" Fiber Gypsum	1		2	1¾	1¼	4		10	
RH135	½" Fiber Gypsum	1		2¾	1¾	1¼	4		11	
F142	19/32" Plywood		1¾			¾	4		7	
F143	15/32" Plywood		1½			1½	4		7	
F143	19/32" Plywood		1¾			1½	4		7	
F150	15/32" Plywood		1½			1¼	4		7	
F150	19/32" Plywood		1¾			1¼	4		7	
F160	⅜" EG/LWP		1½			1¼	4		7	
F144	¼" CBU	1		2		1¼	4		8	
F144	½" CBU	1		4		1¼	4		10	
F144	¼" Fiber-Cement	1		2		1¼	4		8	
F144	7/16" Fiber-Cement	1		2½		1¼	4		9	
F146	¼" Coated-Glass Mat Gypsum	1		1½		1¼	4		8	
F146	½" Coated-Glass Mat Gypsum	1		2		1¼	4		8	
F170	¼" Fiber Gypsum	1		1½		1¼	4		8	
F170	⅜" Fiber Gypsum	1		2		1¼	4		8	
F170	½" Fiber Gypsum	1		2¾		1¼	4		9	
F175	½" Cementitious-Coated Foam	1		½		1¼	4		7	

*Note: For presentation purposes, total assembly weights are rounded to the nearest pound, and individual component weights are rounded to the nearest ¼ pound. However, non-rounded values were used for all calculations.

Table 5: Backer Board and Wood Underlayment Panel Methods (Continued)

Method	Backer Board or Wood Underlayment	Weight (lbs. per sq. ft.)								
		Mortar Under Board	Plywood Underlayment	Backer Board	Radiant System Mortar	Tile Bond Coat	Ceramic Tile	Stone	Total Assembly* (Tile)	Total Assembly* (Stone)
F149	19/32" Plywood		1¾			1¼	4		7	
F151	¼" Coated-Glass Mat Gypsum	1		1½		1¼	4		8	
F151	½" Coated-Glass Mat Gypsum	1		2		1¼	4		8	
F155	19/32" Plywood		1¾			1¼	4		7	
F250S	15/32" Plywood, ¼" CBU	1	1½	2		1¼		6		11
F250S	19/32" Plywood, ¼" CBU	1	1¾	2		1¼		6		12
F250S	15/32" Plywood, ½" CBU	1	1½	4		1¼		6		13
F250S	19/32" Plywood, ½" CBU	1	1¾	4		1¼		6		14
F250S	15/32" Plywood, ¼" Cementitious-Coated Foam	1	1½	½		1¼		6		10
F250S	19/32" Plywood, ¼" Cementitious-Coated Foam	1	1¾	½		1¼		6		10
F250S	15/32" Plywood, ½" Cementitious-Coated Foam	1	1½	½		1¼		6		10
F250S	19/32" Plywood, ½" Cementitious-Coated Foam	1	1¾	½		1¼		6		10
F250S	15/32" Plywood, ¼" Coated-Glass Mat Gypsum	1	1½	1½		1¼		6		11
F250S	19/32" Plywood, ¼" Coated-Glass Mat Gypsum	1	1¾	1½		1¼		6		11
F250S	15/32" Plywood, ½" Coated-Glass Mat Gypsum	1	1½	2		1¼		6		11
F250S	19/32" Plywood, ½" Coated-Glass Mat Gypsum	1	1¾	2		1¼		6		12
F250S	15/32" Plywood, ¼" Fiber-Cement	1	1½	2		1¼		6		11
F250S	19/32" Plywood, ¼" Fiber-Cement	1	1¾	2		1¼		6		12
F250S	15/32" Plywood, 7/16" Fiber-Cement	1	1½	2½		1¼		6		12
F250S	19/32" Plywood, 7/16" Fiber-Cement	1	1¾	2½		1¼		6		12
F250S	15/32" Plywood, ¼" Fiber Gypsum	1	1½	1½		1¼		6		11
F250S	19/32" Plywood, ¼" Fiber Gypsum	1	1¾	1½		1¼		6		11
F250S	15/32" Plywood, ⅜" Fiber Gypsum	1	1½	2		1¼		6		11
F250S	19/32" Plywood, ⅜" Fiber Gypsum	1	1¾	2		1¼		6		12
F250S	15/32" Plywood, ½" Fiber Gypsum	1	1½	2¾		1¼		6		12
F250S	19/32" Plywood, ½" Fiber Gypsum	1	1¾	2¾		1¼		6		13

*Note: For presentation purposes, total assembly weights are rounded to the nearest pound, and individual component weights are rounded to the nearest ¼ pound. However, non-rounded values were used for all calculations.

Table 6: Membrane and Cork Underlayment Methods

Method	Membrane Type	Weight (lbs. per sq. ft.)							
		Plywood	Membrane Bond Coat	Membrane	Tile or Stone Bond Coat	Ceramic Tile	Stone	Total Assembly* (Tile)	Total Assembly* (Stone)
F122/ F122S	Waterproofing		½	¼	1¼	4	6	6	8
F122A/ F122AS	Waterproofing		½	¼	1¼	4	6	6	8
F125/ F125S	Crack Isolation		½	¼	1¼	4	6	6	8
F128	Uncoupling		½	¼	1¾	4		6	
F135	½" Cork		¼	½	1¼	4		6	
F136	Bonded Sound Reduction		¼	½	1¼	4		6	
F147	Uncoupling	1	½	¼	1¾	4		7	
F148	Uncoupling		½	¼	1¾	4		6	
F152	Generic	1	½	¼	1¼	4		7	

*Note: For presentation purposes, total assembly weights are rounded to the nearest pound, and individual component weights are rounded to the nearest ¼ pound. However, non-rounded values were used for all calculations.

APPENDIX C. MATERIAL AND INSTALLATION SPECIFICATIONS, TEST METHODS, AND GUIDES REFERENCED IN THIS *HANDBOOK*

All standards and specifications are revised or updated periodically. The current status can be confirmed by writing the particular authority. For the latest edition of American National Standards ANSI A108, A118, A136.1, or A137.1, contact the Tile Council of North America at www.TCNAtile.com.

American Concrete Institute (ACI)

38800 Country Club Drive
Farmington Hills, MI 48331
www.concrete.org

ACI Specifications:

117-2010	Specifications for Tolerances for Concrete Construction and Materials and Commentary

American National Standards Institute (ANSI)

1899 L Street, NW, 11th floor
Washington, DC 20036
www.ansi.org

ANSI Installation Specifications:

A108.01-2018	General Requirements: Subsurfaces and Preparations by Other Trades
A108.02-2019	General Requirements: Materials, Environmental, and Workmanship
A108.1A-2017	Installation of Ceramic Tile in the Wet-Set Method with Portland Cement Mortar
A108.1B-2017	Installation of Ceramic Tile on a Cured Portland Cement Mortar Setting Bed with Dry-Set or Latex-Portland Cement Mortar
A108.1C-1999 (R2016)	Contractors' Option: Installation of Ceramic Tile in the Wet-Set Method with Portland Cement Mortar or Installation of Ceramic Tile on a Cured Portland Cement Mortar Setting Bed with Dry-Set or Latex-Portland Cement Mortar
A108.4-2019	Installation of Ceramic Tile with Organic Adhesives or Water Cleanable Tile-Setting Epoxy Adhesive
A108.5-2020	Installation of Ceramic Tile with Dry-Set Portland Cement Mortar or Latex-Portland Cement Mortar
A108.6-1999 (R2019)	Installation of Ceramic Tile with Chemical Resistant, Water Cleanable Tile-Setting and -Grouting Epoxy
A108.8-1999 (R2019)	Installation of Ceramic Tile with Chemical Resistant Furan Resin Mortar and Grout
A108.9-1999 (R2019)	Installation of Ceramic Tile with Modified Epoxy Emulsion Mortar/grout
A108.10-2017	Installation of Grout in Tilework
A108.11-2018	Interior Installation of Cementitious Backer Units
A108.12-1999 (R2019)	Installation of Ceramic Tile with EGP (Exterior Glue Plywood) Latex-Portland Cement Mortar
A108.13-2005 (R2016)	Installation of Load Bearing, Bonded, Waterproof Membranes for Thin-Set Ceramic Tile and Dimension Stone
A108.14-2020	Installation of Paper-Faced Glass Mosaic Tile
A108.15-2019	Alternative Method: Installation of Paper-Faced Mounted Glass Mosaic Tile
A108.16-2020	Installation of Paper-Faced, Back-Mounted, or Clear Film Face-Mounted Glass Mosaic Tile
A108.17-2005 (R2016)	Installation of Crack Isolation Membranes for Thin-Set Ceramic Tile and Dimension Stone
A108.19-2020	Interior Installation of Gauged Porcelain Tiles and Gauged Porcelain Tile Panels/Slabs by the Thin-Bed Method Bonded with Modified Dry-Set Cement Mortar or Improved Modified Dry-Set Cement Mortar
A108.20-2020	Exterior Installation of Vertical and Overhead Gauged Porcelain Tiles and Gauged Porcelain Tile Panels/Slabs by the Thin-Bed Method Bonded with Improved Modified Dry-Set Cement Mortar

ANSI Material Specifications:

A118.1-2019	Dry-Set Cement Mortar
A118.3-2013	Chemical Resistant, Water Cleanable Tile-Setting and -Grouting Epoxy and Water Cleanable Tile-Setting Epoxy Adhesive
A118.4-2019	Modified Dry-Set Cement Mortar
A118.5-1999 (R2016)	Chemical Resistant Furan Mortars and Grouts for Tile Installation
A118.6-2019	Standard Cement Grouts for Tile Installation
A118.7-2019	High Performance Cement Grouts for Tile Installation
A118.8-1999 (R2016)	Modified Epoxy Emulsion Mortar/grout
A118.9-2019	Cementitious Backer Units
A118.10-2014 (R2019)	Load Bearing, Bonded, Waterproof Membranes for Thin-Set Ceramic Tile and Dimension Stone Installation
A118.11-2017	EGP (Exterior Glue Plywood) Modified Dry-Set Cement Mortar
A118.12-2014 (R2019)	Crack Isolation Membranes for Thin-Set Ceramic Tile and Dimension Stone Installation
A118.13-2014 (R2019)	Bonded Sound Reduction Membranes for Thin-Set Ceramic Tile Installation
A118.15-2019	Improved Modified Dry-Set Cement Mortar

A136.1-2020	Organic Adhesives for Installation of Ceramic Tile
A137.1-2019	Ceramic Tile
A137.2-2019	Glass Tile
A137.3-2017	Gauged Porcelain Tile and Tile Panel/Slabs

ANSI Test Methods:

A326.3-2017	Dynamic Coefficient of Friction (DCOF)

ASTM International

100 Barr Harbor Drive
West Conshohocken, PA 19428
www.astm.org

ASTM Material Specifications/Test Methods:

C97/C97M-18	Standard Test Methods for Absorption and Bulk Specific Gravity of Dimension Stone
C99/C99M-18	Standard Test Method for Modulus of Rupture of Dimension Stone
C119-20	Standard Terminology Relating to Dimension Stone
C120-19	Test Methods for Flexure Testing of Structural and Roofing Slate
C144-18	Standard Specification for Aggregate for Masonry Mortar
C150/C150M-20	Standard Specification for Portland Cement
C170/C170M-17	Standard Test Method for Compressive Strength of Dimension Stone
C206-14	Standard Specification for Finishing Hydrated Lime
C207-18	Standard Specification for Hydrated Lime for Masonry Purposes
C241/C241M-20	Standard Test Method for Abrasion Resistance of Stone Subjected to Foot Traffic
C373-18	Standard Test Method for Determination of Water Absorption and Associated Properties by Vacuum Method for Pressed Ceramic Tiles and Glass Tiles and Boil Method for Extruded Ceramic Tiles and Non-tile Fired Ceramic Whiteware Products
C424-93 (2020)	Standard Test Method for Crazing Resistance of Fired Glazed Whitewares by Autoclave Treatment
C472-20	Standard Test Methods for Physical Testing of Gypsum, Gypsum Plasters and Gypsum Concrete
C482-20	Standard Test Method for Bond Strength of Ceramic Tile to Portland Cement Paste
C484-20	Standard Test Method for Thermal Shock Resistance of Glazed Ceramic Tile
C503/C503M-15	Standard Specification for Marble Dimension Stone
C568/C568M-15	Standard Specification for Limestone Dimension Stone
C578-19	Standard Specification for Rigid, Cellular Polystyrene Thermal Insulation
C615/C615M-18e1	Standard Specification for Granite Dimension Stone
C616/C616M-15	Standard Specification for Quartz-Based Dimension Stone
C627-18	Standard Test Method for Evaluating Ceramic Floor Tile Installation Systems Using the Robinson-Type Floor Tester
C629/C629M-15	Standard Specification for Slate Dimension Stone
C634-13e1	Standard Terminology Relating to Building and Environmental Acoustics
C645-18	Standard Specification for Nonstructural Steel Framing Members
C650-20	Standard Test Method for Determination of Resistance to Chemical Substances
C880/C880M-18	Standard Test Method for Flexural Strength of Dimension Stone
C919-19	Standard Practice for Use of Sealants in Acoustical Applications
C920-18	Standard Specification for Elastomeric Joint Sealants
C1026-13 (2018)	Standard Test Method for Measuring the Resistance of Ceramic and Glass Tile to Freeze-Thaw Cycling
C1027-19	Standard Test Method for Determining Visible Abrasion Resistance of Glazed Ceramic Tile
C1178/C1178M-18	Standard Specification for Coated Glass Mat Water-Resistant Gypsum Backing Panel
C1193-16	Standard Guide for Use of Joint Sealants
C1201/C1201M-20	Standard Test Method for Structural Performance of Exterior Dimension Stone Cladding Systems by Uniform Static Air Pressure Difference
C1242-20	Standard Guide for Selection, Design, and Installation of Dimension Stone Attachment Systems
C1243-93 (2015) e2	Standard Test Method for Relative Resistance to Deep Abrasive Wear of Unglazed Ceramic Tile by Rotating Disc
C1278/C1278M-17	Standard Specification for Fiber-Reinforced Gypsum Panel
C1288-17	Standard Specification for Fiber-Cement Interior Substrate Sheets
C1325-19	Standard Specification for Fiber-Mat Reinforced Cementitious Backer Units
C1352-15	Test Method for Flexural Modulus of Elasticity of Dimension Stone
C1353-20e	Standard Test Method for Abrasion Resistance of Dimension Stone Subjected to Foot Traffic Using a Rotary Platform, Double-Head Abraser
C1354/C1354M-15	Standard Test Method for Strength of Individual Stone Anchorages in Dimension Stone
C1378-20	Standard Test Method for Determination of Resistance to Staining
C1396/C1396M-17	Standard Specification for Gypsum Board

C1496-18	Standard Guide for Assessment and Maintenance of Exterior Dimension Stone Masonry Walls and Facades
C1515-20	Standard Guide for Cleaning of Exterior Dimension Stone, Vertical and Horizontal Surfaces, New or Existing
C1526-19	Standard Specification for Serpentine Dimension Stone
C1527/C1527M-11 (2018)	Standard Specification for Travertine Dimension Stone
C1528-20	Standard Guide for Selection of Dimension Stone for Exterior Use
C1658/C1658M-19e1	Standard Specification for Glass Mat Gypsum Panels
C1721-15	Standard Guide for Petrographic Examination of Dimension Stone
C1799-19	Standard Guide to Dimension Stone Test Specimen Sampling and Preparation
D4068-17	Standard Specification for Chlorinated Polyethylene (CPE) Sheeting for Concealed Water-Containment Membrane
D4551-17	Standard Specification for Poly(Vinyl Chloride) (PVC) Plastic Flexible Concealed Water-Containment Membrane
E90-09 (2016)	Standard Test Method for Laboratory Measurement of Airborne Sound Transmission Loss of Building Partitions and Elements
E96/E96M-16	Standard Test Methods for Water Vapor Transmission of Materials
E119-20	Standard Test Methods for Fire Tests of Building Construction and Materials
E336-20	Standard Test Method for Measurement of Airborne Sound Attenuation Between Rooms in Buildings
E413-16	Classification for Rating Sound Insulation
E492-09 (2016) e1	Standard Test Method for Laboratory Measurement of Impact Sound Transmission Through Floor-Ceiling Assemblies Using the Tapping Machine
E548-94e1 (withdrawn no date)	Standard Guide for General Criteria Used for Evaluating Laboratory Competence
E596-96 (2016)	Standard Test Method for Laboratory Measurement of Noise Reduction of Sound-Isolating Enclosures
E597 (withdrawn 1999)	Practice for Determining A Single-Number Rating of Airborne Sound Isolation for Use in Multiunit Building Specifications
E717 (withdrawn 2008)	Standard Guide for Preparation of the Accreditation Annex of Acoustical Test Standards
E989-18	Standard Classification for Determination of Single-Number Metrics for Impact Noise
E1007-19	Standard Test Method for Field Measurement of Tapping Machine Impact Sound Transmission Through Floor-Ceiling Assemblies and Associated Support Structures
E1155-20	Standard Test Method for Determining FF Floor Flatness and FL Floor Levelness Numbers
E2179-03 (2016)	Standard Test Method for Laboratory Measurement of the Effectiveness of Floor Coverings in Reducing Impact Sound Transmission Through Concrete Floors
F2329-15	Standard Specification for Zinc Coating, Hot-Dip, Requirements for Application to Carbon and Alloy Steel Bolts, Screws, Washers, Nuts, and Special Threaded Fasteners

Pool & Hot Tub Alliance (formerly the Association of Pool and Spa Professionals)

2111 Eisenhower Avenue
Alexandria, VA 22314
www.phta.org

APSP Specifications:

ANSI/APSP/ICC-1 2014	American National Standard for Public Swimming Pools
APSP-2 1999	Standard for Public Spas
ANSI/APSP/ICC-3 2014	American National Standard for Permanently Installed Residential Spas
ANSI/APSP/ICC-4 2012	American National Standard for Above-ground/On-ground Residential Swimming Pools
ANSI/APSP/ICC-5 2011	American National Standard for Residential In-ground Swimming Pools
ANSI/APSP/ICC-11 2019	American National Standard for Water Quality for Public Pools and Spas

Gypsum Association

6525 Belcrest Road, Suite 480
Hyattsville, MD 20782
www.gypsum.org

Gypsum Association Specifications:

GA-216-2018	Application and Finishing of Gypsum Panel Products

International Code Council (ICC)

500 New Jersey Avenue, NW, 6th Floor
Washington, DC 20001
www.iccsafe.org

ICC Specifications:

International Building Code (IBC)
International Residential Code (IRC)

International Organization for Standardization (ISO)

Chemin de Blandonnet 8
CP 401
1214 Vernier, Geneva
Switzerland
www.iso.org

ISO Material Specifications:

13006:2018	Ceramic tiles—Definitions, classification, characteristics and marking
13007-1:2014	Ceramic tiles—Grouts and adhesives—Part 1: Terms, definitions and specifications for adhesives
13007-2:2013	Ceramic Tiles—Grouts and adhesives—Part 2: Test methods for adhesives
13007-3:2010	Ceramic tiles—Grouts and adhesives—Part 3: Terms, definitions and specifications for grouts
13007-4:2013	Ceramic Tiles—Grouts and adhesives—Part 4: Test methods for grouts

Natural Stone Institute (NSI)

380 E. Lorain Street
Oberlin, OH 44074
www.naturalstoneinstitute.org

NSI Specifications:

Dimension Stone Design Manual

National Fire Protection Association (NFPA)

1 Batterymarch Park
Quincy, MA 02169
www.nfpa.org

NFPA Specifications:

NFPA99-2021	Standard for Health Care Facilities

METHOD LOCATOR BY METHOD NUMBER

METHOD	PAGE # TILE	PAGE # STONE	METHOD	PAGE # TILE	PAGE # STONE	METHOD	PAGE # TILE	PAGE # STONE	METHOD	PAGE # TILE	PAGE # STONE
B411	226	396	EJ171L	437	437	F149	174		RW800	293	
B412	230	400	EJ171M	437	437	F150	158		RW811-825	294	
B413	224	394	F101	56	300	F151	176		S151	285	
B414	238	408	F102	58	302	F152	178		SR613	274	
B415	242	412	F103	60	304	F155	180		SR614	276	
B417	284		F103B	62	306	F160	160		TR418	260	
B419	232	402	F104	64	308	F170	166		TR420	261	
B420	244	414	F105	66	310	F175	168		TR611	292	
B421	250	420	F111	68	312	F180	182		TR711	288	
B421C	254		F112	70	314	F185	184		TR712	290	
B422	252	422	F113	72	316	F200	86	320	TR713	291	
B422C	256		F113A	74	318	F200A	88	322	W201	186	360
B425	234	404	F114	76		F205	90	324	W202E	188	362
B426	246	416	F115	78		F205A	92	326	W202I	196	370
B430	236	406	F115A	80		F250		358	W211	192	366
B431	248	418	F116E	82		P601MB	278		W215	194	368
B440	228	398	F116O	84		P601TB	280		W221	198	372
B441	240	410	F121	94	328	P602	282		W222	200	374
C311	262	424	F122	96	330	R612	286		W223	202	
C312	264	425	F122A	98	332	RH110	116	336	W231/W241	204	376
C315	266	426	F125	100	334	RH110A	118	338	W242	206	
C511	268		F128	102		RH111	122	342	W243	208	378
C512	270		F131	104		RH111A	124	344	W244C	210	380
C513	272		F132	106		RH112	126	346	W244E	190	364
EJ171	434	434	F133	108		RH112A	128	348	W244F	212	382
EJ171A	435	435	F134	110		RH115	130		W245	214	384
EJ171B	435	435	F135	112		RH115A	132		W246	216	386
EJ171C	434	434	F136	114		RH116	134	350	W247	218	388
EJ171D	434	434	F141	150	356	RH116A	136	352	W248	220	390
EJ171E	434	434	F142	154		RH117	120	340	W260	222	392
EJ171F	436	436	F143	156		RH122	138		WS610	292	
EJ171G	435	435	F144	162		RH123	140				
EJ171H	435	435	F145	152		RH130	144				
EJ171I	436	436	F146	164		RH135	146				
EJ171J	436	436	F147	170		RH140	148				
EJ171K	436	436	F148	172		RH141	142	354			

METHOD LOCATOR BY APPLICATION (TILE)

**Interior floor methods—including radiant heat methods—marked with an asterisk (*) may be suitable for exterior applications. Consult Environmental Exposure Classifications section in each method for details.*

METHOD LOCATOR BY APPLICATION (STONE)

**Interior floor methods—including radiant heat methods—marked with an asterisk (*) may be suitable for exterior applications. Consult Environmental Exposure Classifications section in each method for details.*

FLOORS, INTERIOR OVER ON-GROUND CONCRETE SUBSTRATE

Mortar Bed Methods

Thin Bed Methods

Underlayment Methods

FLOORS, INTERIOR OVER ON-GROUND CONCRETE SUBSTRATE, WITH RADIANT HEAT

with Electric Radiant Heat

with Hydronic Radiant Heat

FLOORS, INTERIOR OVER PLYWOOD SUBSTRATE (JOISTS MAX 16" O.C.)

FLOORS, INTERIOR OVER PLYWOOD SUBSTRATE (JOISTS 16" O.C.), WITH RADIANT HEAT

MOVEMENT JOINTS

over Concrete Substrate

over Masonry Substrate

over Wood Substrate

with Backer Board

Other Movement Joint Types

RADIANT HEAT, FLOORS WITH

See Floors, Interior over Above-Ground Concrete Substrate, with Radiant Heat; Floors, Interior over On-Ground Concrete Substrate, with Radiant Heat; Floors, Interior over Plywood Substrate (Joists max. 16" o.c.), with Radiant Heat; and Floors, Interior over plywood substrate (Joists max 19.2" o.c.), with Radiant Heat

SHOWERS

See Bathtubs and Showers

WALLS, EXTERIOR OVER MASONRY OR CONCRETE SUBSTRATE

WALLS, EXTERIOR OVER WOOD OR METAL STUD-FRAMED WALLS

**Interior wall methods marked with an asterisk (*) may be suitable for exterior applications. Consult Environmental Exposure Classifications section in each method for details.*

WALLS, INTERIOR OVER MASONRY OR CONCRETE SUBSTRATE

Mortar Bed Methods

Other Methods

Thin Bed Methods

WALLS, INTERIOR OVER WOOD OR METAL STUD-FRAMED WALLS

Backer Board Methods

Gypsum Board Methods

Mortar Bed Methods

Other Methods

WALLS, INTERIOR, WHERE FLATTENING OR TRUEING OF SUBSTRATE IS REQUIRED

2021 TCNA MEMBER DIRECTORY

REGULAR MEMBERS

A.C. Products Company (A QEP Company)
330-698-1105
www.acproductsco.com

ALCESA
52-55-58454502
www.alcesa.com

American Olean
888-268-8453
www.americanolean.com

American Wonder Porcelain
972-746-2200
www.wonderporcelain.com

ArteVivo Mosaico
52-55-53699713
www.artevivo-mosaico.com

Atlas Concorde USA, Inc.
615-661-7200
www.atlasconcordeusa.com

Cesantoni
52-47-89854200
www.cesantoni.com.mx

Crossville, Inc.
931-484-2110
www.crossvilleinc.com

Daltile
800-933-8453
www.daltile.com

Daltile Mexico
52-81-8124-8124
www.daltile.com.mx

Del Conca USA, Inc.
865-657-3550
www.delconcausa.com

Epro Tile, Inc.
866-818-3776
www.eprotile.com

Estudio Cerámico
52-55-53004576
www.estudioceramico.es

Fiandre USA
800-828-9074
www.granitifiandre.com

Florida Tile, Inc.
800-352-8453
www.floridatile.com

Iberica Tiles
52-55-24513500

Interceramic
214-503-5500
www.interceramicusa.com

Iris US
800-323-9906
www.irisus.com

Ironrock
800-325-3945
www.ironrock.com

Jeffrey Court, Inc.
951-340-3383
www.jeffreycourt.com

Ken Mason Tile (also known as BCIA)
562-432-7574
www.kmt-bcia.com

Landmark Ceramics
931-325-5700
www.lcusa.com

Lamosa USA
(214) 647-0250
www.lamosausa.com

Marazzi
214-398-1411
www.marazzitile.com

Mediterranea LLC
305-718-5091
www.mediterranea-usa.com

Metropolitan Ceramics
800-325-3945
www.metroceramics.com

Milestone
877-FLORIM1
www.florimusa.com

Mohawk Ceramic
800-266-4295
www.mohawkflooring.com

Mosaicos Venecianos de Mexico
52-77-73296640
www.m-v-m.mx

Nitropiso
52-44-48320100
www.nitropiso.com.mx

Porcelanite
52-55-52469912
www.porcelanite.com

Portobello America, Inc.
954-809-5668
www.portobelloamerica.com

Pratt & Larson Ceramics
503-231-9464
www.prattandlarson.com

Quarry Tile Company
509-536-2812
www.quarrytile.com

Ragno
214-398-1411
www.ragnousa.com

Seneca Tiles, Inc.
800-426-4335
www.senecatiles.com

Sonoma Tilemakers
707-837-8177
www.sonomatilemakers.com

StonePeak Ceramics, Inc.
312-506-2800
www.stonepeakceramics.com

Summitville Tiles, Inc.
330-223-1511
www.summitville.com

Vitromex USA
800-848-4146
www.vitromex.com

ASSOCIATE MEMBERS

Alorhum
52-81-88816300

Colorobbia Mexico
52-81-80420050

Covia Corp.
800-243-9004

Ferro Mexicana
52-55-50907200

Grupo Industrial Trébol
52-81-81262380

Gruppo Pro Ceramic
52-81-81540810

IMERYS-North America Ceramics/K-T Clay
770-594-0660

Old Hickory Clay Company
270-247-3042

Precision H2O
800-425-2098

Slcer MX
52-81-11120087

Silicatos Especiales
52-55-57559400

Spinks Clay Company-Lhoist North America
731-642-5414

Torrecid Mexico
52-81-81545757

Vetriceramici de Mexico
52-81-40008400

Visoft USA LLC
404-509-5665

Zinc Nacional
52-81-83426861

ASSOCIATE ART/STUDIO MEMBERS

Allison Eden Studios
212-243-9944

American Bullnose Company
714-630-6042

American Restoration Tile, Inc.
501-455-1000

AMosaic
231-375-8037

Appomattox Tile Art Company
804-732-8810

Artfind Tile
330-264-7706

ARTO Brick & Tile
310-768-8500

B.A. Schmidt Arts & Enterprises, Inc.
847-432-5679

Brick House Tile Co. (dba Cider Press Tile)
603-357-2884

California Art Tile
858-689-9596

Carly Quinn Designs
520-624-4117

Casa Ceramica, Inc.
561-659-7909

Cercan Mosaic
905-851-7923

Chadwick's Surfaces International, Inc.
847-680-3222

Clay Decór, LLC
607-654-7428

Coatings By Sandberg
714-538-0888

CR Studio 4, Inc.
951-296-2270

Creative Edge Master Shop
641-472-8145

Delta Brick & Climate Company Ltd.
970-749-3370

Dunis Studios
512-722-3637

Elon Tile & Stone
203-628-7179

Fine Art Tileworks
907-347-8987

Images in Tile USA, Inc.
(dba Bison Coating & Supply)
417-206-0252

Julian Glass Inc.
248-224-1963

Lowitz & Company
773-784-2628

Malibu Ceramic Works
310-455-2485

McIntyre Tile Company, Inc.
707-433-8866

Mediterra Tile
877-325-5510

Mission Tile West
626-799-4595

MTM Wizard Enterprise LLC
323-756-8430

Quemere Designs
914-934-8366

Rookwood Pottery Co.
513-381-2510

STAR TILE
213-479-6740

StoneImpressions
858-274-3400

Syzygy Tileworks
575-388-5472

Terra Firma, Ltd.
803-643-9399

Terra Green Ceramics, Inc.
765-935-4760

Tile Artisans Digital Imaging LLC
800-601-4199

Tile By Design, LLC
260-918-1784

Tile District, Inc.
310-489-6101

Tile Optima
571-224-5821

Wakei & Company, Ltd.
805-577-1187

Whitehill Enterprises, LLC
417-483-0566

Woodchuck Creations
800-416-0082

ASSOCIATE INSTALLATION MEMBERS

ACG Materials
405-366-9500

AcoustiCORK/Amorim Cork Composites
800-558-3206

Adhesivos del Bajio ADE1000
52-477-7150105

Adhesivos Fixol
52-443-3133159

Aqua Mix
800-366-6877

ARDEX Americas
724-203-5000

Bexel Internacional
52-81-81300200

Blanke Corporation
800-787-5055

Boral
210-503-2634

Bostik, Inc.
800-726-7845

C-Cure
800-895-2874

Cement Pro LLC
951-300-0300

CEMIX
866-GO-CEMIX

ColorFast Industries, Inc.
817-546-4910

Compotite Corporation
800-221-1056

Comp-U-Floor Software - AYA Associates, Inc.
800-766-0330

Cortag Tools USA
305-777-2024

Creaprac
52-81-83389399

Crest
52-81-80475000

Custom Building Products
800-272-8786

Donnelly Distribution Raimondi USA
800-625-6686

Dural USA, Inc.
800-278-1956

DYNA Metro, Inc.
905-761-3309

FILA Chemicals USA Corporation
800-833-6444

Fin Pan, Inc.
800-833-6444

Fortifiber Building Systems Group
800-773-4777

Georgia-Pacific DensShield® Tile Backer
800-225-6119

GranQuartz L.P.
800-458-6222

Guru USA LLC
305-779-3388

Infinity Drain Ltd.
516-767-6786

James Hardie Building Products, Inc.
888-542-7343

Jamo, Inc.
800-826-6852

Kerneos
757-284-3200

Lackmond Products, Inc.
770-919-2100

LATICRETE International, Inc.
800-243-4788

LATICRETE , SA de CV
52-81-83905005

Leggett & Platt Flooring Products
800-475-0010

MAPEI Americas
888-876-2734

Merkrete
866-516-0061

Mezcla Brava
52-22-32753632

More Surface Care, LLC
844-404-6673

MP Global Products - SnapStone
888-379-9695

NAC Products, Inc.
800-633-4622

National Gypsum Company
704-365-7319

NIASA
52-55-53545680

Noble Company
800-878-5788

North American Adhesives
800-747-2722

NUE Tile, LLC
800-686-9158

Panel Rey
52-81-83450055

Park Industries
800-328-2309

Pearl Abrasive Company
562-927-5561

Perdura
52-55-56464151

Pliteq Inc.
416-449-0049

Profilitec Corporation
855-290-9591

Proflex
877-577-6353

Progress Profiles America, Inc.
973-584-2650

Protecto Wrap Company
800-759-9727

PSC, Inc.
509-334-4410

QT Sound Insulation
866-326-5712

REALSA Recubrimientos y Adhesivos de la Laguna
52-18-717506465

Red Tools
52-55-55986500

Reilly Foam Corporation
561-842-8090

Roberts Consolidated Industries, Inc.
866-435-8665

Rubi Tools USA
305-715-9892

Schluter-Systems
800-472-4588

SGM, Inc. (Southern Grouts and Mortars)
800-641-9247

Siena Tile & Stone Installation Products
951-737-7447

Sika Mexicana
52-44-21866587

Sound Seal
413-789-1770

Superior Adhesives & Chemicals
763-202-5374

TEC
800-832-9002

ASSOCIATE INSTALLATION MEMBERS (cont'd)

TEXRITE
800-669-0115

TI-PROBOARD
800-833-6444

Tile Redi, Ltd.
800-232-6156

TileWare Products
828-322-9273

Trimaco, LLC
919-949-1141

Ultimate RB
503-472-4691

UNIBLOCK
52-81-82892100

United States Gypsum Company (USG)
800-USG-4YOU

Wacker Polymers
610-336-2700

wedi Corporation
877-933-9334

X Curb Industries/Rapid Recess LLC
970-852-0505

AFFILIATED EQUIPMENT MANUFACTURERS

Ferrari & Cigarini North America
52-81-83903970

SACMI de Mexico
52-81-87487200

SITI-B&T Group
52-81-8372-9879

System Norte America
1-800-288-1100

AFFILIATED EQUIPMENT MANUFACTURERS

AlysEdwards Tile & Stone
714-917-6720

Artistic Tile
201-864-7000

Chameleon Power, Inc.
248-380-9080

Creative Metals
800-394-8145

ExactaTiles, Inc.
972-323-1063

GBI Tile + Stone
949-567-1880

Hirsch Glass Corp.
732-329-8988

International Wholesale Tile, LLC
772-223-5151

Leap Tools, Inc.
647-352-5320

Lunada Bay Tile
310-257-1300

Maticad USA, Inc.
215-475-6835

New Ravenna
757-442-3379

Pacific Solutions
800-400-4927

Questech Corporation
802-773-1228

Rainbow Inc.
931-552-7783

RFMS, Inc.
800-701-7367

Stonexchange, Inc.
305-513-9795

US Bullnosing, Inc.
954-567-0404

Vidrepur Internacional
55-55-930447

AFFILIATED CERAMIC CONSTRUCTION MATERIALS

Ladrillera Mecanizada
01-800-0022776

AFFILIATED IMPORTER-DISTRIBUTOR

Anatolia Tile & Stone Inc.
905-771-3800

Anthology
888-461-3520

Deutsche Steinzeug America, Inc.
770-442-5500

Marble Systems/Country Floors
703-204-1818

GOLD SPONSORS

ARDEX tile & stone Installation Systems

CREST PREMIUM TILE SETTING SOLUTIONS

ecore™

Perma BASE® BRAND Cement Board By National Gypsum

JamesHardie®

Noble Company Simple Solutions... Proven Performance

The Tile Shop WWW.TILESHOP.COM

SILVER SPONSORS

Armaly proplus® America's Best Soft Tools™

PROGRESS® PROFILES PROFILES & SYSTEMS

RUBI

TEC®

BRONZE SPONSORS

AMERICA'S FLOOR SOURCE

HYDRO-BLOK™ The BETTER Shower System

RETAIL LEAD MANAGEMENT

RUSSO TRADING COMPANY

SGM INSTALLATION SYSTEMS FOR CERAMIC TILE & DIMENSION STONE.

TRALAMA The Trade Labor Marketplace

texrite™ by cemix

UNIVERSITY CTS Ceramic Tile and Stone

wedi®

2021

CTEF FINANCIAL SPONSORS

ELITE SPONSORS

- CTDA
- Ceramics of Italy
- Coverings® THE GLOBAL TILE & STONE EXPERIENCE
- National Tile Contractors Association NTCA
- TCNA Tile Council of North America, Inc.
- Tile of Spain

SIGNATURE SPONSORS

- AMERICAN OLEAN® PROVEN IN TILE™
- CROSSVILLE®
- CORTAG REVOLUTION TOOLS
- D.W. Sanders Tile & Stone Contracting
- CUSTOM® Custom Building Products
- daltile® IMAGINE WHAT'S POSSIBLE™
- floridatile
- Interceramic® CRAFTED FOR LIFE
- LATICRETE®
- MAPEI® ADHESIVES · SEALANTS · CHEMICAL PRODUCTS FOR BUILDING
- MARAZZI® DARE TO BE BOLD™
- Schluter® Systems

PLATINUM SPONSORS

- LAMOSA USA

WHY TILE .COM
SMART.
BEAUTIFUL.
RESOURCE.
If clean air is important to you and your family, tile is the natural choice.
Endless finishes await, from timeless elegance to top trends.
Consumer Information
Searchable Image Database
Commercial Guide and Case Studies
13 Style Innovators and Their Tile Inspirations
Trending articles
Ceramic Tile's Answer to the Plastic Problem

CARAMEL SERIES
BRINGING OUR NEW PORCELAIN PORTAFOLIO TO THE NEXT LEVEL
Choose Perfection
LAMOSA USA Collection
LAMOSA USA
Proud Sponsor of the Ceramic Tile Education Foundation (CTEF)

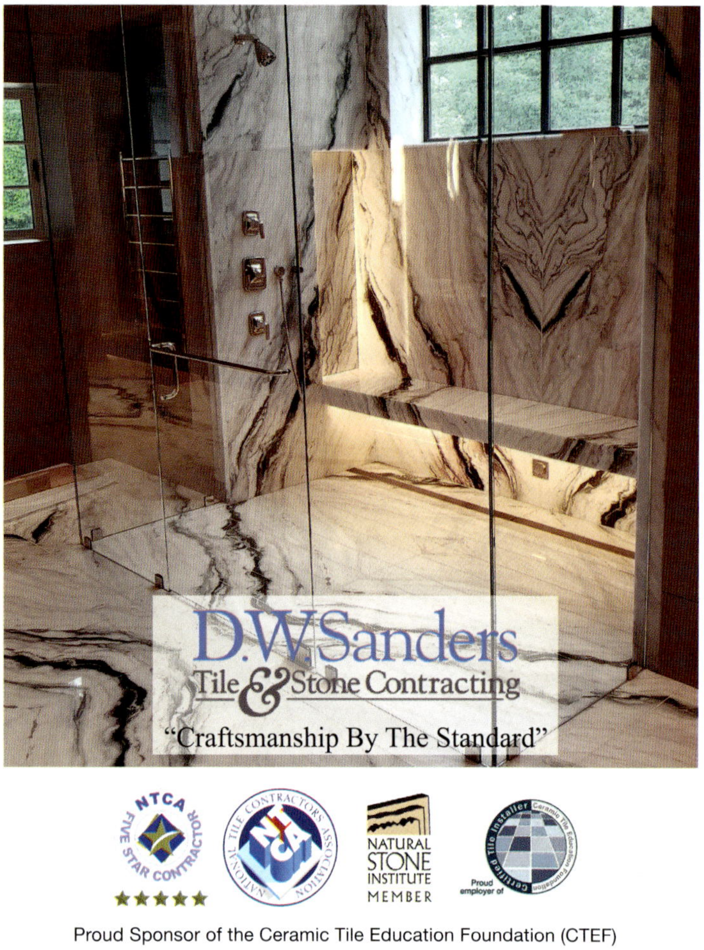
D.W.Sanders
Tile & Stone Contracting
"Craftsmanship By The Standard"
NTCA FIVE STAR CONTRACTOR
NATIONAL TILE CONTRACTORS ASSOCIATION
NATURAL STONE INSTITUTE MEMBER
Proud employer of Certified Tile Installer
Proud Sponsor of the Ceramic Tile Education Foundation (CTEF)

@WhyTile
@WhyTile
@WhyTile
WhyTileIdeas
WHY TILE .COM
Follow the inspiration.

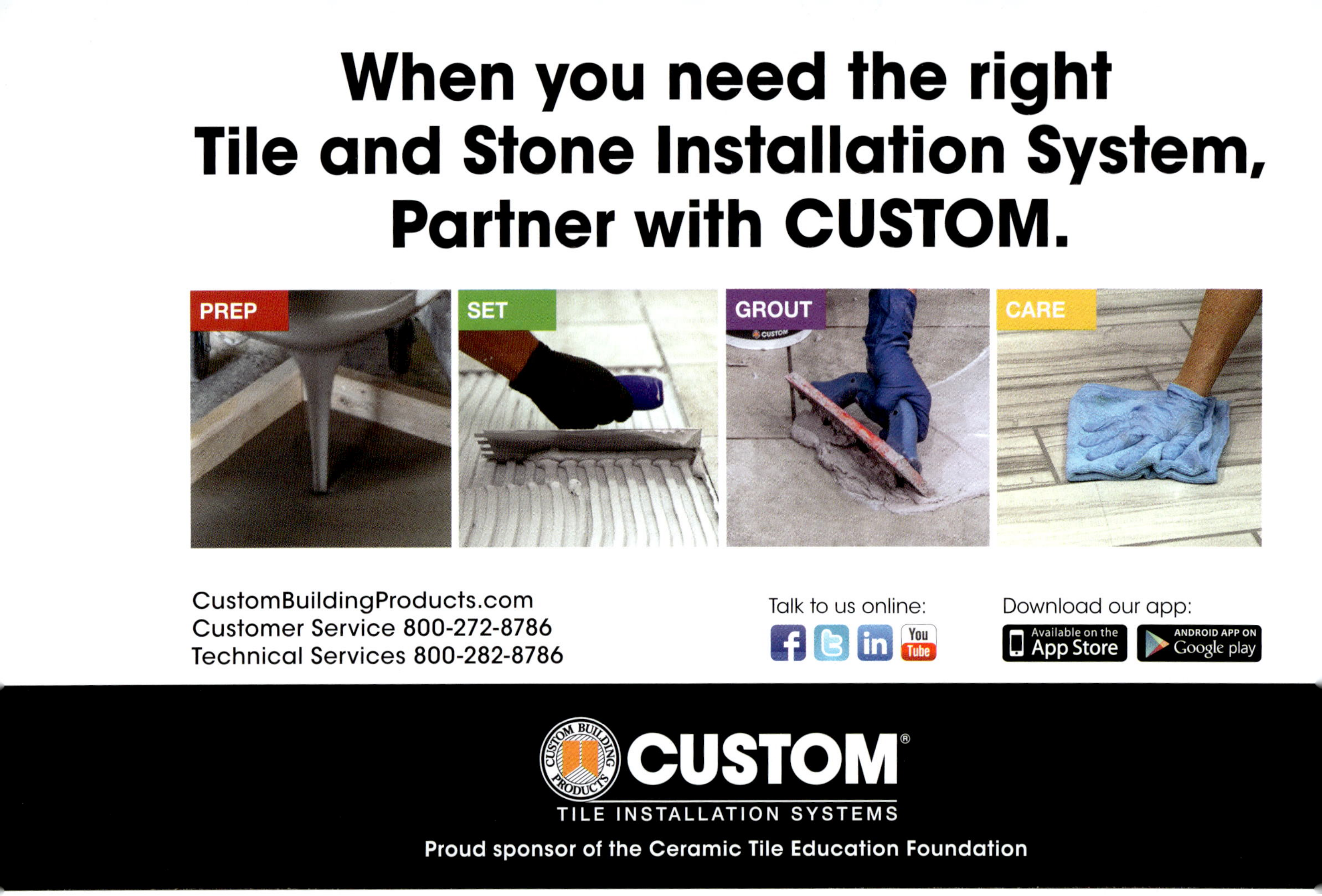
When you need the right
Tile and Stone Installation System,
Partner with CUSTOM.
PREP
SET
GROUT
CARE
CustomBuildingProducts.com
Customer Service 800-272-8786
Technical Services 800-282-8786
Talk to us online:
Download our app:
Available on the App Store
ANDROID APP ON Google play
CUSTOM BUILDING PRODUCTS
CUSTOM®
TILE INSTALLATION SYSTEMS
Proud sponsor of the Ceramic Tile Education Foundation

Effective SOLUTIONS
for tile installation
MADE IN BRAZIL
BASIC 2
Tile leveling system
ZAPP G2
Electric wet saw
up to 61"
1,5 HP
MOTOR
BREAKING POWER
2210
lbs
MASTER PLUS
For tile up to 71"
@cortagusa
info@cortagtools.com
Proud Partner and Sponsor of the
CERAMIC TILE EDUCATION FOUNDATION (CTEF)
CORTAG
REVOLUTION TOOLS